Gerald Doyle

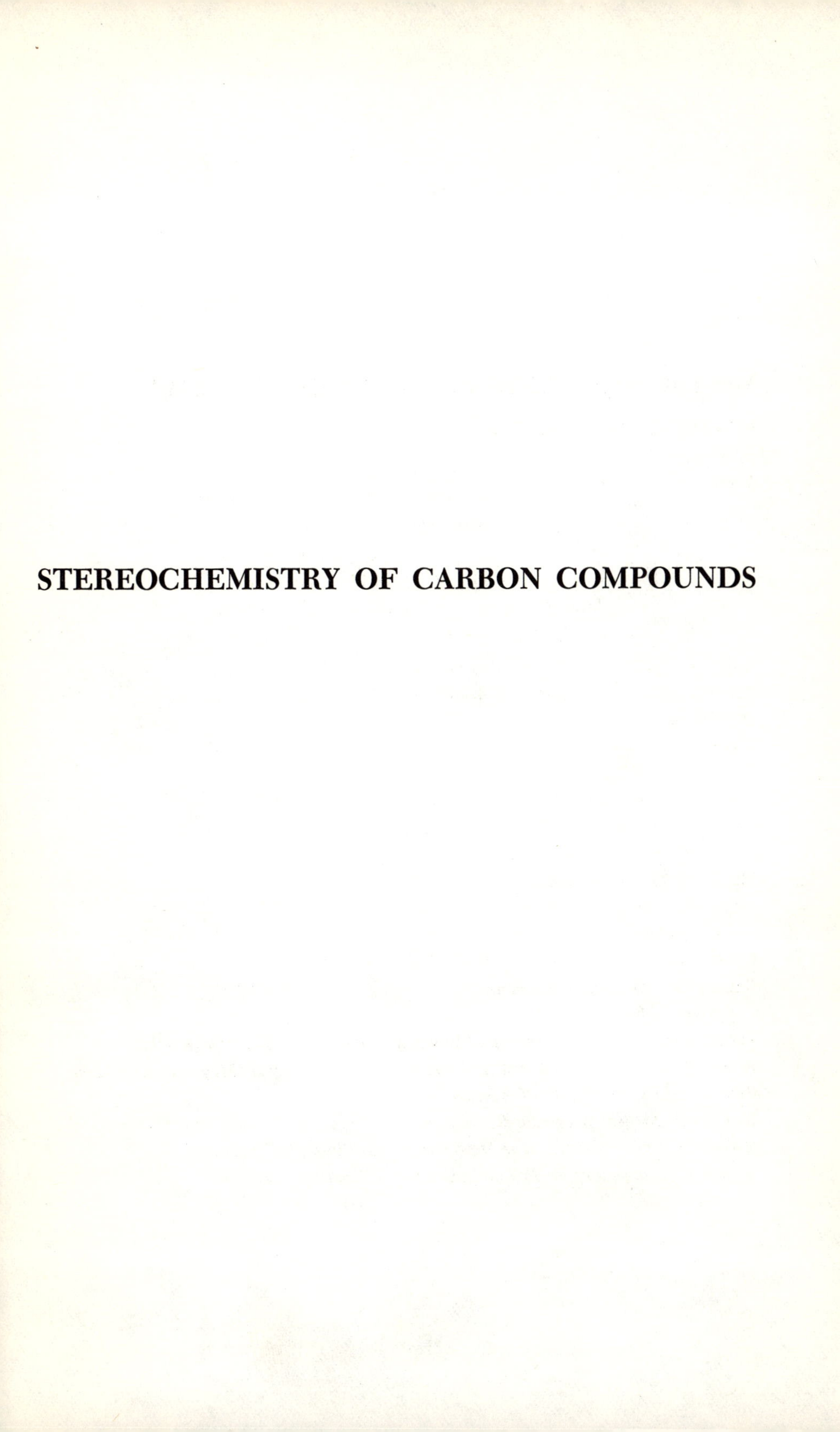

STEREOCHEMISTRY OF CARBON COMPOUNDS

McGRAW-HILL SERIES IN ADVANCED CHEMISTRY

BAIR *Introduction to Chemical Instrumentation*
BENSON *The Foundations of Chemical Kinetics*
DAVIDSON *Statistical Mechanics*
DAVYDOV (*Trans.* Kasha and Oppenheimer) *Theory of Molecular Excitons*
DEAN *Flame Photometry*
DJERASSI *Optical Rotatory Dispersion*
ELIEL *Stereochemistry of Carbon Compounds*
FITTS *Nonequilibrium Thermodynamics*
HELFFERICH *Ion Exchange*
HILL *Statistical Mechanics*
HINE *Physical Organic Chemistry*
KIRKWOOD AND OPPENHEIM *Chemical Thermodynamics*
KOSOWER *Molecular Biochemistry*
LAITINEN *Chemical Analysis*
PITZER AND BREWER (*Revision of* Lewis and Randall) *Thermodynamics*
POPLE, SCHNEIDER, AND BERNSTEIN *High-resolution Nuclear Magnetic Resonance*
PRYOR *Mechanisms of Sulfur Reactions*
ROBERTS *Nuclear Magnetic Resonance*
ROSSOTTI AND ROSSOTTI *The Determination of Stability Constants*
WIBERG *Laboratory Technique in Organic Chemistry*

STEREOCHEMISTRY OF CARBON COMPOUNDS

ERNEST L. ELIEL
Professor of Chemistry
University of Notre Dame

McGRAW-HILL BOOK COMPANY, INC. 1962
New York San Francisco Toronto London

STEREOCHEMISTRY OF CARBON COMPOUNDS

 Library of Congress Catalog Card Number 61-14354

19177

IV

To the Memory of

LOUIS PASTEUR

PREFACE

The subject of stereochemistry is as old as organic chemistry itself. The discovery of optical rotation by Biot antedates Wöhler's famous urea synthesis, and the classical stereochemical research of Pasteur was contemporary with Kekulé's equally classical work on molecular structure. Despite the venerability of the subject, there has been a marked resurgence of interest in it since the end of World War II. The ascertainment of absolute configuration, the elucidation of the configuration of numerous important natural products, the stereoselective synthesis of many of these compounds, and the shaping of stereoregular polymers of distinctively useful physical properties are among the many examples of recent progress in the field. Conformational analysis has led to the systematic interpretation of much chemical knowledge as well as to the prediction of new facts. Last but not least, the years since 1940 have seen remarkable advances in the development and availability of physical instrumentation, with the result that such tools as ultraviolet, infrared, and nuclear magnetic resonance spectroscopy and, most recently, optical rotatory dispersion measurements are now making all-important contributions to the solution of stereochemical problems.

In view of this almost explosive development, it is dismaying that there has been no text in the field of stereochemistry for at least 10 years and that students seeking up-to-date knowledge are forced to proceed directly from the somewhat sketchy treatment of stereochemistry in advanced textbooks to the excellent, but necessarily specialized and unavoidably not comprehensive, information available in monographs and reviews. Although stereochemistry is used extensively in textbooks on reaction mechanism, its basic principles are generally taken for granted in such books.

Under the circumstances, the task of writing a new textbook in stereochemistry seemed timely. This book is designed to present the essentials of the subject to the beginning graduate student as well as to the advanced undergraduate who has had some prior acquaintance with both organic and physical chemistry. Perhaps the book may serve as a text in some of the more theoretically oriented courses in advanced organic chemistry, possibly in conjunction with a companion volume on reaction mechanism. In any case, this book should help the student to acquire the basic knowledge in the field of stereochemistry essential for his later progress. In addition, it may be of interest to those workers in the field who received their graduate training prior to 1950 and who wish to find a comprehensive treatment of modern stereochemistry under one cover. In order to make it possible for the student to follow this book regardless of his previous acquaintance with stereochemistry, the material has been developed from quite elementary concepts, especially in the early chapters. The reader who is already grounded in the subject will pass over these sections rapidly soon to encounter (it is hoped) more challenging topics.

This book is a text, not a treatise, and no pretense is made of complete coverage of the field (which would require a work of encyclopedic proportions). The choice of topics and the extent to which they are dealt with have necessarily been arbitrary and are in part a reflection of the author's personal interests. Advance apologies are tendered to any of my colleagues who find their own work slighted or their favored subjects discussed inadequately. In general, the highly selected literature documentation extends through late 1960. However, it has been possible to include a number of salient articles appearing in 1961. Pertinent monographs and reviews are listed at the end of each chapter.

Since the subjects of stereochemistry and reaction mechanisms are so intimately entwined, it needs to be stressed that this is not a text on organic reaction mechanisms—a field in which other up-to-date textbooks are available. However, since the dividing line between mechanisms and stereochemistry is not well marked, it had to be drawn arbitrarily. Thus, the stereochemical aspects of the fundamental organic processes (such as substitution, addition, and elimination) and of the principal organic intermediates (carbonium ions, radicals, carbanions, and carbenes) have been covered, but no effort has been made to deal with the stereochemistry of organic reactions at large. The area of steric effects in organic chemistry—in which there exists an excellent recent treatise—is also not covered as such in any detail.

A book of this magnitude cannot be produced without the help of many people and the availability of numerous review articles and monographs. Among the latter, I want to mention especially the series "Progress in Stereochemistry," one of whose editors, Professor W. Klyne, kindly made available to me the list of chapters of Volume 3 in advance of publication. These chapters have been included among the general references, even though, with one exception, I had not seen them at the time of writing.

I wish to take this opportunity to express my gratitude to numerous colleagues who provided helpful comments on smaller or larger portions of the manuscript. Professors Carl Niemann, San-ichiro Mizushima, J. Sicher, and Carl Djerassi made a number of detailed suggestions concerning Chapters 4, 6, 9, and 14, respectively. Professor Albert W. Burgstahler read and commented on Chapters 1 to 8, and Professors Jerome Berson and Kurt Mislow reviewed and critically commented on the entire work. I want to express my special appreciation to Professor Mislow, whose detailed and constructive criticism extended far beyond the call of duty of a reviewer, and who also helped with a number of references and supplied several manuscripts in advance of publication. My wife, Eva, and my niece, Kay Eliel, helped with the compilation of the subject index. My wife and a number of my coworkers assisted me with the verification of references and the reading of galley proof. Mr. Thomas J. Brett, Jr. helped with the checking of the entire page proof.

Despite all this generous cooperation, I am solely and exclusively responsible for the contents of this book and for any and all of its errors and shortcomings.

The first part of this book was written while I was a National Science Foundation Senior Research Fellow, and thanks are due to the Foundation for its financial support and to the members of the Division of Chemistry and Chemical Engineering of the California Institute of Technology for their inspiration and hospitality during the fall semester of 1958.

Ernest L. Eliel

CONTENTS

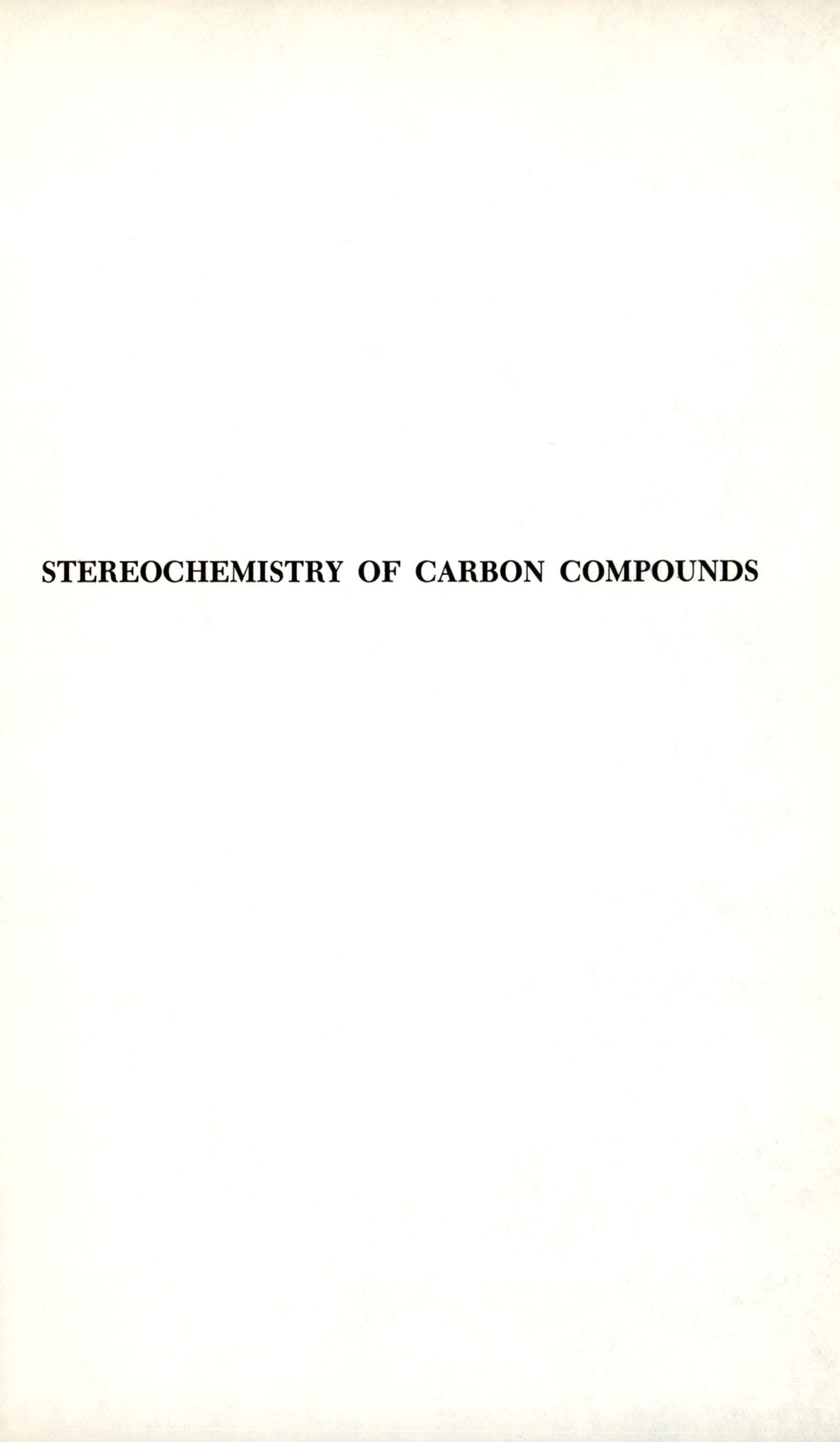

STEREOCHEMISTRY OF CARBON COMPOUNDS

Chapter 1

INTRODUCTION

1-1. Scope

Until a few years ago, stereochemistry was concerned almost exclusively with the subject of stereoisomerism. "Stereoisomers" are isomeric compounds of identical structure† but differing in the arrangement of the atoms in three-dimensional space. (The concept of spatial arrangement, or "configuration," will be dealt with in more detail in Chap. 5.) Today, the scope of stereochemistry extends considerably beyond the static description of molecular geometry and of the physical properties related to such geometry; stereochemistry is concerned also with the relationships in space between the different atoms and groups in a molecule during chemical reactions and the way in which chemical equilibria and rates of reaction are affected by those spatial relationships. While this aspect of stereochemistry borders on the study of reaction mechanism, we shall not deal here with reaction mechanisms as such. For a consideration of these, the reader is referred to other advanced texts.[11-13]

Despite the increasing importance of the dynamic aspects of stereochemistry, it is well to start the discussion by a consideration of static molecules. Since our knowledge of molecular geometry is closely linked to the development of structural organic theory, on the one hand, and to an understanding of the rotation of the plane of polarized light by organic molecules on the other, it is appropriate to consider briefly the history of these ideas.

1-2. History

Ordinary light may be considered as an electromagnetic vibration of a range of different wavelengths vibrating in many different planes at right angles to the direction of propagation of the light ray.‡ Monochromatic light, such as that emitted by a sodium lamp, is of discrete wavelength but

Note: All references above 9 are listed in General References at the end of the chapter.

† The structure of a molecule is completely defined by a statement of the number and kind of atoms in the molecule and of the linkages between them. The subject of structural isomerism is dealt with elegantly and thoroughly by Wheland[10] (chap. 2).

‡ The wave nature of light was recognized by Hooke in 1665 and by Huygens in 1678. The electromagnetic nature of the vibration was established much later (by Maxwell in 1873) but is not essential to the present argument.

still vibrates in an infinite number of planes. In 1808, the French physicist Malus discovered[1] that light reflected from opaque or transparent bodies at a certain angle is endowed with special properties which may be ascribed to the fact that the vibrations are all in one plane, called the "plane of polarization." Light of this kind is said to be "polarized." Malus also found that the two rays produced by the phenomenon of double refraction in Iceland spar (crystalline calcium carbonate) are polarized at planes perpendicular to each other. In fact, one of the most convenient ways of producing polarized light is to pass ordinary light through a prism constructed by cementing together, by means of Canada balsam, two pieces of Iceland spar cut at specified angles. Such a device, called a "Nicol prism" after its discoverer (Nicol, 1828), allows only one of the polarized rays to pass through (the other is reflected). The polarization of the ray of monochromatic light transmitted through a Nicol prism is readily detected by viewing it through a second Nicol prism: If the plane of polarization of the second prism is parallel to that of the first, it transmits the polarized ray with undiminished intensity, but if the plane of polarization of the second Nicol is perpendicular to that of the first (i.e., the Nicol prisms are "crossed"), the ray polarized by the first prism fails to pass through the second.

In 1813 the French physicist Biot, following an earlier observation[2] by his colleague Arago, discovered that a quartz plate cut at right angles to its crystal axis rotates the plane of a ray of polarized light through an angle proportional to the thickness of the plate. Some quartz crystals turn the plane of polarization to the right, others to the left. Two years later,[3] Biot laid the foundation of organic stereochemistry when he found that a similar rotation of the plane of polarized light is produced by certain organic liquids, such as turpentine, as well as solutions of certain organic compounds, such as sugar, camphor, tartaric acid. Biot recognized a difference between the rotation produced by quartz and that produced by organic materials: The former is a property of the crystal and depends on the direction in which the crystal is viewed, whereas the latter is a property of the individual molecules and may be observed even in solution or in the liquid or gaseous state where the molecules are arranged in random fashion.

In 1801, the French mineralogist Hauÿ had noticed that certain quartz crystals exhibit the phenomenon of hemihedrism. For the present purpose, hemihedrism may be defined as the absence of a plane, center, or alternating axis of symmetry (see Chap. 2) in the crystal. In crystals presenting hemihedrism, there are faces that make such crystals non-superimposable with their mirror images (Fig. 1-1). In 1821, Sir John Herschel, British astronomer, observed[4] that all quartz crystals having the odd faces inclined in one direction rotate the plane of polarized light in one and the same direction, whereas the mirror-image crystals, whose odd faces are inclined in the opposite direction, also rotate the plane of polarized light but in the opposite direction.

It was, however, left to the genius of Louis Pasteur to extend this correla-

[1] E. L. Malus, *Mém. soc. d'Arcueil,* **2,** 143 (1809).

[2] D. F. Arago, *Mém. classe sciences math. phys. de l'Institut Imp. France,* **12,** 115 (1811).

[3] J. B. Biot, *Bull. soc. philomath. Paris,* 190 (1815); 125 (1816); *Mém. acad. roy. sci. inst. France,* **2,** 41, 114 (1817).

[4] J. F. W. Herschel, *Trans. Cambridge Phil. Soc.,* **1,** 43 (1821).

tion from the realm of crystals to the realm of molecules. In 1860, following some extensive observations (1848–1853) on optically active compounds which will be presented elsewhere in this book, Pasteur realized that optical activity is caused by an asymmetric grouping of the atoms in the optically active molecule and that molecules of the same substance rotating the plane of polarized light to the right and to the left are related to each other as is an object to its mirror image.[5]

About the same time, in 1858, the German chemist Kekulé had laid the foundation of modern structural organic chemistry when he recognized that carbon has a valence of 4 in all organic compounds and that complex structures can be built up by linking together carbon (and other polyvalent atoms) in chains.[6] It was not long until the concept of the quadrivalence of carbon and the suggestion that optical activity is caused by molecular asymmetry were combined and a new idea emerged, namely, that the four valences of carbon are directed in three-dimensional space toward the corners of a tetrahedron. As happens so often with scientific ideas for which the time is

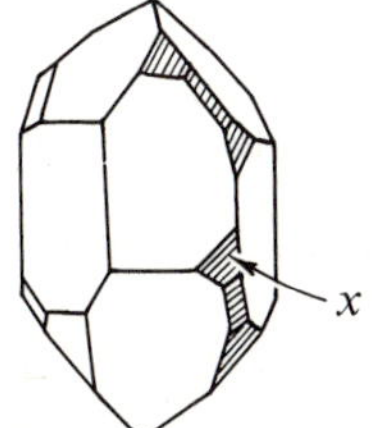

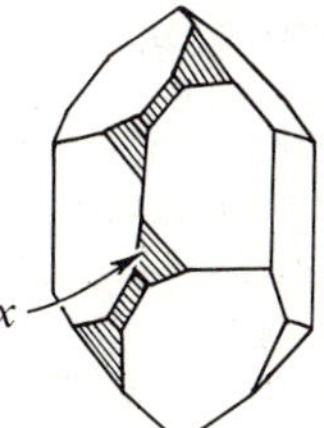

Fig. 1-1. Hemihedral quartz crystals. X: asymmetric face. (*From L. F. Fieser and M. Fieser, "Organic Chemistry," 3d ed., Reinhold Publishing Corporation, New York,* 1956. *By permission of the publishers.*)

ripe, the concept of the tetrahedral carbon originated simultaneously (in 1874) in the minds of two chemists, van't Hoff,[7] who was then working for his Ph.D. degree† in Utrecht (Holland), and Le Bel,[8] working in Paris.‡ A com-

† However, for his Ph.D. thesis van't Hoff presented a rather routine investigation on cyanoacetic and malonic acids. No doubt he feared that his revolutionary work on the relation of optical activity to chemical constitution might not be received well by the examiners. That this fear may have been justified is evidenced by the acrid reception which van't Hoff's theory received at the hands of at least one of the older German chemists, Hermann Kolbe. (For a translation of Kolbe's rather devastating critique, see Ref. 10, p. 197.) Strange as it may seem, Kolbe's criticism may actually have furthered van't Hoff's ideas in that it brought the work of the young investigator to the attention of his older colleagues!

‡ Le Bel and van't Hoff had been working side by side in Wurtz's laboratory in Paris during the academic year preceding publication of their pioneering work. Nevertheless, it appears that they developed the concept of tetrahedral carbon independently and did not even discuss it with each other; see the van't Hoff Memorial Lecture by J. Walker, *J. Chem. Soc.,* **103,** 1127 (1913).

---

[5] L. Pasteur, two lectures delivered before the Société Chimique de Paris, Jan. 20 and Feb. 3, 1860; cf. Ref. 14.

[6] A. Kekulé, *Ann.,* **106,** 154 (1858).

[7] J. H. van't Hoff, *Bull. soc. chim. France,* [2]**23,** 295 (1875); cf. Ref. 14. (The original version appeared in Dutch in 1874.)

[8] J. A. Le Bel, *Bull. soc. chim. France,* [2]**22,** 337 (1874); cf. Ref. 14.

mon (although, as will be seen later, not a necessary) cause for the existence of optical activity is non-identity of the four atoms or groups around a single carbon atom in the active molecule. In such a case, two different arrangements may exist, as shown in Fig. 1-2, which are non-superimposable but

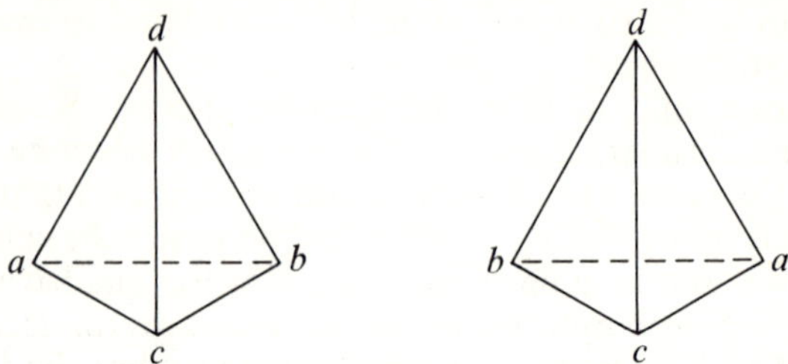

Fig. 1-2. Representation of isomeric arrangements of Cabcd.

which are mirror images of each other. According to van't Hoff and Le Bel, one of these arrangements corresponds to the isomer that rotates the plane of polarized light in one direction, whereas the isomer of opposite rotation corresponds to the other arrangement. When two or more of the atoms or groups attached to the tetrahedra in Fig. 1-2 become identical, the two forms become superimposable and are no longer distinct. Molecules of this type do not rotate the plane of polarized light.

The idea that the valencies of carbon are directed toward the corners of a tetrahedron, originally suggested to explain optical activity, is now firmly supported both by more direct physical measurements (e.g., electron diffraction) and by theoretical quantum-mechanical considerations.

1-3. Optical and Geometrical Isomerism

We thus recognize the existence and cause of one type of stereoisomerism (i.e., isomerism in space) which manifests itself by the rotation that the isomers impart to the plane of polarized light and is caused in many (though not all) instances by the attachment of four different atoms or groups to at least one of the carbon atoms in the molecules, as shown in Fig. 1-2. This type of isomerism, because of its manner of manifestation, is called "optical isomerism." There is another type of spatial isomerism, also recognized by van't Hoff and Le Bel but put on a firm experimental basis mainly through later work by the German chemist Wislicenus,[9] namely, the isomerism caused by different arrangements of groups around ethylenic double bonds (Fig. 1-3).

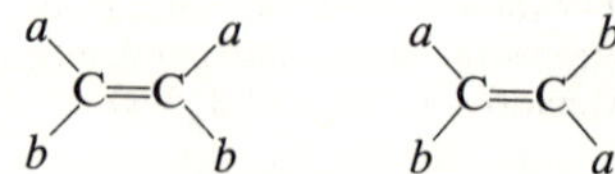

Fig. 1-3. Geometrical isomerism in substituted ethylenes.

[9] J. Wislicenus, *Abhandl. sächs. Ges. Wiss.*, **14,** I (1887); cf. Ref. 14 and *Chem. Zentr.*, **58,** 1005 (1887).

A similar type of isomerism is found in ring compounds, the ring taking the place of the rigid double bond (Fig. 1-4). This kind of isomerism has been called "geometric (or geometrical) isomerism."

Whereas the type of geometrical isomerism occurring in ethylenes is not normally associated with optical activity,† since the plane of the double bond

Fig. 1-4. Geometrical isomerism in ring compounds.

(plane of the paper) is also a plane of symmetry, the type of geometrical isomerism observed in cyclic compounds is often associated with optical activity. This is seen by a comparison of *cis*- and *trans*-crotonic acid (Fig. 1-5) on one hand and *cis*- and *trans*-2-methylcyclopropanecarboxylic acid (Fig. 1-5) on the other. The crotonic acids are geometrical isomers devoid of optical activity. The 2-methylcyclopropanecarboxylic acids are also geometrical isomers, but in addition they have non-superimposable mirror images, and therefore both the cis and the trans compounds exist as a pair of isomers that rotate the plane of polarized light to an equal extent in opposite directions.

It is evident, then, that the division of stereoisomerism into optical and geometrical isomerism is not a clear-cut one and that the two types of isomerism do, in fact, merge in certain cases. Nevertheless, for the purpose of organization, we shall, in this book, deal first with optical isomerism, then with the dual type of isomerism found in ring compounds, and lastly with purely geometrical isomerism.

H_3C CO_2H / C=C / H H — cis

H_3C $C(H_2)$ CO_2H / C—C / H H — cis

HO_2C $C(H_2)$ CH_3 / C—C / H H — cis (mirror image)

H CO_2H / C=C / H_3C H — trans

H $C(H_2)$ CO_2H / C—C / H_3C H — trans

HO_2C $C(H_2)$ H / C—C / H CH_3 — trans (mirror image)

Fig. 1-5. Crotonic acids and 2-methylcyclopropanecarboxylic acids.

† Unless there is some other site in the molecule giving rise to optical isomerism as well, e.g., in $CHCl{=}CH{-}\overset{*}{C}H(CH_3)C_2H_5$, where there is the possibility of geometrical isomerism about the double bond and of optical isomerism involving the starred carbon.

General References

[10] G. W. Wheland, "Advanced Organic Chemistry," 3d ed., John Wiley & Sons, Inc., New York, 1960, chaps. 6–8.

[11] J. Hine, "Physical Organic Chemistry," 2d ed., McGraw-Hill Book Company, Inc., New York, 1962.

[12] E. S. Gould, "Mechanism and Structure in Organic Chemistry," Holt, Rinehart and Winston, Inc., New York, 1959.

[13] C. K. Ingold, "Structure and Mechanism in Organic Chemistry," Cornell University Press, Ithaca, N.Y., 1953.

[14] G. M. Richardson, ed., "The Foundations of Stereochemistry," American Book Company, New York, 1901.

[15] L. F. Fieser and M. Fieser, "Advanced Organic Chemistry," Reinhold Publishing Corporation, New York, 1961, chap. 3.

Chapter 2

OPTICAL ISOMERISM

2-1. Polarimetry

Optical isomerism manifests itself by the rotation that certain molecules impart to the plane of polarized light when in the gaseous, liquid, or molten state or in solution.† This rotation is observed and measured by a rather simple instrument, known as a polarimeter. This is not the place to deal in any detail with the experimental aspects of polarimetry.[4] Suffice it to say that polarimetric readings are obtained by matching light intensities in the field of the instrument; if this is done visually using a sodium lamp ($\lambda = 589$ mμ) as a light source, observations are good to about $\pm 0.005°$. Attachments are available to permit photoelectric matching of the polarimeter; this speeds up the procedure and also makes possible accurate determinations of optical rotation at wavelengths longer and shorter than that of the sodium D line where the human eye becomes less sensitive. When photoelectric recording devices are combined with monochromators attached to a polychromatic light source, a continuous reading of optical rotation over a wide range of wavelengths becomes possible. This technique will be discussed in Chap. 14.

The observed angle of rotation of an optically active liquid, gas, or solution is denoted by the symbol α. In the use of a polarimeter, it is important to realize that no immediate distinction can be made between rotations of $\alpha \pm 180n°$ (n any integer), for if the plane of polarization is rotated in the field by 180°, the new plane will appear to coincide with the old one. Thus, for example, no distinction would appear between rotations of $+50°$, $+230°$, $+410°$, or $-130°$. In order to make the distinction, it is necessary to measure the rotation at at least one other concentration. Since optical rotation is proportional to concentration (see below), if solutions of the above rotations were diluted to one-tenth of their original concentrations, their rotations would become $+5°$, $+23°$, $+41°$, and $-13°$, values which are all clearly distinct. Readings taken at two different concentrations almost always determine α unequivocally. For a pure liquid, the required dilution may be effected with racemic material.

Note: All references above 2 are listed in General References at the end of the chapter.

† As already mentioned in Chap. 1, certain solids, such as quartz, may rotate the plane of polarized light as a result of their peculiar crystal structure. This type of rotation, which is unrelated to molecular properties, will not be considered further in this book.

Factors that affect the magnitude of optical rotation, in addition to the nature of the sample, are sample thickness (i.e., cell length), sample concentration (or density, in the case of a pure liquid), solvent, temperature, and wavelength. The relation of α to cell length (l) is linear and that to concentration (c) is approximately linear, so that $\alpha = [\alpha] \cdot l \cdot c$, where $[\alpha]$ is a proportionality constant depending on the nature of the sample, temperature, solvent, and wavelength of light. When l is measured in decimeters and c in grams per milliliter, $[\alpha]$ is called "specific rotation":

$$[\alpha] = \frac{\alpha}{l\,(\text{dm.})\; c\,(\text{g./ml.})}$$

The dependence on wavelength and temperature is usually indicated by subscripts and superscripts, respectively; thus $[\alpha]_D^{25}$ means the specific rotation of a substance at 25°C. measured at the wavelength of the sodium D line.† Solvent dependence, as well as a minor dependence of $[\alpha]$ on concentration (which is not entirely taken into account by the concentration term in the formula), is denoted by information in parentheses appended to the value of the specific rotation; thus $[\alpha]_{546}^{20}$ $-76.3 \pm 0.3°$ ($c = 5.77$ g./100 ml., ethanol) denotes a specific rotation measured with light of wavelength 546 mμ at 20° in absolute ethanol at a concentration of 5.77 g./100 ml.‡ When the rotation of a pure liquid is cited, no concentration term is, of course, called for, since the density is constant at the given temperature; the word "neat" is often used in the parentheses to specify that the measurement refers to a pure liquid. Thus $[\alpha]_D^{25}$ $+40°$ (neat) refers to the specific rotation of a pure liquid measured with sodium D light at 25°C. If the density of this particular liquid is 1.10 at 25°, an alternative way of reporting this rotation would be by the observed rotation: α_D^{25} $+44°$ (neat, $l = 1$ dm.). Note that, since observed rotation is dependent on cell length, the cell length must now be stated also.

The relation between specific rotation and structure will be discussed at some length in Chap. 14. Suffice it to say at this point that some of the changes of specific rotation associated with temperature, solvent, and concentration changes are related to changes in intermolecular hydrogen bonding and/or the degree of association and dissociation. For example, a sample of atrolactic acid, $C_6H_5C(CH_3)OHCO_2H$, which is dextrorotatory (i.e., rotates the plane of polarized light to the right) in benzene is levorotatory (i.e., rotates to the left) in ether.[1] Presumably in benzene there are strong intermolecular association forces (which would be expected to vary with concentration as well as temperature) whereas in ether there may be strong hydrogen bonding between the acidic hydrogen of the acid and the ether oxygen of the solvent.

Since optical rotatory power is a property of molecules, if two substances

† Because of the temperature dependence of both c and α, most polarimeter cells are constructed so that they can be readily thermostated.

‡ The concentration of solutions in polarimetry is usually expressed in grams per 100 milliliters even though the formula as given above requires it to be expressed in grams per milliliter. Often the formula above is reserved for pure liquids and a formula $[\alpha] = 100\alpha/l \cdot c'$ is employed for solutions, c' now being expressed in grams per 100 milliliters.

[1] E. Eliel, unpublished observation.

have unequal molecular weights but are alike with respect to the power of rotating the plane of polarized light, the substance of smaller molecular weight has the larger specific rotation, simply because it has more molecules per unit weight. (Specific rotation, it may be recalled, is the rotation produced by 1 g. of material in 1 ml. of liquid in a 1-dm. tube.) In order to compensate for the effect of differing molecular weights (which, from the theoretical point of view, is quite unimportant) one defines a new term "molecular rotation" obtained by multiplying specific rotation by molecular weight (thus compensating for the above-indicated disadvantage of larger molecules) and then dividing by 100. (The division by 100 is included arbitrarily in the definition of molecular rotation in order to keep its numerical values manageably small.) Molecular rotation is expressed by the symbols $[M]$ or $[\phi]$. Thus

$$[M] = [\phi] = \frac{[\alpha] \cdot \text{mol. wt.}}{100} = \frac{\alpha}{l\,(\text{dm.})\; c\,(\text{moles}/100\ \text{ml.})}$$

2-2. Molecular Dissymmetry

Van't Hoff and Le Bel related the phenomenon of optical rotation to the presence of asymmetrically substituted carbon atoms (henceforth called, for short, "asymmetric atoms") in the molecule.† They realized, however, that there could exist optically active compounds having no asymmetric atoms (cf. Chaps. 6, 7, and 11) and that in some instances compounds having two or more asymmetric carbon atoms cannot be obtained optically active (Sec. 3-3). Thus it is desirable to go back to the earlier idea of Pasteur, namely, that optical activity is a consequence of *molecular* dissymmetry. In fact, the necessary and sufficient condition for a molecule to show optical activity is that such a molecule not be superimposable with its mirror image.‡ A molecule superimposable with its mirror image cannot be optically active, whereas any molecule not superimposable with its mirror image (allowing for internal rotation) necessarily shows optical activity.¶

† Other kinds of atoms, such as silicon, nitrogen, phosphorus, and sulfur, may give rise to asymmetry. For example, quaternary ammonium salts of the type $RR'R''R'''N^+X^-$ and analogous phosphonium salts, sulfoxides, $RR'SO$, and trialkylsilanes, $RR'R''SiH$, have been obtained optically active. Compounds of this type will not be treated explicitly in this book; for further information on some of them the reader is referred to pp. 400–443 in Ref. 5. The stereochemistry of inorganic coordination compounds has been summarized by F. Basolo, B. P. Block, and Th. D. O'Brien in chaps. 8–10 of J. C. Bailar, ed., "Chemistry of the Coordination Compounds," Reinhold Publishing Corporation, New York, 1956; by R. S. Nyholm in W. Klyne, ed., "Progress in Stereochemistry," vol. 1, Academic Press, Inc., New York, 1954, chap. 9; and by R. J. Gillespie and R. S. Nyholm in W. Klyne and P. B. D. de la Mare, eds., *ibid.,* vol. 2, 1958, chap. 8.

‡ To make the condition sufficient, it may be necessary in some cases to allow the molecule to undergo internal rotation about single bonds before or after the imaginary reflection; cf. Ref. 2 and the discussion at the end of this section.

¶ Some caution is required in translating this statement from the molecular realm to the macroscopic realm of the observable. A macroscopic assembly of large numbers of molecules is not necessarily optically active, even though each one of the molecules is. For it may happen that, in such an assembly, there are approximately equal numbers of molecules rotating the plane of polarized light in one direction and in the other, and the result (if the rotations are of equal magnitude) is that no net over-all rotation is observed. Such an assembly is known as a "racemic modification." This matter is of paramount importance, inasmuch as optical rotation is always observed in an assembly of large numbers of molecules, never in an individual molecule or a few molecules. Chapter 4 deals extensively with this question.

According to group theory, the symmetry properties of a given molecule tell us whether or not the molecule is superimposable with its mirror image. A molecule that has a plane of symmetry, a center of symmetry, or an alternating axis of symmetry† is superimposable with its mirror image, but a molecule that has no element of symmetry, or one that has only a simple axis

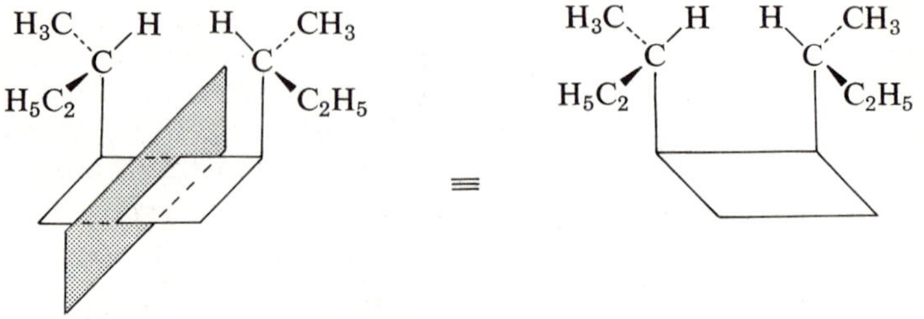

Fig. 2-1. Molecule with plane of symmetry.

of symmetry (and none of the above-mentioned three elements), is not superimposable with its mirror image. It follows that there are usually two ways of deciding whether a given molecule is optically active. One is to construct a model of the molecule and then to build a mirror image of this model; if the model and its image are superimposable, the molecule is not optically active, otherwise it is. Or one may look for symmetry elements in the molecule; if it has a plane (cf. Fig. 2-1), center (cf. Fig. 2-2), or alternating axis (cf. Fig. 2-3) of symmetry, it is not active, but if it has none of these elements, it is active, even though it may have a simple axis of symmetry (Fig. 2-4).‡

The examples shown in Figs. 2-1 to 2-4 are deliberately chosen from molecules having two recurrent structural elements, namely, a cyclobutane ring with two or four *sec*-butyl, $CH_3—\overset{|}{C}H—C_2H_5$, substituents. Taken individually, the *sec*-butyl substituent has an asymmetric carbon. Nevertheless, of the four structures shown in Figs. 2-1 to 2-4, only one (Fig. 2-4) shows optical

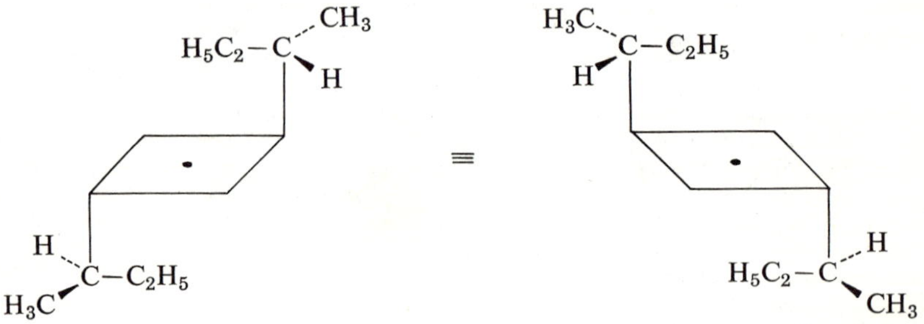

Fig. 2-2. Molecule with center of symmetry.

† The alternating axis of symmetry is actually sufficient inasmuch as a plane of symmetry is equivalent to a onefold alternating axis and a center of symmetry is equivalent to a twofold alternating axis.

‡ An exception is pointed out at the end of this section.

activity. The molecule shown on the left in Fig. 2-1 has a plane of symmetry† bisecting the ring in the manner shown. Its mirror image (Fig. 2-1, right) is superimposable with the original molecule by a simple translation to the left. The molecule shown in Fig. 2-2 has a center or point of symmetry‡ at the center of the ring. Its mirror image (Fig. 2-2, right) may be superimposed with

Fig. 2-3. Molecule with alternating axis of symmetry (fourfold).

the original molecule by turning the ring upside down. The molecule in Fig. 2-3 has a fourfold alternating axis of symmetry,¶ passing through the center of the ring and at right angles to it. Its mirror image (Fig. 2-3, right) is superimposable with the original molecule by turning the ring upside down and rotating it 90° around the axis. Finally, the molecule in Fig. 2-4 has a twofold (simple) axis of symmetry.§ It is not superimposable with its mirror image (Fig. 2-4, right), and both the molecules shown in Fig. 2-4 are optically active.

Molecules of the type shown in Fig. 2-4 which are non-superimposable mirror images are called "enantiomorphs" or "enantiomers" (also known as

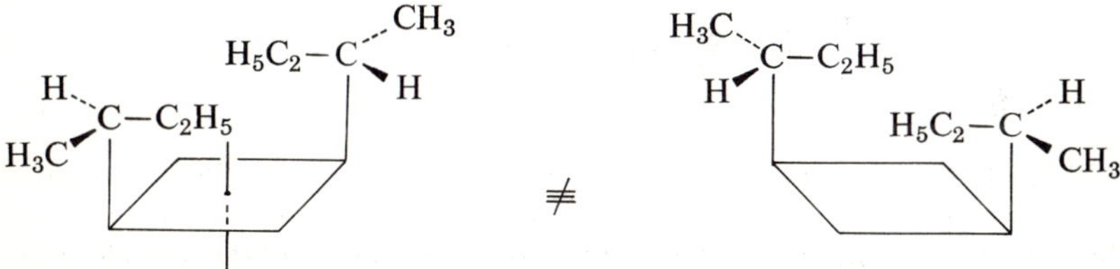

Fig. 2-4. Molecule with simple axis of symmetry (twofold) (optically active).

† A plane of symmetry is a plane such that, if a line from any element on one side of the plane is drawn perpendicular to the plane and the line extended to an equal distance on the other side of the plane, an identical element will be found at the end of the line. Putting it more simply, one half of the molecule is the mirror image of the other half, the plane of symmetry being the plane of the (imaginary) mirror.

‡ A center of symmetry is a point such that, if a line is drawn from any element to this point and then extended an equal distance beyond the point, another identical element will be found at the end of the line.

¶ An n-fold alternating axis of symmetry is an axis such that, when the structure possessing this axis is rotated around the axis by an angle of $2\pi/n$ and then reflected across a plane at right angles to the axis, another, identical structure results.

§ An n-fold axis of symmetry is an axis such that, when a structure possessing this axis is rotated by an angle of $2\pi/n$ around the axis, another, identical structure results.

"optical antipodes" or "antimers"); they rotate the plane of polarized light in opposite directions but have the same absolute value of the specific rotation.

The reader should construct models of the molecules shown in Figs. 2-1 to 2-4 and convince himself both of the presence of the indicated elements of symmetry and of the superimposability (or non-superimposability) of the appropriate models with their respective mirror images.

Wheland[6] has drawn attention to the fact that, since a molecule may be optically active even though it has a simple axis of symmetry, it is not correct to say that optically active molecules lack the elements of symmetry. Therefore, the terms "optically active" and "asymmetric" as applied to molecules should not properly be used in the same context. A new term is called for to denote the absence of an alternating (but not necessarily of a simple) axis of symmetry; this term is "dissymmetric." A molecule that is dissymmetric lacks an alternating axis of symmetry and is usually optically active (see, however, below). A dissymmetric (and therefore usually optically active) molecule may or may not be asymmetric (i.e., lack a simple axis of symmetry). A symmetric molecule is one which possesses an alternating axis of symmetry and is therefore optically inactive. Table 2-1 summarizes these relationships.

Table 2-1
Symmetry Designations

Term	Alternating axis	Simple axis	Optical activity
Symmetric	Present[a]	May or may not be present	Inactive
Dissymmetric	Absent	May or may not be present	Usually active
Asymmetric	Absent	Absent	Usually active

[a] Onefold alternating axis corresponds to plane of symmetry; twofold alternating axis corresponds to center of symmetry.

Although it is generally true that dissymmetric molecules are optically active, a molecule, (*dextro*)-menthyl (*levo*)-menthyl 2,6,2′,6′-tetranitro-4,4′-diphenate (Fig. 2-5), has been synthesized[2a] which is *inactive though devoid of any symmetry element.* [The molecule cannot exist in the form in which the two phenyl rings are coplanar because of the steric interference of the bulky nitro groups in this form (cf. Chap. 6). This excluded form is the only arrangement of the molecule which possesses a plane—or other element—of symmetry.] The lack of activity of this molecule is attributed[2b] to the fact that the molecule and its mirror image, though not actually superimposable, can be readily made superimposable by rotation around the bonds linking the carboxylate groups to the rings. Other dissymmetric but inactive molecules of this type might, in principle, be found.

[2] (*a*) K. Mislow and R. Bolstad, *J. Am. Chem. Soc.*, **77**, 6712 (1955). (*b*) K. Mislow, *Science*, **120**, 232 (1954); *Trans. N.Y. Acad. Sci.*, **19**, 298 (1957).

2-3. Molecular Models

In trying to visualize the symmetry properties of the molecules depicted in Figs. 2-1 to 2-5, the reader may have felt the need for three-dimensional representations of these molecules. The only completely adequate way to visualize molecules in space is by means of three-dimensional models. Since the need for such models will recur throughout this book and, in fact, whenever one considers a problem in stereochemistry, a brief description of the available molecular models is called for.

There are essentially three types of models: those that merely represent the geometric relationship of the atoms in the molecule without being accurate as to interatomic distances; those that show interatomic distances correctly to scale without, however, showing the sizes of the atoms themselves; and those that are to scale with respect to both atomic and interatomic dimensions and therefore allow one to construct a true scale model of the molecule.

The first type of models—those showing geometric relationships only—are by far the least expensive. They are usually models of the ball-and-stick type (affectionately called "tinkertoys" by some chemists). They are very satisfac-

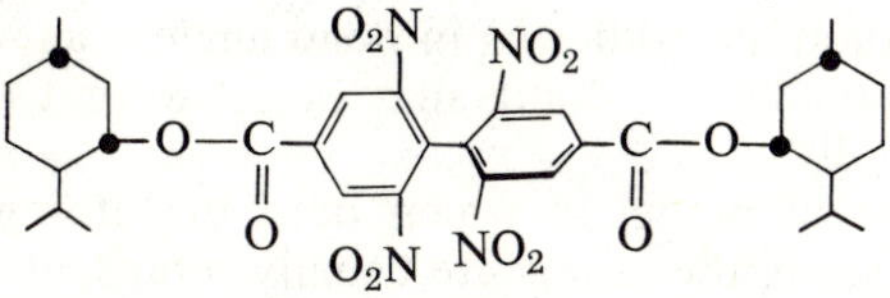

O_2N NO_2 O—C C—O O O O_2N NO_2

Fig. 2-5. (*dextro*)-Menthyl (*levo*)-menthyl 2,6,2′,6′-tetranitro-4,4′-diphenate.

tory for depicting over-all molecular geometry as well as symmetry relationships and allow one to decide the number and kind of stereoisomers that may exist for a given structure. No practicing organic chemist should be without a set or two of these models. Sets are available using colored wooden spheres for atoms and wooden pegs for bonds† (with stout springs available for strained structures, such as the cyclobutane rings in Figs. 2-1 to 2-4) or using a smaller size of colored wooden spheres for atoms and stout springs for all bonds.‡ The latter sets are more compact but make a less rigid and therefore less easily manipulated model.

Where it is desirable to measure interatomic distances in a model, the model must be to scale as far as bond distances are concerned. The "Barton models," "Dreiding models," and "Cenco-Petersen models" are of this type. In the Barton models¶ (scale 1 A. ≡ 10 cm.) the carbon atoms are metal tetrahedra or (for doubly bonded atoms) triangular prisms and the bonds are rigid

† Available in the United States from E. H. Sargent and Co., 4647 W. Foster Avenue, Chicago 30, Ill.

‡ Available in the United States from Central Scientific Co., 1700 Irving Park Road, Chicago 13, Ill.

¶ Available in the United States from Wilkens-Anderson Co., 4525 West Division Street, Chicago 51, Ill. The "bonds" in these models are rather flexible and tend to snap at the constricted (grooved) part.

metal rods which are grooved near the ends, the grooves being held in a predetermined place inside the atom models by means of screws in the latter. A less expensive wooden version of the Barton models is also available;† in these models the interatomic distances are not accurately to scale and the models are not suitable for quantitative measurements, but they are very good for class demonstrations because they are much less cluttered than the ball-and-stick models and show angular relationships more clearly.

In the Dreiding type‡ the atoms and bonds combined are constructed of four metal rods, two solid and two hollow, joined together at the center of the carbon atom at the correct tetrahedral angle. To join two atoms, the solid bond of one is inserted into the hollow bond of the other. The length of the bonds and the position of the stops inside the hollow bonds are such that the C—H and C—C distances are represented to scale by the unattached bonds and by two joined bonds, respectively. A carbon atom consisting of three coplanar rods at 120° angles to each other is used for double-bonded structures. Some hetero atoms are also available. These models are excellent for desk manipulation, although too small for demonstration in large classes (1 A. ≡ 2.5 cm.).

The Cenco-Petersen molecular models¶ are sturdy, large-scale models (1 A. ≡ 5 cm.) in which the bonds are screwed into the spheres representing atoms. A variety of atoms (differing in bond angles), as well as bonds of different lengths, are available. Bonds may be set so that they rotate freely or they may be set rigidly.

In the third type of model in which both the atomic and the bonding dimensions are to scale, the atoms are usually represented as spheres sawed off at right angles to the direction of the bonds and equipped with snap locks that can be joined by snap fasteners. These models are particularly desirable when one wishes to know whether two atoms in a molecule can approach each other very closely, whether there is strain in a molecule due to overcrowding of certain atoms, etc. Both a small-scale version ("Fisher-Hirschfelder-Taylor models," 1 A. ≡ 1 cm.)§ and a somewhat larger-size version ("Stuart-Briegleb models," 1 A. ≡ 1.5 cm.)‖ are available. The so-called "Catalin models"# are a variation of the Fisher-Hirschfelder-Taylor models in which the atoms (made of a phenolic resin) are joined by hard-rubber pegs instead of metal snaps. This imparts a somewhat greater flexibility to the models and allows the representation of moderately strained structures. Another way of increasing the flexibility of the models is used in the "Courtauld atomic models"†† (scale 1 A. ≡ 0.8 in.). In these models the snap-fastener

† From the Bennett Lumber and Manufacturing Co., Zeeland, Mich., U.S.A.

‡ Available in the United States from G.M. Instrument Co., 511 S. Prairie Street, Greenville, Ill.

¶ Available in the United States from Central Scientific Co., 1700 Irving Park Road, Chicago 13, Ill.

§ Available in the United States from Fisher Scientific Co., 717 Forbes Ave., Pittsburgh 19, Pa.

‖ Available in the United States from Arthur S. LaPine and Co., 6001 South Knox Avenue, Chicago 29, Ill.

\# Available in the United States from Arthur F. Smith Co., 311 Alexander Street, Rochester, N.Y.

†† Available in the United States from The Ealing Corporation, 33 University Road, Cambridge, Mass.

socket is equipped with a spring which allows some lateral motion, and the atoms are separated by a rubber spacing collar. Another type of highly flexible models is the "Godfrey models"† (scale 1 A. ≡ 1.65 cm.) in which pliability is achieved by making the atoms out of flexible polyvinyl chloride and the fasteners (bonds) out of polyethylene. Even cyclopropane may be represented by the Godfrey models. In the Courtauld and the Godfrey models the radii of the spherical surfaces of the atomic models are proportional to the actual van der Waals radii of the atoms, whereas in the Fisher-Hirschfelder-Taylor, Stuart-Briegleb, and Catalin models the radius of the atomic spheres is about 80% of the van der Waals radius. For example, in the Catalin models where the scale is 1 cm. ≡ 1 A. in the bond lengths, the radius of the nitrogen-atom model is 1.20 cm., corresponding to a van der Waals radius of 1.5 Å. Thus, the apparent crowding in these models is somewhat less severe than in the actual molecules. This is to compensate for the fact that the bond angles in the models (other than the Courtauld and Godfrey models) are rigid whereas the bond angles in the actual molecules can be deformed relatively readily.

Detailed, illustrated literature about these types of models may be obtained by writing to the suppliers.

General References

[3] F. Ebel, Die Tetraedertheorie, in K. Freudenberg, ed., "Stereochemie," Franz Deuticke, Leipzig, 1932, pp. 525–552.

[4] W. Heller, Polarimetry, in A. Weissberger, ed., "Technique of Organic Chemistry," vol. I, pt. 3, 3d ed., Interscience Publishers, Inc., New York, 1960, pp. 2147–2333.

[5] R. L. Shriner and R. Adams, Optical Isomerism, in H. Gilman, ed., "Organic Chemistry," vol. I, 2d ed., John Wiley & Sons, Inc., New York, 1943, pp. 281–304.

[6] G. W. Wheland, "Advanced Organic Chemistry," 3d ed., John Wiley & Sons, Inc., New York, 1960, chap. 6

[7] R. L. Bent, "Stereoisomerism," *J. Chem. Educ.*, **30**, 328 (1953).

† Available from the Will Corporation, Rochester, N.Y., U.S.A.

Chapter 3

OPTICAL ISOMERISM DUE TO ASYMMETRIC CARBON ATOMS

3-1. Compounds with One Asymmetric Carbon Atom

The simplest source of dissymmetry in an organic molecule is a carbon atom to which four other atoms or groups, each different from the other, are attached. Denoting these atoms or groups by a, b, c, and d, the general symbolism for such a carbon atom becomes Cabcd. One of the oldest known examples is lactic acid, $CH_3\overset{*}{C}HOHCO_2H$, in which the starred carbon atom is asymmetric, being substituted by a hydrogen atom, hydroxyl group, methyl group, and carboxyl group. It has long been known that the lactic acid produced in the living muscle when it performs work ("sarcolactic acid") is dextrorotatory, whereas part of the lactic acid formed by fermentation of lactose in the souring of milk ("fermentation lactic acid") is levorotatory. Synthetic lactic acid, as will be shown in Chap. 4, is a mixture of equal amounts of the dextrorotatory and levorotatory forms and does not affect the plane of polarized light.

Molecular models of (+)-lactic acid (dextrorotatory) and (−)-lactic acid (levorotatory) are shown in Figs. 3-1 and 3-2, along with three-dimensional line drawings of the compounds.† From the information given so far, it cannot be deduced which model corresponds to the dextrorotatory and which to the levorotatory form, but the representations in Figs. 3-1 and 3-2 are correct in this respect. The problem of which arrangement of the atoms in space corresponds to which isomer will be taken up in Chap. 5 under the heading of "configuration."

Since paper and blackboards are two-dimensional, it is inconvenient to represent molecules in their three-dimensional reality, and two-dimensional projection is frequently resorted to. Emil Fischer[1] in 1891 proposed a type of projection formula which is still in use. In this representation, the mole-

Note: All references above 17 are listed in the General References at the end of the chapter.

† The beginner in the field is strongly urged to build his own molecular models from the less costly sets described in Sec. 2-3. It is impossible for the novice to gain an understanding of stereochemistry without frequent reference to three-dimensional models.

[1] E. Fischer, *Ber.*, **24**, 2683 (1891).

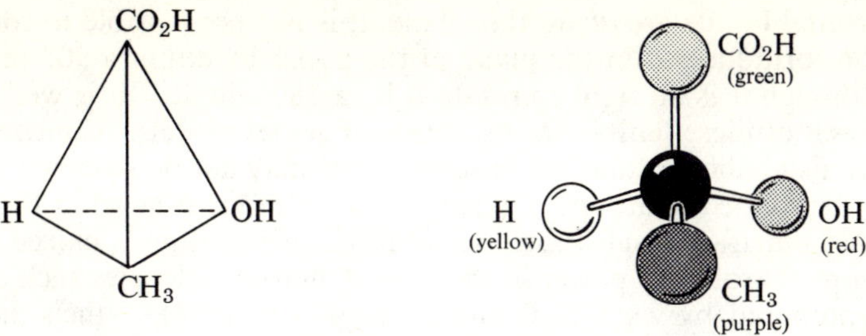

Fig. 3-1. (+)-Lactic acid. (The asymmetric carbon atom is at the center of the tetrahedron. To avoid cluttering the drawing, it is never shown.)†

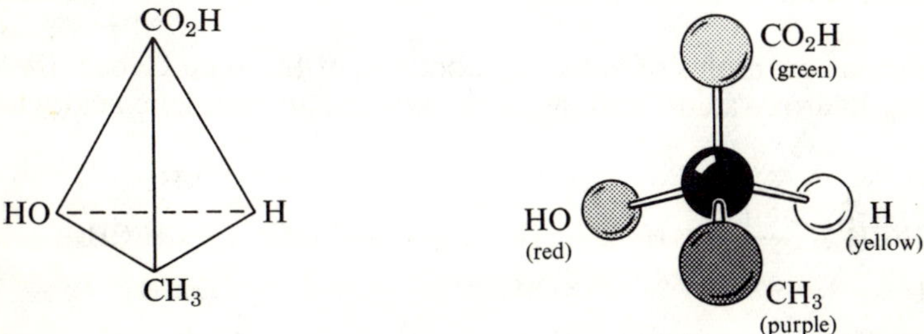

Fig. 3-2. (−)-Lactic acid.†

cule is so oriented that the asymmetric carbon is in the plane of the projection, the groups at the top and bottom are inclined equally below the plane of projection, and the groups on the left and right are similarly inclined equally above the plane of projection. The molecule is then projected in the shape of a cross, as shown in Fig. 3-3. Unfortunately, the orientation of the tetrahedron used for projection in Fig. 3-3 is different from the way in which the tetrahedron is ordinarily shown if represented three-dimensionally (Fig. 3-2); this is one of the several vicissitudes of stereochemical representation to which the newcomer has to become accustomed. Several other points in connection with Fig. 3-3 deserve attention. The projection formula, being two-dimensional, may never be lifted out of the projection plane and turned over. And, since the vertical bonds are in reality *below* the projection plane whereas

$$\mathrm{H}\overset{CO_2H}{\underset{CH_3}{\Diamond}}\mathrm{OH} \equiv \mathrm{H{-}}\overset{CO_2H}{\underset{CH_3}{\mathrm{C}}}\mathrm{{-}OH} \left[\equiv \mathrm{HO{-}}\overset{CO_2H}{\underset{H}{\mathrm{C}}}\mathrm{{-}CH_3} \equiv \mathrm{H_3C{-}}\overset{OH}{\underset{H}{\mathrm{C}}}\mathrm{{-}CO_2H} \quad \text{etc.} \right]$$

Fig. 3-3. Projection formula of (−)-lactic acid.

† The colors indicated in these figures represent typical colors of balls in ball-and-stick models. Symmetrical groups such as CH_3, OH, and CO_2H are customarily represented by a single ball in the model.

the horizontal bonds are *above* this plane, it is not permissible to rotate the projection formula within the plane of the paper by either a 90° or a 270° angle, although it is all right to rotate it by a 180° angle. It is well to convince oneself of these limitations by means of actual models. Contemplation of models also indicates that, while substituents may not be switched in pairs, they may be rotated in groups of three, as shown in the bracket in Fig. 3-3.

If the asymmetrically substituted tetrahedral carbon atom is indeed responsible for optical rotatory power in the case of simple molecules such as lactic acid (as proposed by van't Hoff and by Le Bel in 1874),[2,3] then the interchange of two groups must lead from one enantiomer to the other.† This prediction was verified experimentally[4] by means of the series of reactions shown in Fig. 3-4.‡ (The round-about transformation of the half-ester to the half-amide was resorted to since direct ammonolysis of the ester was unsuccessful.)

Numerous examples of active compounds of the type Cabcd are known, including hydroxy acids, such as lactic, malic, and mandelic acid; carbinols,

$$\underset{(+)\text{ (dextrorotatory)}}{\text{H—}\overset{\displaystyle CONH_2}{\underset{\displaystyle CO_2H}{C}}\text{—}CH(CH_3)_2} \xrightarrow{CH_2N_2} \underset{(+)}{\text{H—}\overset{\displaystyle CONH_2}{\underset{\displaystyle CO_2CH_3}{C}}\text{—}CH(CH_3)_2} \xrightarrow{HNO_2} \underset{(-)}{\text{H—}\overset{\displaystyle CO_2H}{\underset{\displaystyle CO_2CH_3}{C}}\text{—}CH(CH_3)_2} \xrightarrow{NH_2NH_2}$$

$$\underset{(-)}{\text{H—}\overset{\displaystyle CO_2H}{\underset{\displaystyle CONHNH_2}{C}}\text{—}CH(CH_3)_2} \xrightarrow{HNO_2} \underset{(-)}{\text{H—}\overset{\displaystyle CO_2H}{\underset{\displaystyle CON_3}{C}}\text{—}CH(CH_3)_2} \xrightarrow{NH_3} \underset{(-)\text{ (levorotatory)}}{\text{H—}\overset{\displaystyle CO_2H}{\underset{\displaystyle CONH_2}{C}}\text{—}CH(CH_3)_2}$$

Fig. 3-4. Transformation of one enantiomer into the other.

such as butanol-2 and methlyphenylcarbinol; amino acids, such as alanine, phenylalanine, tryptophan, and many others; halides, such as 2-bromoöctane; active amyl alcohol, hydratropic acid, atrolactic acid, etc. In most of these at least two of the four substituents are carbon-containing groups, such as alkyl or carboxyl groups. However, the compound chloroiodomethanesulfonic acid, $CHClISO_3H$, in which none of the attached groups contains carbon,

† The reader should verify this statement by examination of models as well as by inspection of projection formulas.

‡ The arrangement of the groups around the asymmetric carbon in the (+)- and (−)-isopropylmalonamidic acids, unlike that of the lactic acids shown in Figs. 3-1 and 3-2, is not known. The representations chosen here are an enlightened guess based on an empirical rule that an acid of configuration (*i*) forms a more levorotatory salt when the polarizability order (Chap. 14) of the other three substituents is A > B > C (J. H. Brewster, private communication).

$$\underset{(i)}{\text{A—}\overset{\displaystyle CO_2H}{\underset{\displaystyle C}{+}}\text{—B}}$$

[2] J. A. Le Bel, *Bull. soc. chim. France,* [2]**22,** 337 (1874).

[3] J. H. van't Hoff, *Bull. soc. chim. France,* [2]**23,** 295 (1875). (The original version appeared in Dutch in 1874.)

[4] E. Fischer and F. Brauns, *Ber.,* **47,** 3181 (1914).

has been resolved,[5] and there is some tenuous evidence[6] that fluorochlorobromomethane, CHFClBr, in which a, b, c, and d are all individual atoms rather than groups, may be obtained in optically active form also.

Of particular interest is the fact that molecules in which asymmetry is due to isotopic substitution, such as RCHDR′, have been obtained optically active,[7] as shown in Fig. 3-5. The specific rotation of optically pure ethylbenzene-α-d is at least 0.7 to 0.8°.[8] Other active deuterium compounds with

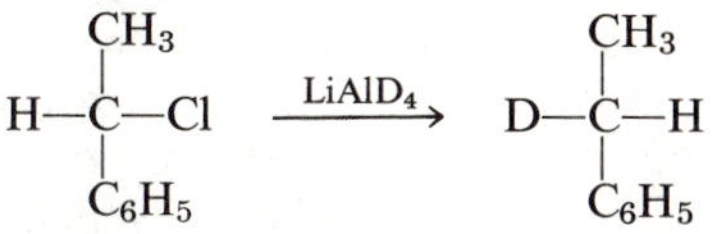

$$\begin{array}{ccc} CH_3 & & CH_3 \\ | & & | \\ H\text{—}C\text{—}Cl & \xrightarrow{LiAlD_4} & D\text{—}C\text{—}H \\ | & & | \\ C_6H_5 & & C_6H_5 \\ [\alpha]_D^{25}\ -49.2^\circ & & [\alpha]_D^{25}\ -0.30^\circ \end{array}$$

Fig. 3-5. Optically active compound of the RCHDR′ type.

rotations of similar magnitude have been synthesized[9] in which it has been shown that the arrangement of the alkyl groups and the hydrogen atom around the asymmetric carbon is opposite in the deuterocarbon from what it was in the halide (as is implied in Fig. 3-5); i.e., the reduction involves inversion of configuration (cf. Chap. 5).

Not only may compounds of type RCHDR′ be obtained optically active but also asymmetric compounds in which the isotope is remote from the asymmetric carbon, such as[10a] $CH_3CHOHCD_3$ and[10b]

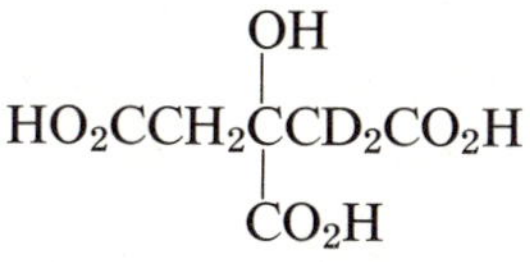

$$\begin{array}{c} OH \\ | \\ HO_2CCH_2CCD_2CO_2H \\ | \\ CO_2H \end{array}$$

One compound of this type, $C_6H_5CHOHC_6D_5$, has even been resolved[10c] by the classical method of crystallization of diastereoisomeric derivatives (Sec. 4-4*b*)! The rotation of these compounds is of the order of a degree also; earlier reports that compounds of this type could not be obtained active[11] (e.g., in the catalytic deuteration of active $C_6H_5CHOHC_6H_{11}$ and $C_2H_5CHOHC{\equiv}CH$) presumably involved accidental racemization of the compounds during their attempted synthesis. The case of the α,α-dideuterocitric acid mentioned above is particularly striking because its specific rota-

[5] W. J. Pope and J. Read, *J. Chem. Soc.*, **105**, 811 (1914).

[6] K. L. Berry and J. M. Sturtevant, *J. Am. Chem. Soc.*, **64**, 1599 (1942).

[7] E. L. Eliel, *J. Am. Chem. Soc.*, **71**, 3970 (1949). See also E. R. Alexander and A. G. Pinkus, *ibid.*, **71**, 1786 (1949).

[8] A. Streitwieser, J. R. Wolfe, and W. D. Schaeffer, *Tetrahedron*, **6**, 338 (1959); H. J. Dauben and L. L. McCoy, *J. Am. Chem. Soc.*, **81**, 5404 (1959).

[9] G. K. Helmkamp and B. F. Rickborn, *J. Org. Chem.*, **22**, 479 (1957).

[10] (*a*) K. Mislow, R. E. O'Brien, and H. Schaefer, *J. Am. Chem. Soc.*, **82**, 5512 (1960). (*b*) C. Martius and G. Schorre, *Ann.*, **570**, 140 (1950). (*c*) Y. Pocker, *Proc. Chem. Soc.*, 140 (1961).

[11] See Ref. 20, pp. 302–304; also R. L. Burwell, F. Hummel, and E. S. Wallis, *J. Org. Chem.*, **1**, 332 (1936).

tion in ammonium molybdate solution is reported[10b] to be over 30°. Presumably a complex is formed which absorbs in the yellow region of the spectrum and gives rise to a Cotton effect (see Chap. 14) near the wavelength of the sodium D line.

3-2. Compounds with Two or More Unequal Asymmetric Carbon Atoms

Many natural products, such as certain carbohydrates, peptides, steroids, terpenes, alkaloids, etc., contain two or more asymmetric carbon atoms, and a thorough understanding of the stereochemistry of systems with more than one asymmetric center is therefore essential. When there are two distinct

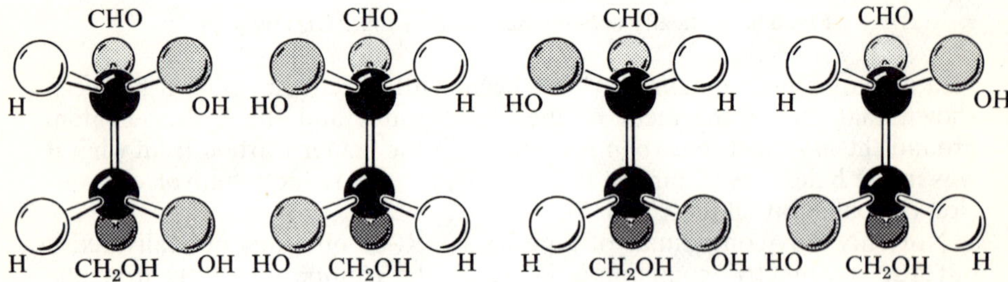

Fig. 3-6. The aldotetroses.

CHO	CHO	CHO	CHO
H—C—OH	HO—C—H	HO—C—H	H—C—OH
H—C—OH	HO—C—H	H—C—OH	HO—C—H
CH_2OH	CH_2OH	CH_2OH	CH_2OH
(−)-Erythrose	(+)-Erythrose	(−)-Threose	(+)-Threose

Fig. 3-7. Projection formulas of the aldotetroses.

asymmetric carbon atoms in a molecule, two arrangements of the groups are possible around each of them, and the total number of possibilities is 2×2 or 4. Adding another asymmetric carbon atom with its two possible arrangements doubles the number of isomers again; thus the number of stereoisomers for a compound with three distinct asymmetric atoms is $2 \times 4 = 2 \times 2 \times 2 = 2^3 = 8$. In general, the number of stereoisomers for a compound with n distinct asymmetric atoms is 2^n.

Examples of compounds with two asymmetric carbon atoms (starred) are the four-carbon sugars (tetroses), $CH_2OH\overset{*}{C}HOH\overset{*}{C}HOHCHO$. Models of the four possible isomers are shown in Fig. 3-6 and the corresponding projection formulas in Fig. 3-7. Also indicated in Fig. 3-7 are the names and signs of rotation of these compounds.

The projection formulas of compounds with two asymmetric atoms are obtained as follows: The model is so oriented that the asymmetric carbon atoms (second and third carbon) lie in the projection plane and the groups on the two sides (hydrogen and hydroxyl) stick out *above* (or in front of) the projection plane. The remaining two groups (aldehyde group and primary alcohol group in the case of the tetroses) then automatically stick out *below* (or behind) the projection plane, and the model is ready for projection. In the case of the aldose sugars, the additional convention is followed that the aldehyde group is to be at the *top* of the formula so projected (No. 1 carbon) and the primary alcohol group at the *bottom* (No. 4 carbon in the case of the tetroses). In an abbreviated writing of the projection formulas, the aldehyde carbon is denoted by a blot or circle, the main carbon chain by a vertical line, and the secondary hydroxyl groups by horizontal lines; the hydrogen atoms are omitted altogether. This notation[12] is shown in Fig. 3-8 for the tetroses.

Inspection of either the models or the projection formulas of (−)- and (+)-erythrose shows that these molecules are mirror images of each other; i.e., (−)- and (+)-erythrose are enantiomers and have identical physical and chemical properties except for the direction in which they rotate the plane of polarized light. Similarly, (−)- and (+)-threose are enantiomers. On the other hand, comparing either of the erythroses with either of the threoses, one

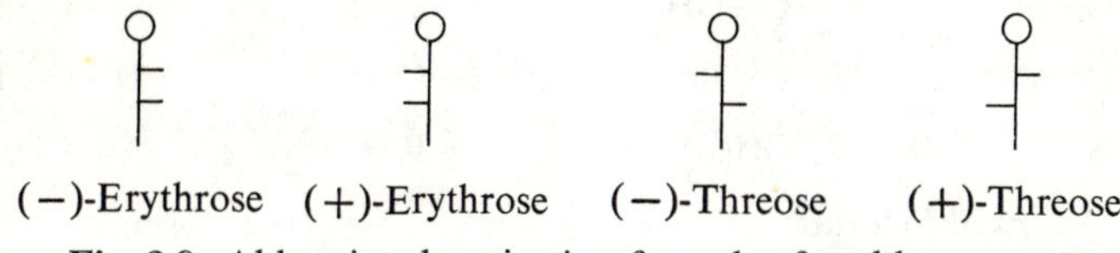

Fig. 3-8. Abbreviated projection formulas for aldotetroses.

finds that, although stereoisomers, they are not mirror images of each other. Such stereoisomers, some or all of which are dissymmetric but which are not mirror images of each other (and therefore not enantiomers), are called "diastereoisomers" (or "diastereomers"). It is evident that, in order to have diastereoisomers corresponding to a given structural formula, there must be at least two asymmetric atoms† in the given structure. If there is only one asymmetric atom,† there will only be two enantiomers.

Interestingly enough, whereas enantiomers have identical properties in a symmetric environment,‡ diastereoisomers may differ widely in both physical and chemical properties—in fact, many diastereoisomers differ among each

† This statement is oversimplified. It was pointed out in Chap. 2 that there are dissymmetric molecules without individual asymmetric atoms. Thus one can have enantiomers, and for that matter diastereoisomers, without asymmetric atoms. Actual cases will be considered in Chaps. 6, 7, and 11. To make the statement in the text universally correct, one should replace the words "asymmetric atoms" by "dissymmetric groupings." For a general definition of "dissymmetric grouping" the reader is referred to Ref. 21*b*, p. 219; specific examples will be pointed out later in this text.

‡ The qualification of the symmetric environment is necessary, for, as will be explained in Chap. 4, enantiomers differ in their reactivity toward dissymmetric reagents and behave differently in a dissymmetric physical environment.

[12] M. A. Rosanoff, *J. Am. Chem. Soc.*, **28**, 114 (1906).

other as much as ordinary (structural) isomers do. This difference between enantiomers and diastereoisomers, while puzzling at first thought, can be readily rationalized upon some reflection. In enantiomers, all intramolecular distances between corresponding groups are the same. Also, any symmetrical reagent approaching the two enantiomers in turn can always be so oriented that the approach is exactly the same in the two cases, i.e., that at a given intermolecular distance all the atoms of the approaching reagent have the same distance from all the atoms in one of the enantiomers as they would have in the other enantiomer. The situation is comparable to that of a man who has perfectly shaped hands and is perfectly ambidextrous; not only do any two fingers of one of his hands bear exactly the same relation to each other as the corresponding fingers of the other hand, but he is able to approach and handle any symmetrical tool (such as a hammer or pair of tweezers) equally well with one hand as with the other.† The same is not true of diastereoisomers. They may be likened to a hypothetical individual two of whose fingers (on one hand only) have been interchanged. Now the distance between fingers (corresponding to groups in the molecule) is no longer the same for the two hands. On the molecular scale, the different dis-

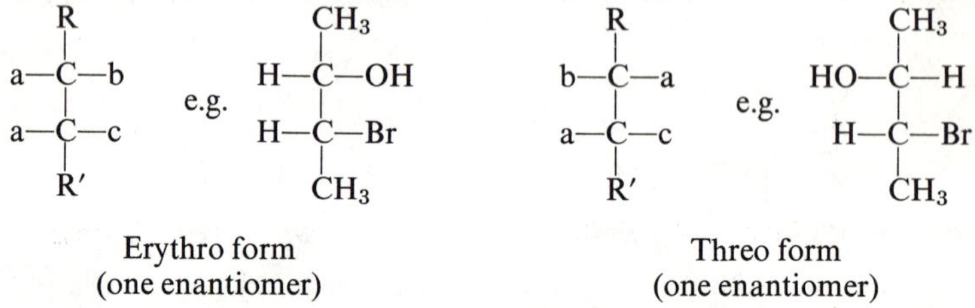

```
     R                 CH3                  R                    CH3
     |                  |                   |                     |
  a—C—b             H—C—OH               b—C—a              HO—C—H
     |       e.g.       |                   |        e.g.         |
  a—C—c             H—C—Br               a—C—c               H—C—Br
     |                  |                   |                     |
     R'                CH3                  R'                   CH3

      Erythro form                           Threo form
   (one enantiomer)                       (one enantiomer)
```

Fig. 3-9. Erythro-threo nomenclature.

tances between corresponding groups in diastereoisomers may produce differences in boiling point, melting point, solubility, spectral properties, etc., in such isomers. The free energy of the two diastereoisomers is also different, since such things as crowding between bulky substituents, hydrogen bonds within the molecule, etc., are not the same. Finally, toward an object (molecule) approaching from the outside, the fingers of the two hands (groups of the two diastereoisomeric molecules) no longer bear the same relationship; i.e., the individual can no longer be perfectly ambidextrous and likewise the diastereoisomeric molecules show differences in reactivity.

Molecules that contain two asymmetric atoms in particular have been used extensively in mechanistic studies, and a special nomenclature and special forms of notation for such systems have sprung up. The nomenclature is derived from the names of the four-carbon sugars, erythrose and threose, and applies to all systems of the type‡ R—Cab—Cac—R′. If the two like groups a in the projection formula are on the same side, as the hydroxyl groups in

† The equivalence of the hands no longer holds when the external agent is dissymmetric; e.g., a left glove will fit only the left hand, not the right hand. This situation, as already mentioned, also has its parallel among molecular enantiomers; cf. Chap. 4.

‡ R—C—C—R′ is the main chain of the molecule.

erythrose, the isomer is called the "erythro" form; if they are on opposite sides, as the hydroxyl groups in threose, the isomer is called the "threo" form. Figure 3-9 indicates the general nomenclature and also gives a particular illustration involving 3-bromo-2-butanol.

As for notation, it might appear at first sight that the Fischer projection formulas, as illustrated in Figs. 3-7 and 3-9, should be adequate for the representation of compounds with two asymmetric atoms. However, in looking at the models from which such projection formulas are derived (e.g., Fig. 3-6), one realizes that these models (and therefore the projections) represent the molecule in the so-called "eclipsed" form (cf. Chap. 6), i.e., in the form in which C_2 and C_3 are so rotated with respect to each other that the groups attached to them approach each other as closely as possible. It turns out (Chap. 6) that in this form crowding between the substituents on C_2 and C_3 is at a maximum and that the real shape of the molecule is not like this at all but more closely corresponds to an arrangement where C_2 and C_3 are rotated with respect to each other by an angle of 60° so that their substituents are now as far apart as possible. In this arrangement the molecule is said to be in the "staggered" form (cf. Chap. 6). In considering reactions of a molecule,

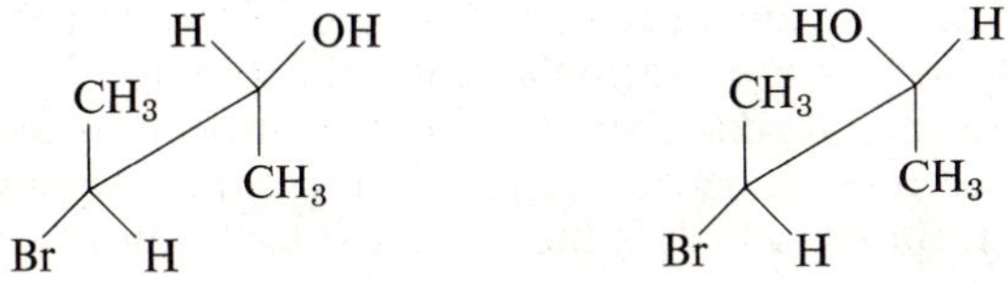

H OH
CH_3
CH_3
Br H

Erythro (one enantiomer)

HO H
CH_3
CH_3
Br H

Threo (one enantiomer)

Fig. 3-10. Sawhorse formulas of the 3-bromo-2-butanols.

it is usually desirable to depict the molecule in its actual staggered form, rather than in the hypothetical eclipsed form shown in the Fischer projection. Of the various ways of representing molecules with two asymmetric atoms in their staggered form, the "sawhorse" representation[13] and the Newman projection formula[14] are, in the author's view, the clearest and most convenient.

In the sawhorse representation, the molecule is simply shown in three dimensions, the bond between the asymmetric carbons being oriented diagonally backward and being exaggerated somewhat in length. Figure 3-10 shows the sawhorse formulas of the 3-bromo-2-butanols whose Fischer projections were given in Fig. 3-9. It should be noted that there are three different ways of staggering C_2 and C_3 with respect to each other; the formulas in Fig. 3-10 are arbitrarily disposed in such a way that the bulky methyl groups point as far away from each other as possible.

In the Newman projection formula, the molecule is viewed from front to back in the direction of the bond linking the asymmetric carbon atoms. These two atoms thus exactly eclipse each other and are represented by two superimposed circles (actually one circle only appears in the drawing). The

[13] See, for example, D. Y. Curtin, *Record Chem. Progr.* (*Kresge-Hooker Sci. Lib.*), **15**, 111 (1954).
[14] M. S. Newman, *J. Chem. Educ.*, **32**, 344 (1955).

bonds and groups attached to the asymmetric carbon atoms are projected into a vertical plane; the bonds thus appear as the spokes of a wheel at angles of 120° for each carbon, the spokes for the rear carbon being displaced by an angle of 60° with respect to the bonds on the front carbon. In order to distinguish the two sets of bonds, the set for the front carbon is drawn to the center of the circle but that for the rear carbon ends at the periphery. Newman projection formulas for the 3-bromo-2-butanols are shown in Fig. 3-11.

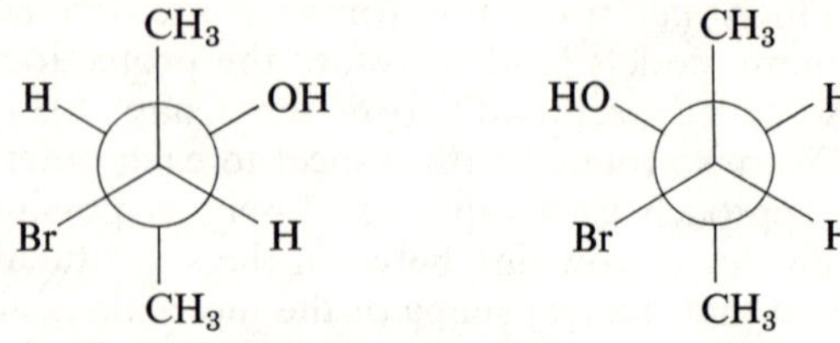

Fig. 3-11. Newman projection for the 3-bromo-2-butanols.

It is desirable that anyone dealing with stereochemical problems be able to translate rapidly one set of formulas into the other. In particular, one should be able to shift readily from the Fischer projection formula (which is the one most often used in books and articles) to either the sawhorse or Newman projection formula (which is the one most useful in assessing the chemical behavior of a molecule, since it more aptly depicts the actual molecular shape). One way of effecting the transformation is to build a model corresponding to the Fischer projection, rotate the model into the staggered form, and then draw the corresponding sawhorse or Newman projection. A quicker way, not requiring models, is shown in Fig. 3-12. Here the Fischer projection is directly translated into an "eclipsed sawhorse" or Newman pro-

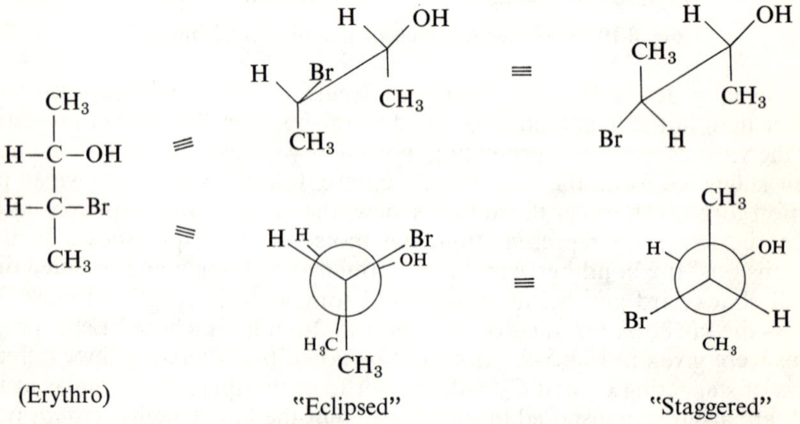

Fig. 3-12. Transformation from Fischer to sawhorse or Newman formula.

jection, and the latter is then rotated by a 180° angle around the C_2—C_3 bond to give the normal sawhorse or Newman projection.

A few further observations regarding the different representations may be helpful. The Fischer formula shows at a glance whether one is dealing with an erythro or threo form, but the sawhorse and Newman formulas do not and usually require at least a mental process of rotation around C_2—C_3 until

Threo Erythro

Fig. 3-13. Flying-wedge formulas of the 3-bromo-2-butanols.

one can decide which is which. Also, the sawhorse formula is quite convenient to represent eclipsed as well as staggered forms, but the Newman projection is awkward for the representation of eclipsed forms (cf. Fig. 3-12).

Yet another way of showing compounds with two asymmetric centers is by means of "flying-wedge" formulas,[15] depicted for erythro- and threo-3-bromo-2-butanol in Fig. 3-13. These formulas, in which the molecule is depicted looking sideways at the bond joining the asymmetric carbon atoms, will not be used in this book.

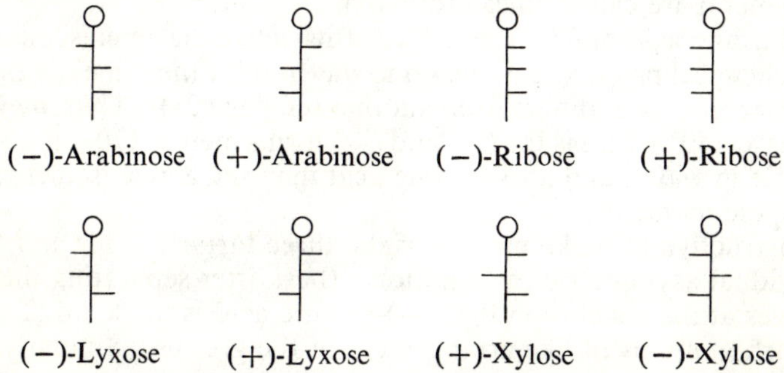

Fig. 3-14. Stereoisomeric aldopentoses.

Much of the early development of stereochemistry was stimulated by investigations of various sugars. It is therefore appropriate to conclude this section by an illustration of compounds with three and four asymmetric carbon atoms, using the aldopentoses and aldohexoses as examples. Figure 3-14 shows the eight stereoisomeric aldopentoses (four pairs of enantiomers), and Fig. 3-15 depicts *one* enantiomer of each of the eight pairs of diastereoisomeric aldohexoses (total number of stereoisomers: 16).†

† The spatial formulas in Figs. 3-8, 3-14, 3-15, and 3-16 are known to represent the correct arrangement in space of the groups attached to the asymmetric carbon atoms; cf. Chap. 5.

[15] For example, D. J. Cram, *J. Am. Chem. Soc.*, **74**, 2129 (1952).

3-3. Compounds Containing Like Asymmetric Carbon Atoms

When a compound contains two or more asymmetric carbon atoms which are alike, meaning that the same atoms or groups are attached to both, the general formula for computing the number of stereoisomers (page 20) no longer applies. A case in point is tartaric acid, $HO_2C\overset{*}{C}HOH\overset{*}{C}HOHCO_2H$. Here the two asymmetric carbon atoms (starred) are alike in that each is linked to a hydrogen atom, a hydroxyl group, a carboxyl group, and a —$CHOHCO_2H$ group. A molecule of this type is sometimes described by the symbolism AA, where A stands for an asymmetric carbon atom. At first sight, four different arrangements appear to be possible; the models and their Fischer projections are shown in Fig. 3-16. The first two formulas do, indeed, represent two active forms, (−)- and (+)-tartaric acid, which are enantiomeric to each other. The third and fourth formulas, however, represent the same molecule, since one can be obtained from the other by a rotation of 180° in the plane of the paper, such a rotation being permitted for a Fischer projection formula (cf. Fig. 3-3). Since the third and fourth formulas are at the same time identical and mirror images of each other, it follows that they represent one and the same, inactive molecule. (This can also be deduced from the fact that the molecule has a plane of symmetry: dotted line.) There is, thus, a tartaric acid diastereoisomeric with the active forms but itself inactive. This acid is called *meso*-tartaric acid, and inactive diastereoisomers of this type in general are called "meso forms."†

meso-Tartaric acid differs from either of its active diastereoisomers in physical and chemical properties in the same way and for the same reason that one active diastereoisomer differs from another (cf. page 21). Thus, *meso*-tartaric acid melts at 140° whereas the (+) and (−) forms melt at 170°; it is less dense, less soluble in water, and a less strong acid than the active isomers and has a higher dipole moment.

It is instructive to make models of the three tartaric acids and to inspect the individual asymmetric carbon atoms (best after separating them). One thus arrives at the conclusion that (−)-tartaric acid is made up of two asymmetric carbon atoms of like arrangement of the groups. Similarly, (+)-tar-

† It is of some interest that the projection of *meso*-tartaric acid shown in Fig. 3-16 as having a plane of symmetry corresponds to the eclipsed form and does not represent a prevalent state of the molecule. Rather, *meso*-tartaric acid exists mainly in the staggered forms shown in (*i*) to (*iii*) below. Of these (*i*) has a center of symmetry, and (*ii*) and (*iii*), though dissymmetric, are mirror images of each other. The absence of activity may be ascribed to the fact that (*ii*) and (*iii*) are present in equal amounts, that they are readily interconverted by internal rotation, and (less important) that during such rotation the molecule passes through the form shown in Fig. 3-16 which does have a plane of symmetry. This subject will be returned to in Chap. 6.

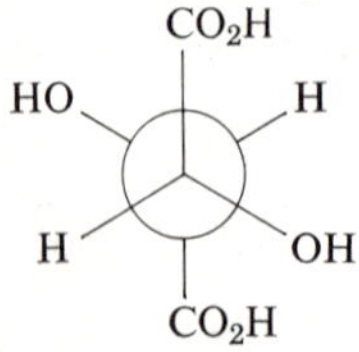

CO_2H / HO, H / H, OH / CO_2H

(*i*)

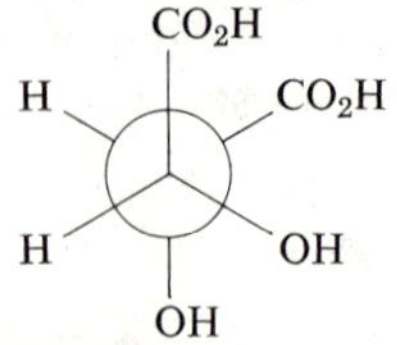

CO_2H / H, CO_2H / H, OH / OH

(*ii*)

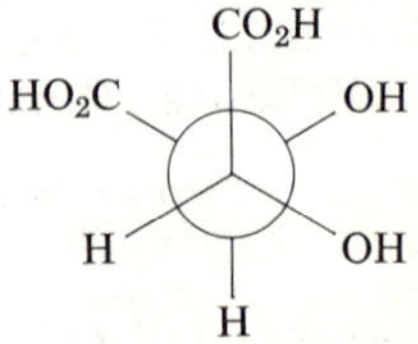

CO_2H / HO_2C, OH / H, OH / H

(*iii*)

taric acid is made up of two asymmetric carbon atoms of like arrangement, the arrangement being opposite to that for the (−) enantiomer. *meso*-Tartaric acid, on the other hand, is made up of two asymmetric carbon atoms of *opposite* arrangement, a fact which can also be gleaned from the projection

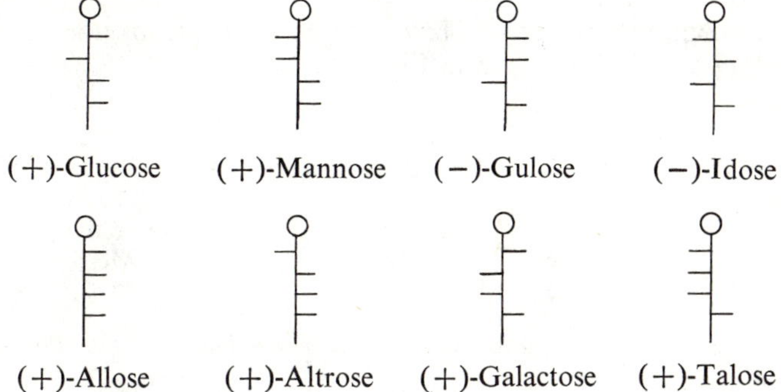

Fig. 3-15. Diastereoisomeric aldohexoses (one member of each pair of enantiomers shown).

formula if one separates the upper half from the lower along the dotted line (Fig. 3-16) and rotates it through a 180° angle. If the arrangement of the substituents about each of the asymmetric carbon atoms in (−)-tartaric acid is called† "S" and that in (+)-tartaric acid is called† "R," then (−)-tartaric

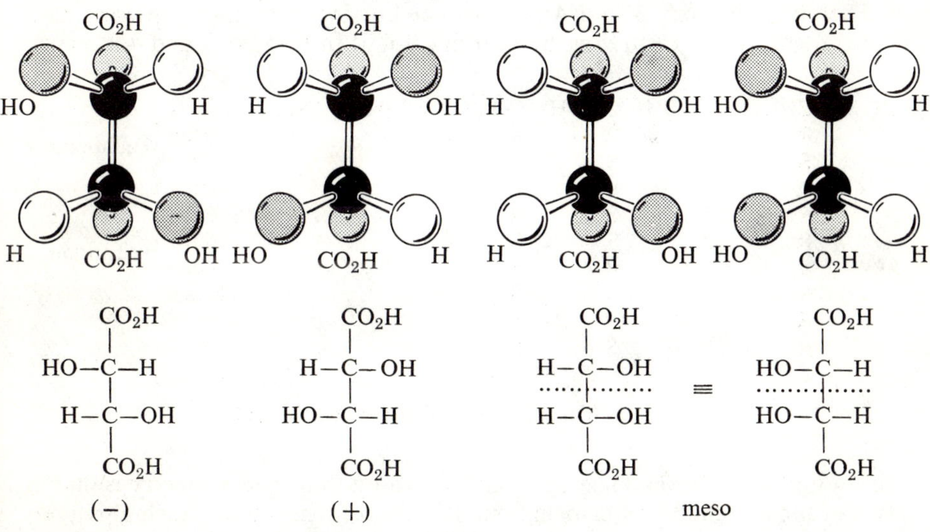

Fig. 3-16. The tartaric acids.

acid has the S,S arrangement, (+)-tartaric acid has the R,R arrangement, and *meso*-tartaric acid has the R,S (or S,R) arrangement.

† At this point, the symbols R and S may be taken to be simply means of denoting that the arrangements of the groups about the two asymmetric carbon atoms are opposite. Actually, the symbols have a more profound meaning which will be explained in Sec. 5-2.

Continuing with this type of symbolism, we may next consider the number of isomers in 2,3,4-trihydroxyglutaric acid, $HO_2C\overset{*}{C}HOH\overset{*}{C}HOH\overset{*}{C}HOHCO_2H$, which may be said to be of type ABA, since the molecule has two asymmetric carbon atoms which are alike (Nos. 2 and 4) and one other one (No. 3) which is asymmetric in some of the stereoisomers but not in all. Each asymmetric atom may, a priori, have the R or S arrangement. Thus one arrives at the possibilities shown in Fig. 3-17.

HO_2C (1)	CHOH (2)	CHOH (3)	CHOH (4)	CO_2H (5)	Isomer No.	Isomer Type	
	R		R		1	Active	enantiomers
	S		S		2	Active	
	R	r*	S		3	Meso	
	R	s*	S		4	Meso	

* Small letters are used because these atoms are "pseudoasymmetric"; see below and Sec. 5-2.

Fig. 3-17. The 2,3,4-trihydroxyglutaric acids.

In isomers 1 and 2 (which are mirror images of each other and therefore represent a pair of enantiomers) the central carbon atom (No. 3) is not asymmetric, since the two groups attached have the same configuration as well as structure. In isomers 3 and 4, the central carbon atom is asymmetric, but the molecule as a whole is not, since it has a plane of symmetry bisecting carbon atom No. 3 (see model or projection formula).

Therefore, isomers 3 and 4 are distinct but inactive; they represent two meso forms. The central carbon atom (No. 3) in isomers 3 and 4 is some-

HO_2C—	CHOH—	CHOH—	CHOH—	CHOH—	CO_2H	No.	Isomer Type	
	R	R	R	R		1	Active	enantiomers
	S	S	S	S		2	Active	
	R	R	R	S		3	Active	enantiomers
	S	S	S	R		4	Active	
	R	R	S	R		5	Active	enantiomers
	S	S	R	S		6	Active	
	R	R	S	S		7	Meso	
	R	S	R	S		8	Meso	
	R	S	S	R		9	Active	enantiomers
	S	R	R	S		10	Active	

Fig. 3-18. The 2,3,4,5-tetrahydroxyadipic acids.

times said to be "pseudoasymmetric," meaning that its asymmetry is due to two of the attached groups being opposite in configuration. Such a pseudoasymmetric atom does not give rise to dissymmetry in the molecule as a whole.

The isomerism of the 2,3,4,5-tetrahydroxyadipic acids (type ABBA), of importance as reference compounds in hexose chemistry, is summarized in Fig. 3-18. There are four pairs of enantiomers and two meso forms. It is instructive to work this out for oneself and to convince oneself that there are no other isomers.

3-4. Compounds with Asymmetric Carbon Atoms in Branched Chains

In the compounds discussed so far, the asymmetric carbons could always be arranged in one main chain. On rare occasions only does one encounter branched dissymmetric molecules, i.e., molecules in which the asymmetric

$$\begin{array}{cc} \begin{array}{c} A^R \\ | \\ H-C-A^R \\ | \\ A^R \end{array} \quad \begin{array}{c} A^S \\ | \\ {}^SA-C-H \\ | \\ A^S \end{array} & \qquad \begin{array}{c} A^R \\ | \\ H-C-A^R \\ | \\ A^S \end{array} \quad \begin{array}{c} A^S \\ | \\ {}^SA-C-H \\ | \\ A^R \end{array} \\ \text{Enantiomers} & \text{Enantiomers} \end{array}$$

Fig. 3-19. CHA$^*_3$ case.

carbon atoms cannot all be aligned in one chain. A general case would be a molecule of the type CA*B*D*E*, where the groups A*, B*, D*, and E* attached to the central carbon atom are all asymmetric and the central carbon atom is asymmetric also. Here there are, then, five asymmetric atoms, and the total number of stereoisomers is 2^5 or 32. Cases where two or more

$$\begin{array}{cc} \begin{array}{c} A^R \\ | \\ H-C-B^R \\ | \\ A^R \end{array} \quad \begin{array}{c} A^S \\ | \\ {}^SB-C-H \\ | \\ A^S \end{array} & \qquad \begin{array}{c} A^R \\ | \\ H-C-B^S \\ | \\ A^R \end{array} \quad \begin{array}{c} A^S \\ | \\ {}^RB-C-H \\ | \\ A^S \end{array} \\ \text{Enantiomers} & \text{Enantiomers} \end{array}$$

$$\begin{array}{cc} \begin{array}{c} A^R \\ | \\ H-C-B^R \\ | \\ A^S \end{array} \quad \begin{array}{c} A^S \\ | \\ {}^SB-C-H \\ | \\ A^R \end{array} & \qquad \begin{array}{c} A^S \\ | \\ H-C-B^R \\ | \\ A^R \end{array} \quad \begin{array}{c} A^R \\ | \\ {}^SB-C-H \\ | \\ A^S \end{array} \\ \text{Enantiomers} & \text{Enantiomers} \end{array}$$

Fig. 3-20. CHA$^*_2$B* case.

of the asymmetric groups attached to the central carbon atom are alike are more complicated, for, depending on the arrangement of the like asymmetric groups, the central carbon atom is or is not itself asymmetric. Thus, in the CHA$^*_3$ case shown in Fig. 3-19, the central carbon atom is never asym-

$$\begin{array}{ccc} \begin{array}{c} A^R \\ | \\ {}^RA-C-A^R \\ | \\ A^R \end{array} \quad \begin{array}{c} A^S \\ | \\ {}^SA-C-A^S \\ | \\ A^S \end{array} & \qquad \begin{array}{c} A^R \\ | \\ {}^RA-C-A^S \\ | \\ A^R \end{array} \quad \begin{array}{c} A^S \\ | \\ {}^RA-C-A^S \\ | \\ A^S \end{array} & \qquad \begin{array}{c} A^R \\ | \\ {}^SA-C-A^S \\ | \\ A^R \end{array} \\ \text{Enantiomers} & \text{Enantiomers} & \text{meso form} \end{array}$$

Fig. 3-21. CA$^*_4$ case.

metric and only two pairs of enantiomers are possible. On the other hand, in the CHA$^*_2$B* case (Fig. 3-20) the central carbon atom *may* become asymmetric and the number of stereoisomers is therefore larger than in the simple ABA case discussed earlier (Fig. 3-17). The CA$^*_4$ case shown in Fig. 3-21 is of special interest. The isomer on the right of the figure is clearly identical with its mirror image (both being $CA_2^RA_2^S$ with the central carbon atom not asymmetric); yet it has neither plane nor center of symmetry. Inspection of the model shows that this molecule has a fourfold alternating axis of symmetry (cf. Chap. 2). A molecule of this type possessing no other element of sym-

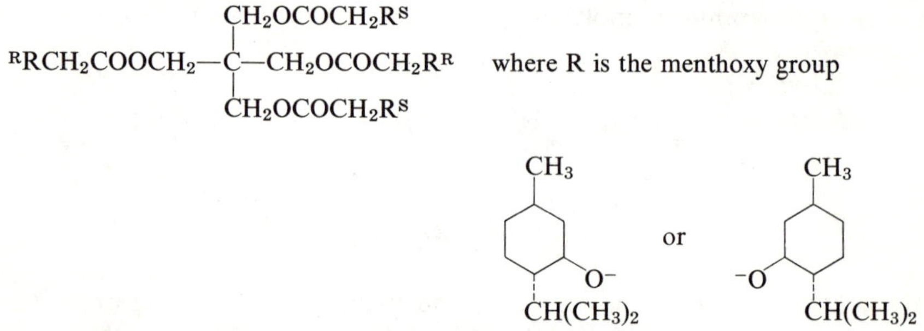

$$\begin{array}{c} CH_2OCOCH_2R^S \\ | \\ {}^RRCH_2COOCH_2—C—CH_2OCOCH_2R^R \\ | \\ CH_2OCOCH_2R^S \end{array}$$

where R is the menthoxy group

CH_3 … O^- … $CH(CH_3)_2$ or CH_3 … ^-O … $CH(CH_3)_2$

Fig. 3-22. Molecule possessing fourfold alternating axis of symmetry.

metry than an alternating axis (Fig. 3-22) has indeed been prepared[16] and shown to be optically inactive.

Mathematical formulas for computing the number of stereoisomers in branched structures have been developed.[17]

General References

[18] G. M. Richardson, ed., "The Foundations of Stereochemistry," American Book Company, New York, 1901.

[19] F. Ebel in K. Freudenberg, ed., "Stereochemie," Franz Deuticke, Leipzig, 1932, pp. 587–602; K. Freudenberg, *ibid.,* pp. 662–668.

[20] R. L. Shriner and R. Adams, Optical Isomerism, in H. Gilman, ed., "Organic Chemistry," 2d ed., John Wiley & Sons, Inc., New York, 1943, pp. 224–240.

[21] (*a*) G. W. Wheland, "Advanced Organic Chemistry," 3d ed., John Wiley & Sons, Inc., New York, 1960; (*b*) *id.,* 2d ed., 1949.

---

[16] G. E. McCasland, R. Horvat, and M. R. Roth, *J. Am. Chem. Soc.,* **81,** 2399 (1959); for an earlier example, see G. E. McCasland and S. Proskow, *ibid.,* **78,** 5646 (1956).

[17] J. K. Senior, *Ber.,* **60B,** 73 (1927).

Chapter 4

RACEMIC MODIFICATIONS

4-1. Nature of Racemic Modifications

It was pointed out in Chap. 2 that the individual molecules of most substances are optically active† if they are dissymmetric. Nevertheless, the substance in bulk may not be optically active, because it may be constituted of approximately equal numbers of dextrorotatory (+) and levorotatory (−) molecules so that the average rotation is zero. Such an assembly of molecules, one half of which are mirror images of the other half, is called a "racemic modification"‡ and is denoted by the symbol (±). The term "racemic modification" evidently does not apply to individual molecules; rather it is a statistical concept which arises when large numbers of molecules are considered. To make this point clear to oneself, one might imagine a lazy Maxwellian demon,¶ sitting between two lakes of one and the same optically active substance, except that the substance in one lake is dextrorotatory, that in the other levorotatory. The demon is then instructed to assemble a mole of the substance in a glass by picking individual molecules at random out of the two lakes. It is clear that the first molecule that he picks will be either dextrorotatory or levorotatory and there is a 50-50 chance that the first two molecules he picks will be of the same rotation. Even after he has picked 20 molecules, there is a fair chance that he will have significantly more of one kind than of the other. But after he has picked 6.02×10^{23} molecules, one can be quite sure that any fluctuation in favor of one kind of molecule or the other will be so minor as to be experimentally undetectable and the material in the glass will not rotate the plane of polarized light.

In this chapter, the formation of racemic modifications will be considered in Sec. 4-2; their properties, especially in the solid state, in Sec. 4-3; and the important process of separating the dextrorotatory and levorotatory molecules, known as "resolution," will be described in Sec. 4-4.

Note: All references above 81 are listed in the General References at the end of the chapter.

† Unfortunately, this statement is of doubtful operational significance, inasmuch as the optical activity of an individual molecule has never been observed and possibly never will be.

‡ The term "*dl* pair" is also frequently used. "Racemic mixture" has been erroneously used in the same context but actually has a more restricted meaning; see below.

¶ A lazy Maxwellian demon is one who accomplishes only what would happen on its own account anyway.

4-2. Formation of Racemic Modifications

a. By Mixing. The most obvious and trivial way of forming a racemic modification is by intimate mixing of exactly equal amounts of the dextrorotatory (+) and levorotatory (−) isomers. This process is associated with an entropy of mixing, since the racemic modification represents a more random state of affairs than the separate enantiomers. The entropy of mixing is calculated by the usual formula $\Delta S = -Rx_1 \ln x_1 - Rx_2 \ln x_2$ (assuming ideal behavior†).

In the case of a racemic modification, $x_1 = x_2 = \frac{1}{2}$, whence $\Delta S = -R \ln \frac{1}{2} = R \ln 2 = 1.4$ cal./deg. mole. The free-energy change in producing the racemic modification from the enantiomers is therefore $\Delta F = -T\,\Delta S = -0.42$ kcal./mole at room temperature.‡ The entropy of mixing is positive because racemization by mixing (like any mixing) leads from a more ordered to a more random (or disordered) state.

The formation of a racemic modification by mixing is sometimes resorted to if racemic material, obtained by synthesis (see below), is to be compared with available dextrorotatory and levorotatory material presumed to be of

```
     CO2H                          CO2H                          CO2H
      |          H—V—Z              |            H—V—Z            |
 Br—C—H      <-----------       H—C—H       ----------->      H—C—Br
      |                             |                             |
     CH3                           CH3                           CH3
     (−)                                                          (+)
```

Fig. 4-1. Bromination of propionic acid.

the same species. As will be seen later, the (±) material will, of course, have the same melting point as that of an intimate mixture of equal amounts of the (+) and (−) isomers of the same chemical species.

b. By Synthesis. Any synthesis of dissymmetric molecules, starting from either symmetric molecules or a racemic modification and using no optically active reagents or catalysts and no asymmetric physical influence (cf. Sec. 4-4*d* and *f*), always produces a racemic modification (i.e., an equal number of the two enantiomeric types of product molecules). This point may be better appreciated by looking at the two common ways of producing an asymmetric carbon in a molecule: by displacement and by addition. The first method is exemplified by the bromination of propionic acid to α-bromopropionic acid by the Hell-Volhard-Zelinsky method (Fig. 4-1). Since each of the two alpha-hydrogens bears the same relationship¶ to the other and to the rest of the

† Ideal behavior in a mixture means that interaction between unlike molecules in the mixture is the same as interaction between the (like) molecules of the individual pure components. For enantiomers, such an assumption is probably excellent for gases and dilute solutions, fair for pure liquids, but not applicable to solids (see below).

‡ Assuming that mixing is a thermoneutral process, i.e., $\Delta H = 0$. This is certainly the case if ideal behavior is encountered. For a possible exception, see A. Ladenburg, *Ber.*, **28**, 163 (1895); also C. J. McGinn, *J. Phys. Chem.*, **65**, 1896 (1961).

¶ In reality, the two relations are of the mirror-image type. Thus in going from one hydrogen in the projection formula to the carboxyl group, one moves up and right, but from the other hydrogen one moves an equal distance up and left. In the absence of any dissymmetric influence, however, the terms right and left are devoid of significance on the molecular scale. This point will be returned to later (Sec. 4-4*g*).

molecule, each is replaced at the same rate as the other and equal numbers of (+) and (−) molecules of α-bromopropionic acid result. The second method is exemplified by the addition of hydrogen cyanide to acetaldehyde to give lactonitrile (Fig. 4-2). Here approach from either side of the carbonyl is equally facile and therefore equal numbers of molecules of the two enantiomeric forms of lactonitrile, $CH_3CHOHCN$, result.

c. By Racemization. Racemization is the process of producing a racemic modification starting with *one* of the pure enantiomers. Since the two enantiomers have the same free energy (Chap. 3), the equilibrium mixture will correspond to a 50-50 composition; i.e., it will be a racemic modification. Since the concentration of the pure enantiomer is reduced to one-half of its original value when racemization occurs, the free-energy change associated with racemization is $\Delta F = RT \ln \frac{1}{2} = -RT \ln 2$. It will be seen that this is equal to the free-energy change associated with forming a racemic modification by mixing. In mixing, one starts with equal amounts of the (+) and (−) forms, in racemization with either form by itself; however, this makes no difference to the free-energy change. Given that racemization is an energetically favored process, the question remains whether a convenient pathway can be

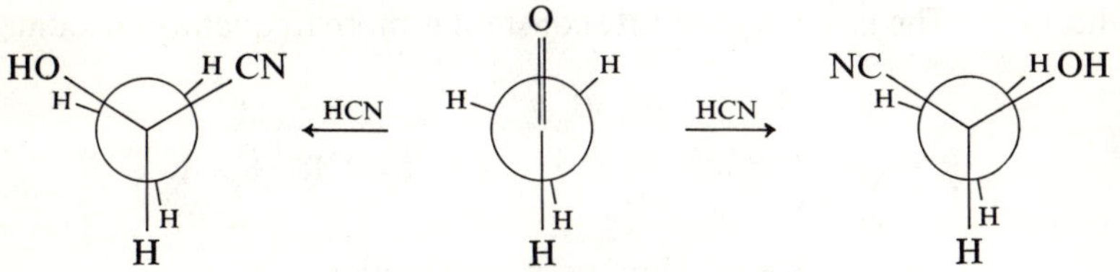

HO H CN H O H NC H H OH H HCN HCN H H H H H H

Fig. 4-2. Addition of hydrogen cyanide to acetaldehyde.

found for equilibrium to be established, or, in other words, whether the activation energy for interconversion of the enantiomers is prohibitively high or not.

If racemization is considered a reversible transformation of the (+) into the (−) form, the kinetic expression for a reversible reaction applies. Let *a* be the initial concentration of the given enantiomer, *x* the amount changed into the opposite enantiomer after time *t*, and *k* the rate constant for both the forward and the reverse reaction (in this particular instance the two rate constants are equal). Then

$$\frac{dx}{dt} = k(a - x) - kx = k(a - 2x)$$

The integrated form of this equation is

$$k = \frac{2.3}{2t} \log \frac{a}{a - 2x}$$

In practice, the concentration of the active species is followed by optical rotation, and since optical rotation is proportional to the concentration of the active species: $\alpha_0 \propto a$ and $\alpha_t \propto (a - 2x)$, where α_0 is the initial rotation and α_t is the rotation after time *t*, the rate constant for racemization is given by

$$k = \frac{2.3}{2t} \log \frac{\alpha_0}{\alpha_t}$$

Some authors, instead of considering racemization a reversible interconversion of the enantiomers, i.e., $(+) \rightleftharpoons (-)$, consider it an irreversible transformation of one of the enantiomers into the racemic modification: $(+) \rightarrow (\pm)$. In this case the kinetics of an irreversible first-order reaction applies: $dy/dt = k'(a - y)$, where a has the same meaning as before and y is the amount of material racemized at time t. Integration gives $k' = (2.3/t) \log a/(a - y)$, and since the initial rotation α_0 is proportional to a and the rotation α_t is proportional to $a - y$ (active material minus racemized material), this may be expressed as

$$k' = \frac{2.3}{t} \log \frac{\alpha_0}{\alpha_t}$$

It should be noted that k' and k are not the same; in fact $k' = 2k$, that is, the rate of racemization is twice the rate of interconversion of the enantiomers. This is reasonable; for each $(-)$ molecule formed from a $(+)$ molecule, *two* molecules are "racemized," since the $(-)$ molecule so formed counterbalances the rotation of another (unchanged) $(+)$ molecule. When reading the literature, it is obviously important to discern whether rate constants for loss of activity refer to interconversion of enantiomers or to formation of the racemic modification. The latter type of rate constant is more frequently encountered.

$$\mathrm{R''{-}\underset{\displaystyle R}{\overset{\displaystyle R'}{C}}{-}R'''} \rightleftharpoons \mathrm{\underset{R''\quad R'''}{\overset{R'}{C}}} + \mathrm{R} \rightleftharpoons \mathrm{R''{-}\underset{\displaystyle R'''}{\overset{\displaystyle R'}{C}}{-}R}$$

Fig. 4-3. Process of racemization.

Whether racemization will take place at a reasonable rate depends on the system under consideration. Several chemical methods for reaching the racemization equilibrium may be distinguished.

i. Thermal Racemization. One general method of racemizing an optically active material is by breaking, temporarily, one of the four bonds to an asymmetric carbon. If, in the subsequent re-formation of the bond, the group separated exchanges places with one of the remaining groups, the dissymmetric molecule is converted to its enantiomer (Fig. 4-3).†

If the bond is to be broken homolytically, i.e., in such a way that one of the electrons of the bond stays with carbon and the other with the group separated, considerable energy (namely, that equivalent to the bond-dissociation energy) must be expended and high temperatures are required. This is apt to lead to chemical changes more deep-seated than simple racemization, and well-authenticated cases of thermal racemization by homolytic cleavage are quite rare. The racemization of α-phenethyl chloride, $C_6H_5CHClCH_3$, upon distillation at atmospheric pressure may be of this type, although this has not been clearly established. Relatively facile thermal racemization may occur in special cases when one enantiomer can be converted to the other by

† It is not immediately obvious why and how such an exchange of places occurs, but in fact it does. This point will be considered in detail in Chap. 13.

stretching and bending bonds rather than by actually breaking them. Such cases are considered in Chap. 6.

ii. By Anion Formation. A second way of bringing about the process depicted in Fig. 4-3 is by heterolytic cleavage, in such a way that the electron pair of the bond broken stays with the R′R″R‴C group. This group is thus converted to an anion. The group R separated without its electrons is usually a proton. In other words, the process involves the temporary separation of a mildly acidic hydrogen. Several examples are shown in Fig. 4-4.

$$\text{(1)}\quad \begin{array}{c} OH \\ | \\ C_6H_5\text{—}C\text{—}CO_2H \\ | \\ H \end{array} \quad \text{via} \quad \begin{array}{c} OH \quad O^- \\ | \quad / \\ C_6H_5\text{—}\underline{C}\text{—}C \\ \quad \backslash\backslash \\ \quad O \end{array} \longleftrightarrow \begin{array}{c} OH \quad O^- \\ | \quad / \\ C_6H_5C\text{=}C \\ \quad \backslash \\ \quad O^- \end{array}$$

Mandelic acid

$$\text{(2)}\quad \begin{array}{c} CH_3 \\ | \\ C_6H_5CO\text{—}C\text{—}C_2H_5 \\ | \\ H \end{array} \quad \text{via} \quad \begin{array}{c} CH_3 \\ | \\ C_6H_5\text{—}C\text{—}\underline{C}\text{—}C_2H_5 \\ \| \\ O \end{array} \longleftrightarrow \begin{array}{c} CH_3 \\ | \\ C_6H_5\text{—}C\text{=}C\text{—}C_2H_5 \\ | \\ O^- \end{array}$$

$$\text{(3)}\quad \begin{array}{c} CH_3 \\ | \\ p\text{-}HO_2CC_6H_4SO_2\text{—}C\text{—}SC_6H_5 \\ | \\ H \end{array} \quad \text{via} \quad \begin{array}{c} CH_3 \\ | \\ p\text{-}\bar{O}_2CC_6H_4SO_2\text{—}\underline{C}\text{—}SC_6H_5 \end{array}$$

$$\text{(4)}\quad \begin{array}{c} OH \\ | \\ C_6H_5\text{—}C\text{—}CO_2H \\ | \\ CH_3 \end{array} \qquad \text{Atrolactic acid, not racemized by base}$$

$$\text{(5)}\quad \begin{array}{c} CH_3 \\ | \\ p\text{-}HO_2CC_6H_4SO_2\text{—}C\text{—}SC_6H_5 \\ | \\ SO_2C_6H_4CH_3\text{-}p \end{array}$$

Not racemized by base

Fig. 4-4. Racemization via anions.

The base used to remove the proton is, in most cases, methoxide or ethoxide. Only those compounds (1 to 3) having an acidic hydrogen attached to the asymmetric carbon are racemized by a base; where there is no acidic hydrogen attached to the asymmetric carbon (4,5) no base-induced racemization occurs. The more labile and acidic the hydrogen, the greater, in general, is the ease of racemization. The hydrogen on the asymmetric carbon in the disulfone $p\text{-}HO_2CC_6H_4SO_2\overset{*}{C}H(CH_3)SO_2C_6H_5$ is so acidic and the compound is therefore converted to the carbanion so readily that all attempts at resolving it have failed.

In a particularly elegant demonstration that racemization by base involves the anion as intermediate, it was shown[1] that the rate of racemization of active phenyl *sec*-butyl ketone in dioxane-D_2O medium using NaOD as base was equal to the rate of deuterium exchange.[1a] Previously it had been shown[2] that the rate of racemization by sodium acetate is equal to the rate of bromination in the presence of the same catalyst. These observations are rationalized readily by assuming that all three processes (racemization, deuterium exchange, halogenation) proceed through a common intermediate—presumably the anion—which is formed in a rate-determining step (Fig. 4-5).

$$C_6H_5COC(CH_3)DC_2H_5 + B:^-$$

$$\text{BD, fast} \uparrow\downarrow \text{Slow}$$

$$(+)\text{-}C_6H_5COCH(CH_3)C_2H_5 + B:^- \underset{\text{Fast}}{\overset{\text{Slow}}{\rightleftharpoons}} BH + C_6H_5CO\bar{C}(CH_3)C_2H_5 \underset{\text{Slow}}{\overset{\text{Fast}}{\rightleftharpoons}} (-)\text{-}C_6H_5COCH(CH_3)C_2H_5 + B:^-$$

$$\text{Fast} \downarrow Br_2$$

$$C_6H_5COCBr(CH_3)C_2H_5 + Br^-$$

Fig. 4-5. Rates of racemization, deuterium exchange, and bromination of phenyl *sec*-butyl ketone in presence of base.

iii. By Cation Formation. Another way of bringing about racemization by the type of process depicted in Fig. 4-3 is to remove (temporarily) the group R with its pair of electrons and to leave $R'R''R'''C^+$ as a carbonium ion. Usually, this type of racemization is brought about by a Lewis acid which

$$(+)\text{-}C_6H_5CHClCH_3 + SbCl_5 \rightleftharpoons C_6H_5\overset{+}{C}HCH_3\ SbCl_6^- \rightleftharpoons (-)\text{-}C_6H_5CHClCH_3 + SbCl_5$$

$$C_6H_5CH(CH_3)C_2H_5 + AlCl_3 \longrightarrow C_6H_5\overset{+}{C}(CH_3)C_2H_5 + AlHCl_3^- \qquad \text{Possible initiating step}$$

$$(+)\text{-}C_6H_5CH(CH_3)C_2H_5 + C_6H_5\overset{+}{C}(CH_3)C_2H_5 \longrightarrow C_6H_5\overset{+}{C}(CH_3)C_2H_5 + (\pm)\text{-}C_6H_5CH(CH_3)C_2H_5$$

Fig. 4-6. Racemization via carbonium ions.

[1] S. K. Hsü, C. K. Ingold, and C. L. Wilson, *J. Chem. Soc.,* 78 (1938).

[1a] See also D. J. Cram, B. Rickborn, C. A. Kingsbury, and P. Haberfield, *J. Am. Chem. Soc.,* **83,** 3678 (1961).

[2] S. K. Hsü and C. L. Wilson, *J. Chem. Soc.,* 623 (1936).

abstracts R:⁻. Examples are the racemization of α-phenethyl chloride by means[3] of antimony pentachloride and of 2-phenylbutane by aluminum chloride (Fig. 4-6).[4] Mercuric chloride, zinc chloride, and stannic chloride are also effective catalysts for the racemization of α-phenethyl chloride, but chlorides devoid of Lewis acid activity, such as lithium chloride and tetramethylammonium chloride, are ineffective.

iv. By Reversible Formation of Stable Inactive Intermediates. The carbonium ions and carbanions involved in the above-described racemizations are

$$(-)\text{-}C_6H_5CHClCH_3 \underset{\text{or HCOOH}}{\overset{\text{Liq. }SO_2}{\rightleftharpoons}} \underset{\text{Inactive}}{C_6H_5CH{=}CH_2} + HCl \rightleftharpoons (+)\text{-}C_6H_5CHClCH_3$$

Fig. 4-7. Racemization of α-phenethyl chloride.

probably true intermediates but of very short half-life. In contrast, this section will be concerned with the reversible formation of symmetric intermediates which are stable entities in their own right. An example is the racemization of α-phenethyl chloride upon dissolution in liquid sulfur dioxide or formic acid; this has been shown to proceed by a dehydrohalogenation-hydrohalogenation process[5] (Fig. 4-7). Racemization of alcohols in the presence of sodium or aluminum alkoxides is another case in point; this reaction requires the presence of a trace of carbonyl compound (usually formed by air oxidation) and

$$(-)\text{-}CH_3\overset{*}{C}HOHCH_2CH_3 \xrightleftharpoons{\substack{OX \\ | \\ CH_3CHCH_2CH_3}} \underset{\text{Inactive}}{CH_3COCH_2CH_3} \rightleftharpoons (+)\text{-}CH_3\overset{*}{C}HOHCH_2CH_3$$

$$(-)\text{-}\overset{C_2H_5}{\underset{CH_3}{>}}CHCH_2OH \overset{a}{\rightleftharpoons} (-)\text{-}\overset{C_2H_5}{\underset{CH_3}{>}}CHCHO \overset{b}{\rightleftharpoons} (+)\text{-}\overset{C_2H_5}{\underset{CH_3}{>}}CHCHO \overset{a}{\rightleftharpoons} (+)\text{-}\overset{C_2H_5}{\underset{CH_3}{>}}CHCH_2OH$$

[a] Alkoxide.
[b] Base.

Fig. 4-8. Racemization of 2-butanol and 2-methyl-1-butanol by alkoxides.

is known to involve a carbinol-carbonyl equilibrium (cf. page 72) (Fig. 4-8). A special case, also shown in Fig. 4-8, concerns the racemization of active amyl alcohol by sodium (actually sodium amyloxide) at 200°; in this case the

[3] K. Bodendorf and H. Böhme, *Ann.*, **516**, 1 (1935).
[4] E. L. Eliel, P. H. Wilken, and F. T. Fang, *J. Org. Chem.*, **22**, 231 (1957); cf. R. L. Burwell and A. D. Shields, *J. Am. Chem. Soc.*, **77**, 2766 (1955).
[5] E. D. Hughes, C. K. Ingold, and A. D. Scott, *Nature*, **138**, 120 (1936).

alcohol is reversibly oxidized to the corresponding aldehyde, which, in turn, is racemized by the anion-forming mechanism described above.[6]

Racemization through reversible formation of a ketene has been observed[7] in the case of 2,4-dichloro-3-phenylcyclobutenone upon heating in chloroform solution (Fig. 4-9). Enolization (which would lead to a cyclobutadiene structure) is ruled out in this case by the observation that no H-D exchange on the cyclobutenone occurs when racemization is effected in CH_3CO_2D as a solvent. When the solvent is ethanol, ketene formation becomes irreversible and an ethyl ester is formed at the same rate at which racemization occurs (Fig. 4-9).

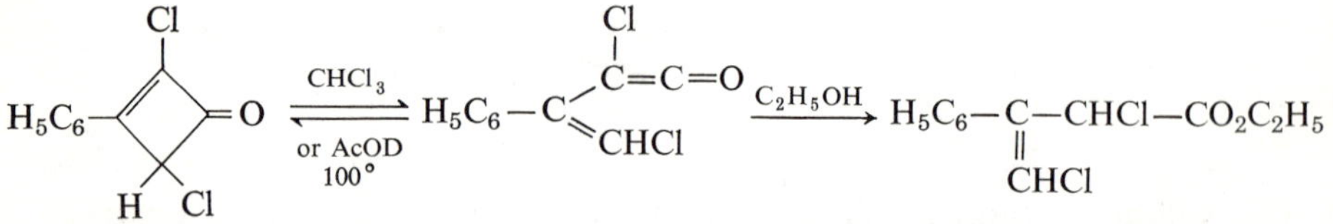

Fig. 4-9. Racemization through ketene intermediate.

v. By Walden Inversion. The racemization of 2-iodoöctane by potassium iodide in refluxing acetone involves a process known as "Walden inversion": $(+)\text{-}CH_3CHIC_6H_{13} + I^- \rightleftharpoons I^- + (-)\text{-}CH_3CHIC_6H_{13}$. This process will be considered in detail in Sec. 5-4*f*.

d. By Chemical Transformation. It is possible to change a dissymmetric molecule to its enantiomeric form without ever breaking any of the bonds leading to the asymmetric atom.† An example has been shown in Fig. 3-4.

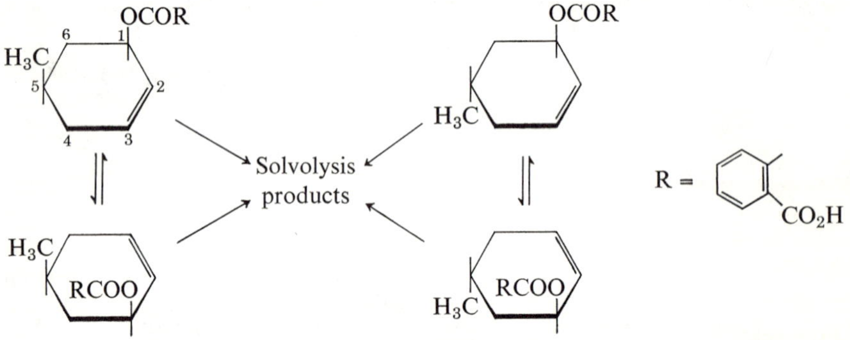

Fig. 4-10. Racemization by rearrangement.

If such a change is brought about in reversible fashion, it will lead to racemization. A case in point[8] is the racemization of 5-methyl-2-cyclohexenyl acid phthalate in aqueous acetone at 100° (Fig. 4-10). Two simultaneous processes take place, namely, solvolysis (to a mixture of 5-methyl-1,3-cyclo-

† More properly we should speak of a "dissymmetric grouping," not an asymmetric atom.

[6] Cf. Ref. 83, p. 866.

[7] E. F. Jenny and J. D. Roberts, *J. Am. Chem. Soc.*, **78**, 2005 (1956).

[8] H. L. Goering and E. F. Silversmith, *J. Am. Chem. Soc.*, **77**, 1129 (1955).

hexadiene, phthalic acid, and 5-methyl-2-cyclohexenols) and rearrangement of the phthalate group from the 1 to the 3 position. Inasmuch as this rearrangement converts the molecule to its mirror image, it leads to racemization without any other observable change. The (−)-*cis*-phthalate is converted to (±)-*cis*-phthalate whereas the (−)-*trans*-phthalate gives (±)-*trans*-phthalate. No interconversion of the *cis* to the *trans* series, and vice versa, is observed. The racemizing rearrangement is believed to involve an ion pair in which the phthalate anion is associated with one or another of the faces of the molecule (Fig. 4-11).† Racemization is evidently due to change of configuration at the No. 5 carbon, even though no bond to this carbon is ever broken. Also configuration at the newly created asymmetric carbon C_3 is opposite to that at C_1 in the starting material.

e. Epimerization, Mutarotation, and Asymmetric Transformation.[91] Although epimerization, mutarotation, and asymmetric transformation do not constitute racemization, they are closely related to racemization and will be taken up in this section.

i. Epimerization. Epimerization is defined as change in configuration (arrangement of the groups) at *one* asymmetric atom in a compound having

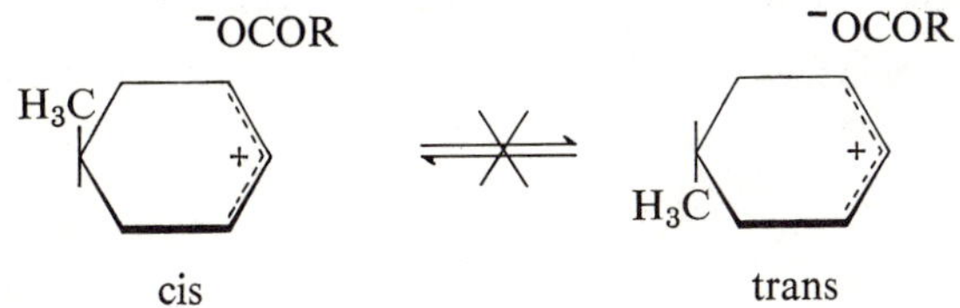

Fig. 4-11. Ion pairs involved in 5-methyl-2-cyclohexenyl acid phthalate rearrangement.

more than one such atom. Except in very unusual circumstances, epimerization of an optically active compound does not involve racemization, since only one of several asymmetric atoms is affected. Rather, epimerization involves interconversion of diastereoisomers. Even a racemic modification may be subjected to epimerization if it contains more than one asymmetric center. In this case, epimerization involves partial conversion to a diastereoisomeric racemic modification. An example is shown in Fig. 4-12. Treatment of (−)-menthone (1) with a base will effect a partial change of configuration at C_4, the asymmetric center adjacent to the carbonyl group (cf. page 35), but will leave the arrangement at C_1 unchanged. The product of the transformation is (+)-isomenthone (2). Because (1) and (2) are diastereoisomers rather than enantiomers, in general, they differ in free energy and therefore equilibrium does not correspond to a 50-50 mixture of the two diastereoisomers. In the case of the menthones, the equilibrium constant[8a] is

† The evidence presented here does not exclude a cyclic transition state of type *i*. However, such a transition state does not readily explain the effect of the solvent dielectric constant on the rate of racemization (Ref. 8) and is excluded on other grounds; cf. H. L. Goering and J. T. Doi, *J. Am. Chem. Soc.*, **82**, 5850 (1960).

(*i*)

[8a] J. Read, G. J. Robertson, and A. M. R. Cook, *J. Chem. Soc.*, 1276 (1927).

close to 2.3, so that equilibrium corresponds to about 70% (1) and 30% (2). Since (3) and (4) have the same free energies as their respective enantiomers, the equilibrium of (3) and (4) is the same as that of (1) and (2), namely, 70% (3) and 30% (4). Since (±)-menthone is a mixture of equal parts of (1) and (3) and (±)-isomenthone is a mixture of equal parts of (2) and (4), the equilibrium of (±)-menthone and (±)-isomenthone is the same (70-30) as that of their pure enantiomers (1) and (2) [or (3) and (4)].

Diastereoisomers that differ in configuration at only one asymmetric center† are sometimes called "epimers"; thus menthone and isomenthone are epimers.‡

ii. Mutarotation and First-order Asymmetric Transformation. In 1846, Dubrunfaut discovered that, when glucose is dissolved in water and the optical activity of the solution observed, there is a gradual change in rotation from an initial value corresponding to $[\alpha]_D^{20}$ of $+111°$ to an equilibrium value of $[\alpha]_D^{20}$ of $+52.5°$. Lowry, in 1899, coined the name "mutarotation" for this

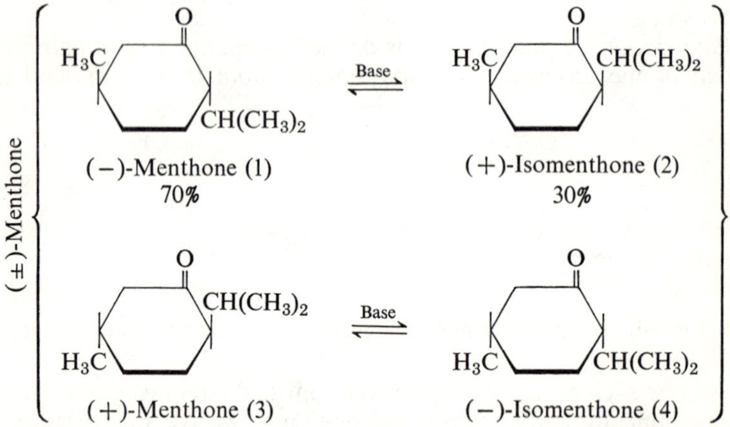

Fig. 4-12. Epimerization of menthone and isomenthone.

phenomenon. Mutarotation, then, is the spontaneous change, with time, in the rotation of freshly prepared solutions of certain optically active substances. Eventually, their rotation reaches an equilibrium value, generally different from zero. Mutarotation may be the result of either a spontaneous epimerization or a spontaneous structural change. In the case of (+)-glucose, mutarotation involves a change of configuration at the No. 1 carbon (the so-called "anomeric center") owing to an opening and reclosing of the hemiacetal

† The term "asymmetric center" is sometimes used to denote a focal point of asymmetry, such as an asymmetrically substituted atom; it is synonymous with "dissymmetric grouping" mentioned earlier.

‡ Originally, the terms epimers and epimerization were coined in the sugar series, epimers being sugars differing in configuration at the No. 2 carbon, such as glucose and mannose (Fig. 3-15). Since this is the center adjacent to the potential aldehyde group in the sugar, epimerization of such sugars may, in principle, be effected by a base. In actual fact, this reaction is very unclean, and epimerization of an aldohexose is best carried out by oxidation to the corresponding gluconic acid, epimerization of the acid by means of pyridine, lactonization, and reduction back to an aldohexose mixture: E. Fischer, *Ber.*, **23**, 799 (1890).

ring (Fig. 4-13). The intermediate open-chain aldehyde form is present in negligibly small concentration. Equilibrium corresponds to 38% of the α and 62% of the β form.

Figures 4-14 and 4-15 illustrate mutarotation due to structural changes. In the gluconolactones (Fig. 4-14) mutarotation is caused by partial hydrolysis to gluconic acid in aqueous solution.[8b] In aniline camphor-10-sulfonate (Fig. 4-15) it is due to isomerization to a ketimine or anil.[8c]

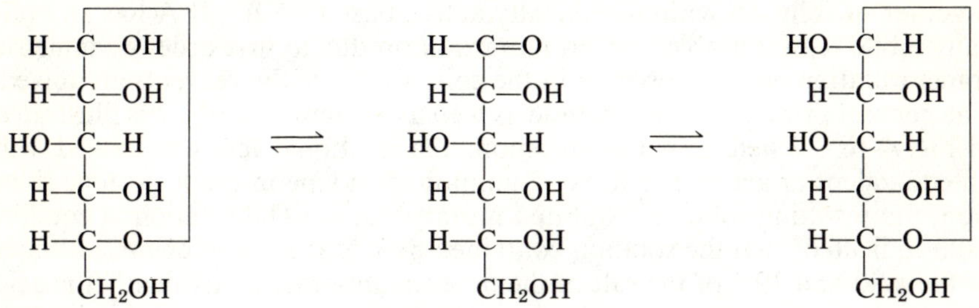

α form $[\alpha]_D^{20}$ +111° ⇌ Equilibrium mixture $[\alpha]_D^{20}$ +52.5° ⇌ β form $[\alpha]_D^{20}$ +19.2°

Fig. 4-13. Mutarotation of (+)-glucose.

In general, the rate of mutarotation depends on temperature, solvent, and catalyst. The mutarotation of glucose is known to be acid-base catalyzed. Temperature and solvent also affect the position of the equilibrium.

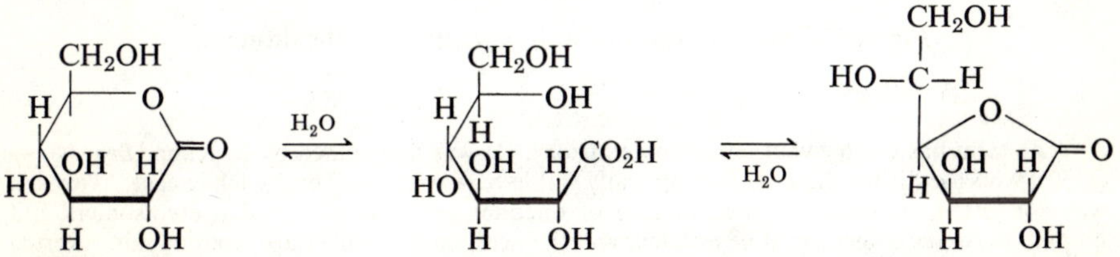

δ-Gluconolactone, $[\alpha]_D^{20}$ +63.5° ⇌ (H_2O) Gluconic acid, $[\alpha]_D^{20}$ −6.7° ⇌ (H_2O) γ-Gluconolactone, $[\alpha]_D^{20}$ +67.8°

(Equilibrium mixture, $[\alpha]_D^{20}$ +12°)

Fig. 4-14. Mutarotation of the gluconolactones.

First-order asymmetric transformation is closely related to mutarotation, but whereas mutarotation is defined phenomenologically, first-order asymmetric transformation is defined by origin as a spontaneous epimerization in solution. Thus, all mutarotations due to configurational (rather than struc-

[8b] K. Rehorst, *Ber.*, **61**, 163 (1928).

[8c] R. S. Schreiber and R. L. Shriner, *J. Am. Chem. Soc.*, **57**, 1306 (1935).

tural) changes involve first-order asymmetric transformations,† and first-order asymmetric transformations usually manifest themselves in mutarotation. The different origin of the two terms is largely a historical one.† In many cases, asymmetric transformation involves not the dissolution of a single pure diastereoisomer but the formation and transformation of a mixture of diastereoisomers in solution. Suppose that a racemic acid, (±)-A, is brought together in solution with an optically active base (−)-B. If A has an optically labile (epimerizable) center, mutarotation due to first-order asymmetric transformation will be observed in the solution, once the reagents are mixed. The general principle of the method as well as an actual example is illustrated in Fig. 4-16. When (±)-chlorobromomethanesulfonic acid was treated with an equivalent of active 1-hydroxy-2-aminohydrindane in dilute acetone solution, the resulting solution exhibited mutarotation. The rotation at equilibrium indicated[9] that the solution contained 81% of the salt of one enantiomer of the acid and 19% of the salt of the other enantiomer.‡ Evidently the acid is very easily converted to its enolate (the α-hydrogen being highly acidic), and therefore epimerization of the two diastereoisomeric salts occurs; one

$CH_2SO_2O^-C_6H_5NH_3^+$ (O=) $\xrightleftharpoons{CHCl_3}$ $C_6H_5N=$ (CH_2SO_2OH) $+H_2O$

Aniline camphor-10-sulfonate $[\alpha]_D^{25}$ +37.5°

2-*N*-(Phenylketimino)-10-camphanesulfonic acid $[\alpha]_D^{25}$ −170.5°

(Equilibrium mixture, $[\alpha]_D^{25}$ +10.7°)

Fig. 4-15. Mutarotation of aniline camphor-10-sulfonate.

† The term has a somewhat unfortunate history. It was first coined by R. Kuhn [*Ber.*, **65**, 49 (1932)] who used it for the salt of an optically stable base with an optically labile acid. This salt was converted, in solution, to a mixture of unequal amounts of two diastereoisomers, but no pure diastereoisomer could be isolated, and the acid, on being liberated from its salt, reverted to the racemic form faster than its rotation could be measured. Kuhn contrasts such "asymmetric transformations of the first kind" with those "of the second kind," where pure diastereoisomers can be isolated and the labile material recovered in an, at least fleetingly, active state. This distinction is not, however, a useful one, since there is no clear-cut division between optically stable and optically labile compounds but rather a gradual transition from one to the other (see especially Chap. 6). The terms will therefore be used here in the sense of M. M. Jamison and E. E. Turner [*J. Chem. Soc.*, 437 (1942)], first-order asymmetric transformations referring to transformations in solution and second-order ones (page 63) referring to cases where one diastereoisomer actually crystallizes. There is also a difficulty in translation; Kuhn's "erster Art" was translated into English as "first-order," whereas the proper meaning is "of the first kind."

‡ The salt of the (−) acid as well as the salt of the (±) acid could be obtained in crystalline form by working in more concentrated solution. By studying the mutarotation of these salts and extrapolating to zero time, the rotation of the pure salts was obtained, and from this the position of the equilibrium mixture could be calculated.

[9] J. Read and A. M. McMath, *J. Chem. Soc.*, 1572 (1925).

salt happens to be more stable than its diastereoisomer by a factor of 4. All attempts to liberate the active acid from its salts failed, for the same labile hydrogen that causes rapid epimerization of the salts also causes rapid racemization of the free acid.

First-order asymmetric transformations are particularly common in the biphenyl series, and further examples will be found in Chap. 6.

4-3. Properties of Racemic Modifications

In the gaseous or liquid state and in solution, a racemic modification is usually an ideal or nearly ideal mixture of equal numbers of enantiomeric molecules.† The usual physical laws applicable to ideal mixtures apply, and since the physical properties of the enantiomers (except toward such asymmetric entities as polarized light) are identical, they are also identical with the

$$2(\pm)\text{-A} + 2(-)\text{-B} \rightarrow \underset{50\%}{(+)\text{-A}\cdot(-)\text{-B}} + \underset{50\%}{(-)\text{-A}\cdot(-)\text{-B}} \rightarrow \underset{(50+x)\%}{(+)\text{-A}\cdot(-)\text{-B}} + \underset{(50-x)\%}{(-)\text{-A}\cdot(-)\text{-B}}$$

$$\underset{50\%}{\text{Br—CHCl—SO}_2\text{OH}} + \underset{50\%}{\text{Cl—CHBr—SO}_2\text{OH}} + 2\text{R—NH}_2 \longrightarrow 2\ \text{(Cl)(Br)}\bar{\text{C}}\text{—SO}_2\text{O}^- \ \text{RNH}_3^+ \ \ \text{H}^+$$

$$\rightleftharpoons \underset{81\%}{\text{Br—CHCl—SO}_2\text{O}^- \ \text{RNH}_3^+} + \underset{19\%}{\text{Cl—CHBr—SO}_2\text{O}^- \ \text{RNH}_3^+}$$

R = 1-hydroxy-2-indanyl (OH), Active

Fig. 4-16. First-order asymmetric transformation.

properties of the racemic modification.[10] Thus racemic modifications have the same boiling points as the pure enantiomers (except for the aforementioned slight deviation from ideality in a few cases); they also have the same refractive index and density in the liquid state[10] and the same infrared spectrum either in the liquid state or in solution.

The same is not true in the solid (crystalline) state.[11] Intercrystalline forces between molecules are highly specific and sensitive to even minor changes in geometry. Therefore, in the solid state, although a molecule of the dextrorotatory form bears the same relation to another molecule of the (+) form as a molecule of the levorotatory form bears to another (−) mole-

† Significant deviations from ideality have been observed in some optically active liquids capable of hydrogen bonding; see, for example, F. B. Thole, *J. Chem. Soc.*, **103**, 19 (1913); C. J. McGinn, *J. Phys. Chem.*, **65**, 1896 (1961).

[10] H. Mauser, *Chem. Ber.*, **90**, 299 (1957).
[11] H. Mauser, *Chem. Ber.*, **90**, 307 (1957).

cule, the interrelation of (−) and (+) molecules is different. As a result, in the solid state, deviations from ideal behavior are generally encountered. The following three cases may arise:

a. Racemic Mixtures. It may happen that in a crystal each enantiomer has greater affinity for molecules of the same kind than for molecules of the other enantiomer. In that case, once a molecule of the (+) form is laid down in the crystal, only (+) molecules will grow on it [and similarly for (−) molecules]. Thus, the macroscopic crystal (or, at least, the unit cell) will correspond to either the (+) or the (−) form. The racemic modification will thus be a gross mixture of crystals of the two forms, and one speaks of a "racemic mixture" (also occasionally called "conglomerate").

Since a racemic mixture is a mixture of crystals of the (+) and (−) forms, its properties are, in most respects, similar to those of the pure enantiomers. In particular, this applies to the X-ray powder diagram and the infrared spectrum in the solid state. However, the melting point of a racemic mixture

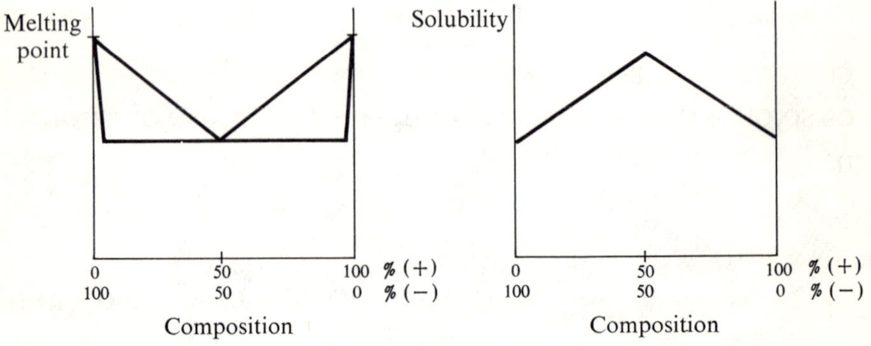

Fig. 4-17. Melting-point and solubility diagrams of racemic mixtures.

(like that of any typical mixture) is lower than that of the pure components and its solubility is higher. Melting point and solubility diagrams of a racemic mixture and the two pure enantiomers are shown in Fig. 4-17. The racemic mixture corresponds to a eutectic which, in this particular case, always occurs at the 50-50 composition mark. An example of a racemic mixture is (±)-sodium ammonium tartrate, provided that the salt is crystallized from water at a temperature below 27°C.

b. Racemic Compounds. A rather more common situation than that described above is that the molecules of one enantiomer have greater affinity for those of the opposite enantiomer than for their own kind. In that case, opposite enantiomers pair up in the unit cell of the crystal which will thus contain equal numbers of (+) and (−) molecules (sometimes just one of each kind). In this case we obtain a true compound in the stoichiometric sense, since any macroscopic crystal upon subdivision (down to the level of the unit cell, though not, of course, down to the molecular level) always gives fragments containing equal numbers of molecules of the two enantiomers. A compound of this type (which exists only in the solid state) is called

a "racemic compound" or a "racemate." Racemic compounds have lower enthalpies than the pure enantiomers. Being true compounds, they differ in most physical properties from the enantiomers; for example, they have different infrared spectra in the solid state, different X-ray powder diagrams, different melting points, and different solubilities. The melting-point diagram for an optically active material that forms a racemic compound is shown in Fig. 4-18; the melting point of the compound lies at a maximum of the curve which may be either higher or lower than the melting point of the pure enantiomers. The corresponding solubility curve is also shown in Fig. 4-18.

A number of cases are known where one and the same racemic modification forms a mixture below a certain temperature and a compound above it, or vice versa. Examples are sodium ammonium tartrate which forms a conglomerate *below* 27°C. but a racemate if crystallized above that temperature, and rubidium tartrate which forms a racemic mixture *above* 40° but a compound below that temperature. In some cases, this change in affinity of like

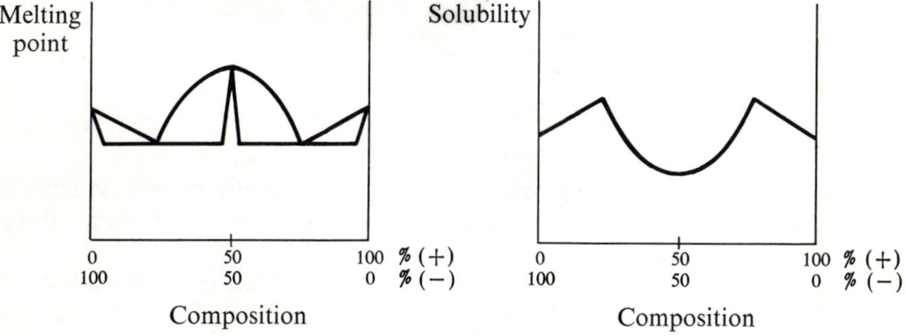

Fig. 4-18. Melting-point and solubility diagrams of racemic compounds. (The compound may melt at a lower point than the pure enantiomers in some cases.)

and unlike molecules to each other with temperature may be accounted for by a change in the number of solvent molecules in the crystal. Thus, sodium ammonium tartrate crystallizes as a mixture (below 27°) with four molecules of water but as a compound (above 27°) with only two. The change in the number of water molecules in the crystal evidently changes the relative ease of fitting like and unlike enantiomers together.

c. Racemic Solid Solutions. In some instances, racemic modifications show nearly ideal behavior even in the solid state, meaning that there is little difference in affinity between molecules of like or opposite configuration. In that case, the arrangement of the molecules in the solid is random and one obtains a "racemic solid solution" or mixed crystal. Such solid solutions are identical with the enantiomers in all respects; even the melting point and solubility of the racemic solid solution either are the same as those of the enantiomers or differ from them only very slightly. A melting-point diagram of a racemic solid solution is shown in Fig. 4-19. The solid horizontal line represents an ideal case; in actual fact the curve may be slightly concave upward or downward (dashed lines). (±)-Camphor oxime is obtained as a racemic solid solution if crystallized above 103°C., although below that temperature it forms a racemic compound.

A racemic compound may be distinguished from a mixture or solid solution by a comparison of its infrared spectrum in the solid state with that of the enantiomers. Only the compound shows a different spectrum. Sometimes the differences are quite profound. For example, the active acid phthalate of *p*-ethylphenylmethylcarbinol (Fig. 4-20) shows evidence of *intra*-

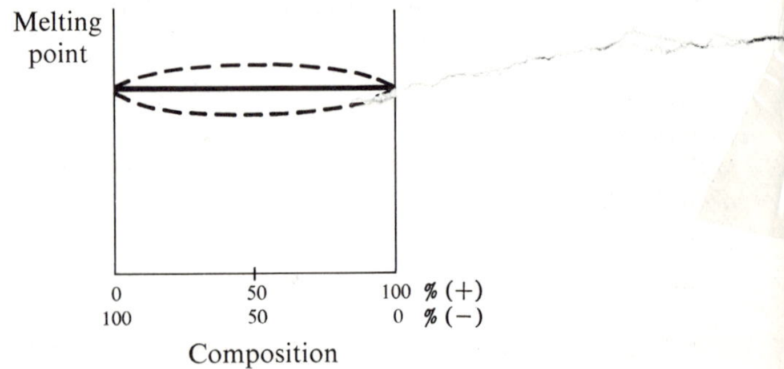

Fig. 4-19. Melting-point diagram of racemic solid solution.

molecular hydrogen bonding only, whereas the racemate shows evidence of strong *inter*molecular hydrogen bonding, presumably between molecules of the (+) and (−) forms.[12]

Another way of distinguishing the three racemic forms is by making use of the melting-point or solubility diagrams.[13] Admixture of one of the pure enantiomers to a racemic mixture increases the melting point, whereas the same process in the case of a compound leads to a depression. In the case of a solid solution, not much change results from such admixture. (For reli-

Fig. 4-20. *p*-Ethylphenylmethylcarbinyl acid phthalate.

able results, it is often necessary to determine the melting points of several different mixtures, which in essence means tracing part of the diagrams of Fig. 4-17, 4-18, or 4-19.) Also, it might be noted that a saturated solution of a racemic mixture or solid solution is saturated also with respect to either enantiomer, but the same is not true for all racemic compounds. Thus, if to a saturated solution of the racemic modification a few crystals of one of the

[12] E. L. Eliel and J. T. Kofron, *J. Am. Chem. Soc.*, **75,** 4585 (1953).
[13] H. W. B. Roozeboom, *Z. physik. Chem.*, **28,** 494 (1899); see also Ref. 11.

enantiomers are added, the crystals will dissolve (and the solution will become optically active) only if one is dealing with a compound.†

Yet another way of distinguishing the racemic forms is by examination of the unit cell through X-ray crystallography.[14] The unit cell of the compound contains equal numbers of enantiomeric molecules, whereas in the mixture or solid solution it contains only molecules of one enantiomer or the other, but not both.

4-4. Resolution of Racemic Modifications

By "resolution" of a racemic modification is meant the separation (not usually in quantitative recovery) of the two enantiomers‡ in the pure state. This process is of considerable practical importance, for, as indicated in Sec. 4-2*b*, synthetic processes usually lead to racemic modifications. On the other hand, many of the chemicals occurring in both plants and animals are found there as pure enantiomers. Any total synthesis of such naturally occurring compounds must, therefore, involve a resolution step.

Of the several methods of resolution which will be described in this section, only resolution via conversion to diastereoisomers (Sec. 4-4*b*) and resolution by biochemical methods (Sec. 4-4*e*) are generally useful. The other methods are included mainly because of their considerable theoretical interest.¶

a. Resolution by Mechanical Separation of Crystals. When one is dealing with a racemic *mixture,* macroscopic crystals of either the (+) or the (−) form are usually present. Provided that the crystals are visually distinct, it should be possible to pick them apart by means of tweezers and thus to effect resolution. The first resolution ever to be brought about was achieved in this way by Louis Pasteur in 1848.[15] Pasteur prepared the sodium ammonium salt of racemic tartaric acid and allowed it to crystallize in large crystals by slow evaporation of the aqueous solution. He then picked apart the two kinds of crystals, making use of the fact that they were hemihedric, i.e., showed dissymmetry in the crystal state. All crystals having the dissymmetric facets disposed in one orientation belonged to one enantiomer, whereas the mirror-image crystals, whose dissymmetric facets were disposed in the opposite orientation, belonged to the other enantiomer (cf. Fig. 1-1). When the two batches of crystals were separately redissolved, the resulting solutions rotated the plane of polarized light to an equal extent in opposite directions. It is fortunate (and perhaps fortuitous) that Pasteur allowed his

† Provided that the compound is less soluble than the enantiomers, as in the case shown in Fig. 4-18.

‡ Sometimes the term is employed even though only one of the enantiomers is obtained pure. When some separation of the enantiomers occurs but neither is obtained completely free of the other, one speaks of "partial resolution."

¶ Sometimes the best route to an optically active compound is synthesis (by conventional means) from optically active precursors rather than resolution. Examples are given in Chap. 5 (Sec. 5-5).

---

[14] For example, K. Pettersson, *Arkiv Kemi,* **7**, 347 (1954).

[15] L. Pasteur, *Ann. chim. et phys.,* [3]**24**, 442 (1848).

solutions to evaporate spontaneously in the cool Parisian climate, for, as indicated in Sec. 4-3, sodium ammonium tartrate crystallizes as a racemic mixture only below 27°; above this temperature it crystallizes as a compound which cannot, of course, be separated mechanically.

The method of mechanical separation is rarely a practical way of separating enantiomers. It is tedious, it cannot be applied to racemic compounds or solid solutions, and it can be applied only to those mixtures in which the crystals of the enantiomers are visually distinct. Although it is now recognized that crystals of dextrorotatory and levorotatory enantiomers always show hemihedrism and are mirror images of each other (as crystals as well as molecules), the distinctive crystal faces are often so poorly developed as to be useless from the practical standpoint.

A more useful variation of mechanical separation is the method of inoculation, originally discovered by Gernez.[16] If a saturated solution of a racemic mixture is carefully inoculated with a pure crystal of one of the enantiomers (with respect to which the solution is supersaturated; cf. Fig. 4-17), the crystal will grow and an appreciable amount of one active form may thus be separated from the racemic mixture. This type of resolution is also applicable to racemic compounds, provided that the compound is more soluble than either of the pure enantiomers. If no crystal of either pure enantiomer is available for inoculation, a crystal of another optically active substance may sometimes serve as a seed. For example, (+)-sodium ammonium tartrate can be crystallized from a solution of the racemic modification not only by inoculation with a crystal of the (+) salt but also by inoculation with (−)-asparagine, $H_2NCOCH_2CH(NH_2)CO_2H$. Sometimes inoculation may even be effected with an optically inactive substance; for example,[17] active asparagine has been crystallized from solutions of the racemic modification by inoculation with crystals of glycine, $H_2NCH_2CO_2H$.† In a few instances, spontaneous crystallization‡ of active material from solutions of the *dl* form has been observed.[18]

Crystallization methods by themselves are rarely practical methods of resolution, but they are often used in a practical way in conjunction with other methods. For example, phenylmethylcarbinyl hydrogen phthalate,

† The success of this experiment has been attributed (Ref. 17) to the possibility that glycine, although a symmetric molecule, might, like quartz, form dissymmetric crystals. In fact, although the ordinary or alpha modification of glycine is symmetric [G. Albrecht and R. B. Corey, *J. Am. Chem. Soc.*, **61**, 1087 (1939)], another crystalline modification, called gamma, exists which is dissymmetric [Y. Iitaka, *Acta Cryst.*, **11**, 225 (1958)].

‡ Surprisingly, K. Vogler and M. Kofler [*Helv. Chim. Acta*, **39**, 1387 (1956)] have reported the resolution of 3,3-diethyl-5-methyl-2,4-diketopiperidine *which forms a racemic solid solution* by this method.

---

[16] M. Gernez, *Compt. rend.*, **63**, 843 (1866). H. E. Zaugg, *J. Am. Chem. Soc.*, **77**, 2910 (1955). See also L. Velluz, G. Amiard, and R. Joly, *Bull. soc. chim. France*, **20**, 342 (1953); L. Velluz and G. Amiard, *ibid.*, **20**, 903 (1953); G. Amiard, *ibid.*, 447 (1956); *Experientia*, **15**, 38 (1959) for a related method.

[17] I. Ostromisslensky, *Ber.*, **41**, 3035 (1908).

[18] (*a*) R. C. Ferreira, *Nature*, **171**, 39 (1953); (*b*) E. Havinga, *Biochim. et Biophys. Acta*, **13**, 171 (1954).

$$C_6H_5\overset{*}{C}H(CH_3)OCOC_6H_4CO_2H\text{-}o$$

after being resolved to the extent of about 95% or so by means of brucine (Sec. 4-4*b*) is dissolved in carbon disulfide and seeded with a small crystal of the racemic compound. In this particular case, the racemic compound is less soluble than the enantiomers (cf. Fig. 4-18) and so most of the excess of it left in the resolved phthalate crystallizes out. The mother liquor is decanted and petroleum ether added to it, whereupon the enantiomeric phthalate crystallizes in turn. In other cases, where a racemic mixture is formed or where the racemic compound is more soluble than the enantiomers, the active form may be purified by crystallization, any residual racemic material remaining in the mother liquor (cf. Fig. 4-17). This is a very common way of further purifying active material which has already been partially resolved to a considerable extent by one of the methods to be discussed below.

b. Resolution by Formation of Diastereoisomers.[94] **Second-order Asymmetric Transformations.** When a racemic modification is allowed to interact with an optically active material to give a derivative (such as a salt), in actual fact two diastereoisomeric derivatives result. For example, in a reaction of a racemic acid (±)-A with an active base, (−)-B, the individual molecules of the acid are either (+) or (−), and, therefore, the individual molecules of the salt formed are either (+)-A·(−)-B or (−)-A·(−)-B. These two types of salt molecules are evidently no longer enantiomers, but diastereoisomers. Therefore, they have different properties and may, in general, be separated on the basis of this difference in properties. Although distillation[19] and chromatographic separation[20, 95] have been employed,† the most efficient method of separating such diastereoisomers is by crystallization. This is because, as already mentioned, crystal structure is apt to be particularly sensitive to minor variations in molecular architecture (such as the difference between diastereoisomers).

Several conditions should be fulfilled by a good resolving agent. First, the compound between the resolving agent and the substance to be resolved should be easily formed and should also be easily broken up, for once one of the diastereoisomers, e.g., (−)-A·(−)-B, is obtained in the pure state, it must be decomposed chemically so that pure (−)-A may be recovered. This condition is generally met by salts, which are usually formed readily by mixing the organic acid and base in a solvent and may be decomposed, following resolution, by treatment with mineral acid (if the organic acid is to be recovered) or mineral base (if the organic base is desired). It is easiest, therefore, to resolve acids (carboxylic or sulfonic) or amines. Other substances, to be resolved, must often be transformed first into an acid, as will be explained further below, or else synthesized from a resolved acid or base

† Gas chromatography in particular may provide a useful separation method for diastereoisomers. It has been applied to the resolution of (±)-camphor through separation of the ketals formed with (−)-2,3-butanediol: J. Casanova and E. J. Corey, *Chemistry & Industry (London)*, 1664 (1961).

[19] M. E. Bailey and H. B. Hass, *J. Am. Chem. Soc.*, **63**, 1969 (1941).
[20] M. M. Jamison and E. E. Turner, *J. Chem. Soc.*, 611 (1942).

(Sec. 5-5). Secondly, the compound between the resolving agent and substance to be resolved must be nicely crystalline,† and there must be an appreciable difference in solubility between (+)-A·(−)-B and (−)-A·(−)-B.‡ Whether this condition is fulfilled depends on both the nature of A and B and the solvent chosen. Unfortunately, resolution is, in this respect, still very much a matter of trial and error, and even in the papers of experienced investigators one is apt to find, from time to time, a statement that a certain compound resisted resolution by any one of a large number of combinations of resolving agents and solvents that were tried. The best hint that can be given for resolution of, let us say, an acid A is to use a base B and a solvent which have been found successful in the past in the resolution of a similar acid A′. The difficulty with this hint lies in the word "similar." For example, it might have been thought that mandelic acid, $C_6H_5CHOHCO_2H$, and atrolactic acid, $C_6H_5C(CH_3)OHCO_2H$, are similar, but whereas the former may be readily resolved by means of the commercially available (−)-ephedrine in aqueous alcoholic solution, the same resolving agent fails for atrolactic acid, and one has to resort to the (not commercially available) α-phenethylamine. Sometimes, even though (+)-A·(−)-B and (−)-A·(−)-B are, individually, nicely crystalline salts of well-differentiated solubility in a given solvent, resolution of (±)-A by (−)-B in the same solvent fails because of formation of a double salt (+)-A·(−)-B·(−)-A·(−)-B (in principle, similar to other types of double salts).¶ This case occurs, for example, in the attempted resolution[21] of β-pipecoline (Fig. 4-21) by means of tartaric acid.§

† It is well to keep in mind that the process of resolution involves crystallization of the desired substance, for example, (−)-A·(−)-B, in the presence of an *equal amount* of an undesirable impurity, (+)-A·(−)-B!

‡ Occasionally, it is possible to crystallize one diastereoisomer from a solution supersaturated with respect to both. For example, when α-phenethyl hydrogen phthalate,

$$C_6H_5\overset{*}{C}H(CH_3)OCOC_6H_4CO_2H\text{-}o$$

is treated with brucine in acetone solution, the salt of the (−)-phthalate with brucine precipitates more readily. Once this is filtered off, however, the filtrate deposits crystals of the (+)-phthalate brucine salt. When crystallization is not carried out carefully, both diastereoisomeric salts may crystallize together, thus spoiling the resolution process.

¶ In other cases, only partial resolution can be achieved because a salt of the type 2(+)-A·(−)-B·(−)-A·(−)-B (or similar type) crystallizes.

§ Because of the possibility of formation of such complexes, one cannot conclude that successful resolution of (±)-A with, say, (−)-B implies that (±)-B can be resolved with, say, (−)-A (Ref. 96). For assume:

$$(\pm)\text{-A} + (-)\text{-B} \rightarrow (+)\text{-A}\cdot(-)\text{-B (in soln.)} + \underline{(-)\text{-A}\cdot(-)\text{-B}}$$

where the latter salt crystallizes out. Also,

$$(\pm)\text{-B} + (-)\text{-A} \rightarrow (+)\text{-B}\cdot(-)\text{-A} + (-)\text{-A}\cdot(-)\text{-B}$$

Now (+)-B·(−)-A has the same solubility as (+)-A·(−)-B (since they are enantiomers), and it would appear at first sight that, if the first resolution succeeds because (−)-A·(−)-B is less soluble than (+)-A·(−)-B, the second resolution should also succeed. The flaw is that (−)-A·(−)-B is a dissymmetric molecule and therefore may form a molecular complex with (+)-B·(−)-A (precluding resolution) although it does not form a complex with (−)-B·(+)-A. Thus, no prediction can, in principle, be made regarding the feasibility of resolving (±)-B with (−)-A (Ref. 96).

[21] A. Ladenburg, *Ber.*, **27**, 75 (1894).

A third condition for a resolving agent is that it be either cheap or readily prepared or else readily and nearly quantitatively recoverable after completion of the resolution. If this condition is not fulfilled, the resolution, at least on a large scale, becomes excessively tedious or expensive. Among the bases, many alkaloids, such as brucine, strychnine, ephedrine, etc., are relatively inexpensive and also readily recoverable after resolution.† Quinine seems to be an exception in that it has a tendency to deteriorate on attempted recovery. Among the acids, (+)-10-camphorsulfonic acid is relatively inexpensive. (+)-Tartaric acid is so cheap that recovery may be forgone. On the other hand, active malic acid is very expensive and, since it is also hard to recover (being quite water-soluble), it is not a useful resolving agent except on a small scale.

A useful basic resolving agent which can be prepared readily in the laboratory is α-phenethylamine, $C_6H_5CH(NH_2)CH_3$. The *dl* form of this material is commercially available (1960). Resolution according to the original procedure[22] required the very expensive (−)-malic acid and was prohibitive on a large scale. An improved procedure[23] used pyroglutamic acid (pyrrolidone-5-carboxylic acid; cf. Fig. 4-23) which can be prepared by a carefully controlled pyrolysis of the relatively cheap (+)-glutamic acid,

$$HO_2CCH_2CH_2CH(NH_2)CO_2H$$

but is difficult to recover from the resolution. Finally, a very simple resolution of α-phenethylamine using the cheap (+)-tartaric acid has been developed,[24] and the active amine may now be considered readily available. The amine may also be resolved via the insoluble complex that the (+)-antipode forms with 2,3,4,6-tetraacetyl-D-glucose.[24a]

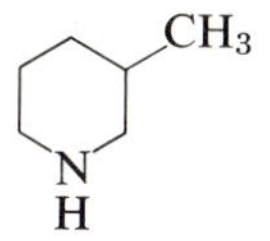

Fig. 4-21. β-Pipecoline.

A fourth, though not absolutely indispensable, condition for a resolving agent is that it should be available in the optically pure state. For, in principle, the substance to be resolved cannot be obtained in a higher state of optical purity than the resolving agent by mere crystallization of diastereoisomers. Suppose that a basic resolving agent (−)-B is 90% optically pure, i.e., that it consists of 90% (−)-B and 10% (±)-B. This means that, of any

† The salt of the alkaloid with the acid resolved by it is decomposed by pouring it, in alcoholic solution, into dilute aqueous hydrochloric acid and filtering or extracting the organic acid liberated. The alkaloid stays in the aqueous-alcoholic layer as hydrochloride and may be liberated and recovered by addition of ammonia.

[22] A. W. Ingersoll, *Org. Syntheses*, Coll. Vol. II, p. 506 (1943), after J. M. Lovén, *Ber.*, **29**, 2313 (1896).

[23] R. J. Dearborn and J. A. Stekol, U.S. Patent 2,528,267 (1950), *Chem. Abstr.*, **45**, 2984c (1951).

[24] W. Theilacker and H. G. Winkler, *Chem. Ber.*, **87**, 690 (1954).

[24a] B. Helferich and W. Portz, *Chem. Ber.*, **86**, 1034 (1953).

100 molecules, 95 are (−)-B and 5 are (+)-B.† If 200 molecules of (±)-(A) are combined with 200 molecules of this base, one will actually have 100 molecules of (+)-A, 100 molecules of (−)-A, 190 molecules of (−)-B, and 10 molecules of (+)-B. Assuming statistical combination, one will then get 95 molecules of (+)-A·(−)-B, 95 molecules of (−)-A·(−)-B, 5 molecules of (+)-A·(+)-B, and 5 molecules of (−)-A·(+)-B. Suppose, further, that (+)-A·(−)·B is the less soluble diastereoisomer and that 3/5 of it or 3/5 × 95 or 57 molecules of this diastereoisomer are recovered after several recrystallizations. Now, since (−)-A·(+)-B is enantiomeric with (+)-A·(−)-B, it has just the same solubility as the latter, and therefore 3/5 of it or 3/5 × 5 or 3 molecules will be found in the final crystallizate along with 57 molecules of (+)-A·(−)-B. After decomposition of the salt, there will be 57 molecules of (+)-A and 3 molecules of (−)-A; i.e., 95% of all molecules will be (+)-A and 5% will be (−)-A. But this means that resolution has proceeded to the extent of 90%,‡ i.e., to the extent of the optical purity of the resolving agent.

In practice, this limitation is not always a serious one. To begin with, one does not always need optically pure materials; i.e., it may not be necessary to achieve complete resolution. Secondly, after extensive resolution has been achieved, the enantiomers can often be further purified by crystallization, as described in Sec. 4-4*a*.

Most naturally occurring resolving agents (such as alkaloids) are readily obtained optically pure, but resolving agents prepared in the laboratory, especially if liquids, such as α-phenethylamine, can be obtained optically pure only with some difficulty.

In the following paragraphs, the resolution of specific types of organic compounds will be discussed. The list of resolving agents is not meant to be exhaustive—only the more common types are mentioned. No specific directions for the resolution of individual compounds will be given here; for these the reader is referred to some of the reference works listed at the end of the chapter (e.g., Ref. 94) and to the original literature.¶

i. Acids. The alkaloids brucine, strychnine, ephedrine, quinine, quinidine, cinchonine, cinchonidine, and morphine have been frequently used to resolve optically active acids. Certain synthetic bases, such as α-phenethylamine, menthylamine (prepared from natural, optically active menthol), and amphetamine are also useful. The formulas of most of these are shown in Fig. 4-22.

ii. Bases. The camphor derivatives camphor-10-sulfonic acid, α-bromocamphor-π-sulfonic acid, hydroxymethylene camphor, and camphoric acid; the menthol derivative menthoxyacetic acid; and the naturally occurring

† The reader should note the general proposition that x% of racemic modification and $(100 - x)$% of an active form, say the (−) form, correspond to $x/2$ molecules of the (+) form and $(100 - x/2)$ molecules of the (−) form. This is, of course, because the racemic modification is only a statistical aggregate and has no meaning on the molecular scale. In the limiting case, when $x = 100$, i.e., one is dealing only with the racemic modification, there are 50 (i.e., $^{100}/_{2}$) molecules of the (+) form and 50 $(100 - {}^{100}/_{2})$ molecules of the (−) form.

‡ Considering 95 molecules of (+)-A and 5 molecules of (−)-A as 90 parts of (+) isomer and 10 parts of (±) isomer.

¶ Resolution is still largely an experimental art. Sometimes procedures in the literature are difficult to repeat, and frequently a skilled experimenter may devise techniques superior to those previously described.

active forms of tartaric acid and malic acid have been used for resolution, among other acids. Also useful are diacetyltartaric acid, obtained by acetylation of tartaric acid; pyrrolidone-5-carboxylic acid, obtained by pyrolysis of naturally occurring optically active glutamic acid; and certain acetyl derivatives of amino acids.[25] Glutamic acid itself, having one free as well as one zwitterionic carboxyl group, may function as an acidic resolving agent; it is used in the commercial resolution of lysine.

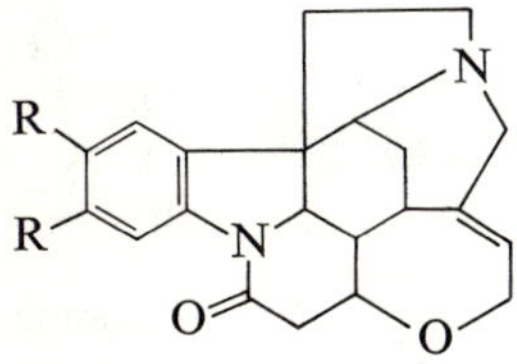

R = H Strychnine
R = CH_3O Brucine

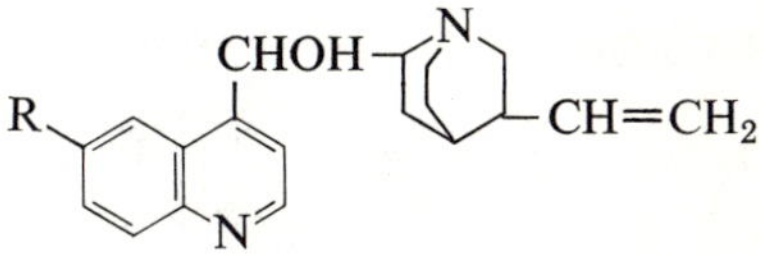

R = H Cinchonine, cinchonidine
R = CH_3O Quinine, quinidine

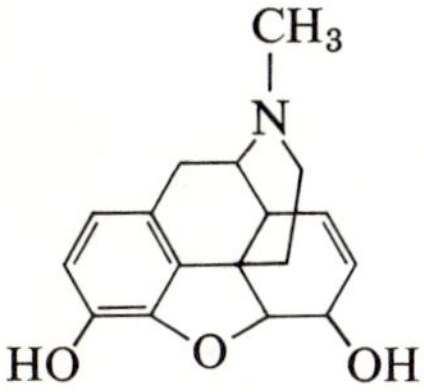

Morphine

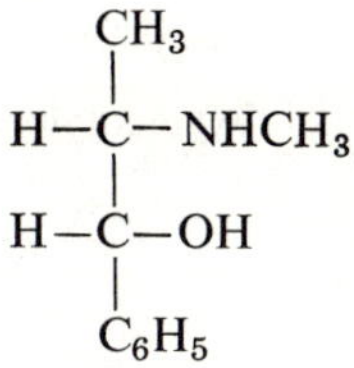

(−)-Ephedrine

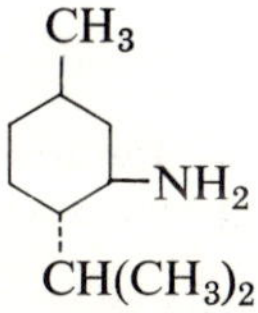

Menthylamine

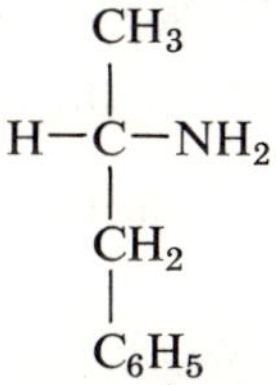

Amphetamine

Fig. 4-22. Formulas of common basic resolving agents.

The formulas of some of these resolving agents are shown in Fig. 4-23.

iii. Amino Acids. Because of their dipolar character, amino acids cannot usually be resolved, as such, using either optically active acids or active bases as resolving agents. A few exceptions are known; thus phenylglycine, $C_6H_5CH(NH_2)CO_2H$, has been resolved using camphor-10-sulfonic acid (which is a very strong acid). Some basic amino acids, which contain a free as well as a zwitterionic amine function, have been resolved by means of

[25] H. D. DeWitt and A. W. Ingersoll, *J. Am. Chem. Soc.*, **73**, 5782 (1951).

H_3C CO_2H
H_3C
H_3C
H CO_2H

Camphoric acid

HNO_3

CH_2SO_2OH O

H_2SO_4, Ac_2O

CH_3 O

Br_2

CH_3 O Br

$H_2SO_4 \cdot SO_3$ or $ClSO_2OH$

CH_3 O Br SO_2OH

Camphor-10-sulfonic acid

Camphor

α-Bromocamphor-π-sulfonic acid

NaOEt
$HCO_2C_5H_{11}$

CH_3 O HOHC

Oxymethylene camphor

CH_3 OH $CH(CH_3)_2$

$ClCH_2CO_2H$

CH_3 OCH_2CO_2H $CH(CH_3)_2$

Menthol

Menthoxyacetic acid

CO_2H
H—C—OR
RO—C—H
CO_2H

R = H Tartaric acid
R = CH_3CO Diacetyltartaric acid

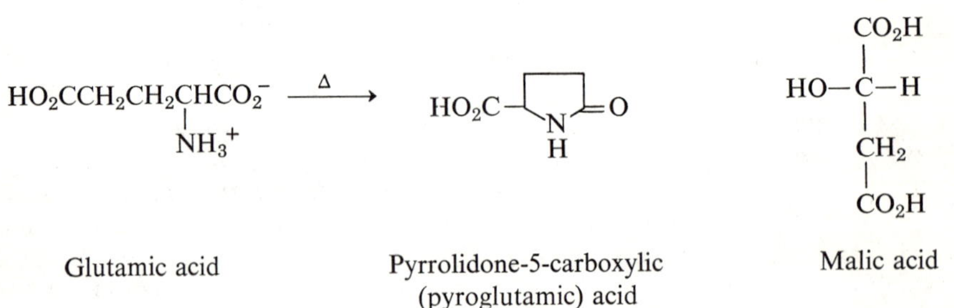

Fig. 4-23. Formulas of common acid resolving agents.

organic acids; thus (+)-tartaric acid has been employed to resolve (±)-histidine (Fig. 4-24), and (±)-lysine may be resolved by means of (+)-glutamic acid.

Commonly, however, amino acids are resolved in the form of their acyl derivatives, which are no longer endowed with zwitterionic properties and may therefore be resolved as typical acids. The classical example is alanine which was resolved in the form of its benzoyl derivative (Fig. 4-24) by Emil Fischer.[26] Crystallization of the brucine salt yielded the benzoyl derivative of (−)-alanine, whereas the strychnine salt gave the derivative of (+)-alanine. In the removal of the benzoyl group from the resolved derivative (by acid hydrolysis), some racemization occurred (note that the hydrogen on the asymmetric carbon is activated by the adjacent carbonyl group; cf. page 35). This difficulty can be largely avoided by subjecting the formyl derivative (Fig. 4-24) to resolution. Removal of the formyl group is effected under very mild conditions and does not lead to racemization of the resolved material.

iv. Alcohols.[86] Alcohols are most often resolved by prior conversion to their acid phthalate or succinate esters, formed by treating the alcohol with

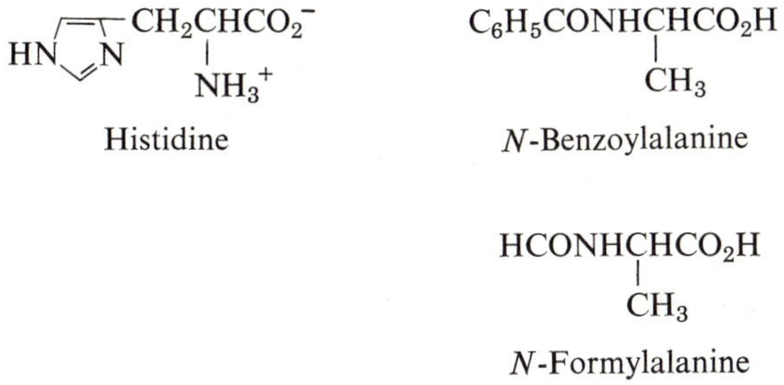

$$\text{HN}\diagdown\!\!\!\diagup\text{N (imidazole ring)}-CH_2\underset{\displaystyle NH_3^+}{\underset{|}{C}}HCO_2^-$$

Histidine

$$C_6H_5CONH\underset{\displaystyle CH_3}{\underset{|}{C}}HCO_2H$$

N-Benzoylalanine

$$HCONH\underset{\displaystyle CH_3}{\underset{|}{C}}HCO_2H$$

N-Formylalanine

Fig. 4-24. Some amino acids and derivatives thereof.

phthalic or succinic anhydride and pyridine. These half-esters are then resolved as typical acids, e.g., by means of the alkaloids brucine and cinchonidine. The pure diastereoisomeric salts, obtained after repeated recrystallization, are decomposed in the usual way (best by dissolution in methanol, pouring into dilute aqueous hydrochloric acid, and extraction of the phthalate precipitated with ether). The half-ester is then either saponified by treatment with hot aqueous sodium hydroxide, or, if there is any danger of racemization by base, the alcohol may be recovered from the half-ester by lithium aluminum hydride reduction. The phthalyl alcohol formed as a by-product in the case of hydride reduction of acid phthalates is very high-boiling so that the resolved alcohol can usually be separated from it by vacuum distillation. The by-product in the reduction of succinates is the very water-soluble 1,4-butanediol. The chemistry of these processes (phthalate case) is summarized in Fig. 4-25.

An alternative method of resolving alcohols and phenols is to convert them to esters of optically active acids. The usefulness of this method is

[26] E. Fischer, *Ber.*, **32**, 2451 (1899).

limited because relatively few esters are satisfactorily crystalline. Tartranilic acid, menthyl isocyanate, and menthoxyacetyl chloride (all derivatives of naturally occurring active substances) are among the reagents that have been used to resolve alcohols; the formulas of these compounds and the derivatives that they form with alcohols are shown in Fig. 4-26. 3β-Acetoxy-Δ^5-etiocholenic acid (Fig. 4-26) forms nicely crystalline esters;[26a] a convenient preparation of this acid has been described.[26b]

Alcohols have also been resolved through the formation of glycosides, using as resolving agent acetobromoglucose.[26c]

v. Aldehydes and Ketones. The need for optically active carbonyl compounds does not arise frequently, and if it does, they are often best obtained by synthesis (cf. Sec. 5-5). A few derivatives of naturally active substances, such as menthylsemicarbazide, menthylhydrazine, and tartramidic acid hydrazide (Fig. 4-27) have been used to resolve carbonyl compounds. The

(±)—R*OH + [phthalic anhydride: CO, O, CO] $\xrightarrow{C_5H_5N}$ (±)— [C_6H_4: CO_2H, CO_2R^*] $\xrightarrow{\text{Brucine}}$

[C_6H_4: CO_2^- H·Brucine$^+$, CO_2R^*] $\xrightarrow{HCl}$ Brucine·HCl + [C_6H_4: CO_2H, CO_2R^*]

(One diastereoisomer purified by crystallization) — Active

$NaOH-H_2O$, Δ — or — $LiAlH_4$

R*OH + [C_6H_4: CO_2Na, CO_2Na] — R*OH + [C_6H_4: CH_2OH, CH_2OH]

Active — Active

Fig. 4-25. Resolution of alcohols.

compounds that these reagents form with carbonyl compounds are all of the hydrazone type. They are sometimes hard to cleave without danger of racemization of the carbonyl moiety, especially if the active center is enolizable, i.e., if it is adjacent to the carbonyl function and bears a hydrogen, as in phenyl *sec*-butyl ketone, $C_6H_5COCH(CH_3)C_2H_5$.

An alternative way of resolving carbonyl compounds, akin to the resolution of alcohols described above, involves their conversion into 4-carboxyphenylsemicarbazones by means of 4-(4-carboxyphenyl)semicarbazide,

$$p\text{-}HO_2CC_6H_4NHCONHNH_2$$

[26a] R. B. Woodward and T. J. Katz, *Tetrahedron,* **5,** 70 (1959).

[26b] C. Djerassi and J. Staunton, *J. Am. Chem. Soc.,* **83,** 736 (1961).

[26c] C. Neuberg, K. P. Jacobson, and J. Wagner, *Fermentforschung,* **10,** 491 (1929); B. Helferich and R. Hiltman, *Ber.,* **70,** 308 (1937).

These semicarbazones, which are acidic, are then resolved by means of brucine and, after resolution, are hydrolyzed back to the (optically active) carbonyl precursor.[26d]

vi. Miscellaneous. Compounds devoid of functional groups, such as saturated hydrocarbons, or possessing only weakly reactive functional groups, such as unsaturated and aromatic hydrocarbons, ethers, alkyl halides, and a variety of sulfur compounds† present a special problem in resolution. Often

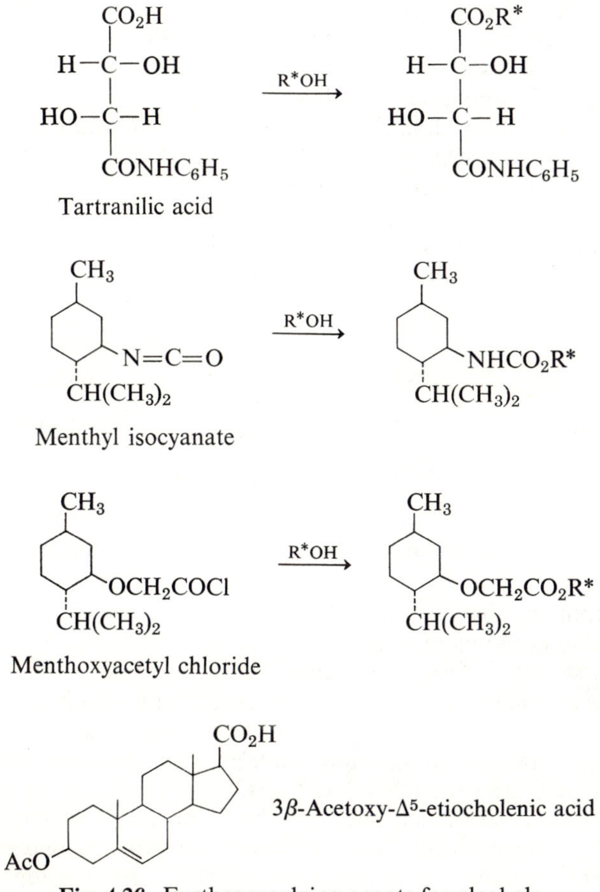

Tartranilic acid

Menthyl isocyanate

Menthoxyacetyl chloride

3β-Acetoxy-Δ^5-etiocholenic acid

Fig. 4-26. Further resolving agents for alcohols.

† It is recognized, of course, that olefins, aromatics, and alkyl halides are perfectly respectable reagents in such reactions as addition, electrophilic substitution, and nucleophilic displacement, respectively. Up to now, however, these processes have not been exploited for purposes of resolution, either because appropriate optically active reagents are not available or because the reactions cannot be reversed at all readily.

[26d] J. K. Shillington, G. S. Denning, W. B. Greenough, T. Hill, and O. B. Ramsay, *J. Am. Chem. Soc.*, **80**, 6551 (1958).

it is better to synthesize such compounds in the active form, starting out with naturally active or resolved starting materials (cf. Sec. 5-5). A few special methods of resolution applicable to such compounds (and some others) are, however, available.

RESOLUTION VIA MOLECULAR COMPLEXES. When a *dl* pair is treated with a dissymmetric reagent with which it can form a crystalline complex compound, it may happen that the resulting two diastereoisomeric complexes are of different solubility and that one will precipitate in preference to the other. Decomposition of the complex by heating, dissolution, chromatography, or chemical treatment will then yield one of the enantiomers of the original substrate. Thus digitonin, a steroidal saponin, has been used to resolve terpineol,[27] (+)-2-naphthylcamphylamine has been used to resolve *N-sec*-butylpicramide,[28] and α-(2,4,5,7-tetranitrofluorenylideneaminoöxy)propionic acid has been used to resolve 1-naphthyl *sec*-butyl ether.[29] The formulas of some of these resolving agents and the substances resolved are shown in Fig. 4-28.

A somewhat different type of resolution has been effected through the so-called clathrate or inclusion complexes. This type of complex is formed

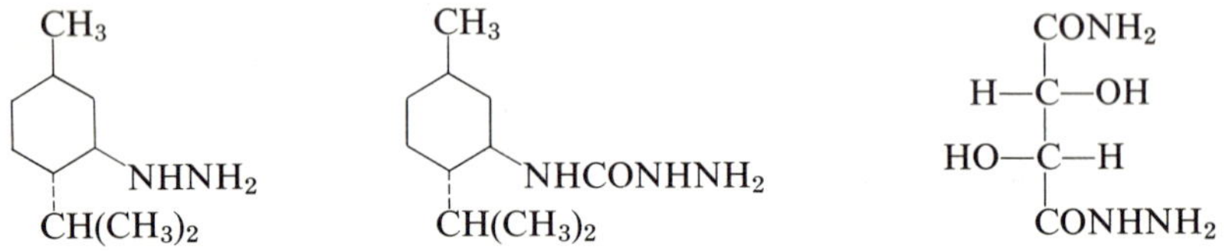

CH_3 / $NHNH_2$ / $CH(CH_3)_2$ — Menthylhydrazine

CH_3 / $NHCONHNH_2$ / $CH(CH_3)_2$ — Menthyl semicarbazide

$CONH_2$ / H—C—OH / HO—C—H / $CONHNH_2$ — Tartramidic acid hydrazide

Fig. 4-27. Resolving agents for carbonyl compounds.

from a component that crystallizes in such a way as to leave a hole into which the other component may fit if it is of suitable size. The formation of the complex seems to depend on relative molecular dimensions rather than on any particular chemical affinity. One compound which has been used in this way is desoxycholic acid (Fig. 4-28), used in the resolution of camphor and the partial resolution of dipentene.[30] The inclusion compounds of desoxycholic acid are called "choleic acids."

A more unusual substance used for the same purpose is tri-*o*-thymotide (Fig. 4-29).[31] This molecule has the shape of a three-bladed propeller and may therefore exist in two enantiomeric forms. In solution it racemizes rapidly by transformation of one form into the other, but, upon crystallization, it may again be obtained optically active through preferential crystal-

[27] A. Windaus, F. Klänhardt, and R. Weinhold, *Z. physiol. Chem.*, **126**, 308 (1923).

[28] R. Weiss and A. Abeles, *Monatsh.*, **59**, 238 (1932).

[29] M. S. Newman, W. B. Lutz, and D. Lednicer, *J. Am. Chem. Soc.*, **77**, 3420 (1955); M. S. Newman and W. B. Lutz, *ibid.*, **78**, 2469 (1956).

[30] H. Sobotka, *Naturwiss.*, **19**, 595 (1931); H. Sobotka and A. Goldberg, *Biochem. J.*, **26**, 905 (1932).

[31] H. M. Powell, *Nature*, **170**, 155 (1952).

lization of one of the enantiomers as a solvate (a kind of spontaneous resolution). When it is crystallized from a solvent possessing an asymmetric carbon atom, such as 2-bromobutane, $CH_3CHBrCH_2CH_3$, the crystals, in addition to containing only one form of the thymotide, also contain a predominance of one enantiomer of the solvent 2-bromobutane, which may thus be partially resolved.

An even more surprising case[32] is the resolution of 2-chlorooöctane, $CH_3CHClC_6H_{13}$-*n*, by means of urea, H_2NCONH_2. Urea is not, of course, an asymmetric molecule but it crystallizes in spiral-shaped asymmetric crys-

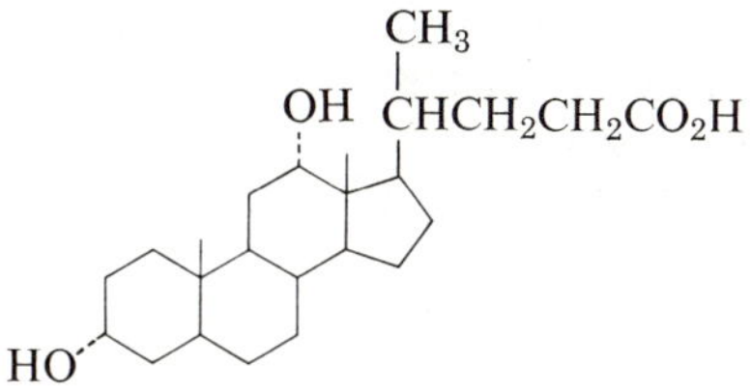

Desoxycholic acid

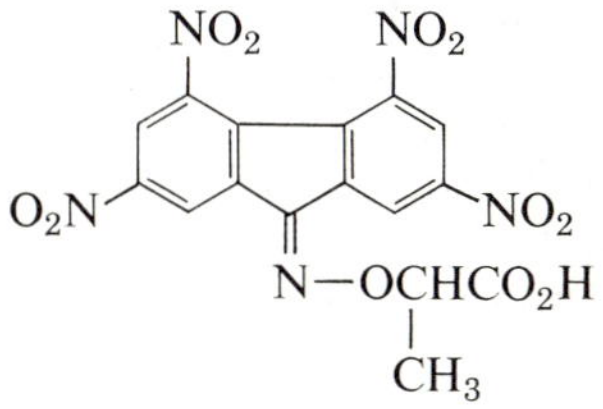

α-(2,4,5,7-Tetranitrofluorenylideneaminoöxy)-propionic acid

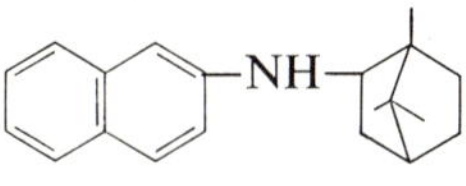

2-Naphthylcamphylamine

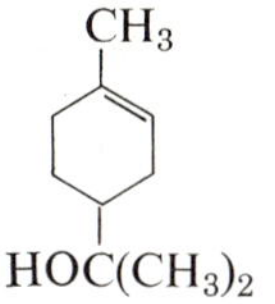

Terpineol

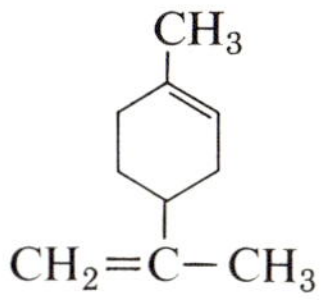

Dipentene

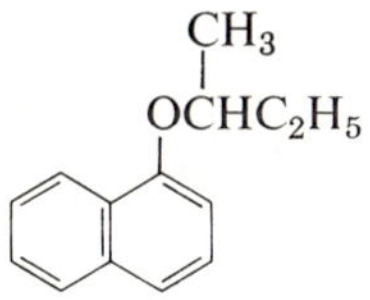

1-Naphthyl *sec*-butyl ether

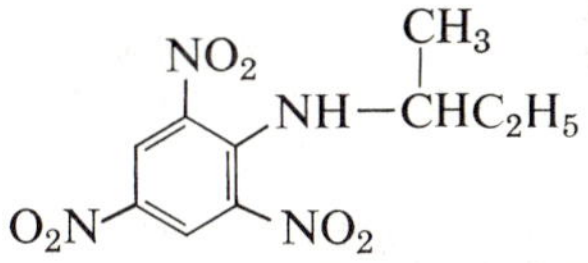

N-*sec*-butyl picramide

Fig. 4-28. Resolution by complex formation; reagents and compounds resolved.

tals. Depending on the "handedness" (hemihedric properties) of the crystal, one or another of the enantiomers of 2-chlorooöctane will fit preferentially into the lattice and will thus be enriched in the inclusion compound. This type of resolution is somewhat hard to effect, since it depends on the preferential formation of one type of hemihedric urea crystal over the other (for which there is no a priori reason) by fortuitous inoculation with only one type of seed followed by undisturbed crystal growth.

[32] W. Schlenk, *Experientia*, 8, 337 (1952).

The above-mentioned inclusion compounds are "lattice inclusion compounds"; i.e., the canals in which the guest molecules are contained are formed by the crystal lattice of the host. There is another type, so-called "molecular inclusion compounds," in which the canals are within one host molecule. This evidently requires a large molecule, and one that has been used in resolution is cyclodextrin.[33] Ethyl mandelate, $C_6H_5CHOHCO_2C_2H_5$, and ethyl phenylchloroacetate, $C_6H_5CHClCO_2C_2H_5$, are among the compounds which have been resolved via their cyclodextrin inclusion compounds.[33]

Some further aspects of resolution by crystallization of diastereoisomers must now be considered. One of the principal difficulties of this method is that generally only one diastereoisomer (the less soluble one) can be obtained from solution in the pure state. The other diastereoisomer is apt to stay in the mother liquor contaminated with a residue of the less soluble material, and decomposition of such a mixture will evidently not give enantiomerically pure material. Thus, the common situation in resolution is that one enantiomer may be obtained pure or nearly so whereas the other is recovered in a far from optically pure state. From the practical point of view this is not always too serious a drawback, since it is rarely necessary to have both enantiomers pure. However, sometimes one *does* want both enantiomers pure, and in any case, if both are obtained pure, this provides a useful check on the optical rotation of the substance (cf. Sec. 4-5).

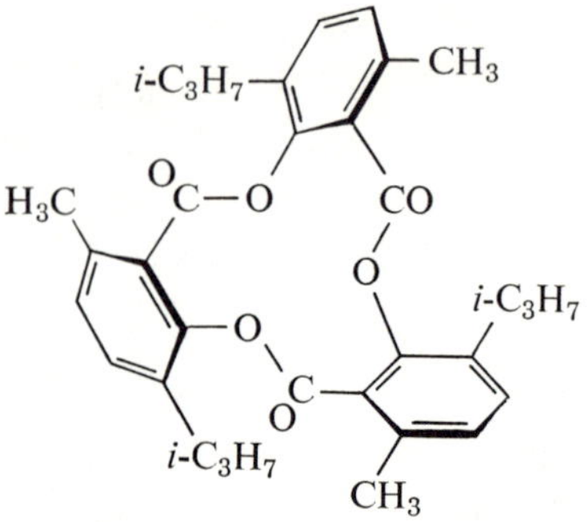

Fig. 4-29. Tri-*o*-thymotide. (*From M. S. Newman, "Steric Effects in Organic Chemistry," John Wiley & Sons, Inc., New York,* 1956. *By permission of the publishers.*)

Sometimes, such as in the case of α-phenethyl hydrogen phthalate, the second diastereoisomer may crystallize from the mother liquor in a state of relatively high purity after the first diastereoisomer has been removed. In other cases, crystallization of the second diastereoisomer can be brought about by distilling off the original solvent and adding a new solvent. Such cases are, however, rare.

One obvious way (first suggested by Marckwald[34]) of obtaining the second enantiomer in a pure state is to use the antipode of the original resolving

[33] F. Cramer, *Angew. Chem.*, **64**, 136 (1952); F. Cramer and W. Dietsche, *Chem. Ber.*, **92**, 378 (1959).

[34] W. Marckwald, *Ber.*, **29**, 43 (1896).

agent. For if (+)-A·(−)-B is less soluble than (−)-A·(−)-B in the resolution of (±)-A by (−)-B, by the same token (−)-A·(+)-B will be less soluble than (+)-A·(+)-B. Thus, (+)-B may be used to complete the resolution of the impure (−)-A recovered from the mother liquors of the resolution of (+)-A [by (−)-B]. The serious drawback of this method is that many resolving agents are natural products and their enantiomers are not usually available. Synthetic resolving agents, such as α-phenethylamine, are useful for the application of Marckwald's method but suffer from the difficulty that both enantiomers of the resolving agent must first be prepared in the pure state.

In most cases, it is necessary to use a second resolving agent to complete the resolution of the impure second enantiomer. To do this, the first resolving agent is removed from the mother liquor of the original resolution by stripping solvent and adding acid (to remove basic resolving agent) or base (to remove acidic resolving agent), and the impure second enantiomer is recovered by filtration or extraction. A new resolution is then carried out in the usual way, by the same method of trial and error as the original resolution, but with the advantage that the material to be resolved is already enriched in the desired enantiomer and that therefore in the second resolution, unlike the original one, the desired diastereoisomer is formed in excess

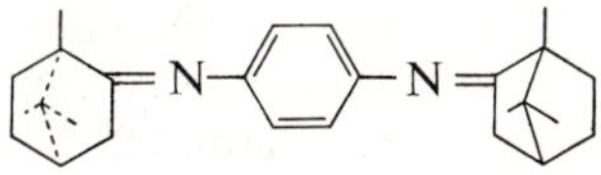

N N

Fig. 4-30. *p*-Phenylene-bis-iminocamphor.

over the undesired one. Quite a number of resolutions have been completed in this way.

Two further variations of the resolution procedure, that devised by Pope and Peachey[35] and that studied by Ingersoll and coworkers,[36] are used only rarely; for a detailed summary the reader is referred elsewhere.[37]

RESOLUTION BY CHROMATOGRAPHY.[95, 97] If an optically active adsorbent, say (−)-Ads., is used for chromatographing a racemate, (±)-X, the individual molecules form adsorbates (−)-Ads.·(−)-X and (−)-Ads.·(+)-X. Since these adsorbates are diastereoisomeric, they are not equally stable; i.e., one enantiomer (the one forming the less strong adsorbate) passes through the column faster than the other. Partial resolution may thus be achieved. An example[38] is the resolution of *p*-phenylene-bis-iminocamphor (Fig. 4-30) on a lactose column. The method is particularly useful to decide whether a given compound is a *dl* pair or an inherently inactive species. If its solution, upon percolation through a column of active adsorbent (assumed to be completely insoluble in the solvent used), becomes active, the material is a *dl* pair. Failure to obtain an active eluate is not conclusive; it means either that the sub-

[35] W. J. Pope and S. J. Peachey, *J. Chem. Soc.,* **75,** 1066 (1899).
[36] For example, A. W. Ingersoll and J. R. Little, *J. Am. Chem. Soc.,* **56,** 2123 (1934).
[37] Ref. 85, pp. 257–258.
[38] G. M. Henderson and H. G. Rule, *Nature,* **141,** 917 (1938); *J. Chem. Soc.,* 1568 (1939).

strate possesses symmetry or that this particular method of resolution failed completely. This type of resolution should be distinguished from resolution by separating diastereoisomers on an ordinary (inactive) column, such as alumina (page 49). It may better be regarded as equilibrium asymmetric transformation (Sec. 4-4*c*).

Instead of using an optically active adsorbent, it is possible to modify the surface of an inactive adsorbent by treatment with an active compound in such a way that the inactive adsorbent will acquire selectivity for one enantiomer over the other. Thus a column of silica gel made by reaction of sodium silicate with (+)-camphorsulfonic acid was found[38a] to adsorb (+)-camphorsulfonic acid in preference to the (−) enantiomer. When a solution of (±)-camphorsulfonic acid was passed through such a column, the first eluate was considerably enriched in the (−) form.

Types of chromatography other than that on a column have been successfully used in resolution. Paper chromatography[39] on No. 4 Whatman filter paper led to separation of (±)-2,3-dihydroxy-β-phenylalanine,

$$2,3\text{-}(HO)_2C_6H_3CH_2CH(NH_2)CO_2H$$

into two distinct spots.[39c] The adsorbent here is, of course, the optically active cellulose. Attempted resolution of racemic amines on optically active ion-exchange resins was initially unsuccessful,[40a] but more recently separations have been claimed, for example, of the enantiomers of (±)-mandelic acid on a resin prepared from chloromethylated styrene–divinylbenzene copolymer treated with (−)-α-phenethylamine.[40b] Gas chromatography has also been adapted to the separation of enantiomers, partial resolution of 2-butanol (among other substances) having been achieved on a stationary phase of (−)-ethyl tartrate.[40c]

OTHER METHODS OF PHYSICAL SEPARATION. Resolution of racemates of optically stable† compounds by crystallization from optically active solvents is usually unsuccessful.[41a] Success may, however, be achieved in cases where the solvent and the substrate to be resolved can associate at two different sites,‡ for example, in the resolution of 2,3-dibromo-1,4-butanediol through crystallization from (+)-diisopropyl tartrate.[41b]

A somewhat unusual method of resolution is dialysis through an optically active membrane,[42] which has been used to separate the enantiomers of tartaric acid.

† As distinct from the optically labile compounds to be described in the next section.

‡ A similar condition—association at more than one site—may be necessary for successful resolution by chromatography; cf. C. E. Dalgliesh, *J. Chem. Soc.*, 3940 (1952), and H. Krebs, J. A. Wagner, and J. Diewald, *Chem. Ber.*, **89**, 1875 (1956).

[38a] R. Curti and U. Colombo, *J. Am. Chem. Soc.*, **74**, 3961 (1952).

[39] (*a*) M. Kotake, T. Sakan, N. Nakamura, and S. Senoh, *J. Am. Chem. Soc.*, **73**, 2973 (1951); (*b*) M. Mason and C. P. Berg, *J. Biol. Chem.*, **195**, 515 (1952); (*c*) C. E. Dalgliesh, *J. Chem. Soc.*, 3940 (1952).

[40] (*a*) J. F. Bunnett and J. L. Marks, *J. Am. Chem. Soc.*, **74**, 5893 (1952); (*b*) S. Tsuboyama and M. Yanagita, *Sci. Papers Inst. Phys. Chem. Research (Tokyo)*, **53**, 245 (1959); (*c*) G. Karagounis and G. Lippold, *Naturwiss.*, **46**, 145 (1959). See, however, G. Goldberg and W. A. Ross, *Chemistry & Industry (London)*, 657 (1962).

[41] (*a*) C. Buchanan and S. H. Graham, *J. Chem. Soc.*, 500 (1950); (*b*) A. Lüttringhaus and D. Berrer, *Tetrahedron Letters*, no. 10, 10 (1959).

[42] V. O. G. Klingmüller and G. Gedenk, *Nature*, **179**, 367 (1957).

SECOND-ORDER ASYMMETRIC TRANSFORMATION.[91] The maximum theoretical yield of pure enantiomer in a resolution based on the original weight of (±) material is obviously 50%, and yields of 25 to 30% are usually considered quite satisfactory. Nevertheless, resolutions have been described in which a nearly 100% yield of one form has been obtained as a crystalline diastereoisomer with the resolving agent. One of the early clear-cut cases is the resolution[43] of 2-(*p*-carboxybenzyl)hydrindanone-1 (Fig. 4-31). When the racemic form was treated with brucine in acetone solution at room temperature, one pure diastereoisomer precipitated in over 90% yield. Upon decomposition with mineral acid, this material gave the (+) enantiomer of the free acid I which, however, lost activity quite rapidly on standing. One is dealing here with a combination of resolution by preferential crystallization of one diastereoisomer and epimerization. The salt (+)-I·(−)-brucine is less soluble in acetone and crystallizes out, but as it does so, more (+)-I·(−)-brucine is formed from (−)-I·(−)-brucine by a process of epimerization involving the enolizable hydrogen marked in Fig. 4-31. The epimerization process itself involves an equilibrium (cf. page 39), but since

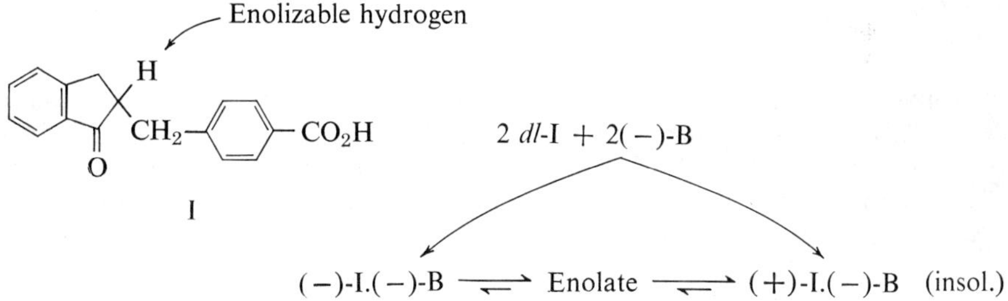

Fig. 4-31. Resolution of 2-(*p*-carboxybenzyl)hydrindanone-1.

one diastereoisomer is continuously removed from the solution by precipitation, the equilibrium between the diastereoisomers is continuously disturbed and eventually the entire material is converted to the less soluble form. This combination of epimerization and precipitation is known as "second-order asymmetric transformation." Second-order asymmetric transformation is involved also in the crystallization of mutarotating sugars such as glucose (cf. page 41). In solution, glucose is present as an equilibrium mixture of the α and β forms (cf. Fig. 4-13). The position of equilibrium depends only slightly on solvent. When a solution of glucose is concentrated, however, the less soluble form crystallizes first, and this disturbs the equilibrium so that more of that form is produced in solution. This will crystallize again, and so on, crystals of entirely one form being obtained in this way. Which form crystallizes preferentially depends on the solvent: Crystallization from ethanol produces the α isomer whereas crystallization from warm pyridine yields the β isomer.

c. Resolution by Equilibrium Asymmetric Transformation. By "equilibrium asymmetric transformation" we mean any change of configuration at one asymmetric atom of a compound having two or more such atoms which pro-

[43] H. Leuchs, *Ber.*, **54**, 830 (1921).

ceeds so as to establish a chemical or physical equilibrium; in other words, this is (cf. Sec. 4-2*e*) an epimerization leading to equilibrium. This term obviously includes first- (Sec. 4-2*e*) and second-order (Sec. 4-4*b*) asymmetric transformations. When equilibrium asymmetric transformation affects an asymmetric center which may subsequently be separated from the rest of the molecule, it constitutes a means of resolution.† The above-mentioned obtention of active 2-(*p*-carboxybenzyl)-1-hydrindanone (Fig. 4-31) is a case in point. Another example is the partial resolution of phenylchloroacetic acid

CH_3 ... $OCOCHClC_6H_5$ ... $CH(CH_3)_2$ $\xrightarrow{\text{Base}}$ CH_3 ... $OCO\bar{C}ClC_6H_5$ ... $CH(CH_3)_2$ $\rightleftharpoons$ CH_3 ... $OCOCHClC_6H_5$ ... $CH(CH_3)_2$

From *dl* acid

Equilibrium mixture
57% ester of (−)-acid
43% ester of (+)-acid

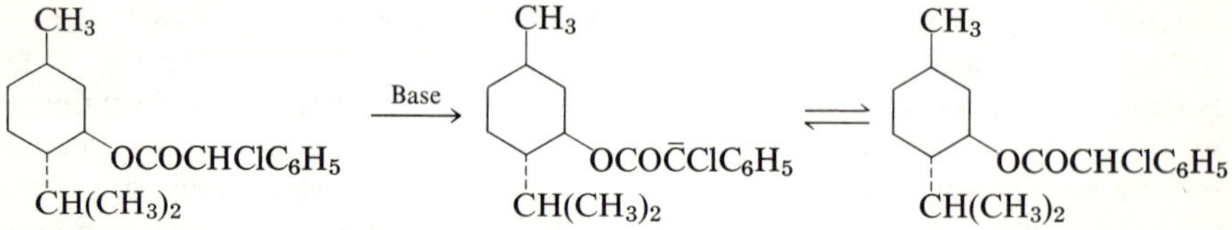

Fig. 4-32. Partial resolution of phenylchloroacetic acid by equilibrium asymmetric transformation.

by conversion to its menthyl ester followed by base-catalyzed epimerization (Fig. 4-32).[43a] The equilibrated ester contained 57% of the (−) acid and 43% of the (+) acid.‡ Equilibrium asymmetric transformations may also take place in solutions of optically active solvents. Thus when *N*,4-dimethyl-*N*-acetyl-2-(*p*-toluenesulfonyl)aniline (Fig. 4-33) was dissolved in (+)-diethyl tartrate, the material which crystallized was slightly dextrorotatory when examined in an optically inactive solvent.[41a] Upon standing, the rotation of the solution decayed to zero, owing to spontaneous racemization. That this

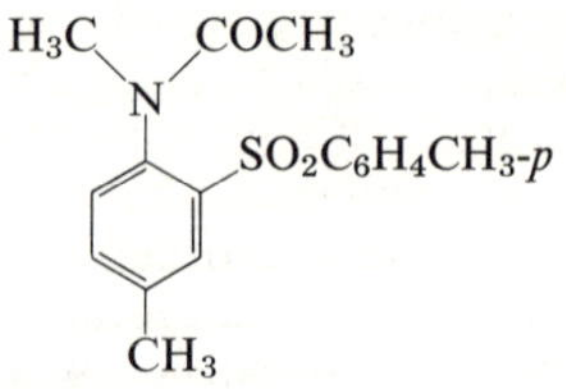

H_3C $COCH_3$
N
$SO_2C_6H_4CH_3$-*p*
CH_3

Fig. 4-33. *N*,4-Dimethyl-*N*-acetyl-2-(*p*-toluenesulfonyl)aniline. (The reason for the activity of this material will be discussed in Chap. 6.)

is an asymmetric transformation rather than a preferential crystallization of one enantiomer is evidenced by the fact that it is not necessary to allow the material to crystallize from the diethyl tartrate solution. Total precipitation

† Elsewhere this has been called "equilibrium method of resolution." We feel that stressing the resolution aspect of this phenomenon in its definition is likely to obscure the underlying physicochemical process.

‡ The acid was not recovered from the partly resolved ester in this instance, although in principle acid hydrolysis could have been used for this purpose.

[43a] A. McKenzie and I. A. Smith, *Ber.*, **58**, 894 (1925).

of the solute by addition of water to the solution also produces active material.

d. Resolution by Kinetic Asymmetric Transformation.[84] By "kinetic asymmetric transformation" we mean the preferential formation, transformation, or destruction of one of two (or of several) stereoisomers in a given reaction because that one isomer is formed or reacts faster than all the others. In terms of chemical kinetics, this means that the free energy of activation for the reaction of the isomer in question is lower than that for all the other isomers. In general, such will be the case if the transition states corresponding to the reactions of the different stereoisomers are diastereoisomeric (not enantiomeric!) and therefore unequal in free energy (cf. Sec. 3-2).†

Three cases of kinetic asymmetric transformation may be distinguished, although all are based on the same principle. Two diastereoisomers may be formed or react at unequal rates without their asymmetric atoms being directly affected in the reaction under consideration. This has sometimes been called the "kinetic method of resolution," properly speaking. A second case arises when a reaction is involved which creates a new asymmetric center in a compound already possessing asymmetry or under the influence of an asymmetric reagent, catalyst, or physical influence. This case is often called "asymmetric synthesis" or "asymmetric induction." Finally a case may arise where enantiomers are destroyed at unequal rates by an asymmetric reagent. This case may be termed "asymmetric destruction."

i. Kinetic Method of Resolution. Diastereoisomers usually differ in free energy, and if they enter into a chemical reaction, the transition states for the isomers also usually differ in energy. Unless these energy differences happen to be the same in the ground and transition states, the activation energies for the diastereoisomers are also different. A case in point is the reaction of *cis*- and *trans*-1,2-cyclopentanediol with lead tetraacetate;[43b] the cis isomer reacts about 3000 times as fast as the trans compound.‡ The differences in ground state here are not likely to be large, and the big rate difference must be caused by differences in transition-state energy levels (Fig. 4-34).¶

The application of this principle to resolution will now be considered, using the case of menthyl mandelate (Fig. 4-35) as an example. When (±)-menthol is esterified with (±)-mandelic acid, a molecule of the (+) acid may be linked to either (+) or (−) alcohol, and similarly for the (−) acid. Therefore there are actually four types of ester molecules, (+)(+), (+)(−), (−)(+), and (−)(−), as shown in Fig. 4-35, which form two pairs

† For a general treatment of the problem, it would be necessary to consider also the differences in free energy of the starting states, inasmuch as activation energy is a difference in energy between initial state and transition state; cf. Sec. 8-5 and Figs. 8-23 and 8-25. This complication does not arise in asymmetric synthesis and asymmetric destruction (Sec. 4-4*d, ii* and *iii* below) where the starting states are either one and the same or enantiomeric and thus equal in free energy.

‡ This is an extreme case; usually the differences are not as large. The cause for these differences will be considered in more detail in Chaps. 6 and 8.

¶ It is recognized that the case discussed here may be somewhat more complicated in that cyclic intermediates may be involved, and the equilibrium constant for the intermediates should also be considered in assessing relative reactivity of stereoisomers. This does not, however, affect the general principle involved.

[43b] R. Criegee, E. Büchner, and W. Walther, *Ber.*, **73**, 571 (1940).

of enantiomers. The two pairs, in turn, are diastereoisomeric (cf. Sec. 3-2). In the esterification, the diastereoisomers may (and probably will) be formed at unequal rates, for, although the ground state is the same for all possible combinations, the diastereoisomeric transition states differ in energy. As a result, if the reaction is interrupted before it is complete, or if insufficient menthol for complete esterification is used, one pair of enantiomers [the (+)(−) and (−)(+), as it happens] is formed in preference over the other. However, since enantiomers have the same free energy, they are formed at the *same* rate. The product, then, is a mixture of (+)(−) and (−)(+) ester (equal amounts) with a lesser proportion of the (+)(+) and (−)(−) isomers (again formed in equal amounts). Total saponification of such a mixture gives back equal amounts of (+) and (−) acid; thus no resolution has been achieved. The situation is different, however, if (−)-menthol is used in the esterification instead of the racemic alcohol. In this case only two diastereoisomers can result, namely, the (+)(−) and the (−)(−) ester,

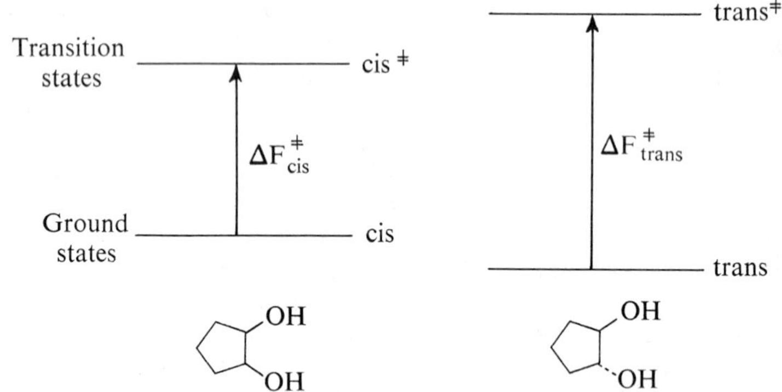

Fig. 4-34. Energetics of lead tetraacetate oxidation of *cis*- and *trans*-1,2-cyclopentanediol.

and from what has been said above, the (+)(−) isomer will predominate when the reaction is stopped before completion. Hydrolysis of the ester mixture will thus return mandelic acid enriched in the (+) isomer, whereas the mandelic acid which remains unesterified will be predominantly (−). Partial resolution may thus be effected.[44]

Several variations of the kinetic method of resolution will now be considered. Instead of bringing together the asymmetric centers (at least two must be involved) in the key reaction, they may be separated. For example, when an equimolar mixture of the esters of (−)-menthol with (+)- and (−)-mandelic acid [obtained by total esterification of the racemic acid with (−)-menthol] is partially hydrolyzed, the acid liberated preferentially is the (+) isomer. The (−) isomer can be recovered by further hydrolysis of the residual ester. In principle, it is not necessary that the reaction effecting resolution should either bring together or separate the asymmetric centers involved. For example, one might envisage resolution of a racemic amino acid by partial acetylation of its (−)-menthyl ester. Since the (−)-menthyl

[44] W. Marckwald and A. McKenzie, *Ber.*, **32**, 2130 (1899).

esters of the (+)- and (−)-amino acid are diastereoisomers, they might be acetylated at different rates, so that one isomer might be enriched in the acylated material, the other in the unacylated residue. However, this particular kind of kinetic resolution does not seem to have been reduced to practice.†

The auxiliary asymmetric center required in the kinetic method may be

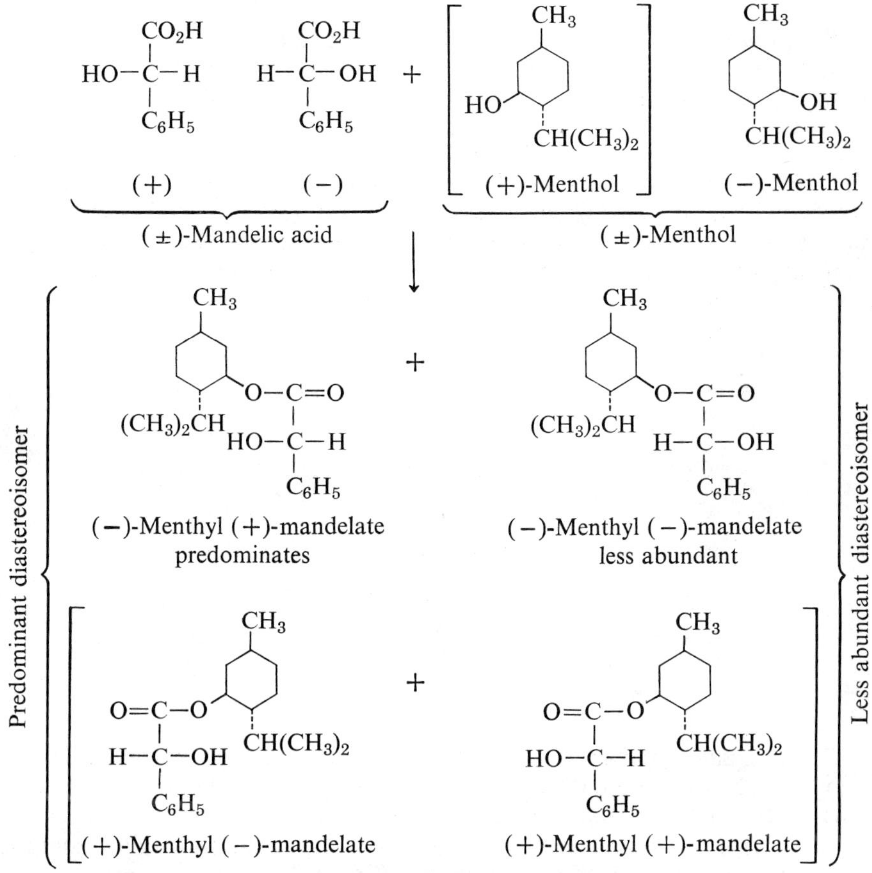

(+) (−) (±)-Mandelic acid + (+)-Menthol (−)-Menthol (±)-Menthol

(−)-Menthyl (+)-mandelate predominates + (−)-Menthyl (−)-mandelate less abundant

(+)-Menthyl (−)-mandelate + (+)-Menthyl (+)-mandelate

Predominant diastereoisomer — Less abundant diastereoisomer

Fig. 4-35. Incomplete esterification of menthol with mandelic acid.

† Practical difficulties may arise because the differences between diastereoisomeric ground states and the corresponding transition states may be the same. In that case (*i*), the activation energies for reaction of the two isomers A and B would be equal.

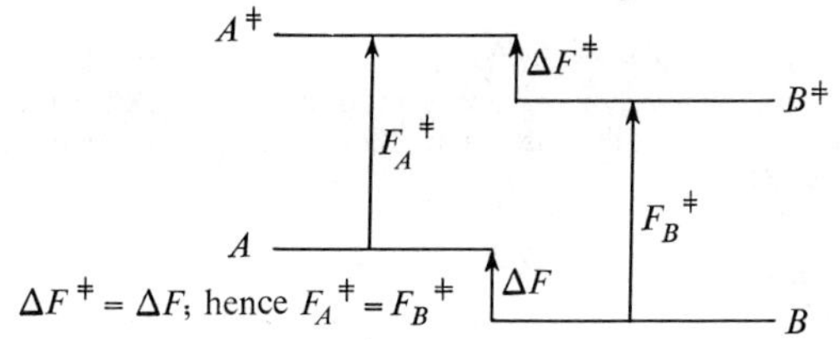

$\Delta F^{\ddagger} = \Delta F$; hence $F_A^{\ddagger} = F_B^{\ddagger}$

(*i*)

present in the catalyst or solvent rather than in the substrate or reagent. Thus incomplete esterification of (±)-methylphenylcarbinol, $C_6H_5CHOHCH_3$, with acetic anhydride in the presence of brucine gives the levorotatory acetate, $C_6H_5CHOAcCH_3$, along with recovered dextrorotatory alcohol.[45] Incomplete hydrolysis of racemic ethyl phenylchloroacetate,

$$C_6H_5CHClCO_2C_2H_5$$

in the presence of cyclodextrin (cf. page 60) produces dextrorotatory phenylchloroacetic acid, $C_6H_5CHClCO_2H$, whereas the recovered ester is levorotatory.[46] Carbonation of racemic 2-butylmagnesium chloride,

$$CH_3CH(MgCl)CH_2CH_3$$

in (+)-2,3-dimethoxybutane, $CH_3CH(OCH_3)CH(OCH_3)CH_3$, gives levorotatory methylethylacetic acid, $CH_3CH(C_2H_5)CO_2H$.[47] Here it seems to be necessary for one active ether molecule to coordinate with the organometallic reagent in two places for kinetic resolution to be effective, for methylethylacetic acid obtained by a method similar to the above in (−)-menthyl methyl ether was inactive.

```
       CN                                          CN
       |                                           |
  HO—C—H                  CHO                  H—C—OH
       |                    |                      |
   H—C—OH      HCN      H—C—OH      HCN        H—C—OH
       |      <----         |      ---->           |
  HO—C—H                HO—C—H                 HO—C—H
       |                    |                      |
  HO—C—H                HO—C—H                 HO—C—H
       |                    |                      |
      CH2OH               CH2OH                  CH2OH

 Gluconononitrile                             Mannononitrile
    (minor)                                       (major)
```

Fig. 4-36. Reaction of (+)-arabinose with hydrogen cyanide.

ii. Asymmetric Synthesis. Asymmetric synthesis or asymmetric induction† is only tenuously different from kinetic resolution. The difference is that the asymmetric atom, instead of being in the molecule to begin with, is introduced in the course of the reaction. An example is the synthesis of gluconononitrile and mannononitrile from (+)-arabinose (Fig. 4-36). The two products are diastereoisomeric, and the transition states leading to them would also be expected to be diastereoisomeric and thus different in free energy. Since the starting state [(+)-arabinose] is the same for the two reactions, the activation energies are expected to differ, and the two products are formed at different

† There appears to be no agreement in the literature on the definition of these terms. In this book they are used interchangeably as meaning production of a new asymmetric atom or entire dissymmetric molecule under conditions where the resulting two stereoisomers are formed in unequal amounts.

[45] R. Wegler, *Ann.*, **498**, 62 (1932).

[46] F. Cramer and W. Dietsche, *Chemistry & Industry* (*London*), 892 (1958); *Chem. Ber.*, **92**, 1739 (1959).

[47] H. L. Cohen and G. F. Wright, *J. Org. Chem.*, **18**, 432 (1953).

rates. In fact, mannonic acid nitrile predominates to such an extent that at one time it was believed to be the only product of the reaction.†

It is well to understand exactly what is involved in asymmetric synthesis. When a new asymmetric center is introduced in a molecule already possessing one or more asymmetric centers, a pair of diastereoisomers can always result, and for reasons already discussed, the two isomers are not usually formed in equal amounts.‡ This is true whether the starting material was originally resolved or whether it is a *dl* pair. Thus, in the reduction of (±)-benzoin, $C_6H_5CHOHCOC_6H_5$, with lithium aluminum hydride, the major product is *meso*-hydrobenzoin, *meso*-$C_6H_5CHOHCHOHC_6H_5$, not its diastereoisomer; the same is true if one starts with (+)- or (−)-benzoin.

The difference in energy between diastereoisomeric transition states and the resulting preference of one diastereoisomeric product over the other is apt to be most marked when the newly created asymmetric center is close to an asymmetric center already in the molecule. For the particular case where the two centers are adjacent to each other and where asymmetry at the new center is created by an addition reaction to a double bond, *Cram's rule* predicts the predominant stereoisomer in the product.[48] The rule may be summarized in a diagram, as in Fig. 4-37, which refers to the particular case of

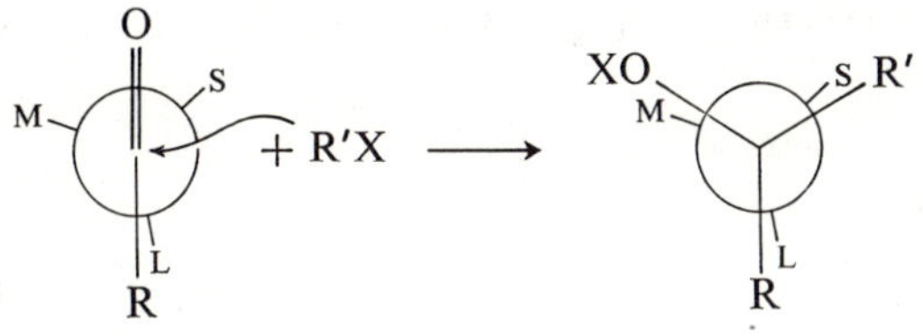

Fig. 4-37. Cram's rule.¶

an addition reaction (e.g., of hydride or an organometallic) to a carbonyl compound SML$\overset{*}{C}$-COR. The rule states that when the asymmetric carbon ($\overset{*}{C}$) is so oriented that the carbonyl function is flanked by the two smaller groups (M and S) attached to $\overset{*}{C}$, the reagent (R′X) preferentially approaches the carbonyl group from the side of the smallest group S.¶ The rule applies only to reactions that are kinetically controlled (cf. Sec. 7-3), i.e., where the product isolated is that formed in a rate-controlled process, not the more stable product formed in a subsequent equilibration.§ Also, the rule specifi-

† It might be noted that, whereas a reaction leading to kinetic resolution must be interrupted short of completion, the same is not true of a reaction leading to asymmetric synthesis.

‡ The introduction of an asymmetric center in a symmetric molecule in the absence of any asymmetric influence of course always leads to a *dl* pair (cf. Sec. 4-2*b*).

¶ S is the smallest group on the asymmetric carbon $\overset{*}{C}$, M the medium-sized group, and L the largest group.

§ For example, in an aluminum isopropoxide (Meerwein-Ponndorf-Verley) reduction of a ketone, the product predicted by Cram's rule is that obtained after a short reaction time. Prolonged reaction leads to formation of the more stable diastereoisomer which is not necessarily that predicted by the rule.

[48] D. J. Cram and F. A. Abd Elhafez, *J. Am. Chem. Soc.*, **74**, 5828 (1952).

cally does not apply to catalytic reduction, and it does not apply to cases where the small group S is a group such as hydroxy, alkoxy, or amino which may complex with the reagent R′X.[49] An example of the application of the rule is the reaction of α-aminobenzyl phenyl ketone (as the hydrochloride) with excess *p*-chlorophenylmagnesium bromide (Fig. 4-38).† The predominant diastereoisomer produced is that predicted by the rule, phenyl being the large group, amino medium, and hydrogen small. The other diastereoisomer may be obtained in predominance by reversing the group already in the molecule and the group introduced in the reaction. In the present case this entails the reaction of α-aminobenzyl *p*-chlorophenyl ketone (as the hydrochloride) with phenylmagnesium bromide (Fig. 4-38).‡

Asymmetric induction of the type described here does not necessarily involve optically active starting materials or products. However, it is obvious from what has been said before that, if the introduction of a new asymmetric center B into a molecule (±)-A gives a predominance of, let us say, (+)-A-(−)-B and (−)-A-(+)-B over the other diastereoisomer (+)-A-(+)-B and

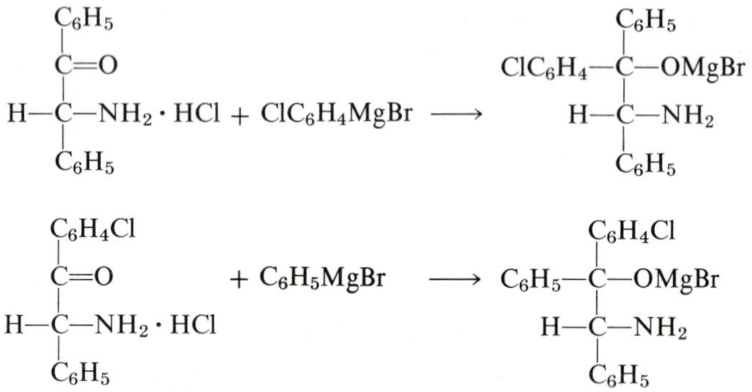

$$\begin{array}{c} C_6H_5 \\ | \\ C{=}O \\ | \\ H{-}C{-}NH_2\cdot HCl \\ | \\ C_6H_5 \end{array} + ClC_6H_4MgBr \longrightarrow \begin{array}{c} C_6H_5 \\ | \\ ClC_6H_4{-}C{-}OMgBr \\ | \\ H{-}C{-}NH_2 \\ | \\ C_6H_5 \end{array}$$

$$\begin{array}{c} C_6H_4Cl \\ | \\ C{=}O \\ | \\ H{-}C{-}NH_2\cdot HCl \\ | \\ C_6H_5 \end{array} + C_6H_5MgBr \longrightarrow \begin{array}{c} C_6H_4Cl \\ | \\ C_6H_5{-}C{-}OMgBr \\ | \\ H{-}C{-}NH_2 \\ | \\ C_6H_5 \end{array}$$

Fig. 4-38. Application of Cram's rule.

(−)-A-(−)-B, then, *if one starts with an already resolved molecule*, for example, (−)-A, one will get a predominance of (−)-A-(+)-B. If now in some way the part of the molecule containing the new asymmetric center B can be separated, this part will be optically active. Originally, the term "asymmetric synthesis" was limited to this particular type of case.[50] Numerous examples have been studied by McKenzie and coworkers. In McKenzie's first example,[51] the phenylglyoxylate ester of (−)-menthol was reduced with aluminum

† Correlation of the Fischer projection formulas used in Fig. 4-38 with the Newman projection formulas used in Fig. 4-37 is left as an exercise for the reader. Cf. Fig. 3-11.

‡ Despite its plausible geometric formulation, Cram's rule is only a *formal* rule based on empirical observation; it does *not* necessarily have any mechanistic implication. In fact, there are cases (Ref. 49) where the rule may lead to a correct prediction even though the transition state is manifestly different from what may be implied in Fig. 4-37.

---

[49] D. J. Cram and K. R. Kopecky, *J. Am. Chem. Soc.*, **81**, 2748 (1959); J. H. Stocker, P. Sidisunthorn, B. M. Benjamin, and C. J. Collins, *ibid.*, **82**, 3913 (1960).

[50] W. Marckwald, *Ber.*, **37**, 349 (1904).

[51] A. McKenzie, *J. Chem. Soc.*, **85**, 1249 (1904).

amalgam to the mandelate ester of (−)-menthol; the ester of (−)-mandelic acid predominated in the product over that of (+)-mandelic acid (Fig. 4-39). In later work the mandelic acid enriched in the (−) isomer was actually recovered from its menthyl ester by hydrolysis. Prelog has recently stated[52] an empirical correlation between the arrangement of the groups (configuration) in the isomer formed predominantly in the type of synthesis studied by

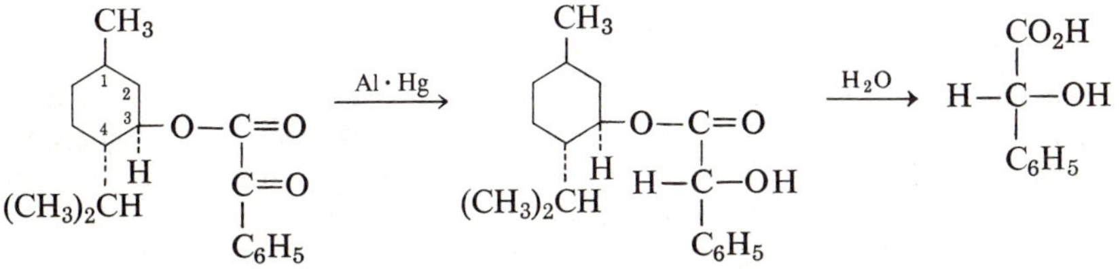

(−)-Menthyl phenylglyoxylate —Al·Hg→ (−)-Menthyl (−)-mandelate (predominant product) —H_2O→ (−)-Mandelic acid (excess)

Fig. 4-39. Asymmetric synthesis of (−)-mandelic acid.

McKenzie and the corresponding arrangement in the optically active alcohol [e.g., (−)-menthol] used as an auxiliary reagent in the synthesis. The correlation, sometimes called *Prelog's rule,*[52] is summarized in Fig. 4-40.† In the statement of the rule, the arrangement of the groups in the starting material must be specified first, since there are several bonds around which the molecule may rotate. By convention, the molecule is so oriented that the two carbonyl groups are antiparallel and that the smallest group (S) in the alcohol portion is eclipsed with the ketone carbonyl.

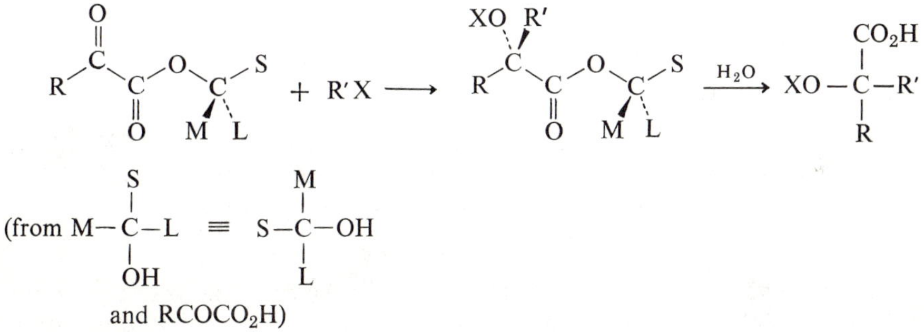

+ R′X ⟶ —H_2O→ XO—C—R′ (CO_2H, R)

(from M—C(S)(OH)—L ≡ S—C(M)(L)—OH and $RCOCO_2H$)

Fig. 4-40. Prelog's rule.

The reagent R′ will then approach the ketone carbonyl group from the side of the smaller of the remaining two groups in the alcohol portion of the molecule, i.e., the side of the medium-sized group M. The case shown in Fig. 4-39 conforms with Prelog's rule if one considers that in the menthol molecule the

† This is also a *formal* representation without necessary mechanistic implications.

[52] V. Prelog, *Helv. Chim. Acta,* **36,** 308 (1953).

small group S is H, the medium-sized group M is the methylene of the ring (at C_2), and the large group L is the methine substituted with the isopropyl group (C_4).† Prelog's rule has been very useful in assigning configurations to a variety of molecules; this aspect will be taken up further in Chap. 5. As in the case of Cram's rule, reversing the order in which the groups are introduced in the molecule reverses their arrangement in the preponderant product. Thus reaction of (−)-menthyl phenylglyoxylate,

$$C_6H_5COCO_2C_{10}H_{19}$$

with methylmagnesium bromide followed by hydrolysis gives predominantly (−)-atrolactic acid, $C_6H_5C(CH_3)OHCO_2H$, but reaction of (−)-menthyl pyruvate, $CH_3COCO_2C_{10}H_{19}$, with phenylmagnesium bromide gives predominantly the stereoisomeric (+)-atrolactic acid.

Several other types of asymmetric synthesis have been rationalized in terms of an optimal steric fit of the reagents involved. Examples are the Meerwein-Ponndorf-Verley reduction[53] of methyl isohexyl ketone,

$$CH_3COCH_2CH_2CH_2CH(CH_3)_2$$

with (+)-2-butanol, $CH_3CHOHC_2H_5$, in the presence of aluminum 2-butoxide to give (+)-methylisohexylcarbinol, $CH_3CHOHCH_2CH_2CH_2CH(CH_3)_2$, and

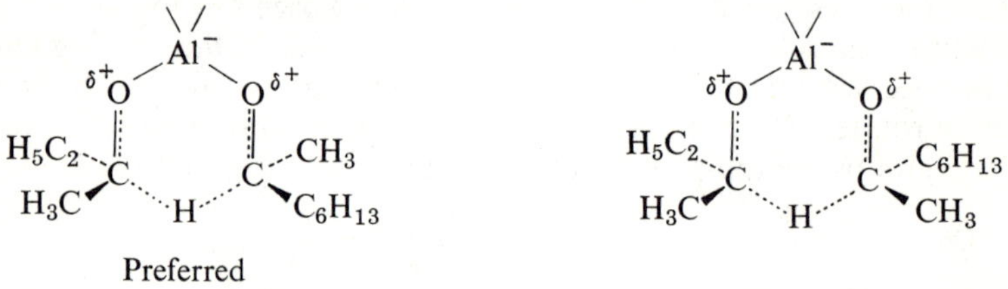

Preferred

Fig. 4-41. Transition states for asymmetric Meerwein-Ponndorf reduction.

2-butanone, $CH_3COC_2H_5$, and the reduction[54] of pinacolone, $CH_3COC(CH_3)_3$, with (+)-2-methyl-1-butylmagnesium chloride, $C_2H_5CH(CH_3)CH_2MgCl$, to give the magnesiochloride derivative of (+)-pinacolyl alcohol,

$$CH_3CHOHC(CH_3)_3$$

and 2-methyl-1-butene, $C_2H_5C(CH_3){=}CH_2$. The probable transition states for these reactions are shown in Figs. 4-41 and 4-42. The preferred transition states are those where the larger groups (ethyl, isohexyl, *t*-butyl) are on opposite sides of the plane of the six-membered ring from each other.‡ An asymmetric synthesis of this type is to be expected only if the asymmetrically

† The reader will probably have to make a model for the case shown in Fig. 4-39 to convince himself that it conforms with the rule.

‡ Unfortunately, examination of the reduction of a wide variety of unsymmetric dialkyl ketones with optically active Grignard reagents has shown that the elegant picture shown in Fig. 4-42 represents an oversimplification: E. P. Burrows, F. J. Welch, and H. S. Mosher, *J. Am. Chem. Soc.*, **82**, 880 (1960).

---

[53] W. von E. Doering and R. W. Young, *J. Am. Chem. Soc.*, **72**, 631 (1950).

[54] H. S. Mosher and E. LaCombe, *J. Am. Chem. Soc.*, **72**, 3994 (1950).

substituted carbon is part of the cyclic transition state. Thus, no asymmetric reduction of pinacolone occurred with 3-methylamylmagnesium chloride, $C_2H_5CH(CH_3)CH_2CH_2MgCl$, even though the Grignard reagent was optically active.[55] In this case the asymmetric center does not form part of the ring in the cyclic transition state, and therefore there is no preference for one configuration of the pinacolyl alcohol formed over the other, the steric fit of the two diastereoisomeric transition states (and therefore their free energy) being the same. An interesting application[56] of this type of asymmetric

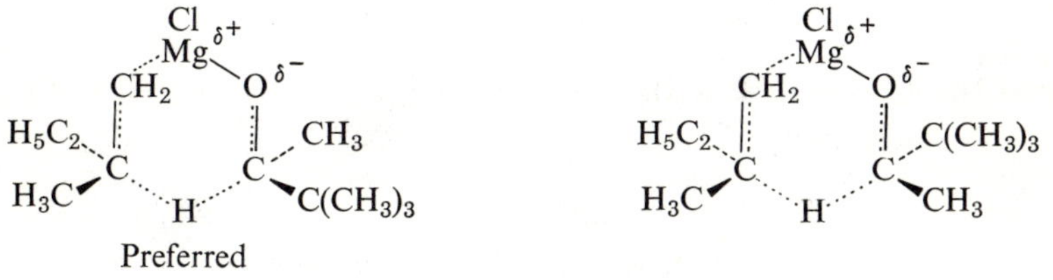

Preferred

Fig. 4-42. Transition states for asymmetric Grignard reduction.

synthesis has been made in the reduction of butyraldehyde, $CH_3CH_2CH_2CHO$, with the magnesiobromide salt† of (+)-2-octanol-2-*d*, $CH_3CDOHC_6H_{13}$, which gave rise to *active* (−)-1-butanol-1-*d*, $CH_3CH_2CH_2CHDOH$, the activity being due to asymmetry at the primary carbon due to hydrogen and deuterium (cf. Sec. 3-1). An asymmetric synthesis somewhat similar to that discovered by McKenzie[51] and rationalized by Prelog,[52] but involving addition to a carbon-carbon rather than to a carbon-oxygen double bond, is the reaction of (−)-menthyl acrylate with diphenyldiazomethane to give, after

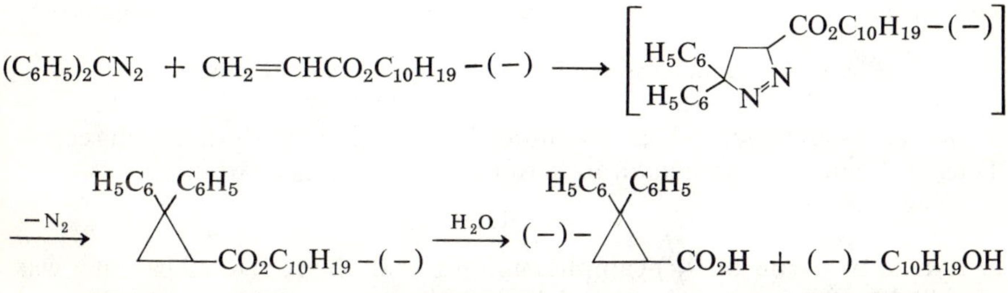

$$(C_6H_5)_2CN_2 + CH_2{=}CHCO_2C_{10}H_{19}-(-) \longrightarrow \left[\text{pyrazoline}\right] \xrightarrow{-N_2} \text{cyclopropane}-CO_2C_{10}H_{19}-(-) \xrightarrow{H_2O} (-)-\text{cyclopropane}-CO_2H + (-)-C_{10}H_{19}OH$$

Fig. 4-43. Addition of diphenyldiazomethane to (−)-menthyl acrylate.

saponification, active 2,2-diphenylcyclopropanecarboxylic acid.[57] The reaction probably proceeds via an active pyrazoline (Fig. 4-43).

Asymmetric synthesis may also occur in the presence of an asymmetric catalyst or solvent rather than an asymmetric reagent. Thus, the synthesis of optically active phenylalanine (25% optically pure) has been claimed by

† This reduction is analogous to a Meerwein-Ponndorf reduction (with —OMgBr instead of —$OAlX_2$) rather than to a reduction with a hindered Grignard reagent.

[55] H. S. Mosher and E. LaCombe, *J. Am. Chem. Soc.*, **72**, 4991 (1950).
[56] A. Streitwieser, *J. Am. Chem. Soc.*, **75**, 5014 (1953).
[57] F. J. Impastato, L. Barash, and H. M. Walborsky, *J. Am. Chem. Soc.*, **81**, 1514 (1959).

catalytic hydrogenation of ethyl α-acetoximino-β-phenylpyruvate, using a palladium catalyst supported on (optically active) silk fibroin followed by hydrolysis[58] (Fig. 4-44).

Active chloromandelic acid has been obtained from chlorobenzaldehyde via the cyanohydrin synthesis in the presence of cyclodextrin[46] (Fig. 4-45). The cyclodextrin appears to act as a solubilizing agent for the aldehyde. The reaction of methyl ethyl ketone with phenylmagnesium bromide[59] in (+)-2,3-dimethoxybutane gives 2-phenyl-2-butanol, $C_6H_5C(CH_3)OHC_2H_5$, of specific rotation $[\alpha]_D^{20}$ + 3.04°.

$$C_6H_5CH_2\text{—}\underset{\underset{NOCOCH_3}{\|}}{C}\text{—}CO_2C_2H_5 + H_2 \xrightarrow[\text{fibroin}]{\text{Pd on silk}} \xrightarrow[\text{ysis}]{\text{Hydrol-}} C_6H_5CH_2\text{—}\underset{\underset{NH_2}{|}}{CH}\text{—}CO_2H$$

$[\alpha]_D^{15}$ + 9.25

Fig. 4-44. Asymmetric hydrogenation.

iii. Asymmetric Destruction. When the asymmetry about one of the atoms (A) in two optically active diastereoisomers, (−)-A-(+)-B and (−)-A-(−)-B, is destroyed,† such destruction could take place at unequal rates in the two isomers, and the resulting product containing only the asymmetry due to B would then be optically active, even though originally the two diastereoisomers had been present in equal amounts. Cases of this type [in which (−)-A is chemically joined to (±)-B] do not seem to have been studied,‡ but very

$$ClC_6H_4CHO + HCN \xrightarrow{\text{Cyclodextrin}} ClC_6H_4CHOHCN \xrightarrow{H_2O} ClC_6H_4CHOHCO_2H$$

o or *p* isomer — Dextrorotatory — Levorotatory

Fig. 4-45. Asymmetric synthesis in the presence of cyclodextrin.

similar cases, in which A is an asymmetric reagent or catalyst, are on record. Thus in the incomplete dehydration of (±)-phenylmethylcarbinol,

$$C_6H_5CHOHCH_3$$

to styrene by means of (+)-camphorsulfonic acid, the recovered carbinol was weakly levorotatory.[60] In the dehydrohalogenation of the (±) form of 1,2,3,4,5,6-hexachlorocyclohexane (cf. Chap. 7), which cannot readily be

† For example, asymmetry of the type R*—CHOH—R′ might be eliminated by oxidation to the ketone or dehydration to the olefin; asymmetry of the type R*—CH(CO_2H)—R′ might be eliminated by decarboxylation, etc. (The asymmetry in the group R* would remain.)

‡ A hypothetical example is the partial oxidation of (−)-menthyl (±)-mandelate,

$$C_6H_5CHOHCO_2C_{10}H_{19}\text{-}(-)$$

The remaining ester, upon total saponification, should give rise to partially active mandelic acid. This would be the reverse of the asymmetric synthesis of active mandelic acid by reduction of (−)-menthyl phenylpyruvate (page 71).

[58] S. Akabori, S. Sakurai, Y. Izumi, and Y. Fujii, *Nature,* **178,** 323 (1956).
[59] N. Allentoff and G. F. Wright, *J. Org. Chem.,* **22,** 1 (1957).
[60] H. Wuyts, *Bull. soc. chim. Belges,* **30,** 30 (1921).

resolved by conventional means, with an insufficient amount of the optically active base brucine (Fig. 4-46), the remaining hexachloride became partly active.[61]

A case in which the catalyst (or catalyst support) supplies the asymmetric influence is provided by the observation[62] that partial decomposition (by dehydration or dehydrogenation) of racemic 2-butanol over nickel deposited on optically active quartz (cf. Sec. 1-2) at 550° causes the residual alcohol to become optically active. A similar example is the incomplete oxidation of 2,2′-dichlorobenzoin, $2\text{-}ClC_6H_4CHOHCOC_6H_4Cl\text{-}2'$, to the corresponding

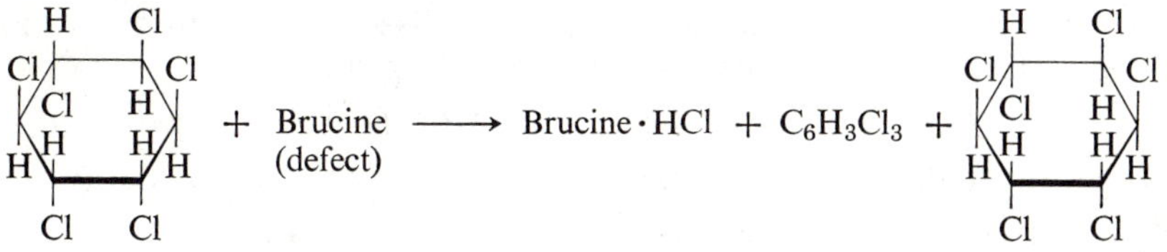

(±) + Brucine (defect) ⟶ Brucine·HCl + $C_6H_3Cl_3$ + Recovered, active

Fig. 4-46. Asymmetric dehydrohalogenation of (±)-1,2,3,4,5,6-hexachlorocyclohexane.

benzil by molecular oxygen in the presence of the catalyst cyclodextrin; the recovered starting material becomes dextrorotatory.[46] A more complicated case,[63] because it involves a substrate with several asymmetric centers, is the partial decarboxylation of (±)-α-carboxycamphor in the presence of the basic catalyst quinine (Fig. 4-47). In this case, not only is the recovered acid active, but so is the camphor obtained, since it still retains two of the original three asymmetric carbon atoms.

e. Biochemical Asymmetric Transformation. Under this heading we shall deal with the production of optically active compounds either by the inter-

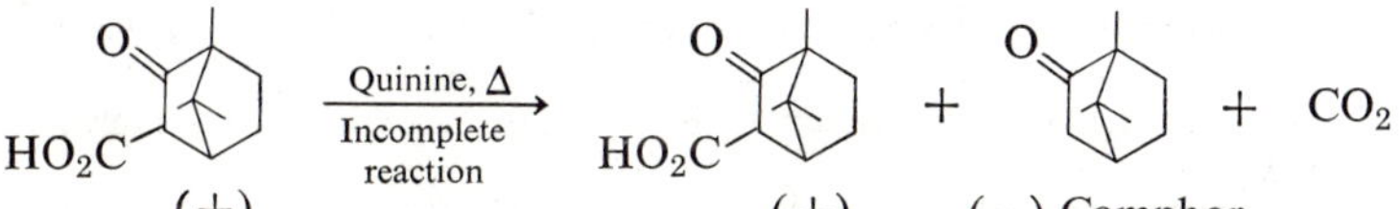

(±) —Quinine, Δ / Incomplete reaction⟶ (+) Recovered + (−)-Camphor + CO_2

Fig. 4-47. Asymmetric decarboxylation of (±)-α-carboxycamphor.

vention of a living organism or by means of one of the catalyst systems, called enzymes, which may be isolated from living organisms. The distinction between biochemical methods of obtaining optically active compounds and other methods is, in many respects, an unfortunate one. In a sense, all the resolutions so far described are biochemical. The alkaloids, terpene derivatives, acids, etc., used in the common methods of resolution are mostly materials of natural origin, and even if they are either strictly synthetic, such as α-phenethylamine, or natural products now accessible synthetically, such

[61] S. J. Cristol, *J. Am. Chem. Soc.*, **71**, 1894 (1949); cf. H. J. Lucas and C. W. Gould, *ibid.*, **64**, 601 (1942).

[62] G. M. Schwab, F. Rost, and L. Rudolph, *Kolloid-Z.*, **68**, 157 (1934).

[63] G. Bredig and K. Fajans, *Ber.*, **41**, 752 (1908); K. Fajans, *Z. physik. Chem.*, **73**, 25 (1910).

as strychnine,[64] somewhere in their synthesis some natural resolving agent must have been employed. Thus α-phenethylamine may be resolved by natural (−)-malic acid or natural (+)-tartaric acid or (−)-pyroglutamic acid obtained by pyrolysis of natural (+)-glutamic acid; and in the synthesis of strychnine, the alkaloid quinidine is used in the resolution of one of the intermediates. Only the mechanical separation method (Sec. 4-4*a*) requires no other optically active agent, but then it requires the active intervention of the most highly developed of all biochemical systems, namely, man!† Another reason why the setting apart of biochemical methods is unfortunate is that it makes a qualitative distinction between enzyme systems and other dissymmetric molecules used in the preparation of dissymmetric products. For example, the reaction of benzaldehyde with hydrogen cyanide,

$$C_6H_5CHO + HCN \rightarrow C_6H_5CHOHCN$$

followed by hydrolysis to mandelic acid, $C_6H_5CHOHCO_2H$, leads to an optically active product if carried out either in the presence of the enzyme emulsin[65] or in the presence of quinine or quinidine.[66] Yet the former process would be called biochemical and the latter an asymmetric synthesis! To be sure, quinine is now available synthetically[67] whereas emulsin is not, but this is not a distinction in principle, since there is no fundamental reason why enzymes should not one day be synthesized also.

In principle, the methods described in this section are merely special cases of kinetic asymmetric transformations (Sec. 4-4*d*). Nevertheless, there is one important quantitative (and practical) difference between the two methods: In the above asymmetric synthesis using quinine or quinidine, the excess of one enantiomer over the other in the product mandelic acid is less than 10%, but, using emulsin, the product is nearly optically pure. In other words, biochemical methods of obtaining active compounds are highly specific and therefore in many cases are of practical importance,‡ whereas other kinetic asymmetric transformations are apt to be of theoretical interest only.¶

† The synthesis of dissymmetric molecules in the absence of other dissymmetric chemical species will be discussed in the next section (Sec. 4-4*f*).

‡ Of the two practical methods of resolution, one—crystallization of diastereoisomers—depends on the high sensitivity of crystal shape to spatial arrangements of the molecules and on the resultant specificity in crystal growth and solubility. The other—the biochemical method—depends on the fact that polymeric molecules containing many asymmetric atoms (the proteins) are involved as catalysts; such molecules might be expected to show considerable specificity toward one enantiomer.

¶ A notable and unique exception is the hydroboration-oxidation of olefins to optically active alcohols by means of di-isopinocampheylborane, the addition product of α-pinene and diborane, which leads to products of optical purity of the order of 83 to 91%: H. C. Brown and G. Zweifel, *J. Am. Chem. Soc.*, **83**, 486 (1961); cf. page 361. Since the α-pinene used in the synthesis was only about 90% optically pure, the "optical yield," i.e., the ratio of optical purity of the product to optical purity of precursor, reagent, or catalyst, is close to 100% in this case. The optical yield of most other chemical asymmetric syntheses does not exceed 20 to 25% at best.

---

[64] R. B. Woodward, M. P. Cava, W. D. Ollis, A. Hunger, H. U. Daenicker, and K. Schenker, *J. Am. Chem. Soc.*, **76**, 4749 (1954).

[65] L. Rosenthaler, *Biochem. Z.*, **14**, 238 (1908).

[66] G. Bredig and P. S. Fiske, *Biochem. Z.*, **46**, 7 (1912).

[67] R. B. Woodward and W. von E. Doering, *J. Am. Chem. Soc.*, **67**, 860 (1945).

The biochemical method of obtaining active compounds, as the method of mechanical separation (Sec. 4-4*a*) and the method of separation by crystallization of diastereoisomers (Sec. 4-4*b*), was discovered by Pasteur.[68] Pasteur noticed that when the ammonium salt of racemic tartaric acid was fermented by means of yeast or *Penicillium glaucum* (a mold), the salt of the natural (dextrorotatory) form was used up preferentially, and after the fermentation had proceeded for some time, the salt of pure (−)-tartaric acid could be isolated from the fermentation broth. This is evidently a case of asymmetric destruction: The penicillium metabolizes the naturally occurring enantiomer preferentially and leaves the unnatural (levorotatory) enantiomer behind. The later work on the synthesis of mandelonitrile in the presence of emulsin (page 76) showed definitely that no living organism or living cell is necessarily involved in this type of asymmetric synthesis but that the active agents are the enzymes (which may function either within or outside the organism).

The above examples of mandelonitrile and tartaric acid illustrate biochemical asymmetric synthesis and biochemical asymmetric destruction. A third and particularly useful method is biochemical kinetic resolution. This method has found widespread application in the preparation of optically active amino

$$\underset{N\text{-acetyl-}dl\text{-alanine}}{\left[\begin{array}{c} CO_2H \\ | \\ H{-}C{-}NHAc \\ | \\ CH_3 \end{array} \quad \begin{array}{c} CO_2H \\ | \\ AcHN{-}C{-}H \\ | \\ CH_3 \end{array}\right]} \xrightarrow[H_2O]{\text{Acylase I}} \underset{(+)\text{-Alanine}}{\begin{array}{c} CO_2H \\ | \\ H_2N{-}C{-}H \\ | \\ CH_3 \end{array}} + \underset{\text{Acetyl derivative of }(-)\text{-alanine}}{\begin{array}{c} CO_2H \\ | \\ H{-}C{-}NHAc \\ | \\ CH_3 \end{array}}$$

Fig. 4-48. Enzymatic resolution of alanine.

acids. For example,[69] when an acylated racemic amino acid in aqueous solution is treated with hog-kidney acylase ("acylase I")† until about half the acyl groups are hydrolyzed, the residual acyl-amino acid is the derivative of the unnatural (so-called "D"; cf. Chap. 5) isomer, whereas the free amino acid obtained in the hydrolysis is the so-called "L" or natural isomer. The unchanged acyl derivative can readily be extracted with ethyl acetate and, in most cases, converted to the free amino acid by non-enzymatic (acid-catalyzed) hydrolysis, whereas the free natural amino acid may be recovered by an ion-exchange process. The method is exemplified in Fig. 4-48 for the case of alanine.

One of the drawbacks of using hog-kidney acylases for resolution is that these enzymes are not stable and must therefore be prepared freshly whenever they are to be used. Fortunately, for some amino acids resolution methods have been elaborated which employ commercially available enzyme preparations. For example,[70] the papain-catalyzed reaction of (±)-acetylphenyl-

† The enzyme seems to be effective for all common amino acids except aspartic acid for which another acylase ("acylase II"), also obtained from kidney, is used.

[68] L. Pasteur, *Compt. rend.*, **46**, 615 (1858).
[69] V. E. Price and J. P. Greenstein, *J. Biol. Chem.*, **175**, 969 (1948); see also Ref. 88.
[70] H. T. Huang and C. Niemann, *J. Am. Chem. Soc.*, **73**, 475 (1951).

alanine with *p*-toluidine gives the *p*-toluide of acetyl-L-phenylalanine and unchanged acetyl-D-phenylalanine which may be readily separated chemically and reconverted individually to D-(+)-phenylalanine and L-(−)-phenylalanine by hydrolysis.

Of particular interest are biochemical methods leading to optically active compounds of the type RR'CHD (cf. Chap. 3). Here may be mentioned the synthesis of (−)-ethanol-1-*d*, CH_3CHDOH, by reduction of acetaldehyde-1-*d*, CH_3CDO, with reduced diphosphopyridine nucleotide in the presence of yeast alcohol dehydrogenase[70a] and the synthesis of optically active tyramine-1-*d*, *p*-$HOC_6H_4CH_2CHDNH_2$, by decarboxylation of (−)-tyrosine, *p*-$HOC_6H_4CH_2CH(NH_2)CO_2H$, in D_2O solution by tyrosine decarboxylase.[70b] Whereas the latter example merely involves the stereospecific conversion of one optically active compound into another (a conversion, however, which it would be difficult to effect stereospecifically by conventional chemical means), the former example actually represents an asymmetric enzymatic synthesis.

While speaking of biochemical discrimination between enantiomers, it is pertinent to point out that such discrimination is also exercised by the human organism. Of the two enantiomeric asparagines, $HO_2CCH(NH_2)CH_2CONH_2$, only the unnatural dextrorotatory isomer is sweet to the taste, and of the monosodium glutamates, only the salt of the natural (+)-glutamic acid (Fig. 4-23) acts as a flavor-enhancing agent. Among drugs there are numerous instances where but one optical antipode of a compound is efficacious. An example is chloramphenicol, *p*-$O_2NC_6H_4CHOHCH(NHCOCHCl_2)CH_2OH$, of whose four stereoisomers only one acts as an antibiotic.[99]

f. Absolute Asymmetric Synthesis. Most of the methods for obtaining optically active compounds discussed so far require other optically active substances as auxiliary devices. These methods, especially the biochemical method, provide rational explanations as to how new dissymmetric molecules may be generated in the presence of the old. They do not, however, explain how optically active molecules originated in the first place.

To some extent, the question of the ultimate origin of optical activity is a metaphysical one, entwined with the related questions of the ultimate origin of matter and of life. There are, however, some purely rational, experimentally demonstrable ways in which optically active substances may be formed in the absence of any other dissymmetric molecules. For example, optically active material may be obtained from a solution of the racemic modification by spontaneous crystallization (Sec. 4-4*a*). This is in no way a mysterious process. It has already been mentioned that all crystals of optically active substances are hemihedric, the crystals of one enantiomer being the mirror images of those of the other. If, in a solution of a racemic modification supersaturated with respect to the enantiomers, a nucleus of one of the enantiomeric crystals begins to form spontaneously and fortuitously, this nucleus is apt to grow by the addition of molecules of its own configuration, and it is quite

[70a] F. A. Loewus, F. H. Westheimer, and B. Vennesland, *J. Am. Chem. Soc.*, **75**, 5018 (1953); H. R. Levy, F. A. Loewus, and B. Vennesland, *ibid.*, **79**, 2949 (1957).

[70b] S. Mandeles, R. Koppelman, and M. E. Hanke, *J. Biol. Chem.*, **209**, 327 (1954); B. Belleau, M. Fang, J. Burba, and J. Moran, *J. Am. Chem. Soc.*, **82**, 5752 (1960).

possible that a macroscopic crystal of this particular enantiomer results before any of the other enantiomer crystallizes. If, through some accident, the mother liquor is at this point separated from the crystal, a partial resolution of the material as between the crystal and the mother liquor will have been effected.

A slight variant of this process has been observed[18b] in the crystallization of methylethylallylanilinium iodide from chloroform solution in sealed glass tubes. The variant here is that the quaternary iodide (whose dissymmetry is due to the asymmetrically substituted nitrogen atom) is subject to spontaneous interconversion of the enantiomers, as shown in Fig. 4-49. Thus, even if one enantiomer crystallizes in preference to the other, due to spontaneous nucleation, the mother liquor will become racemic again after some time, because of the above equilibration process. The crystals isolated from the chloroform solutions after prolonged standing gave rise, in most cases, to optically active solutions when dissolved in water—sometimes dextrorotatory, sometimes levorotatory.

$$\underset{(+)}{H_3C{-}\overset{C_2H_5}{\underset{C_6H_5}{N^+}}{-}CH_2{-}CH{=}CH_2\ I^-} \rightleftharpoons H_3C{-}N\begin{matrix}C_2H_5\\ C_6H_5\end{matrix} + CH_2{=}CHCH_2I$$

$$\Big\Updownarrow$$

$$\underset{(-)}{H_2C{=}CH{-}H_2C{-}\overset{C_2H_5}{\underset{C_6H_5}{N^+}}{-}CH_3\ I^-}$$

Fig. 4-49. Racemization of methylethylallylanilinium iodide.

Another variant of the process is that crystallization of one enantiomer may be induced not by spontaneous nucleation but by seeding. The seed crystal must itself be hemihedric, but it may owe its hemihedrism to crystal dissymmetry rather than molecular dissymmetry. The resolution of solutions of asparagine by seeding with crystals of glycine (p. 48) may be of this type.

Compounds such as quartz and urea which are apt to crystallize in dissymmetric, optically active crystals without themselves possessing molecular dissymmetry may yet induce molecular dissymmetry in other molecules. Several cases, such as resolution by adsorption on active adsorbents (e.g., quartz powder), resolution via inclusion compounds (e.g., with urea), and asymmetric destruction by quartz-supported catalysts (as in the dehydration of 2-butanol) have already been mentioned.

In addition to these possibilities, two methods of so-called "absolute asymmetric synthesis"† have been recognized. These are syntheses of compounds in active form without the intervention of any dissymmetric chemicals. Some sort of physically dissymmetric influence is required in such syntheses, and it

† As distinct from the "partial asymmetric synthesis" discussed in Sec. 4-4*d*.

is necessary that the physical agent in question be essential to the synthesis, rather than accidental to it. For example, experiments[71] on the addition of bromine to methyl cinnamate, $C_6H_5CH{=}CHCO_2CH_3$, to give methyl 2,3-dibromo-3-phenylpropionate, $C_6H_5CHBrCHBrCO_2CH_3$ (two asymmetric carbon atoms), under the influence of visible plane-polarized light passed through a magnetic field were doomed to failure; for although the physical influence is truly dissymmetric, it is not in any way involved in the chemical reaction of bromine addition. A successful absolute asymmetric synthesis[72] was, however, effected in the decomposition of the dimethylamide of α-azidopropionic acid, $CH_3\overset{*}{C}HN_3CON(CH_3)_2$, with circularly polarized light† of wavelength 2800 to 3100 A. The azido compound has an adsorption band in the ultraviolet spectrum at 2900 A. and is photochemically decomposed (with the evolution of nitrogen) by light of that wavelength. Using circularly polarized light of that wavelength introduces an asymmetric influence that is now essential to the reaction, and indeed it was found that the amide recovered after partial (40%) photochemical destruction was in one case levorotatory, α_D $-1.04°$, and in another case dextrorotatory, α_D $+0.78°$, depending on whether the light was levocircularly polarized or dextrocircularly polarized. (These rotations are observed rotations of the liquid in a 1-dm. tube.) This result is close to what had been predicted on theoretical grounds. Whereas this particular case is one of asymmetric *destruction,* cases of absolute asymmetric *synthesis* under the influence of circularly polarized light (e.g., in the photochemical addition of bromine to 2,4,6-trinitrostilbene) have also been recorded,[73] although the observed rotations of the product were very small (less than 0.1°).

Since sunlight reflected by the sea possesses slight circular (or elliptical) polarization, asymmetric syntheses similar to the above could possibly have taken place on the surface of the earth.

Another approach to the problem of absolute asymmetric synthesis is to effect the synthesis under the influence of particulate (alpha or beta) radiation. Since this radiation is itself affected by a kind of handedness or dissymmetry (called "parity"), it might possibly induce dissymmetry in the molecule to be synthesized.[74]

The possibilities for spontaneous formation of optically active material discussed above do not by themselves explain, however, why most naturally occurring compounds, such as carbohydrates, amino acids, terpenes, steroids, alkaloids, antibiotics, etc., are found in nature in only one of the enantiomeric forms. This "configurational specificity" of natural products is very high; thus it appears that all the amino acids occurring in the higher forms of life

† Circularly polarized light is light whose plane of polarization rotates in corkscrew fashion along the line of propagation of the ray (cf. Chap. 14).

71 P. A. Guye and G. Drouginine, *J. chim. phys.*, **7**, 96 (1909).

72 W. Kuhn and E. Knopf, *Z. physik. Chem.*, **7B**, 292 (1930).

73 T. L. Davis and R. Heggie, *J. Am. Chem. Soc.*, **57**, 377 (1935).

74 F. Vester, T. L. V. Ulbricht, and H. Krauch, *Naturwiss.*, **46**, 68 (1959); cf. T. L. V. Ulbricht, *Quart. Revs.* (*London*), **13**, 48 (1959).

in the form of proteins and polypeptides have the so-called "L configuration" (cf. Chap. 5), the arrangement of the atoms about the asymmetric carbon being that shown in Fig. 4-50. (Only a few amino acids occurring in bacteria and fungi have the opposite or "D configuration."[74a]) Propagation of the activity in nature involves enzyme systems made up, in part, of proteins as well as nucleic acids; since these proteins and nucleic acids are high-molecular-weight polymers made up of dissymmetric monomer units, they are highly dissymmetric molecules, and it is perhaps not surprising that the reactions that they catalyze lead almost exclusively to molecules of the same configuration. However, there are still a number of unanswered questions: how the enzymes came to be so stereospecific to begin with, what happens to molecules which are accidentally racemized and to the small but probably finite number of molecules which are synthesized in the wrong configuration, etc. Only speculations on this subject are possible at the present time.[18, 75]

g. Asymmetric Syntheses Involving Symmetric Compounds.[90, 98] In this section will be discussed some reactions of symmetric compounds of the type Cxxyz with dissymmetric reagents in which the two identical substituents (x) behave in a different way, even though the products of the reaction are not, themselves, dissymmetric. Reactions of this type are of importance since

```
        CO2H
         |
  H2N—C—H
         |
         R
```

Fig. 4-50. Natural L-amino acid.

they show that even in a symmetric compound apparently identical groups may be operationally distinguishable.

In order to *demonstrate* that the two identical x groups do, in fact, behave differently it is necessary to label one of them. This has been done,[76] as shown in Fig. 4-51. The optically active lactone I was deuterated in the active methylene position by means of D_2O and then decarboxylated by means of D_2O_2 to citric acid. Because of the dissymmetry of the starting material, the α,α-dideuterocitric acid (II) was labeled exclusively in *one* of the two unequally placed acetic acid groupings. [As it happened, the acid II was also demonstrably optically active (cf. Sec. 3-1), but this is in no way essential to the argument.] The dideuterocitric acid was then degraded to α-ketoglutaric acid (III), using an enzyme preparation made from pigeon-breast homogenate and arsenite. (The arsenite presumably serves to poison enzymes which would bring about further degradation.) Now when one of the antipodes of II (made from active I) was so degraded, all the deuterium was retained in the α-ketoglutaric acid III, indicating that only the *unlabeled* methylene group

[74a] For example, C. M. Stevens, R. P. Gigger, and S. W. Bowne, *J. Biol. Chem.*, **212**, 461 (1955).

[75] W. Langenbeck, "Die Organischen Katalysatoren," 2d ed., Springer Verlag, Berlin, Vienna, 1949, pp. 99–103.

[76] C. Martius and G. Schorre, *Ann.*, **570**, 143 (1950); see also P. E. Wilcox, C. Heidelberger, and V. R. Potter, *J. Am. Chem. Soc.*, **72**, 5019 (1950).

was oxidized by the enzyme. This is not due to an isotope effect, for when the other antipode of II (made from the other antipode of I) was degraded, all the deuterium was lost in III. *Thus which of the methylene groups is oxidized in the enzymatic degradation of II depends on the relative position of these groups with respect to the rest of the molecule.*

It is important to realize that the above result is not dependent on the presence of the deuterium label (although in the absence of the label it would not have been detected). The essential point is that the methylene group in the ring in the enantiomer of lactone I shown in Fig. 4-51 is so located in the citric acid (II) that upon enzymatic degradation it ends up in the α-ketoglutaric acid (III). If the other enantiomer of lactone I had served as starting

$$O{=}C{-}C({=}O){-}CH_2{-}C(CO_2H)(CH_2CO_2H){-}O{-}\ (\text{ring closing } C \text{ and } C \text{ via } O) \xrightarrow[D_2O_2]{D_2O} \left[O{=}C{-}C({=}O){-}CD_2{-}C(CO_2H)(CH_2CO_2H){-}O{-} \right] \xrightarrow[D_2O]{-CO_2}$$

I active

$$DO_2CCD_2{-}C(CO_2H)(OH){-}CH_2CO_2H \xrightarrow[\text{degradation}]{\text{Enzymatic}} HO_2CCD_2CH_2C({=}O)CO_2H$$

II → III

The other* isomer of I gives
$HO_2CC({=}O)CH_2CH_2CO_2H$.

* Assignment of configuration of I is arbitrary. The true configuration does not appear to be known.

Fig. 4-51. Asymmetric degradation of symmetric molecule.

material, its ring methylene group would have ended up in the opposite position in the citric acid (II) and would then have become so oriented on the enzyme that it would have been eliminated (as acetic acid?) rather than retained in the product III. The fact that, although citric acid is a symmetric intermediate in the sequence shown in Fig. 4-51, its methylene groups are not identical either as to origin or as to further fate is of considerable biochemical significance.

Several theories have been developed to explain (or predict) the experimental result discussed above, one of the most interesting being that of Ogston[77] which postulates a three-point contact between enzyme and substrate. It has been pointed out, however, by several investigators[78,79,90] that the above result can be predicted without any specific postulates as to the

[77] A. G. Ogston, *Nature,* **162,** 963 (1948).

[78] P. Schwartz and H. E. Carter, *Proc. Natl. Acad. Sci. U.S.A.,* **40,** 499 (1954).

[79] R. Altschul, P. Bernstein, and S. G. Cohen, *J. Am. Chem. Soc.,* **78,** 5091 (1956); S. G. Cohen and E. Khedouri, *Nature,* **186,** 75 (1960).

nature of the enzyme-substrate interaction. In a molecule Cxxyz (or, for that matter, in any molecule which has two identical atoms or groups attached to the carbon skeleton *but which is devoid of a simple axis of symmetry*†) the two identical groups are not identically located with respect to the rest of the molecule, but the relationship of one group to the rest of the molecule is the mirror image of the relationship of the other group. As long as only symmetric reagents are considered, this is of no consequence, but toward a dissymmetric reagent (such as an enzyme) the two groups may behave differently and react at different rates. This is true even if the reagent or catalyst (enzyme) does not become incorporated in the product but only in the transition state or an intermediate (e.g., page 72), in fact, the product need not be dissymmetric at all (e.g., the case depicted in Fig. 4-51).‡

4-5. Criteria of Optical Purity

By optical purity of a partially resolved material is meant the excess of one enantiomer in the material expressed as a percentage of the total. In a *dl* pair, there is no excess enantiomer and the optical purity is zero; in a completely resolved material, the excess enantiomer is equal in weight to the total material and the optical purity is 100%. It is often desirable to find out whether a given resolution has gone to completion, i.e., whether the enantiomer obtained is really 100% optically pure. Several simple criteria of optical purity have been developed, but none is completely reliable. For example, a crystalline enantiomer is often considered optically pure when its melting point and rotation are unchanged by further crystallization. Reference to Fig. 4-19 shows that such a criterion usually fails when the racemic modification forms a solid solution, since in that case even a *partially* resolved enantiomer may not change in either rotation or melting point by further recrystallization. A resolution is often deemed complete when the diastereoisomeric salt employed (Sec. 4-4*b*) does not change in rotation upon further crystallization. This criterion may also be foiled by certain special types of phase behavior. Finally, resolution is often deemed complete once *both* enantiomers are obtained in a state of equal purity (i.e., with equal and opposite specific rotation), the argument being that, since both were resolved in independent fashion, there is no reason why both should have equal rotations (of opposite signs) unless these rotations are maximal. This approach is not applicable when one of the enantiomers is not readily obtained pure; in that case one cannot tell whether the other one is pure either.

† The reader might note that we are speaking here of molecules devoid of a *simple axis* of symmetry, as distinct from Sec. 2-2 where criteria for optical activity were set up and it was found that molecules lacking an *alternating axis* of symmetry would show mirror-image forms.

‡ The explanation given here is a purely *thermodynamic* one (considering transition-state theory as an extension of thermodynamics); i.e., it says that the two methylene groups in citric acid *may* behave differently toward a dissymmetric reagent and that such behavior is entirely reasonable on the basis of energy differences between diastereoisomeric transition states. This does not mean that such differential behavior will necessarily be observed, i.e., that there is necessarily a *mechanistic* path to implement the potential difference. Ogston's explanation (Ref. 77) may still be useful in providing a picture for such a mechanistic path, thus rationalizing why what is thermodynamically permitted does actually occur.

There are three other ways known of determining optical purity (at least within specifiable experimental limits): an enzymatic method, an isotope dilution method, and a method of relating a compound of unknown optical purity to another one whose purity is known.

The enzymatic method depends on the fact that many enzymes are highly selective for one enantiomer of a *dl* pair.† If one incubates a supposedly pure preparation of the *other* enantiomer with such an enzyme, reaction (which must be detectable by suitable means) would indicate the presence of some of the wrong antipode, owing to incomplete resolution or racemization following complete resolution, whereas the absence of reaction would indicate purity.‡ For example, an L-amino acid oxidase which is almost completely inert to D-amino acids can be isolated from rattlesnakes. A supposedly pure D-amino acid, when treated with this enzyme in a Warburg apparatus in the presence of oxygen, will take up none of this gas, but if L-amino acid is present as an impurity, gas consumption will be registered. The test is supposedly sensitive to 1 part in 1000 of the "wrong" antipode. This method is obviously limited in its application to optically active compounds that are subject to enzymatic reactions.

In the isotope dilution method, the supposedly pure [let us say (−)] enantiomer is mixed with some labeled (radioactively or otherwise) racemic material in solution, and the racemic material is then reisolated (usually by crystallization).¶ Since in solution the racemic material is split up into labeled (+) molecules and (−) molecules, the labeled (−) molecules get commingled with the molecules of the enantiomer whose optical purity is to be determined, but the (+) molecules do not (provided that the enantiomer was pure). One can thus calculate a dilution factor, knowing the weight of the original active material and the weight of the added labeled racemic material, and can thus predict how active the recovered labeled racemic material should be. The predicted activity is then compared with that experimentally found. If the experimental activity is less than that predicted, it indicates that there was some residual (unlabeled) racemic material in the supposedly pure enantiomer. For if one mixes labeled racemic material with unlabeled (−) material, only the (−) molecules in the recovered *dl* pair get diluted isotopically, but if one mixes it with unlabeled racemic material, then *all* molecules in the recovered *dl* pair will be diluted, i.e., the dilution factor will be greater. In fact, a relationship may be established[80]

$$C_{\pm} = aC_0 \frac{a + B}{(2B + a - R)(a + R)}$$

where C_0 is the activity of the added racemic material, $C_{\pm}$ is the activity of

† Not all enzymes are equally stereoselective, and the stereoselectivity of a given enzyme varies with the substrate.

‡ It is assumed that the substrate is *chemically* pure, so that the only contaminant which could possibly be affected by the enzyme is the antipode of the desired compound.

¶ The method is dependent on the possibility of reisolating the racemate in pure form.

[80] J. A. Berson and D. A. Ben-Efraim, *J. Am. Chem. Soc.*, **81**, 4083 (1959); cf. S. Graff, D. Rittenberg, and G. L. Foster, *J. Biol. Chem.*, **133**, 745 (1940).

the recovered racemic material, a is the weight of the added racemic material, B is the weight of the resolved material (whose purity is to be tested) admixed with a, and R is the weight of racemate (if any) in the amount B. By solving the above equation for R, knowing all the other quantities from experiment, one may calculate the amount of racemic contaminant and hence the optical purity which will be

$$100\frac{B-R}{B}\%$$

The third method for determining optical purity is a correlative method. Suppose that the optical purity of a compound Cabde is known. The *minimum* purity of another compound Cabdf may be determined if Cabdf can be converted chemically to Cabde. Cabdf will be *at least* as pure optically as the Cabde prepared from it. (It may be *more* pure, for the reaction converting Cabdf to Cabde may have involved some racemization, and so the Cabde actually obtained may be less pure than the Cabdf starting material.) The following example will illustrate this method. The highest known rotation[81] of α-phenethyl chloride, $C_6H_5CHClCH_3$, is α_D^{25} (neat, 1 dm.) 109°.† When dextrorotatory material of this rotation is allowed to react with allylsodium and the resulting 4-phenyl-1-pentene, $CH_2{=}CHCH_2CH(C_6H_5)CH_3$, hydrogenated to 2-phenylpentane, $CH_3CH_2CH_2CH(C_6H_5)CH_3$, one obtains material of rotation α_D^{25} − 12.96° (neat, 1 dm.).‡ Since the known rotation of optically pure 2-phenylpentane, established in other ways, is α_D^{25} 14.91°, the optical purity of the hydrocarbon obtained from the chloride is 12.96/14.91 × 100 or 86.9%. The *minimum* optical purity of chloride of α_D^{25} 109° is therefore also 86.9%, and it follows that the *maximum* possible rotation of α-phenethyl chloride is 109 × 100/86.9, or 125.4°.

General References

[82] F. Ebel, Optische Antipoden, in K. Freudenberg, ed., "Stereochemie," Franz Deuticke, Leipzig, 1932, pp. 553–586.

[83] Th. Wagner-Jauregg, Racemisierung, in K. Freudenberg, ed., "Stereochemie," Franz Deuticke, Leipzig, 1932, pp. 852–879.

[84] P. D. Ritchie, "Asymmetric Synthesis and Asymmetric Induction," Oxford University Press, London, 1933.

[85] R. Shriner and R. Adams, Optical Isomerism, in H. Gilman, ed., "Organic Chemistry," 2d ed., John Wiley & Sons, Inc., New York, 1943, pp. 240–264, 305–315.

[86] A. W. Ingersoll, The Resolution of Alcohols, in R. Adams, ed., "Organic Reactions," vol. 2, John Wiley & Sons, Inc., New York, 1944, chap. 9.

[87] G. W. Wheland, "Advanced Organic Chemistry," 3d ed., John Wiley & Sons, Inc., New York, 1960, pp. 222–226, 306–357.

[88] J. P. Greenstein, The Resolution of Racemic α-Amino-acids, in M. L. Anson, K. Bailey, and J. T. Edsall, eds., "Advances in Protein Chemistry," vol. 9, Academic Press, Inc., New York, 1954, pp. 121–202.

† This rotation (109°) is obviously the *minimum* rotation of optically pure chloride.
‡ Actually, chloride of α_D^{25} + 77.89° gave 2-phenylpentane of α_D^{25} − 9.26°.

[81] R. L. Burwell, A. D. Shields, and H. Hart, *J. Am. Chem. Soc.*, **76**, 908 (1954).

[89] W. Theilacker, Methoden zur Herstellung optisch aktiver aus inaktiven Verbindungen, in Houben-Weyl, "Methoden der Organischen Chemie," vol. IV, pt. 2, 4th ed., E. Müller, ed., Georg Thieme Verlag, Stuttgart, 1955, pp. 509–538.

[90] H. Hirschmann, The Structural Basis for the Differentiation of Identical Groups in Asymmetric Reactions, in S. Graff, ed., "Essays in Biochemistry," John Wiley & Sons, Inc., New York, 1956, pp. 156–174.

[91] M. M. Harris, The Study of Optically Labile Compounds, in W. Klyne and P. B. D. de la Mare, eds., "Progress in Stereochemistry," vol. 2, Academic Press, Inc., New York, 1958, chap. 5.

[92] E. E. Turner and M. M. Harris, "Organic Chemistry," Longmans, Green & Co., Inc., New York, 1952, chap. 29.

[93] E. E. Turner and M. M. Harris, Asymmetric Transformation and Asymmetric Induction, *Quart. Revs.* (*London*), **1**, 299 (1947).

[94] Méthodes de dédoublement, in L. Velluz, ed., "Substances naturelles de synthèse," vol. IX, Masson et Cie, Paris, 1954, pp. 119–174. (In this work the preparation of a number of resolving agents is described.)

[95] E. Lederer and M. Lederer, The Separation of Stereoisomers, in "Chromatography," Elsevier Publishing Company, Amsterdam, 1957, chap. 39, pp. 420–426.

[96] K. Mislow, Stereoisomerism, in "Comprehensive Biochemistry," Elsevier Publishing Company, Amsterdam, 1962.

[97] H. Krebs, J. Diewald, R. Rasche, and J. A. Wagner, Die Trennung von Racematen auf chromatographischem Wege, "Forschungsber. d. Wirtschafts- und Verkehrsministeriums Nordrhein-Westfalen," Westdeutscher Verlag, Cologne, Germany, 1956.

[98] H. R. Levy, P. Talalay, and B. Vennesland, Steric Course of Enzymatic Reactions at Meso Carbon Atoms: Application of Hydrogen Isotopes, in P. B. D. de la Mare and W. Klyne, eds. "Progress in Stereochemistry," vol. 3, Academic Press, Inc., New York, 1962, chap. 8.

[99] A. H. Beckett, Stereochemical Factors in Biological Activity, in E. Jucker, ed., "Progress in Drug Research," vol. 1, Birkhäuser Verlag, Basel, Switzerland, 1959, pp. 455–530.

Chapter 5

CONFIGURATION

5-1. Definition

By "configuration" is meant the arrangement in space of the atoms or groups around the dissymmetric or rigid part of a molecule—in the simplest case, around an asymmetric carbon. For example, the formulas for (+)- and (−)-lactic acid (Figs. 3-1 and 3-2) represent molecules of identical structure but different configuration. The question may be asked why one needs a new concept "configuration" to distinguish the two lactic acids if one already has a distinctive term, namely, "rotation." This is like asking why one needs two structural formulas to distinguish *n*-pentane and neo-pentane when the two compounds are already distinguished by their boiling points. Rotation is an experimental property, measured in a polarimeter. Configuration is a theoretical concept related to molecular architecture, expressed by a (three-dimensional or projection) formula. While configuration is less tangible than rotation, it is more fundamental. Thus when (+)-lactic acid (Fig. 3-1) is dissolved in aqueous sodium hydroxide, the resulting solution of sodium lactate is levorotatory and would have to be called (−)-sodium lactate. Nevertheless, the arrangement of the groups about the asymmetric carbon is not changed by the abstraction of the ionizable hydrogen; (−)-sodium lactate still has the molecular architecture represented by Fig. 3-1 (with CO_2^- instead of CO_2H).

In a compound with several asymmetric atoms,† configuration is not specified completely until it is specified at each atom. Thus (−)-tartaric acid and *meso*-tartaric acid (Fig. 3-16) have the same configuration at one of their asymmetric carbon atoms but differ at the other.

Cis-trans isomers (Figs. 1-3 and 1-4) also differ in configuration, namely, in the arrangement of the groups about the (rigid) double bond or ring. No additional complication is introduced in cases where both optical and geometrical isomerism coexist (Fig. 1-5). By specifying the configuration at each asymmetric atom and pertinent double bond, one specifies also their relative configuration; in other words, one specifies with which optical isomer one is dealing and to which geometrical configuration this corresponds.

Note: All references above 56 are listed in the General References at the end of the chapter.

† More properly, we should speak about "dissymmetric groupings"; see footnote on page 21.

In this chapter we shall deal only with the configuration of optical isomers. The configuration of geometrical isomers will be taken up in Chaps. 7 and 12.

5-2. Notation

The configuration of a molecule is completely specified either by an appropriate three-dimensional model or formula drawing or by the corresponding projection formula. Fischer projection formulas (Sec. 3-1, Fig. 3-3) are preferred for this particular purpose. Thus it is known (by methods to be discussed in Sec. 5-4) that Fig. 3-1 represents the true arrangement of the groups in (+)-lactic acid and Fig. 3-2 that in (−)-lactic acid. However, just as there are names as well as structural formulas to specify structure, it is desirable to have symbols as well as three-dimensional or projection formulas to specify configuration. Regrettably, while there have long been internationally accepted systems to name structures, there is not yet an internationally accepted system of designating configuration. For the time being, the only *completely safe* way of specifying configuration is by means of appropriate formulas.

Nevertheless, several unofficial systems of configurational nomenclature have been proposed. The oldest one† is applicable to molecules of the type RCHXR′, where R—C—R′ constitutes the main chain of the molecule in the sense of the International Union of Chemistry (I.U.C.) nomenclature. The molecule is so oriented that the No. 1 carbon of the main chain is at the top in a Fischer projection formula.‡ Then, if X is on the right, the molecule is called "D"; if X is on the left, it is called "L."¶ Thus the (−)-lactic acid shown in Fig. 3-3 is D-lactic acid, because X (in this case OH) is on the right when the molecule is oriented as specified above. Other examples are shown in Fig. 5-1. Rotation as well as configuration is indicated in the name.

This system works well for compounds RCHXR′ when X is a hetero atom. It may be applied also to cases where X is an alkyl group, if it is remembered that R—C—R′ is the main chain; note the case of hydratropic acid (and active amyl alcohol) in Fig. 5-2. It is not directly applicable to compounds of type RR′CXR″ but can be extended to such compounds where X is a hetero atom if it is again specified that R—C—R′ is the main chain and that D means that X is to the right and the (small) alkyl group R″ to the left. (See the cases

† One of the earliest, though not wholly successful, attempts to use symbols to specify configuration is that of E. Fischer, *Ber.*, **24**, 2683 (1891).

‡ Another statement of the rule is to say that the most highly oxidized carbon is to be at the top. Since this ($CO_2H > CHO > CH_2OH$) is in any case the No. 1 carbon in the I.U.C. system, there is no contradiction between the two statements.

¶ The older literature used the lower-case symbols *d* and *l*. However, since in many places these symbols had been used for rotation (synonymously with "+" and "−") a great deal of confusion resulted. For this reason the symbols *d* and *l* are not used at all in this text (except in "*dl* pair" where there is no ambiguity). The first suggestion to use symbols different from *d* and *l* for configuration seems to be that of Rosanoff (Ref. 1) who employed the Greek letters δ and λ, but these have been superseded by the capital letters D and L which are used here.

---

[1] M. A. Rosanoff, *J. Am. Chem. Soc.*, **28**, 114 (1906); see also A. Wohl and K. Freudenberg, *Ber.*, **56**, 309 (1923).

of atrolactic acid and 2-phenyl-2-butanol in Fig. 5-2.) The system is not applicable without further operational definitions to compounds of type RR′CR″R‴, that is, compounds in which the asymmetric atom is a quaternary carbon.

It should be noted here that configurational nomenclature is conventional rather than based on genetic relationships. The fact that two compounds

```
    CHO                  CO2H              CH3
    |                    |                 |
 H—C—OH            H2N—C—H           H—C—Cl
    |                    |                 |
    CH2OH                CH3               C6H5
```

D-(+)-Glyceraldehyde | L-(+)-Alanine | D-(−)-α-Phenethyl chloride

```
    CH3                  CO2H              CO2H
    |                    |                 |
 H—C—OCH3           H—C—OH            H—C—OH
    |                    |                 |
    C2H5                 CH2CO2H           C6H5
```

D-(+)-2-Butyl methyl ether | D-(+)-Malic acid | D-(−)-Mandelic acid

Fig. 5-1. Configurational nomenclature. Use of D and L.

have the same configurational symbol means merely that their projection formulas are to be written in a certain specified way; it does *not* mean that compounds bearing the same symbol are interrelated (by chemical transformation or common precursor) in preference to compounds of opposite symbol. For example, the Hofmann bromamide reaction as applied to L-hydratropic acid (Fig. 5-3) is known to proceed without change of configuration (see Sec.

```
    CO2H            CO2H              C2H5             CH2OH           CO2H
    |               |                 |                |               |
 H—C—CH3     H3C—C—OH         H3C—C—OH         H—C—CH3        Cl—C—CH3
    |               |                 |                |               |
    C6H5            C6H5              C6H5             C2H5            C6H5
```

D-(−)-Hydratropic acid | D-(−)-Atrolactic acid | D-(−)-2-Phenyl-2-butanol | D-(+)-2-Methyl-1-butanol | L-(+)-α-Phenyl-α-chloropropionic acid

Fig. 5-2. Configurational nomenclature. Use of D and L. Extensions.

5-4*f*), i.e., by direct replacement of —CO_2H by —NH_2. Nevertheless the nomenclature convention is such (cf. Fig. 5-3) that the product must be labeled D-α-phenethylamine. An even more obvious example that genetic relationships are useless for configurational notation is the conversion of (+)-isopropylmalonamidic acid into (−)-isopropylmalonamidic acid discussed in Chap. 3 (Fig. 3-4). If nomenclature followed a genetic relationship, the two enantiomers should have the same configurational symbol, which is absurd, since they are obviously of opposite configuration.

Difficulties in the use of the D-L nomenclature arise when it is to be applied to compounds with more than one asymmetric atom. For example, in (−)-threose (Fig. 3-7) the hydroxyl group on the lower (No. 3) asymmetric carbon atom is on the right, but that on the upper (No. 2) asymmetric carbon atom is on the left when the formula is written in the conventional way (i.e., with carbons 2 and 3 in front, and 1 and 4 eclipsed pointing toward the rear). Should this compound be called D or L? An additional convention has been developed according to which in the sugars the asymmetric carbon next to the primary alcohol group (*not* the aldehyde group) is the one that specifies convention. Therefore the (−)-threose shown in Fig. 3-7 is D-(−)-threose and the (+)-threose is L-(+)-threose.† The natural (+)-glucose shown in Fig. 3-15 is similarly called D-(+)-glucose.‡ To show that one is dealing here with an extension of the original configurational nomenclature, the symbol D_G is occasionally used instead of D for the sugars; thus (+)-glucose becomes D_G-(+)-glucose. A similar extension has been used for the amino acids. (−)-Serine (Fig. 5-4) is evidently L-(−)-serine, but for (−)-threonine (Fig. 5-4) the situation is not so clear, since there are two asym-

$$\begin{array}{ccccc} CO_2H & & NH_2 & & CH_3 \\ H_3C-\overset{|}{\underset{|}{C}}-H & \xrightarrow{\text{NaOBr on amide}} & H_3C-\overset{|}{\underset{|}{C}}-H & \equiv & H-\overset{|}{\underset{|}{C}}-NH_2 \\ C_6H_5 & & C_6H_5 & & C_6H_5 \\ \text{L-}(+) & & & & \text{D-}(-) \end{array}$$

Fig. 5-3. Hofmann bromamide reaction of hydratropic acid.

metric carbons, one related to the sugars (the alcoholic one) and one related to the amino acids. Conventionally, one lets the latter specify the symbol, and to denote this one uses L_S instead of simply L; thus (−)-threonine becomes L_S-(−)-threonine.¶

Unfortunately, however, this system cannot be generalized. Thus there seems to be no agreement whether the (natural) (+)-tartaric acid shown in Fig. 3-16 should be called "L-(+)-tartaric acid," using the bottom asymmetric carbon as reference, or "D-(+)-tartaric acid," using the top asymmetric carbon as reference. Different authorities in the field have adopted

† This example makes it quite plain, if indeed it is not plain already, that the symbols D and L have nothing to do with the sign of rotation.

‡ The reader might note that the configurational symbol D as here used spells out the configuration of the bottom asymmetric carbon only. The configuration of the other three asymmetric carbons is not specified by the symbol. (In fact, all the aldohexoses shown in Fig. 3-15 have the D configuration.) As long as common names are used, this is not so serious, since *the name itself* specifies the diastereoisomer with which one is dealing (cf. Fig. 3-15), and the configurational symbol serves merely to distinguish the two enantiomers. However, if one were to use I.U.C. names, all the eight aldohexoses in Fig. 3-15 would be D-hexane-2,3,4,5,6-pentol-1-als (or hemiacetals thereof; cf. Fig. 4-13), and, in order to make the names unique, it would be necessary, somehow, to spell out the configuration at the other three asymmetric carbon atoms as well.

¶ The subscript G in D_G stands for glyceraldehyde, the simplest of the sugars, whereas the subscript S in L_S stands for serine, the simplest of the amino acids containing also a hydroxyl group.

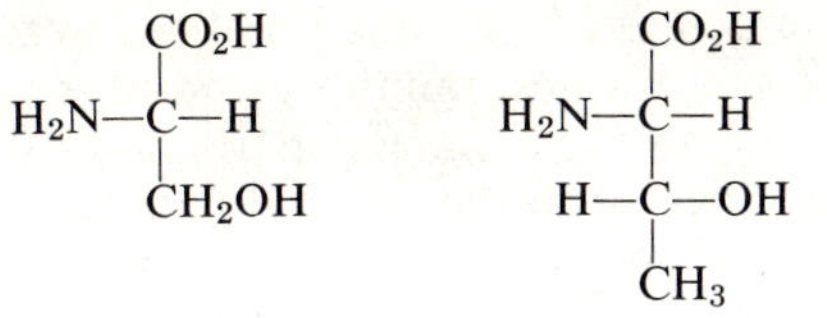

$$\begin{array}{c} CO_2H \\ | \\ H_2N-C-H \\ | \\ CH_2OH \end{array} \qquad \begin{array}{c} CO_2H \\ | \\ H_2N-C-H \\ | \\ H-C-OH \\ | \\ CH_3 \end{array}$$

L-(−)-Serine Ls-(−)-threonine

Fig. 5-4. Amino-acid configurational nomenclature.

(and defended) either usage; it seems to this author that this problem cannot be solved without some additional arbitrary convention and simply serves to point out the inadequacy of naming a compound having two asymmetric atoms with one configurational symbol, even if the two centers are alike, as they are in (+)-tartaric acid.†

Useful extensions of the D-L nomenclature to compounds containing more than one asymmetric carbon and to compounds possessing asymmetric carbons in rings have nevertheless been proposed (Fig. 5-5).[2,3] These systems specify that each asymmetric carbon in a compound possessing more than one is to be given a configurational symbol, that this symbol is to be D when the functional group is to the right of the main chain numbered from the top (according to the I.U.C. system) and L if it is to the left, and that it is to pre-

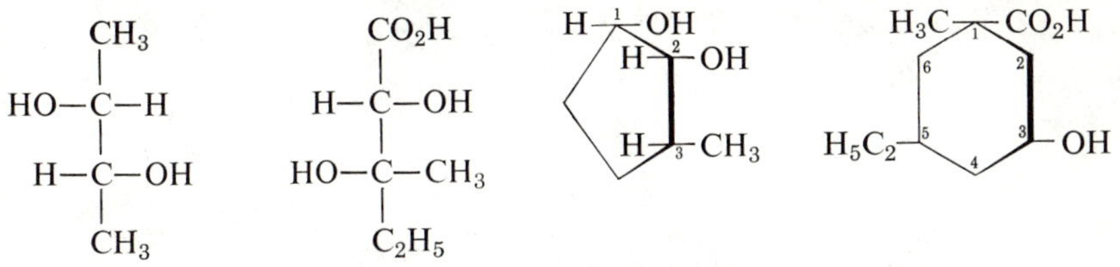

$$\begin{array}{c} CH_3 \\ | \\ HO-C-H \\ | \\ H-C-OH \\ | \\ CH_3 \end{array} \qquad \begin{array}{c} CO_2H \\ | \\ H-C-OH \\ | \\ HO-C-CH_3 \\ | \\ C_2H_5 \end{array}$$

(−)-Butane-2L, 3D-diol

2D-Hydroxy-3D-methyl-3L-hydroxy-pentanoic acid

3D-Methyl-cyclopentane-1D, 2D-diol

5L-Ethyl-3D-hydroxy-1L-methyl-cyclohexane-1D-carboxylic acid

Fig. 5-5. D-L configurational notation for compounds with several asymmetric atoms.

† The protagonists of the L-(+)-tartaric acid name argue that it should be named like a sugar, with the lowermost asymmetric carbon specifying the symbol. The protagonists of the D-(+)-tartaric acid name say that, since it is not in fact a sugar, it should be looked at as R—CHOH—R′, where R = $CHOHCO_2H$ and R′ = CO_2H. Since the carbon adjacent to the asymmetric atom under consideration is more highly oxidized in R′, R′ is placed at the top, and since the No. 2 carbon then has the hydroxyl on the right, it should be D. Usually two extraneous arguments are also brought in: The D school argues that it is better to call dextrorotatory tartaric acid "D"; the L school argues that since (+)-tartaric acid is obtained by oxidation of the two terminal carbons of L-(+)-threose (Fig. 3-7) it should be called "L." These latter two arguments are evidently not pertinent to the problem, for, as we have already seen, neither sign of rotation nor genetic relationships are germane to configurational symbolism.

[2] W. Klyne, *Chemistry & Industry* (*London*), 1022 (1951).

[3] G. E. McCasland, "A New General System for the Naming of Stereoisomers," Chemical Abstracts, Columbus, Ohio, 1950.

cede, immediately, the names of the functional group so placed. Rings are oriented with the edge of the lowest numbers toward the observer. Examples (named according to Klyne[2]) are shown in Fig. 5-5; for a more thorough treatment the reader is referred to the original papers.

The systems mentioned so far are based on Fischer projection formulas. This has certain disadvantages; thus before a unique name can be established for a compound one must specify how its projection formula is to be oriented (cf. Chap. 3). Also it often happens [see the case of the (−)-2,3-butanediol in Fig. 5-5] that two asymmetric carbons having manifestly the same configuration are given opposite configurational symbols. These difficulties are avoided by a system[4] which is based on the actual three-dimensional formula of the compound to be named. In order to avoid confusion with systems based on projection formulas, different symbols are employed, namely, R (for *rectus,* Latin for right) and S (for *sinister,* Latin for left). In order to name a compound Xabcd (X being the asymmetric atom) the groups a, b, c, and d are first arranged in a priority sequence according to the sequence rule below. Let us assume that this sequence is a before b before c before d. The molecule is then viewed from the side *remote* from d. If a → b → c now traces a right-handed (clockwise) turn, the configuration is R,

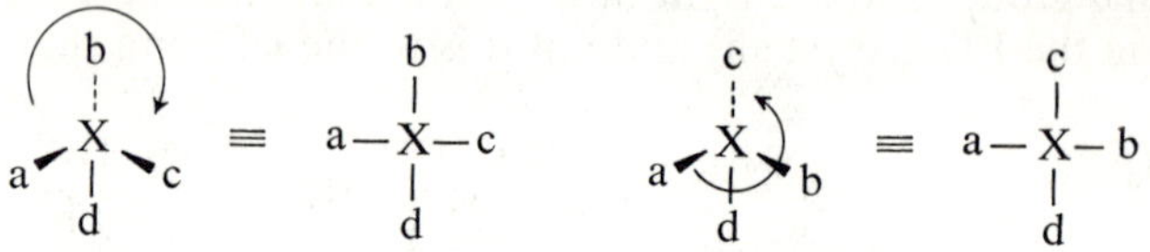

(R) (Fischer) (S) (Fischer)

Fig. 5-6. R-S configurational nomenclature.

but if a → b → c traces a left-handed (counterclockwise) turn, the configuration is S (cf. Fig. 5-6). It is fairly easy to obtain the required three-dimensional arrangement of the groups starting with a Fischer projection formula (Fig. 5-6) so oriented that the appropriate group d (see below) is at the bottom (cf. Sec. 3-1; note that any Fischer projection may be transformed into an equivalent projection by exchanging two pairs of substituents or by exchanging three substituents in order, as shown in Fig. 3-3). In the three-dimensional representation in Fig. 5-6, the model corresponding to the Fischer projection is slightly tilted so that d comes forward into the plane of the paper. The model is then looked at from the top to see whether a → b → c is clockwise (R) or counterclockwise (S).

The *sequence rules* for arranging a, b, c, and d are as follows:

1. The groups are arranged in order of *decreasing* atomic number of the atom directly attached to the asymmetric atom.

2. If two atoms attached to the asymmetric atom are the same, their respective states of substitution are considered. The atom substituted with other

[4] R. S. Cahn and C. K. Ingold, *J. Chem. Soc.,* 612 (1951); R. S. Cahn, C. K. Ingold, and V. Prelog, *Experientia,* **12**, 81 (1956). See also A. P. Terentiev and V. M. Potapov, *Tetrahedron,* **1**, 119 (1957).

atoms of *higher* atomic number takes precedence, or if two atoms are equivalent in that respect, that with *more* substituents of high atomic number comes first. If the second atom out affords no choice, one goes on to the third, etc., always following along branches with atoms of highest atomic number.

3. For $=\mathrm{N}$, substitute $\left\langle \begin{array}{l}\mathrm{N}\\ \mathrm{N}\end{array}\right.$; for $\equiv\mathrm{N}$, substitute $\left\langle \begin{array}{l}\mathrm{N}\\ \mathrm{N}\\ \mathrm{N}\end{array}\right.$; etc. (True $\left\langle \begin{array}{l}\mathrm{N}\\ \mathrm{N}\end{array}\right.$ has precedence over $=\mathrm{N}$.) Thus phenyl corresponds to $\mathrm{C}\left\langle \begin{array}{l}\mathrm{CH}\\ \mathrm{CH}\\ \mathrm{C}\end{array}\right.$.

4. A missing substituent on a tetracovalent atom is, if necessary, replaced by an imaginary atom of atomic number zero.

$$\begin{array}{ccc} & \mathrm{CHO} & \\ & | & \\ \mathrm{H}- & \mathrm{C} & -\mathrm{OH} \\ & | & \\ & \mathrm{CH_2OH} & \end{array} \equiv \begin{array}{ccc} & \mathrm{CHO} & \\ & | & \\ \mathrm{HO}- & \mathrm{C} & -\mathrm{CH_2OH} \\ & | & \\ & \mathrm{H} & \end{array}$$

(R)-Glyceraldehyde

$$\begin{array}{ccc} & \mathrm{CO_2H} & \\ & | & \\ \mathrm{H}- & \mathrm{C} & -\mathrm{OH} \\ & | & \\ & \mathrm{CH_3} & \end{array} \equiv \begin{array}{ccc} & \mathrm{CO_2H} & \\ & | & \\ \mathrm{HO}- & \mathrm{C} & -\mathrm{CH_3} \\ & | & \\ & \mathrm{H} & \end{array}$$

(R)-Lactic acid

$$\begin{array}{ccc} & \mathrm{CH_3} & \\ & | & \\ \mathrm{H}- & \mathrm{C} & -\mathrm{Br} \\ & | & \\ & \mathrm{C_2H_5} & \end{array} \equiv \begin{array}{ccc} & \mathrm{CH_3} & \\ & | & \\ \mathrm{Br}- & \mathrm{C} & -\mathrm{C_2H_5} \\ & | & \\ & \mathrm{H} & \end{array}$$

2(S)-Bromobutane

$$\begin{array}{ccc} & \mathrm{CH_3} & \\ & | & \\ \mathrm{H}- & \mathrm{C} & -\mathrm{NH_2} \\ & | & \\ & \mathrm{C_6H_5} & \end{array} \equiv \begin{array}{ccc} & \mathrm{CH_3} & \\ & | & \\ \mathrm{H_2N}- & \mathrm{C} & -\mathrm{C_6H_5} \\ & | & \\ & \mathrm{H} & \end{array}$$

(S)-α-Phenethylamine

$$\begin{array}{ccc} & \mathrm{CO_2H} & \\ & | & \\ \mathrm{H}- & \mathrm{C} & -\mathrm{CH_3} \\ & | & \\ & \mathrm{C_6H_5} & \end{array} \equiv \begin{array}{ccc} & \mathrm{CO_2H} & \\ & | & \\ \mathrm{H_3C}- & \mathrm{C} & -\mathrm{C_6H_5} \\ & | & \\ & \mathrm{H} & \end{array}$$

(R)-Hydratropic acid

$$\begin{array}{ccc} & \mathrm{CO_2H} & \\ & | & \\ \mathrm{H_3C}- & \mathrm{C} & -\mathrm{OH} \\ & | & \\ & \mathrm{C_6H_5} & \end{array} \equiv \begin{array}{ccc} & \mathrm{CO_2H} & \\ & | & \\ \mathrm{HO}- & \mathrm{C} & -\mathrm{C_6H_5} \\ & | & \\ & \mathrm{CH_3} & \end{array}$$

(R)-Atrolactic acid

Fig. 5-7. R-S configurational designation of some common dissymmetric compounds.

5. In a case such as XabHD, the isotope of higher mass number precedes that of lower mass number. In a case such as Xaa′bc, where a and a′ are stereoisomeric, the *cis*-a group precedes the *trans*-a group and the (R)-a group precedes the (S)-a group. If the atom under consideration is pseudoasymmetric (Sec. 3-3), the prefixes (r) and (s) are used instead of (R) and (S) (see Fig. 3-17).

By applying these rules to some common substituents, one obtains the following sequence (group of highest priority first): I, Br, Cl, SO_2R, SOR, SR, SH, F, OCOR, OR, OH, NO_2, NHCOR, NR_2, NHR, NH_2, CCl_3, COCl, CO_2R, CO_2H, $CONH_2$, COR, CHO, CR_2OH, CHOHR, CH_2OH, C_6H_5, CR_3, CHR_2, CH_2R, CH_3, D, H.

Figure 5-7 shows the common Fischer projection formula, Fischer projection formula suitable for naming, and symbol for some simple substances with one asymmetric carbon atom.

When there is more than one asymmetric atom, a symbol is attached to each. Ring compounds are handled similarly to acyclic ones. Examples are shown in Fig. 5-8.

The system has a few disadvantages; thus it is not readily compatible with common nomenclature (see arabinose in Fig. 5-8), and there is necessarily a certain arbitrariness in the sequence rule which leads to some unusual results. Thus the two asymmetric carbon atoms in (+)-tartaric acid (Fig. 5-8) are

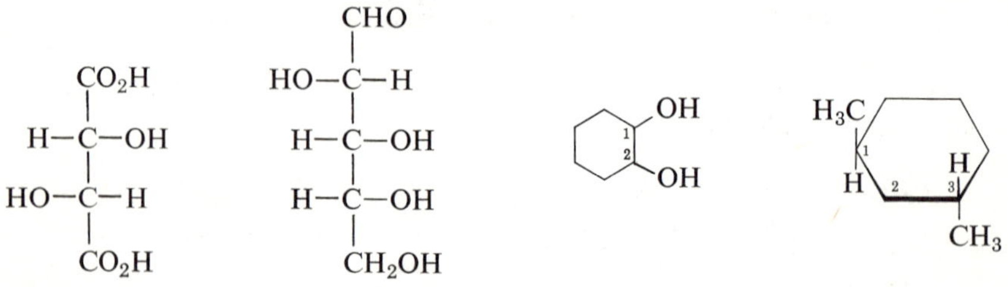

2(R),3(R)-Dihydroxysuccinic acid or (R)-(+)-tartaric acid

2(S),3(R),4(R),5-Tetrahydroxypentanal [D-(−)-arabinose]

Cyclohexanediol-1(R), 2(S) or *meso*-cyclohexanediol-1,2

1(S),3(S)-Dimethylcyclohexane

Fig. 5-8. R-S configurational notation of compounds with several asymmetric atoms or with rings.

both R, as one would expect, but the entirely similarly situated top and bottom asymmetric atoms in (−)-arabinose (Fig. 5-8) have opposite configurational symbols (because the sequence rule specifies —CHO → RCHOH → CH_2OH). However, the system has the great advantage of being universally applicable and unambiguous, qualities which make it particularly suitable for systematic use in abstracting and cataloguing.[5]

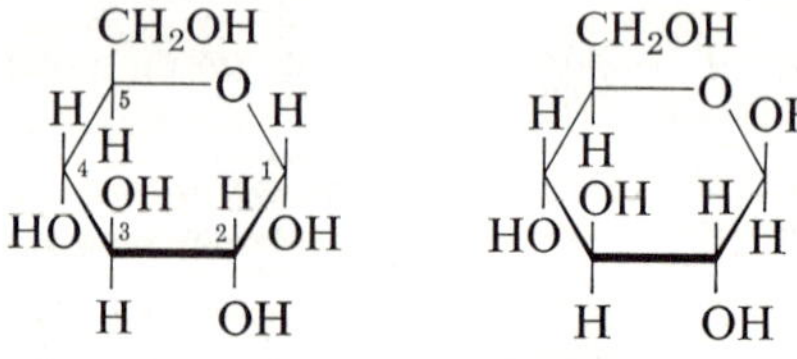

α-D-Glucose

β-D-Glucose

Fig. 5-9. Anomers of glucose.

A few other systems of configurational symbolism are used with particular types of compounds, especially alicyclic compounds, such as the steroids (cf. Chap. 10). Thus the anomeric center (cf. Sec. 4-2, page 40) in the pyranose sugars is usually given the symbol α or β, according to the following convention: When the six-membered pyrane ring is so oriented that the ring oxygen

[5] See also A. Feldman, *J. Org. Chem.*, **24**, 1556 (1959).

is at the right rear and the anomeric carbon on the right side, the α form is the one which has the anomeric hydroxyl below the plane of the ring and the β form that which has it above that plane, as shown in Fig. 5-9. [It is left as an exercise for the reader to establish that the configuration of the remaining asymmetric carbon atoms (C_{2-5}) shown in the cyclic formulas in Fig. 5-9 is equivalent to that shown in the open-chain aldehyde formula of glucose shown in Fig. 3-15. The equivalence can readily be established with the aid of models.]

5-3. Absolute Configuration

In Chap. 3 it was asserted that the three-dimensional formula of (−)-lactic acid shown in Fig. 3-2 corresponds to the real arrangement of the groups in the actual molecule. This section (and, to some extent, the next one) will deal with the problem of how one knows this to be true.

The determination of the absolute configuration (meaning actual arrangement in space) of even a single molecule has been very difficult to achieve; in fact, the problem was solved[6] only in 1951, over 75 years after it was recognized by van't Hoff and Le Bel. The ordinary chemical method of proving structure by converting an unknown compound into a known and usually simpler one and assuming the "principle of minimum structural change"[7] is useless for determining *absolute* configuration, although it does succeed in correlating configurations of different compounds (Sec. 5-4). The determination of absolute configuration requires physical methods, and the techniques of X-ray and electron diffraction come to mind, since they have been used many times to locate directly the atoms in a molecule. Unfortunately, X rays and electron beams are not dissymmetric, and therefore it is not to be hoped that they would show any difference between enantiomers. Figure 5-10 shows the diffraction of X rays by a molecule A—B and its mirror image B—A. A ray hitting A and B is diffracted, and the diffracted waves produce an interference pattern at I, the location of the photographic plate. If B and A are interchanged, the wave diffracted by B now has to traverse the longer path, but the interference pattern at I is the same because the *dif-*

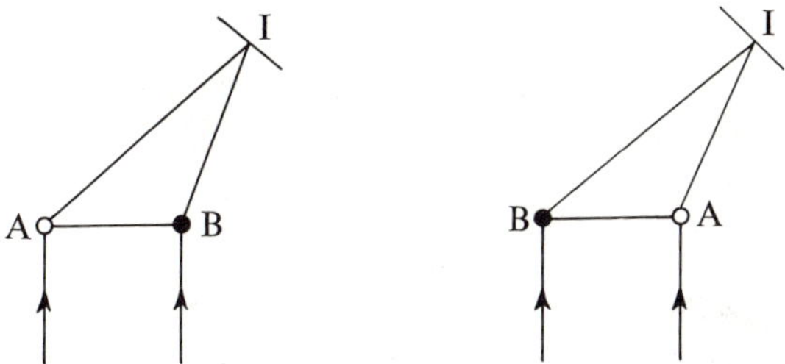

Fig. 5-10. Determination of absolute configuration.

[6] J. M. Bijvoet, A. F. Peerdeman, and A. J. van Bommel, *Nature*, **168**, 271 (1951); see also R. Pepinsky, *Rec. Chem. Progr.* (*Kresge-Hooker Sci. Lib.*), **17**, 145 (1956).
[7] Ref. 61, p. 49.

ference between the paths AI and BI remains unaltered. To produce a difference between the patterns, it is necessary in some way to distinguish the ray refracted by A and that refracted by B. For example, if a *phase lag* could be introduced in the process of diffraction, say at A, then, in the case on the left (Fig. 5-10) the already slower ray diffracted by A (slower because $AI > BI$) would be even further slowed by the phase lag, and the phase difference between the two interfering rays would be *increased.* But in the case on the right (Fig. 5-10), the ray slowed by the phase lag introduced at A is the originally faster one (since now $AI < BI$) and so the phase difference between the two rays is *decreased.* It follows that the two interference patterns will no longer be the same, and it may then be decided which pattern corresponds to A—B and which to B—A.

The means of introducing a phase lag upon refraction at A (but not at B) is to use X rays whose wavelength is near an absorption edge (for X rays) of the atom A. In the original work, sodium rubidium tartrate was used in conjunction with zirconium K_α X rays. The rays refracted by the rubidium suffer a phase lag. The result found was that the double salt of (+)-tartaric acid has the configuration shown in Fig. 5-11.† The same method was ap-

(+)-Tartaric acid sodium rubidium salt	(−)-Isoleucine hydrobromide
CO_2Rb	CO_2H
H—C—OH	H—C—$NH_3^+Br^-$
HO—C—H	H—C—CH_3
CO_2Na	C_2H_5

Fig. 5-11. Compounds whose absolute configuration has been determined.

plied[8] to isoleucine hydrobromide, using uranium L_α radiation which is on the absorption edge of the bromine atom.

In at least one case the X-ray method of determining absolute configuration has been applied to a relatively complex alkaloid.[8a]

Another way in which absolute configuration has been determined is by theoretically deducing the arrangement of groups in space which would produce the observed rotation. Calculations of this type have been developed by W. Kuhn, by Eyring, and by Kirkwood[9] and their coworkers and culmi-

† Before 1951, it had been necessary to assign an arbitrary configuration to *one* molecule with which others were then correlated. The molecule chosen was (+)-glyceraldehyde which was arbitrarily assigned the D configuration shown in Fig. 5-1. Now the relative configurations of (+)-glyceraldehyde and (+)-tartaric acid have long been known (cf. Sec. 5-4), and therefore the true configuration of (+)-glyceraldehyde can be deduced from the now known configuration of (+)-tartaric acid. Fortunately (and fortuitously) it turns out that the previously arbitrarily assigned configuration of (+)-glyceraldehyde is, indeed, the correct one. Therefore configurations determined relative to (+)-glyceraldehyde prior to 1951, are also, as a matter of luck, correct in the absolute sense.

[8] J. Trommel and J. M. Bijvoet, *Acta Cryst.,* **7,** 703 (1954).
[8a] M. Przybylska and L. Marion, *Can. J. Chem.,* **37,** 1843 (1959).
[9] W. W. Wood, W. Fickett, and J. G. Kirkwood, *J. Chem. Phys.,* **20,** 561 (1952).

nated in the prediction of the configuration of 2,3-epoxybutane and 1,2-dichloropropane (see also Chap. 14). The relative configuration (Sec. 5-4) of 2,3-epoxybutane and of tartaric acid is known and the calculated configuration of the former is in agreement with that deduced from the X-ray data for the latter.

5-4. Relative Configuration

The determination of absolute configuration is a difficult and time-consuming task, to be undertaken only by a skilled X-ray crystallographer, and, in any case, is limited to suitably crystalline solids containing appropriate atoms. It is fortunate, therefore, that other means of determining configuration are available. These methods are correlative; i.e., they correlate the configuration of one compound with that of another. However, since these methods, at least in principle, allow the configuration of any active compound to be correlated, either directly or via relays, with that of (+)- or (−)-tartaric acid or (−)- or (+)-isoleucine, of known absolute configurations, correlative methods can be used to determine absolute configuration indirectly.

Several methods of correlating configuration are now recognized, not all of them equally reliable. They are (1) chemical interconversion not affecting bonds to the asymmetric atom, (2) correlation via diastereoisomers, (3) correlation via quasi-racemates, (4) the method of optical comparison, (5) correlation by asymmetric synthesis, and (6) chemical conversion affecting bonds to the asymmetric atom in a predictable way. These methods are dealt with in this section. Rotatory dispersion, which is a refinement of method 4 above, will be dealt with separately in Chap. 14.

a. Chemical Interconversion Not Affecting Bonds to the Asymmetric Atom. The principle of this method is simple: If, let us say, (−)-Cuvxy is converted into Cuvxz *without ever breaking the C—y bond* (or, for that matter, the C—u, C—v, or C—x bond), then, if the product is *levorotatory* Cuvxz (z in place of y), (−)-Cuvxz has the same configuration as (−)-Cuvxy. But if the product is *dextrorotatory* Cuvxz, then (+)-Cuvxz and (−)-Cuvxy have the same configuration.† The method is readily illustrated by the examples of Fig. 5-12 correlating (+)-tartaric acid with (+)-malic acid, (+)-isoserine, (−)-glyceric acid, (+)-glyceraldehyde, and (−)-lactic acid; Fig. 5-13 correlating (+)-lactic acid with (among other compounds) (−)-α-phenethyl alcohol and (−)-mandelic acid; and Fig. 5-14 correlating (+)-alanine and (+)-valine.

b. Chemical Correlation Involving Diastereoisomers. One inherent limitation of the method discussed above is that the atoms immediately attached to the asymmetric carbon atom must be the same in the compounds to be correlated. Thus one can correlate RCHOHR′ (where R and R′ are alkyl

† The statement "two compounds (−)-Cuvxy and (−)-Cuvxz have the same configuration" is not always clear. Its meaning is fairly obvious when, as in the above case, only one group is changed, but it may be quite obscure or even ambiguous when several groups around the asymmetric carbon are altered. The difficulty here is, of course, the semantic equivalent of that involved in configurational notation (Sec. 5-2). A clear and unambiguous statement is to say "given that (−)-Cuvxy has configuration (*i*), (−)-Cuvxz will have configuration (*ii*)."

x
|
u—C—v
|
y

(*i*)

x
|
u—C—v
|
z

(*ii*)

groups) with, let us say, $CH_2OHCHOHCHO$, but one cannot correlate RCHOHR′ with RCHClR′ or $RCH(NH_2)CO_2H$ or RR′CHR″ (R″ being another alkyl group) or $RR'C(OH)CO_2H$, to give a few examples. At first sight it would seem hopeless to effect a completely conclusive correlation of compounds of such divergent types. Nevertheless, a conclusive method is, in principle, available.[10] It requires an optically active compound having at least two asymmetric atoms (or dissymmetric groupings), for example, RCabCxyR′, in which each of the asymmetric atoms can be cleaved off without a bond leading to this atom being broken. Thus in the above case one must be able to degrade the molecule to obtain both active RCabR″ and

CO_2H–CH(OH)–CH(OH)–CO_2H (H—C—OH over HO—C—H)
(+)-Tartaric acid

→ (1) $SOCl_2$ on monoacetyl dimethyl ester (2) Zn, HCl →

CO_2H–(H—C—OH)–CH_2–CO_2H
(+)-Malic acid

→ NaOBr on half-amide →

CO_2H–(H—C—OH)–CH_2NH_2
(+)-Isoserine

→ HNO_2 →

CO_2H–(H—C—OH)–CH_2OH
(−)-Glyceric acid

← HgO ←

CHO–(H—C—OH)–CH_2OH
(+)-Glyceraldehyde

(+)-Isoserine ↓ NOBr

CO_2H–(H—C—OH)–CH_2Br
(−)-Bromolactic acid

→ Na · Hg →

CO_2H–(H—C—OH)–CH_3
(−)-Lactic acid

Fig. 5-12. Determination of configuration of (+)-glyceraldehyde, (−)-lactic acid, (+)-malic acid, and other compounds with respect to (+)-tartaric acid.

R′CxyR‴. If the absolute configuration of RCabR″ is known or can be established by the method discussed earlier and if the *relative* configuration of the two asymmetric atoms in the diastereoisomer RCabCxyR′ can be established,† one can obviously deduce the absolute configuration of the

† The nature of the problem involved in establishing relative configuration of asymmetric atoms in diastereoisomers can be gleaned from the examples given here. A more systematic treatment will be found in Chap. 6 for acyclic diastereoisomers and in Chap. 7 for cyclic diastereoisomers. See also Chap. 12.

[10] K. Freudenberg, W. F. Bruce, and E. Gauf, *Ann.*, **510**, 206 (1934).

R′Cxy— center and, from there, the configuration of R′CxyR‴. In principle, this method imposes no restrictions on the nature of a, b, x, and y.

One of the most elegant applications of this method is the determination of the configuration of a tertiary asymmetric carbon of type RCHR′R″ (to which other tertiary asymmetric carbons could obviously be related by the method discussed in Sec. 5-4*a*). The correlation is summarized in Fig. 5-15. Shikimic acid, a natural, optically active product, is hydrogenated to dihydroshikimic acid which readily forms a δ-lactone, proving[11] the carboxyl and

CH_3 / H—C—OH / CO_2H —(1) EtOH, H^+; (2) CH_3I, Ag_2O→ CH_3 / H—C—OCH_3 / CO_2Et —$BrMg(CH_2)_5MgBr$→ CH_3 / H—C—OCH_3 / C(OH) cyclohexyl

(+)-Lactic acid (−) (+)

↓ (1) Chugaev (2) H_2, cat.

CH_3 / H—C—OH / C_6H_5 —H_2, cat.→ CH_3 / H—C—OH / cyclohexyl —K, CH_3I→ CH_3 / H—C—OCH_3 / cyclohexyl

(−)-Phenylmethylcarbinol (+) (+)

↓ C_2H_5I, Ag_2O

CH_3 / H—C—OC_2H_5 / C_6H_5 ←(1) TsCl; (2) $LiAlH_4$— CH_2OH / H—C—OC_2H_5 / C_6H_5 ←(1) C_2H_5I, Ag_2O; (2) $LiAlH_4$— CO_2H / H—C—OH / C_6H_5

(−) (−) (−)-Mandelic acid

Fig. 5-13. Determination of configuration of (−)-mandelic acid, (−)-phenylmethylcarbinol, and other compounds with respect to (+)-lactic acid.

4-hydroxyl groups in dihydroshikimic acid to be cis.† Shikimic acid has also been degraded to (+)-desoxygluconic acid lactone by the path indicated in Fig. 5-15.[12] This lactone (or the corresponding acid) had previously been

† In 1,4-trans-disubstituted compounds, the substituents are too far apart for bridging to be possible; cf. Chaps. 7 and 8. That the lactone was δ (1,4) and not γ (1,3) was inferred from its lack of reaction with lead tetraacetate, implying the absence of vicinal free hydroxyl groups.

[11] H. O. L. Fischer and G. Dangschat, *Helv. Chim. Acta,* **18,** 1206 (1935).
[12] H. O. L. Fischer and G. Dangschat, *Helv. Chim. Acta,* **20,** 705 (1937).

related to D-(+)-glucose by methods involving, in principle, oxidation of the aldehyde function of glucose to acid and reduction of the No. 2 carbon from the alcohol to the methylene stage. It follows that carbons No. 3, 4, and 5 in desoxygluconic acid have the same configuration as the corresponding carbon atoms in glucose and that therefore shikimic acid (and, therefore, dihydroshikimic acid) has the configuration shown in Fig. 5-15. Freudenberg and coworkers[13] succeeded in degrading dihydroshikimic acid in such a way that the tertiary asymmetric carbon was isolated and, eventually, all other asymmetric atoms destroyed (Fig. 5-15). In this way, the configuration of

CO_2H / $H_2N—C—H$ / CH_3 (+)-Alanine $\xrightarrow[\text{(2) } CH_3MgI]{\text{(1) EtOH, HCl}}$ $(CH_3)_2COH$ / $H_2N—C—H$ / CH_3 (+) (as hydrochloride) $\xrightarrow[\text{(2) } H_2\text{, Pd}]{\text{(1) } PCl_5}$

$CH(CH_3)_2$ / $H_2N—C—H$ / CH_3 (−)-2-Amino-3-methylbutane (as hydrochloride)

CO_2H / $H_2N—C—H$ / $CH(CH_3)_2$ (+)-Valine (as formyl derivative) $\xrightarrow[\text{(3) Na, EtOH}]{\text{(1) EtOH, HCl; (2) } Ac_2O\text{, NaOAc}}$

CH_2OH / $H_2N—C—H$ / $CH(CH_3)_2$ (+) (as hydrochloride) $\xrightarrow[\text{(2) } H_2\text{, Pd}]{\text{(1) HBr}}$ CH_3 / $H_2N—C—H$ / $CH(CH_3)_2$ (+)-2-Amino-3-methylbutane (as hydrochloride)

Fig. 5-14. Correlation of configuration of (+)-alanine and (+)-valine. [The reader should note that (+)-alanine is transformed into (−)-2-amino-3-methylbutane, whereas (+)-valine is transformed into the (+) enantiomer. Nevertheless, the nature of the transformations is such that they prove (+)-alanine and (+)-valine to have the same (L or S) configuration.]

shikimic acid was correlated with that of (+)-trimethyl β-carboxyadipate and (+)-methylethylpropylmethane. This establishes the absolute configuration of the asymmetric carbon atom in these compounds and any others having a tertiary asymmetric carbon which may be related to them.[14]

[13] K. Freudenberg, H. Meisenheimer, J. T. Lane, and E. Plankenhorn, *Ann.*, **543**, 162 (1940); K. Freudenberg and J. Geiger, *Ann.*, **575**, 145 (1951); K. Freudenberg and W. Hohmann, *Ann.*, **584**, 54 (1953).

[14] See also D. S. Noyce and D. B. Denney, *J. Am. Chem. Soc.*, **76**, 768 (1954); D. S. Noyce and J. H. Canfield, *ibid.*, **76**, 3630 (1954).

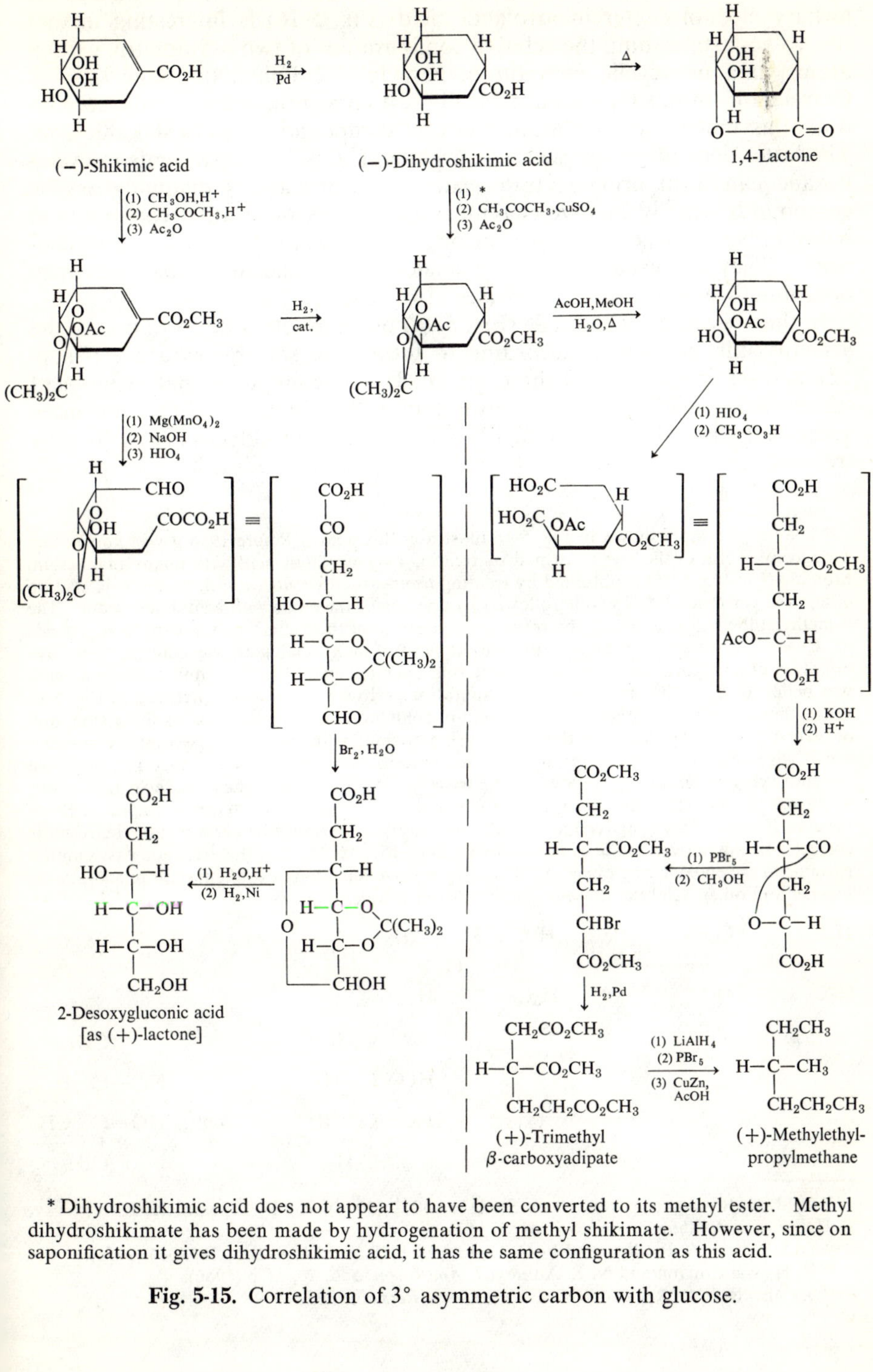

(−)-Shikimic acid $\xrightarrow{H_2,\ Pd}$ (−)-Dihydroshikimic acid $\xrightarrow{\Delta}$ 1,4-Lactone

(−)-Shikimic acid: (1) CH_3OH, H^+ (2) CH_3COCH_3, H^+ (3) Ac_2O

(−)-Dihydroshikimic acid: (1) * (2) $CH_3COCH_3, CuSO_4$ (3) Ac_2O

$\xrightarrow{H_2,\ cat.}$ $\xrightarrow{AcOH, MeOH,\ H_2O, \Delta}$

(1) $Mg(MnO_4)_2$ (2) NaOH (3) HIO_4

(1) HIO_4 (2) CH_3CO_3H

Br_2, H_2O

(1) KOH (2) H^+

(1) H_2O, H^+ (2) H_2, Ni

(1) PBr_5 (2) CH_3OH

H_2, Pd

(1) $LiAlH_4$ (2) PBr_5 (3) CuZn, AcOH

2-Desoxygluconic acid [as (+)-lactone]

(+)-Trimethyl β-carboxyadipate

(+)-Methylethylpropylmethane

* Dihydroshikimic acid does not appear to have been converted to its methyl ester. Methyl dihydroshikimate has been made by hydrogenation of methyl shikimate. However, since on saponification it gives dihydroshikimic acid, it has the same configuration as this acid.

Fig. 5-15. Correlation of 3° asymmetric carbon with glucose.

The correlation[15] of the secondary alcohol center in mandelic acid and the tertiary alcohol center in atrolactic acid (Fig. 5-16) is interesting in that it involves establishing the relative configuration of two asymmetric carbon atoms in a non-cyclic compound, namely, 1,2-diphenylpropane-1,2-diol. Correlations of this type do not usually rest on as firm a basis as the correlation of asymmetric carbon atoms in cyclic compounds, such as shikimic acid. The latter depend on simple steric arguments (cis 1,4 substituents in cyclohexane can form bridges; trans cannot). Correlations involving acyclic compounds usually involve making assumptions of relative configuration based either on spectral properties or on considerations of reaction mechanism. They are usually based on analogy, and, like most analogies, may occasionally give a false answer† (see also Chap. 6).

In the present correlation[15a] (Fig. 5-16) osmium tetroxide oxidation of the α-methylstilbene, which, according to ultraviolet spectral evidence (Chap. 12), is trans, gives the (±)-threo-glycol (I), this configurational assignment (threo) being made because osmium tetroxide oxidation in other instances gives cis hydroxylation (Chap. 12). The same threo-glycol is obtained by treating (±)-mandelamide with methylmagnesium iodide to give phenyl-

† The very example shown in Fig. 5-16 illustrates this point. Before 1956 it was known that the racemic form of the glycol formed by treating (−)-atrolactic acid with phenylmagnesium bromide (I in Fig. 5-16) is obtained by treating *trans*-α-methylstilbene with perbenzoic acid, to obtain the corresponding epoxide followed by opening of the oxide with acidulated water. The α-methylstilbene was assigned the trans configuration on theoretically sound spectral grounds (cf. Chap. 12); the epoxidation reaction, on the basis of many analogies, was believed to involve retention of configuration; and the epoxide ring opening, again based on numerous analogies, was believed to involve inversion of configuration. However, the work pictured in Fig. 5-16 shows that either the assumption of retention of configuration in epoxidation or the assumption of inversion of configuration in the epoxide ring opening is incorrect (or else that the assumption of cis addition of hydroxyl groups in osmium tetroxide oxidation is incorrect, in which case the entire correlation in Fig. 5-16 would be wrong). It is not easy in such instances to discover where exactly the error has been committed, and even if no inconsistency is found (perhaps because the answer was arrived at in only one way!), one cannot be certain that the result is correct. In the above case, it is now known (Refs. 15*a*, 16–18) that the erroneous assumption pertains to the epoxide ring opening. This may involve retention of configuration in some cases, inversion in others, and the present instance is one of retention:

$$(H_3C)(H_5C_6)C{=}C(C_6H_5)(H) \xrightarrow{PhCO_3H} (H_3C)(H_5C_6)C\overset{O}{—}C(C_6H_5)(H) \xrightarrow[H^+]{H_2O}$$

$$\begin{array}{c} C_6H_5 \\ | \\ HO—C—H \\ | \\ H_3C—C—OH \\ | \\ C_6H_5 \end{array} \quad , \quad \textit{not} \quad \begin{array}{c} C_6H_5 \\ | \\ HO—C—H \\ | \\ HO—C—CH_3 \\ | \\ C_6H_5 \end{array}$$

[15] (*a*) J. H. Brewster, *J. Am. Chem. Soc.*, **78**, 4061 (1956); (*b*) D. J. Cram, K. R. Kopecky, F. Hauck, and A. Langemann, *ibid.*, **81**, 5754 (1959).

[16] D. Y. Curtin, A. Bradley, and Y. G. Hendrickson, *J. Am. Chem. Soc.*, **78**, 4064 (1956).

[17] H. H. Wassermann and N. E. Aubrey, *J. Am. Chem. Soc.*, **78**, 1726 (1956).

[18] See also R. E. Parker and N. S. Isaacs, *Chem. Revs.*, **59**, 737 (1959).

$$\underset{\substack{(-)\text{-}\\ \text{or }(\pm)\text{-}}\ \text{Mandelic acid}}{H\text{—}\overset{CO_2H}{\underset{C_6H_5}{C}}\text{—}OH} \xrightarrow[(2)\ NH_3]{(1)\ CH_3OH,\ H^+} \underset{(-)\text{ or }(\pm)}{H\text{—}\overset{CONH_2}{\underset{C_6H_5}{C}}\text{—}OH} \xrightarrow{CH_3MgI}$$

$$\underset{(-)\text{ or }(\pm)}{H\text{—}\overset{COCH_3}{\underset{C_6H_5}{C}}\text{—}OH} \xrightarrow{C_6H_5MgBr} \underset{\substack{(+)\text{-}\\ \text{or }(\pm)\text{-}}\ threo\text{-I}}{\begin{array}{c} C_6H_5 \\ HO\text{—}C\text{—}CH_3 \\ H\text{—}C\text{—}OH \\ C_6H_5 \end{array}}$$

$$\underset{trans\text{-}\alpha\text{-Methylstilbene}}{\begin{array}{c} H_3C \quad C_6H_5 \\ C \\ \| \\ C \\ H_5C_6 \quad H \end{array}} \xrightarrow[\substack{(2)\ KOH,\ mannitol \\ or \\ (1)\ C_6H_5CO_3H \\ (2)\ H_2O,\ H^+}]{(1)\ OsO_4} \underset{(\pm)\text{-}threo\text{-I}}{\begin{array}{c} OH \\ H_3C\text{—}C\text{—}C_6H_5 \\ H_5C_6\text{—}C\text{—}H \\ OH \end{array}} \equiv \begin{array}{c} C_6H_5 \\ HO\text{—}C\text{—}CH_3 \\ H\text{—}C\text{—}OH \\ C_6H_5 \end{array}$$

$$\underset{(+)\text{-Atrolactic acid}}{HO\text{—}\overset{CO_2H}{\underset{C_6H_5}{C}}\text{—}CH_3} \xrightarrow[(2)\ NH_3]{(1)\ CH_3OH,\ H^+} \underset{(-)}{HO\text{—}\overset{CONH_2}{\underset{C_6H_5}{C}}\text{—}CH_3} \xrightarrow{C_6H_5MgBr}$$

$$\underset{(-)}{HO\text{—}\overset{COC_6H_5}{\underset{C_6H_5}{C}}\text{—}CH_3} \xrightarrow{NaBH_4} \underset{(-)\text{-}threo\text{-I}}{\begin{array}{c} C_6H_5 \\ H\text{—}C\text{—}OH \\ HO\text{—}C\text{—}CH_3 \\ C_6H_5 \end{array}}$$

Fig. 5-16. Configurational correlation of (−)-mandelic and (+)-atrolactic acid.

acetylcarbinol which is then treated with phenylmagnesium bromide.† Application of the same series of reactions to (−)-mandelic acid and its amide (the enantiomers shown in Fig. 5-16) gives a dextrorotatory glycol. This must be an enantiomer of the threo form, inasmuch as the preference of one diastereoisomer over another (as, for example, embodied in Cram's rule) is the same for the *dl* pair and the active form (p. 70). Now the configuration of (−)-mandelic acid is known, and so that of the (+)-glycol I must be as shown in Fig. 5-16. It remains to convert active atrolactic acid into the same glycol I, and this was done[19] by converting (+)-atrolactic acid via the amide to (−)-phenylmethylbenzoylcarbinol which, upon sodium borohydride reduction,[15b] gave crystalline (−)-I enantiomeric (i.e., equal in melting point and opposite in rotation) with that obtained from (−)-mandelic acid. These transformations are depicted in Fig. 5-16, and the configuration of (+)-atrolactic acid is therefore as shown.

A problem related to the above, although not falling into exactly the same category, is the determination of the configuration of (+)-glucose by Emil Fischer (1884–1894).[20] Work by Kiliani[21] had indicated that glucose is a 2,3,4,5,6-pentahydroxyhexanal-1.‡ Fischer's work served to indicate which of the 16 stereoisomers of this structure (eight shown in Fig. 3-15) is (+)-glucose.

Before presenting Fischer's proof, it is necessary to state that it does not reveal the absolute configuration of (+)-glucose but only the relative configuration of the four asymmetric carbons. Fischer arbitrarily assumed that the bottom carbon in (+)-glucose as written in the conventional projection formula (Chap. 3) has the hydroxyl to the right. Later work[6] fortuitously showed this assumption to be correct.

The naturally occurring (+)-arabinose, when subjected to a cyanohydrin synthesis, hydrolyzed to a lactone, and reduced,¶ gives a mixture of (−)-glucose and (−)-mannose, enantiomeric with natural glucose and mannose. It follows that carbons No. 3, 4, and 5 in natural (+)-glucose and (+)-mannose have the same configuration, which is opposite to the configuration of the asymmetric carbons in (+)-arabinose. Oxidation of (+)-arabinose gives rise to an *optically active* 2,3,4-trihydroxyglutaric acid. This shows that carbon No. 2 in arabinose has the hydroxyl on the opposite side in the projection formula as carbon No. 4; otherwise a meso (inactive) form of the diacid would necessarily have resulted. Since the bottom asymmetric carbon in (+)-arabinose is, by Fischer's convention, to be written with the hydroxyl

† The reader should convince himself that the reaction of phenylacetylcarbinol with phenylmagnesium bromide to give the threo form of the glycol I follows Cram's rule; cf. Fig. 4-37.

‡ We now know that this "open-chain form" of glucose is present to only a very slight extent in solution, the major part of the substance being in the form of two diastereoisomeric 1,5 cyclic hemiacetals; cf. Fig. 4-13.

¶ It is not possible, within the scope of this book, to discuss in any detail the synthetic and degradative procedures commonly used in sugar chemistry. For an excellent summary of these, the reader is referred to Ref. 60.

[19] A. McKenzie and A. Ritchie, *Ber.*, **70**, 23 (1937).

[20] For a summary, see C. S. Hudson, *J. Chem. Educ.*, **18**, 353 (1941).

[21] H. Kiliani, *Ber.*, **19**, 221 (1886).

group on the left [its configuration being opposite to that of (+)-glucose], the top asymmetric carbon must have the hydroxyl on the right and the partial formulas of (+)-arabinose, (−)-glucose, and (−)-mannose are then as in Fig. 5-17.

The next piece of evidence comes from the fact that *neither* glucose *nor* mannose upon nitric acid oxidation gives a *meso*-2,3,4,5-tetrahydroxyadipic acid, but both give *active* acids. Inspection of Fig. 5-17 (left formula) shows that this means that the hydroxyl groups on C_3 and C_4 must be on opposite sides. For if they were on the same side, then either glucose or mannose—whichever has the hydroxyls on C_2 and C_5 on the same side—would have given a meso acid. The configuration of (+)-arabinose is thus established, inasmuch as the remaining hydroxyl at C_3 (Fig. 5-17) (corresponding to the C_4 hydroxyl in glucose and mannose) is on the left. This leaves the question of the configuration of glucose and mannose at C_2. By fortunate circumstance, there occurs in nature another aldohexose, (+)-gulose, which upon oxidation gives the same 2,3,4,5-tetrahydroxyadipic acid as (+)-glucose.

```
       CHO
      2|
       CHOH                              CHO                          CO2H
      3|                                2|                             |
    H—C—OH     (1) HCN             H—C—OH                        H—C—OH
      4|      <-----------            3|                             |
       CHOH    (2) H2O                 CHOH        HNO3              CHOH
      5|       (3) Na·Hg, H+          4|          ----->             |
   HO—C—H                         HO—C—H                        HO—C—H
       |                               |                             |
       CH2OH                           CH2OH                         CO2H

L-(−)-Glucose and                L-(+)-Arabinose                Active acid
  L-(−)-Mannose
```

Fig. 5-17. Configuration of glucose, first stage.

Contemplation of Fig. 5-18 indicates that this means that (+)-glucose has the top hydroxyl (at C_2) on the right. The diastereoisomeric 2,3,4,5-tetrahydroxyadipic acid shown at the bottom in Fig. 5-18 has an axis of symmetry and therefore cannot be obtained by oxidation of two different sugars; this acid (mannosaccharic acid) therefore corresponds in configuration to (+)-mannose, whereas the acid (glucosaccharic acid) shown at the top is reasonably obtained from two different aldohexoses and therefore corresponds in configuration to (+)-glucose. The configurations of (+)-glucose and (+)-mannose shown in Fig. 5-18 now follow logically.

The problem of the absolute configuration of (+)-glucose could reasonably be solved by carving out from the molecule either carbons No. 2 and 3 or carbons No. 3 and 4 and relating them to tartaric acid whose absolute configuration is known. If the formula for (+)-glucose shown in Fig. 5-18 is correct in the absolute sense, isolation of the top two asymmetric carbons should give (R)-(+)-tartaric acid (cf. Figs. 5-8 and 5-11) whereas carbons 3 and 4 should give the (S)-(−) enantiomer. Correlations of this type have actually been effected and serve to prove that the absolute configuration of (+)-glucose arbitrarily assumed by E. Fischer (shown in Fig. 5-18) is indeed correct.

Yet another important correlation of configurations which has been brought about by the method discussed here is that of the asymmetric carbon atoms in (+)-glyceraldehyde (Fig. 5-1) and (−)-serine (Fig. 5-4).[22] The key compound in this correlation is glucosamine (Fig. 5-19).

c. The Method of Quasi-racemates.[23] The methods so far discussed are in most cases of unquestioned validity, but they are also tedious, time-consuming, and of limited applicability. Fortunately, some other methods are available which, while not as reliable as those discussed in Sec. 5-4*a* and *b*, will nevertheless in many cases give the correct answer with much less effort.

```
    CHO                 CH2OH                  CO2H                   CHO
     |                    |                      |                     2|
HO—C—H                H—C—OH                 H—C—OH                 H—C—OH
     |                    |                      |                     3|
HO—C—H        ≡      HO—C—H     —HNO3→     HO—C—H     ←HNO3—    HO—C—H
     |                    |                      |                     4|
 H—C—OH               H—C—OH                 H—C—OH                 H—C—OH
     |                    |                      |                     5|
HO—C—H                H—C—OH                 H—C—OH                 H—C—OH
     |                    |                      |                      |
    CH2OH                CHO                   CO2H                   CH2OH

L-(+)-Gulose                              Glucosaccharic         D-(+)-Glucose*
                                              acid

                      but

    CHO                    CO2H                   CH2OH
     |                       |                       |
HO—C—H                 HO—C—H                  HO—C—H
     |                       |                       |
HO—C—H     —HNO3→      HO—C—H     ←HNO3—      HO—C—H
     |                       |                       |
 H—C—OH                 H—C—OH                  H—C—OH
     |                       |                       |
 H—C—OH                 H—C—OH                  H—C—OH
     |                       |                       |
    CH2OH                  CO2H                    CHO

D-(+)-Mannose**     Mannosaccharic acid     D-(+)-Mannose**
```

* Configurations at C_3, C_4, and C_5 are opposite to those in L-(−)-glucose established in Fig. 5-17.

** The formulas on the extreme right and left are identical, except that one is turned by 180° with respect to the other.

Fig. 5-18. Configuration of (+)-glucose, second stage.

One of these methods is the method of quasi-racemic compounds (also called quasi-racemates† or partial racemates).[23, 24] The method was sug-

† Use of yet another term, "pseudoracemate," is to be discouraged since the same term has been employed as a synonym for "racemic solid solution" (Sec. 4-3). Because of recent extensions of the method to cases of solid solution (rather than compound) formation, a new term, "method of thermal analysis," is now sometimes used.

[22] M. L. Wolfrom, R. U. Lemieux, and S. M. Olin, *J. Am. Chem. Soc.*, **71**, 2870 (1949).

[23] A. Fredga, in A. Tiselius and K. O. Pederson, eds., "The Svedberg" Memorial Volume, Almqvist and Wiksells, Uppsala, Sweden, 1944, pp. 261–273; A. Fredga, *Tetrahedron*, **8**, 126 (1960).

[24] J. Timmermans, *Rec. trav. chim.*, **48**, 890 (1928).

gested by early observations[25] that (+)-chlorosuccinic acid and (−)-bromosuccinic acid form a molecular compound in the solid state. Detailed

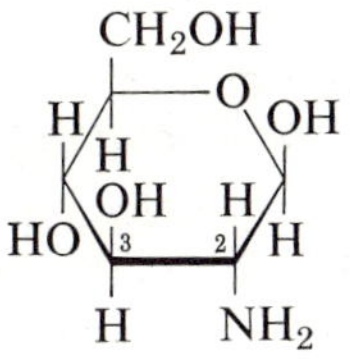

Steps in configurational correlation

(1) C_3 correlated with glyceraldehyde
(2) C_2 correlated with serine
(3) C_3 and C_2 correlated with each other

Fig. 5-19. (+)-Glucosamine.

investigation of this phenomenon employing substances of known configuration showed that, in general, compound formation (as evidenced by melting-point diagrams) occurs between chemically similar optically active species only when they are of opposite configuration. Species of like configuration, instead, show the phase diagram of simple mixtures (see, however, below). Thus the above-mentioned observation indicates that (+)-chlorosuccinic and (−)-bromosuccinic acid are of opposite configuration—a conclusion which could not readily have been reached on the basis of the methods discussed earlier.

The method of quasi-racemates has somewhat severe limitations. The compounds compared must be chemically very similar, or else the result is not reliable.† Even then, an answer is obtained only if, let us say, (−)-A and (+)-B form a compound and (−)-A and (−)-B form a mixture, not if both pairs form a mixture (a very common case) or if both should form a compound.‡ (For cases of solid solutions, see below.) It is necessary to have available *both* enantiomers of at least one of the two components involved in the comparison in order to check this point.

An interesting application[23] of the method is the correlation of configuration of (−)-malic acid and (−)-methylsuccinic acid (Fig. 5-20). Since hydroxyl and methyl are quite unequal in size, the comparison could not be carried out directly. Therefore, as shown in Fig. 5-20, (S)-(−)-malic acid was converted into the dextrorotatory ethyl xanthate derivative (I), which was compared, by the method of quasi-racemates, with the dextrorotatory ethyl thionecarbonate derivative of (+)-2-thiolsuccinic acid (II). The phase diagrams (Fig. 5-21) clearly show that (+)-I and (+)-II are of opposite configuration. The relative configuration of (−)-malic acid and (+)-thiolsuccinic acid (Fig. 5-20) follows. (+)-Thiolsuccinic acid, in turn, has been correlated by the method of partial racemates with (−)-methylsuccinic acid.[23]

† One may look at a quasi-racemate as a true racemate in which one of the groups attached to the asymmetric carbon is *slightly* modified chemically. The modification should be so slight that the preferential steric fit of molecules of the opposite configuration in the crystal of the racemic compound (cf. Sec. 4-3) is not disturbed.

‡ However, J. E. Ricci [*Tetrahedron,* **18**, 605 (1962)] has proposed an extension of the quasi-racemate method: Compounds of opposite configuration depress each other's melting point more than compounds of the same configuration, since compound formation results in a negative deviation from ideality.

[25] M. Centnerzwer, *Z. physik. Chem.,* **29**, 715 (1899).

The phase diagram of mixtures of (+)-thiolsuccinic acid and (−)-methylsuccinic acid is similar to *A* (Fig. 5-21), indicating compound formation, whereas (−)-thiolsuccinic acid and (−)-methylsuccinic acid only form a mixture (cf. Fig. 5-21, *B*).

The method of quasi-racemates has been extended[26] by the postulate that two similar substances, (+)-A and (−)-B, have the *same* configuration when they form a solid solution and (+)-A and (+)-B [or (−)-A and (−)-B] show the phase behavior of a simple mixture. Thus (+)-2-chloro-2-phenylacetamide forms a solid solution with (+)-hydratropamide whereas its levorotatory enantiomer forms only mixtures with (+)-hydratropamide, as shown in phase diagrams in Fig. 5-22. It has been concluded[26] that (+)-2-chloro-2-phenylacetamide and (+)-hydratropamide have the same configuration, the correlation being as shown in Fig. 5-23.

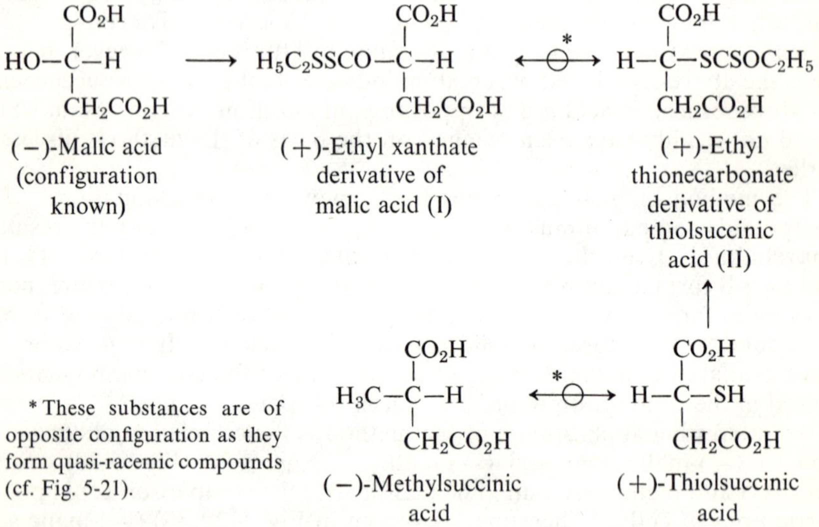

$$\begin{array}{c} CO_2H \\ | \\ HO{-}C{-}H \\ | \\ CH_2CO_2H \end{array} \longrightarrow \begin{array}{c} CO_2H \\ | \\ H_5C_2SSCO{-}C{-}H \\ | \\ CH_2CO_2H \end{array} \overset{*}{\longleftrightarrow} \begin{array}{c} CO_2H \\ | \\ H{-}C{-}SCSOC_2H_5 \\ | \\ CH_2CO_2H \end{array}$$

(−)-Malic acid (configuration known) — (+)-Ethyl xanthate derivative of malic acid (I) — (+)-Ethyl thionecarbonate derivative of thiolsuccinic acid (II)

$$\begin{array}{c} CO_2H \\ | \\ H_3C{-}C{-}H \\ | \\ CH_2CO_2H \end{array} \overset{*}{\longleftrightarrow} \begin{array}{c} CO_2H \\ | \\ H{-}C{-}SH \\ | \\ CH_2CO_2H \end{array}$$

(−)-Methylsuccinic acid — (+)-Thiolsuccinic acid (↑ to II)

* These substances are of opposite configuration as they form quasi-racemic compounds (cf. Fig. 5-21).

Fig. 5-20. Correlation of configuration of malic acid and methylsuccinic acid.

In summary, if A and B form a compound and A and the enantiomer of B (or B and the enantiomer of A) form a mixture (or a solid solution), then A and B are of opposite configuration. If A and B form a solid solution and A and the enantiomer of B form a mixture (or a compound), then A and B are of the same configuration.

Instead of deducing compound formation from melting-point behavior, it may be deduced from solid-state infrared spectra[27,28] or X-ray powder diagrams.[28-30] Just as true racemic compounds differ from the enantiomers

[26] K. Mislow and M. Heffler, *J. Am. Chem. Soc.*, **74**, 3668 (1952).
[27] A. Rosenberg and L. Schotte, *Arkiv Kemi,* **8**, 143 (1955).
[28] S. Gronowitz, *Arkiv Kemi,* **11**, 361 (1957).
[29] K. Pettersson, *Arkiv Kemi,* **7**, 347 (1954).
[30] S. Gronowitz and S. Larsson, *Arkiv Kemi,* **8**, 567 (1955).

in solid-state infrared spectra and X-ray patterns (Sec. 4-3), so do quasi-racemic compounds have infrared spectra and X-ray patterns different from those of simple mixtures of their components. Thus if (−)-A and (−)-B intimately mixed together give infrared spectra and X-ray patterns which are simply summations of the spectra or patterns of (−)-A and (−)-B taken individually, whereas (−)-A and (+)-B give new infrared bands and distinct patterns, it may be concluded that (−)-A and (+)-B have formed a compound and are therefore of opposite configuration.

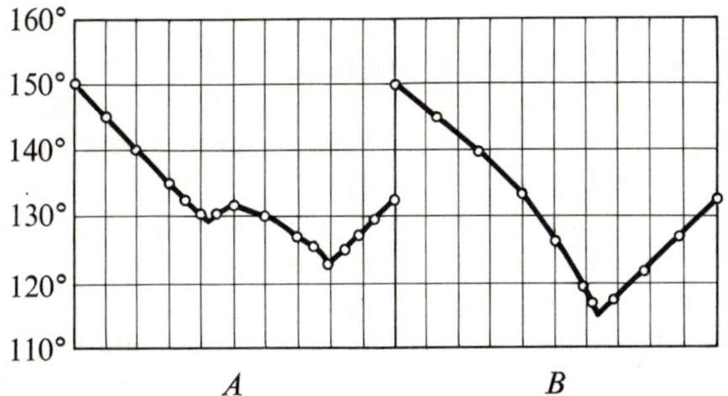

Fig. 5-21. Phase diagrams for mixtures of the dextrorotatory ethyl xanthate derivative of malic acid (I) with the (*A*) dextrorotatory and (*B*) levorotatory ethyl thionecarbonate derivative of thiolsuccinic acid (II). (*From "The Svedberg" Memorial Volume, Almquist & Wiksells, Uppsala, Sweden,* 1944. *By permission of Professor Arne Fredga.*)

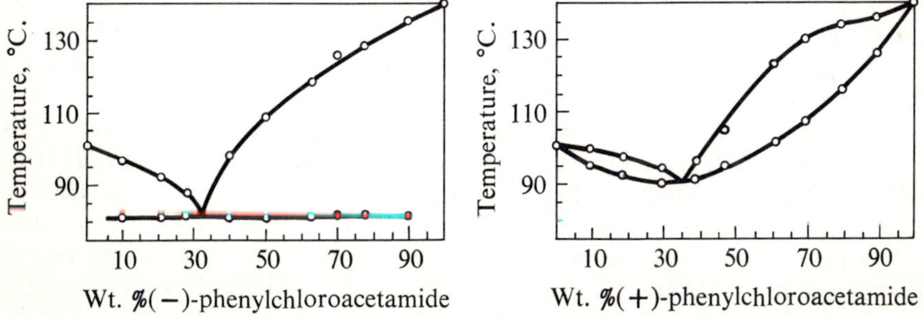

Fig. 5-22. Phase diagrams for mixtures of (−)-and (+)-2-chloro-2-phenylacetamide and (+)-hydratropamide. [*From K. Mislow and M. Heffler, J. Am. Chem. Soc.,* **74,** 3668 (1952). *By permission of the authors and editor.*]

$$\begin{array}{c} CONH_2 \\ | \\ H_3C{-}C{-}H \\ | \\ C_6H_5 \end{array} \qquad\qquad \begin{array}{c} CONH_2 \\ | \\ Cl{-}C{-}H \\ | \\ C_6H_5 \end{array}$$

(+)-Hydratropamide (+)-2-Chloro-2-phenylacetamide

Fig. 5-23. Configurations of hydratropamide and 2-chloro-2-phenylacetamide.

A method which is conceptually somewhat related to the method of quasi-racemates but which has not yet been extensively used is the determination of configuration by adsorption affinity. It was indicated in Chap. 4 that an inactive adsorbent treated with an active compound may acquire some stereochemical specificity. This specificity will be such as to favor adsorption of a compound of similar configuration to that with which the adsorbent was pretreated. Thus[30a] a quinine-treated silica gel adsorbs quinine in preference to its epimer quinidine (cf. Fig. 4-22), and it also adsorbs cinchonidine in preference to cinchonine. This is in agreement with the known fact that cinchonidine has the same stereochemistry as quinine whereas cinchonine is epimeric at the ring carbon adjacent to the carbinol function.

d. The Method of Optical Comparison. It would obviously be desirable if absolute configuration could be deduced on theoretical grounds from the sign of rotation. Some success has been encountered along this line (cf. Chap. 14). Such success, however, is of relatively recent date, and exact methods are not at all easy to apply to molecules of any appreciable degree of complexity. It is fortunate, therefore, that more simple and reasonably reliable methods based on rotation are available to assign *relative* configuration. These methods utilize two simple empirical rules, namely, Freudenberg's "rule of shift"[31] and van't Hoff's "rule of optical superposition."[32]

The rule of shift (or "displacement rule") states that, when two similarly constituted dissymmetric compounds are chemically changed in the same way and the change produces a considerable shift in optical rotation in the same direction, then the two compounds probably have the same configuration. The rule has been applied, for example, to determining the configuration of atrolactic acid.[33] In Table 5-1 are given the *molecular* rotations of (−)-mandelic and (−)-atrolactic acids measured at 578 mμ. For most of the derivatives indicated, the shift for the two acids is in the positive direction. Inconsistencies are observed for the acyl derivatives of the ethyl esters (two derivatives), but the weight of the evidence (five derivatives) indicates that (−)-atrolactic acid and (−)-mandelic acid have the same configuration. This conclusion is in agreement with that derived by more exact methods (Fig. 5-16). It is usually necessary to compare a considerable number of derivatives before any conclusions drawn from the rule of shift may be considered reliable.

The rule of optical superposition[32] in its original form states that two or more asymmetric carbons in an active compound make independent contributions to the total molecular rotation. It is now known, however, that the validity of the rule is severely limited by what is called "vicinal action," meaning that if the asymmetric carbon atoms are close to each other, there are apt to be interactions which change the contribution to the total rotation

[30a] A. H. Beckett and P. Anderson, *Nature,* **179,** 1074 (1957).

[31] K. Freudenberg, W. Kuhn, and I. Bumann, *Ber.,* **63,** 2380 (1930). See K. Freudenberg, F. Brauns, and H. Siegel, *Ber.,* **56,** 193 (1923); see also K. Freudenberg, *Ber.,* **66,** 177 (1933); *Monatsh.,* **85,** 537 (1954).

[32] J. H. van't Hoff, "Die Lagerung der Atome im Raume," Vieweg-Verlag, Brunswick, Germany, 1893, p. 119; P. A. Guye and M. Gautier, *Compt. rend.,* **119,** 740, 953 (1894).

[33] K. Freudenberg, J. Todd, and R. Seidler, *Ann.,* **501,** 199 (1933).

Table 5-1
Molecular Rotation[a] of (−)-Mandelic and (−)-Atrolactic Acids and Some of Their Derivatives at 578 mμ (in Angular Degrees)

Compound	Rotation of derivative of		Direction of shift
	Atrolactic acid	Mandelic acid	
Free acid	−86	−240	
Ethyl ester	−59	−210	+
Amide	+9	−137	++
Benzoyl—ethyl ester	−67	−404	+ and −−
Acetyl—ethyl ester	−40	−280	+ and −
Amide acetone derivative	+103	−181	++ and +
Methyl ether methyl ester	+110	−197	++ and +
Methyl ether dimethylamide	+395	+290	++

[a] From Ref. 33.

from what it would be if each asymmetric carbon existed by itself. A more restricted but at the same time more generally valid statement of the rule is:[34] "Like functional groups in like surroundings make like contributions to the optical rotation." Fortunately, vicinal action decreases as the asymmetric carbons become more remote from each other, and by the time they are separated by three to five saturated atoms, the rule of optical superposition holds quite well.

One of the highly successful applications of the optical superposition rule is the so-called "method of molecular-rotation differences"[35] in the steroids.† Steroids have many asymmetric carbon atoms (cf. Chap. 10); if one focuses attention on any one of them, one may say, applying the van't Hoff rule, that the molecular rotation of the steroid is A + X or A − X, depending on the configuration of the carbon in question, where X is the contribution of that particular carbon and A is the contribution of the rest of the molecule. The values of A and X may be known from other steroids of known configuration, and therefore, comparing the observed molecular rotation of the unknown steroid with the two calculated values A + X and A − X, one may tell the configuration at the carbon atom under consideration. An example[35] of the application of this method is in the elucidation of the configuration at C_6 of Lieberman's compound A31,[36] an allopregnan-3α,6-diol-20-one (Fig. 5-24).

† One reason for this is that some of the more important asymmetric carbons (C_3, C_7, C_{11}, C_{17}) are quite far apart; another is that the steroid system is quite rigid. In flexible carbon chains, optical rotation is affected by conformation (cf. Chaps. 6 and 14) among other factors.

[34] P. de Mayo, "Mono- and Sesquiterpenoids," Interscience Publishers, Inc., New York, 1959, p. 22.

[35] For a summary, see D. H. R. Barton and W. Klyne, *Chemistry & Industry (London)*, 755 (1948).

[36] S. Lieberman, K. Dobriner, B. R. Hill, L. F. Fieser, and C. P. Rhodes, *J. Biol. Chem.*, **172**, 263 (1948).

The molecular rotation of compound A31 is +351°; that of the known allopregnan-3α-ol-20-one is +309°. Hence the contribution of the hydroxyl group at C_6 (using the C_6 hydrogen compound as reference) is 351 − 309 = +42°. From a consideration of the molecular rotations of a wide variety of 6α- and 6β-hydroxy-substituted allopregnanes of known configuration in comparison with C_6-unsubstituted allopregnanes, one knows that the contribution of 6α-OH is ca. +55° whereas that of 6β-OH is ca. −50°.† It follows that compound A31 has the α configuration at C_6 and is therefore allopregnan-3α,6α-diol-20-one.

CH_3 | H_3C CO, H_3C, HO, H, OH

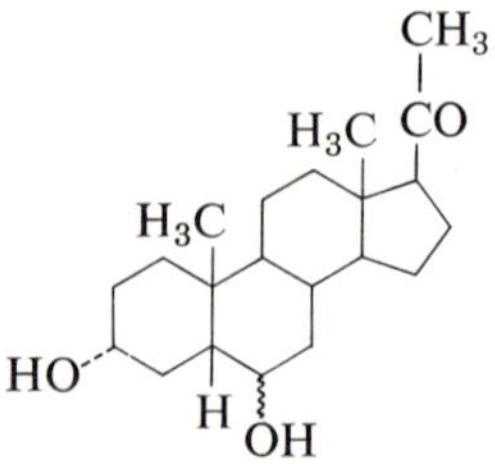

Fig. 5-24. Allopregnan-3α,6-diol-20-one.

The method of molecular-rotation differences has also been used in the field of polycyclic terpenes[37] and more recently[38] in the correlation of the stereochemistry of one group of polycyclic compounds (e.g., triterpenes, diterpenes) with another (e.g., steroids). These correlations have been confirmed by other methods (see Sec. 5-4*a* and *e*).

One of the oldest applications of the optical superposition rule is to the anomeric center (C_1, cf. p. 40) in the carbohydrates[39] by what are sometimes called the "rules of isorotation." These rules may be stated as follows:

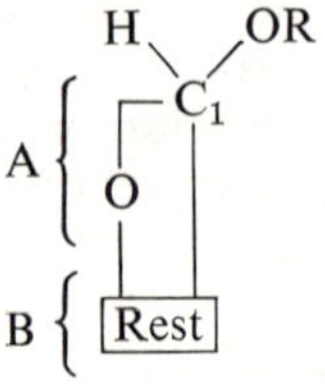

H, OR, C_1, A, O, B, Rest

Fig. 5-25. Molecular-rotation contribution in aldoses.

1. The difference between the molecular rotation of the α and β forms (Fig. 5-9) of all the aldoses and of all their derivatives in which the added substance is not joined directly to the end asymmetric carbon is a nearly constant quantity.

† The contributions are *not* equal in magnitude (and opposite in direction) because of the arbitrary choice of a methylene group at C_6 as the zero of the scale.

[37] D. H. R. Barton and E. R. H. Jones, *J. Chem. Soc.*, 659 (1944).
[38] W. Klyne, *J. Chem. Soc.*, 2916 (1952).
[39] C. S. Hudson, *J. Am. Chem. Soc.*, **31**, 66 (1909).

2. The α and β forms of those derivatives (e.g., glycosides, etc.) of any aldo sugar in which only the end asymmetric carbon is affected have molecular rotations whose sum is equal to the sum for the α- and β-aldoses.

Another way of stating both rules is to say that the rotation of any aldose derivative is the algebraic sum of A and B (Fig. 5-25), where A is the contribution of the anomeric center and B is the contribution of the rest of the molecule.

The reader may wonder why the isorotation rules hold at all, since vicinal action between adjacent hydroxyls is obviously of great importance in the sugars. There are three reasons why the method nevertheless seems to work. One is that the superposition rule is applied more rigorously than in the steroid field. Instead of saying, to use the two-carbon analogy, that the rotation of $RCHOH(CH_2)_nCHOHR'$ is the sum of the rotations of

$$RCHOH(CH_2)_nCH_2R'$$

and $RCH_2(CH_2)_nCHOHR'$ of analogous configuration, one now says that the rotation of $RCHOH(CH_2)_nCHOHR'$ is made up of two parts, one, A, corresponding to the asymmetric carbon on the left and the other, B, corresponding to the asymmetric carbon on the right. The four stereoisomers corresponding to the above structure thus have rotations A + B, A − B, −A + B, and −A − B. The second reason is that, although there is still vicinal action, it may be constant in a series of compounds. For example, (+)-glucose and (+)-galactose (Fig. 3-15) both have the hydroxyl on the No. 2 carbon in the same configuration and therefore any interaction of this hydroxyl with the hydroxyl on the No. 1 carbon in the cyclic hemiacetal form should be constant. The data in Table 5-2 bear this out: The molecular-rotation contribution for (+)-glucose and (+)-galactose at C_1 is about the same, but that for (+)-mannose, which differs in configuration at C_2, is different. (+)-Talose has the same configuration as (+)-mannose at C_2 (cf. Fig. 3-15), and its molecular-rotation contribution at C_1 is indeed similar. Thirdly, speaking qualitatively, the contribution of C_1 to the total rotation is so large that vicinal action plays a relatively minor part and one may say that all the α-D sugars investigated have a positive rotation contribution at C_1. Conversely, a positive contribution of C_1 to the rotation of an anomer

Table 5-2
Molecular-rotation Contribution of the Anomeric Center in Aldoses

Aldose	M_D^{25}		Contribution[a]	
	α form	β form	of C_1	of C_{2-5}
(+)-Glucose	+202	+34	84	118
(+)-Galactose	+272	+95	88.5	183.5
(+)-Mannose	+ 53	−31	42	11
(+)-Talose	+122	+24	49	73

[a] The contribution of C_1 is $\frac{1}{2}(M_\alpha - M_\beta)$, that of C_{2-5} $\frac{1}{2}(M_\alpha + M_\beta)$.

or glycoside of a D-aldose would indicate that the configuration at C_1 is α, a negative contribution that it is β.

A number of other empirical rules of configurational assignment based on rotation have been enunciated from time to time, and a summary and generalization of a number of such rules have been published.[39a] However, these rules will not be reviewed here because a more systematic treatment of the subject is now available which will be presented later (Sec. 14-1).

Methods of configurational assignment based on optical rotatory dispersion are extensions of methods based on optical rotation at a single wavelength. Because of the importance that such methods have assumed, they will also be dealt with separately in Sec. 14-2.

e. Configurational Assignment Based on Asymmetric Synthesis. Prelog's rule[40] (cf. Fig. 4-40) has been used successfully[41] to deduce the configuration of a number of hydroxy acids and alcohols. The rule may be applied in one of two ways. Thus, an ester of a keto acid and an alcohol of known configuration may be reduced (or treated with a Grignard reagent, depending on the result desired) to give the ester of the appropriate hydroxy acid. Prelog's rule indicates the configuration of the diastereoisomer which is obtained preferentially. This is then hydrolyzed quantitatively,† and the rotation of the acid liberated (whose configuration is now known by the rule) is measured. Figure 5-26 indicates application[41a] of this method to (−)-atrolactic acid obtained by reaction of the phenylglyoxylate ester of (−)-menthol with methylmagnesium iodide.‡ A second application concerns the determination of the (unknown) configuration of a given alcohol. The alcohol is esterified with phenylglyoxylic acid and treated with methylmagnesium iodide. From the sign of rotation of the atrolactic acid (configuration now known!) obtained after hydrolysis of the resulting ester, the configuration of

Phenylglyoxylate of (−)-menthol + CH_3MgI ⟶ $R{-}O{-}C(=O){-}C(OH)(CH_3)(C_6H_5)$ $\xrightarrow{H_2O}$ $H_3C{-}C(CO_2H)(C_6H_5){-}OH$ (−)-Atrolactic acid

Fig. 5-26. Configuration of (−)-atrolactic acid by correlation with (−)-menthol.

† Quantitative hydrolysis is essential, since one diastereoisomer may hydrolyze faster than the other. If the isomer formed in minor proportion is hydrolyzed faster, an erroneous conclusion may be drawn if hydrolysis is stopped before it is complete.

‡ The reader should construct models to convince himself that the transformation shown in Fig. 5-26 is in accordance with Prelog's rule.

[39a] A. K. Bose and B. G. Chatterjee, *J. Org. Chem.*, **23**, 1425 (1958).

[40] V. Prelog, *Bull. soc. chim. France*, 987 (1956).

[41] (*a*) V. Prelog and H. L. Meier, *Helv. Chim. Acta*, **36**, 320 (1953); (*b*) W. G. Dauben, D. F. Dickel, O. Jeger, and V. Prelog, *ibid.*, **36**, 325 (1953); (*c*) A. J. Birch, J. W. Clark-Lewis, and A. V. Robertson, *J. Chem. Soc.*, 3586 (1957).

the original alcohol can be deduced.† Application[41b] of this method to androstan-17β-ol, shown in Fig. 5-27, served to establish the configuration of this steroid alcohol at C_{17}.

The configuration of the carbinol carbon in 7α- and 7β-cholestanol and the triterpene alcohols α-amyrin, dihydrolanosterol, and euphol has been deduced similarly,[41b] as has the configuration of the flavanols catechin and epicatechin.[41c]

An extension of the method to the addition of diphenyldiazomethane to menthyl acrylate has been proposed.[42]

Kinetic resolution (Sec. 4-4*d, i*), as obtained in the esterification of an active alcohol with (±)-α-phenylbutyric anhydride, may also be used in assignment of configuration.[42a] Since one enantiomer of the α-phenylbutyric acid, $C_6H_5CH(C_2H_5)CO_2H$, enters preferentially into the ester, the other enantiomer ends up preferentially as the free acid. From the rotation of this acid, the configuration of the alcohol may be deduced.

f. Chemical Transformations Involving the Asymmetric Center in a Predictable Way. It was mentioned in Sec. 5-4*a* that only those chemical transfor-

Phenylglyoxylate of androstan-17β-ol + CH_3MgI ⟶ $RO{-}C(=O){-}C(CH_3)(OH)C_6H_5$ $\xrightarrow{H_2O}$ $HO{-}C(CH_3)(CO_2H)C_6H_5$

(+)-Atrolactic acid

Fig. 5-27. Configuration of androstan-17β-ol relative to (+)-atrolactic acid.

mations serve to establish relative configuration in which the bonds to the asymmetric carbon are not broken. This restriction could obviously be removed if one could tell exactly what happens when one substituent on the asymmetric carbon is replaced by another by a breaking and remaking of the bond, i.e., whether the new substituent takes the place of the old (Fig. 5-28, *A*) or whether more deep-seated changes take place (Fig. 5-28, *B*). In the former case (*A*) one speaks of replacement of the group with "retention of configuration," whereas the latter case (*B*) involves replacement of the group y by z with "inversion of configuration." The curled arrow used in the transformation *B* denotes the inversion. Originally it was assumed that all dis-

† Reduction of the phenylglyoxylate with lithium aluminum hydride to optically active styrene glycol, $C_6H_5CHOHCH_2OH$, of known configuration should also lend itself to the application of Prelog's rule but gives unreliable results: J. A. Berson and M. A. Greenbaum, *J. Am. Chem. Soc.*, **81**, 6456 (1959).

[42] F. J. Impastato, L. Barash, and H. M. Walborsky, *J. Am. Chem. Soc.*, **81**, 1514 (1959).
[42a] A. Horeau, *Tetrahedron Letters*, 506 (1961).

placements proceed with retention of configuration. However, work by Walden[43] summarized in Fig. 5-29 showed that this could not be so.

It is quite evident that the reaction of (−)-malic acid with phosphorus pentachloride to give (+)-chlorosuccinic acid followed by reaction with silver oxide to give the enantiomeric (+)-malic acid cannot have involved retention of configuration in both steps. Similarly, *either* the reaction of (−)-chlorosuccinic acid with potassium hydroxide to give (+)-malic acid *or* its reaction with moist silver oxide to give (−)-malic acid (but not both) must involve inversion of configuration. Thus, Walden's work proves that replacements of

(Inversion) v—C(u)(x)—z ←[−y, +z; B]— v—C(u)(y)—x —[−y, +z; A]→ v—C(u)(z)—x (Retention)

Fig. 5-28. Replacement with retention and with inversion.

the type shown in Fig. 5-28, where a bond to the asymmetric carbon is broken, may involve either retention or inversion of configuration, and it seemed for a long time that such reactions would therefore be useless for configurational correlation. From a completely rigorous point of view, this is still true; nevertheless, the understanding of reaction mechanisms has progressed to such an extent in recent years that one can often predict the steric course (retention or inversion) of reactions of the type shown in Fig. 5-28 with a considerable degree of confidence, so that such reactions may now be used in the correlation of configuration in a number of instances.

(−)-$HO_2C—CH_2—CHCl—CO_2H$
(−)-Chlorosuccinic acid

AgOH ↙ ↖ KOH / PCl_5 ↘

(−)-$HO_2C—CH_2—CHOH—CO_2H$
(−)-Malic acid

(+)-$HO_2C—CH_2—CHOH—CO_2H$
(+)-Malic acid

KOH ↖ / PCl_5 ↘ ↗ AgOH

(+)-$HO_2C—CH_2—CHCl—CO_2H$
(+)-Chlorosuccinic acid

Fig. 5-29. Walden inversion.

To the extent that the stereochemisty of the displacement reaction depends on considerations of mechanism, it can be dealt with here only very briefly; for further details the reader is referred to textbooks on reaction mechanism (cf. Chap. 1), especially to the excellent review of the subject by Ingold.[62] The utilization of reactions involving changes at the asymmetric carbon in configurational correlations proceeds, in general, along the following lines: The reaction to be used is studied in two or three instances where the relative configurations of the starting material and product are known. It is then

[43] P. Walden, *Ber.*, **29**, 133 (1896); *ibid.*, **30**, 3146 (1897).

assumed that the configurational changes so observed occur in other systems as well. In some cases, where stoichiometrically similar reactions may proceed through mechanistically different paths involving different stereochemical changes, it is necessary to set up auxiliary criteria (e.g., kinetic ones) to recognize the type of mechanism involved in the particular case under study. Two examples will be taken up to make the point clearer: nucleophilic displacement† at saturated carbon and rearrangement reactions involving saturated centers.

i. The Nucleophilic Displacement Reaction. As shown in Fig. 5-29, displacement reactions of this type may involve either retention or inversion of configuration. The question of when they involve retention and when inversion proved elusive. Light was first thrown on this subject by the work of Phillips,[44] summarized in Fig. 5-30, which shows definitely that the reaction of 2-octyl tosylate with acetate ions involves inversion of configuration, since

$$\underset{\substack{(-)\text{-2-Octyl}\\ \text{acetate}}}{AcO{-}\overset{\overset{CH_3}{|}}{\underset{\underset{C_6H_{13}}{|}}{C}}{-}H} \xleftarrow{Ac_2O} \underset{(-)\text{-2-Octanol}}{HO{-}\overset{\overset{CH_3}{|}}{\underset{\underset{C_6H_{13}}{|}}{C}}{-}H} \xrightarrow{ArSOCl^*} \underset{\substack{(-)\text{-2-Octyl}\\ p\text{-toluenesul-}\\ \text{finate}}}{ArSOO{-}\overset{\overset{CH_3}{|}}{\underset{\underset{C_6H_{13}}{|}}{C}}{-}H} \xrightarrow{Ox.}$$

$$\underset{\substack{(-)\text{-2-Octyl}\\ p\text{-toluenesul-}\\ \text{fonate}}}{ArSO_2O{-}\overset{\overset{CH_3}{|}}{\underset{\underset{C_6H_{13}}{|}}{C}}{-}H} \xrightarrow[\circ]{OAc^-} \underset{\substack{(+)\text{-2-Octyl}\\ \text{acetate}}}{H{-}\overset{\overset{CH_3}{|}}{\underset{\underset{C_6H_{13}}{|}}{C}}{-}OAc}$$

* Ar = p-$CH_3C_6H_4$—.

Fig. 5-30. Walden inversion in the reaction of 2-octyl tosylate with acetate.

neither in the reaction of 2-octanol with *p*-toluenesulfinyl chloride, nor in the oxidation of the sulfinate to a sulfonate, nor in the reaction of 2-octanol with acetic anhydride is the carbon-oxygen bond broken, so that these latter reactions must necessarily lead to products whose configurations are the same as those of their precursors. These results were later extended,[45] by analogy, to the reaction of 2-octyl tosylate with halide ions and were thus utilized in the assignment of configuration of the 2-octyl halides. The correctness of this argument has been put on a firmer basis by the work summarized in Fig. 5-31.[46] The reaction of 2-octyl iodide with iodide ion leads to racemization, as already mentioned in Sec. 4-2. In the light of what was said previously,

† For terminology, see Ref. 62.

[44] H. Phillips, *J. Chem. Soc.*, **127**, 2552 (1925); for earlier work in this area, see *id., ibid.*, **123**, 44 (1923).

[45] A. J. H. Houssa, J. Kenyon, and H. Phillips, *J. Chem. Soc.*, 1700 (1929).

[46] E. D. Hughes, F. Juliusburger, S. Masterman, B. Topley, and J. Weiss, *J. Chem. Soc.*, 1525 (1935).

it seemed likely that the displacement of iodide by iodide ion in 2-iodoöctane (Sec. 4-2*c*, *v*) would be a process of the inversion type. Proof of this assumption came from the fact that the rate of radioactive exchange (i.e., replacement of iodine by radioiodine in the 2-iodoöctane) was equal to the rate of inversion (or one-half the rate of racemization), indicating that each act of displacement involves inversion at the same time. The mechanism for this sort of process is pictured in Fig. 5-32: The incoming group approaches from the

$$\mathrm{H{-}\overset{\displaystyle CH_3}{\underset{\displaystyle C_6H_{13}}{C}}{-}I} + \mathrm{I^{*-}} \longrightarrow \mathrm{{}^{*}I{-}\overset{\displaystyle CH_3}{\underset{\displaystyle C_6H_{13}}{C}}{-}H} + \mathrm{I^-}$$

Fig. 5-31. Reaction of optically active 2-iodoöctane with radioactive iodide ion. (In actual fact, the racemization and the radioactive exchange were carried out in two separate experiments and the rates were then compared. This does not affect the argument.)

side opposite to that of the leaving group and turns over the other three groups like the ribs of an umbrella in a storm.

The question still remains why not *all* the transformations discovered by Walden (Fig. 5-29) involve inversion of configuration. The complication here is that, in addition to the inversion or "S_N2" mechanism of nucleophilic displacement at saturated carbons, there are at least three other mechanisms: the so-called "S_N1" mechanism, which usually involves substantial racemization; the "S_Ni" mechanism, which may involve retention; and

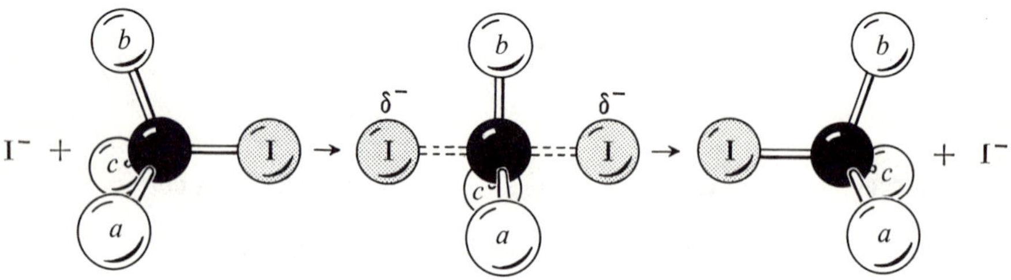

Transition state

Fig. 5-32. Walden-inversion mechanism. (*From L. F. Fieser and M. Fieser, "Organic Chemistry," 3d ed., Reinhold Publishing Corporation, New York,* 1956. *By permission of the publishers.*)

processes involving "neighboring-group participation" which also involve retention of configuration. It is evident that one or both of the latter mechanisms must be involved in some of the transformations depicted in Fig. 5-29. Fortunately, Ingold has developed an empirical rule allowing one to decide which type of nucleophilic displacement reactions may be expected, with some degree of confidence, to involve inversion of configuration. This rule, the so-called "S_N2 rule," states that substitution by the *bimolecular* (S_N2) mechanism regularly involves inversion of configuration, independently of all constitutional details. An application of the rule[47] is in the correlation of configuration of (+)-α-phenethyl chloride and (−)-α-phenethylamine (Fig. 5-33). The reaction of the (+)-chloride with sodium azide gives α-phenethyl azide, and since the kinetics of this reaction was shown to be bimolecular, it may be concluded that inversion of configuration occurred. Further hydrogenation of this azide (without severing a bond to the asymmetric carbon) gave (−)-α-phenethylamine which has therefore the opposite configuration from the starting (+)-chloride.

ii. Intramolecular Rearrangements at Saturated Carbons. This section will deal with a series of rearrangements of the type depicted in Fig. 5-34. In all cases investigated in detail it has been found that when the rearranging group

$$\begin{array}{c}CH_3\\|\\Cl-C-H\\|\\C_6H_5\end{array}\xrightarrow[\text{Bimolecular kinetics}]{NaN_3}\begin{array}{c}CH_3\\|\\H-C-N_3\\|\\C_6H_5\end{array}\xrightarrow[Pt]{H_2}\begin{array}{c}CH_3\\|\\H-C-NH_2\\|\\C_6H_5\end{array}$$

(+)-α-Phenethyl chloride (−)-α-Phenethylamine

Fig. 5-33. Correlation of configuration by application of the S_N2 rule.

Cxyz is asymmetric at its point of attachment to the No. 1 carbon, it will migrate with complete or nearly complete retention of configuration. The rearrangements thus become a convenient means of correlating configuration, especially when X is nitrogen or oxygen, when they serve to correlate compounds of the type $-\overset{|}{\underset{|}{C}}-$Cxyz with $>$N-Cxyz and —O—Cxyz, respectively. Examples[48-50] involving rearrangement to nitrogen (Hofmann, Curtius, Lossen, Schmidt, and Beckmann rearrangements) are shown in Fig. 5-35. Figure 5-36 shows a similar rearrangement to oxygen (Baeyer-Villiger rearrangement).[51] In all these cases, the configurations of the starting materials and

$$\text{zyxC}-\text{C}-\text{X} \longrightarrow \text{C}-\text{X}-\text{Cxyz}$$

Fig. 5-34. Intramolecular 1,2 rearrangements.

products are known by methods described elsewhere in this chapter, and the transformations shown thus serve to establish the stereochemical course of the rearrangements. In contrast, Fig. 5-37 shows a case[52] where the Baeyer-Villiger rearrangement was used to correlate the previously unknown configurations of (+)-α-phenyl-α-methylbutyric acid and (−)-2-phenyl-2-butanol,

[47] P. Brewster, F. Hiron, E. D. Hughes, C. K. Ingold, and P. A. D. Rao, *Nature,* **166,** 179 (1950).
[48] C. L. Arcus and J. Kenyon, *J. Chem. Soc.,* 916 (1939).
[49] J. Kenyon and D. P. Young, *J. Chem. Soc.,* 263 (1941).
[50] A. Campbell and J. Kenyon, *J. Chem. Soc.,* 25 (1946).
[51] K. Mislow and J. Brenner, *J. Am. Chem. Soc.,* **75,** 2318 (1953).
[52] D. J. Cram and J. Allinger, *J. Am. Chem. Soc.,* **76,** 4516 (1954).

CO_2H
H_3C—C—H
C_6H_5
(+)-Hydratropic acid (configuration known to be as shown)

(1) PCl_3, (2) NH_3 → H_3C—C($CONH_2$)(C_6H_5)—H → NaOBr →

(1) PCl_3, (2) NaN_3 → H_3C—C(CON_3)(C_6H_5)—H → (1) Δ, (2) H_2O, H^+ →

(1) PCl_3, (2) NH_2OH, (3) C_6H_5COCl → H_3C—C($CONHOCOC_6H_5$)(C_6H_5)—H → (1) CH_3OK, (2) Δ →

NaN_3, H_2SO_4 →

NH_2
H_3C—C—H ≡ H—C(CH_3)(C_6H_5)—NH_2
C_6H_5
(−)-α-Phenethylamine (configuration known to be as shown)

(1) PCl_3, (2) $(CH_3)_2Cd$ ↓

H_3C—C($COCH_3$)(C_6H_5)—H → NH_2OH → H_3C—C(C(CH_3)=NOH)(C_6H_5)—H → H_2SO_4 → H_3C—C($NHCOCH_3$)(C_6H_5)—H ← $(CH_3CO)_2O$ ← (−)-α-Phenethylamine

Fig. 5-35. Retention of configuration of the migrating group in the Hofmann, Curtius, Lossen, Schmidt, and Beckmann rearrangements.

H_3C—C(CO_2H)(C_6H_5)—H → (1) PCl_3, (2) $(CH_3)_2Cd$ → H_3C—C($COCH_3$)(C_6H_5)—H → $C_6H_5CO_3H$ →
(+)-Hydratropic acid; (+)

H_3C—C($OCOCH_3$)(C_6H_5)—H → KOH → H_3C—C(OH)(C_6H_5)—H ≡ H—C(CH_3)(C_6H_5)—OH
(−); (−)-Phenylmethylcarbinol (configuration known to be as shown)

Fig. 5-36. Retention of configuration in the migrating group in the Baeyer-Villiger rearrangement.

and Fig. 5-38 shows a correlation of (−)*endo*-norbornanecarboxylic acid with (+)-*endo*-norbornylamine accomplished by means of the Schmidt reaction.[53]

In all the reactions discussed so far, the migration terminus is electrophilic. In the example,[54] shown in Fig. 5-39 (Stevens rearrangement), on the other hand, a nucleophilic migration terminus is involved. Thus it appears that in all intramolecular migrations of this type the migrating group retains its configuration, regardless of the polarity of the migration terminus.

[53] J. Berson and D. A. Ben-Efraim, *J. Am. Chem. Soc.*, **81**, 4094 (1959).
[54] J. H. Brewster and M. W. Kline, *J. Am. Chem. Soc.*, **74**, 5179 (1952).

$$\underset{\text{(+)-}\alpha\text{-Phenyl-}\alpha\text{-methylbutyric acid}}{H_5C_2\text{—}\overset{CO_2H}{\underset{C_6H_5}{C}}\text{—}CH_3} \xrightarrow[\text{(2) }(CH_3)_2Cd]{\text{(1) }SOCl_2} \underset{(+)}{H_5C_2\text{—}\overset{COCH_3}{\underset{C_6H_5}{C}}\text{—}CH_3} \xrightarrow[LiAlH_4]{C_6H_5CO_3H}$$

$$\underset{\text{(−)-2-Phenyl-2-butanol}}{H_5C_2\text{—}\overset{OH}{\underset{C_6H_5}{C}}\text{—}CH_3 \equiv H_3C\text{—}\overset{C_2H_5}{\underset{C_6H_5}{C}}\text{—}OH}$$

Fig. 5-37. Correlation of configuration of (+)-α-phenyl-α-methylbutyric acid and (−)-2-phenyl-2-butanol by means of the Baeyer-Villiger rearrangement.

$$\underset{\text{(−)-}endo\text{-Norbornanecarboxylic acid}}{\text{[norbornane]}\cdots CO_2H} \xrightarrow{HN_3} \underset{\text{(+)-}endo\text{-Norbornylamine}}{\text{[norbornane]}\cdots NH_2}$$

Fig. 5-38. Correlation of configuration of (−)-*endo*-norbornanecarboxylic acid and (+)-*endo*-norbornylamine by the Schmidt reaction.

$$\underset{\text{(−)-}\alpha\text{-Phenethylamine (configuration known to be as shown)}}{H\text{—}\overset{CH_3}{\underset{C_6H_5}{C}}\text{—}NH_2} \xrightarrow[\text{(2) }C_6H_5COCH_2Br]{\text{(1) }CH_2O,\ HCO_2H} \underset{(-)}{H\text{—}\overset{H_3C}{\underset{C_6H_5}{C}}\text{—}\overset{CH_2COC_6H_5}{N}(CH_3)_2} \xrightarrow{NaOH}$$

$$\underset{\text{(+) (mixture of diastereoisomers)}}{H\text{—}\overset{CH_3}{\underset{C_6H_5}{C}}\text{—}\overset{*}{C}H(NMe_2)COC_6H_5} \xrightarrow[AcOH]{Zn} \underset{(-)}{H\text{—}\overset{CH_3}{\underset{C_6H_5}{C}}\text{—}CH_2COC_6H_5} \xrightarrow[\text{(2) }SOCl_2]{\text{(1) }NH_2OH}$$

$$H\text{—}\overset{CH_3}{\underset{C_6H_5}{C}}\text{—}CH_2CONHC_6H_5 \xrightarrow[H^+]{H_2O} \underset{\text{(−)-3-Phenylbutyric acid (configuration known to be as shown)}}{H\text{—}\overset{CH_3}{\underset{C_6H_5}{C}}\text{—}CH_2CO_2H \equiv H_3C\text{—}\overset{CH_2CO_2H}{\underset{C_6H_5}{C}}\text{—}H}$$

Fig. 5-39. Retention of configuration of the migrating group in the Stevens rearrangement.

5-5. Synthesis of Optically Active Compounds

The methods of configurational correlation described in Sec. 5-4*a* and *f* may also be, in some cases, convenient for the synthesis of dissymmetric compounds in the active state. As mentioned in Chap. 4, certain types of compounds, such as ethers, halides, hydrocarbons, are not readily resolved and are often better synthesized from optically active precursors. Two examples, shown in Figs. 5-40 and 5-41, must suffice to illustrate this point;

$$\underset{\text{(−)-3-Phenylbutyric acid (by resolution)}}{H_3C\text{—}\underset{C_6H_5}{\overset{CH_2CO_2H}{C}}\text{—}H} \xrightarrow[\text{(2) }PBr_3]{\text{(1) }LiAlH_4} \underset{(-)}{H_3C\text{—}\underset{C_6H_5}{\overset{CH_2CH_2Br}{C}}\text{—}H} \xrightarrow[\text{(2) }H_2O,\ H^+]{\text{(1) Mg}} \underset{\text{(−)-2-Phenylbutane}}{H_3C\text{—}\underset{C_6H_5}{\overset{C_2H_5}{C}}\text{—}H}$$

Fig. 5-40. Synthesis of (−)-2-phenylbutane.

$$\underset{\text{(−)-Phenylmethylcarbinol (by resolution of phthalate)}}{H\text{—}\underset{C_6H_5}{\overset{CH_3}{C}}\text{—}OH} \xrightarrow[C_5H_5N \cdot HCl]{POCl_3} \underset{(+)}{Cl\text{—}\underset{C_6H_5}{\overset{CH_3}{C}}\text{—}H} \xrightarrow{CH_2\text{=}CHCH_2Na}$$

$$\underset{(-)}{H\text{—}\underset{C_6H_5}{\overset{CH_3}{C}}\text{—}CH_2CH\text{=}CH_2} \equiv H_3C\text{—}\underset{C_6H_5}{\overset{CH_2CH\text{=}CH_2}{C}}\text{—}H \xrightarrow[Ni]{H_2} \underset{\text{(−)-2-Phenylpentane}}{H_3C\text{—}\underset{C_6H_5}{\overset{CH_2CH_2CH_3}{C}}\text{—}H}$$

Fig. 5-41. Synthesis of (−)-2-phenylpentane.

they are concerned with the synthesis of (−)-2-phenylbutane[55] and (−)-2-phenylpentane.[56] In the former synthesis, the asymmetric carbon is not touched (cf. Sec. 4-4*a*) whereas the latter involves two reactions of the nucleophilic displacement type (cf. Sec. 5-4*f*).

[55] D. J. Cram, *J. Am. Chem. Soc.,* **74,** 2137 (1952).
[56] R. L. Burwell, A. D. Shields, and H. Hart, *J. Am. Chem. Soc.,* **76,** 908 (1954).

General References

[57] K. Freudenberg, Konfigurative Zusammenhänge optisch aktiver Verbindungen, in "Stereochemie," Franz Deuticke, Leipzig, 1933, pp. 662–720.

[58] Th. Wagner-Jauregg, Sterische Umlagerungen am asymmetrischen Kohlenstoffatom, in "Stereochemie," Franz Deuticke, Leipzig, 1933, pp. 879–912.

[59] R. L. Shriner and R. Adams, Optical Isomerism, in H. Gilman, ed., "Organic Chemistry," 2d ed., John Wiley & Sons, Inc., New York, 1943, pp. 264–281.

[60] M. Wolfrom, Carbohydrates, in H. Gilman, ed., "Organic Chemistry," 2d ed., John Wiley & Sons, Inc., New York, 1943, chap. 20.

[61] G. W. Wheland, "Advanced Organic Chemistry," 3d ed., John Wiley & Sons, Inc., New York, 1960, pp. 357–370, 415–427.

[62] C. K. Ingold, "Structure and Mechanism in Organic Chemistry," Cornell University Press, Ithaca, N. Y., 1953, especially chaps. 7 and 9.

[63] J. A. Mills and W. Klyne, The Correlation of Configurations, in W. Klyne, ed., "Progress in Stereochemistry," vol. I, Butterworth & Co. (Publishers) Ltd., London, 1954, chap. 5.

[64] W. Klyne, Optical Rotation, in E. A. Braude and F. C. Nachod, eds., "Determination of Organic Structures by Physical Methods," Academic Press, Inc., New York, 1955, chap. 3.

[65] E. L. Eliel, Substitution at Saturated Carbon Atoms, in M. Newman, ed., "Steric Effects in Organic Chemistry," John Wiley & Sons, Inc., New York, 1956, chap. 2.

[66] D. J. Cram, Intramolecular Rearrangements, in M. Newman, ed., "Steric Effects in Organic Chemistry," John Wiley & Sons, Inc., New York, 1956, chap. 5.

[67] M. L. Wolfrom, Optical Activity and Configurational Relations in Carbon Compounds, *Rec. Chem. Progr.* (*Kresge-Hooker Sci. Lib.*), **16**, 121 (1955).

[68] W. Klyne, Stereochemical Correlations, in J. K. Grant and W. Klyne, eds., "Steric Aspects of the Chemistry and Biochemistry of Natural Products," Cambridge University Press, London, 1960.

Chapter 6

CONFORMATION AND REACTIVITY IN ACYCLIC COMPOUNDS

6-1. The Meaning of Conformation

It was mentioned in Chap. 1 that compounds containing double bonds, if suitably substituted, may exist as cis and trans isomers, because rotation about the double bond is prohibited. In saturated acyclic systems such isomers are not usually encountered, and it was believed at one time that rotation about most carbon-carbon (or other) single bonds was entirely free. This is not, however, a justified conclusion; for even if there is some hindrance to rotation about single bonds, the resulting stereoisomers are not isolable unless the hindrance is substantial enough to prevent their rapid interconversion. (This point will be returned to in Sec. 6-4.)

The simplest molecule for which one may discuss rotation around a carbon-carbon single bond is ethane. It was shown[1-3] in 1936 that calculations of the enthalpy and entropy of ethane by means of statistical mechanics give poor agreement with the experimental values unless it is assumed that there is a barrier of about 3 kcal./mole to free rotation about the carbon-carbon bond. The barrier seems to be occasioned by the fact that, as the methyl groups rotate with respect to each other, their hydrogen atoms become alternately staggered and eclipsed (Fig. 6-1). The potential energy of the molecule varies with the angle of rotation between two designated hydrogen atoms on adjacent carbons, as shown in Fig. 6-2. To be more precise, the abscissa in Fig. 6-2 denotes the angle between two planes, one defined by a designated C_1—H bond and the C_1—C_2 bond, and the other defined by the C_1—C_2 bond and a designated C_2—H bond. This angle is called the "dihedral angle" or "angle of torsion."

The term "conformation" is used to denote any one of the infinite number of momentary arrangements of the atoms in space that result from rotation about single bonds. Thus any point on the curve shown in Fig. 6-2 corre-

Note: All references above 79 are listed in the General References at the end of the chapter.

[1] J. D. Kemp and K. S. Pitzer, *J. Chem. Phys.*, 4, 749 (1936).

[2] See also E. Teller and B. Topley, *J. Chem. Soc.*, 876 (1935).

[3] See Refs. 82 and 84 for a discussion of earlier indications that rotation about single bonds is not free.

sponds to some conformation of the ethane molecule. Of these, only three—the three staggered conformations—correspond to energy minima and are therefore stable. In ethane these three conformations are the same.

Calculations of the thermodynamic properties of butane show that the potential function has the shape shown in Fig. 6-3. The high barrier *A*, estimated at 4.4 to 6.1 kcal./mole, corresponds to eclipsing of the two bulky

"Staggered" "Eclipsed"

Fig. 6-1. Ethane.

terminal methyl groups. A second barrier *B*, due to eclipsing of the two methyl groups with two hydrogen atoms (plus a hydrogen-hydrogen eclipsing), is only slightly greater (ca. 3.5 kcal./mole) than the barrier in ethane. (In propane, where only one H-CH_3 interaction is involved, the barrier was found to be 3.3 kcal./mole.) The three potential valleys in butane are not of equal energy either. At an angle of rotation of 180° the methyl groups are

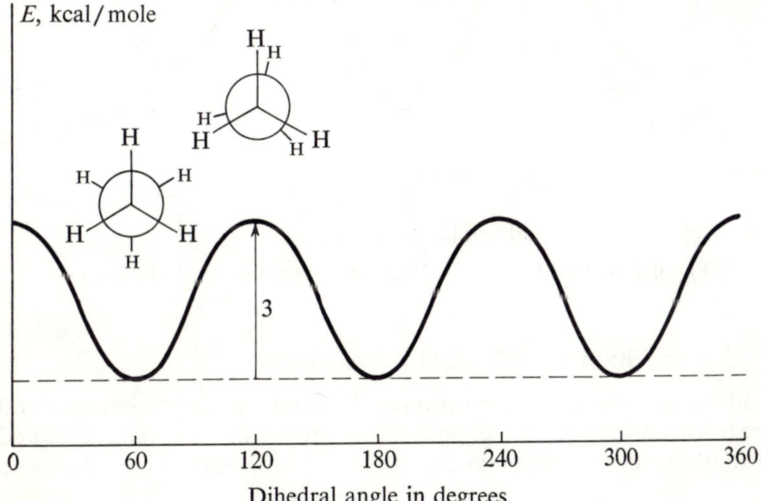

Fig. 6-2. Potential energy of ethane as function of angle of torsion (dihedral angle).

as far apart as they can be, and the system is most stable. When the methyl groups are rotated 60° with respect to each other, some van der Waals interaction occurs and the system is less stable by about 0.8 to 0.9 kcal./mole, but the conformation still corresponds to a potential energy minimum. Such conformations of minimum energy are sometimes called "conformational isomers" or "conformers." According to Fig. 6-3, butane has three conformers, two of which are mirror images of each other and are known as the "gauche" or "skew" conformations and one, more stable than either gauche conformation, which is known as the "trans" or "anti" conformation.† These conformational isomers are shown in Fig. 6-4.

The cause for the hindrance to rotation in ethane and butane, i.e., for the difference in energy between staggered and eclipsed conformations, is not yet completely understood. The interaction is not due simply to van der Waals repulsion forces, for such forces are far too small to account for the observed differences in potential energy. It is now recognized that the differences are in some way associated with interactions of the electron clouds in the C—H (or C—C) bonds, and various possible causes for such interactions have been proposed.[4, 89]

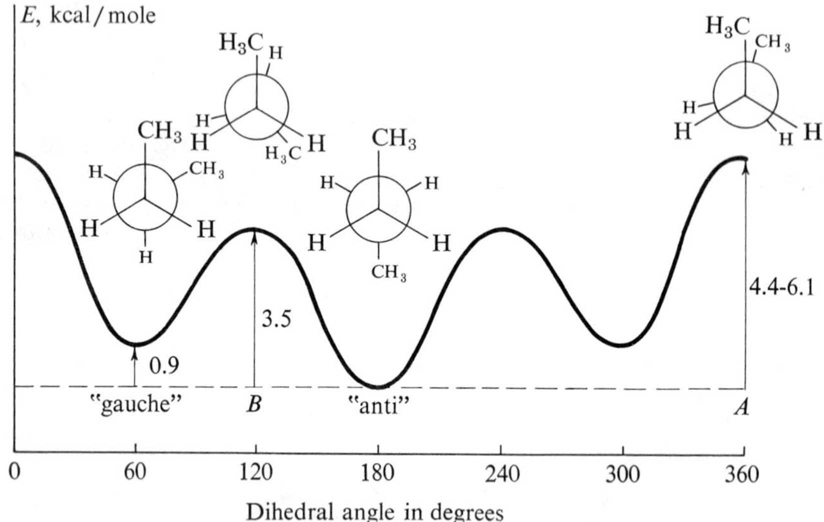

Fig. 6-3. Potential energy of butane as function of dihedral angle.

6-2. Conformation and Physical Properties[82]

In addition to the already mentioned thermodynamic properties of enthalpy and entropy and the related experimental properties of specific heat and heat of formation, several other physically measurable entities are dependent on

† The more common term "trans" is used in both the conformational and configurational sense. This leads to an unfortunate ambiguity which has given rise to much confusion. Therefore, in this text, the term "anti" will be used; cf. E. L. Eliel, *J. Chem. Educ.*, **37**, 126 (1960). For a more detailed conformational nomenclature, see W. Klyne and V. Prelog, *Experientia*, **16**, 521 (1960).

[4] E. N. Lassettre and L. B. Dean, *J. Chem. Phys.*, **17**, 317 (1949); E. B. Wilson, *Proc. Natl. Acad. Sci. U.S.A.*, **43**, 816 (1957); L. Pauling, *ibid.*, **44**, 211 (1958); H. Eyring, G. H. Stewart, and R. P. Smith, *ibid.*, **44**, 259 (1958). Cf. G. H. Stewart and H. Eyring, *J. Chem. Educ.*, **35**, 550 (1958).

conformation, and, in turn, their measurement may be used to determine conformation. Among these are spectral properties—especially infrared, Raman, microwave, and nuclear magnetic resonance spectra—measurements of dipole moments, and data derived from electron diffraction and X-ray diffraction studies.

a. Dipole Moments. Inspection of the anti and gauche forms of 1,2-dibromoethane (Fig. 6-5) shows that these conformations should differ in dipole moment. In the anti form, the C—Br bond dipoles are antiparallel and the dipole moment is zero, whereas in the gauche form it is finite. Since

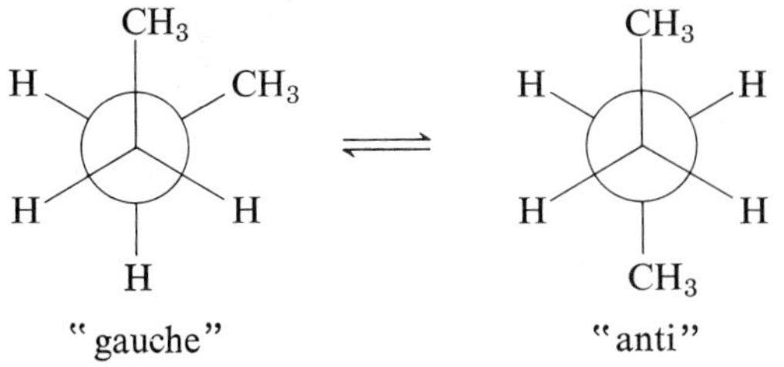

CH_3 H CH_3 H H H ⇌ CH_3 H H H H CH_3

"gauche" "anti"

Fig. 6-4. Stable conformations of *n*-butane.

the actual dipole moment of 1,2-dibromoethane[5] is approximately 1 debye unit (D), the molecule cannot exist entirely in the anti form. Moreover, the dipole moment varies with temperature; thus the molecule cannot be entirely in the gauche form either. There are two ways of explaining the experimental result: Either the anti and gauche forms shown in Fig. 6-5 are in equilibrium and the equilibrium constant varies with temperature (this would make the situation similar to that previously discussed for *n*-butane, Fig. 6-4) or the molecule exists in either form but librates slowly around the carbon-carbon bond, without, however, passing into the other form.† In any case, the temper-

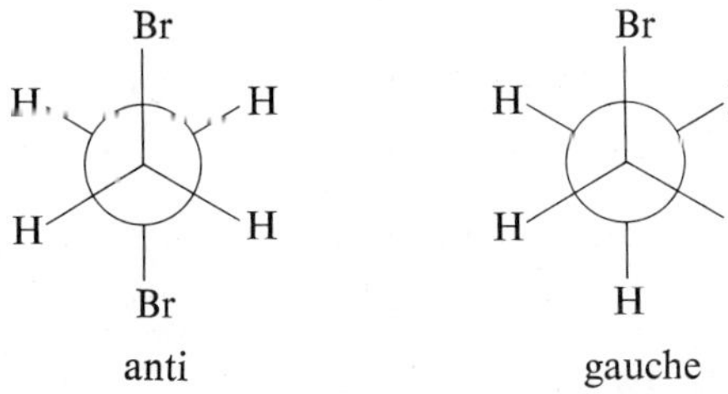

Br H H H H Br Br H Br H H H

anti gauche

Fig. 6-5. 1,2-Dibromoethane.

† By "libration" is meant a torsional oscillation (twisting to and fro) about the bond in question. Although the anti isomer in its resting form has no dipole moment, it would have a dipole moment if it librated, even though the mean position of the libration corresponds to a situation of zero dipole. This is because the over-all dipole moment is *not* the average of the instantaneous dipole moments during libration but rather the square root of the sum of their squares. As the frequency of libration may be expected to vary with temperature, so would the dipole moment vary.

[5] C. P. Smyth and S. E. Kamerling, *J. Am. Chem. Soc.*, **53**, 2988 (1931); cf. Ref. 82, p. 8.

ature variation of the dipole moment is incompatible with the assumption of free rotation. The dipole moments of the diastereoisomeric stilbene dichlorides[6] (*dl* pair and meso form, Fig. 6-6) are also incompatible with free rotation; if rotation about the central carbon-carbon bond were free, the individual bond moments should average out to the same value for the two isomers. This does not follow if the substance exists as discrete, readily interconvertible conformational isomers, as shown in Fig. 6-6. The most favored conformational isomer of the meso form is that (I) in which the bulky phenyl groups are anti to each other; in this conformation the repulsion of the carbon-chlorine dipoles is also minimized. Since this particular conformer has zero dipole moment, the over-all dipole moment of the meso form is

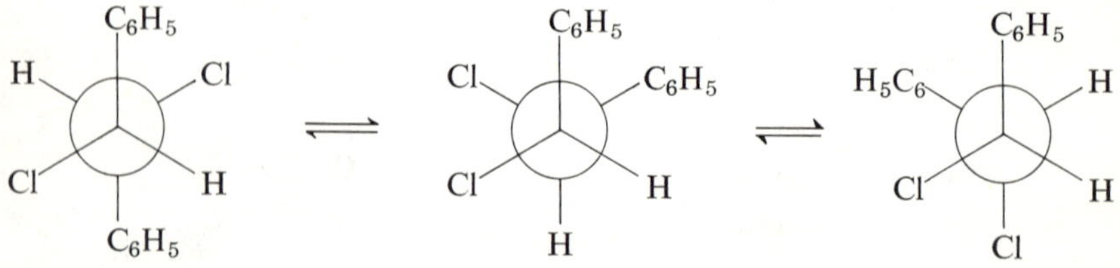

meso form, $\mu = 1.27D$

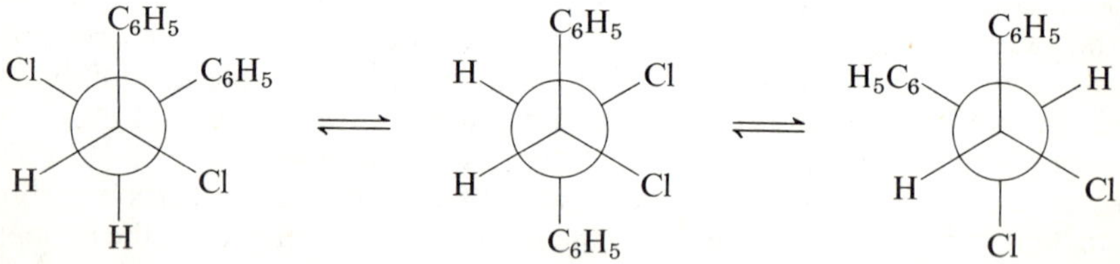

dl-pair (one enantiomer shown), $\mu = 2.75D$

Fig. 6-6. Dipole moments of the stilbene dichlorides.

relatively low. In contrast, in the D (or L) isomer, that conformational isomer (II) in which the carbon-chlorine dipoles are antiparallel (and which has no dipole moment) requires the bulky phenyl groups to be gauche and therefore contributes relatively little. The preferred conformer (III) with the bulky phenyl groups anti to each other has, of course, an appreciable dipole moment and therefore the over-all dipole moment is relatively large.†

b. Infrared and Raman Spectra. Dipole-moment evidence definitely rules out free rotation, but it does not unequivocally prove the existence of several

† Once the principles involved here are recognized, the argument may be reversed and used to decide, on the basis of dipole measurement, which isomer corresponds to the *dl* pair and which to the meso form.

[6] A. Weissberger and R. Sängewald, *Z. physik. Chem.*, **B9**, 133 (1930).

preferred conformations in equilibrium with each other. Librations about one conformation might explain the temperature dependence of dipole moments and other results presented above. More definite evidence for the existence of several conformational isomers, at least in liquid and gaseous substances, comes from infrared and Raman spectra.

The maximum possible number of normal vibrations (and therefore infrared or Raman lines) in a non-linear molecule of N atoms is $3N - 6$. However, a molecule such as 1,2-dibromoethane shows more than the allowed number of lines in the liquid and gaseous states,† although not in the solid state. All the lines found in the solid are also found in the liquid or gas, but additional lines appear in the latter states. Moreover, the spectrum of the solid material discloses that the molecule has a center of symmetry, as evidenced, for example, by the lack of coincidences of infrared and Raman lines below 1200 cm.$^{-1}$ (cf. Ref. 82, p. 28). It appears, therefore, that solid 1,2-dibromoethane is in the anti form (Fig. 6-5) but that, on liquefaction, some molecules turn over into the gauche form (Fig. 6-5) so that the liquid material shows the spectral lines corresponding to both forms.

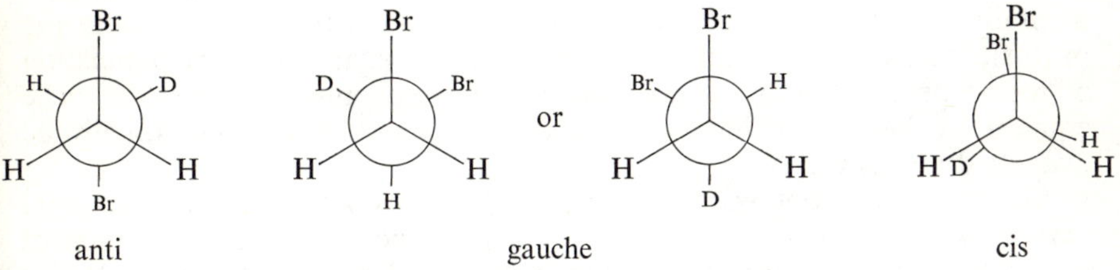

Br, H, D, H, H, Br — anti

Br, D, Br, H, H, H — or — Br, Br, H, H, H, D — gauche

Br, Br, H, D, H, H — cis

Fig. 6-7. Conformations of 1,2-dibromoethane-*d*.

A priori, it might have appeared that the form devoid of a center of symmetry is the cis (Fig. 6-7) rather than the gauche form. This possibility, however, has been disposed of by an ingenious argument,[7] based on the infrared spectrum of monodeuterated ethylene dibromide, $CH_2BrCHDBr$. The anti, as well as the cis, form of this molecule has unique conformations, shown in Fig. 6-7. (The mirror images of these conformations are also present but are, of course, spectrally identical.) The gauche form, on the other hand, presents two distinct conformations (Fig. 6-7) which might be expected to have slightly different spectra. In fact, the extra lines that are found in the spectrum of liquid (but not solid) 1,2-dibromoethane are split when the material is deuterated. Hence these lines must be due to the gauche and not the cis conformation.[8a]

In principle, the position of the equilibrium of the anti and gauche forms of 1,2-dibromoethane (Fig. 6-5) can be calculated from either the observed

† Taken by itself, the appearance of excess lines is not very significant because of the possibility of overtones and combination tones.

[7] J. T. Neu and W. D. Gwinn, *J. Chem. Phys.*, **18**, 1642 (1950).

[8a] See also S. Mizushima, I. Nakagawa, I. Ichishima, and T. Miyazawa, *J. Chem. Phys.*, **22**, 1614 (1954).

dipole moment or the intensity of the spectral lines. In practice, however, there are some difficulties. In calculations based on dipole moment, it is necessary to assign moments to the individual conformational isomers on the basis of analogy with simpler molecules (e.g., ethyl bromide may be used to calculate the bond moments for 1,2-dibromoethane). Corrections should be made for bond-moment interactions and for the fact that the molecules carry out torsional motions about their resting positions. This detracts from the accuracy of the determinations. In calculations based on infrared studies, one uses Beer's law: $\ln (I_0/I) = E \cdot N \cdot l$, where I_0 and I are the intensities of the incident and transmitted radiation, respectively, l is the thickness of the sample, N is the number of molecules per milliliter, and E is the molecular extinction coefficient. If one looks at two bands, one due to the anti, or trans, conformation and one due to the gauche conformation, one has $\ln (I_0/I^t) = E^t \cdot N^t \cdot l$ and $\ln (I_0/I^g) = E^g \cdot N^g \cdot l$, whence

$$N^t/N^g = (E^g/E^t) \ln (I_0^t/I^t)/\ln (I_0^g/I^g)$$

the superscripts t and g referring to trans and gauche, respectively. The logarithmic terms (transmissions) in these expressions can be evaluated experimentally, but the molecular extinction coefficients (corresponding to the intrinsic intensities of the anti and gauche bands, regardless of the population of the respective conformation) can be obtained only in rare cases. The equilibrium between conformational isomers can therefore not usually be evaluated in this way.[8b]

Another way of determining energy differences between conformational isomers is by a study of the temperature dependence of either dipole moment or the intensity of appropriate infrared bands. Application of van't Hoff's equation then gives the enthalpy difference, ΔH, between the conformational isomers. According to van't Hoff's equation,

$$\ln \frac{K_{T_2}}{K_{T_1}} = \frac{\Delta H}{R}\left(\frac{1}{T_1} - \frac{1}{T_2}\right)$$

where K_{T_2} and K_{T_1} are the equilibrium constants between the two isomers at absolute temperatures T_2 and T_1, respectively. But, as explained earlier, for infrared spectral data the equilibrium constant $K = N^t/2N^g$ may be put in the form $K = (E^g/2E^t) \times (D^t/D^g)$, where D^t and D^g stand for the optical density at the wavelengths of the trans and gauche absorption bands, respectively.† By taking the ratio of K's at two different temperatures, the unknown extinction coefficients (which are usually independent of temperature) cancel out, and the final equation is

$$\frac{\Delta H}{R}\left(\frac{1}{T_1} - \frac{1}{T_2}\right) = \ln\left(\frac{D^t}{D^g}\right)_{T_2} - \ln\left(\frac{D^t}{D^g}\right)_{T_1}$$

† $D = \log I_0/I$. The reason for the (statistical) factor of 2 is that the anti form is in equilibrium with *two* gauche forms (cf. Fig. 6-3).

[8b] See, however, S. Mizushima, T. Shimanouchi, K. Kuratani, and T. Miyazawa, *J. Am. Chem. Soc.*, **74**, 1378 (1952).

The energy differences between conformational isomers obtained by the methods so far discussed (thermodynamic properties, dipole moment, infrared, and Raman studies) for 1,2-dihaloethanes are as follows:[8a] dichloride, gas, 1.0 to 1.3 kcal./mole; liquid, 0.0 kcal./mole; dibromide, gas, 1.4 to 1.8 kcal./mole; liquid, 0.73 to 0.76 kcal./mole.† It is interesting that the population of the gauche conformation is greater in the liquid than in the gaseous state. Similarly, it has been found that the gauche form is more important in polar than in non-polar solvents. This is because of the fact that the gauche form has a considerable dipole moment whereas the anti form has nearly none. Solvation (either by polar solvent molecules, or, in the pure liquid phase, by other dihaloethane molecules) is known to reduce the potential energy of a dipole and therefore stabilizes the gauche form relative to the anti. The magnitude of this stabilization may be approached by calculation.[8c]

Evidence presented so far indicates that anti isomers tend to be preferred over gauche, although both are usually present to an appreciable extent at equilibrium. (An enthalpy difference of 1 kcal./mole corresponds to about 72% of the anti isomer at room temperature.‡) Two factors have so far been shown to contribute to this state of affairs, namely, steric interaction, as in butane, and dipole repulsion, as in 1,2-dichloroethane, for which the energy difference between the anti and gauche forms is larger than for *n*-butane, even though the steric interaction should, if anything, be smaller. A third factor to be considered in cases where it may play a role is hydrogen bonding.[9] This usually manifests itself in the infrared spectrum. Thus, when the spectrum of ethylene glycol is recorded in dilute carbon tetrachloride solution (thus avoiding complications due to intermolecular hydrogen bonds) using a lithium fluoride prism (affording high resolution in the short wavelength end of the infrared spectrum), two bands appear, one at 3644 $cm.^{-1}$ and one at 3612 $cm.^{-1}$. The former band corresponds to the O—H stretching frequency due to unbonded hydroxyl seen in dilute solutions of monohydric alcohols, whereas the latter has been assigned to intramolecularly bonded hydroxyl stretching, —O—H ··· O. Now ethylene glycol may exist in two distinct stable conformations (Fig. 6-8), the anti conformation and the gauche conformation (of which there are two enantiomers). Only in the latter are the hydroxyl groups close enough together to give rise to an intramolecular bond. The fact that absorption due to this bond is strong in the spectrum of ethylene glycol means that a considerable fraction of the molecules are in the gauche form, despite the steric and dipolar repulsion of the hydroxyl groups. Evidently this repulsion is more than outweighed by the energy gained in the formation of the hydrogen bond—a conclusion which

† Values of 1.39 and 2.42 kcal./mole have also been recorded.

‡ The statistical factor of two favoring the gauche form is taken into consideration in the 72% figure.

[8c] I. Watanabe, S. Mizushima, and Y. Masiko, *Sci. Papers Inst. Phys. Chem. Research* (*Tokyo*), **40**, 425 (1943).

[9] (*a*) S. Mizushima, T. Shimanouchi, T. Miyazawa, K. Abe, and M. Yasumi, *J. Chem. Phys.*, **19**, 1477 (1951); (*b*) L. P. Kuhn, *J. Am. Chem. Soc.*, **74**, 2492 (1952); (*c*) *ibid.*, **76**, 4323 (1954); (*d*) *ibid.*, **80**, 5950 (1958); (*e*) M. Kuhn, W. Lüttke, and R. Mecke, *Z. anal. Chem.*, **107**, 106 (1959).

does not surprise one, since hydrogen bonds may be as strong as 5 kcal./mole.

From the difference in absorption frequency of the unbonded and bonded hydroxyl groups, one may estimate the strength of the hydrogen bond: the larger the difference, the stronger the bond. Table 6-1 shows this difference

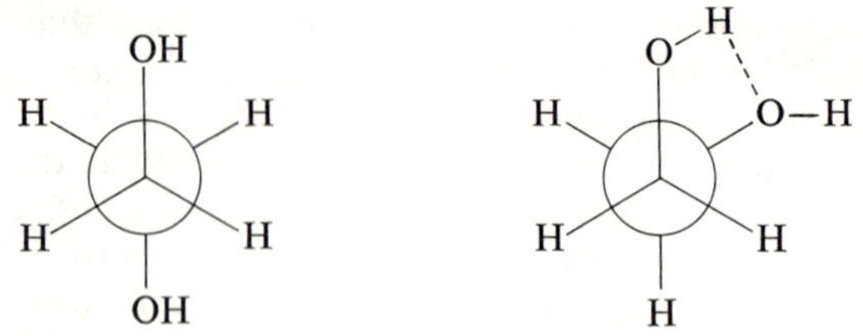

anti (unbonded) gauche (bonded)

Fig. 6-8. Conformational isomers of ethylene glycol.

in the case of a series of glycols of the general formula RCHOHCHOHR for both the meso and the *dl* isomer. The conformations of the two types of diastereoisomeric glycols are shown in Fig. 6-9. The following conclusions may be reached:

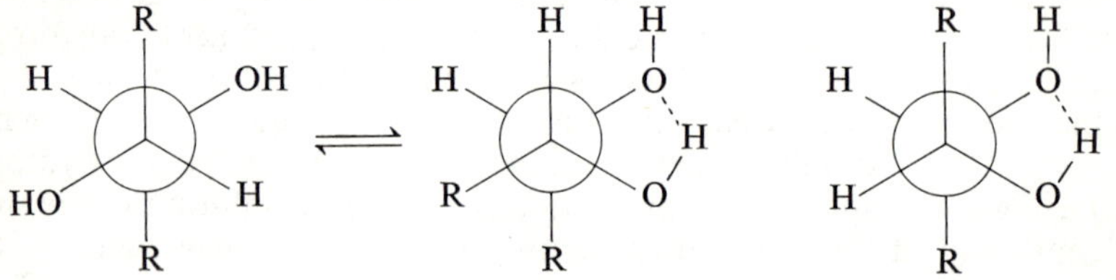

anti gauche *dl*-form, R groups anti

meso form (one enantiomer shown)

Fig. 6-9. 1,2-Dialkylethylene glycols.

Table 6-1

Difference (in cm.$^{-1}$) between Stretching Frequencies Due to Unbonded and Intramolecularly Bonded Hydroxyl in RCHOHCHOHR[a]

R	H	CH_3	n-C_5H_{11}	$(CH_3)_2CH$	$(CH_3)_3C$
(meso)	32	42	43	55[b]	—[c]
(*dl*)	32	49	53	81	94

[a] Data from Ref. 9*d*.
[b] Weak band for bonded species.
[c] No band observed for bonded species.

1. When R is methyl or *n*-amyl, both the meso and the *dl* forms exist to a considerable extent in the conformation in which the hydroxyl groups are gauche to each other, even though, in the meso form, this requires the alkyl groups to be gauche also.

2. When R is isopropyl, and even more when R is *t*-butyl, the conformational isomer of the meso form in which the bulky R groups are gauche contributes little or nothing, and therefore the infrared band due to bonded hydroxyl is weak or absent.

3. As the R groups become bulkier, the hydrogen bond becomes stronger,† presumably because the hydroxyl groups (one in the case of the meso form, both in the *dl* form) bend away from the bulky gauche alkyl group on the adjacent carbon and thus toward each other[9d] (Fig. 6-10).‡

Compounds in which the gauche conformation appears to be more stable in the liquid state than the anti even though hydrogen bonding plays no part are 1,1,2,2-tetrabromoethane and the corresponding tetrachloro compound.[10]

c. Microwave Spectra.[88] Microwave spectral data have been used mainly to arrive at estimates of the energy barriers (potential maxima, correspond-

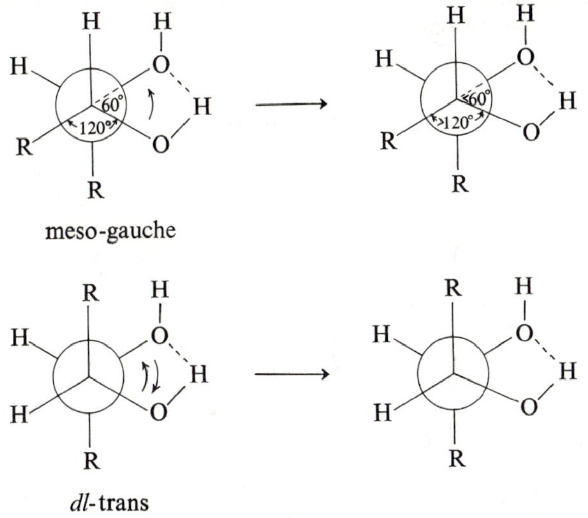

Fig. 6-10. Deformation of RCHOHCHOHR with bulky R group.

ing to eclipsed conformations, cf. Fig. 6-2) between different conformations. Height of the barrier may be deduced from either the frequency or, less easily, the intensity of the rotational absorbtion bands. Some barriers deduced by microwave and other measurements are shown in Table 6-2.

† One must distinguish between the *strength* of the hydrogen bond and the *population* of the hydrogen-bonded conformation. In the meso series, as the size of the alkyl groups is increased, the hydrogen bond in the gauche conformation becomes stronger but the fraction of molecules in the gauche conformation becomes less.

‡ The reader should have clearly in mind the difference between configuration and conformation. The *dl* form and meso form differ in *configuration,* i.e., in the arrangement of the groups about the asymmetric carbon atoms (best seen in Fischer projection formulas). The gauche and anti isomers of the meso configuration differ in *conformation,* i.e., in the rotational arrangement around the carbon-carbon bond. Conformational isomers are readily interconverted at room temperature by rotation; configurational isomers are not.

[10] R. E. Kagarise, *J. Chem. Phys.*, **24,** 300 (1956).

The near constancy of the barrier in such series as CH_3CX_3, CH_3SiX_3, and CH_3COX, where X may be H, F, Cl, or Br, indicates that electrostatic or steric effects of the conventional kind do not seem to affect barrier height appreciably. This supports the statement made earlier that van der Waals forces are not responsible for the barriers. There is, however, a systematic change of barrier height with the symmetry properties of the bond under consideration. Molecules with an sp^2 hybridized carbon (planar) attached to a methyl group have lower barriers than those with an sp^3 (tetrahedral) carbon: compare ethane with acetaldehyde. In cases where there is a sixfold symmetry axis along the pivotal bond (e.g., nitromethane) the barrier is smaller by several orders of magnitude than when the symmetry axis is threefold. It is of interest that the difference in barrier between methanol and methyl mercaptan on one hand and dimethyl ether and dimethyl sulfide on the other is quite large whereas (as mentioned earlier) there is little difference between ethane and propane.

Table 6-2
Potential Barriers
(kcal./mole)[a]

CH_3-CH_3	2.75	CH_3-OH	1.07	$CH_3-CH{=}CH_2$	1.98
CH_3-CH_2F	3.30	CH_3-OCH_3	2.72	$CH_3-CH{=}C{=}CH_2$	1.59
CH_3-CHF_2	3.18	CH_3-SH	1.26	$CH_3-CH{=}CHF$	2.20
		CH_3-SCH_3	2.13		
CH_3-CH_2Cl	3.56	CH_3-NH_2	1.94	$CH_3-CF{=}CH_2$	2.62
CH_3-CH_2Br	3.57	CH_3-CHO	1.15	CH_3CH-CH_2 O (bridging)	2.56
CH_3-SiH_3	1.70	CH_3-COF	1.08		
CH_3-SiH_2F	1.56	CH_3-COCl	1.35	$H_3SiCH{=}CH_2$	1.50
CH_3-SiHF_2	1.28	CH_3-COCH_3	0.78	CH_3-NO_2	0.006
$CH_3-SiH_2CH_3$	1.65	CH_3-CO_2H	0.48	CH_3-BF_2	0.014
		$CH_3-CO_2CH_3$	1.17		

[a] From Refs. 11 and 88.

d. Nuclear Magnetic Resonance Spectra.[90] If a molecule exists in two interconvertible conformations, both approximately equally populated, it may show, depending on the frequency of interconversion, either the nuclear magnetic resonance spectra corresponding to the two individual conformations or an average spectrum. If one has two equally populated conformations A and B in equilibrium in a substance and one heats the substance until a given pair of resonance lines due to A and B just coalesces (or if, originally, there was only one set of lines, if one cools the substance until the resonance lines just begin to split), then the rate of interconversion of A and B at the temperature of incipient coalescence of splitting will be $k = \frac{1}{2}\sqrt{2}\,\pi\,\delta_{AB}$, where δ_{AB} is the chemical shift or distance of the lines under observation in cycles per second at a temperature low enough that the lines are as far resolved as possible. Whenever, at a given temperature, the rate of interconversion of

[11] E. B. Wilson, *Proc. Natl. Acad. Sci. U.S.A.*, **43**, 816 (1957); L. Pierce, *J. Chem. Phys.*, **34**, 498 (1961); J. M. O'Reilly and L. Pierce, *ibid.*, **34**, 1176 (1961).

two equally populated conformations is less than the above value of k, two distinct sets of resonance lines are found corresponding to the two conformations; but if the rate of interconversion is greater than k, then only one set of lines is seen at an intermediate position.

An example is afforded by the behavior of 1,1-difluoro-1,2-dibromo-2,2-dichloroethane (Fig. 6-11).[12] At room temperature or slightly below, this compound shows only one sharp fluorine resonance line, indicating that the two fluorine atoms are in identical environments. This means either that the molecule is all in form I (for in form II, or its enantiomer III, the two fluorines should be distinct) or else that the molecule rotates rapidly enough about the carbon-carbon bond at room temperature to make the fluorine resonance lines coalesce. That the latter explanation is the correct one has been shown by observing the resonance spectrum at low temperatures. At −30°, the spectral line begins to broaden, at −60° it is split into three lines, and at −80° *five* sharp lines may be clearly discerned. Four of these are relatively weak and have been assigned to conformations II and III (which are spectrally and energetically equivalent) whereas the fifth, strong line is due to conformation I. By measurement of the area of the lines, it may be

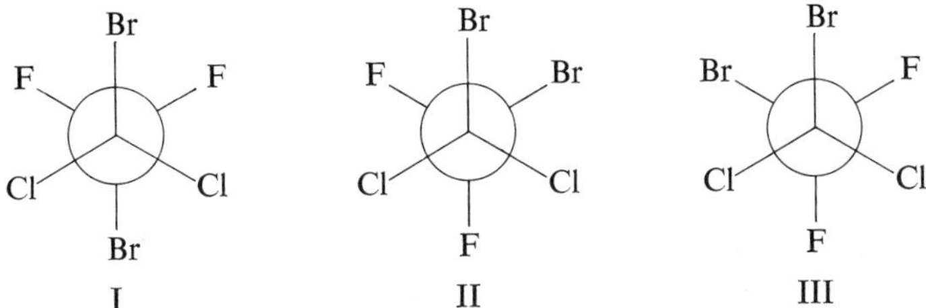

Fig. 6-11. Conformations of 1,1-difluoro-1,2-dibromo-2,2-dichloroethane.

concluded that at −80° the molecule exists preferentially (but by no means exclusively) in conformation I (in which the most bulky atoms, namely, bromine, are anti to each other). Also it may be concluded, using the formula given above and the magnitude of the chemical shifts involved, that at −80° the rate of interconversion of the isomers is much less than 200 times per second whereas at 0°C. it is much greater than 200 times per second.

In a discussion of NMR spectra as related to the stereochemistry of acyclic systems, consideration of compounds of type CabcCXXY is also of interest. In these compounds two apparently identical atoms or groups X are attached to a carbon next to one which is asymmetric, as for example,[12] the methylene hydrogens in methyl 2,3-dibromo-2-methylpropionate (Fig. 6-12). The noteworthy feature of the spectra of such compounds is that the apparently similar groups X are in fact different and will give rise to different chemical shifts. Thus, for example, the left proton in the leftmost conformation in Fig. 6-12 and the right proton in the rightmost conformation are both

[12] P. M. Nair and J. D. Roberts, *J. Am. Chem. Soc.*, **79**, 4565 (1957); cf. J. D. Roberts, "Nuclear Magnetic Resonance," McGraw-Hill Book Company, Inc., New York, 1959, pp. 58, 71.

staggered between Br and CH_3; nevertheless, because of differences in the rest of the two structures, these protons are not equivalent. This conclusion is independent of whether rotation is rapid and whether or not residence time in the three conformations shown is or is not the same.

e. X-ray and Electron Diffraction. Electron diffraction patterns of molecules containing relatively heavy atoms, such as 1,2-dibromoethane, in the vapor phase should give an indication of the position of such heavy atoms and therefore of the conformation or conformations of the molecule. The method has been explored especially by Hassel and his school and by Mizushima and coworkers. Results indicate that the 1,2-dihaloethanes exist preferentially in the anti conformation in the vapor phase, but, unfortunately, give no precise information as to how large a fraction, if any, of the molecules exists in the gauche conformation. Electron diffraction measurements on *n*-butane, using the sector-microphotometer method, have, however, served to indicate that 60 ± 15% of the molecules exist in the anti conformation at 14°C.[12a]

X-ray diffraction has been used to indicate conformation of molecules in the solid state. Once a substance crystallizes, interconversion of various con-

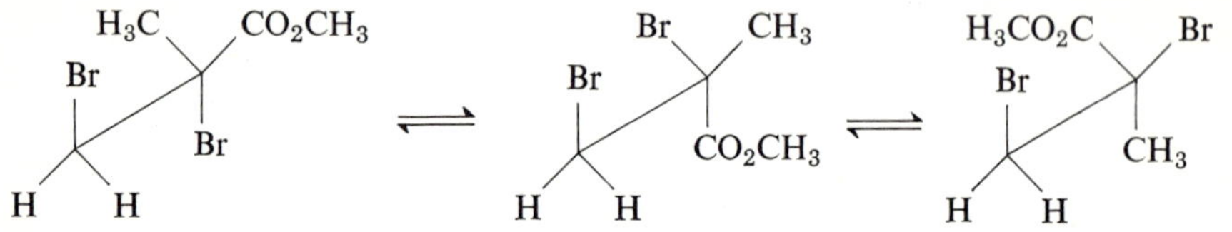

H₃C CO₂CH₃ Br Br H H ⇌ Br CH₃ Br CO₂CH₃ H H ⇌ H₃CO₂C Br Br CH₃ H H

Fig. 6-12. Staggered rotational conformations of methyl 2,3-dibromo-2-methylpropionate. (*From J. D. Roberts, "Nuclear Magnetic Resonance," copyright,* 1959, *McGraw-Hill Book Company, Inc.*)

formations is stopped, and the crystals correspond to one discrete conformation or another. Thus the X-ray diffraction pattern[13] of 1,2-dichloroethane indicates that the material crystallizes exclusively in the anti form.† Where two or more configurations of a molecule exist, it is, in principle, possible that the material should crystallize in several of these, and, in fact, molecules of this type do sometimes show polymorphism. In at least one case—1,1,2,2-tetrabromoethane—it has been shown[10] that the molecule can crystallize in both the gauche and the anti forms.

The preferred conformations of the 2,3-dibromobutanes (Fig. 6-13, R = CH_3) and the stilbene dibromides (Fig. 6-13, R = C_6H_5) have been determined by electron diffraction[14] and X-ray diffraction,[15] respectively. In the

† It is interesting that ethylene chlorohydrin, in contrast, crystallizes in the gauche form, presumably because of intramolecular hydrogen bonding. (This conclusion was reached by infrared and Raman studies, not from X-ray diffraction.)

[12a] R. A. Bonham and L. S. Bartell, *J. Am. Chem. Soc.*, **81**, 3491 (1959).
[13] M. E. Milberg and W. N. Lipscomb, *Acta Cryst.*, **4**, 369 (1951).
[14] D. P. Stevenson and V. Schomaker, *J. Am. Chem. Soc.*, **61**, 3173 (1939).
[15] J. D. McCullough, *J. Am. Chem. Soc.*, **62**, 480 (1940).

case of the stilbene dibromides ($R = C_6H_5$), the meso form crystallizes in conformation I where both the phenyl groups and the bromine atoms are anti (bromine-bromine distance, 4.50 A.) so that both steric and dipolar repulsions are at a minimum; the *dl* form, however, crystallizes in conformation IV in which the bromines are approximately gauche (bromine-bromine distance, 3.85 A., equal to the distance of the combined van der Waals radii) and in which the steric repulsion of the bulky phenyl groups is at a minimum, even though the steric and dipolar repulsions of the bromine atoms are not. In contrast, both meso and *dl*-2,3-dibromobutane ($R = CH_3$) exist in the conformations (I, VI) with the bromine atoms anti (bromine-bromine distance, 4.60 A.). In this case the steric and dipolar repulsions of the bromine atoms determine the stable conformation, even though, in the case of the *dl* isomer, this means that the methyl groups (less bulky than phenyl) are forced to be gauche.

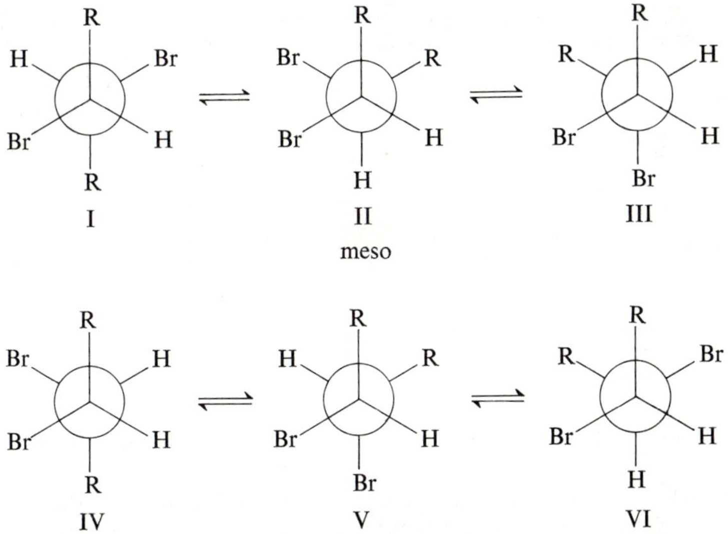

I II III

meso

IV V VI

dl (one enantiomer only shown)

Fig. 6-13. Conformations of meso and *dl* forms of RCHBrCHBrR.

6-3. Conformational Effects on Stability and Reactivity

In assessing the effect of conformation on physical properties (Sec. 6-2), a comparison of the properties of diastereoisomers was frequently used. Similarly, effects of conformational factors on chemical stability and reactivity, especially in acyclic molecules, manifest themselves most often as differences in behavior between diastereoisomers. In the following discussion, relative stability of stereoisomers will be taken up first, since it involves differences in the ground states of the molecules only. Differences in reactivity will then be considered; they are more subtle because they depend on both the differences in ground-state energies and differences in energy of the appropriate transition states. Fortunately, the ground-state differences are usually small compared with the transition-state differences, and attention may usually be focused on the latter when one discusses the relative reactivity of diastereoisomers.

a. Differences in Stability of Diastereoisomers. The free energies of acyclic diastereoisomers usually differ, although not generally by very much. In general, meso forms are more stable than *dl* pairs† and, more tenuously, erythro isomers are apt to be more stable than threo isomers. This may be understood by reference to Fig. 6-14, in which a meso isomer and its active diastereoisomer are depicted in their most stable conformations.‡ L denotes the largest substituent in the two isomers, M the medium-sized substituent, and S the small substituent. The gauche interactions in the meso form are 2L-M + 2L-S + 2M-S, whereas in the active form they are 2L-M + 2L-S + M-M + S-S. The difference between the two forms is clearly 2M-S − (M-M + S-S). Now it is a general principle that the crossed steric interactions between two groups of unequal size are less than the sum of the interactions between the groups of like size,¶ i.e., (M-M + S-S) > 2M-S, *provided*

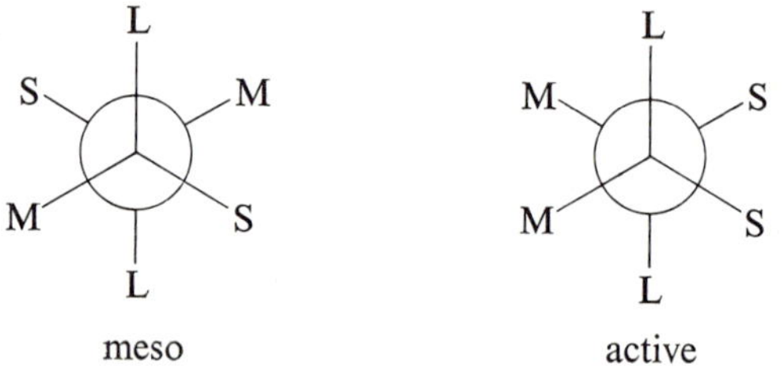

meso active

Fig. 6-14. Most stable conformation of meso and active isomers.

that the interactions are purely steric in origin. With this qualification, then, it follows that the meso isomer is more stable than the active isomer. Moreover, to the extent that Fig. 6-14 may be imagined to represent an erythro form on the left and a threo form on the right (replacing one of the M groups by some similar group M′ of like relative steric requirement), the erythro isomer is apt to be more stable than the threo.

Exceptions are likely to occur when there are specific interactions between groups, such as hydrogen bonding. For example, in 2,3-butanediol,

† Strictly speaking, the meso isomer should be compared with *one* of the active forms. The *dl* pair is always more stable than the active form (except, possibly, in the solid state) by virtue of the entropy of mixing; cf. Sec. 4-2.

‡ In general, the most stable conformations of two compounds determine their relative stability. It is possible to take into account the remaining, less stable, conformations if one so desires; this matter will be considered in Chap. 8.

¶ Thus it is, of course, well known that the best way of packing a box with spheres of two sizes is to alternate the larger and smaller spheres throughout. There is some question whether the analogy of substituents with spheres holds for all types of substituents, but it probably does hold for most. A case in point is butane where the anti isomer (Fig. 6-4) with four H-CH_3 and two H-H interactions is more stable than the gauche with one CH_3-CH_3, two H-CH_3, and three H-H interactions.

$CH_3CHOHCHOHCH_3$, the large group L (Fig. 6-14) is methyl; the medium-sized group M, hydroxyl; and the small group S, hydrogen. But in this case there is a specific attractive interaction of intramolecular hydrogen bonding between the medium-sized (hydroxyl) groups which is apt to be more important than their steric repulsion. This favors conformations in which the two hydroxyl groups are skew. Reference to Fig. 6-9 indicates that in the active isomer the hydroxyl (M) groups are skew as the methyl (L) groups are anti, but in the meso form this is not the case. Therefore in this instance the active form is expected to be the more stable.

Regrettably little experimental information is available on the relative stability of diastereoisomers. In the case of the stilbene dichlorides, $C_6H_5CHClCHClC_6H_5$, it appears that the equilibrium favors the meso isomer as predicted, although no accurate product analyses are available.[16] In the case of the corresponding dibromides, the meso isomer is more stable than the *dl* by at least 1.4 kcal./mole.[17] In α,α'-dimethylsuccinic acid, $HO_2CCH(CH_3)CH(CH_3)CO_2H$, also, the meso form predominates† at equilibrium over the racemic form.[17a] The situation is different with tartaric acid which, upon heating with aqueous sodium hydroxide and acidification, gives a mixture of two parts of the *dl* and one part of the meso form.[18] This again is in agreement with prediction (case of intramolecular hydrogen bond), even though the experimental result is only tentative, the over-all recovery of products being only about 50%.‡

b. The Relative Reactivity of Diastereoisomers. In considering transition states involving reactions of diastereoisomers, two factors need to be taken into account, namely, the conformational requirements of the groups involved in the reaction and the conformational requirements of the non-reactive groups. The former factor is sometimes called the "stereoelectronic" factor and the latter the steric factor. These factors may be illustrated with an example from Chap. 5, namely, the Walden inversion or S_N2 reaction. The stereoelectronic factor here requires that the incoming group approach from the opposite side of the leaving group so that it may engage the rear of the bond orbital previously engaged in the bonding of the leaving group (Fig. 6-15). It is clear that this factor determines stereochemistry (inversion), and it is also clear whence it has its name (it is concerned with the position in space of the electrons involved in bonding in the transition state). It may also, occasionally, be concerned with rate in the sense that a compound com-

† However, with the anhydride or imide, the *dl* isomer predominates at equilibrium. This is because, in order to form the anhydride or imide ring, the carboxyl groups must be eclipsed and, in the meso isomer, this leads also to eclipsing of the methyl groups, an unfavorable situation. In the *dl* anhydride or *dl* imide, on the other hand, the methyl groups are at a 120° angle.

‡ It is perhaps a little surprising that considerations of *intra*molecular hydrogen bonding correctly predict the position of equilibrium in a hydroxylic solvent, such as water, in which *inter*molecular bonding to solvent might have been expected to be strong enough to swamp any effects of stabilization due to intramolecular interaction.

[16] Th. Zincke, *Ber.*, **10**, 999 (1877).

[17] R. E. Buckles, W. E. Steinmetz, and N. G. Wheeler, *J. Am. Chem. Soc.*, **72**, 2496 (1950).

[17a] R. P. Linstead and M. Whalley, *J. Chem. Soc.*, 3722 (1954).

[18] J. Coops and P. E. Verkade, *Rec. trav. chim.*, **44**, 987 (1925).

pletely incapable of fulfilling the stereoelectronic requirements of a reaction cannot undergo this reaction, an example for the S_N2 reaction being apocamphyl chloride (Fig. 6-15; see also Chap. 13). However, the stereoelectronic requirement by itself would not be able to lead to the prediction that methyl bromide (Fig. 6-15) undergoes displacement by iodide much more rapidly than *t*-butyl bromide.[19] This fact may, however, be explained by the steric factor: In methyl bromide the hydrogen atoms attached to the methyl group are sufficiently far away from the incoming iodine and outgoing bromine in the transition state to offer little or no resistance to the reaction, but in

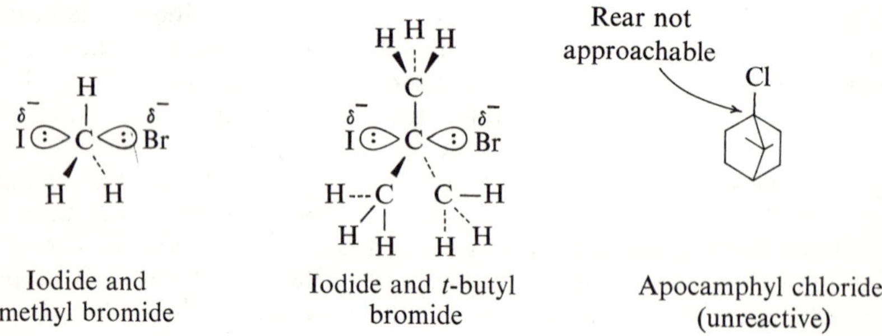

Iodide and methyl bromide — Iodide and *t*-butyl bromide — Apocamphyl chloride (unreactive)

Fig. 6-15. Stereoelectronic and steric aspects of S_N2 reaction.

t-butyl bromide, the much bulkier methyl groups attached to the central carbon atom do produce considerable interference.

i. Ionic Elimination. In ionic elimination reactions, especially bimolecular ones, the stereoelectronic requirement is that the groups to be eliminated be conformationally anti and not gauche or eclipsed. The reason for this may be that the bond orbitals of the two leaving groups must be in one and the same plane (which later becomes the plane of the pi bond of the resulting olefin), with the electrons of the orbital left behind being located at the rear

Fig. 6-16. Stereoelectronic requirement of ionic elimination.

of those departing with the anion (Fig. 6-16). Of the numerous available examples, only two will be cited, namely, the iodide-induced elimination of bromine from 2,3-dibromobutane[20] (Fig. 6-17) and the base-induced dehydrohalogenation of 1-bromo-1,2-diphenylpropane[21] (Fig. 6-18). In both instances one diastereoisomer of the starting material gives the trans olefin whereas the

[19] L. Fowden, E. D. Hughes, and C. K. Ingold, *J. Chem. Soc.*, 3187 (1955).
[20] S. Winstein, D. Pressman, and W. G. Young, *J. Am. Chem. Soc.*, **61**, 1645 (1939).
[21] D. J. Cram, F. D. Greene, and C. H. Depuy, *J. Am. Chem. Soc.*, **78**, 790 (1956).

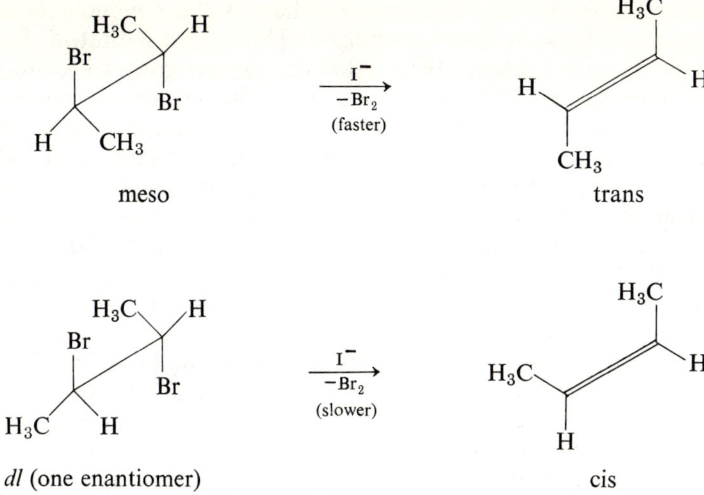

Fig. 6-17. Debromination of *meso*- and *dl*-2,3-dibromobutane.

other gives the cis olefin, despite the fact that the latter is appreciably less stable.

The fact that *cis*-α-methylstilbene is formed from *erythro*-1-bromo-1,2-diphenylpropane more slowly than *trans*-α-methylstilbene is formed from the threo isomer is an indication of the steric effect, since the transition state in which the two bulky phenyl groups are cis is the less stable. (The ground states also differ in stability, but the effect is minor and may be neglected in assessing the relative activation energies.) A number of instances of similar steric effects in both equilibrium and rate-controlled processes are known.

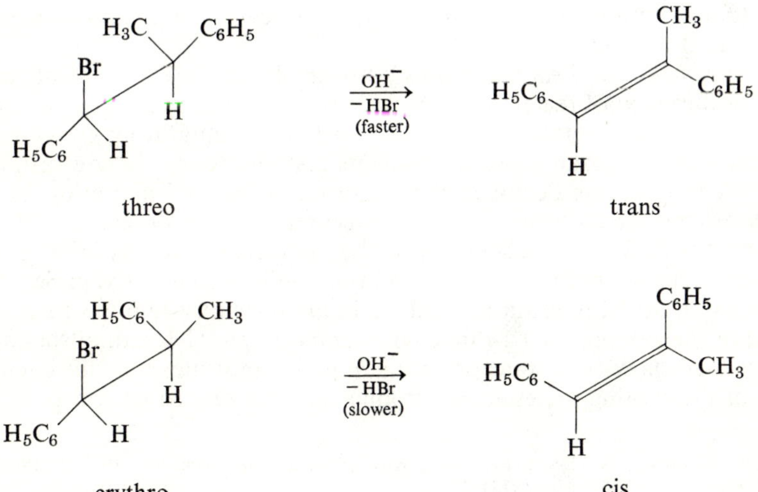

Fig. 6-18. Dehydrobromination of *threo*- and *erythro*-1-bromo-1,2-diphenylpropane.

Thus, for example, (−)-2,3-butanediol enhances the conductivity of boric acid more than does the meso isomer.[22] The reason is that in the borate complex of the active isomer (Fig. 6-19) the methyl groups are anti, but in the borate complex of the meso isomer, the methyl groups are gauche. (The same is true in the starting materials, if they are intramolecularly hydrogen-bonded. One must conclude, from the conductivity data, that either the effect in the borate complexes is more important, or, because of intermolecular bonding to the solvent, water, the meso glycol is more stable than the active isomer. In the former case the free-energy levels would be as in Fig. 6-20*a;* in the latter case, as in Fig. 6-20*b;* in either case the free-energy change for conversion of the active diol to the borate complex is more favorable than for the meso isomer.)

Similar considerations as in Fig. 6-20 apply to a number of rate-controlled processes, except that for ΔF one substitutes the activation energy $\Delta F^{\ddagger}$. The process involving the sterically less favored transition state is slower (Fig. 6-17; rate of formation of *cis*-2-butene is one-half that of formation of the trans isomer), may fail altogether under given conditions (Fig. 6-21),[23] or may

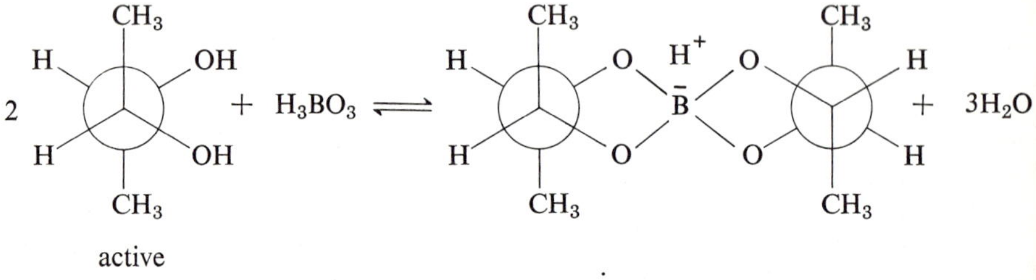

CH3 H OH 2 + H_3BO_3 ⇌ H O H^+ O H B + $3H_2O$ H OH H O O H CH3 CH3 CH3

active

Fig. 6-19. Borate complexes of the 2,3-butanediols.

yield to an alternative process involving a sterically less encumbered transition state (Fig. 6-22).[24]

ii. Intramolecular Rearrangements. Intramolecular rearrangements constitute another type of reaction in which conformational effects play an important role. It was pointed out in Sec. 5-4 that the migrating *group* in most intramolecular rearrangements retains its configuration. In contrast, inversion of configuration occurs at the migration *origin* and migration *terminus* whenever the stereochemistry at these centers can be observed. We shall here deal with the stereochemistry at the migration terminus (cf. Fig. 6-23). The stereochemistry at the migration origin will be considered in Sec. 10-3*c*.

Inversion at the migration terminus in an acyclic system has been established in the reaction of 1,1-diphenyl-2-aminopropanol-1 with nitrous acid.[25] Figure 6-24 shows this transformation and also indicates how the configurations of the starting material and product were correlated (cf. Chap. 5).

[22] J. W. Knowlton, N. C. Schieltz, and D. MacMillan, *J. Am. Chem. Soc.,* **68,** 208 (1946).
[23] P. Pfeiffer, *Ber.,* **45,** 1816 (1912).
[24] E. P. Kohler, W. D. Peterson, and C. L. Bickel, *J. Am. Chem. Soc.,* **56,** 2000 (1934).
[25] H. I. Bernstein and F. C. Whitmore, *J. Am. Chem. Soc.,* **61,** 1324 (1939).

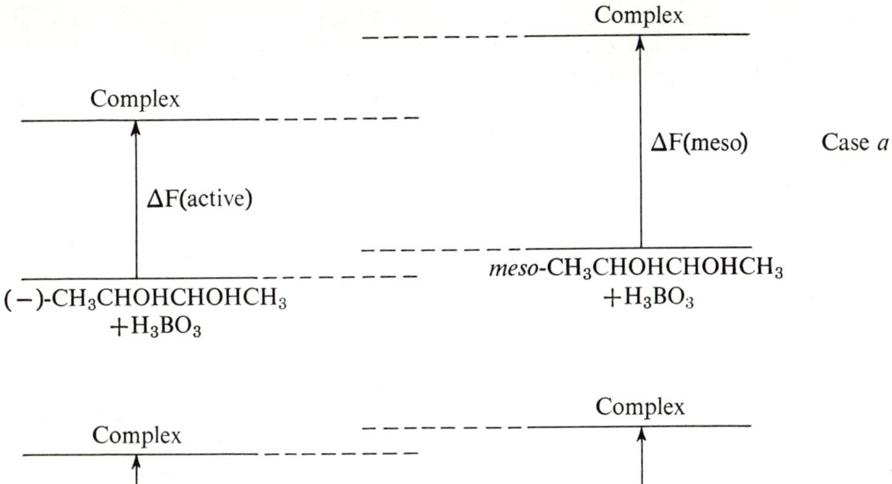

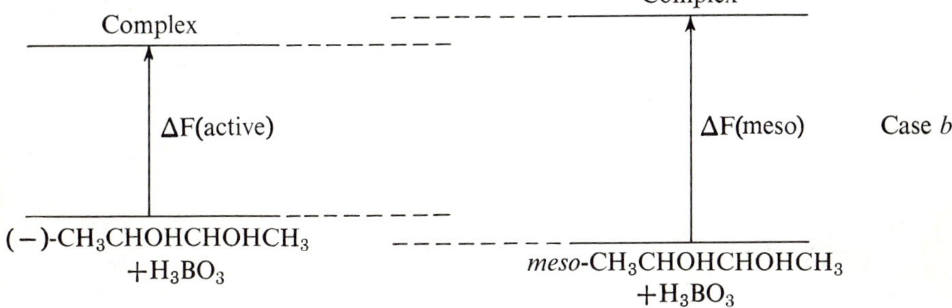

Fig. 6-20. Two possibilities for free-energy diagram of conversion of the 2,3-butanediols to their borate complexes.

H_5C_6, Cl, Cl, H, H, C_6H_5 $+ C_5H_5N \xrightarrow{200^\circ}$ H_5C_6, H, Cl, C_6H_5 $+ C_5H_5\overset{+}{N}H\ Cl^-$

dl (one enantiomer shown)

Cl, C_6H_5, Cl, H, H, C_6H_5 $+ C_5H_5N \xrightarrow{200^\circ}$ No reaction

meso

Fig. 6-21. Dehydrochlorination of the stilbene dichlorides.

In amino alcohols containing two different aryl groups, such as ArAr'COHCH(NH_2)CH_3, because of rotation about the central carbon-carbon bond, either aryl group may in principle be placed in the position favorable to migration, namely, at the rear of the carbon atom from which the

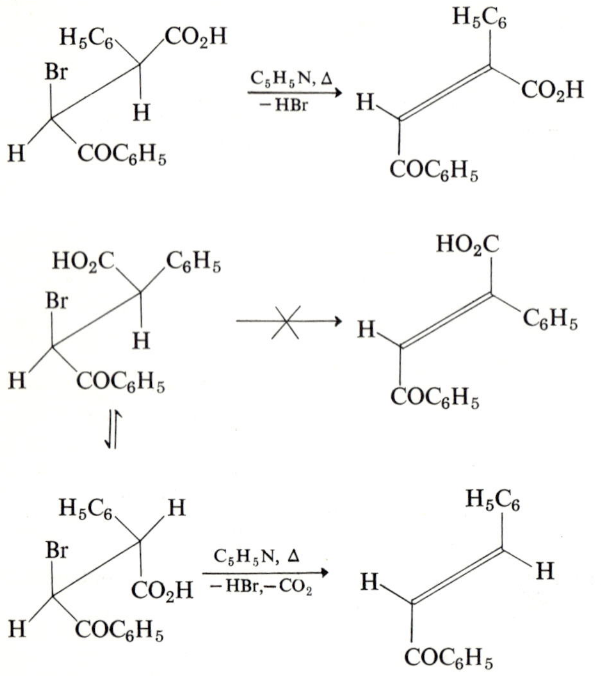
Dehydrohalogenation
Steric hindrance in product and transition state due to "cis effect"
Decarboxylative debromination

Fig. 6-22. Different course of reaction of the diastereoisomeric α-phenyl-β-bromo-β-benzoylpropionic acids with hot pyridine.

diazotized amino group departs. It is clear from Fig. 6-25, however, that, in the diastereoisomer shown, the conformation in which Ar (rather than Ar') is placed anti to the departing group (NH_2 or, more properly, N_2^+) is preferred, because in this conformation the two largest groups at C_1, Ar and Ar',

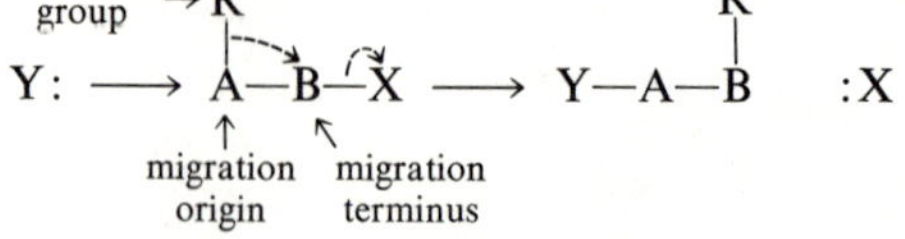
migrating group
migration origin
migration terminus

Fig. 6-23. Saturated rearrangements.

are placed gauche to the smallest group H at C_2. The other conformation shown (which would lead to the migration of Ar') is less favored because both Ar and Ar' are gauche to CH_3. It is therefore not surprising that in a number of such cases where Ar (or Ar') is phenyl and Ar' (or Ar) is α-naphthyl, *p*-anisyl, *p*-tolyl, or *p*-chlorophenyl the group Ar migrates in preference to Ar′ in the diastereoisomer shown.[83]

The case depicted in Fig. 6-25 is concerned with a migration which is

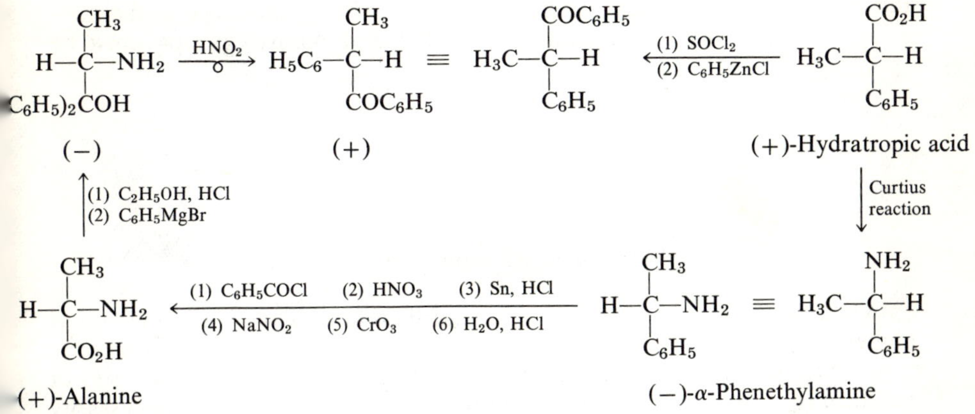

CH_3 / H—C—NH_2 / $(C_6H_5)_2COH$ (−) $\xrightarrow{HNO_2}$ CH_3 / H_5C_6—C—H / COC_6H_5 (+) ≡ COC_6H_5 / H_3C—C—H / C_6H_5 $\xleftarrow{(1)\ SOCl_2\quad (2)\ C_6H_5ZnCl}$ CO_2H / H_3C—C—H / C_6H_5 (+)-Hydratropic acid

(−) ↑ (1) C_2H_5OH, HCl (2) C_6H_5MgBr

(+)-Hydratropic acid ↓ Curtius reaction

CH_3 / H—C—NH_2 / CO_2H (+)-Alanine $\xleftarrow{(1)\ C_6H_5COCl\quad (2)\ HNO_3\quad (3)\ Sn,\ HCl\quad (4)\ NaNO_2\quad (5)\ CrO_3\quad (6)\ H_2O,\ HCl}$ CH_3 / H—C—NH_2 / C_6H_5 ≡ NH_2 / H_3C—C—H / C_6H_5 (−)-α-Phenethylamine

Fig. 6-24. Stereochemistry of reaction of 1,1-diphenyl-2-aminopropanol-1 with nitrous acid.

stereochemically anti. There are, however, also migrations for which the preferred stereochemistry is either cis or at least gauche; an example is the N-O migration in amino alcohols for which the transition state is depicted

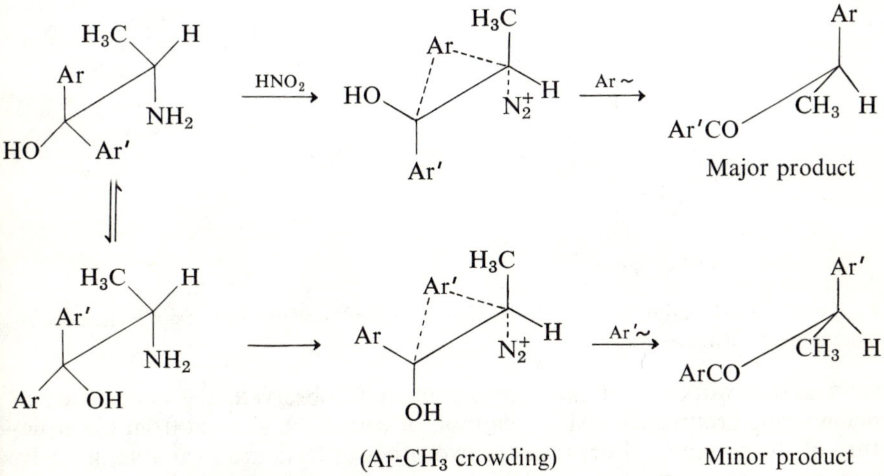

H_3C, H, Ar, NH_2, HO, Ar′ $\xrightarrow{HNO_2}$ H_3C, Ar, HO, N_2^+, H, Ar′ $\xrightarrow{Ar\sim}$ Ar, Ar′CO, CH_3, H — Major product

⇅

H_3C, H, Ar′, NH_2, Ar, OH ⟶ H_3C, Ar′, Ar, N_2^+, H, OH (Ar-CH_3 crowding) $\xrightarrow{Ar'\sim}$ Ar′, ArCO, CH_3, H — Minor product

Fig. 6-25. Rearrangement of ArAr′COHCH(NH_2)CH_3.

in Fig. 6-26. A significant difference of conformational origin is found in *N*-benzoylnorephedrine and *N*-benzoylnor-ψ-ephedrine (Fig. 6-27; see also Chap. 15). The latter diastereoisomer, upon treatment with alcoholic hydrogen chloride, is immediately converted into the hydrochloride of the *O*-benzoyl derivative, whereas the former isomer is unchanged under these conditions.[26] The difference is clearly due to the fact that in the ψ-ephedrine derivative, when the benzamido and hydroxyl groups are gauche, the methyl and phenyl groups are anti, whereas in the ephedrine derivative, in the transition state for N-O migration, the phenyl and methyl groups must be

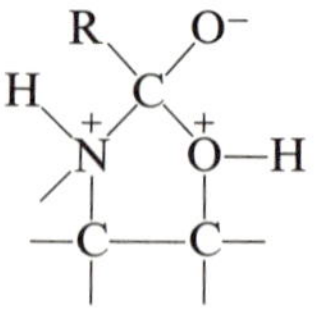

Fig. 6-26. Intermediate in N→O migration.

crowded together in gauche positions and therefore the activation energy is higher.

iii. Neighboring-group Participation.[27] A group adjacent to the site of reaction (cf. Fig. 6-23) may participate in the reaction without necessarily giving rise to rearrangement. Such participation is known as "neighboring-group participation" and has been studied especially extensively by Winstein and coworkers.

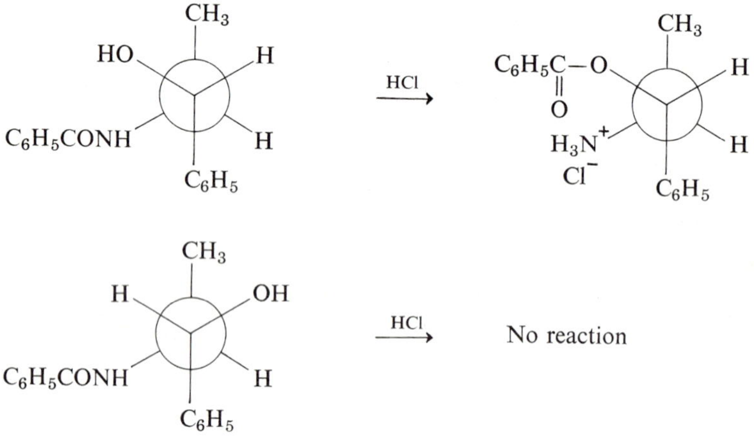

Fig. 6-27. N→O migration in *N*-benzoylnor-ψ-ephedrine and its absence in *N*-benzoylnorephedrine.

stein and coworkers. If no rearrangement is observed, the occurrence of neighboring-group participation cannot, of course, be gleaned from the structure of the product. Fortunately, two other criteria are available, namely, enhancement of rate over what might normally be expected,† in cases where

† It is not always easy to decide what the rate ought to be in the absence of participation. This problem (cf. Ref. 28) cannot be entered into here.

[26] G. Fodor, V. Bruckner, J. Kiss, and G. Óhegyi, *J. Org. Chem.*, **14**, 337 (1949).
[27] W. Lwowski, *Angew. Chem.*, **70**, 483 (1958).
[28] A. Streitwieser, *J. Am. Chem. Soc.*, **78**, 4935 (1956).

participation occurs before the transition state is reached, and a stereochemical criterion—retention of configuration—which manifests itself in suitably substituted systems, regardless of whether participation occurs before the transition state is reached or after.

One of the classical cases of neighboring-group participation as manifested by stereochemistry is the reaction[29] of the 3-bromo-2-butanols with hydrogen bromide (Fig. 6-28). It was found that the active threo isomer, upon treatment with hydrogen bromide, gave rise to *dl*-2,3-dibromobutane (devoid of the meso isomer but totally inactive), whereas the erythro isomer gave rise only to the meso dibromide, free of the *dl* isomer. The explanation, in terms of a cyclic intermediate or transition state ("bromonium ion") which may be attacked at either carbon 2 or carbon 3, is shown in Fig. 6-28.

The correctness of the interpretation of the stereochemical results shown in Fig. 6-28 has been amply confirmed by subsequent work on the acetolysis of

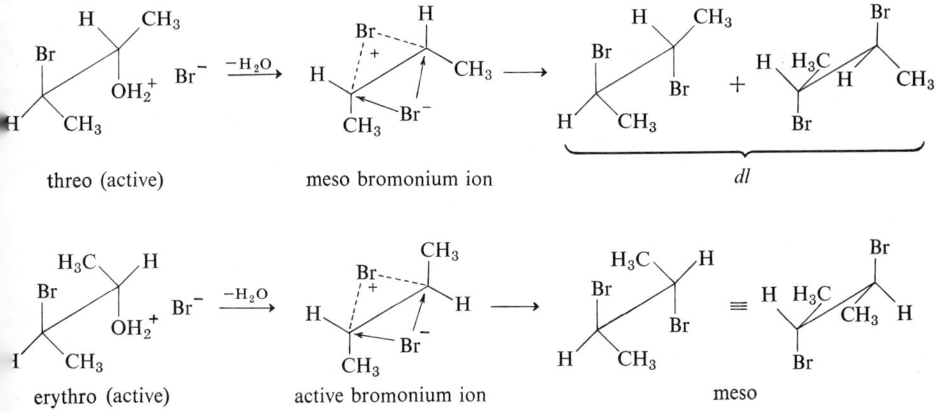

threo (active) $\xrightarrow{-H_2O}$ meso bromonium ion $\longrightarrow$ *dl*

erythro (active) $\xrightarrow{-H_2O}$ active bromonium ion $\longrightarrow$ meso

Fig. 6-28. Reaction of 3-bromo-2-butanols with hydrogen bromide.

the 2-phenyl-3-pentyl *p*-toluenesulfonates and the 3-phenyl-2-pentyl *p*-toluenesulfonates.[30] In these systems, the participating group is phenyl, and, because of the lack of symmetry of the system, all acetolysis products are optically active, there being no meso forms of either the products or intermediates. Nevertheless, the steric course of the substitution reaction shown in Fig. 6-29 is analogous to that shown in Fig. 6-28. The active *threo*-2-phenyl-3-pentyl tosylate (I) upon acetolysis yields, via the cis phenonium ion A, a mixture of active *threo*-2-phenyl-3-pentyl acetate (V) of retained configuration and active *threo*-3-phenyl-2-pentyl acetate (VI). The corresponding active erythro tosylate (II) similarly yields active erythro acetate of retained structure and configuration (VII) as well as active *erythro*-3-phenyl-2-pentyl acetate (VIII), both formed via phenonium ion B. Active *threo*-3-phenyl-2-pentyl tosylate (III) yields a phenonium ion enantiomeric with A and thence the active threo products IX

[29] S. Winstein and H. J. Lucas, *J. Am. Chem. Soc.*, **61**, 1576, 2845 (1939).
[30] D. J. Cram, *J. Am. Chem. Soc.*, **71**, 3875 (1949).

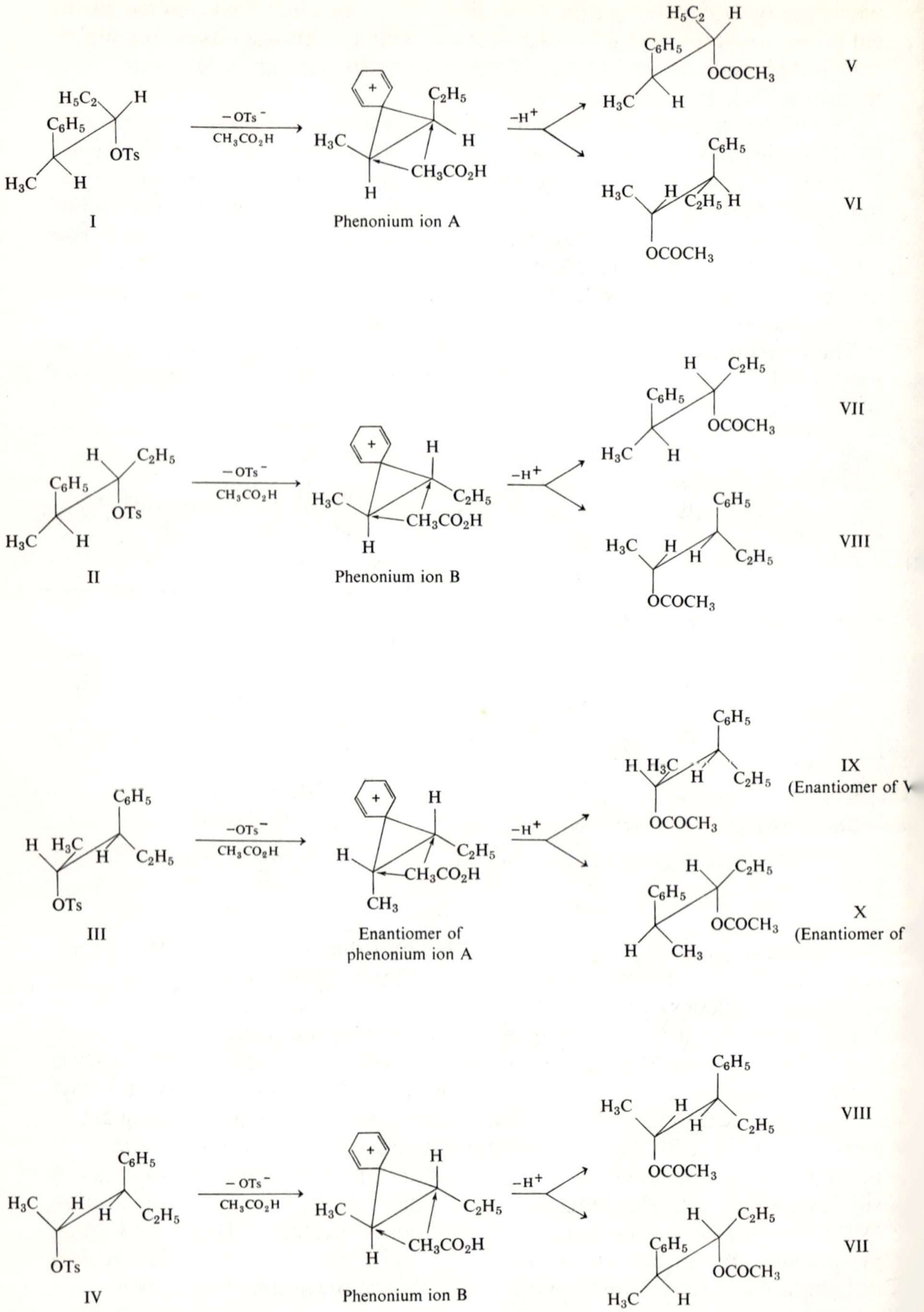

Fig. 6-29. Acetolysis of 3-phenyl-2-pentyl tosylate and 2-phenyl-3-pentyl tosylate.

and X, enantiomeric with VI and V.† Finally, *erythro*-3-phenyl-2-pentyl tosylate‡ (IV) yields the same phenonium ion B as does its positional isomer II, and therefore the solvolysis products VII and VIII from IV are the same as those from II.

Since in phenonium ion A the two alkyl groups are cis whereas in B they are trans (Fig. 6-29), one might expect the erythro isomers II and IV to solvolyze faster than the threo isomers I and III. The difference was measured in the case of the corresponding 3-phenyl-2-butyl tosylates (corresponding to I and II with methyl instead of ethyl), and it was found[31] that solvolysis of the threo isomer was faster than that of the erythro by a factor of only 1.2. Presumably phenyl participation occurs only past the transition state.

iv. Molecular Elimination. It was mentioned above that bimolecular ionic eliminations, such as the base-induced eliminations of halides and tosylates as well as the Hofmann elimination of quaternary hydroxides, usually involve anti stereochemistry. Other groups of elimination reactions are known, however, in which the stereochemistry of the departing groups is cis. In this group is the pyrolysis of xanthates, acetates, and amine oxides. Examples illustrating amine oxide and xanthate pyrolysis are shown in Fig. 6-30. The reason for preferred cis elimination is believed to lie in the fact that the transition state for the reaction is cyclic, as shown in Fig. 6-31.

c. The Relation of Conformation to Reactivity in Individual Compounds. The Curtin-Hammett Principle. So far we have assessed conformational factors on reactivity by looking at diastereoisomers. The question may, however, be raised as to how the reactivity of an individual molecule depends on its existing to a greater or lesser extent in one or another of several possible conformations. The answer, in brief, is as follows: As far as reaction rate is concerned, the population of the various conformations matters a great deal, but as far as reaction product is concerned, it is usually quite immaterial.

Reactivity of a single compound which may exist in different conformations in a given reaction may be expressed by the equation $k = \sum_i N_i k_i$, where k is the empirical rate constant for the reaction in question, N_i is the mole fraction of the ith conformational isomer, and k_i is the specific reaction rate for that isomer. Although this equation is of fundamental importance in considering the reactivity of a wide variety of compounds, it seems to have been exploited almost exclusively in the cyclohexane series. Derivation and further consideration of this equation are therefore postponed to Chap. 8.

More commonly, cases have been studied where different conformations of one and the same compound appear to give rise to different products. An

† That enantiomers of VI and V rather than VI and V were obtained is a fortuitous result of the choice of starting material III. If the enantiomer of III had been used as starting material, its acetolysis products would have been V and VI formed via the same phenonium ion A obtained from I.

‡ In applying the erythro-threo nomenclature to this series of compounds, one converts them into conformations in which the methyl and ethyl groups are eclipsed and then follows the example of Fig. 3-9 with R and R′ methyl and ethyl, a equal to hydrogen, and b and c equal to phenyl and tosylate or acetate, respectively.

[31] S. Winstein, B. K. Morse, E. Grunwald, K. C. Schreiber, and J. Corse, *J. Am. Chem. Soc.*, **74**, 1113 (1952).

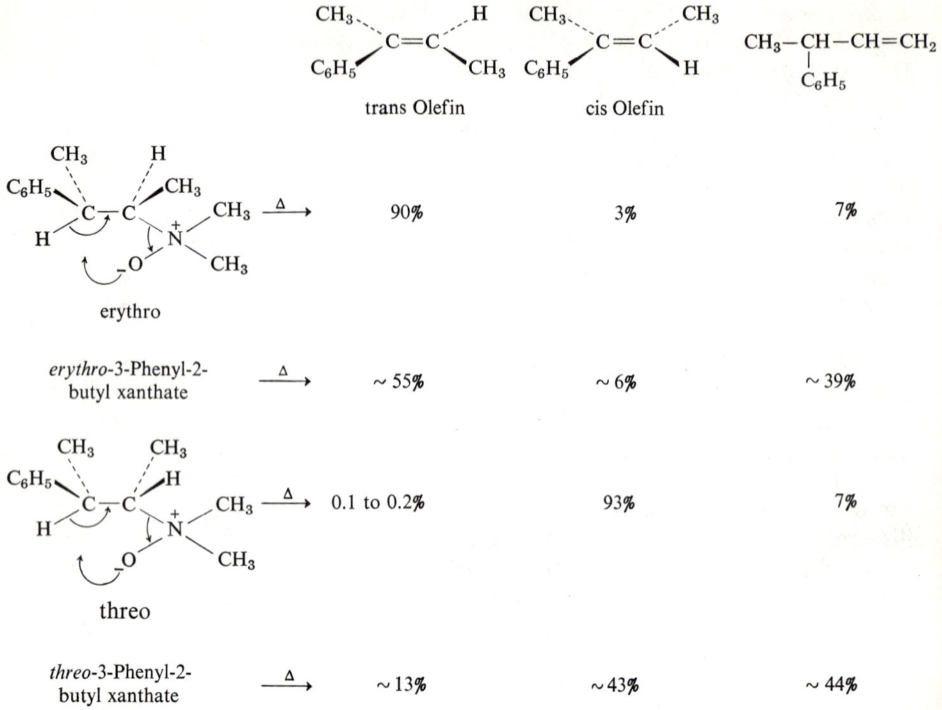

	trans Olefin	cis Olefin	$CH_3{-}CH(C_6H_5){-}CH{=}CH_2$
erythro (amine oxide) $\xrightarrow{\Delta}$	90%	3%	7%
erythro-3-Phenyl-2-butyl xanthate $\xrightarrow{\Delta}$	~55%	~6%	~39%
threo (amine oxide) $\xrightarrow{\Delta}$	0.1 to 0.2%	93%	7%
threo-3-Phenyl-2-butyl xanthate $\xrightarrow{\Delta}$	~13%	~43%	~44%

Fig. 6-30. Steric course of pyrolysis of 3-phenyl-2-butyl xanthate and the corresponding amine oxide. (*From M. S. Newman, "Steric Effects in Organic Chemistry," John Wiley & Sons, Inc., New York, 1956. By permission of the publishers.*)

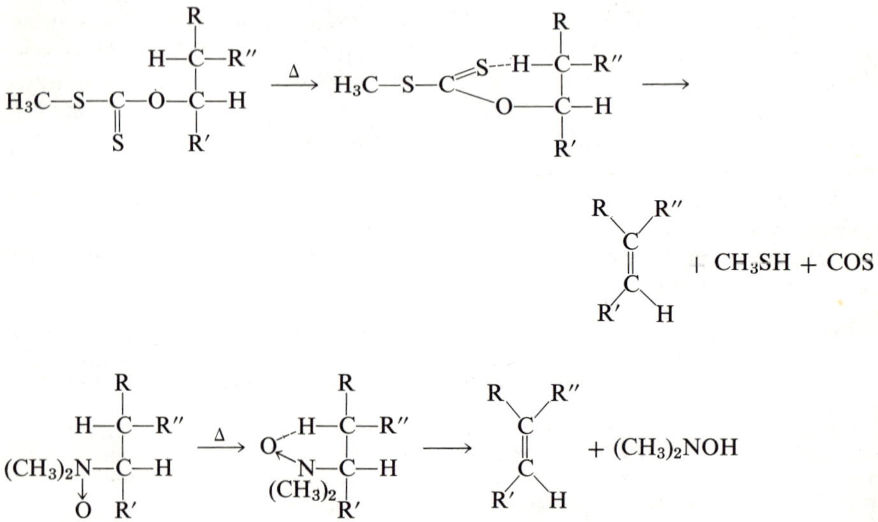

$$H_3C{-}S{-}C({=}S){-}O{-}C(H)(R'){-}C(R)(H)(R'') \xrightarrow{\Delta} \longrightarrow RR''C{=}CR'H + CH_3SH + COS$$

$$(CH_3)_2N(\rightarrow O){-}C(H)(R'){-}C(R)(H)(R'') \xrightarrow{\Delta} \longrightarrow RR''C{=}CR'H + (CH_3)_2NOH$$

Fig. 6-31. Course of xanthate and amine oxide pyrolysis.

example is the base-induced dehydrochlorination of 1,2-diphenyl-1-chloroethane to *cis*- or *trans*-stilbene, shown in Fig. 6-32. It would appear at first sight that the observed predominance of *trans*-stilbene in reactions of this type is caused by the predominance of the anti conformation of the starting material over the gauche conformation. [It has already been shown (cf. Fig. 6-18) that the elimination reaction proceeds readily only when the atoms to be eliminated—hydrogen and chlorine in the present case—are anti.] It has been pointed out by Hammett and Curtin, however, that, provided that the activation energy of the reaction is large compared with the barrier to rotation (which is usually quite low), the proportion of the products in no way reflects the relative population of the ground-state conformations but depends only on the activation energies of the processes leading to these products.[83] This principle, which we shall call the Curtin-Hammett principle, is an

trans-Stilbene $\xleftarrow[\text{fast}]{-HCl}$ anti

gauche $\xrightarrow[\text{slow}]{-HCl}$ *cis*-Stilbene

Fig. 6-32. Dehydrochlorination of 1,2-diphenyl-1-chloroethane.

extension of a more general proposition, according to which the ratio of two products formed from one starting material depends only on the difference in the free-energy levels of the two transition states and is in no way related to the free-energy level of the ground state.† This is shown in Fig. 6-33. Since the ratio of the products P is equal to the ratio of their rates of formation which, in turn, is equal to the ratio of their specific rates of formation (the starting material being the same for both products) we have

$$P = \frac{k_1}{k_2} = \frac{e^{-\Delta F_1^\ddagger}}{e^{-\Delta F_2^\ddagger}} = e^{-(\Delta F_1^\ddagger - \Delta F_2^\ddagger)} = e^{-[(F_1^\ddagger - F) - (F_2^\ddagger - F)]} = e^{(F_2^\ddagger - F_1^\ddagger)}$$

i.e., the product ratio depends only on the difference in the transition-state

† Assuming the reaction is first-order in the starting material or, if not first-order, involves the same kinetic expression for the formation of the two products.

energy levels. The more complex case covered by the Curtin-Hammett principle is depicted in Fig. 6-34. Since it is true for both the molecules in conformation A and conformation B that the product ratio is independent of the ground-state energy level and dependent only on the transition-state energy levels, it must also be true for the system as a whole. The fact that molecules in conformation A may have to change to conformation B before

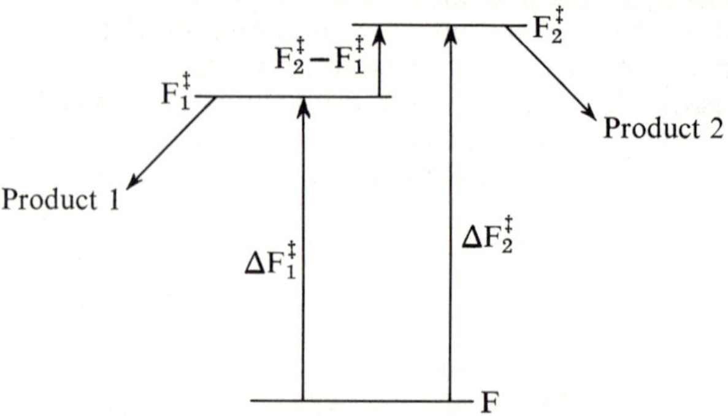

Fig. 6-33. Energetics of formation of two products from common starting material.

traversing the transition state leading to product 2 (and vice versa for molecules in conformation B going to product 1) is in no way relevant, for, inasmuch as transition-state theory is an extension of thermodynamics, its conclusions depend only on the initial and transition-state levels and not on the path traversed between the two. (A somewhat different deduction of the Curtin-Hammett principle will be given in Chap. 8.)

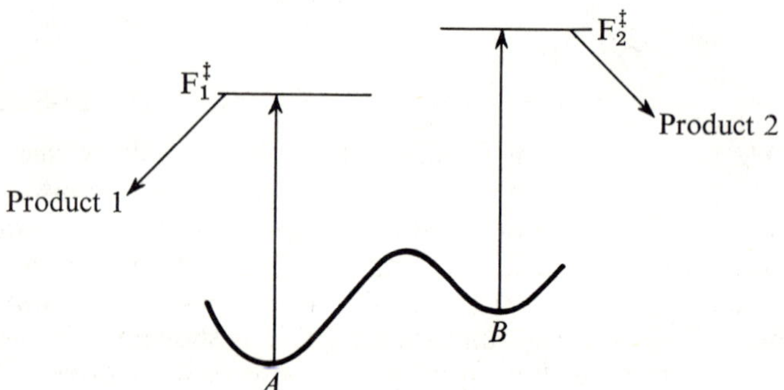

Fig. 6-34. The Curtin-Hammett principle.

A few examples may now be explained on the basis of the principle. Phenylbenzylcarbinyl triethylbenzoate, upon treatment with potassium *t*-butoxide in *t*-butyl alcohol at the boiling point, yields about 100 times as much *trans*-stilbene as *cis*-stilbene (Fig. 6-35),[32] indicating a difference in activation energy of about 3.2 kcal./mole for the formation of the two stil-

[32] D. Y. Curtin and D. B. Kellom, *J. Am. Chem. Soc.*, **75**, 6011 (1953).

bene isomers. Since *cis*- and *trans*-stilbene differ in stability by about 5.7 kcal./mole, the observed difference in activation energy is reasonable. The two transition states should reflect the differences in product stabilities to the extent that whatever unfavorable interactions exist in *cis*-stilbene should exist to a somewhat lesser extent in the transition state leading to it. This type of interaction has been termed[83] the "cis effect."

The cis effect is also observed in the dehydrobromination of 2-bromobutane which leads to *trans*-2-butene and *cis*-2-butene in a ratio of about 6:1.[33] This example demonstrates a *steric* effect in the transition state favoring the more stable trans product and should be clearly distinguished from the case of debromination of *meso*- and *dl*-2,3-dibromobutane (different starting materials!) cited on p. 140 (Fig. 6-17), in which a *stereoelectronic* effect is

$$C_6H_5CH_2-CH(OCOR)C_6H_5 + (CH_3)_3CO^-K^+$$

OCOR
1
100
$(CH_3)_3CO^-$ H C_6H_5 H OCOR H C_6H_5
$(CH_3)_3CO^-$ H C_6H_5 H OCOR H_5C_6 H
C_2H_5 R = H_5C_2 C_2H_5
H H C_6H_5 H_5C_6
H H_5C_6 C_6H_5 H

Fig. 6-35. Elimination from phenylbenzylcarbinyl triethylbenzoate.

operative so that the meso dibromide gives only *trans*-2-butene and the *dl* dibromide gives only *cis*-2-butene.

The explanation given for the preferential migration of Ar as compared with Ar′ for the case shown in Fig. 6-25 would seem to be contrary to the Curtin-Hammett principle, inasmuch as this explanation was based on the population of ground-state conformations. It must be pointed out, therefore, that the principle cannot be expected to apply to this case. The essential step in the deamination (amine–nitrous acid) reaction is the loss of nitrogen from a diazonium salt intermediate RN_2^+. All evidence points to the fact that this step has a very low activation energy and that the assumption embodied in the Curtin-Hammett principle, namely, that the activation energy of the proc-

[33] H. J. Lucas, T. P. Simpson, and J. M. Carter, *J. Am. Chem. Soc.*, **47**, 1462 (1925).

esses studied is large compared with the barrier to internal rotation (cf. Fig. 6-34), can thus not be made in this case. If one makes the extreme opposite assumption, namely, that the activation energy in deamination is very small compared with the barrier height, then the conclusion would follow that the

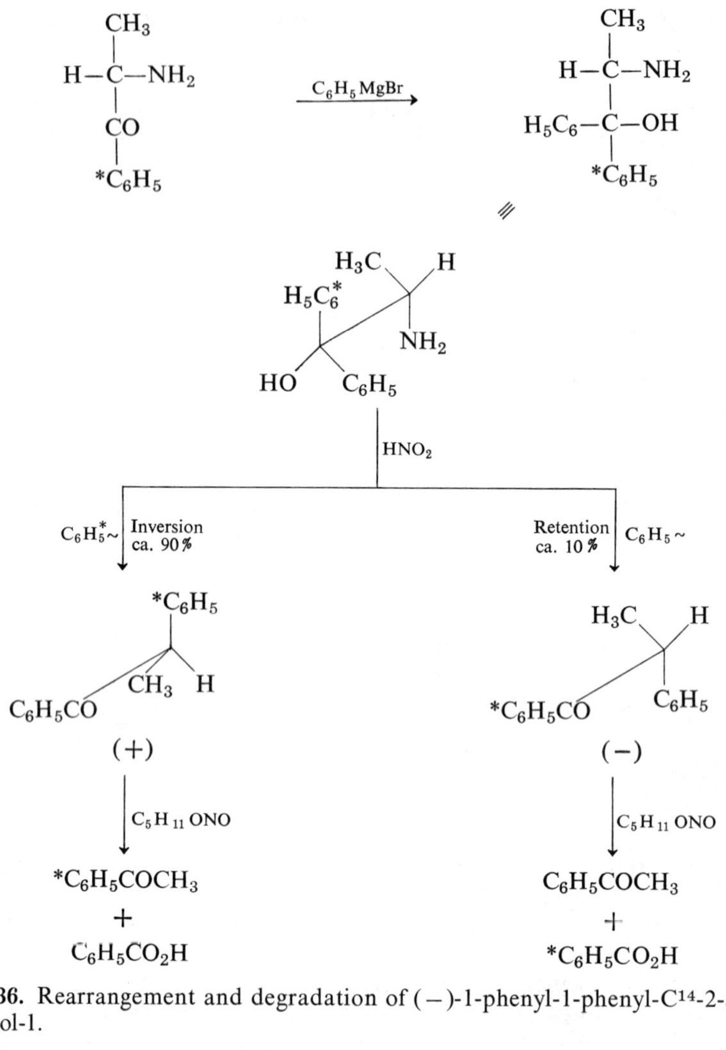

Fig. 6-36. Rearrangement and degradation of (−)-1-phenyl-1-phenyl-C^{14}-2-amino-propanol-1.

ratio of the products is equal to the ratio of the populations of the starting states.

Actually, when Ar is phenyl and Ar′ is anisyl (Fig. 6-25), one observes 12% anisyl migration along with 88% of the expected phenyl migration,[34] and it

[34] D. Y. Curtin and M. C. Crew, *J. Am. Chem. Soc.,* **77,** 354 (1955).

might be expected that this ratio reflects the population of the two starting conformations.

This interpretation is not, however, correct. When stereospecifically labeled 1-phenyl-1-phenyl-C^{14}-2-aminopropanol-1 (Fig. 6-36; the specific labeling follows from the method of preparation and the application of Cram's rule) is deaminated,[35] the labeled phenyl group migrates to the extent of 90% and the unlabeled to the extent of 10%. By using optically active as well as labeled starting material, it can be shown[35] that, while migration of the labeled group occurs with inversion of configuration, *migration of the unlabeled group entails retention of configuration.* This result rules out anti migration of the unlabeled phenyl group starting from a less populated conformation, as implied in the bottom half of Fig. 6-25. A plausible explanation[35] which

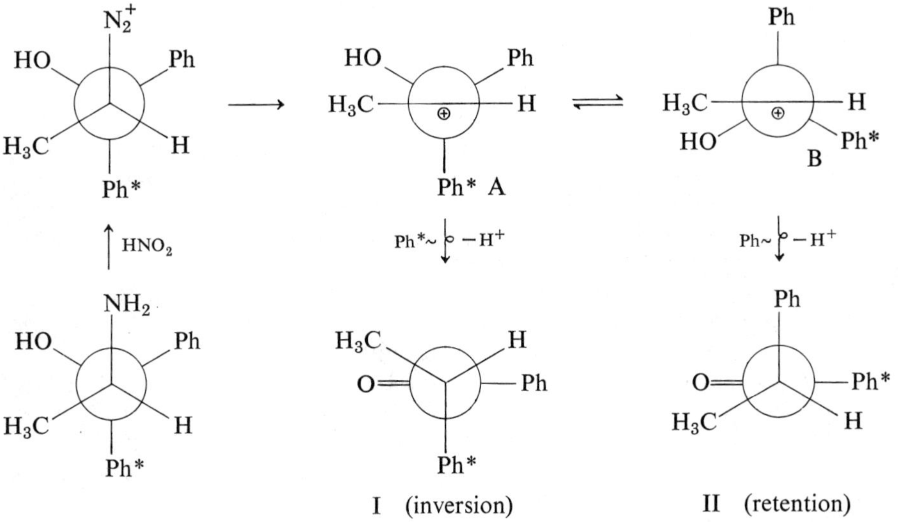

Fig. 6-37 Mechanism of deamination of (−)-1-phenyl-1-phenyl-C^{14}-2-aminopropanol-1.

accounts for both this result and the earlier mentioned observations of Curtin and coworkers is embodied in Fig. 6-37. The open carbonium ion A, formed from the amine via the diazonium cation, is converted to product I by migration of the labeled phenyl group which finds itself in the plane of the vacant *p* orbital of the carbonium ion (cf. Chap. 13). In competition with this migration, however, is rotation of the ion A about the C_1—C_2 bond to give ion B. In this ion it is the unlabeled phenyl group that is favorably located for migration to give product II. Inspection of the formulas also shows that I is formed with inversion of configuration at the amine carbon, but II is formed with retention.

[35] B. M. Benjamin, H. J. Schaeffer, and C. J. Collins, *J. Am. Chem. Soc.*, **79**, 6160 (1957); cf. A. McKenzie, R. Roger, and G. O. Wills, *J. Chem. Soc.*, 779 (1926). See also ref. 35*a*.

[35a] J. C. Martin and W. G. Bentrude, *J. Org. Chem.*, **24**, 1902 (1959).

The interesting conclusion is therefore reached that rotation about the single bond (A $\rightleftharpoons$ B) and migration of phenyl proceed at comparable rates; in fact, the relative rates of the two processes can be calculated from the experimental data.[35a] The deamination of amino alcohols of the type

$$ArAr'COHCH(NH_2)R$$

falls therefore into neither of the two extreme categories mentioned earlier, one in which the activation energy is high compared with the rotational barrier (and to which the Curtin-Hammett principle applies) and the other in which the activation energy is much lower than the barrier (where product composition reflects only population of the ground-state conformations). Rather, this particular reaction occupies an intermediate position where the reaction studied and rotation are about equally fast.

6-4. Optical Isomerism Due to Restricted Rotation ("Atropisomerism")

Conformational isomers, such as the anti and gauche forms of butane (Fig. 6-4), are discrete entities—similar to, let us say, the keto and enol forms of acetoacetic ester—and might, in principle, be isolable. In fact, however, they usually cannot be isolated because of low energy barriers between them. Thus, even when both isomers are present to an appreciable extent at equilibrium, such as in butane (where about two-thirds of the molecules are in the anti form at room temperature and one-third in the gauche form), any attempt to separate them would be frustrated by their tendency to revert to the equilibrium mixture with extreme rapidity.

Nevertheless, isomers of this type have been known since the early 1920s; their discovery[36] antedates by several years the coining of the term "conformation"[37] and the discovery of the rotational barrier in ethane.[1-3] The early examples were all derivatives of biphenyl, and their rotational isomerism is often called "biphenyl isomerism."

The more general term "atropisomerism," taken from the German literature, will be used to denote any kind of stereoisomerism due to restricted rotation about single bonds where the isomers can actually be isolated. It must be emphasized that the transition from conformational isomerism to atropisomerism is not a sharp one. At room temperature a barrier of 16 to 20 kcal. between rotational isomers is usually required to permit their isolation, but at lower temperatures a lower barrier would obviously suffice, and at higher temperatures compounds which are considered "atropisomers" at room temperature may become readily interconvertible.

a. Biphenyl Isomerism. The discovery of biphenyl isomerism is one of those accidents of scientific history where a completely erroneous theory stimulated a highly worthwhile experiment. In the early 1920s it was believed, following a suggestion originally made by Kaufler,[38] that biphenyl was a

[36] G. H. Christie and J. Kenner, *J. Chem. Soc.*, **121**, 614 (1922).

[37] W. N. Haworth, "The Constitution of the Sugars," Edward Arnold & Co., London, 1929, p. 90.

[38] F. Kaufler, *Ann.*, **351**, 151 (1907): *Ber.*, **40**, 3250 (1907).

folded, clam-shaped molecule.† Reasoning on the basis of this theory—which was later shown to be untenable—and on the basis of some experimental observations‡—which were later shown to be erroneous—regarding the occurrence of cis-trans isomerism in 6,6′-dinitrodiphenic acid (Fig. 6-38, I), Christie and Kenner attempted the resolution of this acid and succeeded in obtaining the optically active forms. They realized that any shape of the biphenyl molecule other than a planar one would account for the resolvability of compounds I as well as II (which they also resolved) but on the whole preferred to interpret their results in terms of Kaufler's theory. When this theory was later discredited,¶ another explanation for the resolvability of compounds I and II (Fig. 6-38) and the non-resolvability of III was called for. It was soon suggested[39] that molecules such as I and II cannot readily exist in planar form because of the steric interference of the bulky ortho substituents in this form, and any non-planar form of these molecules is devoid of a plane (or center, or alternating axis) of symmetry. Figure 6-39 represents the situation in terms of energetics. The (+) and (−) forms of I represent distinct conformations which, being enantiomers, must have the same free energies. The planar form of I is a maximum-energy conformation which constitutes the transition state for interconversion of the enantiomers, i.e., for

I $R = NO_2$, $R' = H$
II $R = R' = NO_2$
III $R = H$, $R' = NO_2$

Fig. 6-38. Diphenic acid derivatives studied by Christie and Kenner.

racemization. The more crowded the ortho substituents are in the planar form, the higher the energy maximum is and the more difficult the enantiomers are to racemize. As the ortho substituents are made smaller, the energy barrier between the enantiomers becomes lower and the substance becomes more and more easily racemized.

A molecule such as 6,6′-dinitrodiphenic acid can be racemized in two ways (Fig. 6-39), namely, by forcing a nitro group past a nitro group and a carboxyl group past a carboxyl group (transition state B) and by forcing nitro

† The facile rearrangement of hydrazobenzene to benzidine was one piece of evidence cited in support of Kaufler's theory.

‡ J. Schmidt and A. Kämpf [*Ber.*, **36**, 3745 (1903)] claimed to have proved the structure of 6,6′-dinitrodiphenic acid; when J. Kenner and W. V. Stubbings [*J. Chem. Soc.*, **119**, 593 (1921)] later isolated what they believed to be a second form of the same acid, they suggested that the acid could exist in cis and trans forms.

¶ Partly as a result of further work by G. H. Christie, A. Holderness, and J. Kenner [*J. Chem. Soc.*, 671 (1926)] who showed that the supposed 6,6′-dinitrodiphenic acid of Schmidt and Kämpf was really 4,6′-dinitrodiphenic acid and that the cis-trans isomerism of compound I (Fig. 6-38) (but *not* its optical isomerism) was therefore spurious. The early history of biphenyl isomerism is outlined in detail in Refs. 85 and 91*a*.

[39] F. Bell and J. Kenyon, *Chemistry & Industry* (*London*), **45**, 864 (1926); W. H. Mills, *ibid.*, p. 884; cf. also E. E. Turner and R. J. W. LeFèvre, *ibid.*, p. 831.

past carboxyl on both sides (transition state A). From what has been said before, the transition state (A) in which the groups of unequal size pass each other would be expected to be of lower energy.

Most of the available evidence of biphenyl isomerism is in agreement with the simple picture summarized in Fig. 6-39.† Tetra-ortho-substituted bi-

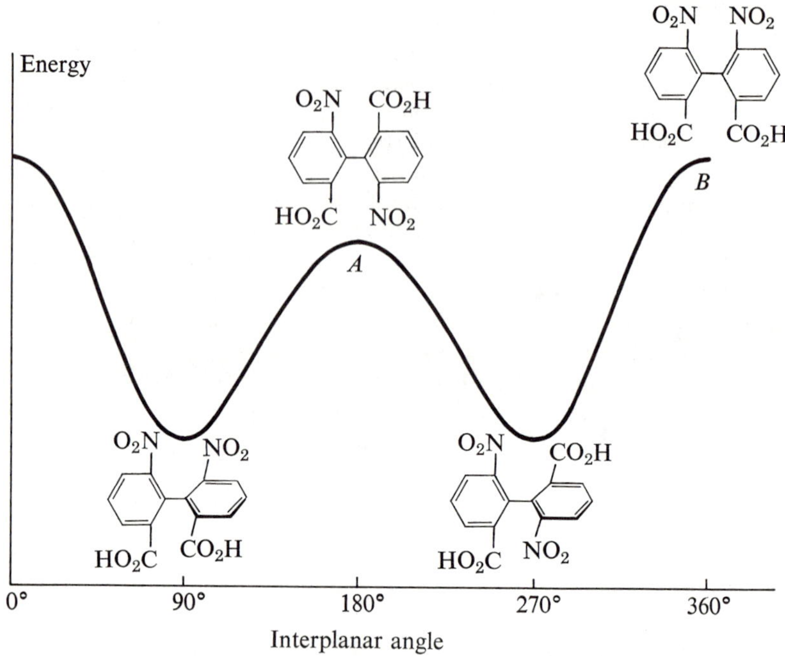

Fig. 6-39. Energy profile for 6,6′-dinitrodiphenic acid.

phenyls of the type shown in Fig. 6-40 can usually be resolved. In the transition state for racemization (Fig. 6-40) two pairs of bulky groups (a-c and b-d) must slip past each other, which is usually difficult, so that the enantiomers once obtained are usually stable. However, it is clear that the

Fig. 6-40. Tetra-ortho-substituted biphenyls.

energy barrier to racemization must depend on the bulk of the ortho substituents; if these are very small, e.g., fluorine or methoxyl (Fig. 6-41),

† Since the extensive studies on biphenyl isomerism—many of them by R. Adams and his school—have been reviewed in two comprehensive summaries (Refs. 85 and 86), some of the documentation will be omitted in the remainder of this section and in the next.

racemization becomes so fast that resolution fails. When one pair of small ortho substituents and one pair of medium-sized ortho groups are juxtaposed (Fig. 6-42), resolution becomes possible but the enantiomers are racemized on warming. When at least two of the four ortho substituents are large, or if all four are at least medium-sized, optically stable enantiomers can be obtained (Fig. 6-43).

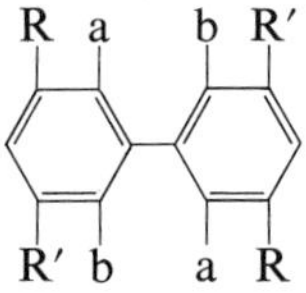

R a b R′

R′ b a R

I $a = b = F$, $R = CO_2H$, $R' = Cl$
II $a = CH_3O$, $b = F$, $R = H$, $R' = CO_2H$
III $a = b = CH_3O$, $R = CO_2H$, $R' = H$
IV $a = b = CH_3O$, $R = NH_2$, $R' = H$

Fig. 6-41. Non-resolvable tetra-ortho-substituted biphenyls.

Before proceeding further in the discussion, it must be pointed out that, regardless of the bulk of the ortho substituents, resolution is of course impossible when the conformation of the biphenyl in which the two rings are at right angles to each other has itself a plane of symmetry. Thus molecules of the type shown in Fig. 6-44, *A,* are not resolvable because of symmetry,

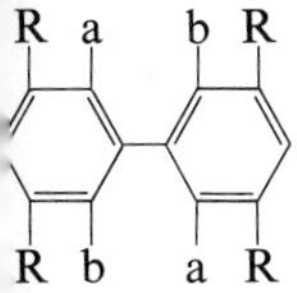

R a b R

R b a R

I $a = F$, $b = CO_2H$, $R = H$ racemized by 10 minutes' boiling in acetic anhydride
II $a = F$, $b = NH_2$, $R = CH_3$ racemized by 30 minutes' boiling in acetic acid
III $a = CH_3O$, $b = CO_2H$, $R = H$ half-racemized after 78 minutes' boiling in acetic acid

Fig. 6-42. Moderately optically stable tetra-ortho-substituted biphenyls.

but the addition of a substituent in the meta position destroying the symmetry (Fig. 6-44, *B*) will lead to resolvability if the ortho substituents are bulky enough.

Tri ortho-substituted biphenyls (Fig. 6-45) are, in general, more easily racemized than tetra-substituted ones, for the transition state for racemiza-

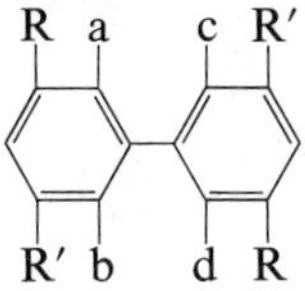

R a c R′

R′ b d R

I $a = d = F$, $b = c = NO_2$, $R = CO_2H$, $R' = CH_3$
II $a = NO_2$, $b = CO_2H$, $c = F$, $d = OCH_3$, $R = R' = H$
III $a = d = Cl$, $b = c = CO_2H$, $R = R' = H$

Fig. 6-43. Optically stable tetra-ortho-substituted biphenyls.

tion (Fig. 6-39) now involves one R-H and one R-R interaction instead of two R-R interactions. Compounds of this type can usually be racemized on heating, and from the ease of racemization (expressed in terms of half-life) the resistance of various groups to pass through the planar transition state can be gauged. For molecules of the type shown in Fig. 6-45 with $a = CO_2H$, $b = NO_2$, $R = H$, and c the group indicated below, the half-lives of racemization are as follows: c = CH_3, 179 min. at 118°; NO_2, 125 min. at 118°; CO_2H, 91 min. at 118°; OCH_3, 9.4 min. at 25°. With a = CO_2H, b = NO_2, and R = CH_3, the following half-lives were observed: c = Br, 3240 min. at 118°; Cl, 154 min. at 118°; F, too short for resolution. The capacity of the ortho substituents to interfere with passage through the planar transition state

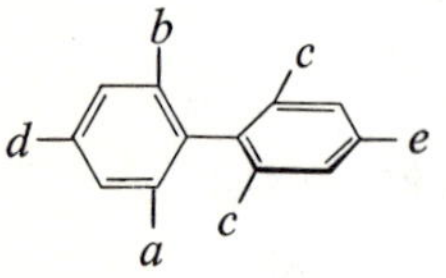

e.g.
$a = CO_2H$
$b = NO_2$
$c = CH_3O$
$d = e = H$

A (non-resolvable)

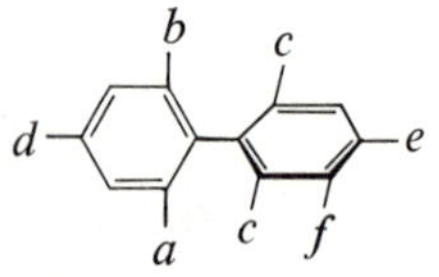

e.g.
$a = c = d = e = NO_2$
$b = H$
$f = CO_2H$

B (resolvable)

Fig. 6-44. Symmetry properties of tetra-ortho-substituted biphenyls.

thus seems to be Br $\gg$ CH_3 > Cl > NO_2 > CO_2H $\gg$ OCH_3 > F. It might be noted that this order roughly parallels the order of size of the groups (as determined by X-ray crystallographic measurements of van der Waals radii), whereas there is no parallel between group interference and polar properties.

In di-ortho-substituted biphenyls of the type shown in Fig. 6-46 racemiza-

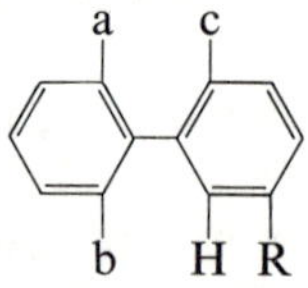

R may be H

Fig. 6-45. Tri-ortho-substituted biphenyls.

tion involves a transition state in which the ortho substituents a and b need to slip past only hydrogen atoms. Compounds of this type which have been resolved are listed in Fig. 6-46. They are all racemized readily on heating. When there is only one ortho substituent, racemization is yet more facile. No compounds of this type have been obtained in optically active form, but

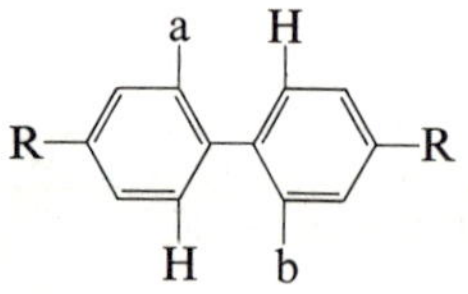

I a = b = SO_3H, R = H
II a = b = $SO_2OC_6H_5$, R = NH_2
III a = $N(CH_3)_2$, b = $N(CH_3)_3$, R = H
IV a = CO_2H, b = $COH(C_6H_5)_2$, R = H
V a = b = Br, R = CO_2H
VI a = b = I, R = CO_2H

Fig. 6-46. Di-ortho-substituted biphenyls.

the (+)-camphorsulfonate of 3′-bromobiphenyl-2-trimethylarsonium iodide (Fig. 6-47) shows mutarotation, indicating that a first-order asymmetric transformation (cf. page 41) takes place in solution. Presumably the (+)-camphorsulfonate of one of the two enantiomers (Fig. 6-39) is more stable than that of the other, and the original mixture of equal parts of the (+)-biphenyl (+)-camphorsulfonate and (−)-biphenyl (+)-camphorsulfonate is equilibrated by passage over the low energy barrier (planar form of the biphenyl) to a mixture containing one of the diastereoisomers in excess. Liberation of the free biphenyl in active form was not achieved because of its high rate of racemization. It is clear, however, that this molecule is not planar but rather that the compound exists as a mixture of enantiomers which are interconverted readily by passage over a relatively low energy barrier (cf. Fig. 6-39).

Perhaps the most conclusive evidence that the interference of the ortho substituents in the transition state for racemization (Fig. 6-39) is steric in origin comes from the elegant work of Westheimer and coworkers.[40] These investi-

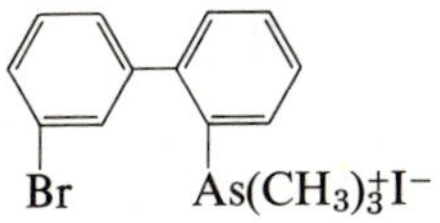

Br $As(CH_3)_3^+I^-$

Fig. 6-47. 3′-Bromobiphenyl-2-trimethylarsonium iodide.

gators were able to estimate correctly the enthalpy of activation for racemization of three different di-ortho-substituted biphenyls from known data on van der Waals radii and stretching- and bending-force constants of various bonds. It turns out that the undeformed planar transition state for racemization drawn in Fig. 6-46 is not the true transition state because of the severe van der Waals interpenetration energy between the groups a and H corresponding to this geometry. The interference can be minimized by compressing the C—H (ortho) bond, bending the ortho hydrogens and bromines away from each other, stretching the interannular bond, deforming the angle between the benzene rings and the interannular bond, and deforming the

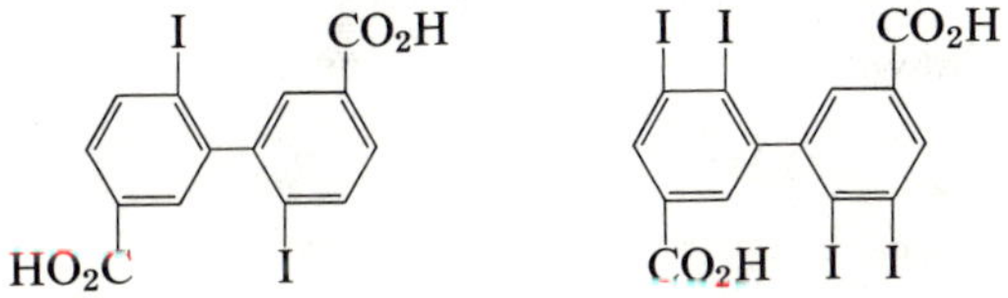

I CO_2H I I CO_2H

HO_2C I CO_2H I I

I II

Fig. 6-48. Substituted 2,2′-diiodobiphenyls.

benzene rings themselves. After all these deformations are allowed for and the over-all energy to bring them about, plus the remaining van der Waals interpenetration energy, is minimized, the following striking agreement between calculated and found enthalpies of activation for racemization is obtained: 2,2′-dibromo-4,4′-dicarboxybiphenyl (Fig. 6-46, V), calculated, 18.2 kcal./mole, found,[40a] 19.0; 2,2′-diiodo-5,5′-dicarboxybiphenyl (Fig. 6-48,

[40] F. H. Westheimer and J. E. Mayer, *J. Chem. Phys.*, **14,** 733 (1946); F. H. Westheimer, *ibid.*, **15,** 252 (1947); M. Rieger and F. H. Westheimer, *J. Am. Chem. Soc.*, **72,** 19 (1950); cf. F. H. Westheimer, Calculation of the Magnitude of Steric Effects, in M. S. Newman, ed., "Steric Effects in Organic Chemistry," John Wiley & Sons, Inc., New York, 1956, chap. 12.

[40a] M. M. Harris, *Proc. Chem. Soc.*, 367 (1959).

I), calculated, 21.4 to 23.6 kcal./mole, found, 21.0; and 2,2′,3,3′-tetraiodo-5,5′-dicarboxybiphenyl (Fig. 6-48, II), calculated, 28.6 to 33.1 kcal./mole, found, 27.3.†

There are, to be sure, some minor difficulties in attributing the energy barrier to racemization *exclusively* to steric factors. Thus, of the three compounds shown in Fig. 6-49, since $CH_3 > NO_2 > CO_2H$ (page 160), I should be the most stable and II and III should be of about equal stability. Actually, however, III was more stable than II which, in turn, was more stable than I toward racemization. In similar fashion, since nitro is larger than carboxyl, 6-nitrodiphenic acid (Fig. 6-50, I) should be more easily racemized than 2,2′-dinitro-

CH_3 CO_2H / H NO_2 — I; CH_3 H / CO_2H NO_2 — II; CH_3 H / NO_2 CO_2H — III

Fig. 6-49. Change of ortho-group arrangement.

6-carboxybiphenyl (Fig. 6-50, II). Actually, I is less easily racemized in butanol although it is racemized more easily than II in acetic acid and much more easily than II in the form of the sodium salt. Comparisons of rates of racemization of sodium salts (in water) with those of the corresponding free acids (in organic solvents) have been made in other instances, but the results are erratic: Sometimes the free acids racemize more readily and sometimes the salts. Similarly the effects of substituents in the 4′ (para) position (Table 6-3) on the rate of racemization of 2-nitro-6-carboxy-2′-methoxybiphenyl (Fig. 6-51) are not well understood.‡ It is of considerable interest that, at least in the case of 6-nitrodiphenic acid (I, Fig. 6-50), the retarding

CO_2H CO_2H / NO_2 H — I; CO_2H NO_2 / NO_2 H — II

Fig. 6-50. Biphenyls whose relative ease of racemization is solvent-sensitive.

effect of nitro groups in the 4 and 4′ positions manifests itself exclusively through a change in *entropy* of activation; the activation enthalpy is unaffected by such nitro substituents.[40b]

† The much higher enthalpy of activation of II (Fig. 6-48) as compared with I is due to the "buttressing effect"; see below.

‡ An explanation of some of these anomalies by M. Calvin, *J. Org. Chem.*, 4, 256 (1939) (cf. also L. N. Ferguson, "Electron Structures of Organic Molecules," Prentice-Hall, Inc., Englewood Cliffs, N. J., 1952, pp. 109–110) does not seem to have been generally accepted.

[40b] J. W. Brooks, M. M. Harris, and K. E. Howlett, *J. Chem. Soc.*, 1934 (1957).

Whereas the variations in half-life caused by substituents in the 4′ position are all within a factor of about 10, corresponding to relatively minor changes of activation energy of up to 1.5 kcal./mole, much more striking changes are brought about by substituents in the 3′ position (Table 6-3). These changes are due to the so-called "buttressing effect." The above-mentioned calculations indicate that, in the transition state for racemization, the groups in the ortho position bend away from each other. This means that the methoxy group in the compound shown in Fig. 6-51 is bent toward the 3′ position during racemization. If this motion is impeded by another

Table 6-3
Half-life Periods (in Minutes) of Substituted 2-Nitro-6-carboxy-2′-methoxybiphenyls[a]

Position of substituent	Nitro	Bromo	Chloro	Methyl	Methoxy
3′	1905	827	711	331	98
4′	115	25	12	2.6	3.6
5′	35	32	31	11.5	10.8

[a] At 25°C. the 3′ and 5′ compounds were dissolved in ethyl alcohol for racemization studies; the 4′, in acetone. The half-life of the parent compound is 9.4 min. in ethanol.

substituent in the 3′ position, racemization is slowed down by a factor of between 10 and 2000, depending on the size of the 3′ substituent.† The buttressing effect can be taken into account in the calculations of the enthalpy of racemization, as in the case of 2,2′,3,3′-tetraiodo-5,5′-dicarboxybiphenyl (Fig. 6-48, II); the agreement of the calculated enthalpy of racemization with the experimental value strengthens one's conviction that the effect of the 3′ substituent is indeed due to buttressing. Buttressing of hydrogen may account for the much smaller retardation of racemization by 5′ substituents (Table 6-3).

The fact that 2,2′,6,6′-tetracarboxy-4,4′-diphenylbipyridyl (Fig. 6-52) racemizes with ease, whereas biphenyls substituted with ortho substituents of

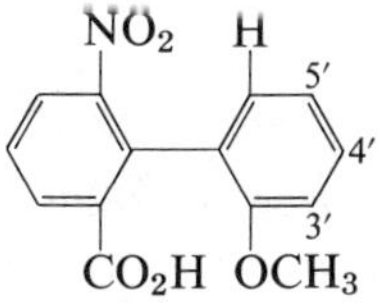

Fig. 6-51. 2-Nitro-6-carboxy-2′-methoxybiphenyls.

comparable size are quite difficult to racemize, may be ascribed to the absence of a hydrogen atom on the pyridine nitrogen and the resultant complete lack of buttressing of the adjacent carboxyl group, provided that there is no buttressing due to the lone pair of electrons on the nitrogen. An alternative possibility is that the effective size of two of the carboxyl groups is diminished through hydrogen bonding to the ring nitrogen.

† The effective order of size for buttressing, $NO_2 > Br > Cl > CH_3 > OCH_3$, is not the same as the previously mentioned (page 160) order of interference in the ortho position.

The presence of a six-membered aromatic ring is in no way required for resolvability; the bipyrryl shown in Fig. 6-53 has been resolved and is of high optical stability.

The effect of bridging the 2,2′ positions in a potentially optically active biphenyl has been studied in molecules of the type shown in Fig. 6-54. When $n = 1$, one has a 4,5-disubstituted fluorene in which the two benzene rings are coplanar and the molecule is therefore not resolvable. With $n = 2$, one has a 9,10-dihydrophenanthrene; molecules of this type, shown in Fig. 6-55, have been resolved,[41] but their optical stability is low, the half-lives being 3 to 5 hr. in boiling toluene, whereas corresponding unbridged biphenyls are

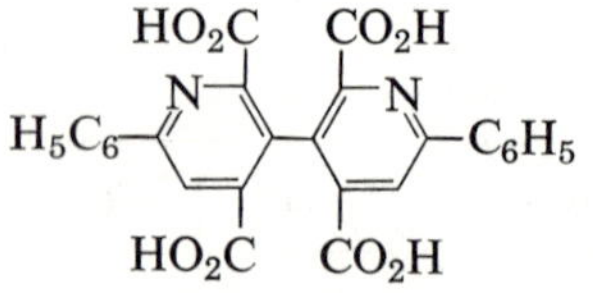

HO_2C CO_2H N N H_5C_6 C_6H_5 HO_2C CO_2H

Fig. 6-52. 2,2′,6,6′-Tetracarboxy-4,4′-diphenylbipyridyl.

completely optically stable. Optical stability increases when the bridge is lengthened to three atoms, as shown in Fig. 6-56; compound I has a half-life[42] of 3 to 4 hr. at 160°, despite the fact that the substituents in the 6 and 6′ positions are methoxyl groups (rather small in size; cf. Fig. 6-41), and compound II is resolvable[43] despite the fact that it has no 6,6′ substituents. (Its half-life is reported to be 80 min. at 32.5° in cyclohexane solution, but the compound is optically stable for at least 200 hr. in the crystalline state, indicating again that the conformation in the crystal is presumably frozen.) It must be pointed out, however, that for compound II (Fig. 6-56) to racemize, a methylene group must slip past another methylene group whereas in a

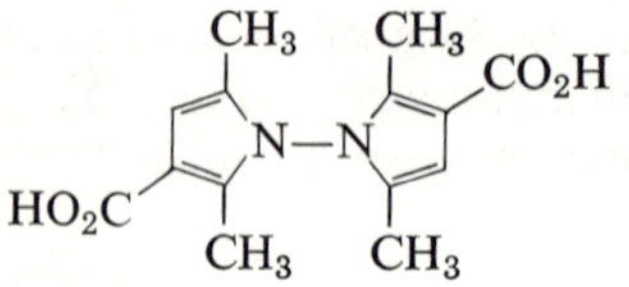

CH_3 CH_3 CO_2H N—N HO_2C CH_3 CH_3

Fig. 6-53. Resolvable bipyrryl.

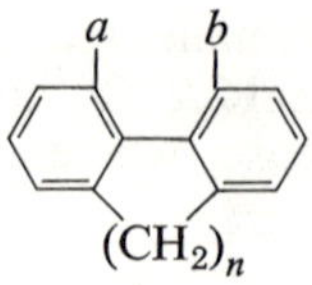

a b $(CH_2)_n$

Fig. 6-54. 2,2′-Bridged biphenyls.

(non-resolvable) non-bridged analog, such as 2,2′-dimethylbiphenyl, the potential antipodes are interconverted by two methyls slipping past two hydrogens which, as indicated earlier, is less difficult. Compounds in which *both* the 2,2′ and 6,6′ positions are spanned by three-atom bridges have also been studied.[43a]

A bridge of four atoms is found in compound III, Fig. 6-56, which has been resolved but is extremely optically labile (half-life, 1.4 min, at 23°).[43b]

[41] D. M. Hall and E. E. Turner, *J. Chem. Soc.*, 1242 (1955); W. L. F. Armarego and E. E. Turner, *ibid.*, 3668 (1956).

[42] G. H. Beaven, D. M. Hall, M. S. Lesslie, and E. E. Turner, *J. Chem. Soc.*, 854 (1952).

[43] D. C. Iffland and H. Siegel, *J. Am. Chem. Soc.*, **80**, 1947 (1958).

[43a] K. Mislow and M. A. W. Glass, *J. Am. Chem. Soc.*, **83**, 2780 (1961).

[43b] S. R. Ahmed and D. M. Hall, *J. Chem. Soc.*, 3383 (1959).

Another compound of this type will be discussed in Chap. 8; it is also quite labile. A compound similar to II (Fig. 6-56) but with a five-atom bridge has also been synthesized[43c] and is found to have somewhat greater optical stability than the lower homologs.

In concluding this section on optically active biphenyls, it might be pointed out that several such molecules occur in nature. An example is 4,5,6,4′,5′,6′-hexahydroxydiphenic acid which is found in bound form (as a glycoside) in certain tannins.[43d]

Fig. 6-55. Resolvable biphenyls with two-atom ortho bridge. (The second ring in the naphthalene system acts as a rather effective bulky ortho substituent; for further examples, see Ref. 86.)

b. Isomerism in Polyphenyls. In polyphenyls with bulky ortho substituents, where there is restricted rotation about more than one interannular bond, cases of diastereoisomerism may arise. In Chap. 3 diastereoisomerism was ascribed to the presence of more than one asymmetric atom in a molecule, but it was pointed out at the same time that this was an oversimplification, and that rather than speak about asymmetric atoms one should speak about "dissymmetric groupings." In biphenyl isomerism, for the first time, we are

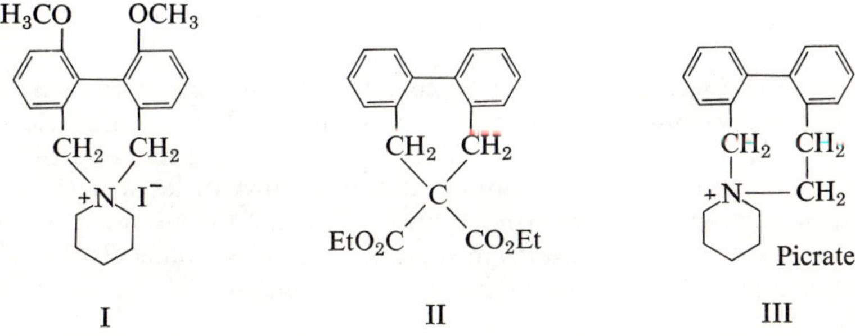

Fig. 6-56. Resolvable biphenyls with three- or four-atom ortho bridges.

encountering molecules that are dissymmetric without possessing asymmetric atoms. The "dissymmetric grouping" in this case is the hindered biphenyl system so substituted as to be devoid of a plane of symmetry. In a *para*-terphenyl (Fig. 6-57) there are two such dissymmetric groupings; in the general case these molecules may exist as two diastereoisomeric pairs of enantiomers

[43c] K. Mislow, S. Hyden, and H. Schaefer, *J. Am. Chem. Soc.*, **84**, 1449 (1962).

[43d] O. T. Schmidt and H. H. Grünewald, *Ann.*, **603**, 183 (1957).

(four optical isomers in all), but in particular instances this number may be reduced by the occurrence of meso forms. The situation, in other words, is analogous to that discussed in Chap. 3 in connection with dissymmetry due to asymmetric atoms.

Case I in Fig. 6-57 is the most general with two *dl* pairs. These may be denoted "cis" (X atoms on the end rings on the same side) and "trans" (X atoms on opposite sides). In case II (Fig. 6-57) the trans isomer has a center of symmetry and is inactive; therefore this molecule exists as a *dl* pair and a meso form. In case III the cis isomer is a meso form because it has a plane of symmetry (bisecting the middle ring), and the trans isomer is *dl*. Finally, case IV presents cis-trans isomerism, but both isomers have a center or a plane of symmetry and are therefore inactive. Molecules of the type shown in Fig. 6-57 have, in fact, been obtained as cis-trans isomers, and in one case (Fig. 6-58) one of the diastereoisomers (presumably the cis isomer) was resolved whereas the other (presumably the trans) resisted resolution.

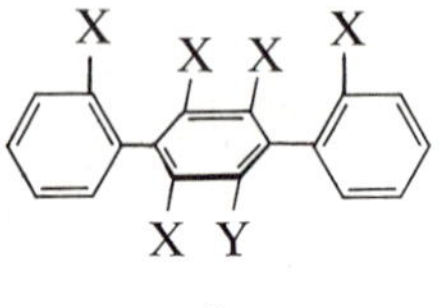

X X X X
X Y
I

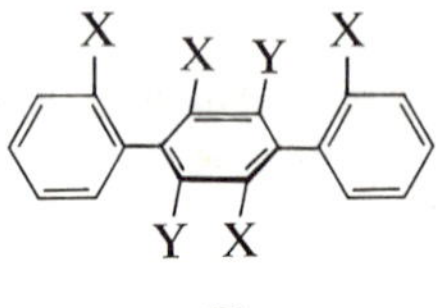

X X Y X
Y X
II

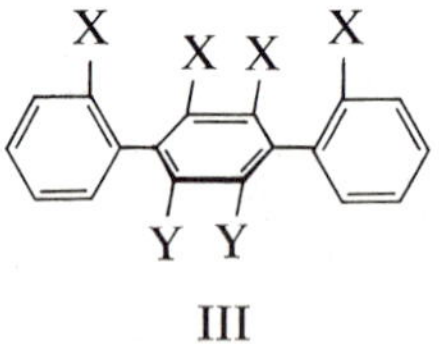

X X X X
Y Y
III

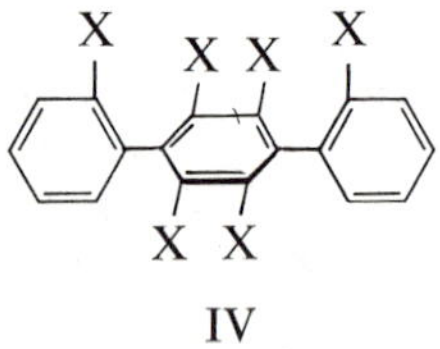

X X X X
X X
IV

Fig. 6-57. Types of dissymmetric *p*-terphenyls. (One isomer only shown.)

c. The Configuration of Active Biphenyls. In biphenyl isomerism, for the first time, we encounter optically active molecules without asymmetric atoms. Although, in principle, this does not change the configurational problem (cf. Chap. 5), namely, which enantiomer, dextrorotatory or levorotatory, corresponds to a given three-dimensional arrangement of the atoms, it does introduce some special complications in the configurational nomenclature (cf. Sec. 5-2) and especially in the experimental determination of the configuration (cf. Sec. 5-4).

The nomenclature problem, while elusive in the older (D,L) system, proved relatively easy to solve in the R-S system.[44] If one looks at a biphenyl in the conformation in which the two rings are perpendicular along the axis of the bond joining the rings and then projects the four ortho substituents on a plane at right angles to this bond, one obtains a representation of the biphenyl which is very similar to a Fischer projection formula (Fig. 6-59).

[44] R. S. Cahn, C. K. Ingold, and V. Prelog, *Experientia,* **12,** 81 (1956).

(For the sake of uniformity, the reader may wish to place the front benzene ring horizontally and the rear ring vertically, but this is of no consequence.) The molecule is then named in the form in which it is projected in Fig. 6-59, as if this projection represented an asymmetric carbon. There is, however, a complication: Because of the different nature of biphenyl isomerism as compared with isomerism due to an asymmetric atom, two or even more of the atoms or groups projected may be the same (e.g., in the case shown in Fig. 6-59 where there are two nitro and two carboxyl groups). At first sight it might appear that the sequence rule would get one around this difficulty; however, a little reflection shows that this is not always the case. For exam-

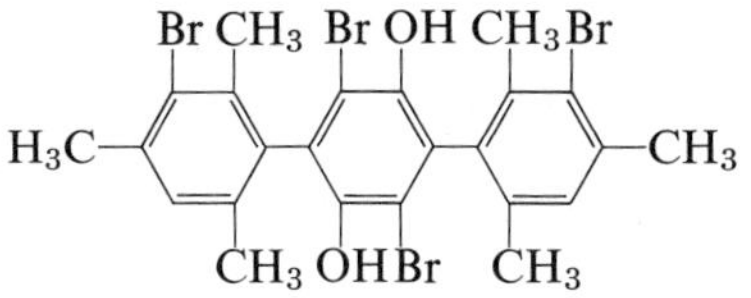

Br CH_3 Br OH CH_3 Br

H_3C— —CH_3

CH_3 OH Br CH_3

Fig. 6-58. Non-resolvable trans form and resolved cis form of a *p*-terphenyl.

ple, in the molecule shown in Fig. 6-59, going around the front ring starting at the carbon atom next to the front nitro group [C → CH → CH → CH → C(CO_2H) → C(Ar)] presents exactly the same sequence as going around the rear ring starting at the carbon next to the rear nitro group.† Therefore the sequence rule is insufficient to determine precedence in biphenyls. A new rule had to be established, namely, *near groups precede far groups.* This rule takes precedence over the sequence rule. Thus, in the molecule shown in Fig. 6-59, the horizontal (near) groups automatically take precedence over the vertical (far) groups. Since, moreover, NO_2 precedes CO_2H (cf. Sec. 5-2), the rear CO_2H group is the group of lowest precedence and is placed at the

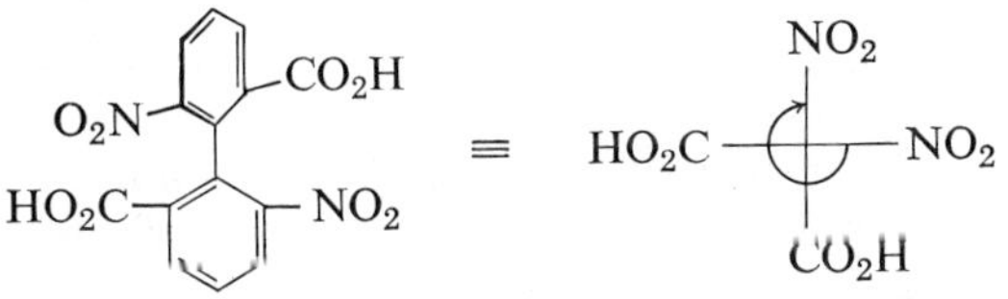

CO_2H O_2N HO_2C NO_2 ≡ NO_2 HO_2C NO_2 CO_2H

Fig. 6-59. Projection formula of a biphenyl.

bottom. The remaining groups are then in order: front NO_2, front CO_2H, rear NO_2. (Note that the near CO_2H group takes precedence over the far NO_2 group.) Since these groups describe a right-handed turn (see arrow in Fig. 6-59), the configurational symbol is R.

Because of the proximity rule, it is actually easier to name dissymmetric biphenyls than other dissymmetric molecules; the proximity rule immediately separates one pair of substituents from the other, and the sequence rule

† The same thing cannot happen in a molecule whose dissymmetry is due to an asymmetric atom, or else the atom would not be asymmetric!

needs to be invoked less frequently than in other cases.† Some more examples will illustrate the nomenclature (Fig. 6-60).

The third and fourth examples in Fig. 6-60 are cases where, because of the nature and position of the substituents, it is *not* immaterial from which end the molecule is viewed. The rule "Near groups precede far groups" in this case is interpreted to mean that the ring that has substituents off the main axis of the molecule *nearest to an external observer* is to be placed in front. Thus, whenever there is a 3 substituent, the ring in which it occurs is placed in front and the 3,5 substituents rather than the 2,6 substituents determine the sequence. (When both rings have substituents in the 3 position, the sequence rule is applied to determine which ring is to be placed in front.)

The problem of determining the configuration of dissymmetric biphenyls is a difficult one whose solution has taken much ingenuity.[45] Although

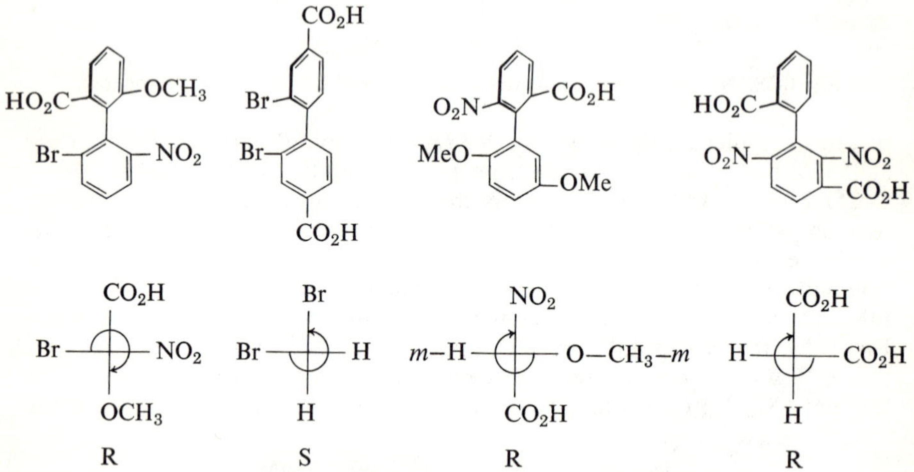

Fig. 6-60. Configurational nomenclature of biphenyls.

the X-ray method of determining absolute configuration (Sec. 5-3) is, in principle, applicable to biphenyls, it has not yet been so applied.‡

Configurational assignment by calculation of optical rotation from fundamental principles and comparison with experimental values (Chap. 14) has been attempted in the biphenyl series,[46, 47] but the results arrived at in the

† The reader may be worried about the arbitrariness of the decision as to which ring is placed in front, thus taking precedence by the proximity rule. Consideration of models allays this worry: the configurational symbol is independent of the orientation of the molecule, as it should be.

‡ For an application of the optical rotatory dispersion method, see Chap. 14.

[45] For a review of the entire problem and all the approaches to its solution, see K. Mislow, *Angew. Chem.*, **70**, 683 (1958).

[46] W. Kuhn and K. Bein, *Z. physik. Chem.*, **24B**, 335 (1934); W. Kuhn and R. Rometsch, *Helv. Chim. Acta*, **27**, 1080, 1346 (1944); W. Kuhn, *Z. Elektrochem.*, **68**, 28 (1958).

[47] D. D. Fitts, M. Siegel, and K. Mislow, *J. Am. Chem. Soc.*, **80**, 480 (1958), using the method of J. G. Kirkwood, *J. Chem. Phys.*, **5**, 479 (1937).

two calculations are inconsistent with each other, and the earlier results are at variance with the experimentally determined configurations of optically active biphenyls discussed below.

Most of the methods used to correlate configurations of molecules with asymmetric atoms (Sec. 5-4) cannot, by their nature, be used to correlate the configurations of such molecules with the configuration of an active biphenyl. An exception is the asymmetric transformation method (Sec. 5-4*e*). If one carries out a reaction involving a dissymmetric biphenyl (say B) and another optically active molecule, let us say (−)-A, the two transition states (−)-A·(+)-B and (−)-A·(−)-B are diastereoisomeric and thus the reactions involving (+)-B and (−)-B will, in general, proceed at different rates; i.e., there will be an experimental distinction in such a reaction between (+)-B and (−)-B, allowing one to say which reacts via the transition state of lower energy. *If* (and only if) one can, in addition, say, on the basis of reasonable molecular models, which of the configurations of B will enter into the transition state of lower energy (or lesser steric interference) with (−)-A, a correlation of the sign of rotation of B and its configuration becomes possible.

CH_3–C(H)(OH)–R (S)-(+), $R = (CH_3)_3C$ or n-C_6H_{13} + O_2N/O_2N biphenyl CH_2–C=O–CH_2 (*dl*) $\xrightarrow[\text{reaction}]{\text{incomplete}}$ CH_3–C(=O)–R + O_2N/O_2N biphenyl CH_2–CHOH–CH_2 (R)-(−)

Fig. 6-61. Asymmetric Meerwein-Ponndorf-Verley reduction used in establishing biphenyl configuration.

Several successful applications of this principle have been made[48–52] of which two will be described here in some detail.

In one of the approaches,[48] an asymmetric Meerwein-Ponndorf-Verley reduction (cf. Chap. 4, page 72) was employed, using (S)-(+)-pinacolyl alcohol (methyl-*t*-butylcarbinol) or (S)-(+)-2-octanol (Fig. 6-61) as reducing agents and the biphenyl ketone shown in Fig. 6-61 as the compound to be reduced. It should be noted that the carbinol carbon in the product alcohol is *not* asymmetric and that it does not matter which side of the ketone group is approached by hydrogen. (In this respect, the reduction is different from the one shown in Fig. 4-41.) However, it *does* matter in which configuration the biphenyl system of the product ends up, for the *t*-butyl group of pinacolyl alcohol (or the *n*-hexyl group in 2-octanol) is much more crowded in the transition state leading to hydride transfer to the S enantiomer of the ketone

[48] P. Newman, P. Rutkin, and K. Mislow, *J. Am. Chem. Soc.,* **80,** 465 (1958).
[49] J. A. Berson and M. A. Greenbaum, *J. Am. Chem. Soc.,* **80,** 445 (1958).
[50] K. Mislow and F. A. McGinn, *J. Am. Chem. Soc.,* **80,** 6036 (1958).
[51] J. A. Berson and M. A. Greenbaum, *J. Am. Chem. Soc.,* **80,** 653 (1958).
[52] K. Mislow, V. Prelog, and H. Scherrer, *Helv. Chim. Acta,* **41,** 1410 (1958).

than in hydride transfer to the R enantiomer. Therefore, if the reduction is terminated short of completion, a kinetic resolution of the ketone occurs such that the S enantiomer is enriched in the ketone left behind, whereas the R enantiomer predominates in the alcohol reduction product. Experimentally it was found that the alcohol produced was levorotatory whereas the residual ketone was dextrorotatory. Therefore the (−) alcohol and (−) ketone shown in Fig. 6-61 have the R configuration, and the (+) alcohol and (+) ketone have the S configuration.† An analogous approach has been used[50] to establish the configuration of a binaphthyl similar to the biphenyl shown in Fig. 6-61. The results agree with calculation.[47]

An entirely different approach[49] is based on a reaction in which a biphenyl system is formed from a compound having asymmetric atoms of the conventional type of known configuration. When thebaine (Fig. 6-62) is treated with phenylmagnesium bromide[53] the product is a mixture of two phenyldihydrothebaines. The absolute configuration of thebaine—whose dissymmetry is due to the presence of asymmetric atoms—is known[54] to be as shown in Fig. 6-62. The phenyldihydrothebaines that result in the Grignard treatment of thebaine have two dissymmetric groupings, namely, the biphenyl system and the asymmetric benzylic carbon. Since further degradation of both phenyldihydrothebaines, resulting in elimination of the asymmetry at the benzylic carbon, yields identical, optically active substituted biphenyls, it follows that the two phenyldihydrothebaines differ only in configuration at the benzylic carbon *and have the same configuration of the biphenyl system.* It is seen, then, that the thebaine-phenyldihydrothebaine transformation is highly stereospecific as far as the configuration of the newly created biphenyl system is concerned: One of the two configurations of this system is obtained to the exclusion of the other. The thebaine-phenyldihydrothebaine rearrangement proceeds via an intermediate or transition state similar to that shown in Fig. 6-62 (center). It can readily be seen with models that, in view of the fact that the five-membered pyrrolidine ring prefers to be planar or nearly so, the intermediate (or related transition state) A is much preferred over B and therefore the biphenyl system in the two dextrorotatory phenyldihydrothebaines obtained in the reaction has the S configuration.

Other determinations[51, 52] of biphenyl configuration are based on an extension of Prelog's rule (Sec. 5-4*e*). A phenylglyoxylate of an optically active hydroxybiphenyl, $C_6H_5COCO_2R$ (R = biphenyl part of molecule), is treated with methylmagnesium iodide to give the corresponding atrolactate which is then hydrolyzed to atrolactic acid, $C_6H_5C(CH_3)OHCO_2H$. Because of the presence of the active biphenyl in the Grignard addition, an asymmetric synthesis occurs, and the atrolactic acid is optically active. Its sign of rota-

† It is very difficult to show the difference between the two transition states in a two-dimensional drawing. The reader is strongly urged to make scale models of (S)-(+)-pinacolyl alcohol and the ketone shown in Fig. 6-61 and convince himself that the less encumbered transition state of hydride transfer leads, indeed, to the R form of the biphenylic alcohol.

[53] Cf. L. Small, L. J. Sargent, and J. A. Bralley, *J. Org. Chem.,* **12,** 839 (1947); for interpretation of the chemistry, see R. Robinson, *Nature,* **160,** 815 (1947).

[54] J. Kalvoda, P. Buchschacher, and O. Jeger, *Helv. Chim. Acta,* **38,** 1847 (1955); H. Corrodi and E. Hardegger, *ibid.,* **38,** 2038 (1955).

tion is measured, and its configuration is thus known (cf. Fig. 5-16). Knowing the configuration of the atrolactic acid formed in preponderance, one can tell from which side of the keto group of the phenylglyoxylate the Grignard addition occurred predominantly. This is evidently the less shielded face of the keto group, and model consideration may then allow one to decide which configuration the biphenyl part of the phenylglyoxylic ester must have had in order to make this particular face of the ketone function more accessible than the other face. Both phenols of the type shown in Fig. 6-62 and alcohols of the type shown in Fig. 6-61 have been used in these experiments.

Once the absolute configurations of a few biphenyls were established by

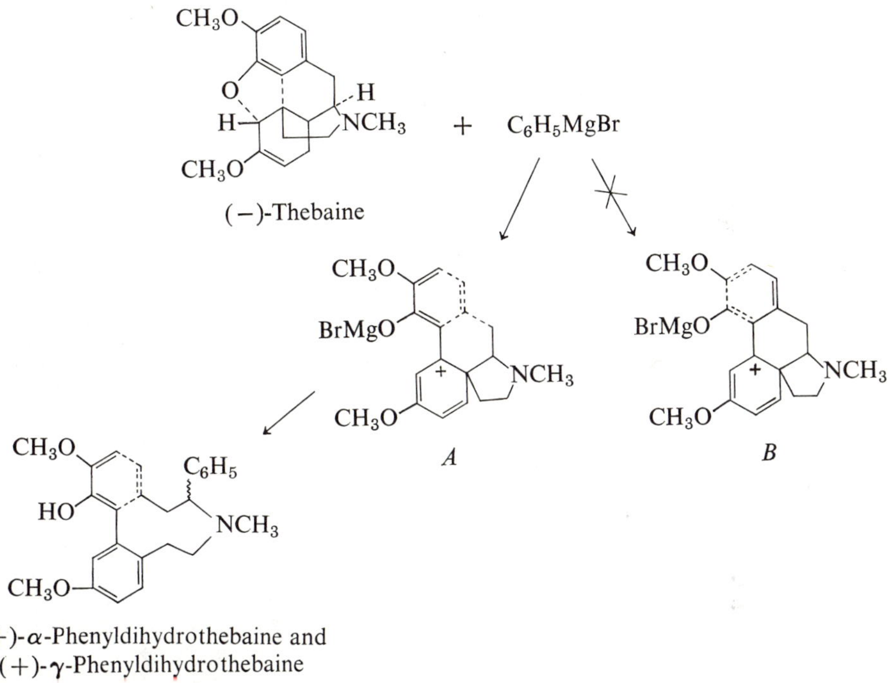

CH_3O … H … NCH_3 + C_6H_5MgBr

CH_3O

(−)-Thebaine

CH_3O / BrMgO / NCH_3 / CH_3O — *A*

CH_3O / BrMgO / NCH_3 / CH_3O — *B*

CH_3O / HO / C_6H_5 / NCH_3 / CH_3O

S(+)-α-Phenyldihydrothebaine and
S(+)-γ-Phenyldihydrothebaine

Fig. 6-62. Stereochemistry of the thebaine-phenyldihydrothebaine reaction.

correlations of the above type, configurations of other biphenyls could be determined by correlation methods of the type discussed in Sec. 5-4. Four methods which have been successfully used are the method of chemical transformation[55] (either without affecting the dissymmetric grouping at all or affecting it in a predictable way; see below), the method of quasi-racemates,[56] a variation of the Freudenberg displacement rule,[47, 57] and optical rotatory dispersion (cf. Chap. 14).[58]

[55] F. A. McGinn, A. K. Lazarus, M. Siegel, J. E. Ricci, and K. Mislow, *J. Am. Chem. Soc.*, **80,** 476 (1958).

[56] M. Siegel and K. Mislow, *J. Am. Chem. Soc.*, **80,** 473 (1958).

[57] See also K. Mislow and P. A. Grasemann, *J. Org. Chem.*, **23,** 2027 (1958).

[58] K. Mislow, M. A. W. Glass, R. E. O'Brien, P. Rutkin, D. H. Steinberg, and C. Djerassi, *J. Am. Chem. Soc.*, **82,** 4740 (1960).

O_2N, CO_2H, O_2N, CO_2H

(S)-(−)

(1) MeOH, H^+
(2) $LiAlH_4$
(3) HBr

O_2N, CH_2Br, O_2N, CH_2Br

(S)-(−)

(1) KCN
(2) H_2O, H^+

O_2N, CH_2, CO, O_2N, CH_2

(S)-(+)

(1) $NaBH_4$, $AlCl_3$
(2) H_2/Pd

H_2N, CH_3, H_3C, NH_2

(R)-(+)

H_2N, CH_3, H_2N, CH_3

(S)-(−)

(1) $NaNO_2$, H^+
(2) Cu_2Cl_2

Cl, CH_3, Cl, CH_3

(S)-(−)

H_3C, Cl, Cl, CH_3

(R)-(+)

(1) $NaNO_2$, H^-
(2) $Cu_2(CN)_2$

(1) $LiAlH_4$
(2) HBr
(3) $NaBH_4$, $AlCl_3$

NC, CH_3, H_3C, CN

(R)-(+)

H_3C, CN, H_3C, CN

(S)-(−)

(1) PCl_5, $ArSO_2NH_2$
(2) C_5H_5N

H_3C, CO_2H, H_3C, CO_2H

(S)-(+)

HO_2C, Cl, Cl, CO_2H

(R)-(+)

Fig. 6-63. Configurational correlation of biphenyls.

A number of biphenyls which have been correlated chemically[55] are shown in Fig. 6-63.† (For a more extensive table, see Ref. 45.)

Thermal analysis has been used to correlate, among other compounds, (S)-(−)-6,6′-dichloro-2,2′-diphenic acid and (S)-(+)-6,6′-dimethyl-2,2′-diphenic acid, which are found to form a solid solution [whereas the two (+) enantiomers form a quasi-racemic compound and are therefore of opposite configuration]. A variation of the displacement rule has been suggested[47] (and successfully tested) according to which "a symmetrically substituted hindered biaryl has the S (resp., R) configuration if, in going from an open to a bridged system (cf. Fig. 6-54), the optical activity suffers a marked shift in the positive (resp., negative) direction." This rule does not hold when the bridge contains a carbonyl function.[50]

d. Other Types of Asymmetry Due to Restricted Rotation. It has been emphasized before that even a molecule as simple as *n*-butane may exist in dissymmetric (gauche) conformations. If the molecule were frozen in that configuration, it would become resolvable. It is not necessary to have a biphenyl system in a molecule for the barrier to rotation to become high enough to prevent rapid racemization. Thus the substituted stilbenes shown in Fig. 6-64 are capable of resolution[61] because the phenyl ring and the

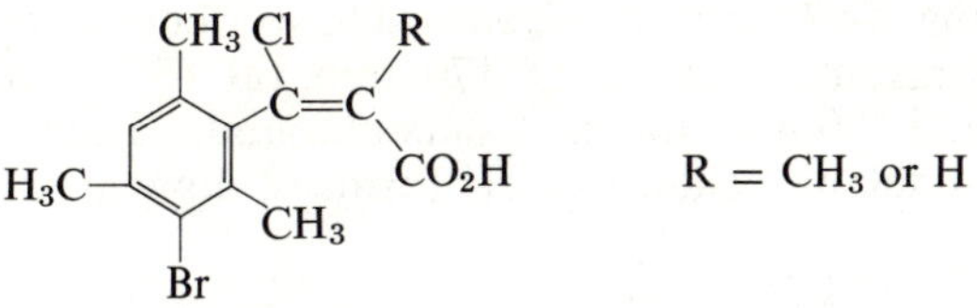

CH_3 Cl R
C=C
H_3C CO_2H R = CH_3 or H
CH_3
Br

Fig. 6-64. Resolvable stilbenes.

ethylenic double bond cannot readily become coplanar. The molecule is optically stable in boiling butanol but becomes racemizable under these conditions (half-life 200 min.) when the α-methyl group is replaced by hydrogen, presumably because of a lesser buttressing effect. Substitution of a second bromine in the ring introduces a plane of symmetry in the non-coplanar molecule and leads to loss of activity.

One of the early examples[62] of activity due to restricted rotation in a molecule other than a biphenyl is shown in Fig. 6-65. The groups attached to the amino nitrogen do not find room in the plane of the naphthalene nucleus

† It is noteworthy that the Sandmeyer reaction (replacement of —NH_2 by —Cl or —CN) proceeds without racemization. Other reactions which have been shown to proceed without racemization in the biphenyl series, although an ortho substituent is replaced, are the Curtius (Ref. 59*a*), Hofmann (Ref. 59), and Wolff rearrangements (Ref. 60).

[59] (*a*) F. Bell, *J. Chem. Soc.*, 835 (1934); (*b*) E. S. Wallis and W. W. Moyer, *J. Am. Chem. Soc.*, **55**, 2598 (1933).

[60] J. F. Lane and E. S. Wallis, *J. Org. Chem.*, **6**, 443 (1941).

[61] R. Adams and M. W. Miller, *J. Am. Chem. Soc.*, **62**, 53 (1940); R. Adams, A. W. Anderson, and M. W. Miller, *ibid.*, **63**, 1589 (1941).

[62] W. H. Mills and K. A. C. Elliot, *J. Chem. Soc.*, 1291 (1928).

and cannot readily pass through that plane.† The molecule is rather easily racemized ($t_{1/2} = 17$ min. at 15°); an analogous molecule without the nitro group could not be resolved. Derivatives of this type have also been investigated extensively by Adams and coworkers;[63] of particular interest is the effect of a substituent in the 4 position of the naphthalene system on the half-life of racemization[64] (Fig. 6-66, I; the methyl group in the 2 position makes the molecule resolvable, taking the place of the nitro group in the 8 position). The large difference between an electron-withdrawing group, such as nitro,

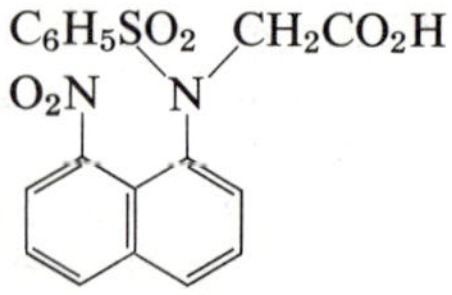

$C_6H_5SO_2$ CH_2CO_2H
O_2N N

Fig. 6-65. Resolvable naphthylamine derivative.

and an electron-donating group, such as hydroxyl, is ascribed to resonance stabilization of the coplanar transition state for racemization, as shown in Fig. 6-66, II. This matter will be returned to in the next section.

Although diphenic acid itself is not resolvable, the meta-bridged diphenic acids shown in Fig. 6-67 can be resolved and have moderate optical stability,[65] the half-life of racemization being 170 min. (at 43° in dioxane) for the C_8 compound and 120 min. for the C_{10} compound. The rapid interconversion of the non-coplanar, enantiomeric forms of diphenic acid undoubtedly

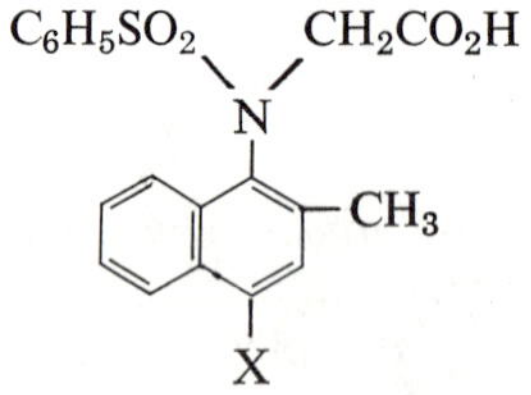

I: $C_6H_5SO_2$, CH_2CO_2H, N, CH_3, X

X	$t_{1/2}$ hr.*
NO_2	0.42
OH	8.7

*in dimethylformamide at 118°

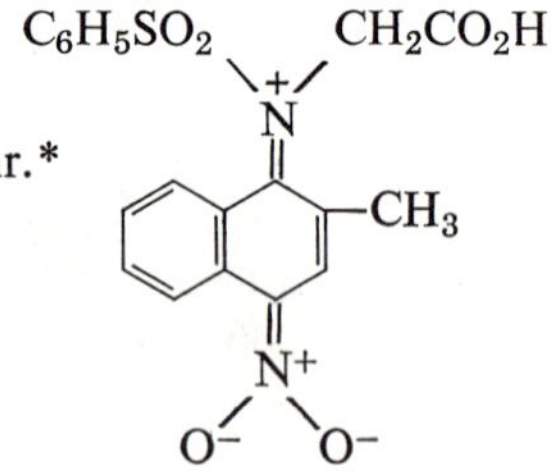

II: $C_6H_5SO_2$, CH_2CO_2H, $\overset{+}{N}$, CH_3, N^+, O^-, O^-

Fig. 6-66. Effect of substituents in the 4 position on half-life of racemization of 1-naphthylamine derivatives.

involves a passage through a planar transition state in which carboxyl slips past hydrogen. In the meta-bridged acids this is not possible, and racemiza-

† The alternative possibility that activity is due to the asymmetrically substituted nitrogen atom has been ruled out; cf. Chap. 13.

[63] R. Adams and J. S. Dix, *J. Am. Chem. Soc.*, **80**, 4579 (1958).

[64] R. Adams and K. V. Y. Sundstrom, *J. Am. Chem. Soc.*, **76**, 5474 (1954); R. Adams and H. H. Gibbs, *ibid.*, **79**, 170 (1957).

[65] R. Adams and N. Kornblum, *J. Am. Chem. Soc.*, **63**, 188 (1941).

tion must involve slippage of carboxyl past carboxyl. The activation energy for this process is not very sensitive to the length of the bridge.

A somewhat similar type of restricted rotation is found in the so-called[66] "ansa compounds" shown in Fig. 6-68 (ansa = Latin for handle). Provided that the bridge is small enough, the phenyl ring cannot swivel through

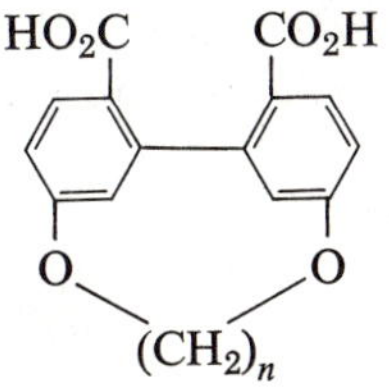

Fig. 6-67. Meta-bridged diphenic acids.

the alicyclic ring, and the molecule is resolvable. Compounds I and II with two ortho substituents and a 10-membered methylene bridge are optically stable. When there is only one substituent, activity disappears in the corresponding compound III, but compound V with an eight-membered methylene bridge can be resolved and is optically stable. Compound IV with its

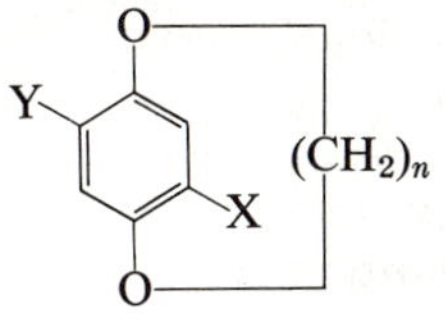

I $X = CO_2H$, $Y = Br$, $n = 10$
II $X = Y = Br$, $n = 10$
III $X = CO_2H$, $Y = H$, $n = 10$
IV $X = CO_2H$, $Y = H$, $n = 9$
V $X = CO_2H$, $Y = H$, $n = 8$

Fig. 6-68. Restricted rotation in ansa compounds.

nine-membered bridge is of intermediate optical stability; it can be resolved but is racemized on heating ($t_{1/2} = 444$ min. at 95.5°).

Yet another, similar type of activity is found in the "paracyclophanes" shown in Fig. 6-69.[67] The compound with $m = n = 2$ is optically stable, that with $m = 3$, $n = 4$ is resolvable but racemizes at 160°, and that with

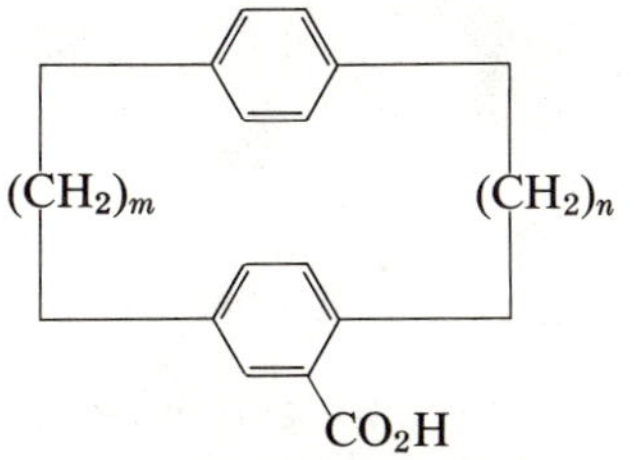

$m = n = 2$
$m = 3$, $n = 4$
$m = n = 4$

Fig. 6-69. Resolvability in paracyclophanes.

[66] A. Lüttringhaus and H. Gralheer, *Ann.*, **550**, 67 (1940); **557**, 108, 112 (1945); A. Lüttringhaus and G. Eyring, *Ann.*, **604**, 111 (1957). For a carbon-bridged analog, see A. T. Blomquist, R. E. Stahl, Y. C. Meinwald, and B. H. Smith, *J. Org. Chem.*, **26**, 1687 (1961).

[67] D. J. Cram and N. L. Allinger, *J. Am. Chem. Soc.*, **77**, 6289 (1955); D. J. Cram, W. J. Wechter, and R. W. Kierstead, *ibid.*, **80**, 3126 (1958).

$m = n = 4$ cannot be resolved. In the smaller members of the series, the phenyl rings are evidently forced out of the coplanar arrangement into a sort of sandwich-like arrangement.

All the molecules presenting conformational enantiomerism discussed so far have two enantiomeric conformations of minimum energy separated by one or two high energy maxima. In contrast, the molecules discussed earlier in this chapter whose stable conformations are not isolable have three energy minima (Fig. 6-3). It is significant, therefore, that two forms of one

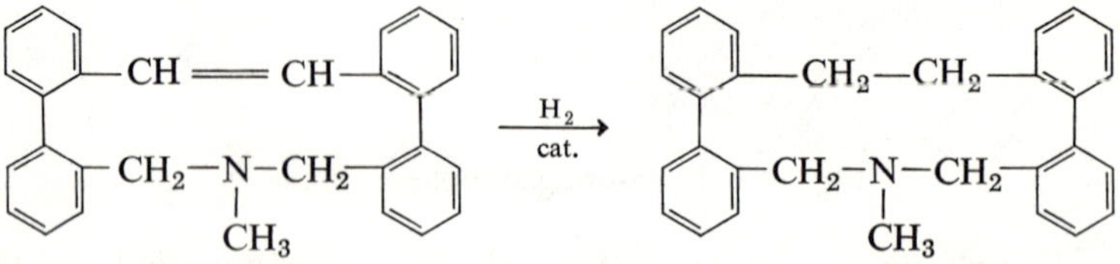

I: cis, m.p. 200° → gauche, m.p. 140°; trans, m.p. 194° → anti, m.p. 180° (II; interconverted by Δ; H_2, cat.)

Fig. 6-70. Isolable conformational isomers of substituted ethane.

such molecule, shown in Fig. 6-70 (II), have finally been isolated.[68] Since the two forms were formed by catalytic hydrogenation (involving cis addition; cf. Chap. 12) of the corresponding cis and trans olefins (Fig. 6-70, I), they probably correspond to the gauche and trans forms of compound II, respectively. It should be noted that the two molecules are geometrical rather than optical isomers, unlike the conformational enantiomers discussed earlier. The conformational isomers are interconverted upon heating, presumably by rotation about the CH_2—CH_2 bond. A similar situation will be discussed in more detail in Chap. 8.

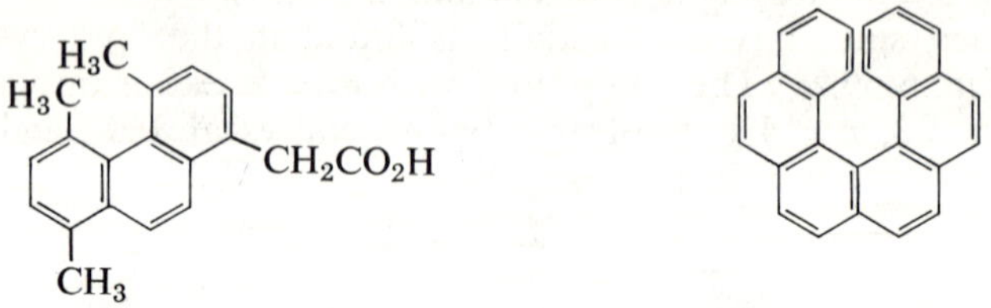

I II

Fig. 6-71. Dissymmetry due to molecular overcrowding.

Still a different kind of molecular dissymmetry is found in the phenanthrene derivatives shown in Fig. 6-71. The 4,5-dimethylsubstituted phenanthrene I can be resolved but is of low optical stability.[69] On the other hand, hexahelicene (II) is very stable toward racemization and has the amazingly large rotation[70]

[68] G. Wittig, G. Koenig, and K. Clauss, *Ann.*, **593**, 127 (1955).
[69] M. S. Newman and A. S. Hussey, *J. Am. Chem. Soc.*, **69**, 3023 (1947).
[70] M. S. Newman and D. Lednicer, *J. Am. Chem. Soc.*, **78**, 4765 (1956).

of $[\alpha]_D^{25} = 3700°$. This molecule is dissymmetric because molecular overcrowding twists the terminal rings out of the coplanar arrangement.

e. Effect of Atropisomerism on Physical Properties.[80, 91b] Assuming that the Arrhenius PZ factor in the equation $k = PZe^{-E/RT}$ is normal or slightly below normal, as is usually the case for thermal racemization of biphenyls, a compound will have isolable rotational isomers (i.e., be resolvable) if the barrier to rotation is greater than about 20 kcal./mole.[71] A barrier of at least 16 kcal./mole[72] may lead to such fleeting indications of resolvability as mutarotation of diastereoisomeric salts (first-order asymmetric transformation; cf. Sec. 4-2*e*). With barriers much smaller than that, resolution will not succeed. Nevertheless there is tangible evidence of lack of coplanarity of the rings even in biphenyls that cannot be resolved. The most striking evidence for this comes from ultraviolet spectra. Thus, whereas biphenyl itself has a strong absorption maximum at 249 mμ(ε = 15,000) because of an excited state involving conjugation of the two benzene rings (Fig. 6-72), the introduction of even a single methyl group in the ortho position[73] shifts the absorption maximum to 237 mμ and lowers the extinction coefficient to 10,500. When there are two methyl groups in the ortho and ortho′ positions, although the molecule is not resolvable (cf. Sec. 6-4*a*), the biphenyl absorption band disappears,[74] and the absorption spectrum of the molecule resembles that of

Fig. 6-72. Excited state of biphenyl.

m-xylene. The reason for this is that an ortho-disubstituted biphenyl spends most of its time in a conformation which deviates considerably from the planar, and therefore, by the Franck-Condon principle, the excited state must deviate from the planar also. As a result, the extended resonance shown in Fig. 6-72 is greatly dampened, and the biphenyl band disappears. The suppression of the resonance shown in Fig. 6-72 due to steric causes (in the present case, due to interference of ortho methyl groups with ortho hydrogens) is called "steric inhibition of resonance."

It is of interest that biphenyl itself, although it shows spectral evidence for considerable resonance involving both rings, is *not* actually coplanar but the rings form an angle with each other of approximately 40° in the vapor, as shown by electron diffraction.[75] Nevertheless there is considerable resonance of the type shown in Fig. 6-72. Theoretical considerations[76] indicate that

[71] G. B. Kistiakowsky and W. R. Smith, *J. Am. Chem. Soc.*, **58**, 1043 (1936).

[72] W. H. Mills, *J. Chem. Soc.*, 194 (1943).

[73] E. A. Braude, F. Sondheimer, and W. F. Forbes, *Nature*, **173**, 117 (1954).

[74] M. T. O'Shaughnessy and W. H. Rodebush, *J. Am. Chem. Soc.*, **62**, 2906 (1940). The disappearance of the biphenyl band was first observed by L. W. Pickett, G. F. Walter, and H. France [*ibid.*, **58**, 2296 (1936)] in bimesityl.

[75] I. L. Karle and L. O. Brockway, *J. Am. Chem. Soc.*, **66**, 1974 (1944); O. Bastiansen, *Acta Chem. Scand.*, **3**, 408 (1949).

[76] J. Guy, *J. chim. phys.*, **46**, 469 (1949).

resonance is not severely inhibited until the interplanar angle becomes quite large.

In di-ortho-substituted biphenyls the interplanar angle between the ring varies from 60 to 90°, depending on the substituent.[77] For example, in *o,o′*-dichlorobiphenyl, it is estimated to be between 62 and 74°. Of particular interest is the finding,[78] based on dipole-moment studies, that the rings are so inclined in their equilibrium positions as to place the chlorine atoms on

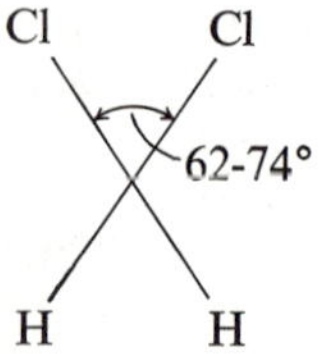

Fig. 6-73. Equilibrium conformation of *o,o′*-dichlorobiphenyl.

the same side, rather than on opposite sides, of the rings (see the end-on view in Fig. 6-73). This has been explained[78] on the basis of attractive London forces between the chlorine atoms.

Another interesting manifestation of steric inhibition of resonance is found in the diphenoquinone derivative shown in Fig. 6-74. This molecule exists as a biradical, as evidenced by its paramagnetism, the quinoid structure being avoided because of excessive crowding.[79]

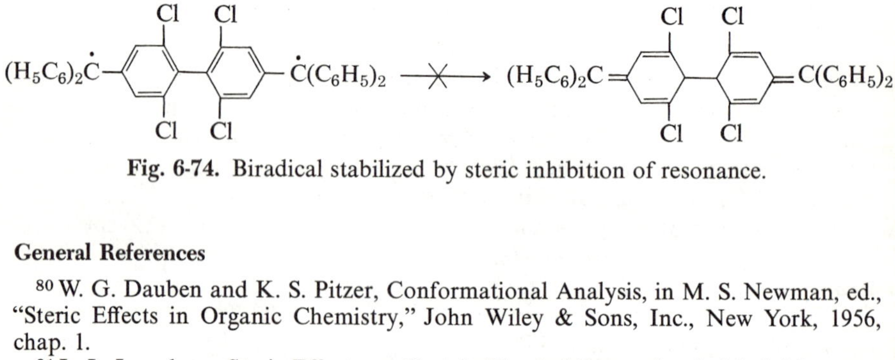

Fig. 6-74. Biradical stabilized by steric inhibition of resonance.

General References

[80] W. G. Dauben and K. S. Pitzer, Conformational Analysis, in M. S. Newman, ed., "Steric Effects in Organic Chemistry," John Wiley & Sons, Inc., New York, 1956, chap. 1.

[81] L. L. Ingraham, Steric Effects on Certain Physical Properties, in M. S. Newman, ed., "Steric Effects in Organic Chemistry," John Wiley & Sons, Inc., New York, 1956, chap. 11.

[82] S. Mizushima, "The Structure of Molecules and Internal Rotation," Academic Press, Inc., New York, 1954.

[83] D. Y. Curtin, Stereochemical Control of Organic Reactions, *Record Chem. Progr.* (*Kresge-Hooker Sci. Lib.*), **15,** 111 (1954).

---

[77] Ref. 81, p. 511.

[78] A. Weissberger, R. Sängewald, and G. C. Hampson, *Trans. Faraday Soc.,* **30,** 1 (1934); G. C. Hampson and A. Weissberger, *J. Am. Chem. Soc.,* **58,** 2111 (1936). See also A. C. Littlejohn and J. W. Smith, *J. Chem. Soc.,* 2552 (1954).

[79] E. Müller and H. Neuhoff, *Ber.,* **72,** 2063 (1939).

[84] F. Ebel, Umlagerung von Äthankörpern, in K. Freudenberg, ed., "Stereochemie," Franz Deuticke, Leipzig, 1932, pp. 825–851.

[84a] R. Kuhn, Atrop-Isomere, in K. Freudenberg, ed., "Stereochemie," Franz Deuticke, Leipzig, 1932, pp. 810–824.

[85] R. Adams and H. C. Yuan, The Stereochemistry of Diphenyls and Analogous Compounds, *Chem. Revs.,* **12,** 261 (1933).

[86] R. L. Shriner and R. Adams, Optical Isomerism, in H. Gilman, ed., "Organic Chemistry," John Wiley & Sons, Inc., New York, 1943, chap. 4, pp. 343–382.

[87] E. E. Turner and M. M. Harris, "Organic Chemistry," Longmans, Green & Co., New York, 1952, chap. 29.

[88] E. B. Wilson, The Problem of Barriers to Internal Rotation in Molecules, in I. Prigogine, ed., "Advances in Chemical Physics," vol. II, Interscience Publishers, Inc., New York, 1959, pp. 367–393.

[89] D. J. Millen, Restricted Rotation about Single Bonds, in P. B. D. de la Mare and W. Klyne, eds., "Progress in Stereochemistry," vol. 3, Academic Press, Inc., New York, 1962, chap. 4.

[90] R. J. Gillespie and R. F. M. White, Nuclear Magnetic Resonance and Stereochemistry, in P. B. D. de la Mare and W. Klyne, eds., "Progress in Stereochemistry," vol. 3, Academic Press, Inc., New York, 1962 chap. 2.

[91] (*a*) E. E. Turner, Configuration and Steric Effects in Conjugated Systems, in G. W. Gray, ed., "Steric Effects in Conjugated Systems," Academic Press, Inc., New York, 1958, pp. 1–7; (*b*) G. H. Beaven, The Study of Steric Effects in Substituted Diphenyls by Ultraviolet Absorption Spectroscopy, *ibid.,* pp. 22–33.

[92] H. H. Lau, Prinzipien der Konformationsanalyse, *Angew. Chem.,* **73,** 423 (1961).

Chapter 7

SOME ASPECTS OF THE STEREOCHEMISTRY OF RING SYSTEMS

In this chapter, those aspects of the stereochemistry of alicyclic and some heterocyclic compounds which are not critically dependent on the conformation of these compounds will be considered. The more strictly conformational aspects of ring stereochemistry will be taken up in Chap. 8 for six-membered rings and in Chap. 9 for rings of other sizes.

7-1. Stereoisomerism of Rings

a. Number and Kind of Stereoisomers. A monosubstituted cyclane, such as cyclopropanecarboxylic acid (Fig. 7-1, I), always has a plane of symmetry. The substituted carbon is not asymmetric because one encounters the same atoms going around the ring in one direction as in the other. Introduction of additional substituents in the ring may lead to dissymmetry; thus 2,2-dimethylcyclopropanecarboxylic acid (Fig. 7-1, II) has (+) and (−) forms. A non-geminally disubstituted odd-membered ring has two asymmetric carbon atoms, and there are therefore two diastereoisomeric pairs of enantiomers (e.g., 2-methylcyclopropanecarboxylic acid, Fig. 7-1, III). When the two substituents are the same, as in cyclopropane-1,2-dicarboxylic acid, there are a *dl* pair and a meso form, just as in tartaric acid (Fig. 3-16). As was explained before (Chap. 2), since the ring defines a plane, one may also speak of cis-trans isomers in cyclic systems. Thus one of the two diastereoisomeric *dl* pairs of 2-methylcyclopropanecarboxylic acid (Fig. 7-1, III*a*) may be called cis and the other (III*b*) trans. Similarly, the *dl* pair of cyclopropane-1,2-dicarboxylic acid (IV*a*) is a trans form and the meso isomer (IV*b*) is a cis form. The configuration of one of the enantiomers of IV*a* is not, however, completely specified by saying that it is trans; one must also specify whether it is R or S (Sec. 5-2). The same is not true of IV*b,* whose configuration is completely specified by saying that it is meso.

Similar considerations apply to all odd-membered rings, no matter whether the substituents are 1,2, 1,3, 1,4, or in any other position relative to each other. Thus cycloheptane-1,4-dicarboxylic acid (Fig. 7-2) exists as meso-cis and *dl*-

Note: All references above 75 are listed in the General References at the end of the chapter.

trans forms. This is true regardless of the fact that the seven-membered ring (or, for that matter, any ring larger than a three-membered one) is not, in fact, planar. It is true even though in large rings, because of the relatively facile rotation about the ring bonds, the cis substituents may not actually be

CO_2H — I

H_3C, CO_2H, H_3C — II, *dl*

H_3C, CO_2H — IIIa, cis-*dl*

CO_2H, H_3C — IIIb, trans-*dl*

CO_2H, HO_2C — IVa, trans-*dl*

HO_2C, CO_2H — IVb, cis-meso

Fig. 7-1. Cyclopropane derivatives.

pointing in the same direction. In fact, it is necessary to *define* cis substituents as substituents that are on the same side of the ring when the ring is molded into a planar form (even though it may not normally assume that form) whereas trans substituents are defined as substituents on opposite faces of the plane of the ring in that particular form.†

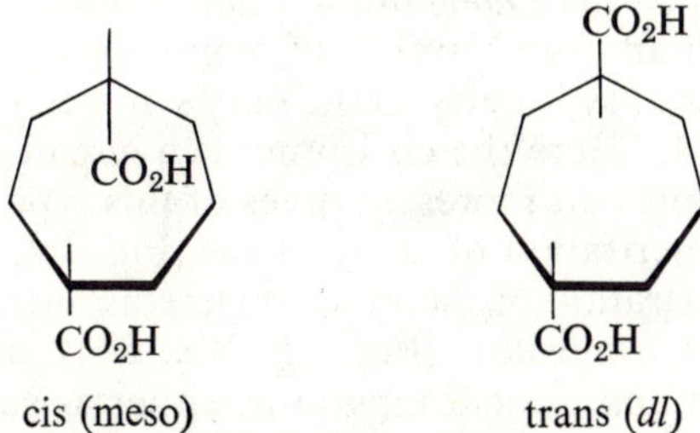

Fig. 7-2. Stereoisomers of cycloheptane-1,4-dicarboxylic acid.

In even-membered rings, there is a slight additional complication. When the substituents are across the ring from each other (on carbons No. 1 and

† The reader may convince himself with models that in a large ring (e.g., 15-membered) conformations exist in which the cis substituents are conformationally anti and the trans substituents are conformationally cis.

$1 + n/2$ in an n-membered ring) the molecule always has a plane of symmetry (provided that the substituents are not in themselves dissymmetric) and therefore no active forms exist, although the molecule may still exhibit cis-trans isomerism. Examples from 1,3-substituted cyclobutanes are shown in Fig. 7-3; similar considerations apply to 1,4-substituted cyclohexanes, 1,5-substituted cycloöctanes, etc.

Comparison of the molecules shown in Figs. 7-2 and 7-3 (II and III) is instructive in illustrating the difference between optical and geometrical isomerism previously mentioned in Chap. 2. The isomers shown in Fig. 7-2 are optical isomers as well as geometrical isomers, the cis and trans forms may be called diastereoisomers of each other, and the cis isomer is a meso form. *This is because at least one isomer in the set* (the trans *dl* pair) *exists in optically active forms.* In contrast, the isomers shown in Fig. 7-3 (II and III) are geometrical isomers but not optical isomers. They are not usually called diastereoisomers of each other (see, however, Ref. 1) and are not called

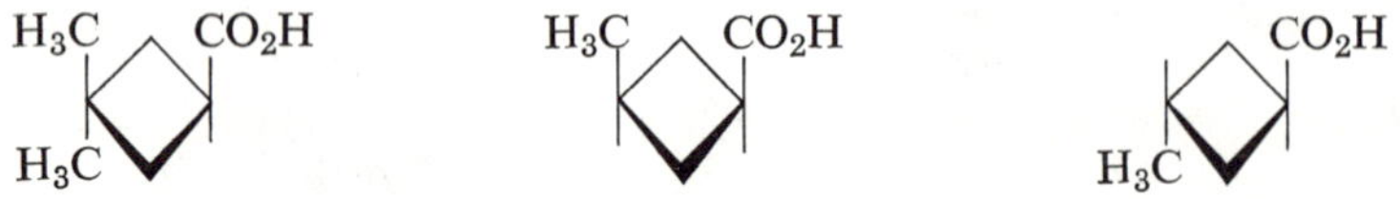

H_3C CO_2H H_3C — I

H_3C CO_2H — IIa (cis)

CO_2H H_3C — IIb (trans)

HO_2C CO_2H — IIIa (cis)

CO_2H HO_2C — IIIb (trans)

Fig. 7-3. Inactive 1,3-substituted cyclobutanes.

meso forms. *This is because none of the isomers in this set is optically active.*

Certain rings with an even number of members equal to or greater than six may have a center of symmetry as the only symmetry element. Examples are shown in Fig. 7-4. Here the cis isomer is a *dl* pair but the trans isomer is a meso form. Important representatives of this type are the diketopiperazines formed by dimerization of amino acids (Fig. 7-4, II).

An interesting illustration of the kind of stereoisomerism discussed in this section is found in the inositols (Fig. 7-5, X = OH), which are widely distributed in the plant and animal kingdoms either as such or in the form of derivatives, and in the hexachlorocyclohexanes (Fig. 7-5, X = Cl), one of which (the gamma isomer) is a widely used insecticide. These compounds have eight cis-trans isomers of which seven are meso forms and one is a *dl* pair. The isomers, with their names or symbols (name for X = OH, symbol for X = Cl), are shown in Fig. 7-5; the reader will readily detect the plane or

[1] G. W. Wheland, "Advanced Organic Chemistry," 3d ed., John Wiley & Sons, Inc., New York, 1960, pp. 221 and 295–298.

center of symmetry in the seven meso forms. It is sometimes said that these compounds present optical isomerism without asymmetric carbons; however, in the one isomer which is resolvable, each carbon atom is asymmetric by virtue of the fact that its substituents are diastereoisomeric.

b. Determination of Configuration. The determination of the configura-

I (meso-trans) [*dl*-cis not shown]

meso-trans [center] II *dl*-cis [no center]

Fig. 7-4. Rings with center of symmetry.

tion of an asymmetric carbon atom in a ring is no different from the corresponding determination in an acyclic compound and has, in fact, already been discussed in connection with the correlation of shikimic acid to the sugars (cf. Fig. 5-15). Somewhere along the line, of course, the ring must be broken (or formed) away from the asymmetric atom. When there is more

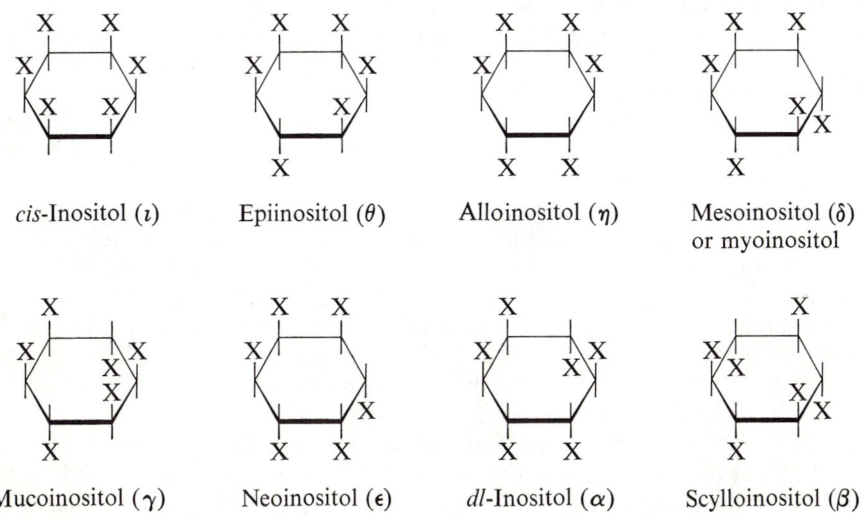

Fig. 7-5. Inositols and hexachlorocyclohexanes.

than one asymmetric atom, in principle, the configuration of each of them could be established separately by the methods already discussed. In fact, this is rarely done; it is usually much easier to determine the configuration of the second atom by correlating it with that of the first. This amounts to determining whether appropriate substituents are cis or trans with respect to each other. An example has already been given for the case of dihydroshikimic acid (Fig. 5-15).

Six methods have been used to determine relative configuration within one molecule, namely, (1) testing resolvability, (2) determining the number of isomers formed upon chemical transformation, (3) establishing the ease of bridging or ring formation, (4) drawing conclusions from mechanistic considerations, (5) drawing conclusions from physical properties, and (6) correlating one compound with another. Only the first two and last of these methods are unequivocal, and the last one only if appropriate precautions are taken.

The method of testing resolvability is straightforward when it is applicable: For example, of the two cycloheptane-1,4-dicarboxylic acids (Fig. 7-2) evidently only one (the trans isomer) is resolvable whereas the other (the cis isomer), being a meso form, is not. If both isomers are available and one can be resolved, or if only one isomer is available and it is resolvable, assignment of configuration is immediate. If both isomers are available but, because of technical difficulties, neither can be resolved, the method obviously fails. If only one isomer is available and it cannot be resolved, no conclusion can be drawn; it may be the meso form, but it is also possible that resolution failed for technical reasons rather than because the compound is not resolvable. The resolvability method has been used to determine the configurations of the hexahydrophthalic acids[2] and epoxyethane-1,2-dicarboxylic acids[3] (Fig. 7-1, IV, O instead of CH_2). In both cases, the trans acid was resolved whereas attempts to resolve the cis isomer failed. This particular method is limited to cases where one isomer is a meso form and the other a *dl* pair. It is not limited to cyclic compounds but is a general method for assigning relative configurations to the asymmetric atoms in diastereoisomers.

An improved way of establishing that a given compound is a meso form is to synthesize the compound under mild conditions (not affecting the asymmetric atoms) from an optically active precursor. If the product is inactive, one can be reasonably sure that it is a meso form (excluding the possibility of racemization of an active form). This method may be applicable even if only the meso form (and not its racemic diastereoisomer) is available. An example is the establishment of configuration of *cis*-hexahydrophthalic acid by saponification, under mild conditions, of its active monomethyl ester.[4] The product was inactive and is therefore meso (Fig. 7-6). Analogous cases in acyclic diastereoisomers have been shown in connection with the establishment of the configuration of glucose (cf. Sec. 5-4*b*).

Configurational assignments may also be made on the basis of the number of chemical transformation products obtained from a given compound. The method is illustrated in Fig. 7-7 for the 2,5-dimethylcyclopentane-1,1-dicarboxylic acids.[5] The meso-cis isomer, upon thermal decarboxylation, gives rise to two discrete and separable meso forms of 2,5-dimethylcyclopentanecarbox-

[2] A. Werner and H. E. Conrad, *Ber.*, **32**, 3046 (1899).
[3] R. Kuhn and F. Ebel, *Ber.*, **58**, 919 (1925).
[4] G. Vavon and P. Peignier, *Bull. soc. chim. France,* [4]**45**, 293 (1929).
[5] J. Wislicenus, *Ber.*, **34**, 2565 (1901).

ylic acid (depending upon which carboxyl group is lost) whereas the *dl*-trans isomer of the dicarboxylic acid gives rise to a single *dl*-trans monocarboxylic acid (regardless of which carboxyl is lost). It might be noted that this method also may be extended to acyclic diastereoisomers and that a conclusive result is obtained only if the meso-cis isomer is available as a starting material and if, indeed, two meso forms of the decarboxylation product can be demonstrated to have been formed. (It is conceivable that, for mechan-

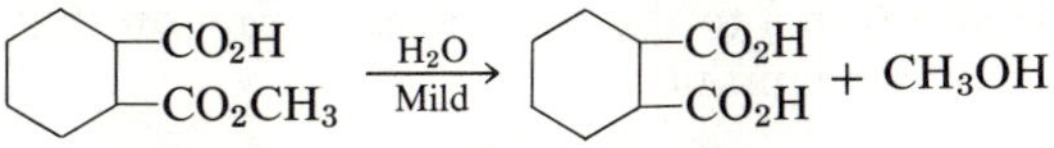

$-CO_2H$, $-CO_2CH_3$ $\xrightarrow[\text{Mild}]{H_2O}$ $-CO_2H$, $-CO_2H$ + CH_3OH

Active Inactive

Fig. 7-6. Proof of meso-cis configuration.

istic reasons, only one of the isomers is formed or that, for technical reasons, only one can be isolated.)

The above methods are applicable only to sets of diastereoisomers that contain meso forms. The method of ring formation, already illustrated in connection with dihydroshikimic acid (Fig. 5-15), is more general, but caution must be exercised in drawing conclusions. For example, it is quite clear from consideration of models that only the cis isomer of cyclopropane-1,2-dicarboxylic acid (Fig. 7-1, IV*b*) can form an intramolecular (cyclic)

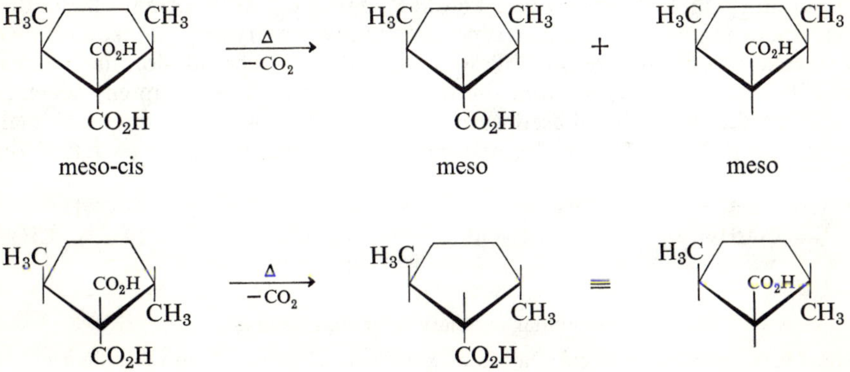

H_3C CH_3 CO_2H CO_2H $\xrightarrow[-CO_2]{\Delta}$ H_3C CH_3 CO_2H + H_3C CH_3 CO_2H

meso-cis meso meso

H_3C CO_2H CH_3 CO_2H $\xrightarrow[-CO_2]{\Delta}$ H_3C CH_3 CO_2H $\equiv$ H_3C CO_2H CH_3

dl-trans *dl*

Fig. 7-7. Proof of configuration of the 2,5-dimethylcyclopentane-1,1-dicarboxylic acids.

anhydride, whereas in the trans isomer (Fig. 7-1, IV*a*) the carboxyl groups are too far apart for anhydride formation. In fact, of the two known acids, only one forms a cyclic anhydride upon treatment with acetyl chloride,[6] and it seems quite safe to assign the cis configuration to this particular isomer. Similarly, it is evident that of the two isomers of 3-hydroxycyclohexanecarboxylic acid (Fig. 7-9) the one that forms a cyclic lactone must be the

[6] E. Buchner, *Ber.*, **23**, 701 (1890).

cis isomer.[7] However, the method is not foolproof; thus, for example, *both* isomers of cyclohexane-1,2-dicarboxylic acid form a cyclic anhydride upon treatment with acetyl chloride, although anhydride formation from the cis isomer appears to be more facile.[8] In fact, closure of a five-membered ring (as exists in the anhydride) across adjacent trans bonds appears entirely reasonable from models in cyclohexane and larger rings (cf. Chaps. 8 and 10), although, as mentioned earlier, 1,3 and 1,4 ring closure can occur only cis in a six-membered ring.† In case of doubt, it is obviously desirable to subject *both* isomers (cis and trans) to the ring-closure reaction that is to be used in the establishment of configuration.

OH, Cl —OH^-, fast→ O

OH, Cl —OH^-, slow→ O (hydride shift)

Fig. 7-8. Reaction of 2-chlorocyclohexanols with alkali.

Sometimes mechanistic considerations may be invoked in configurational assignments. For example, of the two 2-chlorocyclohexanols (Fig. 7-8), only one forms an epoxide upon treatment with mineral base whereas the other is rather sluggishly converted to a ketone.[9] Since epoxide formation is a displacement reaction of the S_N2 type and involves rearward attack, as shown in acyclic cases,[10] it is reasonable to assume that the epoxide-forming chlorohydrin is the trans isomer and the other is the cis. The argument becomes much weaker when it is based on mechanistically less well-understood analogies. In this category is the assumption often made in the past that the

OH, CO_2H —$LiAlH_4$→ OH, CH_2OH —TsCl→ OH, CH_2OTs —$LiAlH_4$→ OH, CH_3

and similar correlation for trans isomers

Fig. 7-9. Configurational correlation of 3-methylcyclohexanol with 3-hydroxycyclohexanecarboxylic acid.

isomer formed predominantly in catalytic hydrogenation in acid medium of an aromatic ring or a ketone (such as a methylcyclohexanone) is the cis isomer. As will be explained in Chap. 8, this assumption is now known to

† The subject of fused rings will be resumed in Chap. 10.

[7] W. H. Perkin and G. Tattersall, *J. Chem. Soc.,* **91,** 480 (1907).

[8] A. Baeyer, *Ann.,* **258,** 216 (1890).

[9] P. D. Bartlett, *J. Am. Chem. Soc.,* **57,** 224 (1935).

[10] For example, S. Winstein and H. J. Lucas, *J. Am. Chem. Soc.,* **61,** 1576 (1939); H. J. Lucas and C. W. Gould, *ibid.,* **63,** 2541 (1941); H. J. Lucas and H. K. Garner, *ibid.,* **70,** 990 (1948).

be sometimes incorrect, and reasoning of this type should be used only with great circumspection and when the analogies being made are quite close and reasonably well understood.

Arguments based on physical properties are on a similar footing. When the rings are small and therefore relatively rigid and the relation of the property to be studied to configuration is well understood, reliable results may be obtained. For example,[11] the first dissociation constant of *cis*-cyclopropane-1,2-dicarboxylic acid corresponds to a pK_a of 3.33 whereas the trans isomer has a pK_a of 3.65; for the second proton, pK_a is 6.47 for the cis isomer and 6.15 for the trans isomer. The greater strength of the cis diacid and lesser acid strength of the corresponding monoanion are to be expected on the basis of the greater proximity of the functional groups in the cis isomer† and could have been used to assign the configurations of the acids. With the cyclopentane-1,2-dicarboxylic acids the situation is already not quite so clear-cut, for although pK_2 is smaller for the trans acid (5.91) than for the cis (6.51), pK_1 is also smaller for the trans isomer (3.89 vs. 4.37). In the cyclohexanedicarboxylic acids the non-planar shape of the ring must definitely be taken into account; these will be considered in Chap. 8.

Dipole moments could, in principle, be used to establish configuration; thus one would predict that *cis*-1,2-dichlorocyclopentane should have a larger dipole moment than the trans isomer. Data of this type are available in six-membered rings where again the situation is more complicated than in the smaller systems; they will be considered in Chap. 8.

Infrared spectra may also be helpful in assigning configuration. For example, *cis*-cyclopentane-1,2-diol in dilute carbon tetrachloride solution shows[12] hydroxyl stretching frequencies due to hydrogen-bonded hydroxyl at 3572 cm.$^{-1}$ as well as due to free hydroxyl at 3633 cm.$^{-1}$, whereas the trans isomer, in which the hydroxyl groups are too far apart for intramolecular bonding, shows only the unbonded absorption at 3620 cm.$^{-1}$.

Of particular value in the assignment of relative configuration is X-ray diffraction. Elucidation of the configuration of the antibiotic fumagillin[12a] provides an elegant example of this method.

Empirical assignments based on differences in density, refractive index, and other physical properties whose origin is not so well understood have been made in some instances but are often of doubtful reliability; proper use of such data will be discussed in Sec. 8-4.

As in the case of acyclic stereoisomers, once the configuration of certain compounds has been established, that of others can be elucidated by a process of chemical correlation. For example, the configuration of *cis*-3-methylcyclohexanol has been established by correlation with the known configuration (see above) of 3-hydroxycyclohexanecarboxylic acid, as shown in

† The first proton is repelled by the adjacent carbonyl dipole; the second is held back by the adjacent carboxylate anion. For a more detailed discussion, see Sec. 12-3*c*.

[11] A. Wasserman, *Helv. Chim. Acta*, **13**, 207, 223 (1930).
[12] L. P. Kuhn, *J. Am. Chem. Soc.*, **74**, 2492 (1952).
[12a] N. J. McCorkindale and J. G. Sime, *Proc. Chem. Soc.*, 331 (1961).

Fig. 7-9.[13] Such correlations involve a complication not present in the configurational correlation of optical isomers (Sec. 5-4*a*). The equilibrium between enantiomers necessarily corresponds to a 50-50 mixture, so that any reaction in an attempted correlation which racemizes the asymmetric atom will immediately be detected by the resultant loss of optical activity. However, the equilibrium between cis-trans isomers may be quite far from 50-50, and if, by mischance, one of the reactions in an attempted correlation leads to epimerization of one of the asymmetric carbons, this fact may escape notice. For example if (and this is a most unlikely eventuality)† the lithium aluminum hydride reduction of *cis*-3-hydroxycyclohexanecarboxylic had involved epimerization of the carboxyl group to the trans position, the 3-methylcyclohexanol obtained as shown in Fig. 7-9 would be the trans isomer to which the cis configuration might then have been erroneously assigned. To avoid this pitfall, it is necessary to undertake the transformations involved in the correlation starting with *both* geometrical isomers. If a process of equilibration by epimerization is involved somewhere along the line, then both epimeric starting materials will be so equilibrated and will necessarily give rise to the same product or mixture of products. The fact, established specifically in the case shown in Fig. 7-9, that the two epimeric starting materials give different (epimeric) products rules out the possibility of an equilibration of this type.

7-2. Stability of Rings

a. As a Function of Ring Size. A number of interesting studies correlating stability of rings with size are available. The meaning of "stability" in this context is subject to a certain amount of confusion. For example, ease of ring closure has occasionally been used as an indicator of stability. However, ease of ring closure is a matter of kinetics and depends on the difference in free energy between the acyclic starting material and the transition state for ring closure; it is not necessarily related to the stability of the product. Therefore this subject will be considered separately (Sec. 7-3). Thermodynamic stability of rings should be measured by the position of equilibrium in a reaction in which an open-chain and a cyclic compound are equilibrated (such as a hydroxy acid–lactone equilibrium) but very little information of this type is available. The most informative studies so far are thermochemical in nature; they are concerned with heats of combustion of cycloalkanes and the differences in enthalpy between them. Several such measurements are summarized in Table 7-1. The heat of combustion per methylene group is high in cyclopropane, drops to a minimum in cyclohexane (which has about the same heat of combustion per methylene group as an open-chain

† Unlikely on two grounds: one that the cis-1,3 isomer is actually the more stable (cf. Chap. 8) and the other that lithium aluminum hydride reduction rarely involves epimerization, e.g., D. S. Noyce and D. B. Denney, *J. Am. Chem. Soc.*, **72**, 5743 (1950). [For an unusual contrary case, see J. A. Berson and M. A. Greenbaum, *ibid.*, **81**, 6456 (1959).]

[13] H. L. Goering and C. Serres, *J. Am. Chem. Soc.*, **74**, 5908 (1952); D. S. Noyce and D. B. Denney, *ibid.*, **74**, 5912 (1952); S. Siegel, *ibid.*, **75**, 1317 (1953).

compound), rises to a maximum in cyclononane, and then drops again to reach the *n*-alkane value at about cyclotetradecane.

V. Prelog and H. C. Brown have classified[18] ring compounds in four categories, namely, "small rings" (three- and four-membered), "common rings" (five-, six-, and seven-membered), "medium rings" (8- to 11-membered†), and "large rings" (12-membered† and larger).

Table 7-1 shows that small and medium rings tend to have anomalously high heats of combustion whereas large rings have "normal" heats of combustion similar to those of appropriate acyclic analogs. Among the common rings, cyclohexane has a "normal" heat of combustion but that of cyclopentane and of cycloheptane is somewhat enhanced.

The high heat of combustion of small rings is appropriately accounted for by the Baeyer strain theory.[19] Baeyer pointed out that the bond angles in small rings (60° in cyclopropane, ca. 90° in cyclobutane, and ca. 108° in

Table 7-1
Heats of Combustion of Cycloalkanes per Methylene Group[a]

n	H_c/n	$(H_c/n) - 157.4$	n	H_c/n	$(H_c/n) - 157.4$
3	166.6	9.2	11	158.4	1.0
4	163.95	6.55	12	157.7	0.3
5	158.7	1.3	13	157.8	0.4
6	157.4	0.0	14	157.4	0.0
7	158.3	0.9	15	157.5	0.1
8	158.6	1.2	16	157.5	0.1
9	158.8	1.4	17	157.2	−0.2
10	158.6	1.2	∞[b]	157.4	0.0

[a] In kilocalories per mole of gaseous cycloalkane, divided by the number of methylene groups, n. Data from Refs. 14–17.

[b] Value per methylene group in an *n*-alkane.

cyclopentane‡) deviate from the normal tetrahedral angle of 109° 28′ and that therefore these rings are strained. The strain (often called angle strain

† Originally (Ref. 18, footnote 21) the 12-membered rings were classified with the medium ones. However, Table 7-1 indicates that their heat of combustion per methylene group is closer to that of the large rings than to that of the 8- to 11-membered rings. Other recent evidence (Chap. 9) also tends to group the 12-membered rings with the larger ones. In view of these new data, Profs. H. C. Brown and V. Prelog have agreed to the reclassification (private communication to the author).

‡ Because of non-coplanarity of the ring carbons (cf. Chap. 9) the angles in cyclobutane and cyclopentane may be slightly smaller.

[14] S. Kaarsemaker and J. Coops, *Rec. trav. chim.*, **71**, 261 (1952).

[15] J. W. Knowlton and F. D. Rossini, *J. Research Natl. Bur. Standards*, **43**, 113 (1949).

[16] R. Spitzer and H. M. Huffman, *J. Am. Chem. Soc.*, **69**, 211 (1947).

[17] H. van Kamp, Ph.D. Dissertation Vrije Universiteit te Amsterdam, Amsterdam, Netherlands, 1957; cf. J. Coops, H. van Kamp, W. A. Lambregts, B. J. Visser, and H. Dekker, *Rec. trav. chim.*, **79**, 1226 (1960).

[18] H. C. Brown, R. S. Fletcher, and R. B. Johannesen, *J. Am. Chem. Soc.*, **73**, 212 (1951).

[19] A. von Baeyer, *Ber.*, **18**, 2277 (1885).

or "Baeyer strain") is defined as ½(109° 28′ − actual bond angle), the factor of ½ being put in because the strain is spread over two bonds. Table 7-2 shows the angle strain for rings of different sizes.

It is seen that the decreasing angle strain in the cyclopropane-cyclobutane-cyclopentane series accounts well for the decreasing heat of combustion shown in Table 7-1. However, the increase in strain postulated for cyclohexane (assumed to be planar and having bond angles of 120°) is not in agreement with its especially low heat of combustion. It is now known[20, 21] that the strain theory does not apply to rings of six members and larger, because such rings are puckered. Cyclohexane, in fact, exists in a completely strain-free chair form (cf. Chap. 8), as evidenced by its "normal" heat of combustion. The high heat of combustion of the medium-sized rings is due not only to angle strain but also to the existence of eclipsed conformations in these rings as well as to crowding of atoms across the rings. This subject will be returned to in Chap. 9. Large rings again have "normal" heats of combustion, as one might expect on the assumption that they begin to resemble open-chain compounds. Again this is contrary to the strain theory which would postulate large negative angle strains and presumably large heats of combustion per methylene group for such rings.

Table 7-2
Angle Strain in Cycloalkanes

n	3	4	5	6	7	15
Strain	24°44′	9°44′	0°44′	(−5°16′)	(−9°51′)	(−23°16′)

Equilibrium studies measuring thermodynamic stability of rings are unfortunately almost non-existent. One of the few exceptions is the hydroxy-aldehyde–hemiacetal equilibrium of the sugars (Fig. 4-13). The predominance of pyranose (six-membered) over furanose (five-membered) rings suggests that the former are more stable. Another example will be discussed in Sec. 7-3*a*.

b. As a Function of the Nature of the Ring. So far we have discussed the stability of saturated alicyclic rings. The question may be raised as to the effect of introducing elements of unsaturation, hetero atoms, or other structural elements (such as phenyl rings) into the cycle. Unfortunately, very little thermochemical information on this point is available. The best that we can do in this section is to take stock of some systems that have been synthesized and are therefore stable enough to exist. Some of these systems are obviously quite strained. In a number of cases, the synthesis of similar systems of still greater strain has been attempted without success, and this fact will be noted. The inference should not be drawn, however, that because the synthesis of a given system has been unsuccessful in a single attempt or even in repeated attempts, the system is necessarily too strained to exist.

In Fig. 7-10 are listed some ring systems containing unsaturation which have been synthesized. A double bond may be introduced even in the small-

[20] For early suggestions, see H. Sachse, *Ber.*, **23**, 1363 (1890); *Z. physik. Chem.*, **10**, 203 (1892).
[21] E. Mohr, *J. prakt. Chem.*, **98**, 315 (1918).

est of rings; cyclopropene (I) and a number of its derivatives have been synthesized[22] despite the large strain that would appear to be involved, and the system is even found in the naturally occurring sterculic acid[23] (Fig. 7-10, II). The supposedly even more strained cyclopropenone has been synthesized in the form of its diphenyl and di-*n*-propyl derivatives[23a] (Fig. 7-10, III); the system appears to be appreciably stabilized by having aromatic character. The smallest known cyclic diene, cyclobutadiene (IV),[23b, c] has been isolated as a silver nitrate complex,[23d] the free hydrocarbon having very fleeting stability at best.[23d] The nickel chloride complex of tetramethylcyclo-

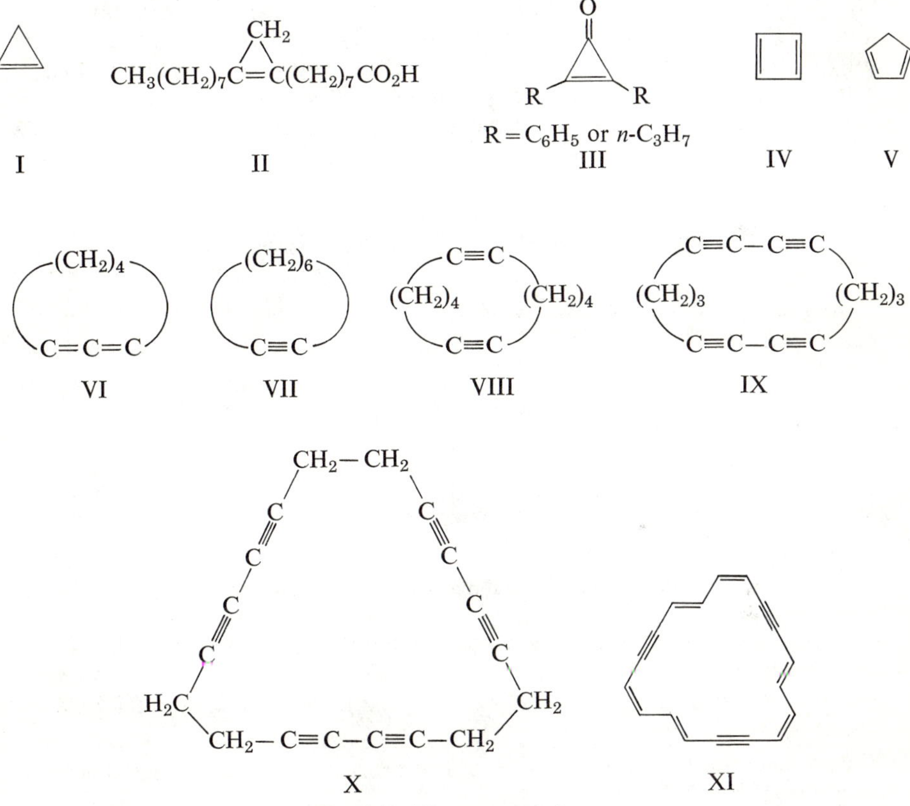

Fig. 7-10. Unsaturated rings.

[22] N. J. Demjanov and M. Dojarenko, *Ber.,* **56,** 2200 (1923).

[23] J. R. Nunn, *J. Chem. Soc.,* 313 (1952).

[23a] R. Breslow, R. Haynie, and J. Mirra, *J. Am. Chem. Soc.,* **81,** 247 (1959); R. Breslow and R. Peterson, *ibid.,* **82,** 4426 (1960); M. E. Volpin, Y. D. Koreshkov, and D. N. Kursanov, *Izvest. Akad. Nauk S.S.S.R., Otdel. Khim. Nauk,* 560 (1959), English translation, p. 535.

[23b] W. Baker and J. F. W. McOmie, Cyclobutadiene and Related Compounds, in D. Ginsburg, ed., "Non-benzenoid Aromatic Compounds," Interscience Publishers, Inc., New York, 1959, pp. 43–105.

[23c] Regarding an unusual suggestion for the possible structure of cyclobutadiene, see W. N. Lipscomb, *Tetrahedron Letters,* no. 18, 20 (1959).

[23d] M. Avram, E. Marica, and C. D. Nenitzescu, *Chem. Ber.,* **92,** 1088 (1959); M. Avram, Gh. Mateescu, I. G. Dinulescu, E. Marica, and C. D. Nenitzescu, *Tetrahedron Letters,* 21 (1961).

butadiene is also known.[23e] Considerably greater stability is encountered in cyclopentadiene (V), although this hydrocarbon dimerizes on standing. Both 1,3- and 1,4-cyclohexadiene are known, but a 1,2-diene (allenic) system was first claimed in 1,2-cycloheptadiene (VI).[24] The smallest cyclic acetylene whose synthesis has been reported[25] is cycloöctyne (VII).† A 12-membered cyclic diyne (VIII)[26] and a 14-membered cyclic tetrayne (IX)[27] have also been obtained, the latter by oxidative dimerization of hepta-1,6-diyne,

$$HC{\equiv}C(CH_2)_3C{\equiv}CH$$

The next lower homolog, hexa-1,5-diyne, $HC{\equiv}CCH_2CH_2C{\equiv}CH$, would not yield the cyclic dimer[27] but did give cyclic trimers (X), tetramers, and pentamers.[28] Treatment of the cyclic hexayne (X) with base converted it to the fully conjugated macrocycle XI which, in turn, upon mild catalytic hydrogenation gave the completely conjugated cyclic polyolefin cycloöctadecanonaene (XI, double instead of triple bonds),[29] a compound of considerable interest because of its possible aromatic character.[29a]

Table 7-3
Heats of Hydrogenation[a] and Relative Stability[b] of Cis and Trans Cycloölefins

Ring size	Heat of hydrogenation		$\Delta\Delta H^0$, kcal./mole	$\Delta\Delta F^0$, kcal./mole
	cis	trans		
8	−22.98	−32.24	−9.26	—
9	−23.62	−26.49	−2.87	−4.04
10	−20.67	−24.01	−3.34	−1.86
11	—	—	0.12[c]	0.67
12	—	—	−0.41[c]	0.49

[a] In acetic acid; Ref. 31.
[b] From Ref. 32.
[c] From temperature dependence of equilibrium constant; from Ref. 32.

A trans-substituted double bond can evidently not be accommodated in a small or common ring, but eight-membered (as well as larger) trans cycloölefins have been synthesized,[30] and even the trans-trans form of 1,5-cyclo-

† Cyclohexyne has been claimed as a reaction intermediate: F. Scardiglia and J. D. Roberts, *Tetrahedron*, **1**, 343 (1957).

[23e] R. Criegee and G. Schröder, *Angew. Chem.*, **71**, 70 (1959).
[24] A. E. Favorskii, *J. Gen. Chem. (U.S.S.R.)*, **6**, 720 (1936); see, however, L. Skattebäl, *Tetrahedron Letters*, 167 (1961).
[25] A. T. Blomquist and L. H. Liu, *J. Am. Chem. Soc.*, **75**, 2153 (1953); cf. N. A. Domnin, *J. Gen. Chem. (U.S.S.R.)*, **8**, 851 (1938).
[26] D. J. Cram and N. L. Allinger, *J. Am. Chem. Soc.*, **78**, 2518 (1956).
[27] F. Sondheimer, Y. Amiel, and R. Wolovsky, *J. Am. Chem. Soc.*, **79**, 6263 (1957).
[28] F. Sondheimer, Y. Amiel, and R. Wolovsky, *J. Am. Chem. Soc.*, **79**, 4247 (1957).
[29] F. Sondheimer and R. Wolovsky, *J. Am. Chem. Soc.*, **81**, 1771 (1959); *Tetrahedron Letters*, no. 3, 3 (1959).
[29a] See also F. Sondheimer, R. Wolovsky, and D. A. Ben-Efraim, *J. Am. Chem. Soc.*, **83**, 1686 (1961).
[30] K. Ziegler and H. Wilms, *Ann.*, **567**, 1 (1950).

öctadiene has been obtained.[30] Heats of hydrogenation,[31] summarized in Table 7-3, indicate greater thermochemical stability of the cis cycloölefins as compared with the trans isomers for 8-, 9-, and 10-membered cycles, the difference being by far the greatest for the eight-membered ring. In the 11- and

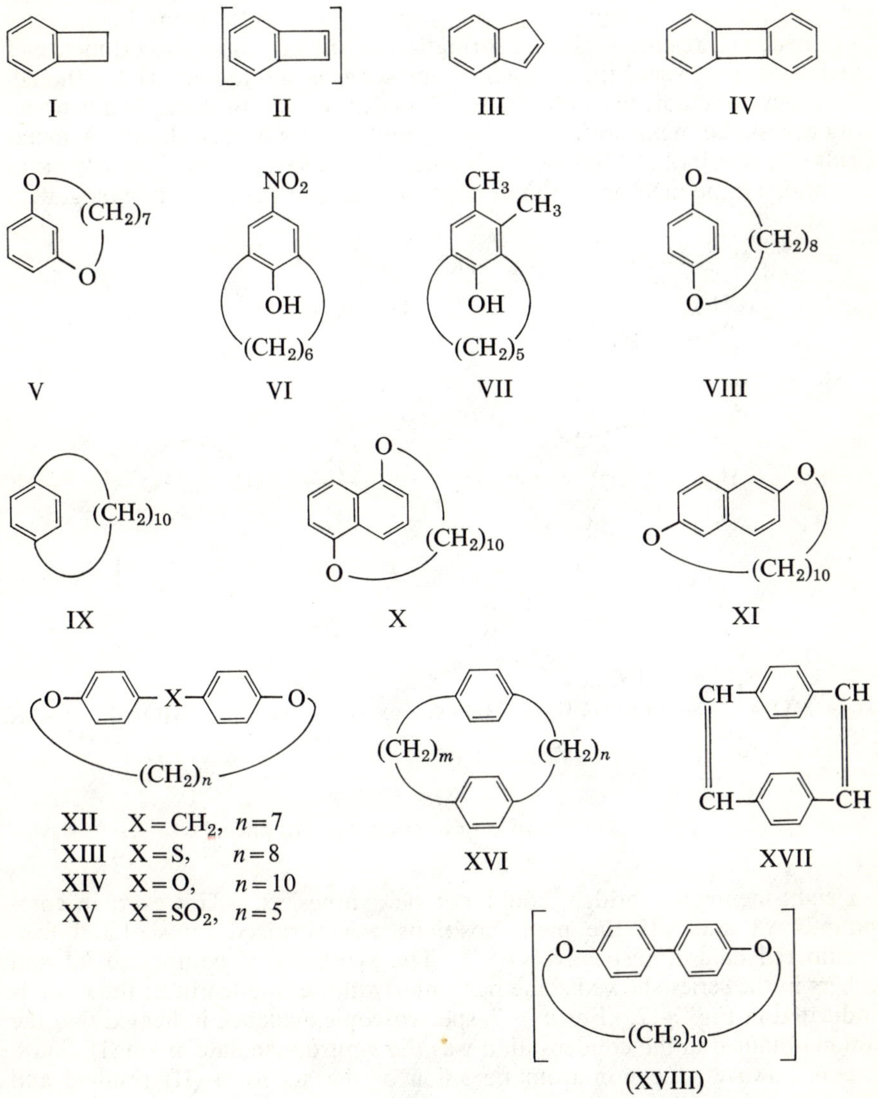
I
II
III
IV
O
$(CH_2)_7$
O
NO_2
OH
$(CH_2)_6$
CH_3
CH_3
OH
$(CH_2)_5$
O
$(CH_2)_8$
O
V
VI
VII
VIII
$(CH_2)_{10}$
O
$(CH_2)_{10}$
O
O
O
$(CH_2)_{10}$
IX
X
XI
O
X
O
$(CH_2)_n$
XII X=CH_2, $n=7$
XIII X=S, $n=8$
XIV X=O, $n=10$
XV X=SO_2, $n=5$
$(CH_2)_m$
$(CH_2)_n$
XVI
CH
CH
CH
CH
XVII
O
O
$(CH_2)_{10}$
(XVIII)

Fig. 7-11. Cycles containing benzene rings.

12-membered rings, there is little difference in enthalpy between cis and trans isomers.[32] Free energy also favors the cis isomer in cyclononene and cyclo-

[31] R. B. Turner and W. R. Meador, *J. Am. Chem. Soc.*, **79**, 4133 (1957).
[32] A. C. Cope, P. T. Moore, and W. R. Moore, *J. Am. Chem. Soc.*, **81**, 3153 (1959).

decene, but there is a slight predominance of the trans isomer at equilibrium for cycloundecene and cyclododecene.[32]

In Fig. 7-11 are shown some cyclic systems containing benzene rings. Among the ortho-bridged benzene rings, benzocyclopropene is not known but benzocyclobutene (I) has been synthesized.[33] The higher homologs (hydrindene, tetralin, etc.) are well known. Attempts to obtain benzocyclobutadiene (II) resulted[33] in the formation of the corresponding dimer, and the lowest member of the cycloalkadiene series is still indene (III), although dibenzocyclobutadiene (biphenylene, IV) is known.[34] Bridging of a benzene ring across the meta positions takes a considerably longer chain. A meta-bridged resorcinol (V) has been obtained[35] with seven methylene groups in the bridge (nine members in the bridge altogether); the lower homolog, with

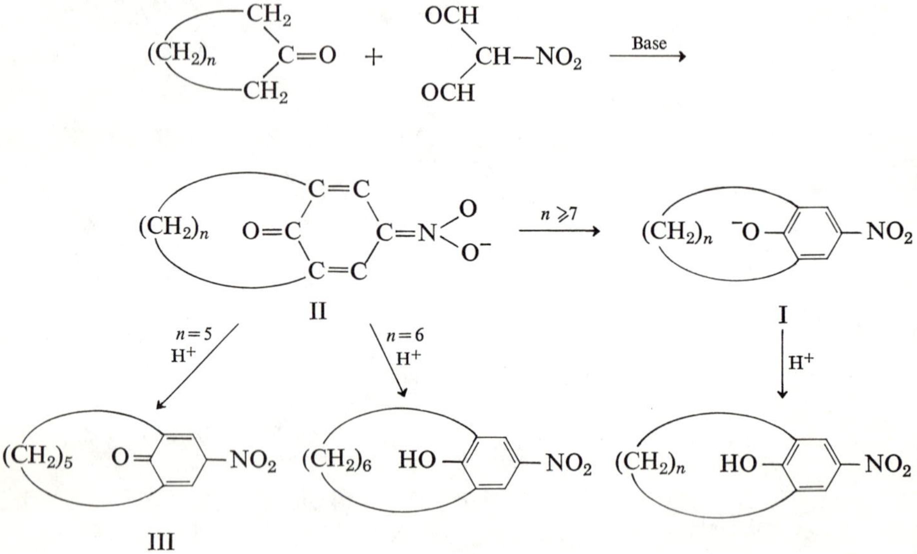

Fig. 7-12. Formation of *m*-bridged *p*-nitrophenols.

an eight-membered bridge, could not be synthesized. However, in compounds VI and VII the meta positions were bridged by six- and five-membered chains, respectively.[36-38] The synthesis of compound VI and others in the series showed some particularly interesting features; the route is indicated in Fig. 7-12. For $n \geqslant 7$, spectroscopic evidence indicated that the anion obtained in the condensation was the *p*-nitrophenolate anion (I). With $n = 6$, however, the non-aromatic anion of the aci form (II) resulted and

[33] M. P. Cava and D. R. Napier, *J. Am. Chem. Soc.*, **78**, 500 (1956). Cf. H. Finkelstein, Doctoral Dissertation, University of Strasbourg, Strasbourg, France, 1909 (via Ref. 23*b*).

[34] W. C. Lothrop, *J. Am. Chem. Soc.*, **63**, 1187 (1941).

[35] A. Lüttringhaus, *Ann.*, **528**, 181 (1936).

[36] V. Prelog and K. Wiesner, *Helv. Chim. Acta*, **30**, 1465 (1947).

[37] V. Prelog, K. Wiesner, W. Ingold, and O. Häfliger, *Helv. Chim. Acta*, **31**, 1325 (1948).

[38] V. Prelog, P. Barman, and M. Zimmerman, *Helv. Chim. Acta*, **33**, 356 (1950).

reverted to the aromatic only upon acidification. Finally with $n = 5$, the aromatic system was not formed even upon acidification, presumably because of excess strain; the product had the spectral properties of a dienone (III).

The smallest bridge which has yet been used to span the para positions in a benzene ring is 10-membered, as indicated in compounds VIII[35] and IX[39], Fig. 7-11. Attempts to span hydroquinone with a heptamethylene bridge to obtain the next lower homolog of VIII were unsuccessful;[35] but a compound similar to IX with a nine-membered carbonyl-containing bridge, $—CO(CH_2)_8—$, has been synthesized in 0.7% yield.[40] Also of interest are the 1,5- and 2,6-bridged naphthalene ansa compounds X and XI.[35] The lower homolog of X containing an octamethylene bridge was not obtained.[35]

Among cycles containing two benzene rings, the ethers shown in Fig. 7-11 (XII-XIV) are of interest because the length of the bridge required to span the rings depends on the nature of the internuclear atom or group.[41] When X = methylene (XII), a heptamethylene bridge can be accommodated; for X = sulfur (XIII), the octamethylene-bridged compound was obtained (synthesis of the heptamethylene-bridged compound was not attempted but that of the hexamethylene-bridged compound failed); but when X = oxygen, the synthesis of the octamethylene-bridged compound failed and only a decamethylene bridge could be accommodated. With the sulfone (XV, $X = SO_2$) a bridge of as few as five methylene groups could be constructed.[42] This variation of the ease of bridging has been ascribed[43] to the differences in bond angle at methylene and sulfur (110 to 112°), oxygen (129°), and sulfone (<90°). However, the assigned C—O—C angle in the ether appears unreasonably large and the C—S—C angle in the sulfone unreasonably small;† it would appear that causes other than differences in bond angle must be partly responsible for the observed variations. Among such causes may be mentioned differences in bending-force constants, including stiffening of the C—X bond by overlap of the *p* electrons of X with the pi electrons of the aromatic ring.

A number of carbocyclic compounds containing two para-bridged benzene rings, called "paracyclophanes" (Fig. 7-11, XVI), have been synthesized by D. J. Cram and coworkers. The smallest members of the series have[44] $m = n = 2$ and $m = 1, n = 7$.[45] The smaller members present quite anomalous spectral properties suggesting strong interactions of the benzene rings with each other [so-called "transannular interaction" (cf. Chap. 9) because it occurs across the large cycle]. In the smallest member ($m = n = 2$) it has been demonstrated by X-ray diffraction studies[46] that the benzene rings are

†Values of 120° and 100° might appear more likely; cf. L. E. Sutton, ed., "Tables of Interatomic Distances and Configuration in Molecules and Ions," The Chemical Society, London, 1958.

[39] D. J. Cram and H. U. Daeniker, *J. Am. Chem. Soc.*, **76**, 2743 (1954).

[40] R. Huisgen, W. Rapp, I. Ugi, H. Walz, and I. Glogger, *Ann.*, **586**, 52 (1954).

[41] A. Lüttringhaus, (*a*) *Ber.*, **72**, 887 (1939); (*b*) *Ann.*, **528**, 223 (1936); (*c*) *Ann.*, **528**, 211 (1936).

[42] A. Lüttringhaus and K. Buchholz, *Ber.*, **72**, 2057 (1939).

[43] A. Lüttringhaus and R. Kohlhaas, *Ber.*, **72**, 907 (1939); R. Kohlhaas and A. Lüttringhaus, *Ber.*, **72**, 897 (1939).

[44] D. J. Cram and H. Steinberg, *J. Am. Chem. Soc.*, **73**, 5691 (1951).

[45] D. J. Cram and M. F. Antar, *J. Am. Chem. Soc.*, **80**, 3103 (1958).

[46] C. J. Brown, *J. Chem. Soc.*, 3265 (1953).

appreciably distorted from their normal planar shape. Other members of this family of compounds are the doubly unsaturated paracyclophane XVII (Fig. 7-11),[47] the meta-bridged analog of XVI ($m = n = 2$),[48] and paracyclophanes containing three[48] and four[49] aromatic rings. Attempts to prepare compound XVIII (Fig. 7-11) have failed,[41c] but the analogous compound with NH instead of O has been prepared[49a] even with a nonamethylene bridge.

In comparing oxygen-containing cycles with carbocycles (e.g., V with VII; VIII with IX; and XVI, $m = 1$, $n = 7$, with XII), it appears that the smallest known carbocycles of this type contain fewer ring members than their oxygen analogs. This difference, if real, may be due to the above-mentioned variation in bond angle, bond bending-force constants, etc.

c. As a Function of Ring Substituents.[50] There are numerous indications in the literature that, for a given ring size, alkyl substituents favor the ring form in an equilibrium involving the opening and closing of a ring, such as

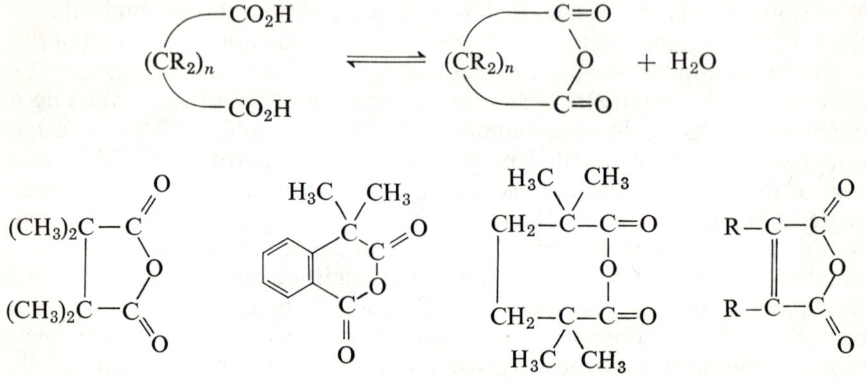

$$(CR_2)_n(CO_2H)_2 \rightleftharpoons (CR_2)_n(C{=}O)_2O + H_2O$$

I II III IV

Fig. 7-13. Dicarboxylic acid–cyclic anhydride equilibria.

the equilibrium between a dicarboxylic acid and the corresponding cyclic anhydride, plus water (Fig. 7-13). Thus tetramethylsuccinic anhydride (Fig. 7-13, I) is formed by hydrolysis of esters of the acid with hydrobromic acid, by heating the acid with concentrated aqueous hydrochloric acid in a sealed tube[51] at 200°, and even by steam-distilling the acid;[52] dimethylhomophthalic

[47] K. C. Dewhirst and D. J. Cram, *J. Am. Chem. Soc.*, **80**, 3115 (1958).

[48] M. Pellegrin, *Rec. trav. chim.*, **18**, 457 (1899); W. Baker, J. F. W. McOmie, and J. M. Norman, *J. Chem. Soc.*, 1114 (1951).

[49] H. Steinberg and D. J. Cram, *J. Am. Chem. Soc.*, **74**, 5388 (1952); E. D. Bergmann and Z. Pelchowicz, *ibid.*, **75**, 4281 (1953).

[49a] G. Wittig and J. E. Grolig, *Chem. Ber.*, **94**, 2148 (1961).

[50] For summaries see (*a*) G. S. Hammond in M. S. Newman, ed., "Steric Effects in Organic Chemistry," John Wiley & Sons, Inc. New York, 1956, pp. 460–470; (*b*) E. L. Eliel, *ibid.*, pp. 117–120; (*c*) C. K. Ingold, "Structure and Mechanism in Organic Chemistry," Cornell University Press, Ithaca, N.Y., 1953, pp. 537–543; (*d*) J. W. Baker, "Tautomerism," D. Van Nostrand Company, Inc., Princeton, N.J., 1934, chap. 10.

[51] K. Auwers and V. Meyer, *Ber.*, **23**, 101, 293 (1890); K. Auwers and L. L. Jackson, *ibid.*, **23**, 1599 (1890).

[52] P. E. Verkade, *Rec. trav. chim.*, **40**, 199 (1921).

anhydride (II) is obtained in the hydrolysis of the imide by aqueous hydrochloric acid;[53] and $\alpha,\alpha,\alpha',\alpha'$-tetramethyladipic anhydride (III, seven-membered ring!) is not affected by hot water or aqueous sodium carbonate.[54] Furthermore, the dialkylmaleic acids exist only in the form of their anhydrides (Fig. 7-13, IV, R = methyl, ethyl, or phenyl) which are formed spontaneously upon acidification of aqueous solutions of the salts of the acids.[55] [In all these cases the unsubstituted homologs (hydrogen instead of alkyl) have a much smaller tendency to form rings.]

By way of explanation of these observations, Ingold and Thorpe[56] suggested long ago that the diminution of the internal angle in a small ring (e.g., to 60° in cyclopropane) leads to a spreading apart of the external angle. This, in turn, relieves steric compression between substituents attached to one and the same carbon (Fig. 7-14), thus favoring the ring form over the open-chain form. This explanation is probably correct for small rings.[57, 58] In common rings with their normal or nearly normal bond angles the theory fails, however, to explain the enhanced ring stability. Evidence against bond-angle spreading is also found in the formation of macrocyclic rings from 2,2-bis(*p*-hydroxyphenyl)propane.[59] The phenolic groups in this compound could be spanned with a decamethylene or octamethylene bridge (Fig. 7-15) with no

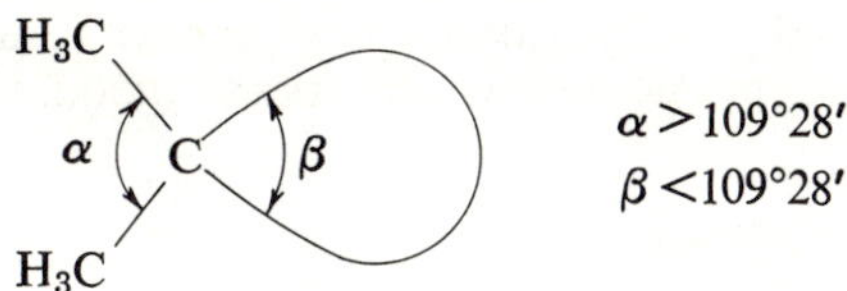

Fig. 7-14. Suggested bond angles in *gem*-dimethyl substituted cycloalkanes. (*Thorpe and Ingold.*[56])

greater ease than those in the parent compound having hydrogen atoms instead of methyl groups (Fig. 7-11, XII); attempts to span them with a hexamethylene bridge failed, as they had with the unmethylated compound. It was argued that, if spreading of the methyl groups was desirable and was favored by a decrease in the phenyl-CR_2-phenyl bond angle, the system shown in Fig. 7-15 might have been spanned with a smaller bridge than its unmethylated homolog.

A theory of the "*gem*-dialkyl effect" which appears to be applicable at least in the case of six-membered rings has recently been proposed[60] in terms of the enthalpies and entropies of open-chain vs. ring compounds. Analysis of a number of specific cases of substituted hexanes on one hand and substituted cyclohexanes on the other discloses that there are fewer extra gauche interac-

[53] S. Gabriel, *Ber.,* **19,** 2363 (1886); **20,** 1198 (1887).

[54] E. H. Farmer and J. Kracovski, *J. Chem. Soc.,* 680 (1927).

[55] R. Anschütz, *Ann.,* **254,** 168 (1889).

[56] R. M. Beesley, C. K. Ingold, and J. F. Thorpe, *J. Chem. Soc.,* **107,** 1080 (1915); C. K. Ingold, *ibid.,* **119,** 305 (1921).

[57] S. Searles, E. F. Lutz, and M. Tamres, *J. Am. Chem. Soc.,* **82,** 2932 (1960).

[58] P. von R. Schleyer, *J. Am. Chem. Soc.,* **83,** 1368 (1961).

[59] A. Lüttringhaus and K. Buchholz, *Ber.,* **73,** 134 (1940).

[60] N. L. Allinger and V. Zalkow, *J. Org. Chem.,* **25,** 701 (1960).

tions due to the alkyl substituents in the cycles than there are in the open chains. This means that, compared with an unsubstituted chain, the substituted chain has a more favorable enthalpy of ring closure. There is, in addition, an entropy effect due to branching based on the fact that branching reduces the rotational entropy of open-chain compounds but cannot, of course, reduce the entropy of the ring compound very much, because the ring compound has little freedom of internal rotation to begin with.[50a] Thus the

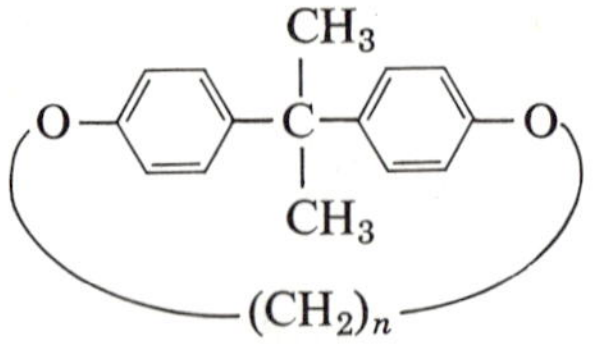

$n = 8, 10$
$n = 6$ could not be made

Fig. 7-15. Macrocycle with *gem*-dimethyl groups.

entropy factor also favors ring closure for the more branched compounds. Since branching both reduces the enthalpy and increases the entropy of ring closure, it decreases the free energy of ring closure and thus leads to an equilibrium more favorable to the ring structure. The effect has been calculated for a number of methyl-substituted cyclohexanes[60] and agreement with experimental thermodynamic data is remarkably good.

7-3. Ease of Ring Formation[61]

a. As a Function of Ring Size. Ease of ring formation is not synonymous with ring stability. Whereas the activation energy for ring closure might be expected to reflect the stability of the ring formed (cf. Table 7-1) to some extent, other factors enter as well. The most important of these involves the probability of having the ends of the ring-forming chain approach each other. This probability decreases as the ring size increases and reflects itself in an unfavorable activation entropy for the formation of medium and large rings. The over-all ease of ring closure thus may be derived from two factors: a monotonous decrease in the ease of having the ends of the ring meet and a strain factor which becomes more favorable to closure as the ring size increases from three- to six-membered, then less favorable as it increases further up to nine-membered, and then more favorable again for larger rings. In the over-all result, ease of ring formation is relatively high for three-membered rings (because of the high probability factor; three atoms are necessarily in the optimum position for ring formation). It drops sharply for the four-membered ring (if four atoms are arranged in the most stable conformation, as in trans-butane, they cannot form a ring; the best conformation for ring formation is the very unfavorable cis conformation). For the five-membered ring there is a sharp rise because of the considerable reduction in the strain factor. The ease of formation of a six-membered ring is less than for a five-membered one because the slight improvement in the strain factor is outweighed by a deterioration

[61] See also Ref. 50*b*, pp. 114–117.

in the distance factor. There is a sharper drop for the seven-membered ring (both strain and distance factor become worse) and an even sharper drop for the eight-membered ring (where non-classical strain sets in). After that, the distance factor has become about constant and the ease of forming larger rings reflects the strain factor: low ease of ring formation for 9- and 10-membered rings (because of non-classical strain, cf. p. 259), improving for 11- and 12-membered rings, and leveling off for still larger cycles.

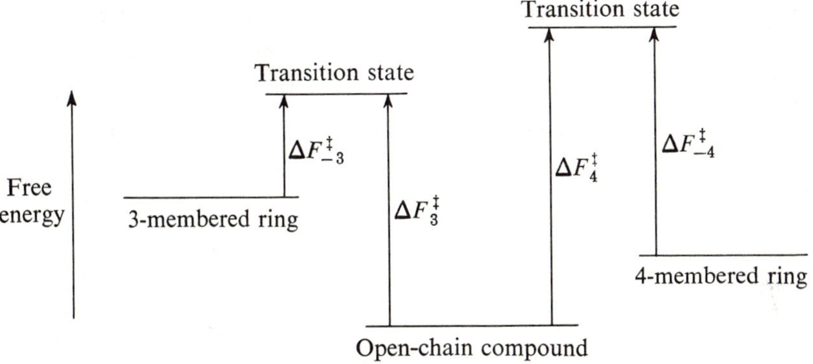

Fig. 7-16. Ease of ring formation vs. ring stability.

The easier closure of three-membered rings as compared with four-membered ones is well illustrated by the ready formation of epoxides from 1,2-halohydrins in contrast to the difficulty in preparing trimethylene oxides from 1,3-halohydrins. Yet the epoxides, once formed, are much more easily opened than the trimethylene oxides, reflecting the lesser stability of the three-membered ring. The energetic situation is summarized in Fig. 7-16.

H_2O + (six-membered cyclic acetal, OH) II $\overset{H^+}{\rightleftharpoons}$ CH_2OH–CHOH–CH_2OH + CH_2O $\overset{H^+}{\rightleftharpoons}$ (five-membered cyclic acetal)–CH_2OH I + H_2O

Fig. 7-17. Reaction of glycerol with formaldehyde.

A similar situation exists with regard to the ease of formation and stability of five- and six-membered rings. The five-membered ring is easier to close but is less stable. A clear-cut illustration of this state of affairs is the reaction[62] of glycerol with formaldehyde to give either a five-membered or a six-membered cyclic acetal (Fig. 7-17). Formation of the five-membered-ring compound 4-hydroxymethyl-1,3-dioxolane (I) is faster, and this product will predominate if the reaction mixture is worked up after a short time. How-

[62] H. Hibbert and N. M. Carter, *J. Am. Chem. Soc.*, **50**, 3120 (1928).

ever, the formation of I is reversible, and if the reaction is allowed to continue for a long period, equilibrium of all three sets of species shown in Fig. 7-17 will be established. If the mixture is worked up at this point, the main product is the six-membered 5-hydroxy-1,3-dioxane (II), the thermodynamically more stable of the two cyclic acetals I and II.†

The dioxolane I is said to be the product of "kinetic control," which means that it is the product of the faster of the two reactions (kinetics being that branch of chemistry which deals with reaction rates). In contrast, the dioxane II is said to be the product of "thermodynamic control," meaning the product predominating at equilibrium (since thermodynamics is the science that deals with chemical equilibria, among other matters).

As explained above, the formation of rings larger than seven-membered is not easy, and the problem has long been a challenge to synthetic organic

$(CH_2)_n(CO_2\frac{Th}{4})_2 \xrightarrow{\Delta} (CH_2)_n\,C{=}O + \frac{Th}{2}CO_3$ (Ruzicka)

$(CH_2)_n(CN)_2 \xrightarrow{LiNEt_2} (CH_2)_{n-1}[C{=}NH][CH{-}CN]$ (Ziegler)

$(CH_2)_n(CO_2Et)_2 \xrightarrow[Xylene]{Na} (CH_2)_n[CHOH][C{=}O]$ (Hansley, Prelog, Stoll)

Fig. 7-18. Methods for synthesizing medium and large rings. (For other methods, see Refs. 77 and 79.)

chemists, especially since some valuable naturally occurring perfumes, such as muskone (3-methylcyclopentadecanone), contain large rings. One of the early methods of synthesizing large rings is due to Ruzicka[63] and consists in

† Although six-membered acetals are usually more stable than five-membered ones, the opposite is usually true of ketals (cf. Ref. 61). The reason for this is shown in Fig. *i:* When both R and R′ are larger than hydrogen (as will be the case in a ketal) an axial interaction (cf. Chap. 8) will necessarily be generated in the six-membered ring, diminishing its stability below that of the corresponding five-membered ring. This is the reason why the acetone derivatives of sugars are 1,2 ketals rather than 1,3 ketals.

H R R′ H O O

(*i*)

[63] For example, L. Ruzicka, M. Stoll, and H. Schinz, *Helv. Chim. Acta,* **9**, 249 (1926); for a comprehensive summary, see Ref. 77.

the pyrolysis of the thorium salts of dibasic acids to give cyclic ketones (Fig. 7-18). The method served to synthesize 8-membered rings as well as rings of 13 members and more (although the yields were low) but fails for rings of 9 to 12 members. Another method, due to Ziegler,[64] involves ring closure of a dinitrile with bases (Fig. 7-18). In order to obtain good yields in this method, it is necessary to operate in very dilute solution so as to avoid linear polymerization of the difunctional nitriles. This so-called "dilution principle" was first recognized and applied by Ruggli[65] and is based on the fact that the rate of an intramolecular ring closure generally depends on the first power of substrate concentration, whereas the rate of an intermolecular dimerization is proportional to the *square* of the concentration of the substrate. Therefore, decreasing substrate concentration decreases rate of dimerization much more than rate of intramolecular ring closure. However, even when the dilution principle is used, the Ziegler ring closure fails for rings of 9, 10, and 11 members. In fact, rings of this size were not accessible at all until it was found[66] that the acyloin synthesis (reaction of a diester with sodium to give an α-hydroxyketone) was particularly suitable for the synthesis of ring compounds, as shown in Fig. 7-18. This synthesis gives high yields for large as well as medium rings and has even been applied to four-[67] and five-membered rings.[68] It is not necessary to use high dilution. The amazing success of this method has been explained[78] as being due to adsorption of the functional groups of the diester on the surface of the sodium, which automatically brings the two ends of the molecule in close proximity.

As a result of the application of the acyloin synthesis to medium and large rings, many functional derivatives of these rings are now readily available and have been extensively studied. The interesting properties of such rings will be returned to in Chap. 9.

b. As a Function of the Nature and Degree of Substitution of the Ring Atoms. Little is known about the effect of introducing hetero atoms into the ring on the ease of ring closure. Inasmuch as some of the difficulty of forming medium-sized rings is due to transannular hydrogen interactions (cf. Chap. 9) one might think that replacement of methylene groups by —NH—, —S—, or —O— would facilitate ring closure by diminishing these interactions. In fact, the introduction of a single hetero atom seems to have little effect: Ring closure of ω-bromoalkylamines, $Br(CH_2)_nNH_2$, is extremely slow for $n = 9$ or 11,[69] and ring closure of ω-chlorosulfides, $RS(CH_2)_nCl$, to cyclic sulfonium salts fails for cycles containing between 8 and 13 atoms.[70] Compounds of the type shown in Fig. 7-19 could be formed readily by the

[64] For example, K. Ziegler, H. Eberle, and H. Ohlinger, *Ann.*, **504,** 94 (1933); see also Ref. 77.

[65] P. Ruggli, *Ann.*, **392,** 92 (1912); **399,** 174 (1913); **412,** 1 (1917).

[66] V. L. Hansley, U.S. Patent 2,228,268 (1941), *Chem. Abstr.*, **35,** 2534 (1941); V. Prelog, L. Frenkiel, M. Kobelt, and P. Barman, *Helv. Chim. Acta*, **30,** 1741 (1947); M. Stoll and J. Hulstkamp, *ibid.*, **30,** 1815 (1947); M. Stoll and A. Rouvé, *ibid.*, **30,** 1822 (1947).

[67] A. C. Cope and E. C. Herrick, *J. Am. Chem. Soc.*, **72,** 983 (1950).

[68] J. C. Sheehan and R. C. Coderre, *J. Am. Chem. Soc.*, **75,** 3997 (1953).

[69] G. Salomon, *Helv. Chim. Acta*, **19,** 743 (1936).

[70] G. M. Bennett and H. Gudgeon, *J. Chem. Soc.*, 1891 (1938).

reaction shown for n between 2 and 10, the rate of ring closure dropping off gradually with increasing n, with no rate minimum appearing for 9- to 11-membered rings.[71] This system, however, contains not only two hetero (oxygen) atoms but also two benzene carbons which are part of the ring, so that transannular hydrogen interactions would be expected to be a great deal less important here than in polymethylene carbocycles.

The effect of alkyl substituents in general is to facilitate ring closure, at least where small and common rings are concerned. (Little information is available along this line on medium and large rings.) This effect (the "Thorpe-

$O(CH_2)_nBr$ / OH → O, O, $(CH_2)_n$

Fig. 7-19. Cyclic ethers of catechol.

Ingold effect") runs parallel to the corresponding effect on ring stability discussed in Sec. 7-2*c*. A very clear-cut example is the effect of substitution on the rate of epoxide formation from chlorohydrins under the influence of base shown in Table 7-4.[72] A number of other examples of the effect may be found in the literature.[50b, 61, 73–75]

Table 7-4
Relative Rate of Ring Closure of Chlorohydrins

$HOCH_2CH_2Cl$	1
$HOCH_2CHClCH_3$	5.5
$CH_3CHOHCH_2Cl$	21
$HOCH_2CCl(CH_3)_2$	248
$(CH_3)_2COHCH_2Cl$	252
$(CH_3)_2COHCHClCH_3$	1360
$CH_3CHOHCCl(CH_3)_2$	2040
$(CH_3)_2COHCCl(CH_3)_2$	11,600

General References

[76] R. A. Raphael in E. H. Rodd, ed., "Chemistry of Carbon Compounds," Elsevier Publishing Company, Amsterdam, 1953, pp. 1–10.

[77] V. Prelog, Bedeutung der Vielgliedrigen Ringverbindungen, in A. R. Todd, ed., "Perspectives in Organic Chemistry," Interscience Publishers, Inc., New York, 1956, pp. 96–116.

[78] V. Prelog, Newer Developments of the Chemistry of Many-membered Ring Compounds, *J. Chem. Soc.*, 420 (1950).

[71] K. Ziegler, A. Lüttringhaus, and K. Wohlgemuth, *Ann.*, **528**, 162 (1936).
[72] H. Nilsson and L. Smith, *Z. physik. Chem.*, **166A**, 136 (1933).
[73] F. G. Bordwell, C. E. Osborne, and R. D. Chapman, *J. Am. Chem. Soc.*, **81**, 2698 (1959).
[74] M. S. Newman and R. J. Harper, *J. Am. Chem. Soc.*, **80**, 6350 (1958).
[75] T. C. Bruice and U. K. Pandit, *J. Am. Chem. Soc.*, **82**, 5858 (1960).

[79] K. Ziegler, Methoden zur Herstellung und Umwandlung grosser Ringsysteme, in E. Müller, ed., "Methoden der Organischen Chemie" (Houben-Weyl), vol. 4-2, Georg Thieme Verlag, Stuttgart, 1955, pp. 729–822.

[80] A. Wassermann, Spannungstheorie und physikalische Eigenschaften ringförmiger Verbindungen, in K. Freudenberg, ed., "Stereochemie," Franz Deuticke, Leipzig, 1932, pp. 781–802.

[81] R. C. Fuson, Alicyclic Compounds and the Theory of Strain, in H. Gilman, ed., "Organic Chemistry," John Wiley & Sons, Inc., New York, 1943, chap. 2.

Chapter 8

THE ACTUAL SHAPE OF SIX-MEMBERED RINGS AND ITS RELATION TO PROPERTIES AND REACTIVITY

8-1. The Shape of Cyclohexane Rings

Data were presented in the previous chapter showing that six-membered rings are, in general, more stable than five-membered ones. This, as was already pointed out, is not in agreement with the relative strains to be expected in planar five- and six-membered rings (cf. Table 7-2). However, it was pointed out by Sachse[1] in 1890 that two puckered, i.e., non-planar, models of cyclohexane could be constructed in which the valence bonds of all the carbons were at the tetrahedral angle (109° 28′) to each other and which were therefore free of angle strain. These are the rigid or chair form (Fig. 8-1, *C*) and a mobile form which can be readily distorted into a variety of shapes, some of which resemble a boat (Fig. 8-1, *B*).†

If one plots the potential energy of cyclohexane as a function of the angles between certain valence bonds, in a way similar to that shown for ethane and *n*-butane in Chap. 6 (Figs. 6-2 and 6-3), one obtains the diagram shown in Fig. 8-2. In the chair form, not only is there no angle strain but there is no bond opposition (eclipsing) strain either, and the potential energy is at a minimum (point *a* in diagram). The reader should convince himself of this

Note: All references above 69 are listed in the general references at the end of the chapter.

† Despite Sachse's remarkable understanding of the three-dimensional geometry of cyclohexane, his ideas found no acceptance whatsoever and were, in fact, completely emasculated in one of the better-known textbooks of his time.[2] This was due in part to the preeminence of the strain theory of the famous Adolph von Baeyer and in part to the fact that Sachse suggested, though only tentatively, that what we now call conformational isomers might be isolable and might, in fact, correspond to certain substances then known—a prediction which proved to be erroneous. Sachse's theory was revived and applied to decalin by E. Mohr [*J. prakt. Chem.*, [2]**98,** 315 (1918)] and became established when the cis-trans isomerism of the decalins predicted by Mohr was experimentally confirmed by W. Hückel [*Ann.*, **441,** 42 (1925)]. The theory of the puckered rings is therefore often referred to as "Sachse-Mohr theory."

---

[1] H. Sachse, *Ber.*, **23,** 1368 (1890); *Z. physik. Chem.*, **10,** 203 (1892).

[2] O. Aschan, "Chemie der Alicyclischen Verbindungen," F. Vieweg u. Sohn, Brunswick, Germany, 1905.

fact with the aid of a model. When the bottom part of the chair is bent upward (or the back bent downward), the chair is transformed into a boat. During the transformation, some angular distortion is required (unlike the case of ethane) because several bonds have to be rotated at the same time; the molecule passes through a form of high potential energy (point *b*). The height of the potential barrier has been determined[3] by the nuclear magnetic resonance method (cf. Chap. 6) and is found to be 10 to 11 kcal./mole for cyclohexane,[3a, d] 10.85 kcal./mole for cyclohexyl bromide,[3b] and 9.9 kcal./mole for perfluorocyclohexane.[3c] These values are free-energy rather than potential-energy or enthalpy values and correspond to the usual potential-energy values only if the entropy of activation for conversion of the conformational isomers is zero—an assumption which, at least in two of the cases

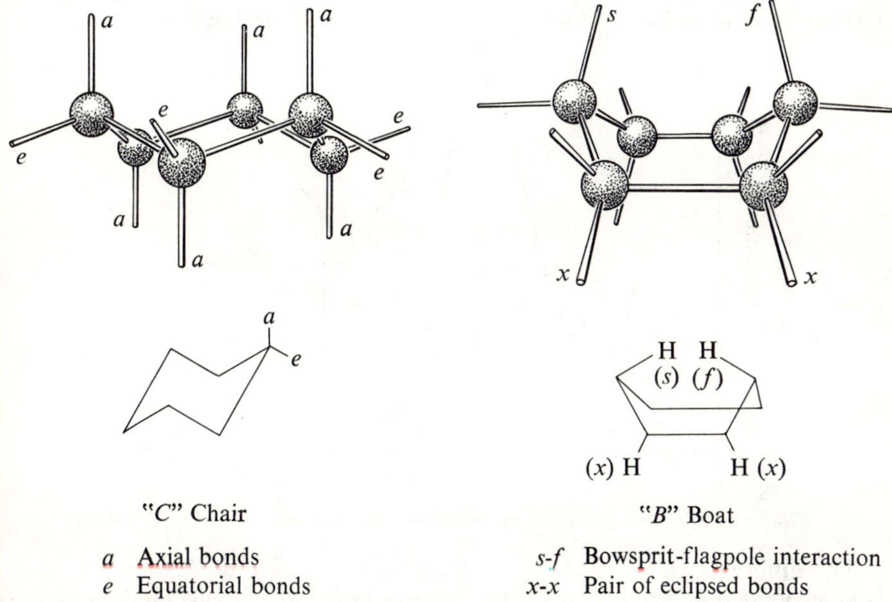

Fig. 8-1. Chair ("*C*") and boat ("*B*") forms of cyclohexane.

that have been studied, is evidently unwarranted.[3c, d] The potential-energy barrier between the rigid and flexible forms (*a* and *c*) is high enough to ensure that these conformations are discrete entities but not high enough to prevent their rapid interconversion at room temperature or to make possible their isolation. In the boat form (point *d*) there is no angle strain, but there is bond opposition strain (of the type encountered in the eclipsed form of ethane; cf. Chap. 6) involving the four pairs of hydrogens at the side of the

[3] (*a*) F. R. Jensen, D. S. Noyce, C. H. Sederholm, and A. J. Berlin, *J. Am. Chem. Soc.*, **84,** 386 (1962); (*b*) L. W. Reeves and K. O. Strømme, *Can. J. Chem.*, **38,** 1241 (1960); (*c*) G. van Dyke Tiers, *Proc. Chem. Soc.*, 389 (1960); (*d*) R. K. Harris and N. Sheppard, *ibid.*, 418 (1961). These authors report an enthalpy barrier of 9.0 ± 0.2 kcal./mole.

boat (one pair shown in Fig. 8-1, *B*, at the bottom), and there is also strain due to the interference of the pair of hydrogens shown at the top of Fig. 8-1, *B*, which are only about 1.8 A. apart (the sum of the van der Waals radii of two hydrogens is 2.4 A.). This is sometimes called the "bowsprit-flagpole interaction."

As a result of these unfavorable interactions, the boat is considerably less stable than the chair. Careful inspection of a model of the flexible form reveals, however, that it is misleading to equate this form with the "boat" form. In distorting the model in such a way as to pass from one boat to another, one obtains forms in which both bowsprit-flagpole interactions and eclipsings of adjacent hydrogens are somewhat alleviated (Fig. 8-3*b*). In the energy diagram (Fig. 8-2) these forms (sometimes called "skew-boats" or "twist forms") correspond to an energy minimum (point *c*). The true boat (point *d*) actually lies at the energy maximum between two skew-boats and has been calculated to be 1.6 kcal./mole less stable than the latter.[4d]

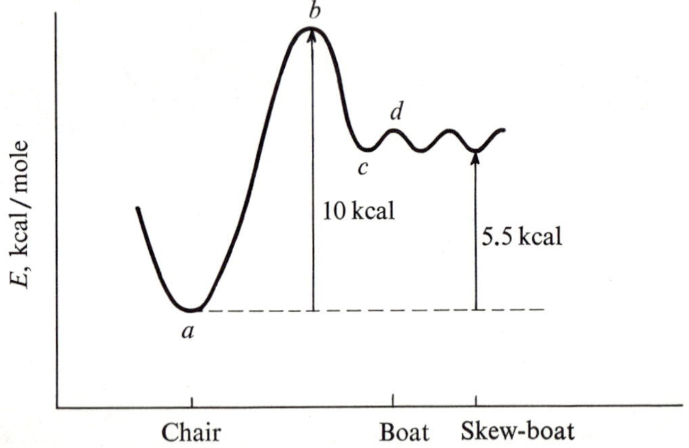

Fig. 8-2. Potential energy of cyclohexane as a function of conformation.

The difference in energy between the chair form and the flexible form has been estimated at 5 to 6 kcal. on the basis of both calculations[4] and measurements.[5] The measurements are based on heats of combustion[5a] or equilibrium determinations[5b] of molecules which for one reason or another must exist in the boat form (see below). Under normal circumstances, however, the enthalpy difference of 5 to 6 kcal./mole between flexible and chair forms ensures that cyclohexanes exist almost exclusively in the chair form at room temperature. In order to put this statement into quantitative terms, it is necessary also to take into account the finding[5b] that the flexible form, because of its greater mobility, has an entropy of 5 e.u. greater than the chair.

[4] (*a*) N. L. Allinger, *J. Am. Chem. Soc.*, **81**, 5727 (1959); (*b*) K. E. Howlett, *J. Chem. Soc.*, 4353 (1957); (*c*) P. Hazebroek and L. J. Oosterhoff, *Discussions Faraday Soc.*, **10**, 87 (1951); (*d*) J. B. Hendrickson, *J. Am. Chem. Soc.*, **83**, 4537 (1961).

[5] (*a*) W. S. Johnson, V. J. Bauer, J. L. Margrave, M. A. Frisch, L. H. Dreger, and W. N. Hubbard, *J. Am. Chem. Soc.*, **83**, 606 (1961); (*b*) N. L. Allinger and L. A. Freiberg, *ibid.*, **82**, 2393 (1960).

Therefore, at 298°K., $\Delta F = \Delta H - T\Delta S = 5500 - 5 \times 298 = 4000$ cal. or 4 kcal./mole. This means, according to Table 8-1, that only about one molecule in a thousand will be in the boat form.

[Table 8-1, computed from the familiar relationship $\Delta F = -2.3RT \log K$, gives a relation between $-\Delta F$ and K at 298°K. (25°C.) which will be found generally useful.]

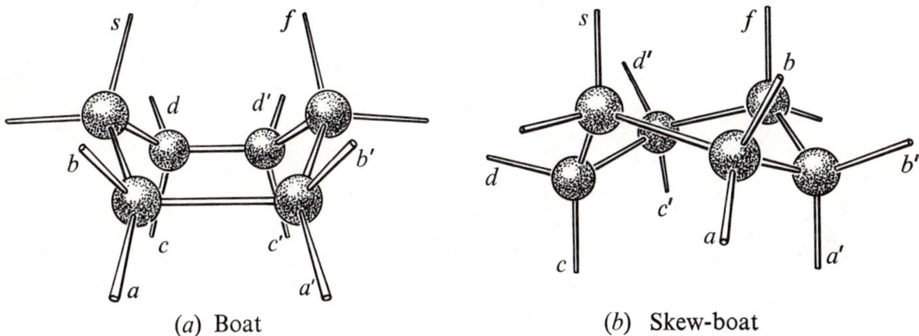

(*a*) Boat (*b*) Skew-boat

Fig. 8-3. Cyclohexane: boat and skew-boat.

The chair shape of a number of cyclohexane derivatives has been confirmed by such physical methods as X-ray diffraction and electron diffraction and has also been deduced from Raman and infrared spectra.[6]

There are, nevertheless, some molecules in which the cyclohexane ring exists in the flexible form. [2.2.2]-Bicyclooctane (Fig. 8-4) represents such a case. This molecule must necessarily exist as a boat. Another case is *trans*-1,3-di-*t*-butylcyclohexane[5b] (Fig. 8-5, I). If this molecule existed as a chair,

Table 8-1
Relation between Standard Free-energy Difference and Equilibrium Constant at 25°C.

K (298°K.)	2	3	4	5	10	20	100	1000	10,000
Population of favored state, %	67	75	80	83	91	95	99	99.9	99.99
$-\Delta F$, kcal./mole	0.41	0.65	0.82	0.95	1.4	1.8	2.7	4.1	5.5

there would be an intolerable interaction between the axial *t*-butyl group and the axial hydrogens, as will be explained in the next section. [The energy difference between the chair and the flexible form mentioned earlier may be obtained by study of the temperature equilibrium of I with its cis isomer (Fig. 8-5, II).]

Yet another molecule in which the cyclohexane ring is boat-shaped is 1,2,2,6,6-pentamethyl-4-hydroxy-4-phenylpiperidine[7] (Fig. 8-6), as may be

[6] Cf. Ref. 76, p. 77, and Ref. 73, footnote 3.

[7] R. E. Lyle, *J. Org. Chem.*, **22**, 1280 (1957); see also J. W. Huffman, *ibid.*, **24**, 447 (1959).

seen from the fact that this molecule gives infrared spectral evidence (cf. Chap. 6) of intramolecular hydrogen bonding.

Even when molecules exist predominantly in the chair form, the flexible form, though sparsely populated, may be of importance in certain reactions. An example is the lactonization of *cis*-4-hydroxycyclohexanecarboxylic acid which must proceed via the boat conformation (Fig. 8-7); the trans isomer does not lactonize.

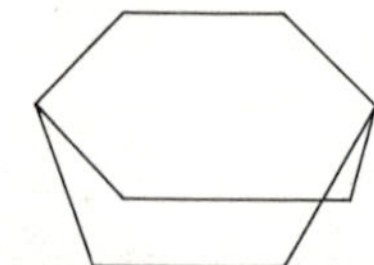

Fig. 8-4. [2.2.2]-Bicyclooöctane.

A number of cases where flexible forms are of importance have been summarized in the literature,[8] and further examples will be presented in Chap. 10.

8-2. Monosubstituted Cyclohexanes

In a chair-shaped cyclohexane ring, there are two possible kinds of bonds,[9, 10] namely, those (Fig. 8-8) pointing up and down, called "axial" (*a*), and those

$(H_3C)_3C$ $C(CH_3)_3$ ⇌ $C(CH_3)_3$ $C(CH_3)_3$ $\underset{\Delta}{\overset{Pd/C}{\rightleftharpoons}}$

I (trans) (bad)

$(H_3C)_3C$ $C(CH_3)_3$

II (cis)

Fig. 8-5. The 1,3-di-*t*-butylcyclohexanes.

[8] J. Levisalles, *Bull. soc. chim. France,* 551 (1960). This excellent summary contains an error: The activation energy for passing from one chair form to the other is *not* the sum of the activation energies for passing from the chair to the flexible form plus that for passing from the flexible form to the other chair. If a molecule has sufficient energy to pass from the chair into the flexible form over the barrier, it will have enough energy to pass over the second barrier (assumed to be no higher than the first) into the other chair form.

[9] (*a*) K. W. F. Kohlrausch, A. W. Reitz, and W. Stockmair, *Z. physik. Chem.,* **B32,** 229 (1936); (*b*) O. Hassel, *Tidsskr. Kjemi, Bergvesen Met.,* **3,** 32 (1943).

[10] C. W. Beckett, K. S. Pitzer, and R. Spitzer, *J. Am. Chem. Soc.,* **69,** 2488 (1947).

(Fig. 8-8) pointing sideways which are called "equatorial" (*e*).† Substituents attached to these bonds are customarily called axial and equatorial substituents, respectively.

Thus, in principle, there may be two isomeric forms of a monosubstituted

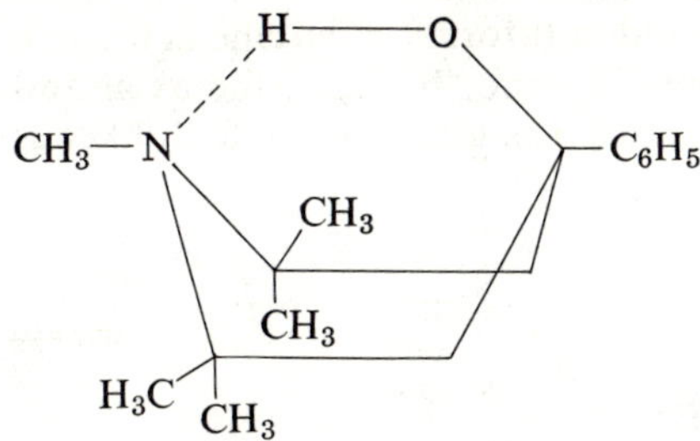

Fig. 8-6. 1,2,2,6,6-Pentamethyl-4-hydroxy-4-phenylpiperidine.

cyclohexane, such as methylcyclohexane, namely, one in which the substituent is equatorial and another in which it is axial. However, reference to Fig. 8-3 shows that the equatorially substituted chair can be readily converted into a flexible form, and this, in turn, can be similarly converted into another

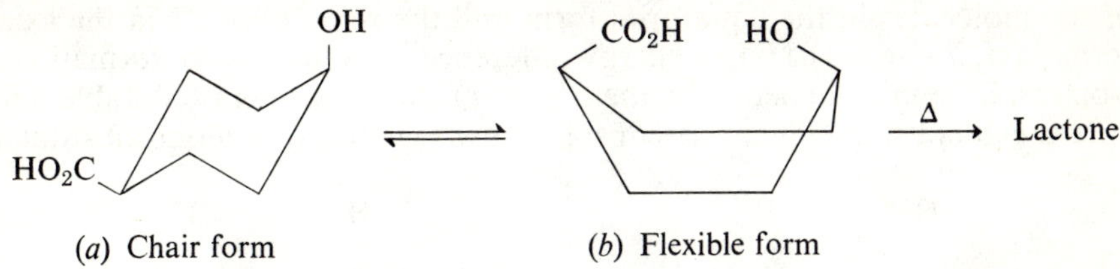

Fig. 8-7. *cis*-4-Hydroxycyclohexanecarboxylic acid.

chair in which the substituent is now axial (Fig. 8-9). This type of transformation is readily illustrated by means of a model. The two isomers *A* and *E* in Fig. 8-9 are conformational isomers, or conformers (cf. Chap. 6). The difference in potential energy between them can be estimated from an inspec-

Fig. 8-8. Axial bonds and equatorial bonds.

† Hassel originally called the equatorial bonds "κ" (Greek for prostrate) and the axial bonds "ε" (Greek for erect); Pitzer and Barton originally used "*p*" (polar) instead of axial. These designations will be found in papers published prior to 1954 when the present nomenclature was agreed on (Ref. 11).

---

11 D. H. R. Barton, O. Hassel, K. S. Pitzer, and V. Prelog, *Science*, **119**, 49 (1953); *Nature*, **172**, 1096 (1953).

tion of the models. The axial isomer, *A*, has two interactions of the type present in the gauche form of butane (Chap. 6). These are shown in heavy lines in Fig. 8-9, and one of them can be seen even more clearly in the Newman formulations for methylcyclohexane (Fig. 8-10). No such interactions are present in the equatorial isomer, *E*. Since the potential-energy difference between the gauche and anti forms of butane is 0.8 to 0.9 kcal./mole (Chap. 6), the corresponding difference between the axial and equatorial forms of methylcyclohexane is twice as great, or 1.6 to 1.8 kcal./mole, the equatorial

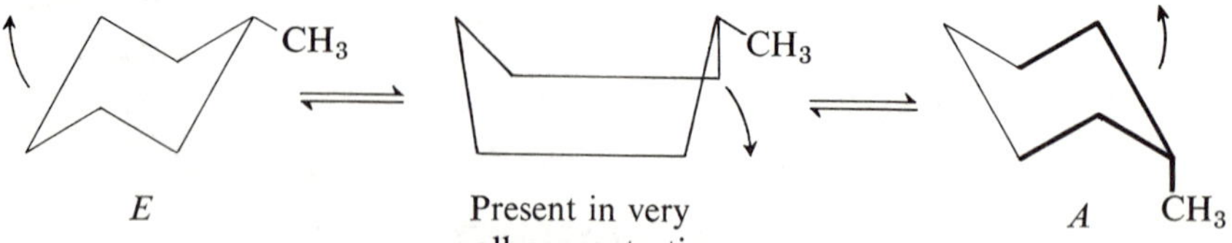

CH_3 CH_3

E Present in very small concentration *A* CH_3

Fig. 8-9. Interconversion of chair forms of methylcyclohexane.

form being the more stable.† The potential energy of methylcyclohexane as a function of conformation is shown in Fig. 8-11. At room temperature, both isomers are present in a rapidly established equilibrium, with about 95% of the molecules in the equatorial form and the remaining 5% in the axial form[12] (cf. Table 8-1).‡ The energy difference between the conformational isomers, of course, depends on the nature of the substituent (cf. Table 8-6).

It is evident that, when one considers any reaction of a monosubstituted

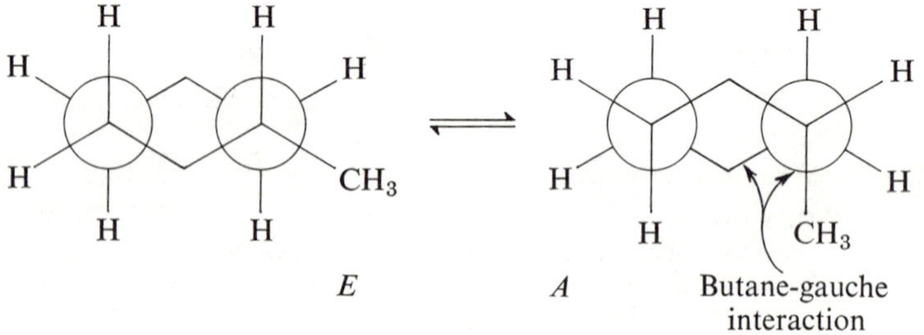

H H H H H H H H H H CH_3 H H H H H H CH_3

E *A* Butane-gauche interaction

Fig. 8-10. Newman formulations of methylcyclohexane.

†A check on this value (Ref. 10) comes from the good agreement of the observed entropy of methylcyclohexane with that calculated using $\Delta E = 1.8$ kcal./mole. See also Table 8-2 and Ref. 14, which support a value of 1.9 kcal./mole.

‡Assuming that the difference in free energy between the isomers is equal to the potential energy. This, in turn, involves two other assumptions. The first is that ΔE or ΔH may be equated with potential-energy differences. This assumption (which is usually satisfactory) involves neglect of the distribution of molecules over various vibrational-energy states, since potential-energy differences can be measured only between specified states and should properly be summed up over all possible pairs of vibrational-energy levels to give ΔE. The second assumption is that $\Delta S = 0$; this is probably true for the conformers of methylcyclohexane. In other instances, mentioned elsewhere in this book, ΔS may not equal zero and must then be taken into account.

[12] See E. L. Eliel and M. Rerick, *J. Am. Chem. Soc.*, **82**, 1367 (1960).

cyclohexane, one must, in fact, consider the reaction of both the equatorial and the axial species, just as, in looking at certain reactions of acetoacetic ester, one must consider both the keto and the enol form. This point, as well as the methods for determining the energy difference between the two species, will be returned to in Sec. 8-6.

8-3. Disubstituted Cyclohexanes

As mentioned in Chap. 7, 1,2-, 1,3-, and 1,4-disubstituted cyclohexanes may each exist in two diastereoisomeric forms, one cis, the other trans. It was pointed out that, for equal substituents, the cis-2, cis-3, cis-4, and trans-4 isomers are meso forms whereas the trans-2 and trans-3 isomers form *dl* pairs. This was derived by considering planar forms of these isomers. Now that we know that the compounds are not, in fact, planar, it is desirable to reconsider this matter and to decide whether the same conclusion follows from three-dimensional formulas.

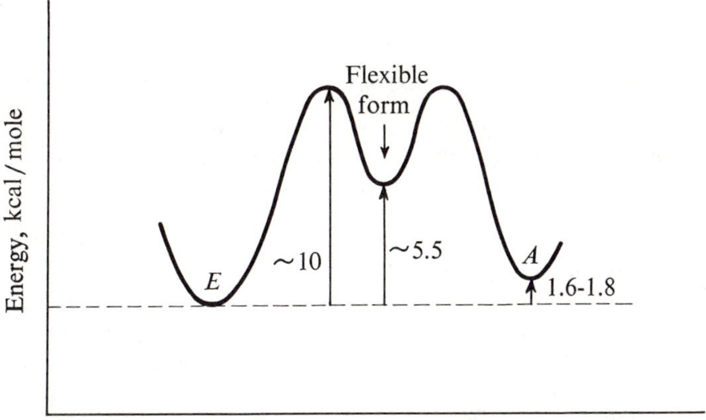

Fig. 8-11. Potential energy of methylcyclohexane as a function of conformation.

trans-1,2-Dimethylcyclohexane is shown in Fig. 8-12 (top). The diaxial and diequatorial forms are readily interconvertible. The diaxial form has four *gauche*-butane interactions (twice as many as the axial form of methylcyclohexane) corresponding to an interaction energy of 4×0.9, or 3.6, kcal./mole. The diequatorial form has only one such interaction (between the methyl substituents) corresponding to an interaction energy of 0.9 kcal./mole. Thus the difference in stability between the conformational isomers is $3.6 - 0.9$, or 2.7, kcal./mole; i.e., the molecule is almost entirely (99%) in the diequatorial form at room temperature. Both forms shown in Fig. 8-12 are dissymmetric and possess a non-identical mirror image which represents the enantiomer. Therefore, *trans*-1,2-dimethylcyclohexane forms a *dl* pair, in agreement with the conclusion reached from planar models.

The situation with *cis*-1,2-dimethylcyclohexane (Fig. 8-12, bottom) is more noteworthy. Each conformational isomer of this compound has one axial and one equatorial substituent. Since the two forms are equally stable, half the molecules exist in one and half in the other. It may be noted that the molecule has no element of symmetry and thus, at first sight, might be capable of optical activity, contrary to what had been concluded in Chap. 7 on the basis of planar models. However, *the two conformational isomers are also mirror images.* This can be seen (best by manipulation of models) by rotating the cis isomer marked *a,e* in Fig. 8-12 by an angle of 120° around a vertical axis. Since the two conformational isomers are present in equal

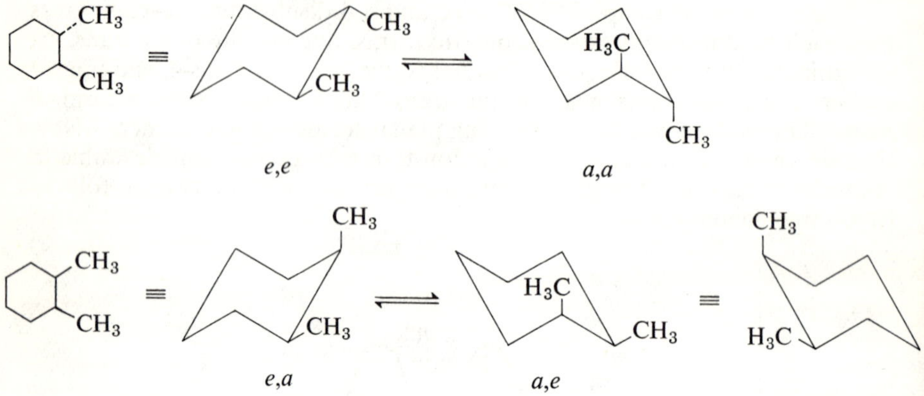

CH_3 CH_3 ≡ CH_3 CH_3 ⇌ H_3C CH_3

e,e *a,a*

CH_3 CH_3 ≡ CH_3 CH_3 ⇌ H_3C CH_3 ≡ CH_3 H_3C

e,a *a,e*

Fig. 8-12. *trans*-1,2-Dimethylcyclohexane (one enantiomer) and *cis*-1,2-dimethylcyclohexane.

amounts, *cis*-1,2-dimethylcyclohexane exists as a *dl* pair, but since the potential barrier between the isomers is too low to allow separation, the *dl* pair is inseparable. The situation resembles that discussed in Chap. 6 for an asymmetrically substituted biphenyl in which the barrier to rotation is too low to allow separation of the enantiomers. It is conceivable that resolution will be effected at some future date by working at very low temperature or by devising a molecule in which the barrier between the two conformational isomers is high enough to allow separation at room temperature. In fact, a

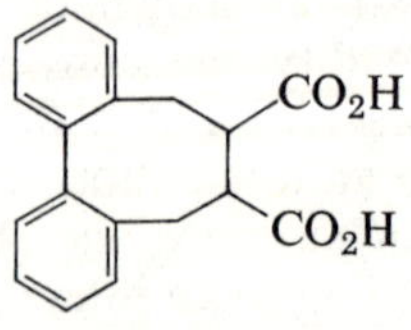

CO_2H CO_2H

Fig. 8-13. *cis*-1,2,3,4-Dibenz-1,3-cycloöctadiene-6,7-dicarboxylic acid.

compound of this type (Fig. 8-13), though containing an eight-membered instead of the above-discussed six-membered ring, has already been obtained optically active.[13] The conformational stability of this compound is enhanced by the built-in biphenyl system with its hindered rotation. It racemizes on standing with a half-life of 85 min. at 31.5°.

[13] L. V. Dvorken, R. B. Smyth, and K. Mislow, *J. Am. Chem. Soc.*, **80**, 486 (1958).

The conclusion that *cis*-1,2-dimethylcyclohexane is an inseparable *dl* pair rather than a meso form is different in principle from that derived from planar models, although it still follows that the substance is optically inactive.

Since *cis*-1,2-dimethylcyclohexane has three butane-gauche interactions and the more stable, diequatorial form of the trans isomer has only one, the two configurational isomers† differ in potential energy by $3 \times 0.9 - 0.9$, or 1.8, kcal./mole. Experimental values for the corresponding enthalpy differences and free-energy differences have been determined;[14] the experimental ΔH (at 25°C.) is 1.87 kcal./mole, and ΔF (at 25°) is 1.61 kcal./mole.

The reader should now be able to extend these concepts to the 1,3- and 1,4-dimethylcyclohexanes by a consideration of models. The 1,3 isomer exists in diastereoisomeric cis and trans forms. In the cis form, either both substituents are equatorial or both are axial; in either case there is a plane of symmetry passing through carbon atoms 2 and 5. Actually, the steric situation in the diaxial isomer is quite unfavorable, since the two methyl groups stick out on the same side of the molecule and crowd each other badly. The resulting steric interaction has been found[14a] to give rise to an unfavorable energy term of 5.5 kcal./mole. In *trans*-1,3-dimethylcyclohexane, one substituent is always equatorial and the other axial. The molecule does not have a plane of symmetry and, unlike the case of the cis-1,2 compound, flipping the ring converts the molecule, not to its enantiomer, but to a species which is superposable with the original molecule. Thus in this case, the (+) and (−) forms are capable of discrete existence, and *trans*-1,3-dimethylcyclohexane is a resolvable molecule.[15] Since there are two butane-gauche interactions in either form of *trans*-1,3-dimethylcyclohexane but no such interactions in the stable (diequatorial) form of the cis isomer, it follows that the cis isomer should be more stable by 1.8 kcal./mole. Confirmatory experimental data are indicated in Table 8-2.‡ In 1,4-dimethylcyclohexane, cis and trans isomers exist; both have a plane of symmetry bisecting carbon atoms 1 and 4 and are therefore inactive. The less stable cis isomer has one equatorial and one axial substituent in either form of the chair, whereas the more

† *cis*- and *trans*-1,2-Dimethylcyclohexane are *configurational* isomers, since they differ in spatial arrangement of the groups around one of the asymmetric carbon atoms (No. 1 or 2). In the trans isomer, one methyl substituent is uppermost (as seen by an observer looking at the ring from above) and the other is lowermost. In the cis isomer, however, both are uppermost or both lowermost. This situation remains unchanged even though one chair form may be transformed into the other chair form. In contrast, the *e,e* and *a,a* forms of a given enantiomer of *trans*-1,2-dimethylcyclohexane are *conformational* isomers, since they are interconvertible by rotation about single bonds. The configurations of the asymmetric carbon atoms are not changed by this rotation. The reader should convince himself of this point by the study of models.

‡ Prior to 1947, it was believed that trans isomers in cyclic systems are always more stable than cis isomers. The above considerations indicate that in cyclohexane, at least, this is true only for the 1,2 and 1,4 isomers; in the 1,3 series, the cis isomer is now known to be more stable.

---

[14] J. E. Kilpatrick, H. G. Werner, C. W. Beckett, K. S. Pitzer, and F. D. Rossini, *J. Research Natl. Bur. Standards,* **39,** 523 (1947); E. J. Prosen, W. H. Johnson, and F. D. Rossini, *ibid.,* **39,** 173 (1947).

[14a] N. L. Allinger and M. A. Miller, *J. Am. Chem. Soc.,* **83,** 2145 (1961).

[15] M. Mousseron and R. Granger, *Bull. soc. chim. France,* [5]**5,** 1618 (1938).

stable trans isomer has two distinct conformational isomers, a (preferred) diequatorial one and a diaxial one. The energy situation is summarized in more detail in Table 8-2. The experimental differences in enthalpy between cis and trans isomers (ΔH) are in good agreement with the calculated differences in interaction energy for the more stable conformational isomers.

The fact that the absolute values of the free-energy differences ΔF in Table 8-2 are always smaller than the enthalpy differences ΔH indicates that there

Table 8-2
Conformations and Energies of the Dimethylcyclohexanes

Isomer	Conformation	No. gauche[a]	Interaction, kcal./mole	ΔH, exptl.,[b] kcal./mole	ΔF, exptl.,[b] kcal./mole
cis-1,2	*e,a* ⇃↾ *a,e*	3	2.7	1.87	1.66
trans-1,2	*e,e*	1	0.9		
	⇃↾ *a,a*	4	3.6		
cis-1,3	*a,a*	4	5.4[c]		
	⇃↾ *e,e*	0	0	−1.96	−1.59
trans-1,3	*e,a* ⇃↾ *a,e*	2	1.8		
cis-1,4	*e,a* ⇃↾ *a,e*	2	1.8	1.90	1.55
trans-1,4	*e,e*	0	0		
	⇃↾ *a,a*	4	3.6		

[a] Number of butane-gauche interactions.

[b] Value for cis isomer minus value for trans isomer at 25° (Ref. 14). ΔF is calculated from experimental ΔH and ΔS data, the latter being shown in Table 8-3.

[c] Includes diaxial methyl–methyl interaction.

is a difference in entropy between the diastereoisomers which always favors the equatorial-axial species and thus offsets, to some extent, the enthalpy difference which favors the diequatorial species. The major part of this entropy difference is trivial in origin, being made up of an entropy-of-mixing term and a symmetry-number term. The entropy of mixing must be considered for any species that consists of *non-identical* conformational isomers† and

† No entropy-of-mixing term enters when the two conformational isomers are superimposable, as in *cis*-1,4-dimethylcyclohexane.

amounts to $-R(x_1 \ln x_1 + x_2 \ln x_2)$, where x_1 and x_2 are the mole fractions of the two non-identical conformational isomers. As already mentioned in Chap. 3, an entropy of mixing of $R \ln 2$ (1.38 cal./deg. mole) must also be taken into account for any *dl* pair, whether separable or inseparable. Symmetry numbers enter into the entropies of molecules that have elements of symmetry, the symmetry number σ being equal to the number of equivalent spatial orientations that a molecule can occupy as a result of simple rotation. The entropy contribution due to symmetry is $-R \ln \sigma$. In the dimethylcyclohexanes only a twofold axis of rotation may occur (in the trans-1,2 and trans-1,4 isomers) leading to a symmetry number of 2 and an entropy contribution of -1.38 cal./deg. mole. Table 8-3 summarizes the entropy situation in the dimethylcyclohexanes. The observed experimental entropy differences between cis and trans isomers are included for comparison.

Table 8-3
Entropy Differences in Dimethylcyclohexanes[a]

Isomer	Symmetry no. σ	Entropy				Exptl. entropy difference
		$-R \ln \sigma$	Due to *dl* forms ($R \ln 2$)	Due to *e,e-a,a* equilibrium[b]	Total, calc'd	
cis-1,2	1	0	1.38	0	1.38[c]	0.72
trans-1,2	2	-1.38	1.38	0.11	0.11	
cis-1,3	1	0	0	ca. 0	0	-1.24
trans-1,3	1	0	1.38	0	1.38	
cis-1,4	1	0	0	0	0	1.19
trans-1,4	2	-1.38	0	0.03	-1.35	

[a] Entropies in cal./deg. mole. Ref. 10 and H. M. Huffman, S. S. Todd, and G. D. Oliver, *J. Am. Chem. Soc.*, **71**, 584 (1949).

[b] $-R(x_1 \ln x_1 + x_2 \ln x_2)$, where x_1 and x_2 are the mole fractions of *e,e* and *a,a* isomers, respectively, computed from $x_1 + x_2 = 1$ and $x_1/x_2 = e^{-\Delta F/RT}$. Although ΔF is not known a priori, it is sufficiently accurate to approximate it by the calculated ΔH of the two conformational isomers (cf. Table 8-2).

[c] This value is considerably in excess of the experimental value because interference with methyl rotation is not taken into account. If this interference is taken into account, better agreement results.

It might be noted that the agreement of calculated with experimental values is good, except in *cis*-1,2-dimethylcyclohexane. For this isomer it has been suggested[10] that there is a diminution of the rotational entropy of the two methyl groups due to steric interference ("cogwheeling"). Surprisingly enough, the same difference is *not* found in *trans*-1,2-dimethylcyclohexane, implying that in this isomer the methyl groups are not as close together as in the cis form, contrary to what would be expected on the basis of a perfect chair. The suggestion has been made[16] that the chair in *trans*-1,2-dimethylcyclohexane is distorted in such a way as to bend the methyl groups away from each other.

[16] E. L. Eliel, *J. Chem. Educ.*, **37**, 126 (1960).

8-4. Physical Properties of Substituted Cyclohexanes

The rule that summarizes the relationship between certain physical properties and conformation was first developed by von Auwers and by Skita[17] and is generally known as the "von Auwers–Skita rule" or "conformational rule." An up-to-date statement[18] of this rule (which has undergone several modifications since it was first enunciated) is that among alicyclic epimers not differing in dipole moment the isomer of highest heat content (enthalpy) has the higher density, index of refraction, and boiling point.† An application of the rule to the dimethylcyclohexanes is shown in Table 8-4. The diastereoisomer of lower enthalpy in each pair (cf. Table 8-2) is starred in Table 8-4. It is clear that this isomer has the lower physical constants in each instance. The reason for this is that, in general, both heat content and boiling point, refractive index, and density are functions of molecular volume: The isomer of greater molecular volume has the smaller heat content and the smaller physical constants. The inverse relationship of molecular volume to density

Table 8-4
Physical Properties of the Dimethylcyclohexanes

Isomer	Conformation	B.p., °C.	n_D^{25}	d_4^{25}
cis-1,2	*e,a*	129.7	1.4336	0.7922
trans-1,2*	*e,e*	123.4	1.4247	0.7720
cis-1,3*	*e,e*	120.1	1.4206	0.7620
trans-1,3	*e,a*	124.5	1.4284	0.7806
cis-1,4	*e,a*	124.3	1.4273	0.7787
trans-1,4*	*e,e*	119.4	1.4185	0.7584

is obvious and that to refractive index follows, inasmuch as the refractive index is a function of density. The relation of molecular volume to heat content and boiling point is less obvious but may be rationalized as follows: Increased molecular volume means increased distance both between nonbonded atoms within the same molecule and also between adjacent molecules in a liquid or gas. The former results in lesser intramolecular crowding and therefore lesser heat content, the latter in diminished van der Waals attraction and therefore lower boiling point.

A few exceptions to the von Auwers–Skita rule are known. For example, it does not apply to the boiling points of the alkylcyclohexanols,† here the *e,e* isomers, while having the lower enthalpy, refractive index, and density, are

† H. van Bekkum, A. van Veen, P. E. Verkade, and B. M. Wepster, *Rec. trav. chim.*, **80**, 1310 (1961), have suggested that, since the rule frequently fails for the boiling point, it should properly be applied only to density and refractive index. Other limitations of the rule are also discussed in their paper.

[17] K. von Auwers, *Ann.*, **420**, 84 (1920); A Skita, *Ber.*, **56**, 1014 (1923).
[18] Cf. N. L. Allinger, *J. Am. Chem. Soc.*, **79**, 3443 (1957).

more strongly hydrogen-bonded than the *e,a* isomers (the axial hydroxyl group being less well-disposed for hydrogen bonding) and therefore boil at a higher temperature.[19] Also, the rule does not apply to isomers differing appreciably in dipole moment. Such isomers follow the "van Arkel rule" or "dipole rule" according to which the isomer of higher dipole moment has the higher physical constants, regardless of heat content.[20]

Evidently the conformational rule may (with some caution) be used to assign configurations to epimers when the more reliable methods discussed in Chap. 7 cannot be readily applied. Physical properties other than those summarized in the conformational rule have been studied as a function of conformation and these, also, may be used to make tentative configurational assignments to epimers. For example, equatorial substituents usually show typical infrared absorption at higher frequencies (shorter wavelengths) than axial substituents. Thus the C—O stretching frequency of a number of

Table 8-5
Infrared Absorption Bands Due to C—X Stretching for Substituents X in Equatorial and Axial Positions in Cyclohexane[a]

Substituent X	Equatorial C—X, cm.$^{-1}$	Axial C—X, cm.$^{-1}$
OH	1037–1044	996–1036
OAc[b]	1013–1022	1025–1031
OMe	1100–1104	1086–1090
Cl	736–856	646–730
Br	682–833	542–692
D	{2155–2162 2171–2177	2114–2138 2139–2164

[a] From Ref. 77.
[b] This is the only substituent of those studied which has a higher absorption frequency when in the axial position.

equatorial steroid alcohols is found around 1040 cm.$^{-1}$, whereas the corresponding stretching frequency for axial alcohols occurs at about 1000 cm.$^{-1}$, at least when the A/B ring junction is trans. Unfortunately, the situation is not quite the same in terpene alcohols nor in simple alkylcyclohexanols, although, at least in the latter, the *shift* in the C—O frequency in going from equatorial to axial hydroxyl substituents is the same as for the steroids. Similar correlations have been made for other types of substituents and are summarized in Table 8-5.

Conclusions as to configuration (and conformation) based on dipole-moment studies or measurements of acid strength are usually more reliable, since they are based on more clear-cut theoretical concepts. The 1,2-dibromocyclohexanes (Fig. 8-14) may be cited as an example of dipole studies.[21]

[19] E. L. Eliel and R. G. Haber, *J. Org. Chem.*, **23**, 2041 (1958).
[20] A. E. van Arkel, *Rec. trav. chim.*, **51**, 1081 (1932); **53**, 246 (1934).
[21] P. Bender, D. L. Flowers, and H. L. Goering, *J. Am. Chem. Soc.*, **77**, 3463 (1955).

The moment for the diaxial isomer is assumed to be zero; the calculated moment for either the diequatorial or the equatorial-axial isomer is 3.09 D.† The observed moments for the two known isomers are 3.12 and 2.11 D, respectively. It is clear that the higher moment corresponds to the cis isomer (*e,a;* cf. Fig. 8-14) and the lower to the trans (*e,e* or *a,a*). One must also assume that the diaxial form makes an appreciable contribution to the trans isomer in order to account for its moment being so much lower than that of the cis compound. Undoubtedly this is because the diequatorial form, although favored sterically (cf. Table 8-2) is destabilized by dipole repulsion.

Turning now to acid-strength measurements, the difference between the first and second ionization constants in cyclohexane-*cis*-1,2-dicarboxylic acid (aqueous solution) corresponds to 2.42 pK units, whereas the corresponding difference for the trans isomer is 1.75 pK units.[22] In general, this difference

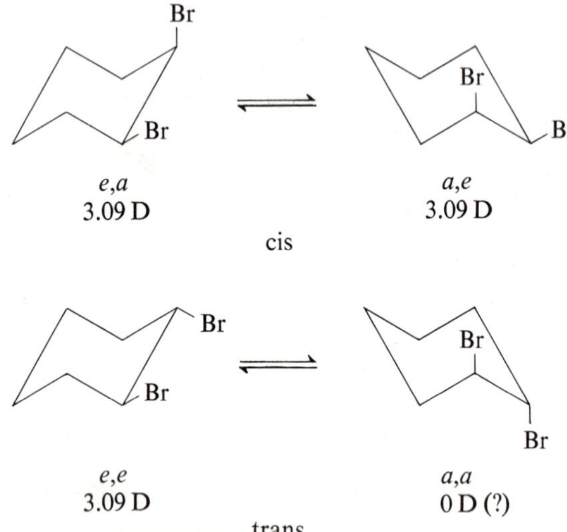

Fig. 8-14. 1,2-Dibromocyclohexanes.

is the greater, the closer together the carboxyl groups are, for the positive C=O dipole of one carboxylic acid group eases the departure of the proton

† It is assumed that the dihedral angle between the C—Br bond is 60° for both the *e,e* and *e,a* isomers. This assumption may be poor since, because of steric and dipolar repulsion, the bromine atoms may tend to rotate away from each other, the cyclohexane chair becoming deformed (Ref. 16). It can be seen in models that such rotation would be relatively facile in the *e,e* isomer, where it leads to unpuckering of the chair, but would be difficult for the *e,a* isomer where it would lead to increased puckering (see the earlier discussion on the entropies of the 1,2-dimethylcyclohexanes). Therefore it is not surprising that the dipole moment of the *e,a* isomer is close to that calculated. The assumption that the *a,a* isomer has zero dipole moment is also questionable since this molecule does not have a center of symmetry. For detailed discussion of this problem, cf. E. Wessels, Ph.D. Dissertation (with E. Havinga), University of Leiden, Holland, 1960.

[22] R. Kuhn and A. Wasserman, *Helv. Chim. Acta,* **11,** 50 (1928).

from a closely proximate group by a field effect (induction through space), but the negative charge of the COO^- ion in the once ionized acid prevents the departure of the second proton, if the other carboxyl group is close by. (See the more detailed discussion in Chap. 12.) It follows, then, that the carboxyl groups in the cis-1,2-dicarboxylic acid are closer together than in the trans isomer—a reasonable conclusion if it is granted that the trans isomer, for reasons of dipole repulsion, exists at least partly in the diaxial form. Conversely, if the configurations of the cyclohexane-1,2-dicarboxylic acids had not been known, they could have been derived from the pK data.

A more subtle difference is found between the acid strengths[23] of *cis*- and *trans*-4-*t*-butylcyclohexanecarboxylic acid (Fig. 8-15).† The trans acid is stronger because the corresponding anion (equatorial COO^-) is more readily solvated than that of the cis acid which, being axial, encounters some steric hindrance to solvation.

Another physical difference between certain equatorial and axial groups lies in their different affinity for adsorbents, especially when the groups are somewhat polar in nature. Thus *trans*-4-*t*-butylcyclohexanol (cf. Fig. 8-20, equatorial hydroxyl) is more strongly adsorbed on alumina than the cis

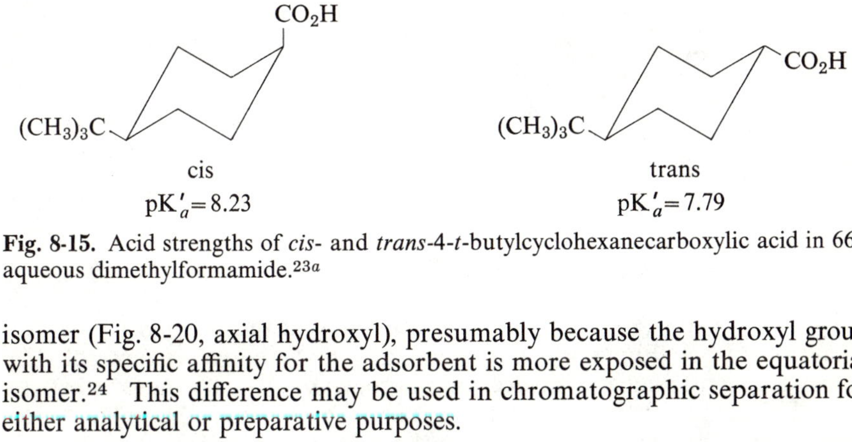

CO_2H

$(CH_3)_3C$

cis

$pK'_a = 8.23$

CO_2H

$(CH_3)_3C$

trans

$pK'_a = 7.79$

Fig. 8-15. Acid strengths of *cis*- and *trans*-4-*t*-butylcyclohexanecarboxylic acid in 66% aqueous dimethylformamide.[23a]

isomer (Fig. 8-20, axial hydroxyl), presumably because the hydroxyl group with its specific affinity for the adsorbent is more exposed in the equatorial isomer.[24] This difference may be used in chromatographic separation for either analytical or preparative purposes.

8-5. Conformation and Chemical Reactivity in Cyclohexane

The relationship between conformation and chemical reactivity was first pointed out by D. H. R. Barton in an epoch-making paper[25] in 1950. In this paper, attention is drawn to the difference in chemical behavior between substituents placed in equatorial and axial positions in cyclohexanes. To avoid

† The size of the *t*-butyl group is such that almost all these molecules exist in the conformation shown and not in the alternative conformation in which *t*-butyl is axial (Ref. 24).

[23] (*a*) R. D. Stolow, *J. Am. Chem. Soc.*, **81**, 5806 (1959); (*b*) M. Tichý, J. Jonáš, and J. Sicher, *Collection Czechoslov. Chem. Commun.*, **24**, 3434 (1959).

[24] S. Winstein and N. J. Holness, *J. Am. Chem. Soc.*, **77**, 5562 (1955).

[25] D. H. R. Barton, *Experientia*, **6**, 316 (1950).

complications from the above-mentioned conformational transformation ("flipping" or "conformational inversion")—which can place the substituent in a monosubstituted cyclohexane in either the equatorial or the axial position—Barton's early examples came mostly from systems which are either

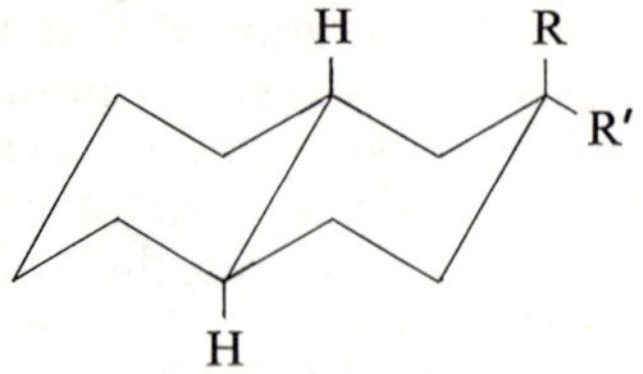

R = H, R′ = OH trans isomer, equatorial hydroxyl
R = OH, R′ = H cis isomer, axial hydroxyl

Fig. 8-16. The two diastereoisomeric *trans*-decalin-2-ols.

rigid or, though mobile, exist almost exclusively in one of the two possible conformations. Examples of the first kind are derivatives of *trans*-decalin (Fig. 8-16) which, as will be explained in Chap. 10, cannot flip. Among these derivatives are many natural products, such as steroids (e.g., Fig. 8-17) and

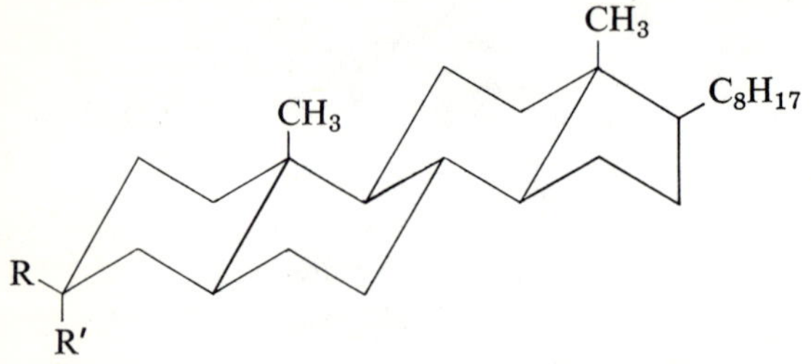

R = OH, R′ = H cholestanol, equatorial hydroxyl

R = H, R′ = OH *epi*-cholestanol, axial hydroxyl

Fig. 8-17. Cholestanol and *epi*-cholestanol.

polycyclic terpenes (e.g., Fig. 8-18). Examples of the second kind are menthol and neomenthol (Fig. 8-19) and *trans*- and *cis*-4-*t*-butylcyclohexanol (Fig. 8-20) which exist almost exclusively in the conformation shown, for otherwise both the methyl and the isopropyl group (Fig. 8-19) or the bulky *t*-butyl group (Fig. 8-20) would be forced into the axial position.

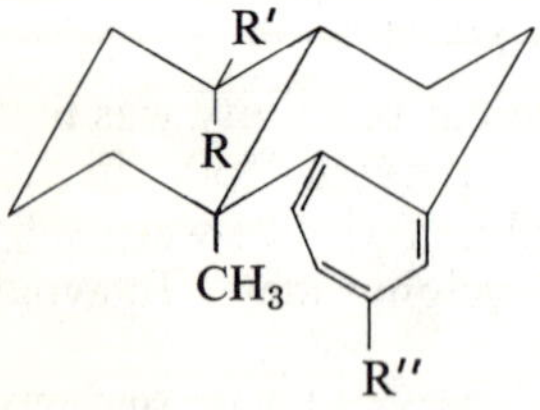

R = CH_3, R′ = CO_2H, R″ = $(CH_3)_2CH$ dehydroabietic acid, equatorial carboxyl

R = CO_2H, R′ = CH_3, R″ = H podocarpic acid, axial carboxyl

Fig. 8-18. Dehydroabietic acid and podocarpic acid.

It has already been mentioned that compounds with equatorial substituents are generally more stable (because of lesser steric crowding) than compounds with corresponding axial substituents, and evidence from thermodynamic measurements supporting this statement has been cited. It follows that in cases where, of two chemically interconvertible epimers, one has an equatorial substituent and the other an axial one, the equatorial isomer will predominate in the equilibrium mixture. Thus the equilibrium mixture of *cis*- and *trans*-4-*t*-butylcyclohexanol, obtained by treating either isomer with

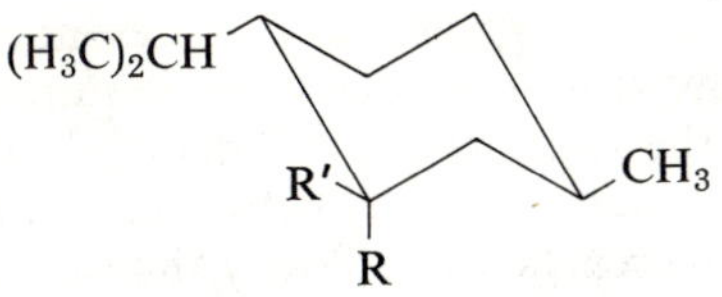

$(H_3C)_2CH$ ... R' ... CH_3 ... R

R = H, R' = OH menthol, equatorial hydroxyl
R = OH, R' = H neomenthol, axial hydroxyl

Fig. 8-19. Menthol and neomenthol.

an aluminum isopropoxide–isopropyl alcohol–acetone mixture, contains four parts of the trans isomer (equatorial hydroxyl) to one part of cis (axial hydroxyl), as shown in Fig. 8-20.[26] Similarly, the equilibrium mixture of carvomenthone and isocarvomenthone (Fig. 8-21) contains about 91% of the

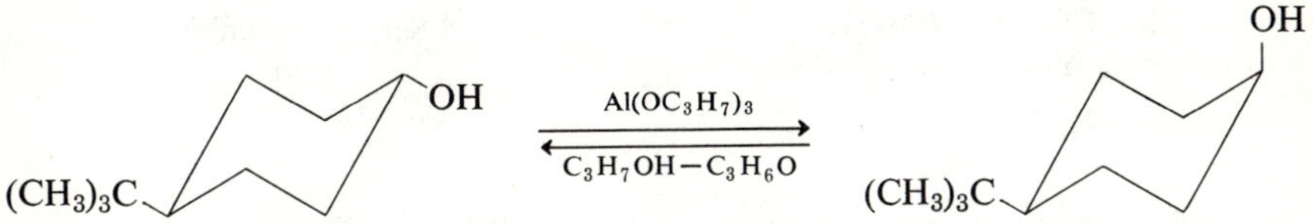

$(CH_3)_3C$ ... OH $\xrightleftharpoons[C_3H_7OH-C_3H_6O]{Al(OC_3H_7)_3}$ $(CH_3)_3C$ ... OH

trans (80%) cis (20%)

Fig. 8-20. 4-*t*-Butylcyclohexanol equilibrium.

former (both methyl and isopropyl equatorial) and 9% of the latter (methyl group, axial).[27] The fact that the alkaline saponification of both yohimbine and corynanthine gives yohimbic acid (Fig. 8-22) (since re-esterification of this acid returns yohimbine and not corynanthine) leads to the assumption

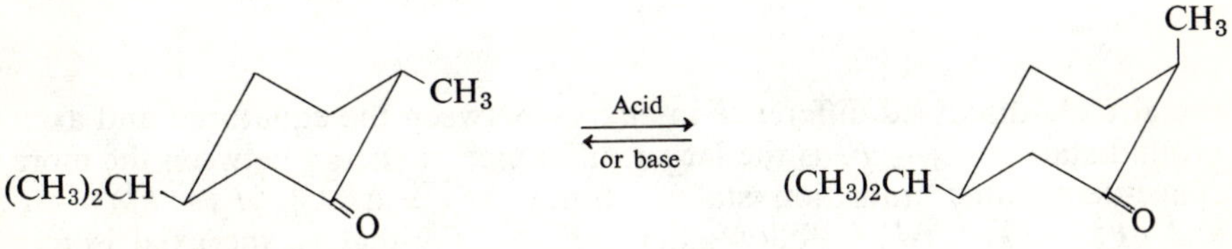

$(CH_3)_2CH$ ... CH_3 ... O $\xrightleftharpoons[\text{or base}]{\text{Acid}}$ $(CH_3)_2CH$ ... CH_3 ... O

Carvomenthone, 80% Isocarvomenthone, 20%

Fig. 8-21. Carvomenthone-isocarvomenthone equilibrium.

that the carbomethoxy group is equatorial in yohimbine and axial in corynanthine; alkali epimerizes corynanthine to yohimbine before the ester group is saponified. (Acid hydrolysis of corynanthine does not produce epimerization

[26] E. L. Eliel and R. S. Ro, *J. Am. Chem. Soc.*, **79**, 5992 (1957).
[27] B. Rickborn, *J. Am. Chem. Soc.*, **84**, 2414 (1962).

and leads to a different acid, corynanthic acid, whose re-esterification returns corynanthine.) In fact, the configuration of the carbomethoxy group in yohimbine and corynanthine has been derived from the behavior of these alkaloids upon saponification.[28]

Axial substituents are in a more crowded environment than equatorial substituents, and this, in general, gives rise to differences in reactivity between the two types. In many reactions, the steric requirements of the transition state are greater than the steric requirements of the ground state. In such reactions, the crowding around the axial substituent produces steric hindrance, and the axial substituent reacts more slowly than the equatorial substituent. The energetics of this situation is depicted in Fig. 8-23. ΔF in this figure is

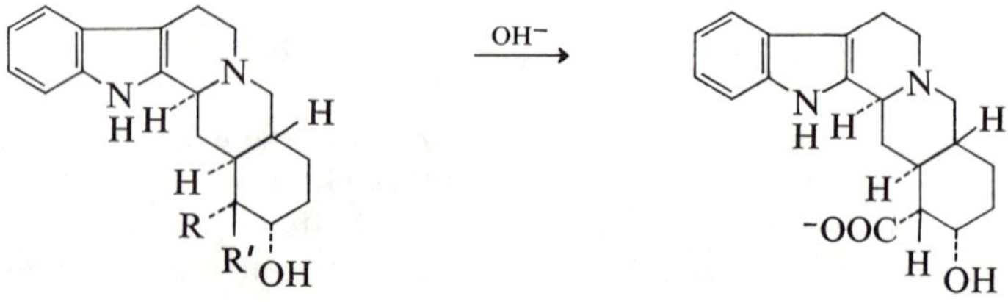

$R = CO_2CH_3$, $R' = H$
Yohimbine

$R = H$, $R' = CO_2CH_3$
Corynanthine

$\xrightarrow{OH^-}$

Salt of yohimbic acid

Fig. 8-22. Saponification of yohimbine and corynanthine. (The reader should construct models of the two alkaloids and convince himself that, for the configuration shown, the carbomethoxy group is, indeed, equatorial in yohimbine and axial in corynanthine. The ability to appreciate at a glance whether a given substituent in a planar formula, in which configurations are indicated, is equatorial or axial can be acquired only by continual practice. In the present instance, the reasoning is as follows: Since the D-E ring junction is trans, the bonds leading to ring D are equatorial with respect to ring E, and the hydrogen atoms at the ring junctions are axial. Hence the carbomethoxy group in yohimbine, being cis to the adjacent axial hydrogen, is equatorial, while in corynanthine it is trans to the adjacent axial hydrogen and therefore axial.)

the already-discussed difference in energy between the equatorial and axial ground states. $F_A^\ddagger - F_E^\ddagger$ is the larger difference in energy between the more space-demanding transition states. Since $\Delta F_A^\ddagger = \Delta F_E^\ddagger + (F_A^\ddagger - F_E^\ddagger) - \Delta F$ and $(F_A^\ddagger - F_E^\ddagger) > \Delta F$, it follows that $\Delta F_A^\ddagger > \Delta F_E^\ddagger$, that is, the axial isomer reacts more slowly.

Examples are found in the saponification of esters of cyclohexanols[29] and of cyclohexanecarboxylic acids[30] (Fig. 8-24). It is known that these reactions

[28] R. C. Cookson, *Chemistry & Industry (London)*, 337 (1953).

[29] G. F. Hennion and F. X. O'Shea, *J. Am. Chem. Soc.*, **80**, 614 (1958). See also N. B. Chapman, R. E. Parker, and P. J. A. Smith, *J. Chem. Soc.*, 3634 (1960).

[30] E. L. Eliel, H. Haubenstock, and R. V. Acharya, *J. Am. Chem. Soc.*, **83**, 2351 (1961); cf. E. A. S. Cavell, N. B. Chapman, and M. D. Johnson, *J. Chem. Soc.*, 1413 (1960).

involve not merely transition states but discrete intermediates[31] of the type

$$\begin{array}{c}\text{OH}\\ |\\ \text{—C—OR}\\ |\\ \text{OH}\end{array}$$

in which the carboxyl carbon has become tetrahedral and has therefore increased in bulk from its original trigonal state. The effect on rate is greater when the cyclohexyl substituent is in the acid part of the ester than

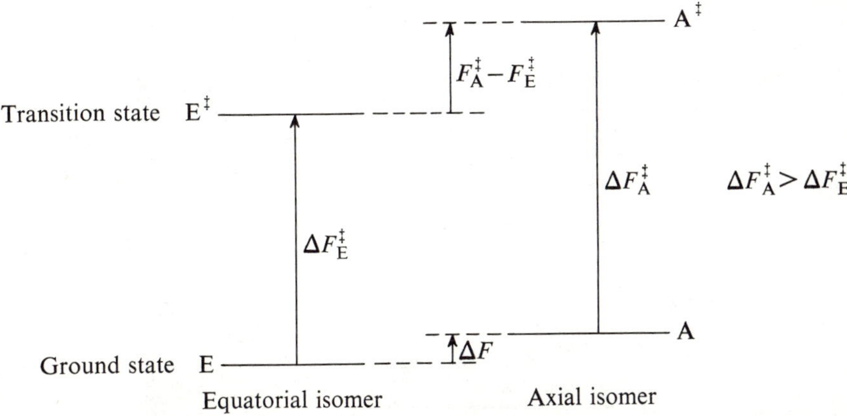

Fig. 8-23. Energy level of ground state and transition state: case of steric hindrance.

when it is in the alcohol part. This is to be expected, since the site of crowding—the C=O group—is closer to the ring in the former case than in the latter.

There are also some reactions in which the steric requirements of the ground state are greater than those of the transition state. In that case, the crowding of the axial substituent is greater in the ground state, that is, ΔF (Fig.

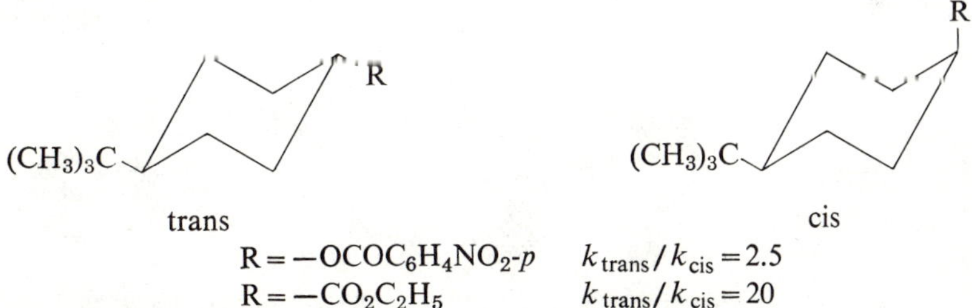

Fig. 8-24. Relative rates of saponification of the 4-*t*-butylcyclohexyl *p*-nitrobenzoates and ethyl 4-*t*-butylcyclohexanecarboxylates.

8-25) is larger than $F_A^‡ - F_E^‡$ and therefore $\Delta F_E^‡$ is larger than $\Delta F_A^‡$. In other words, the more crowded substituent reacts faster. In this case, one speaks of "steric assistance."[32] A case in point is the solvolysis of *cis*- and *trans*-4-

[31] M. L. Bender, *J. Am. Chem. Soc.*, **73**, 1626 (1951).
[32] H. C. Brown, *Science*, **103**, 385 (1946).

t-butylcyclohexyl tosylates (Fig. 8-26).[24] To the extent that the transition state resembles the carbonium ion (Fig. 8-26), the initial difference between the energy levels of the cis and trans tosylate has disappeared and the less stable cis tosylate reacts faster. In fact, it does so by a factor of about 3 to 4. Another case of steric assistance is the chromic acid oxidation of alcohols to ketones.[33] Since an understanding of this case is dependent on a detailed knowledge of steroid structure and geometry, it will be discussed in Chap. 10.

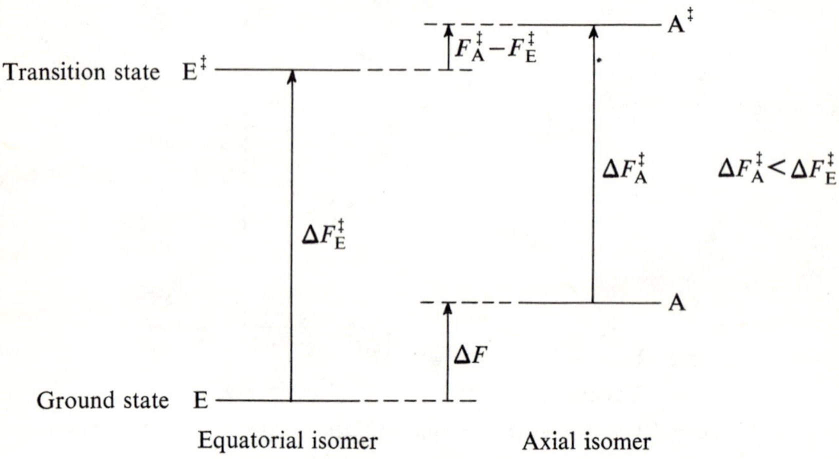

Fig. 8-25. Energy level of ground state and transition state: case of steric assistance.

The situation in the S_N2 reaction is somewhat more complicated than that in solvolysis (presumably largely S_N1), because there may be differences in the energy levels of the transition states as well as of the ground states as one switches from an equatorial to an axial substituent in the substrate. The most general case is shown in Fig. 8-27.

OTs
$(CH_3)_3C$ → $(CH_3)_3C$ + ← $(CH_3)_3C$ O
trans ↓ cis
Solvolysis products

Fig. 8-26. Solvolysis of *cis*- and *trans*-4-*t*-butylcyclohexyl tosylates.

A simplification occurs when the substitution is symmetrical, i.e., when the incoming and outgoing groups are the same, as in the substitution of 4-*t*-butylcyclohexyl iodide by radio-iodide. The energetics of this hypothetical case is shown in Fig. 8-28. The difference in activation energy $\Delta F_E^{\ddagger} - \Delta F_A^{\ddagger}$ between the equatorial and axial iodide is simply the difference

[33] J. Schreiber and A. Eschenmoser, *Helv. Chim. Acta,* **38,** 1529 (1955).

in free energy of the respective equatorial and axial ground states (ΔF). The more complex energetics of an unsymmetric substitution is shown in Fig. 8-29. Depending on the spatial requirements of the incoming and outgoing groups, the equatorial transition state may be either higher or lower than the axial transition state.† In the two cases so far studied—displacement of bromide and of tosylate by thiophenolate—the incoming group is more bulky than

Transition state

Fig. 8-27. S_N2 reaction.

the outgoing group and the difference in activation energy between the two isomers is greater than the difference in ground-state energy. This is the case shown in Fig. 8-29. The measured differences ΔF are ca. 0.7 kcal./mole for the ground states in both the bromide[34] and the tosylate[35] case, whereas the respective differences in activation energies ($\Delta F_E^{\ddagger} - \Delta F_A^{\ddagger}$) are 2.4 and 2.1 kcal./mole, leaving a difference $F_E^{\ddagger} - F_A^{\ddagger}$ of 1.7 and of 1.4 kcal./mole, respectively.

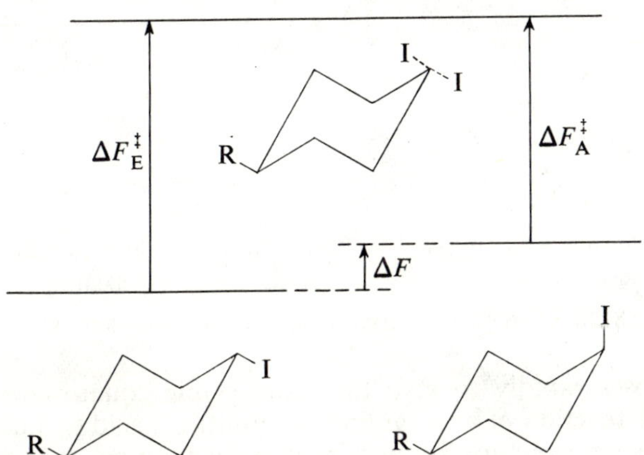

Fig. 8-28. Energetics of reaction of 4-*t*-butylcyclohexyl iodide with iodide ion. $R = (CH_3)_3C$.

† The correctness of this statement follows from the principle of microscopic reversibility. If Y is larger than X, then the transition state for the replacement of equatorial X (Fig. 8-29) will be higher than that for replacement of axial X. However, by the same token, the transition state for the replacement of axial Y will be higher than that for replacement of equatorial Y. If the reaction is measurably reversible, it should be possible to verify this point.

[34] E. L. Eliel and R. G. Haber, *J. Am. Chem. Soc.,* **81,** 1249 (1959).

[35] E. L. Eliel and R. S. Ro, *J. Am. Chem. Soc.,* **79,** 5995 (1957).

tively, to be explained by differences in the energy levels of the transition states.

Another type of substitution reaction that has been investigated extensively in cyclohexyl systems is the amine–nitrous acid reaction. Equatorial amines react cleanly with nitrous acid to give equatorial alcohol, probably via a fleeting diazonium salt intermediate. Because of the clear-cut retention of configuration, this substitution has been called an "S_Ni" (as distinct from S_N1 or S_N2) reaction in the classification of Hughes and Ingold. Retention has been attributed to a cyclic intermediate[36] or to a short-lived solvated ion

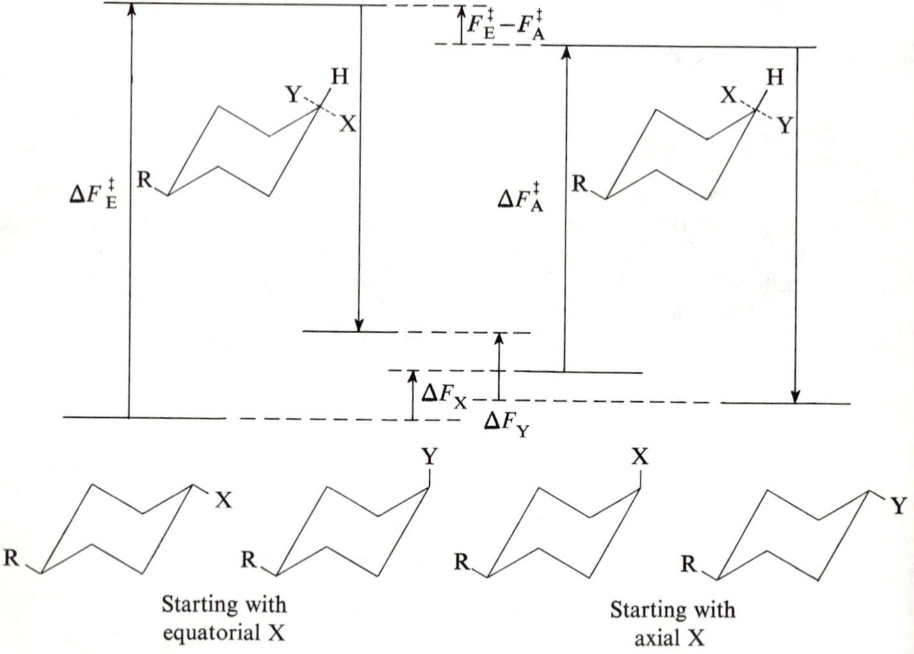

Fig. 8-29. Energetics of unsymmetric S_N2 substitution (Y > X).

which collapses rapidly to give the more stable equatorial isomer (Fig. 8-30).[37] The reaction of axial amines with nitrous acid is much less clear-cut. Since the diazonium group is in an axial position, it is well disposed toward trans diaxial elimination (see below, page 227), and the predominant reaction product is olefin. The alcohol product that is formed at the same time may be predominantly equatorial[37] or exclusively axial[38] and sometimes includes products of rearrangement.

So far, reactions have been discussed in which only simple *steric* factors need to be considered. In the following, attention will be focused on reac-

[36] J. A. Mills, *J. Chem. Soc.*, 260 (1953); A. K. Bose, *Experientia*, **9**, 256 (1953).

[37] W. G. Dauben, R. C. Tweit, and C. Mannerskantz, *J. Am. Chem. Soc.*, **76**, 4420 (1954).

[38] C. W. Shoppee, D. E. Evans, and G. H. R. Summers, *J. Chem. Soc.*, 97 (1957); C. W. Shoppee, R. J. W. Cremlyn, D. E. Evans, and G. H. R. Summers, *ibid.*, 4364 (1957).

tions that have definite "*stereoelectronic*" requirements (cf. Chap. 6), i.e., which proceed best when certain spatial relationships pertain between electrons involved in the bonds formed or broken during the reaction. A case in point is the ionic elimination reaction which is known to proceed best when the atoms or groups to be eliminated are conformationally anti to each other (cf. Chap. 6). In a cyclohexane ring, this relationship pertains for adjacent groups which are both axial (trans) but not for adjacent groups which are both equatorial (although these are also, configurationally speaking, trans) or for adjacent axial and equatorial substituents (cis). Thus in a number of steroid derivatives trans diaxial vicinal dibromides react with iodide ion reasonably rapidly (to give olefin, bromide ion, and iodine), whereas the corresponding trans diequatorial or cis equatorial-axial dibromides react extremely sluggishly.[39, 40] Similarly, in the dehydrohalogenation of the menthyl chlorides by base,[41] neomenthyl chloride (Fig. 8-19, R = Cl, R′ = H) readily

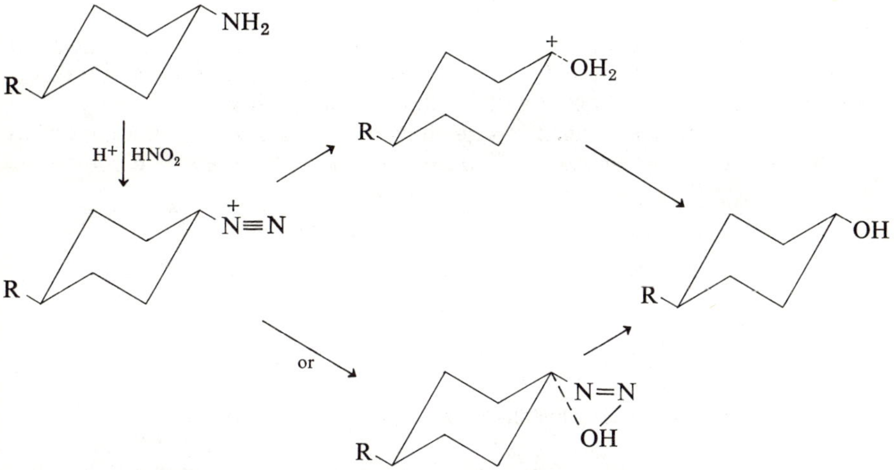

R NH₂
H⁺ HNO₂
R N≡N (+)
R OH₂ (+)
or
R N=N OH
R OH

Fig. 8-30. Reaction of equatorial amine with nitrous acid.

loses the elements of hydrogen chloride to give 3-menthene, whereas menthyl chloride (Fig. 8-19, R = H, R′ = Cl) reacts at a specific rate 200 times smaller, the product being 2-menthene (Fig. 8-31). Evidently, the compound that has adjacent axial chlorine and hydrogen (neomenthyl chloride) reacts readily and follows the usual Saytzeff rule which demands that the double bond be formed in the most highly substituted position. The compound that has equatorial chlorine (menthyl chloride) reacts much more slowly, and when it does react, it does so, apparently, in the conformation (Fig. 8-32) in which the normal chair (Fig. 8-19) is flipped so that all three substituents, including the chlorine, become axial (Fig. 8-32). Because this conformation

[39] D. H. R. Barton and W. J. Rosenfelder, *J. Chem. Soc.*, 1048 (1951).

[40] G. H. Alt and D. H. R. Barton, *J. Chem. Soc.*, 4284 (1954).

[41] W. Hückel, W. Tappe, and G. Legutke, *Ann.*, **543**, 191 (1940); E. D. Hughes, C. K. Ingold, and J. B. Rose, *J. Chem. Soc.*, 3839 (1953).

is so unfavorable, the rate is slow, and since the only axial hydrogen next to the axial chlorine in this conformation is at C_2, the elimination product is 2-menthene, contrary to the Saytzeff rule. Yet another illustration comes from the hexachlorocyclohexanes; of the eight possible isomers (cf. Fig. 7-5) all but the one shown in Fig. 8-33 undergo ready elimination with mineral

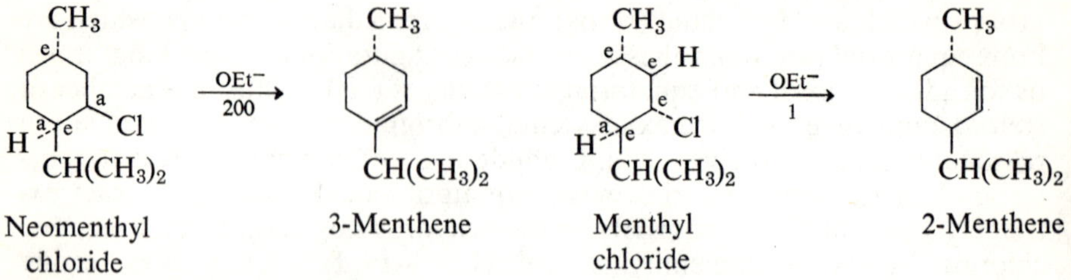

Neomenthyl chloride — OEt⁻, 200 → 3-Menthene; Menthyl chloride — OEt⁻, 1 → 2-Menthene

Fig. 8-31. Dehydrohalogenation of the menthyl chlorides.

base.[42] The β isomer (Fig. 8-33) is the only one that does not have any axial hydrogen next to an axial chlorine in either chair conformation, and it reacts more slowly, by several powers of 10, than any of its isomers.

There are several other reactions in which a trans diaxial or anti relationship of the electron pairs involved is desirable. Among these are molecular

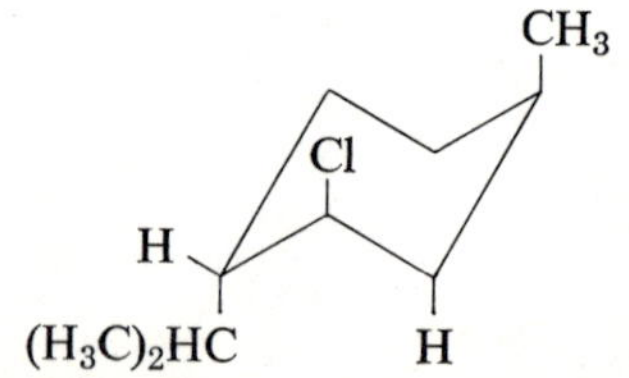

Fig. 8-32. Menthyl chloride: unfavorable conformation.

rearrangements of the concerted type. Thus when *trans*-2-aminocyclohexanol (Fig. 8-34) is treated with nitrous acid, the intermediate diazonium ion rearranges in such a way that the ring bond (C_2—C_3), which is conformationally trans to the departing nitrogen, migrates, and a ring contraction to cyclopentanecarboxaldehyde results.[43] In contrast, the cis isomer (Fig. 8-35)

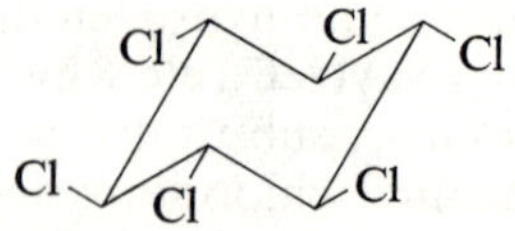

Fig. 8-33. β-Benzene hexachloride.

may involve either an equatorial or an axial diazonium intermediate.[43] In the former, the situation is similar to that pertaining in the trans isomer and

[42] S. J. Cristol, *J. Am. Chem. Soc.*, **69**, 338 (1947).
[43] G. E. McCasland, *J. Am. Chem. Soc.*, **73**, 2293 (1951).

the product is the ring-contracted cyclopentanecarboxaldehyde, but in the latter the axial hydrogen at C_2 is now in the best position for migration and the product is cyclohexanone.

Neighboring-group participation, even if it does not involve rearrangement but only enhancement of reaction rate, requires a trans diaxial arrangement

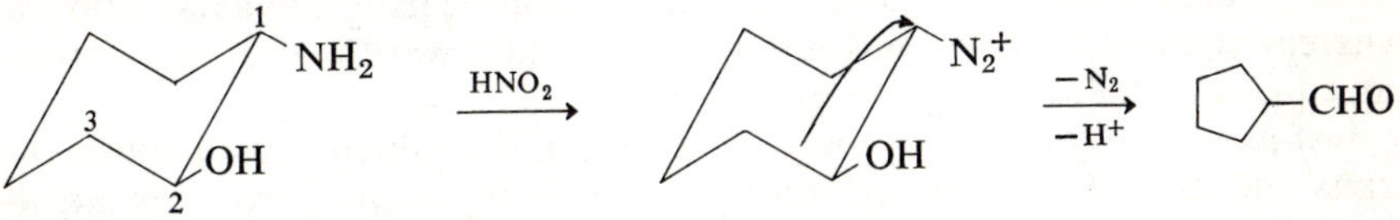

Fig. 8-34. Rearrangement of *trans*-2-aminocyclohexanol.

of the departing and participating groups.[40] Thus neomenthyl tosylate (Fig. 8-19, R = OTs, R′ = H) solvolyzes 170 times faster in acetic acid than does menthyl tosylate (Fig. 8-19, R = H, R′ = OTs).[44] This large enhancement of rate cannot be merely a matter of steric assistance since the rate factor due to this cause is no larger than about 3 to 5.[24] Winstein has suggested

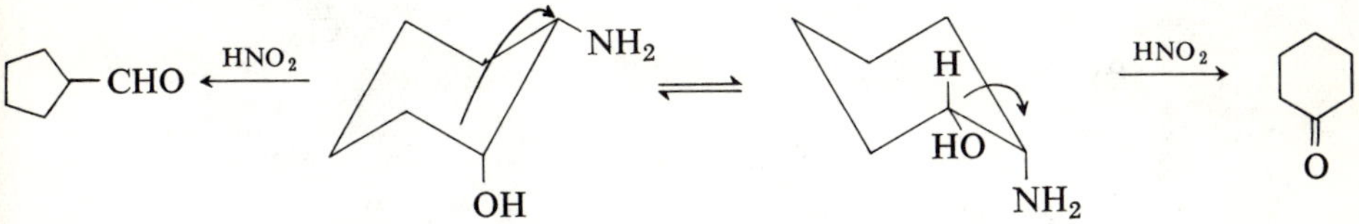

Fig. 8-35. Rearrangement of *cis*-2-aminocyclohexanol.

that the much larger rate enhancement in neomenthyl tosylate is caused by participation of the neighboring tertiary hydrogen (axial) in the departure of the tosylate group (also axial), as shown in Fig. 8-36.†

Related to the matter of neighboring-group participation is the stereochemistry of epoxide ring formation which also proceeds best when the groups involved are both axial. Thus,[77] a diequatorial halohydrin, 2α-chloro-

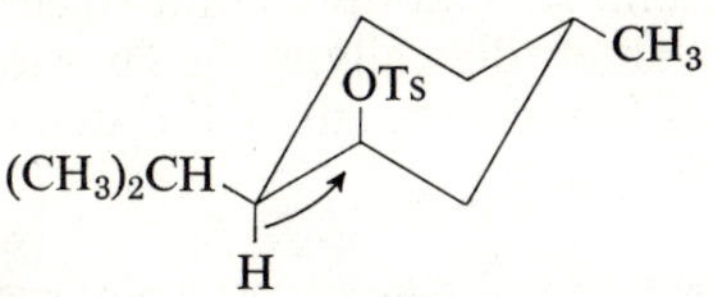

Fig. 8-36. Neighboring-group participation in neomenthyl tosylate solvolysis.

† However, the solvolysis rates of *cis*- and *trans*-2-*t*-butylcyclohexyl tosylate differ by factors of only 2 to 5: H. L. Goering and R. L. Reeves, *J. Am. Chem. Soc.*, **78**, 4931 (1956). There appears to be no neighboring-group participation in the rate-determining step in this case.

[44] S. Winstein, B. K. Morse, E. Grunwald, H. W. Jones, J. Corse, D. Trifan, and H. Marshall, *J. Am. Chem. Soc.*, **74**, 1127 (1952).

cholestan-3β-ol,† is cyclized to the epoxide (2,3β-epoxycholestane) several thousand times more slowly than either the diaxial position isomer 3α-chlorocholestan-2β-ol (which yields the same product) or the diaxial stereoisomer 2β-chlorocholestan-3α-ol (which yields the 2,3α-oxide) (Fig. 8-37). The difference in rate here is evidently not a result of differences in product stability but originates from the fact that the axial alkoxide is favorably located for a rear attack on the carbon bearing the axial halogen, whereas equatorial alkoxide is not well disposed toward rear attack displacing equatorial halogen.

Just as ionic elimination proceeds best when the substituents involved are trans and diaxial, so ionic addition to double bonds also proceeds largely diaxially.[40, 45]‡ Thus addition of bromine to 2-cholestene[40] (ring A shown in Fig. 8-38) gives principally the diaxial 2β,3α-dibromocholestane.[40] Similarly, just as epoxides are formed most easily from diaxial halohydrins, the epoxide ring (which resembles a double bond electronically in some of its behavior) also opens up largely diaxially.[46, 47] Thus the addition of hydrogen bromide to cholestane-2,3α-epoxide gives the diaxial 2β,3α-bromohydrin (Fig. 8-39).[40]

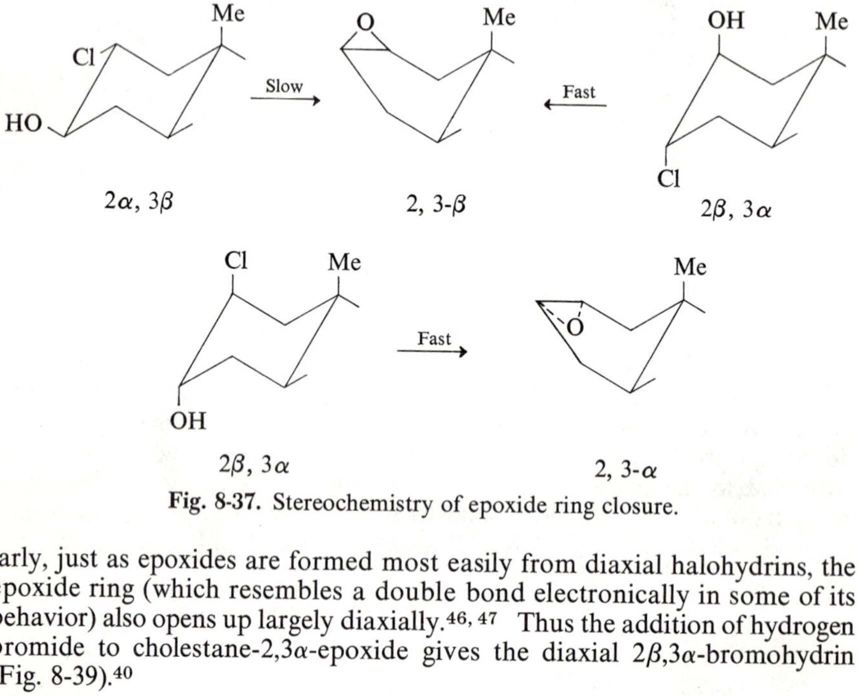

Fig. 8-37. Stereochemistry of epoxide ring closure.

†A substituent that is on the same side as the angular methyl groups in a steroid is called "β" and one on the opposite side "α." These letters thus refer to configuration; they are not based on the equatorial or axial conformation of the substituent. The complete correlation between configuration and conformation in steroids is shown in Chap. 10 (Fig. 10-29).

‡The addition of HCl is an exception which will be considered in Chap. 10.

[45] D. H. R. Barton, A. da S. Campos-Neves, and R. C. Cookson, *J. Chem. Soc.*, 3500 (1956).

[46] A. Fürst and P. A. Plattner, *Abstr. Papers 12th Intern. Congress Pure and Appl. Chem.*, New York, 1951, p. 409.

[47] See also E. L. Eliel in M. Newman, ed., "Steric Effects in Organic Chemistry," John Wiley & Sons, Inc., New York, 1956, pp. 130–134.

Free-radical addition to cyclohexenes also proceeds configurationally trans. Thus the ultraviolet-light-induced addition of hydrogen bromide to 1-bromocyclohexene gives *cis*-1,2-dibromocyclohexane, the product of trans addition.[48] This type of addition will be considered in more detail in Sec. 12-6*d*.

Because of its well-defined and (compared with aliphatic compounds) relatively rigid stereochemistry, the cyclohexane system has been used as a probe for the steric course of a number of reactions other than those already mentioned. Although the answers obtained are of pertinence primarily to the

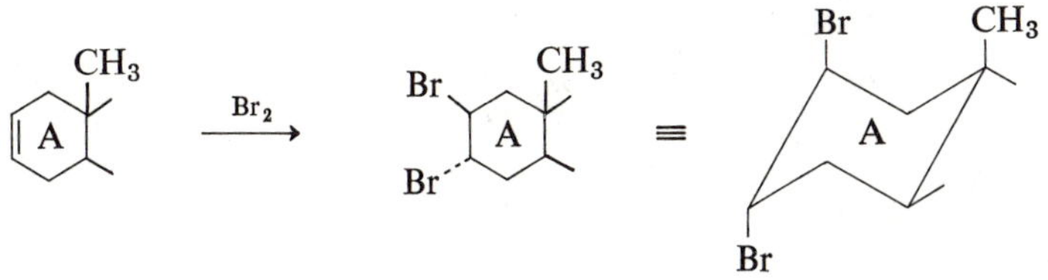

Fig. 8-38. Addition of bromine to 2-cholestene.

mechanistic course of the reaction in question—and therefore belong in a textbook on reaction mechanisms more properly than in one on stereochemistry—a few more cases will be considered here.

In addition to the already-mentioned *ionic* type of elimination reaction, several other mechanistic types are known, and all seem to differ in their stereochemistry. The pyrolysis of acetates, xanthates, and tertiary amine oxides to produce olefins has already been mentioned in Sec. 6-3*b, iv*. Whereas ionic elimination preferentially involves axially situated trans substituents, pyrolytic elimination preferentially involves cis substituents. Thus

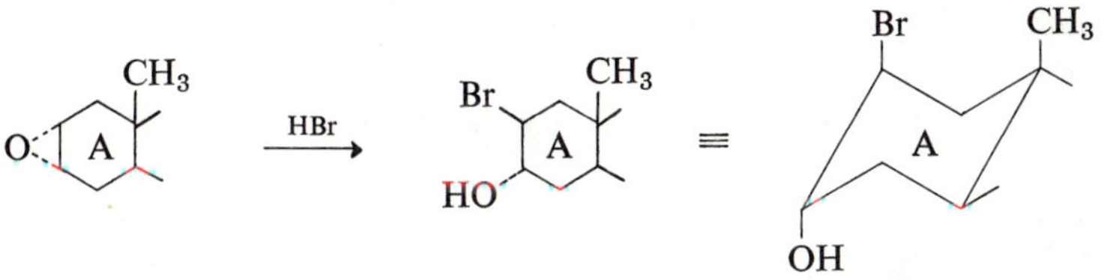

Fig. 8-39. Addition of hydrogen bromide to cholestane-2,3α-epoxide.

the pyrolysis of *trans*-2-phenylcyclohexyl xanthate (Fig. 8-40) produces predominantly the conjugated olefin 1-phenylcyclohexene, whereas the corresponding cis isomer produces almost entirely the unconjugated isomer 3-phenylcyclohexene.[49] Formation of the more stable conjugated olefin in the latter case would involve an unfavorable trans elimination of the xanthate group and the adjacent tertiary hydrogen. Similar results were obtained in the pyrolysis of acetates[49] and amine oxides.[50] The reason for this stereo-

[48] H. L. Goering, P. I. Abell, and B. F. Aycock, *J. Am. Chem. Soc.*, **74**, 3588 (1952).
[49] E. R. Alexander and A. Mudrak, *J. Am. Chem. Soc.*, **72**, 1810 (1950).
[50] A. C. Cope and C. L. Bumgardner, *J. Am. Chem. Soc.*, **79**, 960 (1957).

chemical preference seems to be the following: In the transition state for the pyrolytic elimination (Fig. 8-41), there is a concerted shift of bonding electrons in such a way that acetic acid (or some other molecule or molecules) is split out all at once. In the case of acetate, the transition state contains three partial double bonds (two about to be formed and one about to be broken), and therefore the six atoms or groups involved tend toward coplanarity in the transition state. As will be explained in connection with hydrindane (Chap.

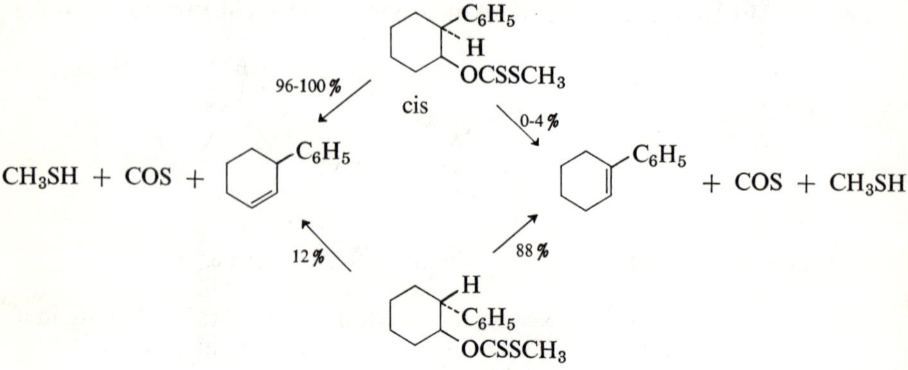

C_6H_5 H $OCSSCH_3$ cis

96-100% 0-4%

CH_3SH + COS + C_6H_5 C_6H_5 + COS + CH_3SH

12% 88%

H C_6H_5 $OCSSCH_3$ trans

Fig. 8-40. Pyrolysis of the 2-phenylcyclohexyl xanthates.

10, page 276), it is easier to force an equatorial and an axial bond into a plane than two equatorial bonds. Therefore an equatorial and an axial substituent (cis to each other) undergo pyrolytic elimination of acetic acid more readily than two equatorial substituents (trans to each other). Similar considerations apply to xanthate. In the case of the amine oxide, approximate coplanarity of the reacting atoms is required because the transition state contains a five-membered ring.

Three other types of elimination mechanism are elimination preceded by

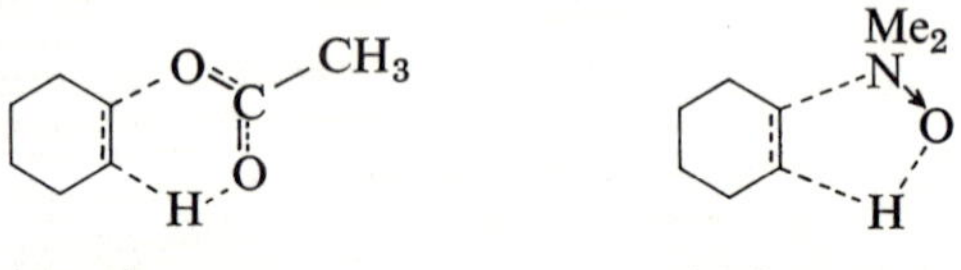

O C CH_3 O H Me_2 N O H

Acetate   Amine oxide

Fig. 8-41. Transition state for acetate and amine oxide pyrolysis.

substitution, elimination involving carbanion intermediates, and the "merged mechanism of substitution and elimination."

cis-1,2-Dibromocyclohexane undergoes bromine elimination to cyclohexene with potassium iodide in methanol at 80° at a rate which is only about eleven times slower than that for the trans isomer,[51] even though the trans isomer can undergo diaxial elimination and the cis isomer cannot. The explanation,

[51] H. L. Goering and H. H. Espy, *J. Am. Chem. Soc.*, 77, 5023 (1955).

based on kinetic study, is that in the cis dibromide the axial bromine atom quite readily undergoes bimolecular substitution by iodide to give the trans iodobromide, which may then flip over into the diaxial conformation and undergo dehydrohalogenation in turn.

The *p*-toluenesulfonates of *cis*- and *trans*-2-*p*-toluenesulfonylcyclohexanol (Fig. 8-42) undergo a base-induced elimination to give, in either case, 1-*p*-

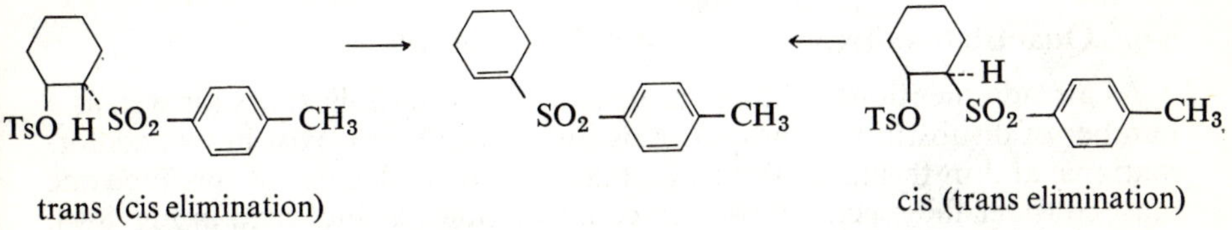

TsO H SO_2— —CH_3 → SO_2— —CH_3 ← TsO H SO_2— —CH_3

trans (cis elimination) cis (trans elimination)

Fig. 8-42. Reaction of the *p*-toluenesulfonates of *cis*- and *trans*-2-*p*-toluenesulfonylcyclohexanol with sodium hydroxide.

toluenesulfonylcyclohexene.[52] In the case of the trans isomer, this reaction involves cis elimination. The reason why the product from this isomer is nevertheless the 1-olefin (and not 3-*p*-toluenesulfonylcyclohexene, which could be formed by anti elimination) appears to be the loosening of the proton at the ring carbon adjacent to the sulfone function, which overcomes the stereochemical handicap sufficiently to lead to cis elimination.

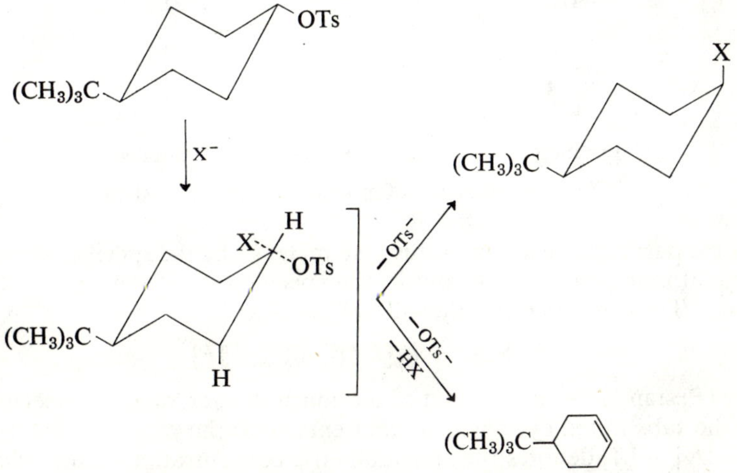

$(CH_3)_3C$ OTs

X^-

$(CH_3)_3C$ X H OTs H

$-OTs^-$ $-OTs^-$ $-HX$

$(CH_3)_3C$ X

$(CH_3)_3C$—

Fig. 8-43. Merged substitution-elimination.

trans-4-*t*-Butylcyclohexyl tosylate (Fig. 8-43) undergoes bimolecular elimination with the bases bromide and thiophenolate, although not with the much stronger base ethoxide. Here, again, direct diaxial elimination is not possible. It has been suggested[53] that bimolecular attack of the equatorial

[52] J. Weinstock, R. G. Pearson, and F. G. Bordwell, *J. Am. Chem. Soc.*, **76**, 4748 (1954).
[53] S. Winstein, D. Darwish, and N. J. Holness, *J. Am. Chem. Soc.*, **78**, 2915 (1956).

tosylate by the highly nucleophilic bromide or thiophenolate ions (although not by the less nucleophilic and more basic ethoxide) gives rise to the *intermediate* shown in Fig. 8-43 which can then go on to either substitution product or (by loss of both the incoming nucleophile and the tosylate group and adjacent hydrogen) elimination product. This process has been called "merged substitution and elimination."

8-6. Quantitative Treatment of Mobile Systems[16]

As already mentioned, simple monosubstituted cyclohexanes (as well as a number of disubstituted and polysubstituted ones) may exist in two conformations, and in trying to assess the reactivity of molecules of this type one must consider the reactivity of both conformations, as indicated in Fig. 8-44. Let $[E]$ be the concentration of the conformational isomer in which the functional group is equatorial and $[A]$ that of the axial isomer and let $K = [E]/[A]$ be the equilibrium constant between the two isomers.† In consider-

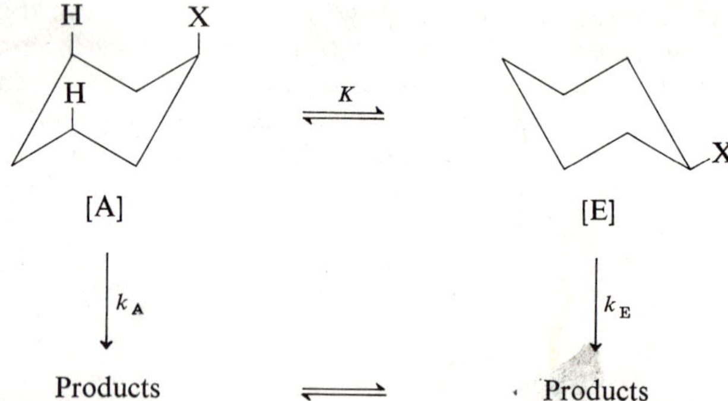

Fig. 8-44. Reaction of mobile cyclohexane system.

ing some particular reaction of this system, let k_E be the specific rate at which the equatorial isomer reacts and k_A the specific rate of reaction of the axial isomer. Then the over-all rate will be

$$\text{Rate} = k_E[E][P] + k_A[A][P] \tag{1}$$

where $[P]$ stands for the product of the concentration of all the species, other than the substituted cyclohexane, that enter into the rate equation.‡ But, if $[C] = [E] + [A]$ denotes the stoichiometric concentration of the substituted cyclohexane and k is the normally reported empirical rate constant, it is also true that

$$\text{Rate} = k[C][P] \tag{2}$$

† It is assumed throughout that equilibrium is maintained during the reaction, i.e., that the rate of interconversion of A and E is fast compared with their rate of reaction.

‡ For a reaction which is first-order in the cyclohexane species, e.g., the solvolysis of cyclohexyl tosylate, $[P] = 1$. An example of a second-order reaction would be the alkaline saponification of ethyl cyclohexanecarboxylate. In this case, $[P] = [OH^-]$.

Equating the rates as given by (1) and by (2), one gets

$$k_E[E] + k_A[A] = k[C] = k([E] + [A]) \tag{3}$$

But $[E] = K[A]$, and making this substitution in (3), one obtains

$$k_E K[A] + k_A[A] = k[A](K + 1) \tag{4}$$

Canceling $[A]$ and regrouping terms, this may be expressed as

$$k = (k_E K + k_A)/(K + 1) \tag{5}$$

Equation (5) derived by Eliel and coworkers[16] is equivalent to relationship (6) derived by Winstein and Holness:[24]

$$k = k_E N_E + k_A N_A \tag{6}$$

where N_E and N_A are the mole fractions of the substituted cyclohexane in the equatorial and axial conformations, respectively. Since $K = N_E/N_A$ and $K + 1 = (N_E + N_A)/N_A = 1/N_A$, the equivalence of relationships (5) and (6) is readily established.

Equation (6) shows at a glance that the specific reaction rate for a substituted cyclohexane depends on the mole fractions in the equatorial and axial forms and on the specific rate at which these individual conformations react. A few illustrations will make this relationship clearer.

The simplest situations arise when only one of the two conformations is reactive. For example, as already mentioned, in bimolecular ionic eliminations, the reacting substituents must be trans diaxial, that is, $k_E = 0$. In that case, $k = k_A N_A$. Thus, in the elimination reaction of *cis*-4-*t*-butylcyclohexyl tosylate (Fig. 8-26) with ethoxide ion, since the bulky *t*-butyl group is equatorial, the tosylate group is entirely axial, that is, $N_A = 1$, and the observed specific rate, $k = 7.1 \times 10^{-3}$ l. mole^{-1}sec.$^{-1}$ at 75°, is equal to k_A. On the other hand, in the trans isomer (Fig. 8-26) the tosylate group is entirely equatorial, and $N_E = 1$, $N_A = 0$; hence for this isomer $k = 0$, that is, the compound will not undergo bimolecular elimination with ethoxide ion, in agreement with observation.[24] Finally, for cyclohexyl tosylate itself, $0 < N_A < 1$, since this compound exists in both conformations shown in Fig. 8-44, and therefore $0 < k < 7.1 \times 10^{-3}$; the experimental value is 2.4×10^{-3} l. mole^{-1}sec.$^{-1}$, and from this value the conformational equilibrium constant K for the tosylate group can be calculated.

The acetylation of cyclohexanol with acetic anhydride in pyridine at 25° presents a case where the reactivity of both conformations must be taken into account.[54] The specific reaction rates of *cis*-4-*t*-butylcyclohexanol (Fig. 8-20, $N_A = 1$, $N_E = 0$) and *trans*-4-*t*-butylcyclohexanol (Fig. 8-20, $N_A = 0$, $N_E = 1$) give $k_A = 2.89 \times 10^{-5}$ l. mole^{-1}sec.$^{-1}$ and $k_E = 10.65 \times 10^{-5}$ l. mole^{-1}sec.$^{-1}$, respectively. For cyclohexanol itself, the specific rate is intermediate, namely, $k = 8.37 \times 10^{-5}$ l. mole^{-1}sec.$^{-1}$.

Equation (5) may be solved for K by transforming it to

$$K = \frac{k_A - k}{k - k_E} \tag{7}$$

[54] E. L. Eliel and C. A. Lukach, *J. Am. Chem. Soc.*, **79**, 5986 (1957).

Equation (7) is a convenient one for computing conformational equilibrium constants. For example, using the rate constants given in the previous paragraph, one may calculate K for the hydroxyl group in cyclohexanol to be 2.40, corresponding to a free-energy difference ($\Delta F = -RT \ln K$) of -0.52 kcal./mole between equatorial and axial hydroxyl. Corresponding energy differences for some other substituents determined by several different methods are shown in Table 8-6.

Table 8-6
Experimental Free-energy Differences between Equatorial and Axial Substituents

Group X (Fig. 8-44)	$-\Delta F$, kcal./mole	Ref.
OH	0.4–0.9	16
OAc	0.4–0.7	56, *a*
O—PNB[i]	0.9	29
OTs[j]	0.6, 0.7, 1.7	35, *a*, 24
OCH_3	0.5–0.7	*a*, *b*
OC_2H_5	1.0	*b*
F	0.25	*c*
Cl	0.3–0.5	*c*, *d*, *e*
Br	0.2–0.7	34, 60, *c*, *d*, *e*, *f*, *g*
I	0.4	*c*
SC_6H_5	0.6	*a*
CO_2H	1.6–1.7	23
$CO_2C_2H_5$	1.1	30
CH_3	1.5–1.9	10, 12, *b*
C_2H_5	2.1	*b*
C_6H_5	ca. 2.6	12
HgBr	ca. 0	*h*

[a] E. L. Eliel and M. Gianni, *Tetrahedron Letters,* 97 (1962).
[b] D. S. Noyce and L. J. Dolby, *J. Org. Chem.,* **26,** 3619 (1961).
[c] A. J. Berlin and F. R. Jensen, *Chemistry & Industry* (*London*), 998 (1960).
[d] L. W. Reeves and K. O. Strømme, *Can. J. Chem.,* **38,** 1241 (1960).
[e] G. Chiurdoglu, L. Kleiner, W. Masschelein, and J. Reisse, *Bull. soc. chim. Belges,* **69,** 143 (1960).
[f] E. L. Eliel, *Chemistry & Industry* (*London*), 568 (1959).
[g] F. R. Jensen and L. H. Gale, *J. Org. Chem.,* **25,** 2075 (1960).
[h] F. R. Jensen and L. H. Gale, *J. Am. Chem. Soc.,* **81,** 6337 (1959).
[i] *p*-Nitrobenzoate.
[j] *p*-Toluenesulfonate.

The accuracy of these values is not very high, perhaps ± 0.2 kcal./mole, at best. Where several interactions occur within one molecule, the values might be expected to be additive. For example, for the *cis*-4-methylcyclohexanol conformational equilibrium shown in Fig. 8-45, one may calculate $\Delta F = 1.8 - 0.5 = 1.3$ kcal./mole, for, whereas 0.5 kcal./mole is gained by changing hydroxyl from the axial to the equatorial conformation, 1.8 kcal./mole is lost by flipping methyl from the equatorial to the axial position. The experimental value, found by putting the experimental rate constant for the acetylation of *cis*-4-methylcyclohexanol (3.76×10^{-5} l. mole^{-1}sec.$^{-1}$) in Eq. (7), is

$K = 0.126$ and $\Delta F = 1.22$ kcal./mole, in good agreement with the calculated values. The agreement is not always so good.[16]

The energy differences listed in Table 8-6 are equal to twice the interaction energy between the group X and an axial hydrogen, since in the axial conformation (Fig. 8-44) there are two such interactions. Interaction energies

Fig. 8-45. *cis*-4-Methylcyclohexanol.

between two groups X and Y both differing from hydrogen, of the type shown in Fig. 8-46, have also been estimated for a few substituents and are listed in Table 8-7.

The equation $k = N_e k_e + N_a k_a$ suggests that the *rate of reaction* of a mobile substituted cyclohexane depends on the relative population of the

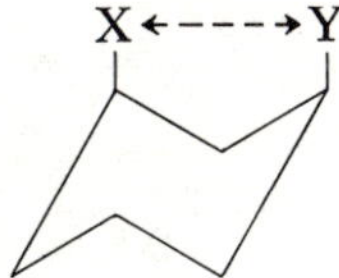

Fig. 8-46. Interaction in 1,3 diaxially substituted cyclohexanes.

equatorial and axial ground states. At first sight, this seems to be in contradiction to the Curtin-Hammett principle discussed in Chap. 6, according to which *product composition,* in a reaction which yields one product from one conformational isomer and another product starting with another conformational isomer of the same substrate, does *not* depend on the population of the

Table 8-7
Interaction Energies between Axial Substituents

Group X	—OH	—OAc	$—CH_3$	$—CH_3$
Group Y	—OH	—OAc	—OH	$—CH_3$
$-\Delta F$, kcal./mole	1.9[a]	2.0[b]	2.2–2.4[c]	3.7[d]

[a] Ref. 55. [b] Ref. 56. [c] Ref. 54; E. L. Eliel and H. Haubenstock, *J. Org. Chem.* **26,** 3504 (1961). [d] Ref. 14*a*.

ground states but only on the relative energies of the respective transition states. In fact, however, there is no contradiction; the Winstein-Holness equation refers to reaction rate whereas the Curtin-Hammett principle refers

[55] S. J. Angyal and D. J. McHugh, *Chemistry & Industry* (*London*), 947 (1955).

[56] R. U. Lemieux and P. Chu, Abstracts, Meeting of American Chemical Society, San Francisco, 1958, p. 31N.

to product composition (which depends on the ratio of the rates of formation of the two products). To illustrate this point further, it is instructive to derive the Curtin-Hammett principle from first principles in similar fashion as the equation $k = (k_e K + k_a)/(K + 1)$ was derived above.

$$P_A \xleftarrow{k_A} A \underset{}{\overset{K}{\rightleftharpoons}} E \xrightarrow{k_E} P_E$$

Fig. 8-47. Two conformers giving rise to different products.

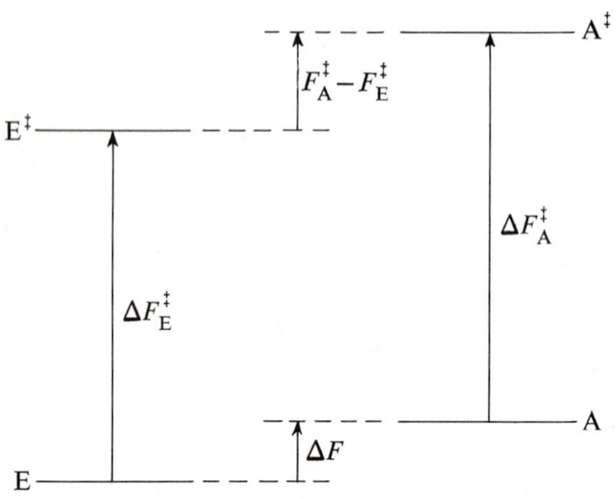

Fig. 8-48. Energy diagram for reaction shown in Fig. 8-47.

Consider a reaction in which conformation A of one of the factors gives rise to product P_A and conformation E gives product P_E, P_A and P_E in this case being chemically distinct entities rather than different conformations of one and the same entity. The situation is schematized in Fig. 8-47. In that case, $dP_A/dt = k_A\,[A][P]$ and $dP_E/dt = k_E\,[E][P]$, where $[P]$ has the same significance as in Eq. (1) above.

Hence
$$\frac{dP_E/dt}{dP_A/dt} = \frac{k_E[E]}{k_A[A]} = k_E K/k_A \tag{8}$$

Using transition-state theory (cf. Fig. 8-48) Eq. (8) can be transformed as follows:

$$k_E = \kappa\frac{kT}{h}e^{-\Delta F_E^{\ddagger}/RT} \qquad k_A = \kappa\frac{kT}{h}e^{-\Delta F_A^{\ddagger}/RT}$$

Also
$$K = e^{\Delta F/RT}$$

Assuming equality of the transmission coefficients and inserting into Eq. (8), one obtains

$$\frac{dP_E/dt}{dP_A/dt} = \frac{e^{-\Delta F_E^{\ddagger}/RT}\;e^{\Delta F/RT}}{e^{-\Delta F_A^{\ddagger}/RT}} = e^{(\Delta F_A^{\ddagger} + \Delta F - \Delta F_E^{\ddagger})/RT}$$

But, according to Fig. 8-48, $\Delta F_A^\ddagger + \Delta F - \Delta F_E^\ddagger = F_A^\ddagger - F_E^\ddagger$; hence

$$\frac{dP_E/dt}{dP_A/dt} = e^{(F_A^\ddagger - F_E^\ddagger)/RT} \qquad (9)$$

i.e., the ratio of the products P_E and P_A [which is equal to the ratio of the reaction rates in Eq. (9)] depends only on the difference in energy of the two transition states, as postulated by Curtin and Hammett.

8-7. Conformational Effects in Six-membered Rings Containing Unsaturation

When a cyclohexane ring contains trigonal carbon atoms (sp^2 hybridization), its shape is distorted from that of a normal chair. A case in point is cyclohexene (Fig. 8-49) in which the ethylenic carbons and the two adjacent allylic carbon atoms (Nos. 3 and 6) are in a plane, whereas the atoms on the other side (Nos. 4 and 5) are normally staggered, as in cyclohexane.† The

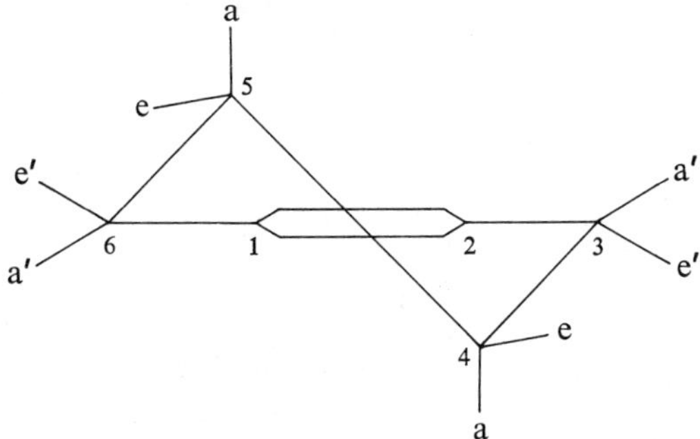

Fig. 8-49. Cyclohexene.

hydrogen (or other) atoms attached to atoms 4 and 5 occupy normal equatorial (*e*) and axial (*a*) positions, but those attached to atoms 3 and 6 are imperfectly staggered and do not take up the normal equatorial and axial positions; these atoms are said to occupy *pseudo*equatorial (*e*′) and *pseudo*-axial (*a*′) positions. This geometry has been confirmed in the case of certain substituted cyclohexenes by means of X-ray diffraction; the prediction that substituents in the *e* and *e*′ position should be more stable than those in *a* and *a*′ positions has also been confirmed.[57] The shape of the alicyclic ring in tetralin is analogous to that in cyclohexene.

More extensive studies have been carried out on the stereochemistry of

† The conformational situation in cyclohexenes seems to have been first appreciated by J. Böeseken and W. J. F. de Rijck van der Gracht, *Rec. trav. chim.*, **56**, 1203 (1937). In this, as in the understanding of many other aspects of cyclane stereochemistry, Böeseken seems to have been well ahead of his time.

[57] Cf. Ref. 79, pp. 38–39.

cyclohexanone (Fig. 8-50). It has been pointed out[58] that, as a result of the trigonal geometry of the carbonyl carbon, an equatorial substituent in the 2 position is nearly eclipsed with the carbonyl oxygen, whereas an axial substituent in this position is staggered with respect to this oxygen atom. It was originally believed[58] that for any kind of alkyl group in the equatorial position next to a ketone function this eclipsing would lead to an interaction energy term called the "2-alkyl ketone effect." Recent work[58a] has shown, however, that no such effect exists for a 2-methyl ketone; the energy difference between an equatorial and axial methyl group in such a ketone is 1.6 kcal./mole, almost the same as that between an equatorial and axial methyl in cyclohex*ane* (Table 8-6). The 2-alkyl ketone effect does, however, play a part in the case of 2-ethyl- and 2-isopropylcyclohexanone, reducing the difference between equatorial and axial alkyl to 1.1 kcal./mole for ethyl and 0.4 kcal./mole for isopropyl.[58a]

Another observation[58] is that on one side of the cyclohexanone molecule (the bottom side in Fig. 8-50), instead of three axial substituents, there are only two. This leads to a decrease in the axial interaction for an axial substituent at C_3 or C_5. For a methyl substituent, this decrease should amount to one CH_3-H axial interaction or 0.9 kcal./mole (cf. Table 8-6). Experimentally a value of ca. 0.6 kcal./mole is indicated for this "3-alkyl ketone effect."[58b]

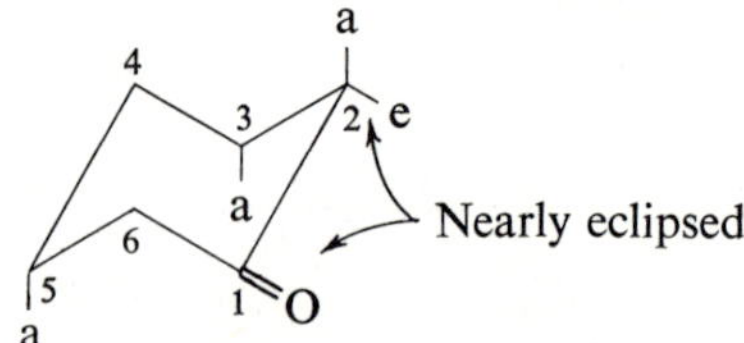

Fig. 8-50. Cyclohexanone.

The special spatial relationship of the carbonyl function in cyclohexanone to the equatorial and axial groups on the adjacent (α-) carbon gives rise to certain peculiarities in the infrared and ultraviolet spectra of these compounds when the substituent is a polar group, such as halogen. In the infrared spectrum, the carbonyl stretching frequency is increased by the nearly parallel dipole of an equatorial α-halogen by about 20 $cm.^{-1}$, whereas the dipole of an axial α-halogen is so oriented that it leaves the carbonyl stretching frequency nearly the same as in the unsubstituted ketone.[59] This fact has been used extensively to assign configuration to halogen substituents α to carbonyl groups. In a simple α-halocyclohexanone, the chair is largely (but not exclusively) in the form in which the halogen is axial,[60, 61] indicating that

[58] P. A. Robins and J. Walker, *J. Chem. Soc.*, 1789 (1955); *Chemistry & Industry* (*London*), 772 (1955); W. Klyne, *Experientia*, **12**, 119 (1956).

[58a] N. L. Allinger and H. M. Blatter, *J. Am. Chem. Soc.*, **83**, 994 (1961).

[58b] N. L. Allinger and L. A. Frieberg, *J. Am. Chem. Soc.*, **84**, 2201 (1962).

[59] R. N. Jones, D. A. Ramsay, F. Herling, and K. Dobriner, *J. Am. Chem. Soc.*, **74**, 2828 (1952).

[60] E. J. Corey, *J. Am. Chem. Soc.*, **75**, 2301 (1953).

[61] For example, N. L. Allinger, J. Allinger, L. A. Freiberg, R. F. Czaja, and N. A. LeBel, *J. Am. Chem. Soc.*, **82**, 5876 (1960), and earlier papers by N. L. Allinger and coworkers.

the dipole repulsion between the carbonyl function and the adjacent equatorial α-halogen (cf. Fig. 8-51) is greater than the steric repulsion between axial halogen and the axial hydrogens. The situation is reversed (Fig. 8-51) when the axial position for the α-halogen involves steric interaction with another axial group larger than hydrogen. As expected for a dipole-dipole interaction, the effect (and therefore the equilibrium shown in Fig. 8-51) is dependent on solvent dielectric.

In the ultraviolet spectrum, the overlap for the pi electrons of the carbonyl group and the sigma electrons of the α-halogen (or other polar α-substituent)

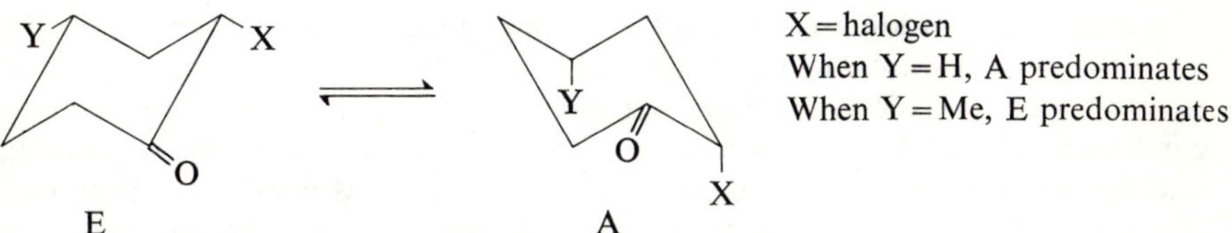

Fig. 8-51. Conformation of α-halocyclohexanones.

is greater for an axial substituent. Therefore, the axial groups produce a shift in carbonyl absorption to longer wavelength (with a concomitant increase in absorption intensity), whereas equatorial groups produce, if anything, a small shift to shorter wavelengths, as shown in Table 8-8.[62]

The stereochemistry of enolization and ketonization has been studied extensively by Corey[63] and Zimmerman.[64] A priori, in enolization, either the axial or the equatorial α-hydrogen may be removed preferentially. Experimentally, it has been shown by deuterium-labeling experiments that,

Table 8-8
Shifts in Wavelength of Ultraviolet Absorption Maximum of Saturated Ketones Produced by α Substituents

Substituent	Shift Δ (mμ)	
	e	*a*
Cl	−7	+22
Br	−5	+28
OH	−12	+17
OAc	−5	+10

when the carbonyl function forms part of the ring system (Fig. 8-52), the axial hydrogen is abstracted preferentially (though not exclusively), despite its greater steric encumbrance. This is believed to be due to stereoelectronic factors. As already mentioned in connection with ultraviolet spectra, the

[62] R. C. Cookson, *J. Chem. Soc.*, 282 (1954).

[63] For example, E. J. Corey and R. A. Sneen, *J. Am. Chem. Soc.*, **78**, 6269 (1956), and earlier papers.

[64] For example, H. E. Zimmerman and T. W. Cutshall, *J. Am. Chem. Soc.*, **81**, 4305 (1959), and earlier papers.

sigma electrons of the axial bond are more favorably located for overlap with the pi electrons of the carbonyl function, which overlap must occur in the formation of the enol or enolate ion where there is a pi bond in the 1,2 position (Fig. 8-52). By the same token, addition to the enol of either a proton or some other cation also occurs preferentially from the axial position. Thus bromination of a variety of ketosteroids usually produces predominantly axial α-bromoketones (provided that the process is kinetically controlled), even though, as mentioned above, there are instances where the equatorial α-bromine substituent is the more stable. Exceptions have been noted, however, in cases where the axial approach is strongly sterically hindered.[64a]

A different situation arises when the ketone function is located exocyclically, as in the system shown in Fig. 8-53. In this case, stereoelectronic considerations play no part, since the bond between the carbonyl function and the ring carbon can always be so rotated that the pi electrons of the carbonyl function and the sigma electrons of the α-hydrogen are oriented parallel.† Therefore the steric (bulk) effect is now controlling; enolization occurs more easily if it involves the equatorial ring hydrogen, and addition to the enol form occurs on the equatorial side of the ring. Examples for both these statements are

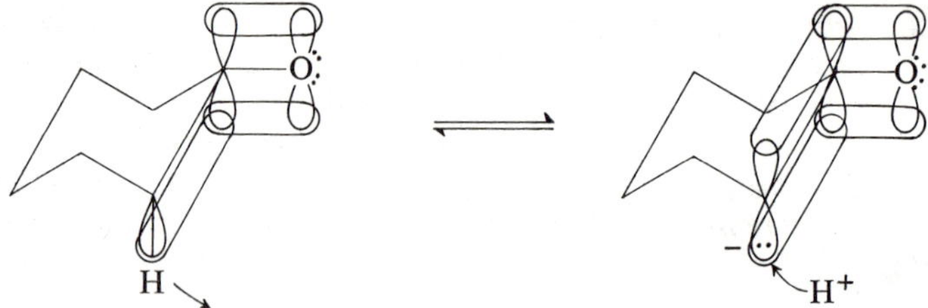

Fig. 8-52. Stereoelectronic situation in enolization and ketonization.

shown in Fig. 8-53. 1,4 Addition of phenylmagnesium bromide to acetylcyclohexene gives, in the first instance, the enolate which then, upon acidification, ketonizes in such a way that the proton approaches equatorially. The product is *cis*-2-phenyl-1-acetylcyclohexane (C), evidently as a result of kinetic control, since thermodynamic control would have produced the more stable trans epimer (T), in which both substituents are equatorial. The equatorial hydrogen in compound C is readily enolized, and, upon bromination, this compound gives 1-bromo-1-acetyl-2-phenylcyclohexane. In contrast, the tertiary *axial* ring hydrogen in the trans epimer (T) is not readily enolized, and bromination of this compound proceeds at the primary position to give 1-bromoacetyl-2-phenylcyclohexane.

Another point of interest about cyclohexanones is that in those cases where the ring is rigid or all in one conformation (such as steroidal ketones and

† The reader should convince himself of the correctness of this statement by means of models, both for the case where the carbonyl substituent is axial and the α-hydrogen equatorial and for the case where the carbonyl is equatorial and the α-hydrogen axial.

[64a] R. Villotti, H. J. Ringold, and C. Djerassi, *J. Am. Chem. Soc.,* **82,** 5693 (1960). See also J. Valls and E. Toromanoff, *Bull. soc. chim. France,* 758 (1961), and Sec. 10-2*b*.

trans-decalones on one hand, 4-*t*-butylcyclohexanone and menthone on the other) addition reactions to the carbonyl function—such as catalytic hydrogenation, chemical reduction, addition of organometallic reagents, addition of cyanide, etc.—may proceed either from the axial side so as to give an equatorial alcohol or from the equatorial side so as to give an axial alcohol.† Investigation as to which of the two reaction paths is preferred has thrown considerable light on the mechanism of the respective carbonyl addition reaction under investigation.

One of the simplest reactions of cyclohexanones is catalytic hydrogenation to the corresponding cyclohexanols. The stereochemistry of this process is

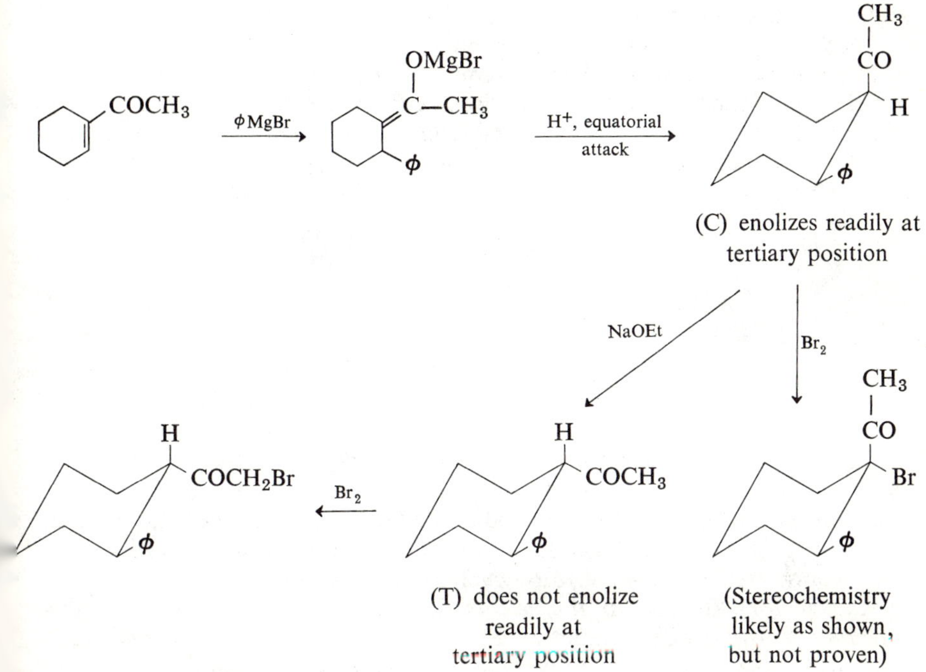

Fig. 8-53. Enolization and ketonization of exocyclic ketones.[65]

governed by the von Auwers–Skita hydrogenation rule (not to be confused with the von Auwers–Skita conformational rule, page 216) which, in somewhat modified form, states that hydrogenation in acid medium produces predominantly the axial alcohol. The rationale of this rule seems to be that the hydrogen must approach the adsorbed ketone molecule from the catalyst surface and since it is easier for the ketone to be adsorbed from the less-

† This statement is, of course, still true in a mobile cyclohexanone system, such as cyclohexanone itself or 4-methylcyclohexanone. In this case, however, since the system may flip over after reaction, it may be difficult or impossible to decide which course the addition actually took.

[65] H. E. Zimmerman, *J. Am. Chem. Soc.*, **79**, 6554 (1957).

hindered equatorial side, the hydrogen approaches also from this side, producing the axial alcohol. The function of the acid medium may be to promote rapid desorption of the alcohol once formed, for if the alcohol sojourns on the catalyst for a prolonged period, it may be dehydrogenated back to ketone and then rehydrogenated again, possibly after changing its orientation with respect to the catalyst surface. If this process is allowed to proceed long enough, it leads to an equilibrium mixture of alcohols rich in the equatorial isomer;[66] it also leads to a slowing down of the hydrogenation process. Empirically it is found that the faster the hydrogenation (i.e., the less chance for equilibration) the more the axial alcohol predominates. The best way to obtain products rich in axial alcohol is to speed up the hydrogenation by adding mineral acid to the acetic acid solvent.

Another type of reduction which produces predominantly the axial alcohol is the Meerwein-Ponndorf-Verley reduction employing aluminum isopropoxide, whose mechanism has been discussed in Chap. 4. The space demand of the hydride ion which is transferred from the isopropoxy group to the carbonyl group of the ketone (Fig. 4-41) is such that it approaches predominantly from the unhindered equatorial side of the cyclohexanone, giving an axial

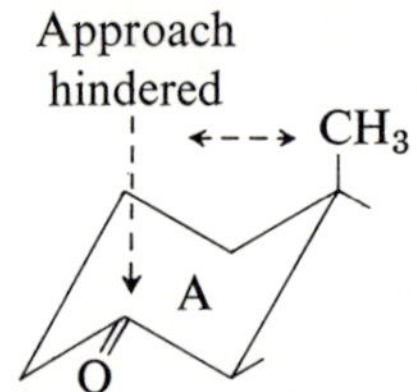

Fig. 8-54. Steric approach control.

cyclohexanol, which is thus the product of kinetic control, i.e., the product formed fastest (cf. page 200). However, on prolonged reaction with aluminum isopropoxide, the substituted cyclohexanol is reoxidized to the ketone by the acetone formed in the Meerwein-Ponndorf-Verley reduction (this process is known as Oppenauer oxidation) which is then again reduced, and so on in a process which leads, eventually, to the establishment of equilibrium between the two epimeric alcohols. When the reaction is allowed to proceed to this point (which takes quite a long time), the equatorial alcohol, being the product of thermodynamic control (cf. page 200), will predominate, as already indicated (Fig. 8-20).[26, 67]

Prediction of the predominant product of a metal hydride reduction is more uncertain, since it depends on the space requirements both of the metal hydride and of the system under study. It appears that, in most cases, the more stable equatorial alcohol predominates among the products, presumably because the space requirement of the approaching hydride ion is quite small.[67] This case has been termed one of "product development control." However, with hindered systems, such as 4-ketosteroids (Chap. 10) in which

[66] R. J. Wicker, *J. Chem. Soc.*, 2165 (1956).
[67] W. G. Dauben, G. J. Fonken, and D. S. Noyce, *J. Am. Chem. Soc.*, **78**, 2579 (1956).

the axial approach of hydride is strongly inhibited by an axial methyl group (see Fig. 8-54), the hydride may be forced to approach equatorially and the predominant product will be the axial alcohol. In this case, one speaks of "steric approach control" of the product.[67]

Data so far discussed in connection with the reactivity of cyclohexanones are kinetic data; i.e., they are concerned with relative rates of reaction. Thermodynamic data are also available in the form of dissociation constants of the cyanohydrins derived from various substituted cyclohexanones;[68] some of these are shown in Table 8-9.

It will be seen that those compounds having only equatorial substituents (Nos. 1 to 3) have similar equilibrium constants. Introduction of an axial

Table 8-9
Relative Dissociation Constants of Cyanohydrins

R, R′, R″, R‴ (ring) — OH, CN or R, R′, R″, R‴ (ring) — CN, OH $\underset{K}{\rightleftharpoons}$ R, R′, R″, R‴ (ring) =O + HCN

No.	Compound	R	R′	R″	R‴	$K_{rel.}$
1	Cyclohexanone	H	H	H	H	1.0
2	3-Methylcyclohexanone	H	Me	H	H	0.93
3	*cis*-3,5-Dimethyl-cyclohexanone	H	Me	H	Me	4.4
4	3,3-Dimethylcyclo-hexanone	Me	Me	H	H	30
5	3,3,5-Trimethyl-cyclohexanone	Me	Me	H	Me	38
6	3,3,5,5-Tetramethyl-cyclohexanone	Me	Me	Me	Me	800

substituent (R = Me) in the 3 position (Nos. 4 and 5) increases the dissociation constant about 30-fold by introducing an unfavorable 1-3 diaxial interaction of either hydroxyl or cyanide with methyl. With two axial substituents (R = R″ = Me, entry 6, Table 8-9), the number of such interactions is doubled and the dissociation constant increases another ca. 25-fold.

Before leaving the subject of cyclohexanone chemistry, it must be mentioned that the equilibrium between the chair and the flexible forms of cyclohexanones is probably not as unfavorable to the latter as it is in the case of corresponding saturated cyclohexanes (Sec. 8-1). The energy difference between the two forms of cyclohexanone has been estimated[4a] as 2.7 kcal./mole, about one-half the value of the corresponding difference for the saturated system. The reason for this is that the eclipsing between the car-

[68] O. H. Wheeler and J. Z. Zabicky, *Can. J. Chem.*, **36**, 656 (1958).

bonyl oxygen and adjacent equatorial hydrogens in the chair may be partially abolished in the flexible form and that, on the other hand, the eclipsing between adjacent cis hydrogens in the boat form of cyclohexane is partly abolished upon introduction of a keto group (see also Sec. 9-4).† As a result, a number of polycyclic cyclohexanone derivatives are now known in which the cyclohexanone ring is in the flexible form. Specific examples will be cited in Chaps. 10 and 14. Among monocyclic cyclohexanones, *cis*-2,4-di-*t*-butylcyclohexanone[58a] and *cis*-2-*t*-butyl-5-methylcyclohexanone[68a] appear to exist extensively or exclusively in the skew-boat form.

8-8. Six-membered Heterocyclic Rings

Because of the similarity in bond length of C—C, C—N, and C—O bonds as well as in the C—C—C, C—N—C, and C—O—C angles, the stereochemistry of six-membered rings containing nitrogen (such as piperidine and many alkaloids) and oxygen (such as dioxane and the pyranose sugars) is similar to that of cyclohexane. It is of interest that physical measurements on piperidine and morpholine in benzene solution[69] suggest that the hydrogen attached to nitrogen in these molecules is axial (Fig. 8-55). If this impli-

H
N:
X
X = CH_2 or O

Fig. 8-55. Piperidine and morpholine.

cation is correct, it must mean that an axial electron pair, unlike an axial hydrogen, interacts unfavorably with the axial hydrogen atoms on the same side of the ring.

General References

[70] W. Hückel, Der gegenwärtige Stand der Spannungstheorie, *Fortschr. Chem., Physik u. physik. Chem.,* **19,** 243 (1927).

[71] A. Wasserman, Spannungstheorie and physikalische Eigenschaften ringförmiger Verbindungen, in K. Freudenberg, ed., "Stereochemie," Franz Deuticke, Leipzig, 1932, pp. 781–802.

[72] S. J. Angyal and J. A. Mills, The Shape and Reactivity of the Cyclohexane Ring, *Rev. Pure Appl. Chem. Aust.,* **2,** 185 (1952).

[73] D. H. R. Barton, The Stereochemistry of Cyclohexane Derivatives, *J. Chem. Soc.,* 1027 (1953).

† Alternatively, if the carbonyl group is at the bowsprit position, the bowsprit-flagpole interaction (p. 206) will be abolished.

[68a] C. Djerassi, E. J. Warawa, J. M. Berdahl, and E. J. Eisenbraun, *J. Am. Chem. Soc.,* **83,** 3334 (1961).

[69] M. Aroney and R. J. W. Le Fèvre, *J. Chem. Soc.,* 3002 (1958).

[74] O. Hassel, Stereochemistry of Cyclohexane, *Quart. Revs.* (*London*), **7,** 221 (1953).

[75] W. Klyne, The Conformations of Six-membered Ring Systems, in "Progress in Stereochemistry," vol. I, Butterworth & Co. (Publishers), Ltd., London, 1954, chap. 2.

[76] S. Mizushima, "The Structure of Molecules and Internal Rotation," Academic Press, Inc., New York, 1954.

[77] D. H. R. Barton, Some Recent Progress in Conformational Analysis, *Experientia, Suppl.* II, 121 (1955).

[78] D. H. R. Barton and R. C. Cookson, The Principles of Conformational Analysis, *Quart. Revs.* (*London*), **10,** 44 (1956).

[79] W. G. Dauben and K. S. Pitzer, Conformational Analysis, in M. S. Newman, ed., "Steric Effects in Organic Chemistry," John Wiley & Sons, Inc., New York, 1956, chap. 1.

[80] D. H. R. Barton, Some Recent Progress in Conformational Analysis, in "Theoretical Organic Chemistry" (Kekulé Symposium), Butterworth & Co. (Publishers), Ltd., London, 1959, pp. 127–143.

[81] H. H. Lau, Prinzipien der Konformationsanalyse, *Angew. Chem.,* **73,** 423 (1961).

[82] H. D. Orloff, The Stereoisomerism of Cyclohexane Derivatives, *Chem. Revs.,* **54,** 347 (1954).

Chapter 9

THE SHAPE OF RINGS OTHER THAN SIX-MEMBERED ONES

This chapter will deal with the more detailed geometry of common and medium-sized rings other than six-membered ones which were discussed in Chap. 8.

9-1. Five-membered Rings

Cyclopropane cannot be anything but planar. In cyclobutane there seems to be some deviation from planarity,[1] even though this increases even further the already substantial angle strain in the planar model. The cause for the deviation has been sought mainly in 1,3 interference of ring carbon atoms (in addition to the bond eclipsing strain mentioned below), but the matter has not been explored in detail as yet. In trimethylene oxide both types of interaction are diminished and the ring is essentially planar.[1a]

Cyclopentane has been studied more extensively. In the planar form this hydrocarbon would have very little angle strain (the internal angle in a regular pentagon, 108°, being very close to the tetrahedral angle of 109°28′), but there would be substantial eclipsing strain between adjacent hydrogen atoms, similar to the strain existing in the eclipsed form of ethane (cf. Chap. 6). It turns out that, by puckering the cyclopentane ring somewhat, one gains more energy through staggering the hydrogen atoms than one loses through increasing the angle strain in the molecule. The decrease in total strain (angle strain plus bond opposition strain) achieved through this puckering is estimated[2] as 4 to 5 kcal./mole. As a result, cyclopentane has the puckered shapes shown in Fig. 9-1 rather than the planar form shown. Experimentally, this fact was first derived from entropy measurements.[3] Actually, the shape

Note: All references above 42 are listed in the General References at the end of the chapter.

[1] J.D. Dunitz and V. Schomaker, *J. Chem. Phys.*, **20**, 1703 (1952); G. W. Rathjens, N. K. Freeman, W. D. Gwinn, and K. S. Pitzer, *J. Am. Chem. Soc.*, **75**, 5634 (1953). See also A. Almenningen, O. Bastiansen, and P. N. Skancke, *Acta Chem. Scand.*, **15**, 711 (1961).

[1a] S. I. Chan, J. Zinn, J. Fernandez, and W. D. Gwinn, *J. Chem. Phys.*, **33**, 1643 (1960).

[2] J. E. Kilpatrick, K. S. Pitzer, and R. Spitzer, *J. Am. Chem. Soc.*, **69**, 2483 (1947).

[3] J. G. Aston, S. C. Schumann, H. L. Fink, and P. M. Doty, *J. Am. Chem. Soc.*, **63**, 2029 (1941); J. G. Aston, H. L. Fink, and S. C. Schumann, *ibid.*, **65**, 341 (1943).

of cyclopentane is not fixed; the individual carbon atoms move up and down at right angles to the average plane of the ring in such a manner as to cause the irregularity or puckering to move around the ring in what has been termed[2] by Pitzer a "pseudorotation." Of the many possible puckered shapes two are especially noteworthy because they are more symmetrical than all the others. One of these (Fig. 9-1*B*) has C_s symmetry (plane of symmetry at right angles to the average plane of the ring). Since this form resembles an open envelope (with its flap pointing upward), it has been called[4] the "envelope form." The other symmetrical form (Fig. 9-1*B*) has C_2 symmetry (twofold axis of symmetry in the average plane of the ring) and has been called[4] the "half-chair form"; it resembles cyclohexene (Fig. 8-49) except that the two olefinic carbons in the latter are replaced by a single methylene group in the C_2 form of cyclopentane. In cyclopentane itself the change from C_s to C_2 and back again via intermediate asymmetric arrangements appears to involve no substantial change in potential energy, but in substituted cyclopentanes (to be discussed below) one form or the other may have greater stability.

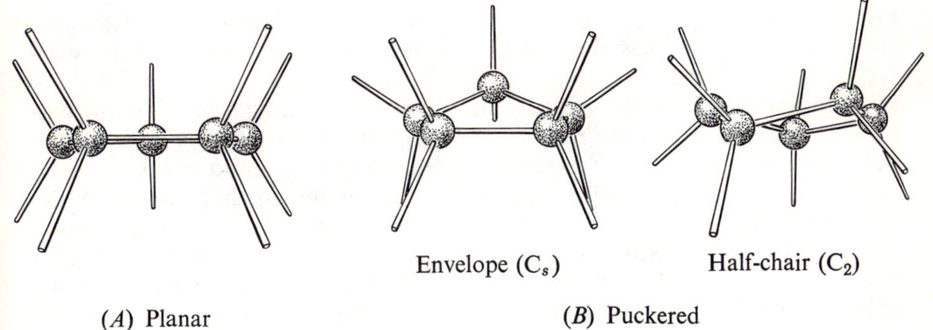

Fig. 9-1. Cyclopentane.

Projection formulas of the envelope and half-chair forms of cyclopentane are shown in Fig. 9-2. The plane of the paper is both the average plane of the ring and the projection plane. Atoms above the projection plane are denoted by a plus sign, below the projection plane by a minus sign, and in the projection plane by 0 (zero). The figures indicate the displacement, in angstroms, of the atoms above or below the plane. It should be realized that the two drawings shown in Fig. 9-2 present but an instantaneous view of the cyclopentane molecule, since each ring carbon is continuously vibrating up or down at right angles to the plane of the paper.

A direct chemical demonstration of the non-planarity of cyclopentane comes from a consideration of *cis*- and *trans*-1,3-dimethylcyclopentane. If the ring were flat, the cis isomer should be no more stable—and possibly, because of crowding across the ring, less stable—than the trans isomer. In

[4] F. V. Brutcher, T. Roberts, S. J. Barr, and N. Pearson, *J. Am. Chem. Soc.*, **81**, 4915 (1959).

actual fact, however, the cis isomer is more stable by about 0.5 kcal./mole.[5,6] This is understandable on the basis of the envelope model, for if the No. 2 carbon is out of the plane of carbons Nos. 1 and 3 (which bear the methyl groups), the cis methyl groups can be placed in a position resembling the equatorial positions in cyclohexane (Fig. 9-3). In the trans isomer, only one methyl group can be placed in this favorable position, in which bond oppositions involving the methyl groups are minimized. It is also reasonable that the enthalpy difference between *cis*- and *trans*-1,3-dimethylcyclopentane (0.54

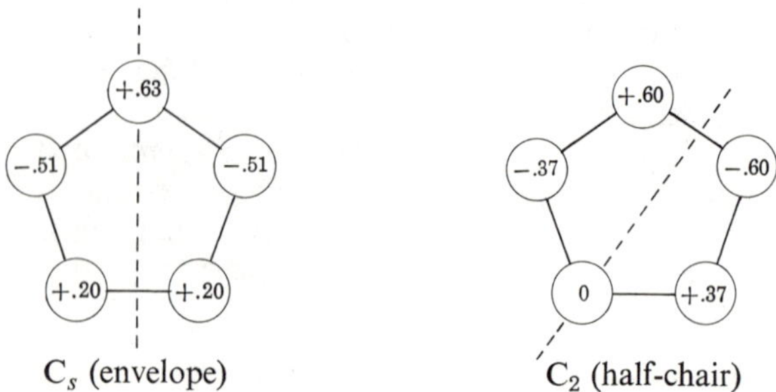

Fig. 9-2. Envelope and half-chair forms of cyclopentane. [Figures indicate displacement of atoms, in angstroms, above (positive) or below (negative) the plane of the paper.]

kcal./mole) is less than the difference between *cis*- and *trans*-1,3-dimethylcyclohexane (1.96 kcal./mole; cf. Table 8-2), since the puckering in the cyclopentane ring is not as great as the complete staggering found in cyclohexane, and therefore the advantage gained by placing the methyl groups in the equatorial positions in cyclopentane is not as great either. Or, putting it differently, the conformational energy minima are deeper in cyclohexanes than in cyclopentanes. It might be noted that, in order to minimize the eclipsing of the methyl groups in *cis*-1,3-dimethylcyclopentane with the

Fig. 9-3. *cis*-1,3-Dimethylcyclopentane.

hydrogens on carbon No. 2 (Fig. 9-3), the molecule must be in the envelope form, as shown. One would therefore expect this conformation to have considerably greater stability than any of the other conformations, contrary to the situation in cyclopentane itself. Similarly, one would expect methylcyclo-

[5] M. B. Epstein, G. M. Barrow, K. S. Pitzer, and F. D. Rossini, *J. Research Natl. Bur. Standards,* **43,** 245 (1949).

[6] J. N. Haresnape, *Chemistry & Industry* (*London*), 1091 (1953); S. F. Birch and R. A. Dean, *J. Chem. Soc.,* 2477 (1953).

pentane to be most stable in the envelope form, with the methyl group at the tip of the flap (Fig. 9-4), since in this form there is a minimum of eclipsing between the methyl substituent and hydrogens on adjacent carbons. Calculation[7] indicates the extra stabilization in this conformation to be 0.9 kcal./mole, whereas a value of 0.75 kcal./mole is suggested by measurements of the entropy of methylcyclopentane.[8]

In cyclopentanone, methylenecyclopentane, and heterocyclic analogs of cyclopentane, such as tetrahydrofuran, pyrrolidine, and tetrahydrothiophene, two pairs of hydrogen-hydrogen eclipsings are removed. These molecules are most stable in the half-chair (C_2) form with the doubly bonded carbon or hetero atom located on the axis of symmetry (Fig. 9-2, atom marked 0). The fact that, in the C_2 form, substituents on this atom are ordinarily nearly eclipsed with substituents on the two neighboring atoms is of no consequence in the molecules under discussion;† apart from this, the C_2 form provides the optimal amount of staggering between the hydrogens attached to the other four carbon atoms. The picture is confirmed by calculations of the enthalpy of methylenecyclopentane, pyrrolidine, and tetrahydrothiophene and by cal-

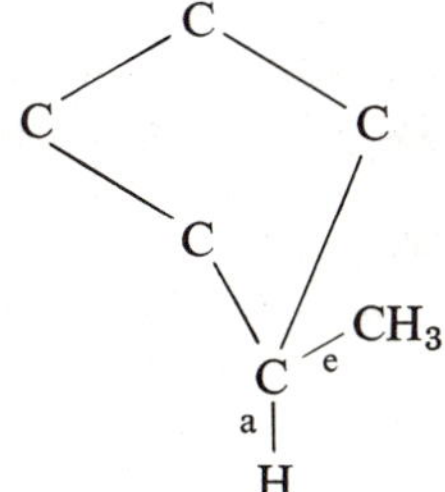

Fig. 9-4. Methylcyclopentane.

culations of the free-energy change in the cyclopentanone–cyclopentanone cyanohydrin equilibrium,[7] the calculated data agreeing quite well with the experimental. Further confirmation for the half-chair form of cyclopentanone comes from measurements of infrared carbonyl stretching frequencies and dipole moments in 2-halocyclopentanones.[4] It has already been pointed out (Sec. 8-7) that the magnitude of the shift in carbonyl frequency in going from a ketone to the corresponding α-haloketone depends on the angle between the carbon-oxygen and carbon-halogen bonds. The dipole moment of an α-haloketone depends also on this angle, of course. The measurements in the 2-halocyclopentanones indicate an angle of 77° between the C=O and C—X bonds, very close to the calculated 78° for the half-chair but incom-

† It is of considerable interest that this seems to be true even in the five-membered heterocycles where one might have expected eclipsing due to the lone pairs on the hetero atoms; cf. Sec. 8-8.

[7] K. S. Pitzer and W. E. Donath, *J. Am. Chem. Soc.*, **81**; 3213 (1959); cf. J. P. McCullough, R. E. Pennington, J. C. Smith, I. A. Hossenlopp, and G. Waddington, *ibid.*, **81**, 5880 (1959).

[8] J. E. Kilpatrick, H. G. Werner, C. W. Beckett, K. S. Pitzer, and F. D. Rossini, *J. Research Natl. Bur. Standards*, **39**, 523 (1947).

patible with the 94° angle expected at the side of the envelope or the 60° angle expected in a planar model.

9-2. Rings Larger than Six-membered

At the time of writing (1961) conformational analysis of rings larger than six-membered is still in its infancy, but it is a subject that is being actively developed. Three approaches are being used: physical measurements (such as X-ray analysis,[9] infrared analysis,[10] and dipole-moment studies[11]), calculations using torsional potential functions, [12, 13, 13a, 47] and chemical (kinetic and equilibrium) studies.[14] The infrared bands of substituted cycloalkanes of this type, unlike those of substituted cyclohexanes (Chap. 8), show little temperature dependence,[10] suggesting either that these molecules are in a fixed conformation or that there are several conformations of approximately the same energy.[14] Calculation of the actual shape of rings larger than six-membered[12, 13] is complicated by two interrelated factors: one that in such rings there may be van der Waals repulsions between hydrogen atoms across the ring in many of the possible conformations (this can be readily seen in scale models) and the other that, as a result of the potential van der Waals and torsional† interactions in such rings, it is often cheaper, energetically, to deform valence angles than to let these interactions stand as they are. The energy required to deform bond angles is very small; taking the carbon-carbon bending-force constant as 0.8×10^{-11} erg radian^{-2} molecule^{-1} gives $E = 0.0175x^2$, where E is the angle deformation energy in kilocalories per mole and x is the deformation angle in degrees.[47] Thus for angle deformations of 1, 5, and 10°, the required energies are 0.018, 0.44, and 1.8 kcal./mole, respectively. It is therefore not surprising that in cyclododecane some of the C—C—C angles are as high as 117° (instead of the normal 109°28′).[9] (In fact, small angular deformations occur even in alkanes, the C—C—C angle in *n*-butane being 112.4°.)[14a]

In cycloheptane, there seems to be agreement[12, 13, 13a] that the stable conformation is a somewhat deformed chair (Fig. 9-5). The difference in energy

† Torsional interactions are interactions resulting from the fact that substituents on adjacent carbon atoms are imperfectly staggered. The maximum torsional interaction corresponding to two adjacent substituents which are eclipsed is 1 kcal., that is, one-third of the total interaction of the eclipsed form of ethane (Chap. 6). This interaction is sometimes called "bond opposition strain" or "Pitzer strain."

---

[9] J. D. Dunitz and V. Prelog, *Angew. Chem.*, **72,** 896 (1960).

[10] For example, (*a*) H. E. Bellis and E. J. Slowinski, *Spectrochim. Acta,* **12,** 1103 (1959); (*b*) E. Billeter and H. H. Günthard, *Helv. Chim. Acta,* **41,** 338,686 (1958); (*c*) G. Chiurdoglu, T. Doehaerd, and B. Tursch, *Chemistry & Industry* (*London*), 1453 (1959); *Bull. soc. chim. France,* 1322 (1960).

[11] (*a*) For example, N. L. Allinger and S. Greenberg, *J. Am. Chem. Soc.,* **81,** 5733 (1959); (*b*) N. L. Allinger and S. Hu, *ibid.,* **83**; 1664 (1961).

[12] N. L. Allinger, *J. Am. Chem. Soc.,* **81,** 5727 (1959).

[13] R. Pauncz and D. Ginsburg, *Tetrahedron,* **9,** 40 (1960).

[13a] J. B. Hendrickson, *J. Am. Chem. Soc.,* **83,** 4537 (1961).

[14] For example, N. L. Allinger, *J. Am. Chem. Soc.,* **81,** 232 (1959).

[14a] R. A. Bonham and L. S. Bartell, *J. Am. Chem. Soc.,* **81,** 3491 (1959).

between the diequatorial *cis*-3,5-dimethylcycloheptanone and the equatorial-axial trans 3,5 isomer is 0.8 to 0.9 kcal./mole, about the same as would be the difference between *cis*- and *trans*-3,5-dimethylcyclohexanone.†[14] In cycloöctane, three conformations have been considered to be the most stable: the "crown" (Fig. 9-6),[10a, 10c, 11, 12] the "boat" (Fig. 9-6),[10a, 12] and the "chair" (Fig. 9-6),[13] with the additional assumption that there is some bond-angle deformation in all these forms. The weight of the present evidence points to the crown form as being by far the most populated conformation in cycloöctane.[9, 11b]

Suggestions have also been made for the conformation of cyclononane,[9, 10c] cyclodecane,[9, 10c, 12] and cyclododecane.[9, 12]‡ The case of cyclo-

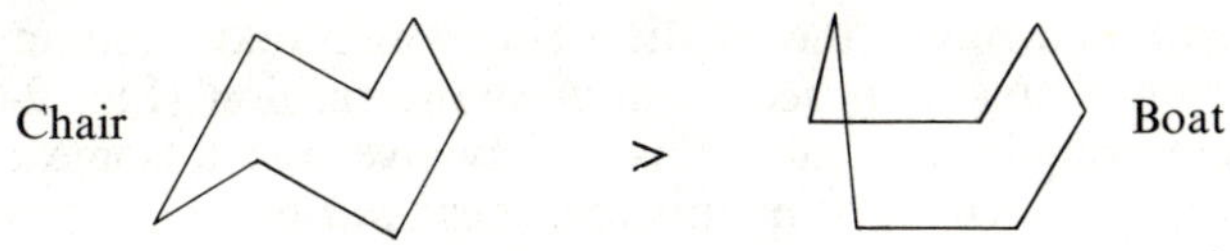

Fig. 9-5. Cycloheptane.

nonylamine hydrobromide[9] is of particular interest in that this molecule appears to exist in two different conformations in the solid state.

9-3. Conformational Effects in Medium-sized Rings. Transannular Effects[46]

Although the precise conformational situation in medium-sized rings is as yet imperfectly understood, some of the chemical consequences of their peculiar geometry have been studied and will be presented in this section and the next. Before entering into this subject, it is well to point out that,

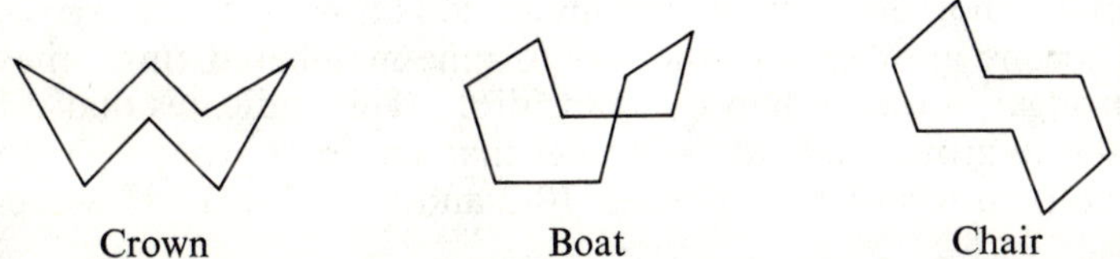

Fig. 9-6. Conformations of cycloöctane.

although one may speak of cis-trans isomerism in medium and even in large rings (cf. Chap. 7), models indicate that, because of the large number of degrees of freedom in such rings, no definite conformational relationship need correspond to the cis and trans configurations. Even in a ring as small

† The normal difference of 1.8 kcal./mole between equatorial and axial methyl must be diminished in this case by about 1.0 kcal./mole because of the 1,3-ketone effect; cf. Sec. 8-7.

‡ Heats of combustion of cycloalkanes have been calculated from torsional potentials for the appropriate conformations.[12] Whereas the calculated values for cycloheptane and cycloöctane are in good agreement with those found experimentally (Table 7-1), the agreement is quite poor for cyclodecane and cyclododecane. This is probably due to neglect of angle-deformation strain, which turns out to be quite important in these cycles.[9]

as an eight-membered one, it is perfectly possible to rotate cis substituents (configurationally speaking) on adjacent carbons into a position where they become conformationally anti with respect to each other. This is a consequence of the fact that, as rings become larger and larger, they become more and more similar to open chains. In hydrocarbon chains, the preferred conformation is a zigzag one, corresponding to an *anti*-butane conformation of any segment of four adjacent carbon atoms. A ring, no matter how large, cannot have a complete zigzag conformation (for the ends of a zigzag never meet). However, it can be seen from models that, whereas in small and common rings the conformation of any four adjacent carbon atoms must correspond to either the cis conformation (four- and five-membered ring) or the gauche conformation (five-, six-, and seven-membered ring†) of butane, the eight-membered ring is the smallest ring which may accommodate four adjacent carbon atoms disposed in *anti*-butane fashion (Fig. 9-6). In the eight-membered ring in particular, this moiety gives rise to some angle strain, but in larger rings, such as 10-membered ones, *anti*-butane segments can be accommodated without angle strain.

A fairly clear-cut demonstration of the change in dihedral angle between adjacent substituents with ring size is found in the infrared spectra[15] and lead tetraacetate cleavage rates[16, 17] of cycloalkane-1,2-diols. The data are summarized in Table 9-1.

The column headed $\Delta\tilde{\nu}$ gives a measure of the strength of the intramolecular hydrogen bond between the adjacent hydroxyl groups (cf. Sec. 6-2*b* and Table 6-1): The greater the distance $\Delta\tilde{\nu}$ between bonded and unbonded hydroxyl stretching frequencies, the stronger is the intramolecular hydrogen bond. In turn, this bond is strongest when the hydroxyl groups are eclipsed or nearly eclipsed; it declines in strength with increasing dihedral angle between the C—OH bonds. The strongest hydrogen bond is found in cyclopentane-*cis*-1,2-diol where the OH groups are nearly eclipsed.‡ In cyclopentane-*trans*-1,2-diol, the hydroxyl groups are too far apart to undergo hydrogen bonding to one another. In the six- to nine-membered rings, the cis diol shows stronger bonding than the trans diol. This indicates that it is easier to bend the cis groups toward each other than the trans groups, a point which will be returned to in Chap. 10 (Sec. 10-2 and Fig. 10-17). However, in the

† The seven-membered ring may, in principle, contain an eclipsed conformation in which the first and fourth carbons are at a 120° angle (*i*). However, this conformation is not apt to be favored over the gauche arrangement (staggered, 60° angle).

‡ Because of the puckering of the cyclopentane ring, the eclipsing is not complete; cf. H. Kwart and W. G. Vosburgh, *J. Am. Chem. Soc.*, **76**, 5400 (1954).

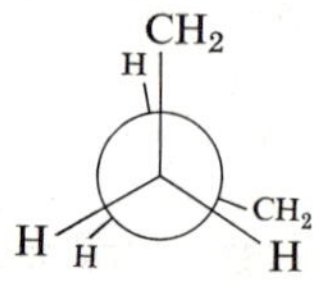

(*i*)

[15] L. P. Kuhn, *J. Am. Chem. Soc.*, **76**, 4323 (1954).

[16] (*a*) R. Criegee, E. Büchner, and W. Walther, *Ber.*, **73**, 571 (1940); (*b*) R. Criegee, L. Kraft, and B. Rank, *Ann.*, **507**, 159 (1933); (*c*) R. Criegee, E. Höger, G. Huber, P. Kruck, F. Marktscheffel, and H. Schellenberger, *Ann.*, **599**, 81 (1956).

[17] (*a*) V. Prelog, K. Schenker, and W. Kung, *Helv. Chim. Acta*, **36**, 471 (1953); (*b*) V. Prelog, K. Schenker, and H. H. Günthard, *ibid.*, **35**, 1598 (1952); (*c*) V. Prelog and M. Speck, *ibid.*, **38**, 1786 (1955). (*d*) Cf. S. J. Angyal and R. J. Young, *J. Am. Chem. Soc.*, **81**, 5467, 5251 (1959).

10-membered ring the difference becomes insignificant, and in the 12-membered ring, the trans hydroxyls actually approach each other more closely than the cis hydroxyls. The reason for this can be seen in Fig. 9-7. In the larger rings, the trans diol may use the more stable *anti*-butane conformation in which the hydroxyl groups may readily approach each other somewhat by a rotation about the 1,2 bond. But even if the trans diol uses the *gauche*-butane conformation, the hydroxyl groups may approach each other readily, since the rotation that brings this about does not eclipse the ring carbons. The cis diol, on the other hand, must use the *gauche*-butane conformation (no hydrogen bonding is possible in the *anti*-butane conformation), and in

Table 9-1
Hydroxyl Stretching Frequencies and Lead Tetraacetate Cleavage Rates in Cycloalkane-1,2-diols

Ring size	Configuration	$\Delta\tilde{\nu}^a$	k^b
5-membered	cis	61	40,000
	trans	. . [c]	12.8
6-membered	cis	38	5.0
	trans	33	0.22
7-membered	cis	44	7.7
	trans	37	1.2
8-membered	cis	51	[d]
	trans	43	[d]
9-membered	cis	49	2.87
	trans	45	20.7
10-membered	cis	44	2.6
	trans	45	100
12-membered	cis	38	1.27
	trans	51	73.6
16-membered	cis	. . [d]	7.84
	trans	50	91.2

[a] Distance, in wave numbers, between unbonded hydroxyl stretching band and intramolecularly bonded hydroxyl stretching band in the infrared.
[b] Rate of cleavage with lead tetraacetate in glacial acetic acid at 20°C. in l. $mole^{-1}$ $min.^{-1}$.
[c] This compound does not have an intramolecular hydrogen bond.
[d] Not determined.

this conformation the hydroxyl groups cannot approach each other readily. For if they did—by rotation about the 1,2 bond—the ring carbons would become eclipsed. In fact, the situation here is very similar to that depicted in Figs. 6-9 and 6-10 for acyclic glycols, the cyclic cis glycol being meso and the trans glycol *dl*.†

Lead tetraacetate cleavage is similar;[17a] the reaction requires a cyclic intermediate (Fig. 9-8) which can be formed most readily when the hydroxyl

† The situation in the 2-aminocycloalkanols is quite similar, except that the crossover occurs between the seven-membered and the eight-membered rings: J. Sicher, M. Horák, and M. Svoboda, *Collection Czechoslov. Chem. Commun.*, **24**, 950 (1959).

groups are eclipsed or nearly eclipsed. However, the crossover point (where the trans diol begins to react faster than the cis) comes earlier here than in the spectral series, being somewhere between cycloheptane-1,2-diol and cyclononane-1,2-diol. The reason may be that formation of the cyclic intermediate in cyclohexane-*cis*-1,2-diol, and presumably also in the larger diols which are in the *gauche*-butane conformation, requires attachment of the lead atom to a crowded axial (or axial-like) position—an unfavorable situation which is avoided in the *anti*-butane conformation of the trans diol.

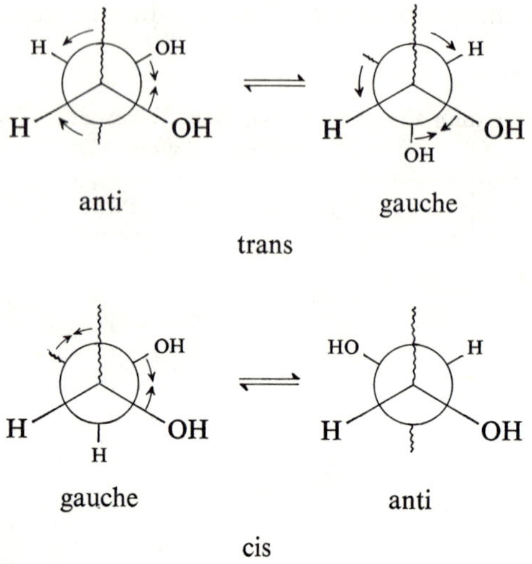

Fig. 9-7. Conformation of large-ring 1,2-diols.

Other reactions in which these conformational differences between rings of different sizes affect the stereochemical course are the pyrolysis of cycloalkylamine oxides[18] and the pyrolysis of cycloalkyltrimethylammonium hydroxides,[17-20] both giving rise to cycloalkenes. The cyclononyl- and cyclodecylamine derivatives give trans olefins, whereas the cyclohexyl- and cycloheptylamine derivatives give only cis olefins (the corresponding trans olefins

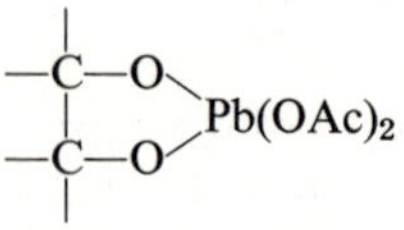

Fig. 9-8. Intermediate in lead tetraacetate cleavage. (An 11-membered-ring intermediate is postulated in Ref. 16*c*.)

[18] (*a*) A. C. Cope, R. A. Pike, and C. F. Spencer, *J. Am. Chem. Soc.*, **75**, 3212 (1953); (*b*) A. C. Cope, D. C. McLean, and N. A. Nelson, *ibid.*, **77**, 1628 (1955).

[19] K. Ziegler and H. Wilms, *Ann.*, **567**, 1 (1950); K. Ziegler, H. Sauer, L. Bruns, H. Froitzheim-Kühlhorn, and J. Schneider, *Ann.*, **589**, 122 (1954).

[20] A. T. Blomquist, L. H. Liu, and J. C. Bohrer, *J. Am. Chem. Soc.*, **74**, 3643 (1952); A. T. Blomquist, R. E. Burge, and A. C. Sucsy, *ibid.*, **74**, 3636 (1952).

being unknown; cf. Chap. 7). In the cycloöctyl series, the amine oxide gives pure *cis*-cycloöctene whereas the quaternary ammonium hydroxide gives a mixture of *cis*- and *trans*-cycloöctene, with the latter predominating. The facts are summarized in Fig. 9-9. In order to explain the facts, one has to keep in mind that pyrolysis of amine oxides requires that the amine oxide and hydrogen to be eliminated be nearly cis-coplanar (Fig. 8-41) whereas the Hofmann elimination reaction requires the hydrogen and trimethylammonium group to be anti-coplanar (Sec. 8-5). Now, as may be seen from Fig. 9-10, cis as well as anti elimination from an *anti*-butane segment gives a trans olefin, whereas cis as well as anti elimination from a *gauche*-butane segment may yield a cis olefin.† The question then arises: Does elimination from cyclic amine oxides and quaternary ammonium salts involve *anti*-butane or *gauche*-butane segments?‡ Obviously, in six- and seven-membered rings, *gauche*-butane segments must be involved, since such rings cannot accommodate *anti*-butane segments. In 9- and 10-membered rings, reasoning after the fact, one sees that *anti*-butane segments are involved, and this is not

$(CH_2)_{n-2}$ ring with CH_2 and $CH{-}NMe_2{\rightarrow}O$ $\xrightarrow[-Me_2NOH]{\Delta}$ $(CH_2)_{n-2}$ ring with $CH{=}CH$ $\xleftarrow[-NMe_3,\ -H_2O]{\Delta}$ $(CH_2)_{n-2}$ ring with CH_2 and $CH{-}\overset{+}{N}Me_3OH^-$

n		Olefin		n
6, 7	→	cis	←	6, 7
8	→ cis	trans + cis	←	8
9, 10	→	trans	←	9-12*

* The 12-membered quaternary ammonium hydroxide gives some cis, along with much trans, olefin.

Fig. 9-9. Pyrolysis of cycloalkyldimethylamine oxides and cycloalkyltrimethylammonium hydroxides.

unreasonable, inasmuch as rings of this size may accommodate *anti*-butane segments readily. The eight-membered ring occupies an intermediate position. It may contain *anti*-butane segments, but only with considerable difficulty. In the amine oxide pyrolysis, such conformations are evidently not favored and only cis olefin results, whereas in the Hofmann elimination, part of the reaction does proceed via a transition state with *anti*-butane geometry.[18b] The greater tendency of the Hofmann elimination to use the

† With a *gauche*-butane segment, the group to be eliminated (X) may occupy two different positions (corresponding to the equatorial and axial positions in cyclohexane). Only one of these (the axial) is suitable for trans elimination, but both are suitable for cis elimination, especially in rings larger than six-membered. (Regarding the restriction in six-membered rings, see Chap. 10.) When the "equatorial" group undergoes cis elimination, the product may be either cis or trans olefin, depending on which hydrogen leaves. It is not clear as yet why the cis olefin is formed in the 8-membered ring but the trans olefin is formed in the 9- and 10-membered rings.

‡ The situation is being oversimplified somewhat, inasmuch as the conformation of the transition state may deviate somewhat from pure anti or pure gauche geometry. In particular, the transition state in amine oxide pyrolysis may be closer to cis geometry than to gauche geometry. This does not change the argument materially, however.

otherwise unfavorable *anti*-butane conformation most probably lies in some unfavorable conformational situation affecting the $N(CH_3)_3$ group in elimination from the *gauche*-butane conformation.† In this connection, it is of interest that the Hofmann elimination of cyclohexyltrimethylammonium

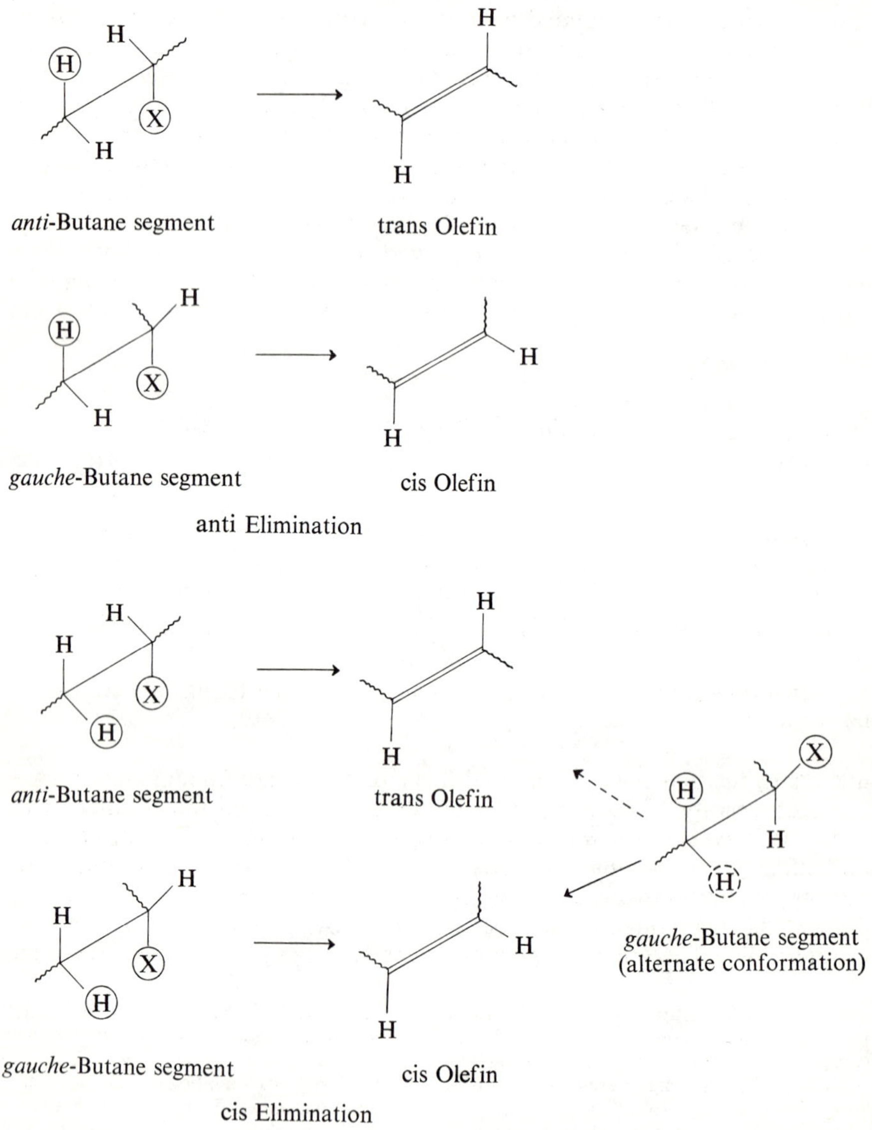

anti-Butane segment

trans Olefin

gauche-Butane segment

cis Olefin

anti Elimination

anti-Butane segment

trans Olefin

gauche-Butane segment

cis Olefin

gauche-Butane segment (alternate conformation)

cis Elimination

Fig. 9-10. Steric course of elimination.

† A much less likely alternative is that some specific factor favors the cis geometry of the ring carbons involved in amine oxide pyrolysis over the gauche arrangement leading to cis olefin in the case of the Hofmann reaction.

hydroxide to cyclohexene[18a] requires the $N(CH_3)_3$ group to assume the very unfavorable axial position, as shown in Fig. 9-11.† If a similarly unfavorable location of the $N(CH_3)_3$ group is involved in the anti elimination of cycloöctyltrimethylammonium hydroxide to give *cis*-cycloöctene, it becomes understandable why an alternative geometric arrangement leading to *trans*-cycloöctene competes successfully, even though it involves other types of strain.

That there is some resistance for 1,2-cis substituents to assume the anti conformation even in 10- and 12-membered rings is evidenced by the results of the pyrolysis of *cis*-2-hydroxy-cycloalkyltrimethylammonium hydroxides.[20a] The cyclodecane derivative undergoes a shift of the trans hydrogen (anti) to give cyclodecanone, resembling in this respect the cyclohexane derivative (cf. Fig. 7-8 for corresponding behavior of the halohydrins). The cyclotridecane and cyclohexadecane derivatives give trans epoxides; the cyclododecane compound is intermediate in giving a mixture of cyclododecanone and cyclododecane-*trans*-epoxide.

It was mentioned in Sec. 9-2 that one of the strain factors in medium-sized cycles is van der Waals repulsion of hydrogen atoms across the ring. Strain

$\overset{+}{N}(CH_3)_3$

$\xrightarrow{\Delta}$ (cyclohexene) $+ N(CH_3)_3 + H_2O$

H---OH⁻

Fig. 9-11. Hofmann elimination of cyclohexyltrimethylammonium hydroxide.

from this source has been termed[43] "transannular strain." Although it is now recognized[9] that this particular type of strain contributes only in a minor way‡ to the instability of medium-sized rings, the fact remains that atoms may approach each other closely across such rings and may thus enter into physical interaction or chemical reaction through what have been called "transannular effects" by Prelog and "proximity effects" by Cope.

† D. Y. Curtin, R. D. Stolow, and W. Maya [*J. Am. Chem. Soc.*, **81,** 3330 (1959)] have shown that under milder conditions than the usual pyrolysis of the nearly dry quaternary ammonium hydroxide, namely, when the cyclohexyltrimethylammonium chloride is boiled with potassium *t*-butoxide in *t*-butyl alcohol at 75°, only 7% of the product is cyclohexene, resulting from elimination of an axial $N(CH_3)_3$ group, the remainder being cyclohexyldimethylamine, resulting from displacement of an *N*-methyl group by *t*-butoxide, mostly in the equatorial conformation of $N(CH_3)_3$.

‡ The total strain in medium-sized rings has been termed[43] "non-classical strain" to distinguish it from classical or angle strain. Although this term is widely used it has turned out to be somewhat of a misnomer, inasmuch as it is now known[9] that angle strain is sometimes a substantial part of the total strain in medium-sized rings. "Non-classical strain" is thus actually a composite of torsional, angle, and (in a minor way) transannular strain.

[20a] M. Svoboda and J. Sicher, *Collection Czechoslov. Chem. Commun.*, **23,** 1540 (1958). See also Ref. 45.

Physical effects of transannular interactions may be seen in spectra. In the infrared spectra of some of the cyclic aminoketones shown in Fig. 9-12, the carbonyl absorption (around 1700 cm.$^{-1}$) disappears on protonation, indicating, presumably, the formation of a transannular bond.[21] Transannular bond formation under these circumstances has been observed in 8-membered, 9-membered, and 10-membered aminoketones (A) which give

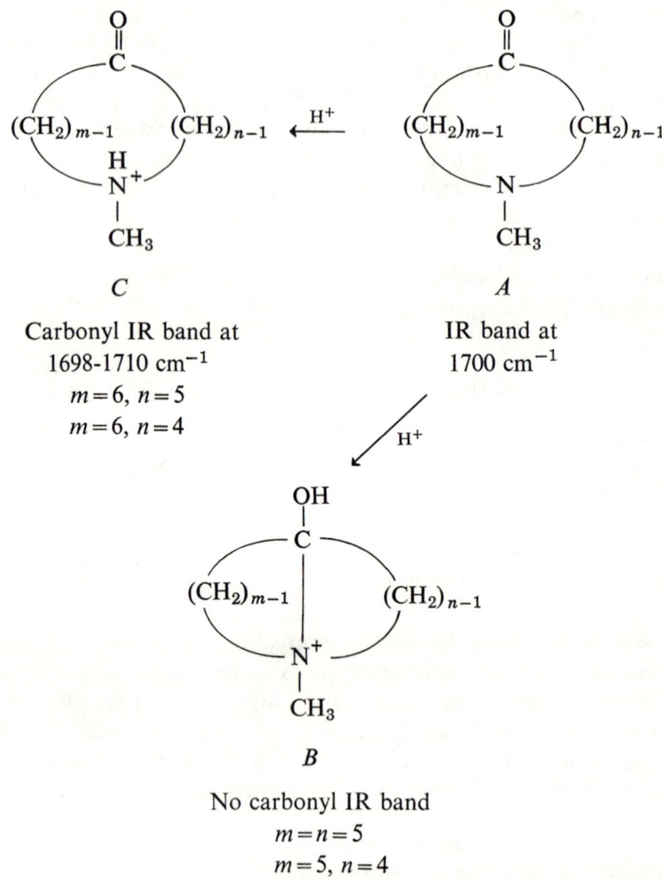

Fig. 9-12. Change of infrared spectrum of cyclic aminoketones upon protonation.

rise to bicyclic 5-5, 5-6, or 6-6 systems (B), respectively, upon protonation. However, in an 11-membered aminoketone (which would have to give rise to a 6-7 bicyclic system) transannular bond formation did not occur upon protonation; instead the simple aminoketone salt C was obtained, as evidenced by the preservation of the carbonyl infrared absorption in strong acid. Similarly, a 10-membered-ring aminoketone which would have given a 7-5

[21] N. J. Leonard, *Record Chem. Progr. (Kresge-Hooker Sci. Lib.)*, **17**, 243 (1956); N. J. Leonard, J. A. Adamcik, C. Djerassi, and O. Halpern, *J. Am. Chem. Soc.*, **80**, 4858 (1958).

(rather than a 6-6) system upon transannular ring closure appeared, instead, to give largely the simply protonated form C.

In those aminoketones where transannular ring closure occurred upon protonation, the transannular interaction was also demonstrated in the free aminoketones by the lowering of the carbonyl absorption frequency below the normal value, by a decrease in the chemical reactivity of the carbonyl function, by an abnormally high dipole moment, by unusual effects of solvent on the base strength of the amine, and by anomalous rotatory dispersion (Chap. 14).[21] Similar transannular interaction has been found in cyclic ketosulfides.[22]

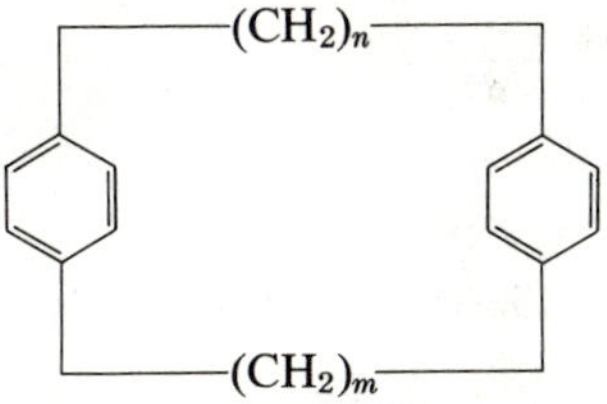

$(CH_2)_n$

$(CH_2)_m$

Fig. 9-13. Paracyclophanes.

Transannular interactions also manifest themselves in ultraviolet spectra in ring compounds having chromophores affected by such interaction. An example is presented by the paracyclophanes (Fig. 9-13; cf. Chap. 6).[23] When n and $m \geqslant 4$, or $n = 1$ and $m = 12$, or $n = 4$, $m = 6$, the spectra are normal or nearly so; i.e., they resemble the spectra of 1,4-dialkylbenzenes. Increasingly anomalous spectra are observed in systems where $n = 3$, $m = 4$; $n = m = 3$; $n = 2$, $m = 3$; and $n = m = 2$; as well as in systems where $n = 3$, $m = 6$; $n = 2$, $m = 4$; or $n = 1$, $m = 7$. Transannular interaction causes a solvent-sensitive absorption band in *trans*- (but not *cis*-) 5-cyclodecenone.[23a]

Unusual transannular ring closures are sometimes observed in medium-sized rings. Examples are shown in Figs. 9-14[24] and 9-15.[25]

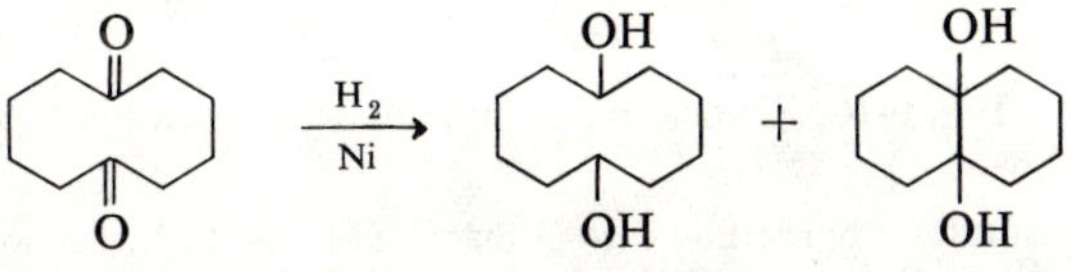

O O $\xrightarrow[\text{Ni}]{H_2}$ OH OH + OH OH

Fig. 9-14. Side reaction involving transannular ring closure.

Of particular interest is the fact, discovered independently by Prelog[26] and by Cope,[27] that groups which may approach each other across the ring can get involved in rearrangements and in neighboring-group participation (cf.

[22] N. J. Leonard, T. W. Milligan, and T. L. Brown, *J. Am. Chem. Soc.*, **82,** 4075 (1960).

[23] Cf. D. J. Cram, *Record Chem. Progr.* (*Kresge-Hooker Sci. Lib.*), **20,** 71 (1959).

[23a] E. M. Kosower, W. D. Closson, H. L. Goering, and J. C. Gross, *J. Am. Chem. Soc.*, **83,** 2013 (1961).

[24] P. A. Plattner and J. Hulstkamp, *Helv. Chim. Acta,* **27,** 211 (1944).

[25] A. C. Cope, R. J. Cotter, and G. G. Roller, *J. Am. Chem. Soc.,* **77,** 3594 (1955).

[26] V. Prelog and K. Schenker, *Helv. Chim. Acta,* **35,** 2044 (1952).

[27] A. C. Cope, S. W. Fenton, and C. F. Spencer, *J. Am. Chem. Soc.,* **74,** 5884 (1952).

Sec. 6-3*b*). Transannular rearrangements have been observed in the hydroxylation of cycloalkenes with performic acid and in the acid-catalyzed opening of epoxycycloalkanes. Reaction of *cis*-cyclooöctene with performic acid gives, among other products, *cis*-1,4-cyclooöctanediol whereas the trans olefin gives the corresponding trans diol.[27] Similarly, *cis*-cyclodecene gives *cis*-cyclodecane-1,6-diol and *trans*-cyclodecene gives *trans*-cyclodecane-1,6-diol.[26] In the cyclononene series *both* olefins apparently give mixtures of *cis*- and *trans*-cyclononane-1,5-diol.[17a] These reactions involve normal epoxidation of the

OTs

$C_6H_5N(C_2H_5)_2$ Δ

+

and
cis Isomer

Fig. 9-15. Transannular ring closure.

olefin followed by a transannular hydride shift in the acid-catalyzed ring opening of the epoxide,[28] as shown in Fig. 9-16. The shift in the cyclodecane system is of the 1,5 rather than of the 1,6 type, suggesting conformation A (Fig. 9-17) rather than B for the transition state. Analogous transannular reactions are found with cycloundecene[29] but not (or only to a minor extent) with cyclohexene, cycloheptene,[30] and cyclododecene.[17c]†

Solvolysis of cycloalkyl *p*-toluenesulfonates labeled[31] with C^{14} at the carbinol carbon or treatment[32] of the corresponding cycloalkylamines with nitrous acid leads to products (olefins and substitution products) having the

HCOOH

HCO_3H H O H_2O OH OH trans

Fig. 9-16. Transannular epoxide ring opening.

† Cyclooöctene constitutes a borderline case in that it gives rise to 1,2- as well as 1,4-diol. No 1,2-diol is found in the C_9-C_{11} rings. Cyclohexene gives 0.30% cyclohexane-1,4-diol: A. C. Cope, H. E. Johnson, and J. S. Stephenson, *J. Am. Chem. Soc.*, **78**, 5599 (1956); and ring opening of cycloheptene oxide yields cycloheptane-*cis*-1,4-diol as a minor product (2.4%); A. C. Cope, T. A. Liss, and G. W. Wood, *J. Am. Chem. Soc.*, **79**, 6287 (1957).

---

[28] A. C. Cope, A. H. Keough, P. E. Peterson, H. E. Simmons, and G. W. Wood, *J. Am. Chem. Soc.*, **79**, 3900 (1957); A. C. Cope, A. Fournier, and H. E. Simmons, *ibid.*, **79**, 3905 (1957). See also A. C. Cope, G. A. Berchtold, P. E. Peterson, and S. H. Sharman, *J. Am. Chem. Soc.*, **82**, 6366 (1960).

[29] V. Prelog and V. Boarland, *Helv. Chim. Acta*, **38**, 1776 (1955).

[30] A. C. Cope and W. N. Baxter, *J. Am. Chem. Soc.*, **76**, 279 (1954).

[31] H. J. Urech and V. Prelog, *Helv. Chim. Acta*, **40**, 477 (1957). See also Ref. 44 for a correction of some of these results.

[32] V. Prelog, H. J. Urech, A. A. Bothner-By, and J. Würsch, *Helv. Chim. Acta*, **38**, 1095 (1955).

label at positions other than C_1 in the ring. The distribution of the label in the products indicates that transannular hydride shifts take place in the course of the reaction (Fig. 9-18). These shifts are most important with 8- to 11-membered rings, but occur to some extent in 7-membered-ring systems and

HCOOH HCOOH
H rather than H
O O
A *B*

Fig. 9-17. Nature of hydride shift in transannular ring opening of cyclodecene oxide.

to a very slight extent in 12-membered-ring systems.[33a] Figure 9-19 shows a transformation in which the hydride shift was demonstrated directly.[33b]

It is perhaps well to point out that proximity effects are not limited to medium-sized rings; they may occur whenever a group disposed toward

$(CH_2)_x$ $(CH_2)_x$ $(CH_2)_x$
H—C—H CH_2 H—C—H CH_2 H—C—H CH_2
14 $\xrightarrow{Y}$ 14 + 14
H—C—H·········→CHX HO—C—H CH_2 H—C—H CHOH
$(CH_2)_y$ $(CH_2)_y$ $(CH_2)_y$

$X = OTs,\ Y = H_2O$
or
$X = NH_2,\ Y = HNO_2$

Fig. 9-18. Transannular hydride shifts in solvolysis and amine–nitrous acid reactions.

participation is in the close neighborhood of the reacting functional group. A dramatic illustration[34] is provided by participation across the five-membered ring in *anti*-7-norbornenyl *p*-toluenesulfonate (Fig. 9-20) which reacts faster than the syn isomer by a factor of 10^7 and faster than the saturated isomer by a factor of 10^{11}. The pi electrons of the double bond are well dis-

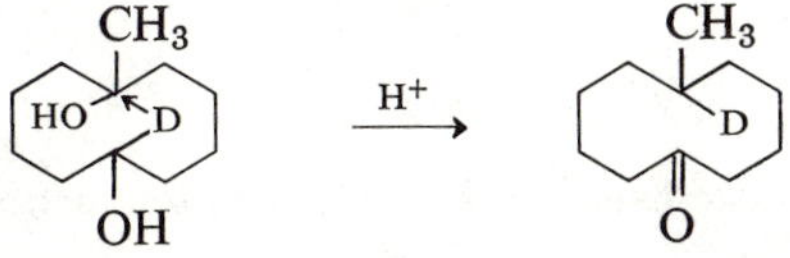

Fig. 9-19. Hydride shift in transannular pinacol rearrangement.

[33] (*a*) See also V. Prelog, *Record Chem. Progr.* (*Kresge-Hooker Sci. Lib.*), **18,** 247 (1957); *Angew. Chem.,* **70,** 145 (1958). (*b*) V. Prelog and W. Küng, *Helv. Chim. Acta,* **39,** 1394 (1956).

[34] S. Winstein, M. Shatavsky, C. Norton, and R. B. Woodward, *J. Am. Chem. Soc.,* **77,** 4183 (1955); S. Winstein and M. Shatavsky, *ibid.,* **78,** 592 (1956); S. Winstein and E. T. Stafford, *ibid.,* **79,** 505 (1957).

posed, sterically, to assist the departure of the tosylate ion. Acetolysis proceeds with retention of configuration, the product being *anti*-7-norbornenyl acetate.

1,3 Participation across a six-membered ring is observed in the solvolysis of cholesteryl *p*-toluenesulfonate (Fig. 9-21). Because of participation, cholesteryl tosylate is methanolyzed[35] and acetolyzed[36] five to six times faster[37] than its epimer, epicholesteryl tosylate, despite the fact that the tosylate function in the latter isomer is axial and should, therefore, other things being

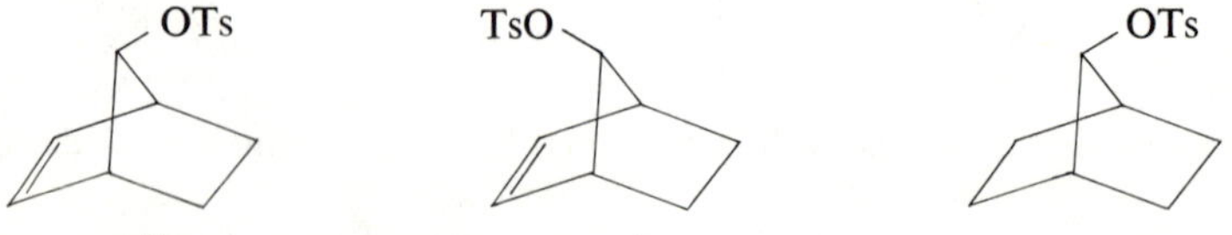

OTs TsO OTs

10^{11} 10^{4} 1

Fig. 9-20. Relative acetolysis rates of 7-norbornyl and 7-norbornenyl derivatives.

equal, be solvolyzed more rapidly (cf. Chap. 8; see Fig. 10-29 for steroid conformation). Methanolysis of cholesteryl tosylate (Fig. 9-21) gives, depending on conditions, cholesteryl methyl ether (retention of configuration) or 3,5-cyclocholestan-6β-yl methyl ether (also called *i*-cholesteryl methyl ether),[38] thus confirming the operation of neighboring-group participation.

1,4 Participation across a six-membered ring (in the boat form) is seen in the methanolysis of *trans*-4-methoxycyclohexyl *p*-toluenesulfonate (Fig. 9-22).[39] The equatorial trans isomer acetolyzes about twice as fast as the

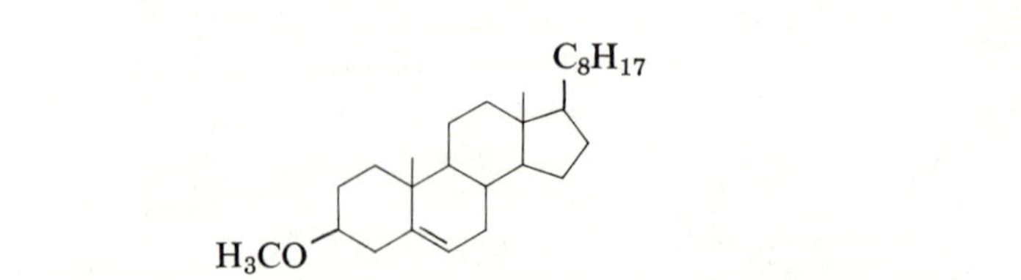

C_8H_{17} C_8H_{17}

CH_3OH / NaOAc CH_3OH

OCH_3 TsO (β)

C_8H_{17}

H_3CO

Fig. 9-21. Methanolysis of cholesteryl *p*-toluenesulfonate.

[35] L. C. King and M. J. Bigelow, *J. Am. Chem. Soc.*, **74**, 6238 (1952).

[36] C. W. Shoppee and D. F. Williams, *J. Chem. Soc.*, 686 (1955).

[37] For earlier indications of the fast solvolysis of cholesteryl tosylate, see W. Stoll, *Z. physiol. Chem.*, **246**, 1 (1937), and S. Winstein and R. Adams, *J. Am. Chem. Soc.*, **70**, 838 (1948).

[38] W. Stoll, *Z. physiol. Chem.*, **207**, 147 (1932); J. H. Beynon, I. M. Heilbron, and F. S. Spring, *J. Chem. Soc.*, 1459 (1937).

[39] D. S. Noyce, B. R. Thomas, and B. N. Bastian, *J. Am. Chem. Soc.*, **82**, 885 (1960); D. S. Noyce and B. N. Bastian, *ibid.*, **82**, 1246 (1960).

partly axial cis isomer, and when the starting material is labeled with tritium in the 1 position, the *trans*-4-methoxycyclohexyl acetate product has the label nearly equally divided between the 1 and 4 positions (Fig. 9-22).

Finally, in the solvolysis of acyclic *p*-bromobenzenesulfonates, participation by a suitable substituent on the carbon chain may occur if that substituent may be brought into spatial proximity of the sulfonate function. Thus participation has been observed[40] in compounds of the type $Ar(CH_2)_nCH_2OBs$ for $n = 1$ and 3.†

9-4. The Concept of I-strain[41, 42]

Many chemical reactions involve changes from sp^3 to sp^2 hybridization, or vice versa, at the site of reaction in the transition state. Examples are dis-

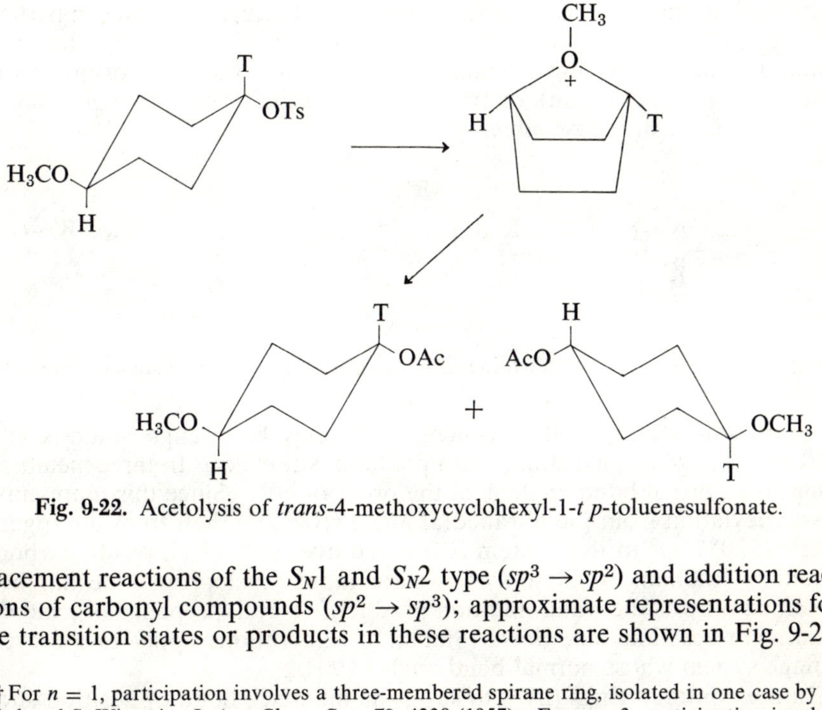

Fig. 9-22. Acetolysis of *trans*-4-methoxycyclohexyl-1-*t* *p*-toluenesulfonate.

placement reactions of the S_N1 and S_N2 type ($sp^3 \rightarrow sp^2$) and addition reactions of carbonyl compounds ($sp^2 \rightarrow sp^3$); approximate representations for the transition states or products in these reactions are shown in Fig. 9-23.

† For $n = 1$, participation involves a three-membered spirane ring, isolated in one case by R. Baird and S. Winstein, *J. Am. Chem. Soc.*, **79,** 4238 (1957). For $n = 3$, participation involves either a five-membered spirane ring or a six-membered hydronaphthalene ring (Ref. 40); such participation occurs only when there are suitably placed methoxyl groups in the aromatic (Ar) ring.

---

40 R. Heck and S. Winstein, *J. Am. Chem. Soc.*, **79,** 3105, 3114 (1957).

41 (*a*) H. C. Brown, R. S. Fletcher, and R. B. Johannesen, *J. Am. Chem. Soc.*, **73,** 212 (1951); (*b*) H. C. Brown, *Record Chem. Progr.* (*Kresge-Hooker Sci. Lib.*), **14,** 83 (1953); (*c*) H. C. Brown, *Bull. soc. chim. France,* 980 (1956); (*d*) H. C. Brown, *J. Chem. Soc.*, 1248 (1956); (*e*) G. A. Russell, *J. Am. Chem. Soc.*, **80,** 4997 (1958); (*f*) H. C. Brown, J. H. Brewster, and H. Shechter, *ibid.*, **76,** 467 (1954).

42 E. L. Eliel in M. S. Newman, ed., "Steric Effects in Organic Chemistry," John Wiley & Sons, Inc., New York, 1956, pp. 121–124, 148–149; M. S. Newman, *ibid.*, pp. 237–239.

Some chemical equilibria, such as the cyanohydrin equilibrium,

$$>C=O + HCN \rightleftharpoons >C\begin{matrix}OH \\ CN\end{matrix}$$

involve similar changes in hybridization. H. C. Brown[41] has pointed out that, if steric strains are greater in the sp^2 (planar) than in the sp^3 (tetrahedral) arrangement of the groups, reactions of the $sp^2 \rightarrow sp^3$ type, such as carbonyl addition reactions, are facilitated and reactions of the $sp^3 \rightarrow sp^2$ type (for example, S_N1 and S_N2 substitutions) are retarded. Contrariwise, if steric strains are greater in the sp^3 (tetrahedral) than in the sp^2 (planar) arrangement, $sp^2 \rightarrow sp^3$ reactions (e.g., carbonyl addition) are hindered and $sp^3 \rightarrow sp^2$ reactions (e.g., nucleophilic substitution) are assisted, *provided always that there are not other effects of a steric or polar nature more important than those mentioned here.*†

In the I-strain concept, these ideas are applied to cyclic systems in particular. In such systems, changes of bond hybridization may produce concomitant changes in angle strain (Baeyer strain), bond opposition strain (torsional or Pitzer strain), or transannular strain. These changes may be either favorable or unfavorable.

R′ R C⁺ R″ (sp^2) $\xleftarrow{S_N1}$ R R′ C–X R″ (sp^3) $\xrightarrow[S_N2]{Y}$ R′ R Y···C···X R″ (sp^2)

R′ R C=O (sp^2) $\xrightarrow{HX}$ R R′ C–OH X (sp^3)

Fig. 9-23. Changes in hybridization of bond orbitals in transition state or product.

In the small (three- and four-membered) rings, bond-angle strain is large, and any change in such strain has a predominant effect. In three-membered rings, the normal bond angle is of the order of 60°. Since this represents a lesser deviation from the tetrahedral angle (109°28′) than from the trigonal angle (120°), sp^3 in this system is favored over sp^2. As a result, carbonyl addition reactions in these systems are rapid (cyclopropanone has been obtained only as the water-addition product, 1,1-cyclopropanediol), and displacement reactions are slow. Similar considerations apply to the cyclobutane system whose normal bond angle is 90°.‡

In the common rings (five- to seven-membered), bond opposition strain seems to be predominant. As already mentioned, in cyclopentane, if considered

† For example, in S_N2 reactions, the steric encumbrance of the incoming group Y is very important but is not explicitly considered in the above argument.

‡ However, as pointed out previously, the effects due to I-strain may be swamped by the operation of other factors. For example, J. D. Roberts and V. C. Chambers [*J. Am. Chem. Soc.*, **73**, 5034 (1951)] found that the solvolysis reactions of cyclobutyl halides and tosylates are fast, probably because ionization proceeds to an ion more stable than the expected cyclobutyl cation. The stability of cyclopropenones (Sec. 7-2*b*) is due to their aromatic character. Even when I-strain does predict rate effects correctly, the operation of other factors is not excluded. Thus the slowness of solvolysis reactions in the cyclopropyl system may be due, in part, to inductive effects.

planar, there are 10 C—H bond oppositions, giving rise to ca. 10 kcal./mole eclipsing strain.†

When the hybridization at any one carbon is changed from sp^3 to sp^2, four bond oppositions are relieved (this can best be seen in models), and the strain is decreased by ca. 4 kcal./mole.‡ This favors the $sp^3 \rightarrow sp^2$ transition, and, correspondingly, displacement reactions in cyclopentane systems are relatively rapid, whereas carbonyl addition reactions are relatively slow. Similar considerations apply to seven-membered rings which are also mobile and beset with bond oppositions of adjacent C—H bonds, although to a somewhat lesser extent than the five-membered rings. In the six-membered ring, however, the rigid (chair) form is free of bond oppositions. Change of hybridization from sp^3 to sp^2 engenders bond oppositions and is, therefore, resisted. Correspondingly, displacement reactions in cyclohexyl systems are slow; addition reactions to cyclohexanone are fast.

Medium-sized rings (8 to 11-membered) are beset with angle deformations, bond opposition, and transannular interactions. Most of these are due to hydrogen-hydrogen interactions, as evidenced by the much greater ease of ring closure of medium-sized rings containing oxygen atoms (bare) as com-

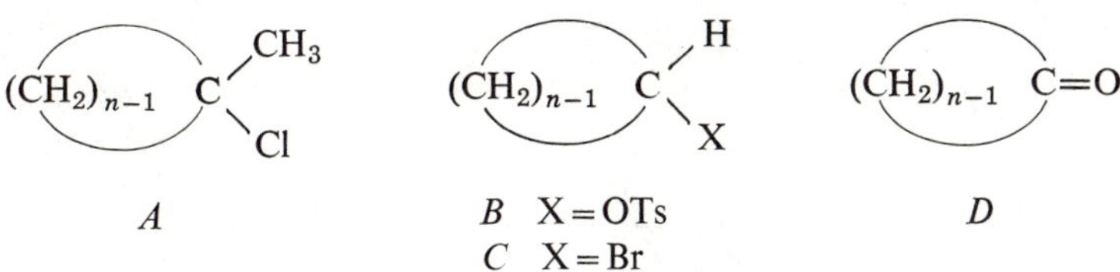

Fig. 9-24. Systems chosen to illustrate I-strain.

pared with carbocyclic rings (Sec. 7-3*b*). The sp^2 configuration, with its lesser number of C—H bonds, is therefore favored over sp^3.[44]

Large rings resemble open chains, and considerations of I-strain in these systems are not important.

A number of reactions showing the effects of I-strain are summarized in Table 9-2. Solvolysis of methylcycloalkyl chlorides (Fig. 9-24, *A*) and of cycloalkyl tosylates (Fig. 9-24, *B*) illustrates I-strain effects in S_N1 reactions. Reaction of cycloalkyl bromides (Fig. 9-24, *C*) with lithium iodide illustrates the effect in S_N2 reactions. The cyanohydrin equilibrium of cycloalkanones (Fig. 9-24, *D*) illustrates the effect in an equilibrium process involving carbonyl compounds, whereas the rate of sodium borohydride reduction of the corresponding ketones to alcohols illustrates the effect in a rate-controlled carbonyl reaction. Less complete (and, in some cases, less convincing) data in support of I-strain come from free-radical reactions,[41c–e] carbanion reactions,[41c, d] and the relative stability of exocyclic and endocyclic double bonds in ring systems.[41f]

† Since the molecule is puckered (Sec. 9-1), some eclipsing strain is converted into angle strain, and the total strain is less. The detailed geometry of the ring must be taken into account when one assesses the reactivity of substituted cyclopentyl halides.

‡ The decrease may be somewhat less, because of the puckering of the ring.

Table 9-2
Predictions and Observations on I-Strain[a]

Ring size	Prediction for $sp^3 \rightarrow sp^2$	*A*	*B*	*C*	Prediction for $sp^2 \rightarrow sp^3$	*D*	*E*
Acyclic		1.00	1.00	1.00		1.00	1.00
3	Difficult[b]		0.00002	0	Facile[b]		
4	Difficult[b]	0.97	8.50	0.0075	Facile[b]		581
5	Facile[c]	43.7	10.5	1.2	Difficult[c]	3.33	15.4
6	Difficult[c]	0.35	0.75	0.015	Facile[c]	70	355
7	Facile[c]	38.0	19.0	0.79	Difficult[c]	0.54	2.25
8	Facile[d]	100	144	0.14	Difficult[d]	0.081	0.172
9	Facile[d]	15.4	129	0.036	Difficult[d]	0.041	0.070
10	Facile[d]	6.22	286	0.050	Difficult[d]	small	0.0291
11	Facile[d]	4.21	30.8		Difficult[d]	0.063	0.0518
12	[e]		2.44	0.0080	[e]	0.226	0.401
13	[e]	1.00	2.63		[e]	0.269	0.427
14	[f]		0.99	0.020	[f]	1.17	
15	[f]	0.64	1.65	0.064	[f]	0.64	0.925
16	[f]				[f]	0.77	
17	[f]	0.67	1.63		[f]	0.58	1.31
18	[f]				[f]	0.70	
19	[f]				[f]	0.70	
20	[f]		1.35		[f]	1.00	

[a] From Refs. 41*c* and *d* and 42.
[b] Indicates that angle strain is incurred ("difficult") or relieved ("facile") in the transition indicated.
[c] Indicates that bond opposition strain is incurred or relieved.
[d] Indicates that non-classical strain is incurred or relieved.
[e] The prediction for 12- and 13-membered rings is uncertain because the change in strain is small.
[f] No appreciable change in strain is predicted for 14-membered and larger rings.
A: Relative rates of solvolysis of methyl cycloalkyl chlorides in 80% ethanol: Ref. 41*a* and H. C. Brown and M. Borkowski, *J. Am. Chem. Soc.,* **74,** 1894 (1952).
B: Relative rates of acetolysis of cycloalkyl tosylates: H. C. Brown and G. Ham, *J. Am. Chem. Soc.,* **78,** 2735 (1956); see also R. Heck and V. Prelog, *Helv. Chim. Acta,* **38,** 1541 (1955).
C: Relative rates of reaction of cycloalkyl bromides with lithium or potassium iodide: P. J. C. Fierens and P. Verschelden, *Bull. soc. chim. Belges,* **61,** 427,609 (1952); L. Schotsmans, P. J. C. Fierens, and T. Verlie, *ibid.,* **68,** 580 (1959).
D: Equilibrium constants (relative) for reactions of cycloalkanones with hydrogen cyanide to give cyanohydrins: V. Prelog and M. Kobelt, *Helv. Chim. Acta,* **32,** 1187 (1949).
E: Relative rates of reduction of cycloalkanones with sodium borohydride: H. C. Brown and K. Ichikawa, *Tetrahedron,* **1,** 221 (1957).
All rates are expressed relative to the acyclic analog taken as unity.

Inspection of Table 9-2 indicates that transformations which involve an increase in I-strain are generally slow while those involving decrease in I-strain are generally fast. This is true in particular in the region of the medium-sized (8- to 11-membered) rings. Some irregularities are observed. The solvolysis of cyclobutyl tosylate has already been mentioned. The S_N1

reactions of cyclohexyl compounds are generally faster and S_N2 reactions of cyclic compounds in general are slower than expected.† Cyclopentanone and cycloheptanone are more reactive than predicted. These irregularities probably indicate that effects other than I-strain may be important in certain cases. It is probably best to consider the I-strain approach as a first approximation, useful in predicting reactivity of cyclic compounds where more detailed information is not available. As higher approximations in terms of more detailed conformational considerations become available (e.g., Chap. 8), the predictions of reactivity will become more certain and more quantitative. In any case, it should be remembered that I-strain and conformational considerations are primarily steric in nature and that polar factors may also have to be considered.

General References

[43] V. Prelog, Bedeutung der vielgliedrigen Ringverbindungen, in A. R. Todd, ed., "Perspectives in Organic Chemistry," Interscience Publishers, Inc., New York, 1956, pp. 96–133.

[44] V. Prelog, Some Newer Developments of the Chemistry of the Medium-sized Ring Compounds, *Bull. soc. chim. France,* 1433 (1960).

[45] J. Sicher, M. Svoboda, J. Jonáš, J. Roček, and F. Mareš, Some Recent Studies in Macrocyclic Stereochemistry, *Bull. soc. chim. France,* 1438 (1960).

[46] J. Sicher, The Stereochemistry of Many-membered Rings, in P. B. D. de la Mare and W. Klyne, eds., "Progress in Stereochemistry," vol. 3, Academic Press, Inc., New York, 1962, chap. 6.

[47] F. H. Westheimer, Calculation of the Magnitude of Steric Effects, in M. S. Newman, ed., "Steric Effects in Organic Chemistry," John Wiley & Sons, Inc., New York, 1956, chap. 12.

† The abnormal slowness of S_N2 displacements in medium and even some large rings (Table 9-2, column *C*) has been ascribed to difficulties in achieving collinearity of the incoming and outgoing groups. Such collinearity is, of course, required for the optimum stereoelectronic situation in the transition state (Fig. 6-15). The difficulty seems to be most pronounced in the 12-membered ring. Although I-strain may play a part in S_N2 reactions, it is evidently not the major factor.

Chapter 10

FUSED RINGS AND BRIDGED RINGS

10-1. Scope

Among compounds containing two (or more) rings, four cases may be distinguished, depending on the number of atoms that are common to two adjacent rings. If the rings have no atoms in common, as in cyclohexylcyclohexane (Fig. 10-1), no new stereochemical features appear, and such systems need not be separately considered. Systems in which two rings have one atom in common are called "spiranes"; these systems present a special type of stereoisomerism and will be considered in Chap. 11. This chapter will be concerned with systems in which the adjacent rings have two or more atoms in common. When there are two atoms (necessarily adjacent) in common between two rings, one speaks of "fused systems," whereas in "bridged systems" the rings are linked through non-adjacent atoms and there are at least three atoms common to the two rings.

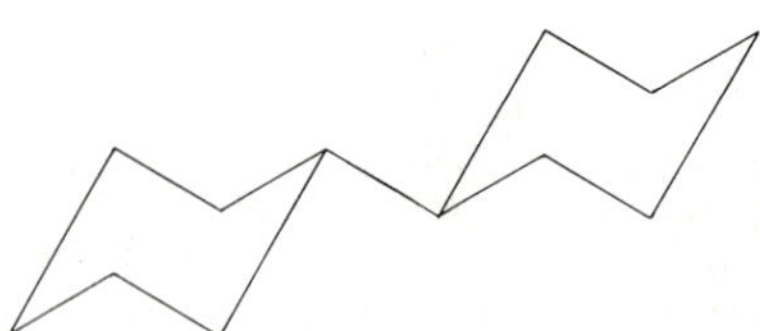

Fig. 10-1. Cyclohexylcyclohexane.

10-2. Fused Systems

a. Bicyclic Systems. The smallest imaginable carbocyclic fused-ring compound is bicyclo[1.1.0]butane (Fig. 10-2, X = H).† A derivative of this compound, ethyl bicyclo[1.1.0]butane-1-carboxylate (Fig. 10-2, X = CO_2Et),

Note: All references above 83 are listed in the General References at the end of the chapter.

† In the following discussion, fused bicyclic systems will sometimes be referred to in terms of the number of atoms contained in each of the rings considered separately. Thus the system shown in Fig. 10-2 would be called a "3-3 system."

has been synthesized by a transannular ring closure of ethyl 3-bromocyclobutane-1-carboxylate.[1]

The bicyclic system containing five atoms, bicyclo[2.1.0]pentane (Fig. 10-3), has been synthesized by pyrolysis of a bicyclic azo compound.[2] The system is surprisingly stable thermally, being rearranged to cyclopentene only upon heating to 330°. The corresponding heterocycle cyclobutene oxide does not appear so far to have been isolated, although rate studies including the oxidation of cyclobutene by peracetic acid are on record.[3] Systems containing three-membered rings fused to five-membered and larger rings are well known. In cyclopentene oxide and cyclohexene oxide—which exemplify

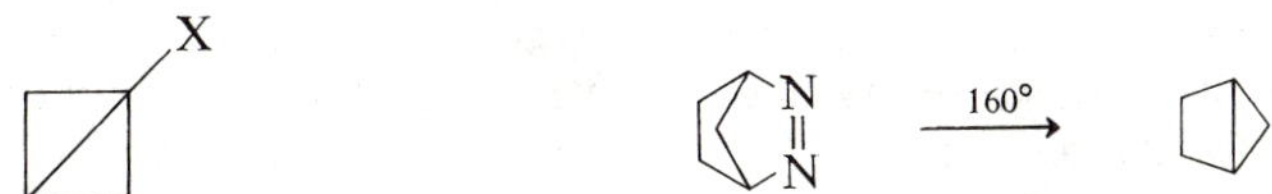

Fig. 10-2. [1.1.0] systems. **Fig. 10-3.** Synthesis of bicyclo[2.1.0]pentane.

such systems—the fusion of the rings is necessarily cis. The cis bonds on adjacent carbon atoms in cyclopentane make nearly a 0° dihedral angle with each other and therefore lend themselves well to the fusion of the planar three-membered ring. The favored conformation of cyclopentene oxide should be an envelope form (Fig. 9-2), with the oxide ring fused to atoms No. 3 and 4. In cyclohexane the situation is less favorable for epoxide formation, inasmuch as the normal disposition of the bonds is staggered rather than eclipsed. Accordingly, the rate of formation of cyclohexene oxide from the trans chlorohydrin and base is less than the corresponding rate of formation of cyclopentene oxide.[4] The six-membered ring in cyclohexene oxide has been shown, by X-ray diffraction data,[5] to be severely deformed from the chair form (Fig. 10-4); it resembles the half-chair of cyclohexene (Fig. 8-49). In-

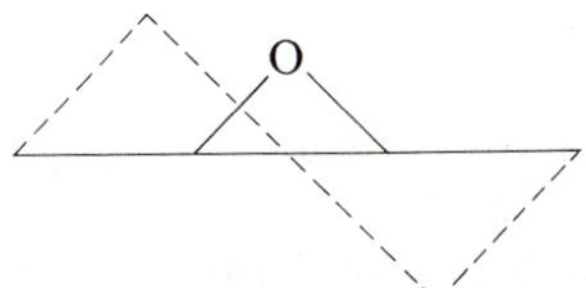

Fig. 10-4. Cyclohexene oxide.

spection of a model of cyclohexane clearly indicates that the cis (*e,a*) bonds may be readily brought into a plane by internal rotation to make the half-chair, whereas attempted rotation of the originally equidistant trans (*e,e*) bonds into a plane causes prohibitive distortion of the model.

[1] K. B. Wiberg and R. P. Ciula, *J. Am. Chem. Soc.*, **81,** 5261 (1959).

[2] R. Criegee and A. Rimmelin, *Chem. Ber.*, **90,** 414 (1957).

[3] J. Böeseken and J. Stuurman, *Rec. trav. chim.*, **56,** 1034 (1937).

[4] F. V. Brutcher and T. Roberts, Abstracts, Meeting of American Chemical Society, Cincinnati, Ohio, 1955, p. 39N.

[5] B. Ottar, *Acta Chem. Scand.*, **1,** 283 (1947).

In cycloheptane, unlike in cyclohexane, it becomes possible to deform the ring in such a way that adjacent trans bonds are nearly parallel. The formation of a cycloheptene trans oxide might therefore appear possible; however, the compound is not known. Only the cis oxide obtained by epoxidation of (cis) cycloheptene has been reported. In the cycloöctane series, where both cis and trans olefins are known (cf. Sec. 7-2*b*), cis and trans epoxides may be formed by oxidation of these olefins,[6] and the same is undoubtedly true in rings larger than eight-membered.

Before leaving the subject of fused three-membered rings, it may be pointed out that the carbocyclic 5-3 and 6-3 systems are found in a series of naturally occurring terpenes derived from thujane and carane (Fig. 10-5), respectively.†

Turning now to fused four-membered-ring systems, the first member of the series (Fig. 10-3) has already been mentioned. The next higher homolog, bicyclo[2.2.0]hexane (Fig. 10-6), is claimed to have been synthesized by the treatment of *cis*-1,4-dibromocyclohexane with sodium.[7a] More recently, this hydrocarbon has been identified[7b] among the photolysis products of bicyclo-[3.2.0]heptanone-3 (Fig. 10-6).

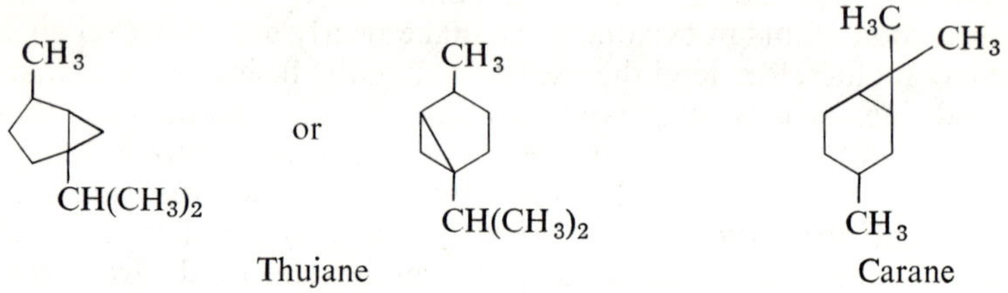

CH_3 / $CH(CH_3)_2$ or CH_3 / $CH(CH_3)_2$ — Thujane

H_3C, CH_3 / CH_3 — Carane

Fig. 10-5. Thujane and carane.

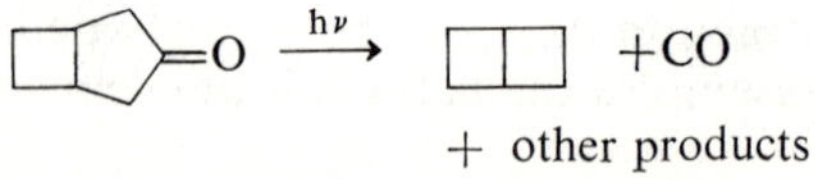

=O $\xrightarrow{h\nu}$ +CO

\+ other products

Fig. 10-6. Synthesis of bicyclo[2.2.0]hexane.

It is of interest, also, that attempts to obtain tetramethylcyclobutadiene by dehalogenation of a suitable vicinal dihalide gave instead a dimer whose properties suggest that it contains the bicyclo[2.2.0]hexane system (Fig. 10-7).[8] The next-higher homologs, bicyclo[3.2.0]heptane[9] (Fig. 10-8) and bicyclo-[4.2.0]octane[10] (Fig. 10-9, *a*), are both known; one method of preparation in-

† For a facile synthesis of the 5-3 system (bicyclo[3.1.0]hexane), see S. Winstein, J. Sonnenberg, and L. de Vries, *J. Am. Chem. Soc.*, **81**, 6523 (1959).

[6] A. C. Cope et al., *J. Am. Chem. Soc.*, **79**, 3900, 3905 (1957).

[7] (*a*) N. D. Zelinskii and K. A. Kozeshkov, *Ber.*, **60B**, 1102 (1927); cf. N. A. Domnin, *J. Gen. Chem., U.S.S.R.*, **26**, 2200 (1956) (English translation, 2459). (*b*) S. Cremer and R. Srinivasan, *Tetrahedron Letters*, no. 21, 24 (1960).

[8] R. Criegee and G. Louis, *Chem. Ber.*, **90**, 417 (1957). See also M. Avram, G. Mateescu, I. G. Dinulescu, E. Marica, and C. D. Nenitzescu, *Tetrahedron Letters*, 21 (1961).

[9] (*a*) A. T. Blomquist and J. Kwiatek, *J. Am. Chem. Soc.*, **73**, 2098 (1951); (*b*) O. L. Chapman and D. J. Pasto, *Chemistry & Industry (London)*, 53 (1961).

[10] W. Reppe, O. Schlichting, K. Klager, and T. Toepel, *Ann.*, **560**, 1 (1948).

volves the 1,2 addition of ketene to an appropriate cyclic conjugated diene (Fig. 10-8; starting with 1,3-cyclohexadiene instead of cyclopentadiene, one obtains hydrocarbon (*a*) shown in Fig. 10-9).[9a] An interesting photochemical synthesis of bicyclo[3.2.0]heptane from cycloheptadiene-1,3 (Fig. 10-8) has also been reported.[9b]

Bicyclo[5.2.0]nonane (Fig. 10-9, *b*) has been obtained in both cis and trans forms,[10a] indicating that the seven-membered ring is the smallest cycle that may span a four-membered ring in trans fashion. In fact, the physical properties of the two hydrocarbons in the 7-4 system suggest that they are of about

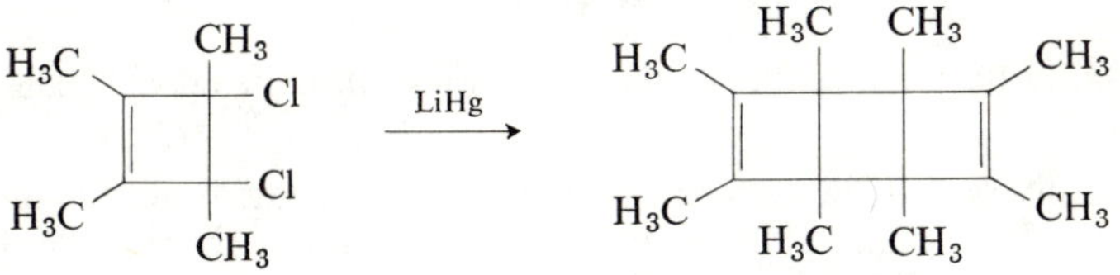

H_3C CH_3 Cl H_3C Cl CH_3 $\xrightarrow{\text{LiHg}}$ H_3C CH_3 H_3C CH_3 H_3C CH_3 H_3C CH_3

Fig. 10-7. Reaction of 1,2,3,4-tetramethyl-3,4-dichlorocyclobutene with lithium amalgam.

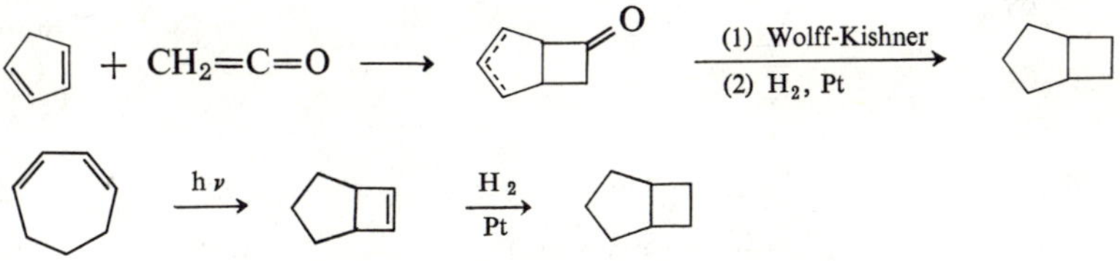

+ $CH_2{=}C{=}O$ ⟶ O $\xrightarrow[\text{(2) } H_2\text{, Pt}]{\text{(1) Wolff-Kishner}}$

$\xrightarrow{h\nu}$ $\xrightarrow[\text{Pt}]{H_2}$

Fig. 10-8. Synthesis of bicyclo[3.2.0]heptane.

cis only (*a*) cis or trans (*b*)

Fig. 10-9. Bicyclo[4.2.0]octane (*a*) and bicyclo[5.2.0]nonane (*b*).

equal heat content if the conformational rule (Sec. 8-4) is applicable. The 4-9 system occurs in the naturally occurring sesquiterpene caryophyllene (Fig. 10-10).

Among fused five-membered-ring systems, we have already encountered the 5-3 and 5-4 combinations. A five-membered ring fused to a five-membered ring is found in bicyclo[3.3.0]octane (Fig. 10-11) which is known in both cis and trans forms.[11] The trans isomer is the harder to make and has the higher heat of combustion by about 6 kcal./mole, indicating the presence

[10a] N. L. Allinger, M. Nakazaki, and V. Zalkow, *J. Am. Chem. Soc.*, **81**, 4074 (1959).
[11] J. W. Barrett and R. P. Linstead, *J. Chem. Soc.*, 611 (1936).

of a substantial amount of strain. This is what one would expect on the basis of consideration of models. It is so much the more surprising, therefore, that two of the heterocyclic analogs of bicyclo[3.3.0]octane, shown in Fig. 10-12, can be formed with equal ease in both the cis and trans fused forms.[12]

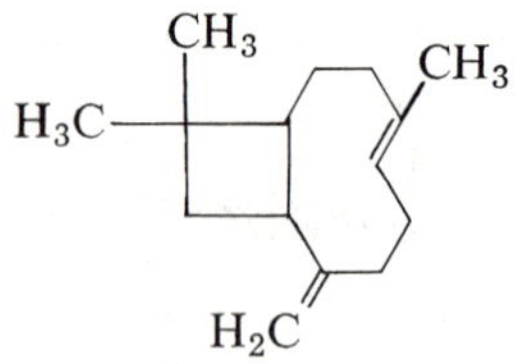

CH_3 CH_3 H_3C H_2C

Fig. 10-10. Caryophyllene. (Note the trans-substituted double bond.)

A naturally occurring material with a fused 5-5 ring system is the vitamin biotin (Fig. 10-13). Since the biotin structure contains three asymmetric carbon atoms (starred), there are eight stereoisomers (or four *dl* pairs) possible, *provided* that the ring fusion may be either cis or trans. Biotin is one of the

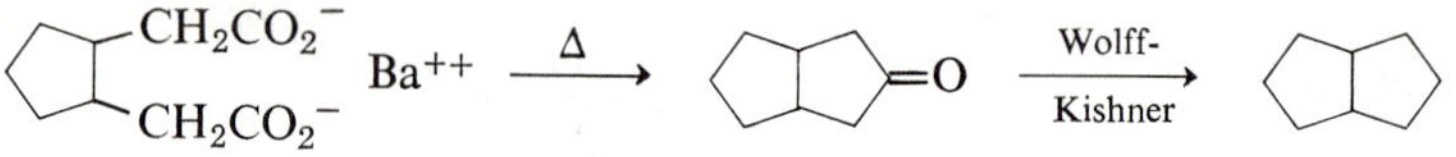

$CH_2CO_2^-$ Ba^{++} $CH_2CO_2^-$ $\xrightarrow{\Delta}$ =O $\xrightarrow[\text{Kishner}]{\text{Wolff-}}$

				Heat of combustion (vapor) kcal/mole
cis	$\xrightarrow{(280°)}$	(72% yield) →	cis	1208.3
trans	$\xrightarrow{(340°)}$	(50% yield) →	trans	1214.3

Fig. 10-11. Bicyclo[3.3.0]octane.

four stereoisomers in which the ring fusion is cis, but synthetic biotin isomers (allobiotin and epiallobiotin) have been obtained in which the ring fusion is trans,[13] and all the possible *dl* pairs are now known.

The 6-5 fused-ring system occurs in the well-known compound hydrindane.

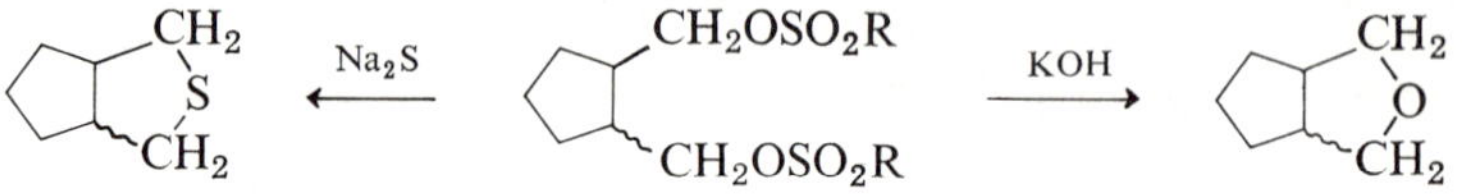

CH_2 S CH_2 $\xleftarrow{Na_2S}$ CH_2OSO_2R CH_2OSO_2R $\xrightarrow{KOH}$ CH_2 O CH_2

cis	←	cis	→	cis
trans	←	trans	→	trans

Fig. 10-12. Synthesis of heterocyclic analogs of bicyclo[3.3.0]octane.

This compound, like its lower homologs, exists in cis and trans forms (Fig. 10-14) but in this case the trans isomer has a slightly smaller heat of com-

[12] L. N. Owen and A. G. Peto, *J. Chem. Soc.*, 2383 (1955).

[13] S. A. Harris, R. Mozingo, D. E. Wolf, A. N. Wilson, and K. Folkers, *J. Am. Chem. Soc.*, **67**, 2102 (1945).

O
‖
C
HN NH
HC* *CH
H_2C $^*CH(CH_2)_4CO_2H$
S

Fig. 10-13. Biotin.

bustion (by 1.04 ± 0.52 kcal./mole in the vapor state).[14a] The configuration of the two hydrindanes was determined by two of the methods mentioned in Sec. 7-1*b;* the salient data[15, 16] are summarized in Fig. 10-15.

trans (*dl*) cis (meso)

Fig. 10-14. Hydrindane.

The optically active (and therefore trans) form of cyclohexane-1,2-diacetic acid, upon Dieckmann cyclization of the ester, is converted to 2-hydrindanone which, being optically active, must be the trans form. Reduction of the *dl* form of this ketone gives a single alcohol, as expected. On the other hand, the meso (cis) form of the diacid gives *meso*-2-hydrindanone which, upon reduction, gives two diastereoisomeric meso-cis alcohols. This hydrindanone

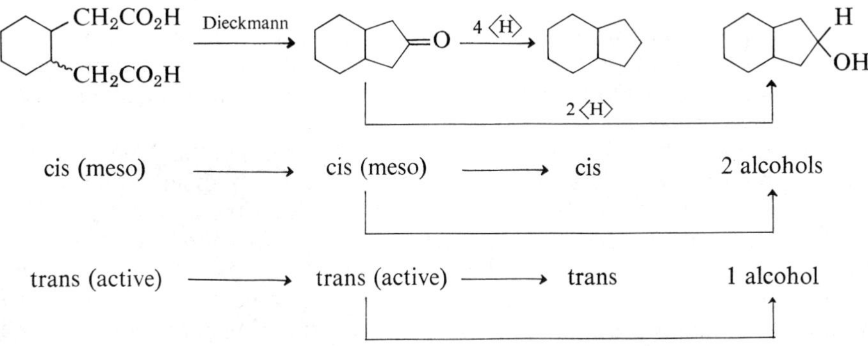

Fig. 10-15. Configurational determination of hydrindanes.

[14] (*a*) C. C. Browne and F. D. Rossini, *J. Phys. Chem.*, **64**, 927 (1960); (*b*) N. L. Allinger and J. L. Coke, *J. Am. Chem. Soc.*, **82**, 2553 (1960).

[15] W. Hückel and H. Friedrich, *Ann.*, **451**, 132 (1926).

[16] W. Hückel, M. Sachs, J. Yantschulewitsch, and F. Nerdel, *Ann.*, **518**, 155 (1935). The configurational assignment of the 1-hydrindanones is not, however, entirely secure.

is therefore the cis isomer. The configuration of the hydrindanes follows from their correlation with the 2-hydrindanones.

The relative thermodynamic (as distinct from thermochemical) stability of the hydrindanes has been found[14b] to be a function of temperature. Below 466°K. the trans isomer is more stable, but above that temperature the cis form predominates at equilibrium. This is because *cis*-hydrindane has a greater entropy (by 2.3 cal./deg. mole) than the trans form and this, at higher temperatures, more than compensates for the lesser enthalpy of the cis isomer. In substituted hydrindanes, such as 1-hydrindanone,[16] and analogous heterocycles, such as hexahydrophthalic anhydride[17] (Fig. 10-16), the cis isomer is the more stable. A priori, one might have expected the trans isomers to be more stable in all these cases, because in them the six-membered ring is fused via the more stable equatorial bonds, whereas in the cis isomers the fusion involves one equatorial and one axial bond. Moreover, as emphasized before, in an undeformed cyclohexane chair the *e,e* trans and *e,a* cis 1,2 bonds are equidistant and the ease of fusing a second ring across these bonds should be the same.

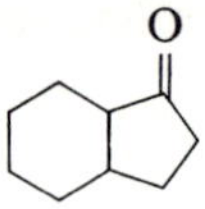

1-Hydrindanone

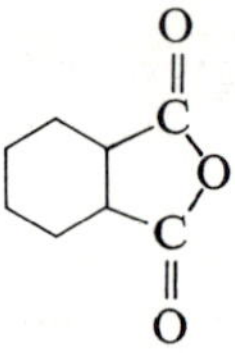

Hexahydrophthalic anhydride

Fig. 10-16. 1-Hydrindanone and hexahydrophthalic anhydride.

The fact, then, that the *trans*-hydrindane system is either only slightly more stable or actually less stable than the *cis*-hydrindane system must mean that there is some distortion in these systems which favors the cis isomer. Two suggestions have been made, namely, that in *cis*-hydrindane the six-membered ring is a boat, with the nearly planar five-membered ring strainlessly fused to the eclipsed bonds at the side of the boat, or that the six-membered ring is a deformed chair.[18] It may be shown with models that the cyclohexane chair in *cis*-hydrindane may be easily distorted in such a way as to bring the bonds linking the five-membered ring closer together. This distortion involves an unpuckering of the chair without producing any violent interactions. An attempt to bring the bonds linking the five-membered ring in *trans*-hydrindane closer together, however, increases the puckering of the chair and is energetically quite unfavorable.[19]

[17] A. von Baeyer, *Ann.*, **258**, 169, 217 (1890).

[18] Cf. E. L. Eliel and C. Pillar, *J. Am. Chem. Soc.*, **77**, 3600 (1955).

[19] See also (*a*) S. J. Angyal and C. G. MacDonald, *J. Chem. Soc.*, 686 (1952); (*b*) L. P. Kuhn, *J. Am. Chem. Soc.*, **74**, 2492 (1952).

A decision between the two possibilities has been made[18] in the case of the 2-oxahydrindane-5,6-diols (Fig. 10-17) through observation of their rates of oxidation with lead tetraacetate and determination of the strength of the intramolecular hydrogen bonds between the hydroxyl groups by infrared measurement (cf. Sec. 6-2*b* and Table 9-1). The prototypes for the chair and boat forms are taken to be the cyclohexane-1,2-diols and cyclopentane-1,2-diols, respectively. Cyclohexane-1,2-diol is chair-shaped, and the bonds linking the hydroxyl groups to the ring make a dihedral angle of about 60° with each other in both the cis and trans isomers (i.e., they are staggered). Accordingly, the two isomers form intramolecular hydrogen bonds of about the same strength,[19b] the shift $\Delta\tilde{\nu}$ between the unbonded and bonded O—H stretching frequency being 39 cm.$^{-1}$ in the cis isomer and 32 cm.$^{-1}$ in the

	k_{cis}/k_{trans}[b]	$\Delta\tilde{\nu}$[c]
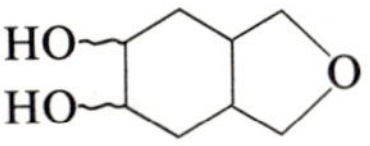 *cis*-2-Oxahydrindane-5,6-diol[a]	21	30–34
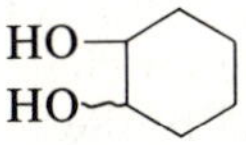 Cyclohexane-1,2-diol	22	32–39
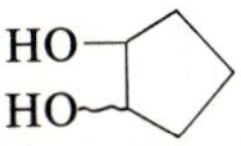 Cyclopentane-1,2-diol	>3000	61[d]

[a] Cis ring junction. Only one of the two possible cis diols was investigated.
[b] Rates of cleavage with lead tetraacetate. The subscripts cis and trans refer to the relative position of the hydroxyl groups.
[c] Infrared shift; see text.
[d] No bonded hydroxyl in trans isomer.

Fig. 10-17. Geometry of 2-oxahydrindane-5,6-diol.

trans.† Also the two isomers react with lead tetraacetate[20] at rates which differ by only a factor of 22.† In cyclopentane-1,2-diol, on the other hand, the dihedral angle is nearly (though not quite; cf. Chap. 9) 0° in the cis isomer and 120° in the trans isomer. Accordingly the cis isomer shows strong intramolecular hydrogen bonding ($\Delta\tilde{\nu} = 61$ cm.$^{-1}$)[19b] whereas the trans isomer shows none, and the difference in rate of tetraacetate cleavage between the two isomers is by a factor of over 3000.[20] The situation in a boat-shaped

† The reason for the slight difference lies in the deformation of the cyclohexane ring from the chair form discussed above: The cis OH groups can approach each other somewhat more readily than the trans OH groups and therefore show stronger hydrogen bonding and a higher rate of glycol cleavage.

[20] R. Criegee, E. Büchner, and W. Walther, *Ber.*, **73**, 571 (1940); R. Criegee, L. Kraft, and B. Rank, *Ann.*, **507**, 159 (1933).

2-oxahydrindane-5,6-diol should be similar to that in cyclopentane-1,2-diol, inasmuch as the dihedral angles at the side of the boat are 0° for the cis isomer and 120° for the trans. In fact, however, the situation with respect to both hydrogen bonding and tetraacetate cleavage in the oxahydrindane derivative resembles that in cyclohexane and differs greatly from that in cyclopentane (cf. Fig. 10-17). It may be concluded that the six-membered ring in the oxahydrindane is a chair (presumably slightly deformed) rather than a boat. This conclusion has recently been supported by NMR studies[20a] of *cis*-hydrindane which suggest that the molecule may be "frozen" in one of the two interconverting chair forms at low temperatures. Approximate measurement of the barrier to chair interconversion gives a value of about 6 kcal./

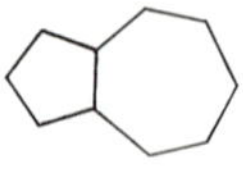

Fig. 10-18. Bicyclo[5.3.0]decane.

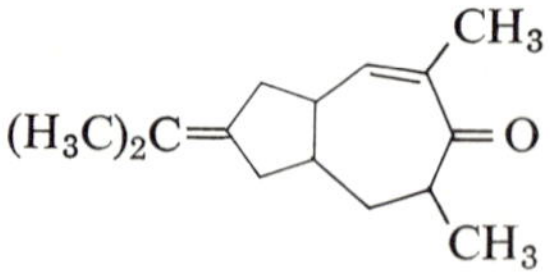

CH_3 $(H_3C)_2C=$ $=O$ CH_3

Fig. 10-19. Vetivone.

mole, appreciably lower than for cyclohexane (Chap. 8); this is reasonable on the basis of models which suggest that the chair distortion in *cis*-hydrindane makes the molecule approach the transition state for chair inversion.

A rather more complex situation with respect to the relative stabilities of the cis and trans isomers arises when there is an angular methyl group in the hydrindane system or when it forms part of a larger system of fused rings.[20b]

The last fused five-membered-ring system to be discussed here is the 5-7 system (perhydroazulene) shown in Fig. 10-18. This system occurs in a number of sesquiterpenes, such as vetivone (Fig. 10-19). The difference in free energy between the cis and trans isomers in perhydroazulene is extremely small. The trans isomer is very slightly more stable than the cis, being favored by 0.3 kcal./mole in enthalpy and 0.3 cal./deg. mole in entropy.[21]

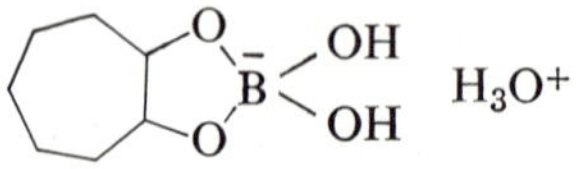

O O B OH OH H_3O^+

Fig. 10-20. Boric acid complex of cycloheptane-1,2-diol.

Earlier evidence for the slight difference in stability between the cis and trans fused 5-7 systems had come from the observation that both *cis*- and *trans*-1,2-cycloheptanediol form cyclic ketals with acetone[22]—in contrast to the situation with the 1,2-cyclohexanediols where only the cis isomer forms an acetone derivative. Also both *cis*- and *trans*-1,2-cycloheptanediol enhance

[20a] W. B. Moniz and J. A. Dixon, *J. Am. Chem. Soc.*, **83**, 1671 (1961).

[20b] N. L. Allinger, R. B. Hermann, and C. Djerassi, *J. Org. Chem.*, **25**, 922 (1960); J. F. Biellmann, D. Francetić, and G. Ourisson, *Tetrahedron Letters*, no. 18, 4 and no. 23, 39 (1960). See also ref. 84, pp. 211–216.

[21] N. L. Allinger and V. B. Zalkow, *J. Am. Chem. Soc.*, **83**, 1144 (1961).

[22] J. Böeseken and H. G. Derx, *Rec. trav. chim.*, **40**, 529 (1921).

the conductivity of boric acid[23] (cf. Fig. 6-19) and must thus be able to form complexes of the type shown in Fig. 10-20; actually the cis diol has the somewhat greater effect, suggesting that it forms the stronger borate complex.

Among the fused six-membered-ring systems of which the 6-3, 6-4, and 6-5 systems have already been discussed, by far the most important compound is the 6-6 compound decalin. It was probably the demonstration by W. Hückel[24] that decalin may exist in cis and trans forms, of which the trans is the more stable, which first convinced chemists that cyclohexane rings in decalin and elsewhere are puckered, as postulated by Sachse and Mohr (cf. Sec. 8-1), and not planar, as postulated by Baeyer.† The counting of butane-gauche interactions in the decalins (cf. Sec. 8-3) reveals three such interactions in the cis isomer‡ (cf. Fig. 10-21) and none in the trans.§ Therefore the calculated difference in heats of formation between the decalins is 3×0.9, or 2.7, kcal./mole, the trans isomer being the more stable.[25] The experimental value, obtained either from temperature dependence of the *cis*-decalin $\rightleftharpoons$ *trans*-decalin equilibrium (established at elevated temperature over a platinum catalyst)[26a] or directly from the difference in heats of combustion,[26b,c] is 2.7 kcal./mole for the liquid phase[26a] or 3.1 kcal./mole for the vapor,[26b] in good agreement with the calculated value.

trans-Decalin (Fig. 10-21) has a center of symmetry (midway between C_9 and C_{10}) and is therefore an optically inactive molecule. It also has a two-fold axis of symmetry (passing horizontally between C_2 and C_3, C_9 and C_{10}, and C_7 and C_6); thus its symmetry number is 2. *cis*-Decalin (Fig. 10-21) is

† Actually, J. Böeseken [*Rec. trav. chim.*, **40**, 553 (1921)] had somewhat earlier concluded that *cis*-cyclohexane-1,2-diol could not be either planar or boat-shaped, since it does not enhance the conductivity of boric acid and forms a cyclic isopropylidene ketal with less readiness than other 1,2-diols in which the hydroxyl groups may become coplanar. He reasoned that the cyclohexane ring was therefore chair-shaped.

‡ In the earlier literature, *cis*-decalin is often depicted as a double boat. However, electron-diffraction studies [O. Bastiansen and O. Hassel, *Nature*, **157**, 765 (1946)] indicate that it exists in the double-chair form (Fig. 10-21). In view of the unfavorable interactions in the boat (Sec. 8-1), this is what one would expect.

§ The reader should convince himself of the correctness of this statement by contemplating appropriate ball-and-stick models. By analogy with 1,2-dimethylcyclohexane (Sec. 8-3), one might have expected two butane-gauche interactions in the trans isomer, considering each ring as a 1,2-disubstituted cyclohexane. However, these interactions (5-10-9-8 and 4-10-9-1 in Fig. 10-21) would then be *all within the other ring*, and such interactions, which also occur in cyclohexane itself, are not counted. By a similar argument, the six butane-gauche interactions in the cis isomer are reduced to four. Furthermore, in counting these four interactions in *cis*-decalin, it will be noted that one (5-10-9-1 in Fig. 10-21) has been counted twice over, so that there are actually only three, viz., 5-10-4-3, 5-10-9-1, and 1-9-8-7 in Fig. 10-21.

---

[23] H. G. Derx, *Rec. trav. chim.*, **41**, 329 (1922).

[24] W. Hückel, R. Mentzel, W. Brinkman, and E. Goth, *Ann.*, **441**, 1 (1925); see also Ref. 15.

[25] R. B. Turner, *J. Am. Chem. Soc.*, **74**, 2118 (1952). Turner uses 0.8 kcal./mole for the butane-gauche interaction and refers to an older value for the difference in heats of combustion of the decalins.

[26] (*a*) N. L. Allinger and J. L. Coke, *J. Am. Chem. Soc.*, **81**, 4080 (1959); (*b*) D. M. Speros and F. D. Rossini, *J. Phys. Chem.*, **64**, 1723 (1960). (*c*) See also W. A. Roth and R. Lassé, *Ann.*, **441**, 48 (1925). (*d*) See also J. P. McCullough, H. L. Finke, J. F. Messerly, S. S. Todd, T. C. Kincheloe, and G. Waddington, *J. Phys. Chem.*, **61**, 1105 (1957) and T. Miyazawa and K. S. Pitzer, *J. Am. Chem. Soc.*, **80**, 60 (1958).

dissymmetric. However, unlike *trans*-decalin—which is rigid because the ring fusion can only be through *e,e,* not through *a,a,* bonds—the cis isomer has two interconvertible chair conformations, similar to *cis*-1,2-dimethylcyclohexane (cf. Table 8-2). Also, just as in the case of the latter, the "flipping" or chair inversion converts the molecule to its mirror image. *cis*-Decalin is therefore a non-resolvable *dl* pair. *cis*-Decalin also has a twofold axis of symmetry, passing through the $C_9—C_{10}$ bond at right angles to the bond in a plane bisecting the dihedral angle between the 9- and 10-hydrogens. Its symmetry number is thus also 2.

Since both *cis*- and *trans*-decalin have a symmetry number of 2, their difference in entropy should result only from the fact that the cis isomer is a *dl* pair (even though non-resolvable) and should favor the cis isomer by $R \ln 2$ or 1.38 cal./deg. mole (cf. Table 8-3). The experimental value[26a] is 0.55 cal./deg. mole, i.e., less than the theoretical, suggesting that at least in the liquid phase there is slightly more ordering of the molecules of the cis isomer than of the trans.[26d]

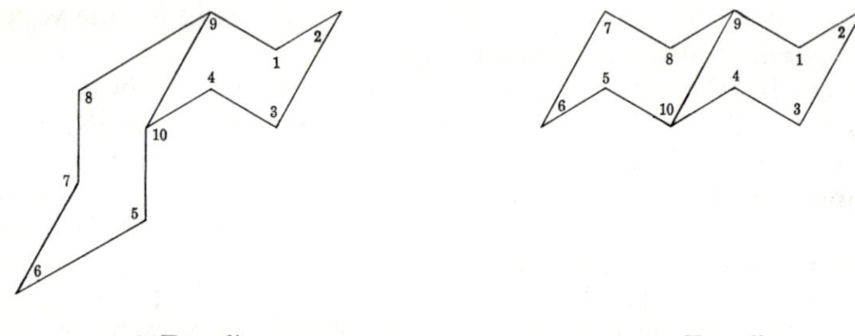

cis-Decalin *trans*-Decalin

Fig. 10-21. *cis*-Decalin and *trans*-decalin.

From the enthalpy and entropy differences between *cis*- and *trans*-decalin their free-energy difference at 250° C. may be calculated to amount to about 2.4 kcal./mole. This is in semiquantitative agreement with the observation[27] that the equilibrium between *cis*- and *trans*-1-decalone favors the trans isomer by between 10:1 and 20:1.†

That *trans*-decalin is a rigid molecule whereas the cis isomer has two interconvertible conformations is dramatically demonstrated by the nuclear magnetic resonance spectra of these compounds.[27a] *cis*-Decalin shows a single narrow resonance line, since, because of the rapid flipping of the molecule from one chair form into the other, all methylene hydrogens become identical and are seen in an average position as far as the NMR spectrum is concerned (cf. Sec. 6-2*d*). *trans*-Decalin, on the other hand, shows a broad and partially

† Strictly quantitative comparisons are not possible, not only because the exact position of the 1-decalone equilibrium is not known, but also because there are complications due to the 2- and 3-ketone effects; cf. Sec. 8-7.

[27] H. E. Zimmerman and A. Mais, *J. Am. Chem. Soc.*, **81**, 3644 (1959).
[27a] J. Musher and R. E. Richards, *Proc. Chem. Soc.*, 230 (1958).

resolved band caused by the non-equivalent equatorial and axial hydrogens and complicated by the spin-spin coupling between them.

The difference in potential energy between the *cis*- and *trans*-9-methyldecalin system (which occurs in the A/B rings of the steroids) is considerably less than that between *cis*- and *trans*-decalin.[25] The angular methyl group in *trans*-9-methyldecalin is axial and gives rise to four butane-gauche interactions (two with respect to either ring; cf. methylcyclohexane, Sec. 8-2). In *cis*-9-methyldecalin, however, the methyl group is axial with respect to one ring only and equatorial with respect to the other, giving rise to two butane-gauche interactions in addition to the three already present in the *cis*-decalin system. The total number of butane-gauche interactions in the cis compound is therefore five, i.e., only one more than in the trans isomer, and the difference in enthalpy should thus be 0.9 kcal./mole. With such a small ΔH, differences in entropy (ΔS) between the isomers may cause ΔF to be either positive or negative.[28a] Accordingly, in some 9-methyldecalin systems the cis isomer is found to be more stable, thermodynamically, whereas in others

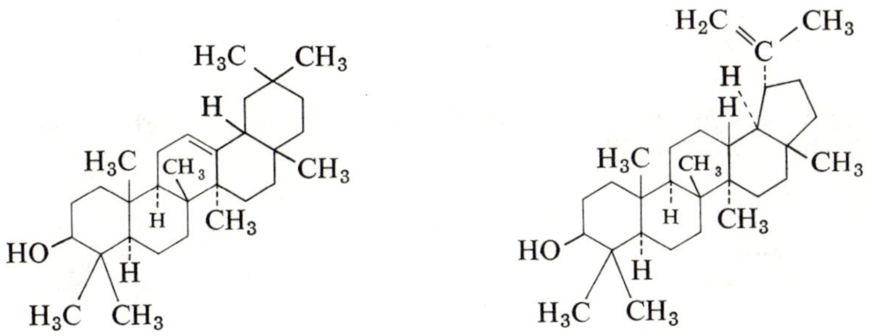

β-Amyrin Lupeol

Fig. 10-22. The triterpenes β-amyrin and lupeol.

the trans form predominates at equilibrium.[28] For the parent compound, 9-methyldecalin (liquid state), the trans isomer has the lower enthalpy, ΔH being 0.55 ± 0.28 kcal./mole from temperature dependence of equilibrium[28f] or 1.39 ± 0.64 kcal./mole from the difference in heats of combustion.[28g] Here, as in the case of the decalins, the entropy favors the cis isomer slightly, although again by less than would be expected on the basis that the cis isomer is a (non-resolvable) *dl* pair.

b. Polycyclic Systems. Although the 6-5 system occurs in the C/D ring of the steroid skeleton[29] and in a few other natural products, such as lupeol (Fig. 10-22), by far the most common polycyclic systems are made up of fused

[28] (*a*) N. L. Allinger, *J. Org. Chem.*, **21**, 915 (1956); (*b*) A. Ross, P. A. S. Smith, and A. S. Dreiding, *ibid.*, **20**, 905 (1955); (*c*) D. Arigoni, J. Kalvoda, H. Heusser, O. Jeger, and L. Ruzicka, *Helv. Chim. Acta*, **38**, 1857 (1955); (*d*) D. N. Jones, J. R. Lewis, C. W. Shoppee, and G. H. R. Summers, *J. Chem. Soc.*, 2876 (1955); (*e*) F. Sondheimer and D. Rosenthal, *J. Am. Chem. Soc.*, **80**, 3995 (1958); (*f*) N. L. Allinger and J. L. Coke, *J. Org. Chem.*, **26**, 2096 (1961); (*g*) W. G. Dauben, O. Rohr, A. Labbauf, and F. D. Rossini, *J. Phys. Chem.*, **64**, 283 (1960).

[29] See Ref. 84, pp. 212–216.

six-membered rings, and this section will be devoted exclusively to such systems. Examples are the steroids,[84] such as cholestanol (Fig. 8-17); numerous alkaloids,[90, 91] e.g., yohimbine (Fig. 8-22); and many di- and triterpenes,[86, 88, 89] exemplified by dehydroabietic acid (Fig. 8-18) and β-amyrin (Fig. 10-22) and lupeol (Fig. 10-22), respectively.

The stereochemistry of even a relatively simple polycyclic system such as perhydrophenanthrene (Fig. 10-23) has not been easy to unravel. Perhydrophenanthrene is a molecule of the ABBA type (cf. Fig. 3-18) and may exist as four *dl* pairs and two meso forms. The nomenclature of these is indicated in Fig. 10-23. The prefixes cis and trans refer to the stereochemistry of the fusion of the individual terminal rings to the center rings, whereas syn and anti are used to denote the orientation of the terminal rings with respect to each other. A heavy dot indicates a hydrogen in front of the plane of the paper whereas the absence of a dot denotes a hydrogen behind the plane of the paper. The stereochemistry of a number of perhydrophenanthrene

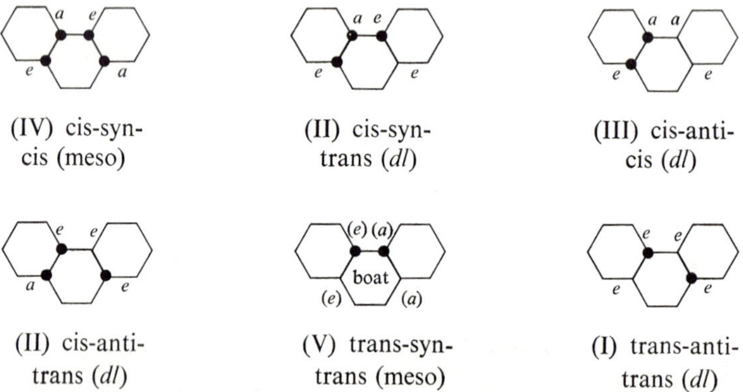

Fig. 10-23. The ten stereoisomeric perhydrophenanthrenes. (In the case of *dl* pairs, only one enantiomer is shown. Roman numerals indicate order of stability.)

derivatives, including several of the ketones and all the perhydrodiphenic acids (Fig. 10-25), has been elucidated by the elegant work of Linstead and coworkers[30] whose arguments will be given here in detail.†

The correlation of several perhydrophenanthrones with the perhydrodiphenic acids is schematized in Fig. 10-24 and requires no comment beyond the observation that epimerization during oxidation must be guarded against if the assumption of equal configuration of a ketone and the corresponding diacid obtained from it by oxidation is to be valid. The configurational assignment of the perhydrodiphenic acids themselves is summarized in Fig. 10-25.

† For the elucidation of the stereochemistry of even more complicated steroid, terpenoid, and alkaloid systems, the reader is referred to the monographs in these areas.[84–91, 93]

[30] R. P. Linstead, W. E. Doering, S. B. Davis, P. Levine, and R. R. Whetstone, *J. Am. Chem. Soc.*, **64**, 1985, 1991, 2003, 2006, 2009, 2014 (1942); *J. Chem. Soc.*, 1425, 1428 (1950).

Catalytic hydrogenation of diphenic acid gives, in addition to a hexahydrodiphenic acid, three of the six possible perhydrodiphenic acids, a major product melting at 289° and minor products melting at 198° and 200°. A further stereoisomer is obtained by epimerizing the diester of the 289° acid

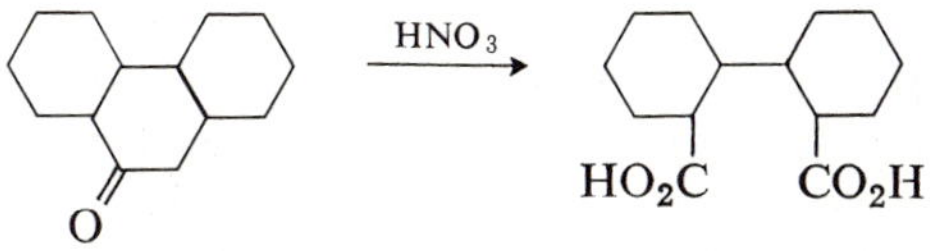

HNO_3

HO_2C CO_2H

O

Fig. 10-24. Correlation of perhydrophenanthrones with perhydrodiphenic acids.

with base, followed by saponification. Similar treatment of the 198° acid gives a fifth isomer melting at 247°. The sixth isomer, melting at 206°, is obtained by epimerization of the half-ester of the 198° acid, followed by saponification.

O_3

HO_2C CO_2H 8%

HO_2C CO_2H *cis*-Hexahydrophthalic acid

HO_2C CO_2H H_2 Pt

$NaOCH_3$ on half-ester

HO_2C CO_2H (meso) m.p. 289° (~50%)

HO_2C CO_2H m.p. 198° (2%)

HO_2C CO_2H m.p. 200° (2%)

$NaOCH_3$ on diester

$NaOCH_3$ on half-ester

$NaOCH_3$ on diester

HO_2C CO_2H m.p. 223° (meso)

HO_2C CO_2H m.p. 206°

HO_2C CO_2H m.p. 247°

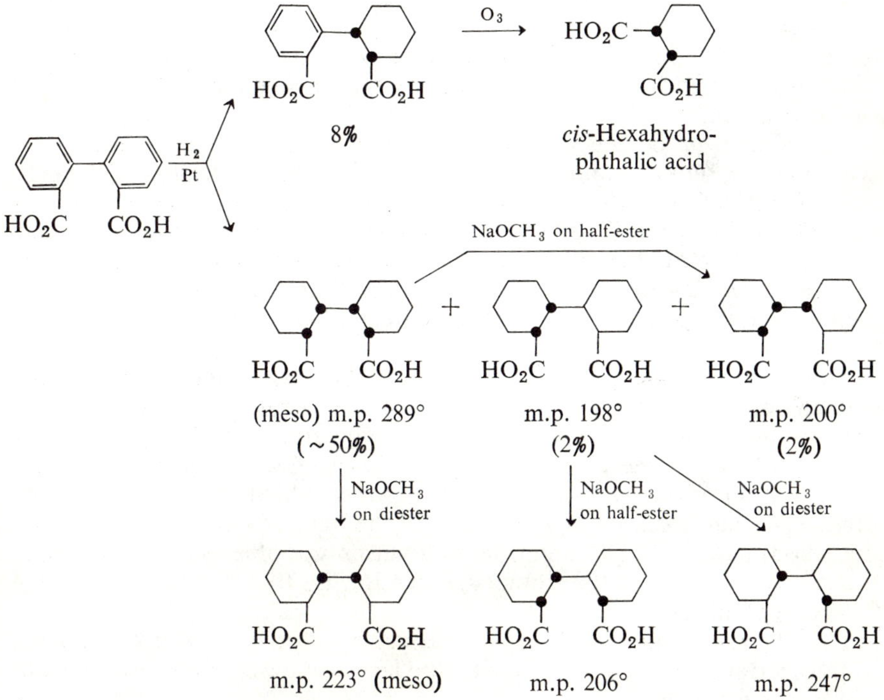

Fig. 10-25. Synthesis and configurational assignment of the perhydrodiphenic acids.

Since epimerization can affect only the asymmetric carbon atoms adjacent to an ester group and not the "backbone" atoms involved in the linkage of the rings, it follows that the 198°, 206°, and 247° acids all have one and the same backbone configuration and that the 289°, 223°, and 200° acids have the opposite backbone configuration. Now it follows from stereochemical considerations that the syn-backbone series is the one that has the two meso forms (possessing planes of symmetry), whereas the anti-backbone series has no meso forms. It was shown experimentally that the 289-223-200° series is the one that contains the meso forms and therefore has the syn backbone, as follows:

1. The 198° and 247° acids were both resolved; therefore the 198-206-247° series cannot contain two meso forms.

2. The monomethyl ester of the 289° acid was resolved but, upon treatment with diazomethane, gave an inactive diester. Hence the diester and its precursor, the 289° acid, are meso forms (cf. page 184).

Since ozonization of the hexahydrodiphenic acid shown in Fig. 10-25 gave *cis*-hexahydrophthalic acid, the phenyl substituent and carboxyl group in the hexahydrodiphenic acid must be cis. Further hydrogenation of this acid gave exclusively the 289° acid, and therefore this acid also has the carboxyl groups and adjacent ring juncture cis. It has already been shown that the 289° acid is syn and meso, and hence it must be the cis-syn-cis isomer, as shown in Fig. 10-25. The configuration of the 200° acid as the cis-syn-trans isomer follows from its method of formation by epimerization of the half-ester of the 289° acid.† The 223° acid must, therefore, be the trans-syn-trans isomer, which is in agreement with the observation that (being a meso form) it could not be resolved.

Assignments in the anti series do not rest on as firm a basis as those in the syn series. It should be noted that, in the syn series, conversion of the half-ester of the 289° acid to the 200° acid and of the diester of the 289° acid to the 223° acid involves, in all cases, a change from a cis acid to a trans acid. By analogy, it may then be tentatively assumed that the (most stable) 247° acid in the anti series is the trans-anti-trans; its (least stable) precursor, melting at 198°, the cis-anti-cis; and the intermediate isomer, melting at 206°, the cis-anti-trans.

Although the configurational assignment of these compounds was originally worked out with a view to elucidating the stereochemistry of catalytic hydrogenation (cf. Chap. 12), the assignment of configuration of derivatives of the perhydrophenanthrenes shown in Fig. 10-23, which was deduced[30] from the configuration of the perhydrodiphenic acids to which they are related (cf. Fig. 10-24), has taken on new meaning in the light of some theoretical considerations adduced by W. Johnson.[31] Johnson was able to predict correctly the relative stability of the isomers shown in Fig. 10-23 on the basis of the following premises:

1. If the ring-fusion bonds are considered equatorial or axial with respect to the central ring, the system having the larger number of equatorial bonds will be the more stable.‡

† It might be noted that the 289° acid can form only *one* half-ester (*dl* pair), regardless of which carboxyl group is esterified.

‡ A priori, there are two ways of labeling a system, since the first bond might be arbitrarily considered to be either equatorial or axial. The proper labeling is that which gives the largest number of equatorial bonds.

[31] W. S. Johnson, *Experientia*, **7**, 315 (1951); *J. Am. Chem. Soc.*, **75**, 1498 (1953).

2. When there are two axial bonds at the ring-fusion points, the system having the axial bonds 1,2 or 1,4 with respect to each other will be more stable than that which has the axial bonds 1,3 with respect to each other. In the latter system there will be a severe interaction of the 1,3-diaxial methylene groups (cf. Fig. 8-46).

3. Regardless of premise 1 or 2, if a ring junction involves two axial bonds either way the system is labeled, the system cannot exist in that form but the central ring in such a system must be boat-shaped (or flexible) and the system will be quite unstable.

In Fig. 10-23 the bonds are labeled *e* (equatorial) or *a* (axial) with respect to the central ring, and the Roman numerals in parentheses below each isomer denote its relative order of stability (I, most stable; V, least stable).

Figure 10-26 shows the five stereoisomeric perhydroanthracenes.[32] It is left as an exercise for the reader to confirm the *dl* or meso nature of these compounds, label the ring fusion bonds *a* or *e*, and confirm the indicated order of stability.

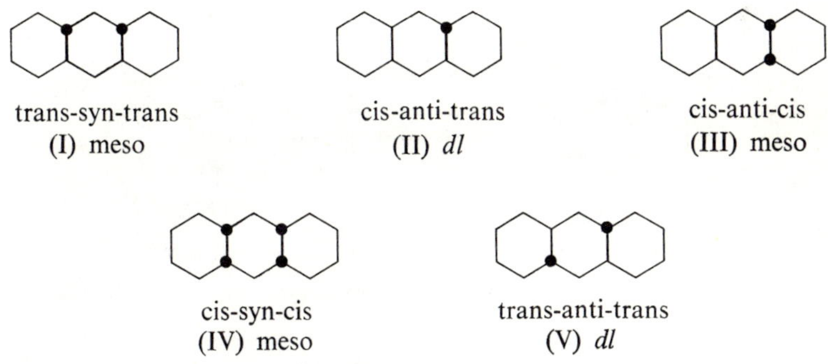

Fig. 10-26. The perhydroanthracenes.

The predicted order of stability of the perhydrophenanthrenes (Fig. 10-23) has been confirmed by equilibration of ketones in at least three cases:[30] The cis-syn-cis system (IV) is converted to the trans-syn-cis (II), the cis-anti-trans (II) is converted to the trans-anti-trans (I), and the cis-syn-trans (II) could not be converted to the trans-syn-trans (V). The last instance, shown in Fig. 10-27, is the most remarkable because it is contrary to the primitive (but not always correct) notion that a system with the maximum number of trans fusions should be the most stable.

An important application of the above ideas on perhydrophenanthrene stability has been in W. Johnson's classical steroid synthesis.[33] A precursor of the alicyclic D ring was formed in the course of this synthesis by catalytic hydrogenation of a phenol, as shown in Fig. 10-28. Since the stereochemistry of the C/D ring fusion (especially as regards C_{14}) is established in this step, it was important to determine the configuration of the two stereoisomers

[32] Cf. R. L. Clarke, *J. Am. Chem. Soc.*, **83**, 965 (1961) and earlier references there cited.
[33] W. S. Johnson, E. R. Rogier, and J. Ackerman, *J. Am. Chem. Soc.*, **78**, 6278, 6322 (1956).

actually formed. Since compound A was stereochemically homogeneous, having the configuration shown at its four asymmetric carbon atoms, production of three new asymmetric atoms in the catalytic hydrogenation could produce, in principle, 2^3, or 8, stereoisomeric alcohols (each new asymmetric atom multiplies by 2 the number of isomers in the precursor molecule). Actually only two isomers were obtained, presumably because the hydrogenation involves predominantly an all-cis approach, as it did with the perhydrodiphenic acids (Fig. 10-25). The two isomers (B and C) differ in configuration of the ring junction rather than just of the hydroxyl group, since, upon oxidation, they gave two distinct ketones D and E. Because of the expected predominant cis approach in the hydrogenation step, it was assumed that both ketones would have a C/D cis fusion and that (if the B-C-D ring system was regarded as a perhydrophenanthrene system) one would be trans-anti-cis and the other trans-syn-cis. Upon treatment with base, one of the ketones (labeled E in Fig. 10-28) was epimerized to a new ketone F but the other ketone (D) remained unchanged. Since the cis-syn-trans system is not epimerized (cf. Fig. 10-27) but the cis-anti-trans system would be expected

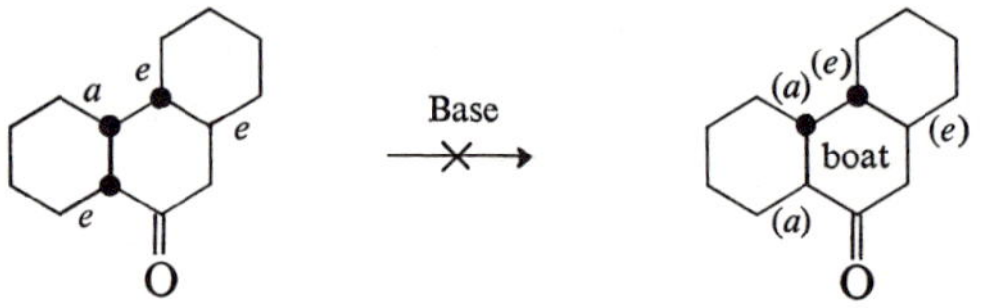

Fig. 10-27. Evidence of instability of the trans-syn-trans perhydrophenanthrene system.

to be epimerized to trans-anti-trans (Fig. 10-23), it may be concluded that ketone D is cis-syn-trans, E is cis-anti-trans, and F is trans-anti-trans.

Although it is not pertinent to the present discussion, it may be added that the configurations of the alcohols B and C were also assigned on the basis of the fact that B, the precursor of D, was less rapidly oxidized to the ketone than was C, the precursor of E. Since equatorial alcohols are generally less readily oxidized than axial alcohols (see below), it may be tentatively concluded that the hydroxyl group in B is equatorial and in C axial. The complete configurations of these alcohols are, therefore, as shown at the bottom of Fig. 10-28. (Note that, since the C/D ring fusion is cis, a bond that is equatorial with respect to ring C is axial with respect to ring D, and vice versa.) If these assignments are correct, B and C are, indeed, products of all-cis addition of hydrogen to the aromatic ring in the precursor A.

As the above example indicates, stereochemical, and especially conformational, considerations are of great importance in the synthesis (as well as structure elucidation) of natural products. Unfortunately, it is not possible even to attempt to cover this area at all adequately within the scope of a textbook. We shall, instead, limit ourselves in this chapter to a consideration of some of the more interesting conformational aspects of the chemistry of *one* class of natural compounds, namely, the steroids. In Chap. 15 we shall return briefly to stereochemical aspects of natural-product synthesis.

Almost all steroids contain the ring system shown in Fig. 10-29. The A/B ring fusion may be either trans (also called 5α or allo) or cis (also called 5β or normal); all other ring junctures are trans. The appropriate conformations are shown in Fig. 10-29, which correctly represents the absolute configuration of the steroids in question. Groups pointing upward (in the direction of the C_{10} and C_{13} methyl groups) are arbitrarily assigned the trivial stereochemical prefix β whereas the groups pointing downward are called α. It

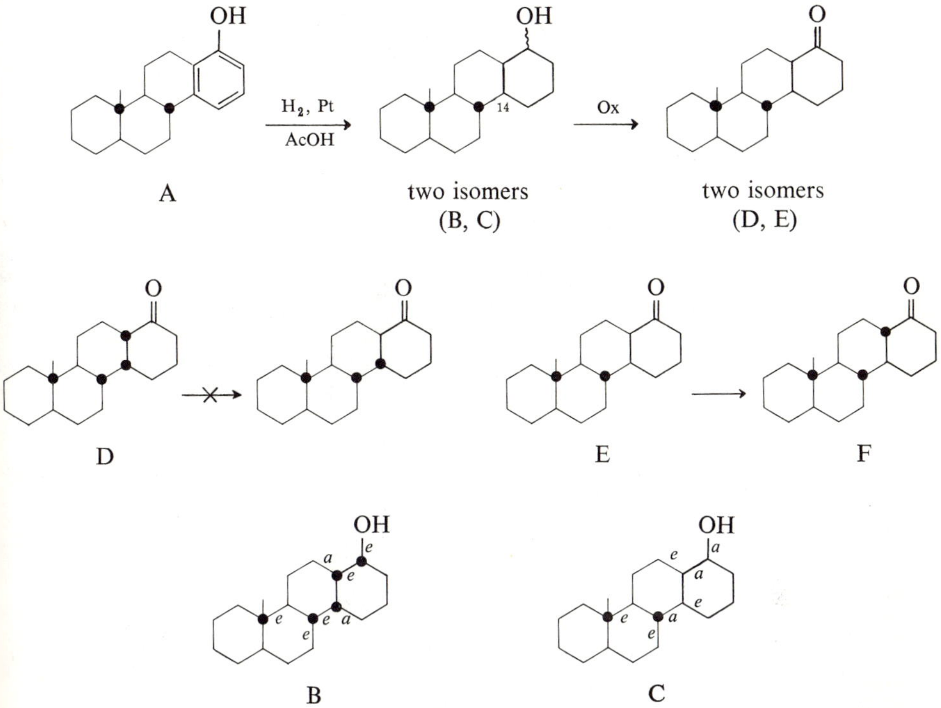

Fig. 10-28. Establishment of stereochemistry of D ring in Johnson's steroid synthesis.

should be noted that the designations α and β have nothing to do with the axial or equatorial nature of the substituent; as Fig. 10-29 shows, an α (or β) substituent may be either axial or equatorial. In the trans series, β substituents at C_2, C_4, C_6, C_8, C_{10}, C_{11}, and C_{13} and α substituents at C_1, C_3, C_5, C_7, C_9, C_{12}, and C_{14} are axial. Contrariwise, α substituents at C_2, C_4, C_6, and C_{11} and β substituents at C_1, C_3, C_7, and C_{12} are equatorial. In the cis series, β substituents at C_1, C_3, C_6, C_8, C_{11}, and C_{13} and α substituents at C_2, C_4, C_7, C_9, C_{12}, and C_{14} are axial, whereas α substituents at C_1, C_3, C_6, and C_{11} and β substituents at C_2, C_4, C_7, and C_{12} are equatorial. The β substituents at C_5 and C_{10} in the cis series are equatorial with respect to one of the rings and axial with respect to the other.

It was pointed out in Chap. 8 (cf. Fig. 8-24) that equatorial ester groups are more easily saponified than axial ester groups because they are less hindered. For the same reason, equatorial alcohols are esterified more readily than axial alcohols.[34] Many examples are found in the steroid literature to illustrate these points; thus cholestan-3β-yl acetate (equatorial $-$OAc) is hydrolyzed more rapidly than its 3α (axial) epimer,[35] and treatment of cholestan-3β,7α-diol with excess ethyl chloroformate leads to the formation of the 3-monocathylate ("cathylate" = R—$OCO_2C_2H_5$), the axial 7α-hydroxyl

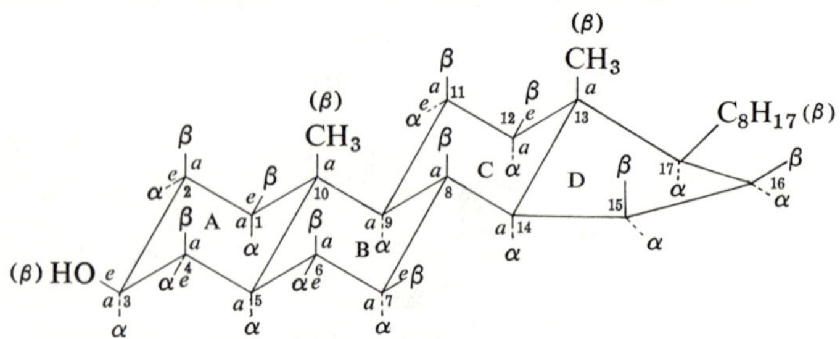

A/B trans: Cholestanol

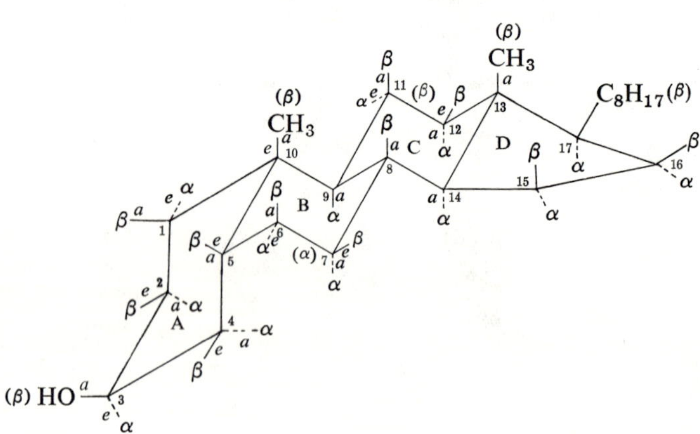

A/B cis: Coprostanol

Fig. 10-29. Representative steroids.

group resisting esterification. (In contrast, the epimeric cholestan-3β,7β-diol in which both hydroxyl groups are equatorial is readily dicathylated.[36]) A further differentiation with respect to steric hindrance may be demonstrated in the steroids: An axial hydroxyl group that is encumbered only by axial hydrogen atoms on the same side of the ring, such as the 3α group in the A/B trans system (Fig. 10-29), is less hindered than an axial hydroxyl group

[34] E. L. Eliel and C. A. Lukach, *J. Am. Chem. Soc.*, **79**, 5986 (1957).

[35] A. Fürst and P. A. Plattner, *Helv. Chim. Acta*, **32**, 275 (1949).

[36] L. F. Fieser, J. E. Herz, M. W. Klohs, M. A. Romero, and T. Utne, *J. Am. Chem. Soc.*, **74**, 3309 (1952).

that stands out parallel with one of the angular methyl groups, such as the 2β group in the same system. Thus cholestan-2β-yl acetate is hydrolyzed more slowly than the -3α-yl acetate.[35] The most hindered position of all is the 11β position which is encumbered by both the angular methyl groups at C_{10} and C_{13} (cf. Fig. 10-29). Little steric hindrance is provided by a vicinal hydroxyl group, as evidenced by the ready *di*cathylation of cholestan-3β,5α,-6α-triol to the 3,6 dicathylate.[36]

An exception to the above generalizations is provided by cholic acid derivatives (Fig. 10-30). Inspection of the three-dimensional formula would lead one to believe that the equatorial 3α-hydroxyl group should be least hindered and the axial 7α-hydroxyl group most hindered because of encumbrance by the C_4-methylene group of the A ring. While it is true that the 3-OH group is most accessible, as evidenced by the formation of a 3-monocathylate from methyl cholate,[36] the next most accessible hydroxyl group is the one at C_7 and not the one at C_{12}. This is shown by the formation of a

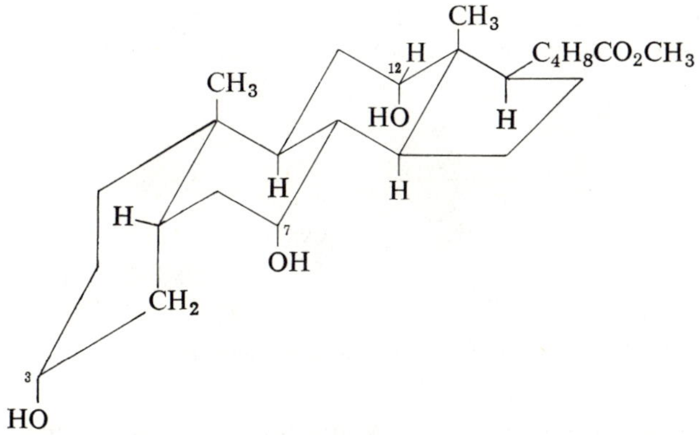

Fig. 10-30. Methyl cholate.

3,7-diacetate upon treatment of methyl cholate with acetic anhydride in dioxane-pyridine.[37] It is believed[38] that the inaccessibility of the 12α-hydroxyl group to acetylation is due to its becoming associated intramolecularly with the ester group in the 17-side chain which may readily curl around so as to approach 12α-OH quite closely. Thus when the 17-side chain is foreshortened, as in methyl etiocholate, the 12α-hydroxy group actually becomes easier to esterify than 7α-hydroxyl.[39]

A very interesting correlation has been made of the conformation of steroid alcohols and their rate of oxidation with chromic acid.[40] Since the rate-determining step in chromic acid oxidation of alcohols is the abstraction of the hydrogen on the carbinol carbon,[41] the observation[42] that axial alcohols

[37] L. F. Fieser, S. Rajagopalan, E. Wilson, and M. Tishler, *J. Am. Chem. Soc.*, **73**, 4133 (1951).
[38] Ref. 84, p. 222.
[39] A. Lardon, *Helv. Chim. Acta*, **30**, 597 (1947).
[40] J. Schreiber and A. Eschenmoser, *Helv. Chim. Acta*, **38**, 1529 (1955).
[41] F. H. Westheimer and N. Nicolaides, *J. Am. Chem. Soc.*, **71**, 25 (1949).
[42] G. Vavon and C. Zaremba, *Bull. soc. chim. France*, [4]**49**, 1853 (1931). See also S. Winstein and N. J. Holness, *J. Am. Chem. Soc.*, **77**, 5562 (1955).

are more readily oxidized than equatorial alcohols was originally rationalized by assuming that in the axial alcohol the carbinol hydrogen, being equatorial, is more accessible than the (axial) carbinol hydrogen in the equatorial alcohol.[43] Observations in the steroid field[40] did not, however, bear out this contention, for the effect of crowding the axial carbinol hydrogen in an equatorial steroid alcohol is to accelerate, rather than to retard, chromic acid oxidation. Thus cholestan-4α-ol (carbinol hydrogen crowded by C_{10} methyl substituent; cf. Fig. 10-29) reacts twice as fast as the 3β-ol, and 5α-pregnan-3,20-dione-11α-ol, in which the carbinol hydrogen is in the highly hindered 11β position, is oxidized seven times as fast as cholestan-3β-ol. A hypothesis in agreement with these results[40] is that oxidation of a sterically crowded alcohol (such as an axial alcohol) is accelerated because of the high ground-state energy level (cf. Fig. 8-25) of the starting material. Since the crowding is relieved in the product ketone as well as in the transition state leading toward it, the transition-state energy levels for the crowded (axial) and uncrowded (equatorial) alcohol do not differ much, and the situation of steric acceleration depicted in Fig. 8-25 therefore results. Accordingly, it was found[40] that the more highly crowded the hydroxyl group is in the alcohol,

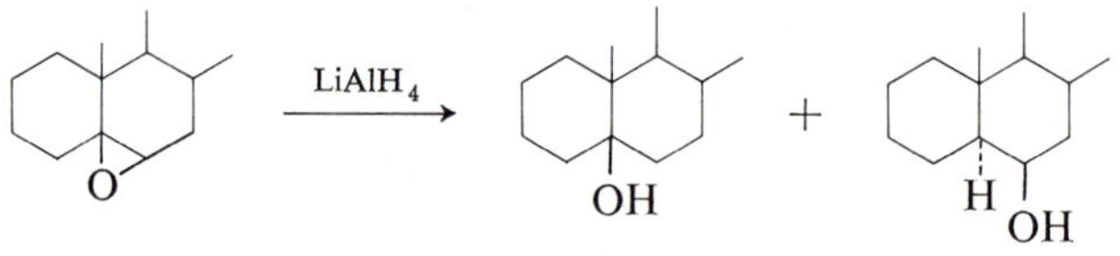

$\xrightarrow{LiAlH_4}$ +

O OH H OH

2 parts 1 part

Fig. 10-31. Reduction of 5,6β-epoxycholestane with lithium aluminum hydride.

the faster the alcohol is oxidized. Cholestan-3α-ol, which has an axial but otherwise not very crowded hydroxyl group, is oxidized only three times faster than its equatorial 3β epimer. On the other hand, the 2β-ol, in which the hydroxyl group is crowded by the methyl substituent at C_{10}, reacts twenty times as fast as the 3β-ol, and the doubly crowded 5α-pregnan-3,20-dione-11β-ol reacts sixty times as fast.

Axial addition to double bonds and axial ring opening of epoxides have already been mentioned in Sec. 8-5 (Figs. 8-38 and 8-39); most of the pertinent examples come from the steroid field. Exceptions to the rule may occur under special circumstances; thus 5β,6β-epoxycholestane upon lithium aluminum hydride reduction yields largely the equatorial 5β alcohol and only a minor amount of the 6β-ol (Fig. 10-31).[44] Formation of the latter would involve axial attack by the hydride at the tertiary C_5 carbon atom, and it is known that nucleophilic displacements at tertiary centers occur only with difficulty.

2,3β-Epoxylanost-8-ene is a disecondary epoxide which undergoes diequatorial ring opening with hydrogen bromide (Fig. 10-32).[44a] The observation

[43] D. H. R. Barton, *Experientia,* **6,** 316 (1950).
[44] A. S. Hallsworth and H. B. Henbest, *J. Chem. Soc.,* 4604 (1957).
[44a] D. H. R. Barton, D. A. Lewis, and J. F. McGhie, *J. Chem. Soc.,* 2907 (1957).

has been rationalized by assuming that the epoxide ring in the lanostane derivative is a distorted boat rather than a distorted chair. The reason for this may be that, in the chair form of the β epoxide (Fig. 10-32), there would be severe interaction of the axial 4β- and 10β-methyl groups with the epoxy oxygen on the same side of the ring.† The α-oxide (in which the oxygen is on the opposite side of the ring from the axial methyl groups) appears to exist in the chair form, for it gives the normal diaxial bromohydrin with hydrogen bromide.

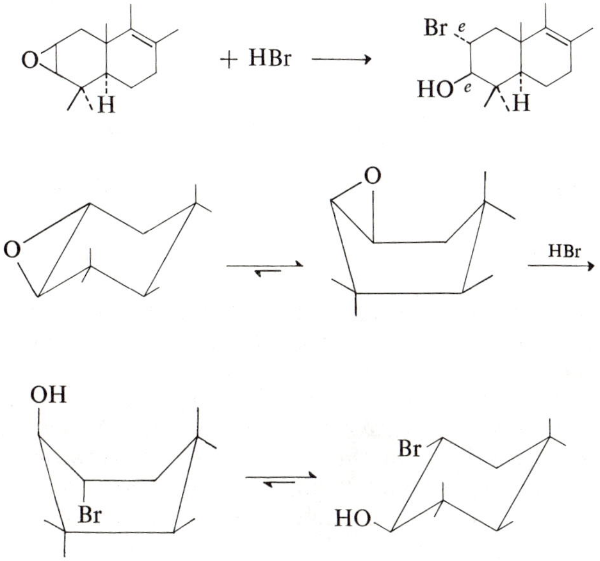

Fig. 10-32. Addition of hydrogen bromide to $2,3\beta$-epoxylanost-8-ene.

† Because of the axial-axial methyl interaction at C_4 and C_{10}, there is considerable strain in the chair conformation in the lanostane system. This strain *by itself* is not sufficient to force the molecule into the boat form to an appreciable extent, unless there are additional interactions, as in the above-mentioned $2,3\beta$-epoxylanost-8-ene or in lanost-8-en-3α-ol (partial formula *i*) in which there would be an additional axial hydroxyl interaction in the chair form: J. W. Huffman, *J. Org. Chem.* **24**, 447 (1959). In ring-A ketones which have 1,3-diaxial methyl interactions—such as lanostanone and 2β-methylcholestan-3-one—the tendency to change over to the boat or flexible conformation should be even greater, since, as mentioned in Sec. 8-7, the difference in energy between the chair and flexible forms of a cyclohexanone is considerably less than for the corresponding cyclohexane. It is therefore not surprising that the introduction of additional CH_3-Br diaxial interactions (as in 2β-bromolanostan-3-one) or of a carbonyl-bromine interaction (as in 2α-bromo-2β-methylcholestan-3-one) should force the molecule into the flexible form; cf. Ref. 44*a* and C. Djerassi, N. Finch, R. C. Cookson, and C. W. Bird, *J. Am. Chem. Soc.*, **82**, 5488 (1960). See also J. S. E. Holker and W. B. Whalley, *Proc. Chem. Soc.*, 464 (1961).

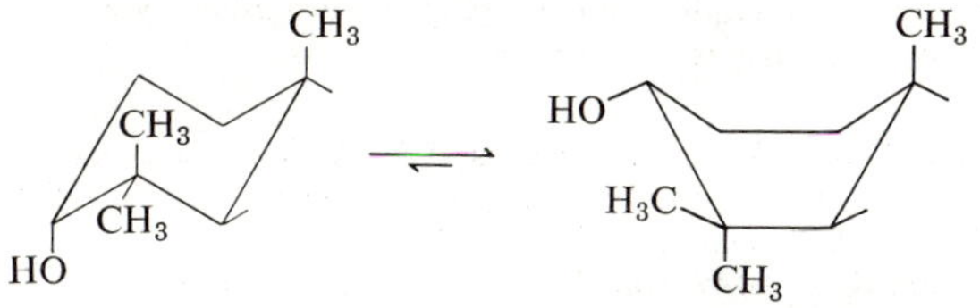

(*i*) (*ii*)

Among olefin reactions, addition of hydrogen chloride to 3-methylcholest-2-ene (Fig. 10-33) gives what appears to be† the equatorial chloride.[45] This has been explained by the assumption that the addition of HCl is non-concerted and that the tertiary carbonium ion intermediate gives the "more stable" equatorial halide.‡ However, since addition of hydrogen bromide also gives an equatorial bromide (spectral evidence) whereas bromine appears to be smaller than methyl (cf. Table 8-6), it is hard to see why the compound with equatorial halide and axial methyl should be the more stable isomer. The picture is further complicated by the observation,[47] based on infrared

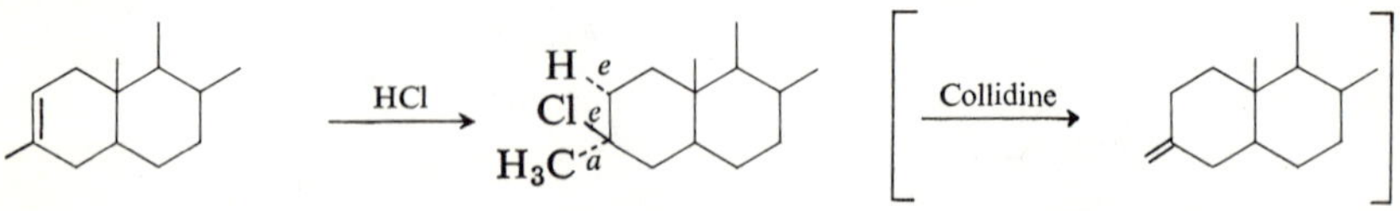

HCl → H e, Cl e, H_3C a ... [Collidine →]

Fig. 10-33. Addition of hydrogen chloride to 3-methylcholest-2-ene.

spectra (cf. Table 8-5), that in the addition of deuterium chloride the deuterium appears in the equatorial (α) position at C_2. This has been explained on the basis of a pi-complex intermediate involving a proton bridge. Such a bridge, in analogy with epoxide formation (see below), should form more readily on the less hindered α side of the ring.

Another conformational point which is well illustrated in the steroids is concerned with the stereochemistry of epoxidation. Since the upper or β face of the steroid system in the A/B trans configuration is somewhat screened by the two angular methyl groups, epoxidation occurs preferentially from the under or α side. Thus cholest-1-ene,[48] cholest-2-ene,[35] cholest-3-

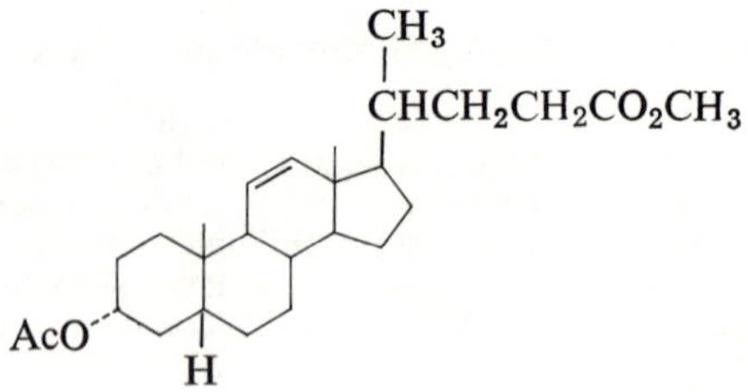

CH_3
$CHCH_2CH_2CO_2CH_3$
AcO
H

Fig. 10-34. Methyl 3α-acetoxy-Δ^{11}-cholenate.

†The equatorial nature of the chlorine was concluded from the position of the C—Cl stretching band in the infrared and from the fact that dehydrohalogenation gave the exocyclic olefin 3-methylenecholestane whereas an axial chloride should have given one of the more stable endocyclic olefins by diaxial elimination.

‡The greater stability of halide in the equatorial position is deduced from the observation[46] that, in a 1-methylcyclohexyl halide, the halogen shows exclusively what seems to be the equatorial C—X stretching frequency in the infrared.

[45] D. H. R. Barton, A. da S. Campos-Neves, and R. C. Cookson, *J. Chem. Soc.*, 3500 (1956).
[46] D. H. R. Barton, J. E. Page, and C. W. Shoppee, *J. Chem. Soc.*, 331 (1956).
[47] Cf. D. H. R. Barton, *Experientia, Suppl.*, **2**, 121 (1955).
[48] H. B. Henbest and R. A. L. Wilson, *J. Chem. Soc.*, 3289 (1956).

ene,[49] and methyl 3α-acetoxy-Δ^{11}-cholenate (Fig. 10-34)[50] all afford exclusively the corresponding α-oxide upon treatment with perbenzoic acid. With cholest-5-ene the α-oxide appears to be the predominating but not the exclusive product.[51] Possibly the slight lack of stereoselectivity here is due to the fact that the methyl group at C_{10} in a $\Delta^{5,6}$ steroid is pseudoaxial (cf. Fig. 8-49) rather than axial. Some β-oxide is also formed from methyl 5-cholestene-3β-carboxylate but the corresponding free acid seems to give only the α-oxide.[52] In 4-cholestene the methyl group at C_{10} is also pseudoaxial, but this olefin appears to give a homogeneous epoxide on epoxidation, presumably the α-oxide,[53] and the 3-methoxy- and 3-acetoxy- derivatives of 4-cholestene are known to give the α-oxide in high yield.[54a] In the epoxidation of steroids having A/B cis fusion and containing the double bond in ring A, the angular methyl groups do not interfere with the approach of the reagent

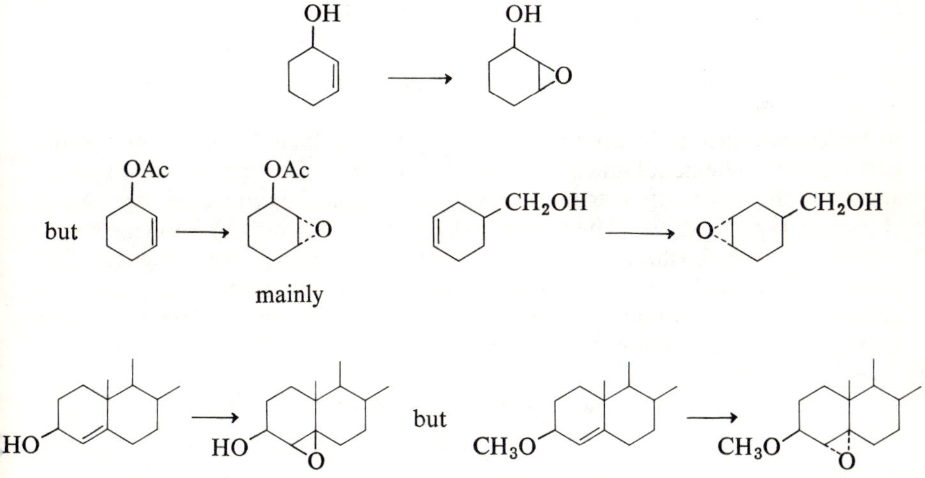

OH → OH O

but OAc → OAc O mainly; CH_2OH → O CH_2OH

HO → HO O but CH_3O → CH_3O O

Fig. 10-35. Effect of allylic hydroxyl group on stereochemistry of epoxidation.

from the β side, but the B ring, especially the C_6-methylene group, interferes with approach from the α side; such steroids, therefore, give β-oxides.[55]

Surprisingly enough, even in simple substituted cyclohexenes, such as 4-hydroxymethylcyclohexene (Fig. 10-35), epoxidation seems to occur mainly from the side trans to the substituent (provided that the substituent is bulky enough to occupy nearly exclusively the equatorial position), even though a model would seem to indicate that the double bond is about equally accessible

[49] A. Fürst and R. Scotoni, *Helv. Chim. Acta*, **36**, 1332 (1953).

[50] T. F. Gallagher and W. P. Long, *J. Biol. Chem.*, **162**, 495 (1946).

[51] H. B. Henbest and T. I. Wrigley, *J. Chem. Soc.*, 4596 (1957). Cf. A. J. Fudge, C. W. Shoppee, and G. H. R. Summers, *ibid.*, 958 (1954).

[52] G. Roberts, C. W. Shoppee, and R. J. Stephenson, *J. Chem. Soc.*, 3178 (1954).

[53] I. M. Heilbron, W. Shaw, and F. S. Spring, *Rec. trav. chim.*, **57**, 529 (1938).

[54] (*a*) H. B. Henbest and R. A. L. Wilson, *J. Chem. Soc.*, 1958 (1957). (*b*) See also P. A. Plattner, H. Heusser, and A. B. Kulkarni, *Helv. Chim. Acta*, **32**, 265 (1949).

[55] C. Djerassi and J. Fishman, *J. Am. Chem. Soc.*, **77**, 4291 (1955).

from either side.[56] An exception to these rules occurs when there is a hydroxyl group in the position allylic to the double bond, as in cholest-4-en-3-ol or 2-cyclohexenol (Fig. 10-35).[54, 56c] Apparently the epoxidation reagent gets associated, in some way, with the hydroxyl group in the molecule prior to attacking the double bond and is, therefore, constrained to approach the latter from the side of the hydroxyl group.

10-3. Bridged Systems

a. Occurrence and Availability. The smallest conceivable bridged system is bicyclo[1.1.1]pentane. This hydrocarbon, or derivatives of it, does not seem

$\xrightarrow[-N_2, +H_2O]{h\nu}$ HO_2C

Fig. 10-36. Synthesis of 1,6,6-trimethylbicyclo[2.1.1]hexane-5-carboxylic acid.

to be known although some heterocyclic analogs have been claimed in the literature.[92] The next higher homolog, bicyclo[2.1.1]hexane, was first synthesized in the form of a trimethyl substituted carboxylic acid derivative, shown in Fig. 10-36.[57a] The parent hydrocarbon is now also known.[57c]

The bicyclo[2.2.1]heptane system is found in a number of terpenes. The hydrocarbon skeletons of these terpenes and their names are shown, in correct absolute configuration, in Fig. 10-37. Derivatives of bicyclo[3.1.1]-heptane also occur naturally (Fig. 10-37).

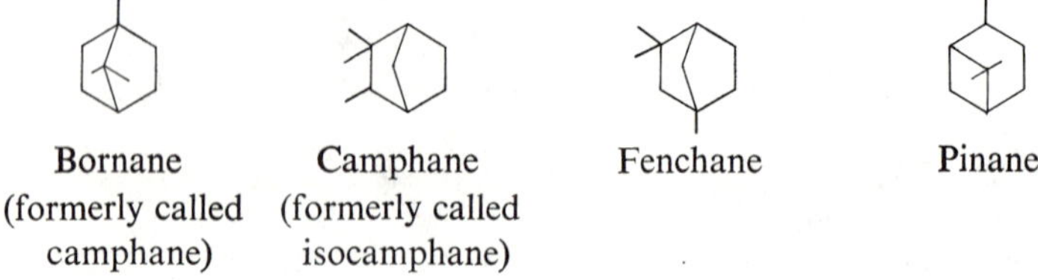

Fig. 10-37. Parent hydrocarbons of bicyclic monoterpenes. (The one-carbon bridge across the six-membered ring is always forward.)

Derivatives of bicyclo[2.2.1]heptane (norbornane) and bicyclo[2.2.2]octane are also readily accessible by Diels-Alder reactions, starting with cyclopentadiene or 1,3-cyclohexadiene (Fig. 10-38).† It should be noted that, with

† The completely unsaturated bicyclo[2.2.2]octatriene system (*i*) has been synthesized by H. E. Zimmerman and R. M. Dauffler [*J. Am. Chem. Soc.*, **82**, 1514 (1960)] and has been called "barrelene" because the pi-electron cloud is barrel-shaped. The system appears to have no aromatic properties.

(*i*)

---

[56] (*a*) H. B. Henbest and B. Nicholls, *J. Chem. Soc.*, 221 (1959); (*b*) H. Hopff and H. Hoffmann, *Helv. Chim. Acta*, **40**, 1585 (1957); (*c*) H. B. Henbest, B. Nicholls, W. R. Jackson, R. A. L. Wilson, N. S. Crossley, N. B. Meyers, and R. S. McElhinney, *Bull. soc. chim. France*, 1365 (1960).

[57] (*a*) L. Horner and E. Spietschka, *Chem. Ber.*, **88**, 934 (1955). (*b*) See also J. Meinwald and P. G. Gassman, *J. Am. Chem. Soc.*, **82**, 2857 (1960). (*c*) K. B. Wiberg, B. R. Lowry, and T. H. Colby, *ibid.*, **83**, 3998 (1961).

cyclopentadiene, the adduct might, a priori, be either the exo product (in which the substituent is on the side of the C_7 bridge, i.e., *outside* the cage formed by carbon atoms 2, 3, 5, and 6) or the endo product (in which the substituent is *inside* the cage formed by carbon atoms 2, 3, 5, and 6). In the examples of the Diels-Alder reaction originally studied (e.g., Fig. 10-39), the endo product was obtained predominantly or exclusively. This empirical finding was generalized in the so-called "Alder rule" according to which addition is predominantly endo; the rule has been rationalized on the basis

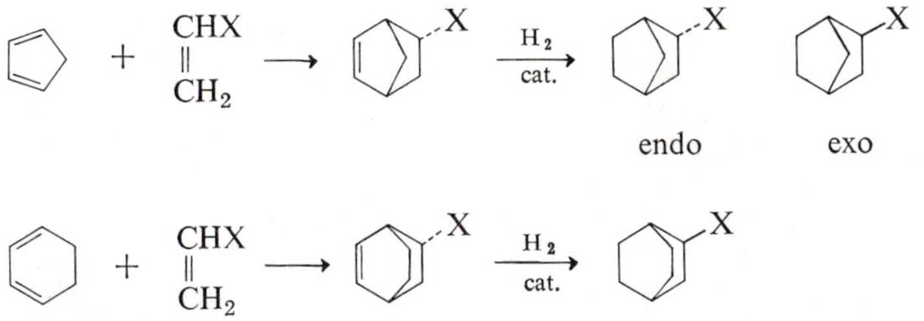

CHX ‖ CH_2 $\xrightarrow[\text{cat.}]{H_2}$ endo exo

CHX ‖ CH_2 $\xrightarrow[\text{cat.}]{H_2}$

Fig. 10-38. Synthesis of bicyclic systems by Diels-Alder reaction.

that endo addition involves maximum overlap of pi electrons in the transition state.[58a, b] Unfortunately, more recent examples[58c] imply that the rule may be of rather limited validity. With cyclohexadiene, it might be noted, although there is exo-endo isomerism in the Diels-Alder adduct, this disappears upon hydrogenation because of the symmetry of the substituted bicyclo[2.2.2]octanes produced. In the case of the Diels-Alder reaction with a 1,2-disubstituted olefinic dienophile, the stereochemistry of the addition, besides being endo, is also cis; i.e., a cis olefin gives a cis-1,2-disubstituted adduct and a trans olefin precursor gives a trans adduct.[58d] This is also exemplified in Fig. 10-39.

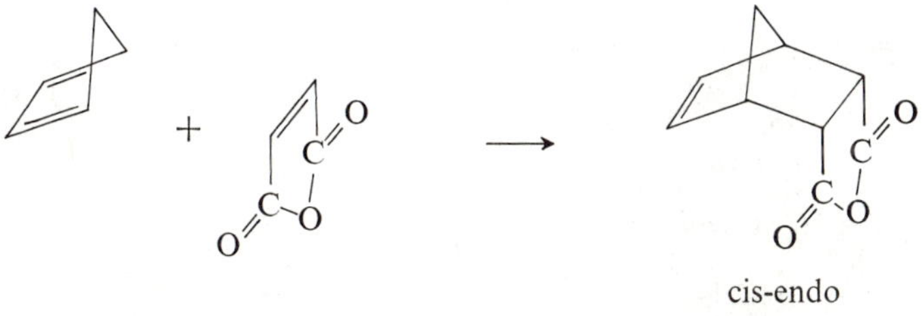

cis-endo

Fig. 10-39. Example of Alder rule and cis addition.

Returning now to the bicyclic bridged hydrocarbons, there are three systems besides those already enumerated which deserve special mention. One of these is the bicyclo[3.3.1]nonane system which, in a ball-and-stick model,

[58] (*a*) M. C. Kloetzel in R. Adams, ed., "Organic Reactions," vol. 4, John Wiley & Sons, Inc., New York, 1948, pp. 10–12; (*b*) K. Alder and G. Stein, *Angew. Chem.*, **50,** 510 (1937). (*c*) For example, J. A. Berson, A. Remanick, and W. A. Mueller, *J. Am. Chem. Soc.*, **82,** 5501 (1960); K. Alder and W. Günzel, *Chem. Ber.*, **93,** 809 (1960); and J. Berson and W. A. Mueller, *Tetrahedron Letters,* 131 (1961). (*d*) For a review of the stereochemistry of the Diels-Alder reaction, see J. G. Martin and R. K. Hill, *Chem. Revs.*, **61**; 537 (1961).

may be constructed by the 1,3 fusion of two cyclohexane chairs free of angle strain (Fig. 10-40). Construction of a scale model of this hydrocarbon shows, however, that there is an intolerable transannular interaction between the axial hydrogens at C_3 and C_7 (which are only ca. 1 A. apart), and it would appear that the molecule must therefore exist in a chair-boat form (Fig. 10-40) even though this still has some fairly bad interactions. The hydrocarbon (Fig. 10-40) has been synthesized[59] and shows the remarkably high

Fig. 10-40. Bicyclo[3.3.1]nonane.

melting point of 145 to 146°, the boiling point being 170°. Heterocyclic analogs of this hydrocarbon are cyclohexane-1,3-dicarboxylic anhydride and the alkaloid pseudopelletierine (Fig. 10-41). In these compounds, because one of the interfering methylene groups is replaced by —O— or —CO—, respectively, the non-bonded interactions in the chair-chair form are reduced.

The second system of interest is obtained by bridging the offending methylene groups C_3 and C_7 in the chair-chair form of bicyclo[3.3.1]nonane

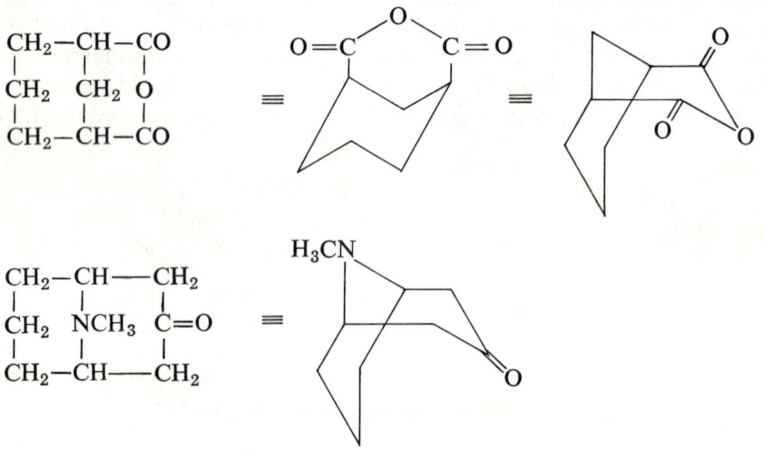

Fig. 10-41. Heterocyclic analogs of bicyclo[3.3.1]nonane.

by another methylene group. This completely removes the non-bonded hydrogen-hydrogen interaction, and the resultant molecule (Fig. 10-42) is a beautifully symmetric structure entirely free of strain. (The reader is strongly urged to build a model of this molecule.) The compound repre-

[59] H. Meerwein and W. Schürmann, *Ann.*, **398**, 196 (1913); H. Meerwein, F. Kiel, G. Klösgen, and E. Schoch, *J. prakt. Chem.* [2]**104**, 161 (1922).

sented in Fig. 10-42 was originally isolated from a high-boiling petroleum fraction.[60] The structure was first assigned[60] on the basis of the melting point (270°) (which is remarkably high for a hydrocarbon of the molecular formula $C_{10}H_{16}$), the high stability, the unusual crystal symmetry (the compound crystallizes in the cubic system), and the X-ray diffraction pattern. Later on, it was confirmed by an elegant if painstaking synthesis.[61] The molecule is now readily accessible[62] by isomerization of the hydrogena-

Fig. 10-42. Adamantane and its synthesis.

tion product of cyclopentadiene dimer (Fig. 10-42). Because the compound simulates the geometrical arrangement of the carbon atoms in a small piece of the diamond crystal lattice, it has been called "adamantane." The well-known ammonia-formaldehyde condensation product urotropine, $(CH_2)_6N_4$, is a heterocyclic analog of adamantane, with nitrogen at the bridgehead positions instead of C—H.

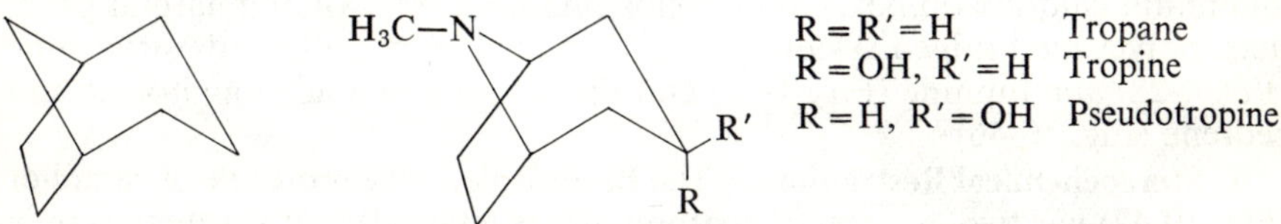

Fig. 10-43. Bicyclo[3.2.1]octane, tropane, tropine, and pseudotropine.

The last system to be mentioned here is bicyclo[3.2.1]octane (Fig. 10-43). The carbocyclic compound is readily accessible,[63] but of more interest is the heterocyclic analog tropane, which constitutes the parent ring system of a

[60] S. Landa and V. Macháček, *Collection Czechoslov. Chem. Commun.* **5**, 1 (1933).
[61] V. Prelog and R. Seiwerth, *Ber.*, **74**, 1644, 1769 (1941).
[62] P. von R. Schleyer and M. M. Donaldson, *J. Am. Chem. Soc.*, **82**, 4645 (1960).
[63] For example, G. Komppa, T. Hirn, W. Rohrmann, and S. Beckmann, *Ann.*, **521**, 242 (1936); J. W. Barrett and R. P. Linstead, *J. Chem. Soc.*, 611 (1936).

series of alkaloids. The alkaloid tropine and its epimer pseudotropine are shown in Fig. 10-43. In the benzoyl derivative of the lower homolog of ψ-tropine (*N*-methyl group missing), called "nor-ψ-tropine," there is facile migration of the benzoyl group from oxygen to nitrogen, and vice versa[64] (Fig. 10-44). Since this can occur only in the boat form, it follows that the boat form of the tropine system is at least not much less stable than the chair form. No N ⇌ O benzoyl migration can, of course, occur in nortropine because even in the boat form the oxygen and nitrogen are not close together.

Adamantane (Fig. 10-42) represents a system in which there are three bridged rings. Other interesting systems containing three or more rings of which at least two are bridged are nortricyclene and the tricycloheptane, quadricycloheptane, and perchloroquadricyclodecane derivatives shown in Fig. 10-45.

The very unusual tricycloheptanes in Fig. 10-45 are made by a self-insertion reaction of the carbene derived from 7,7-dibromo-bicyclo[4.1.0]heptane.[64a] The quadricycloheptane derivative is made photochemically,[65] but the tricyclene, interestingly enough, is formed by a simple, reversible thermal rear-

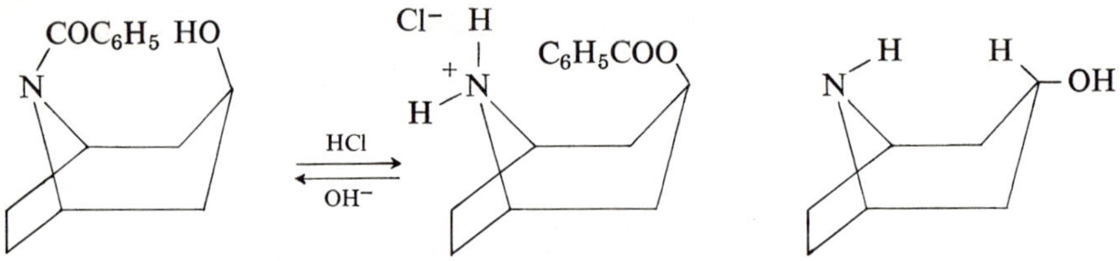

COC_6H_5 HO N

HCl / OH^-

Cl^- H $\overset{+}{N}$ H C_6H_5COO

N H H OH

Nortropine
(no N⇌O migration)

Fig. 10-44. N → O migration in nor-ψ-tropine.

rangement of the thermodynamically less stable† norbornene.[66] The quadricyclodecane structure is assigned to the product from the reaction of aluminum chloride on hexachlorocyclopentadiene.[66a] Among natural products, polycyclic bridged systems are exemplified by the alkaloids strychnine (Fig. 4-22) and quinine (Fig. 4-22) and the sesquiterpenoids longifolene and cedrene (Fig. 10-46).

b. Stereochemical Restrictions. The Bredt Rule. The molecule of camphor (Fig. 10-47) has two asymmetric carbon atoms (starred) and yet there is only

† Electron-diffraction studies by H. Mendel quoted by E. C. Kooyman and G. C. Vegter [*Tetrahedron,* **4,** 382 (1958), footnote 3] indicate that the 2,6 distance in norbornane is only 2.2 A. as compared with the normal value of 2.5 A. in an aliphatic chain.

[64] G. Fodor and K. Nádor, *J. Chem. Soc.,* 721 (1953).
[64a] W. R. Moore, H. R. Ward, and R. F. Merritt, *J. Am. Chem. Soc.,* **83,** 2019 (1961).
[65] S. J. Cristol and R. L. Snell, *J. Am. Chem. Soc.,* **80,** 1950 (1958).
[66] P. von R. Schleyer, *J. Am. Chem. Soc.,* **80,** 1700 (1958).
[66a] E. T. McBee, C. W. Roberts, J. D. Idol, and R. H. Earle, *J. Am. Chem. Soc.,* **78,** 1511 (1956).

one *dl* pair. The reason for this, readily understandable upon inspection of a model, is that the bridge must necessarily be cis and therefore the configurations of the two starred carbons are not independent. A similar situation

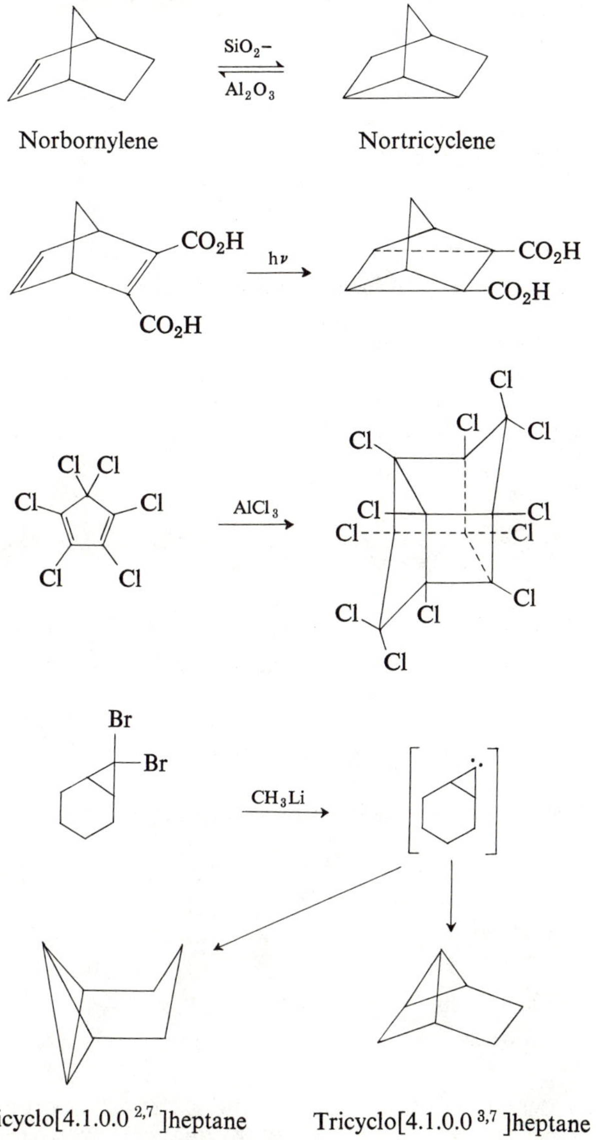

Norbornylene $\underset{Al_2O_3}{\overset{SiO_2-}{\rightleftharpoons}}$ Nortricyclene

CO_2H $\xrightarrow{h\nu}$ CO_2H CO_2H

CO_2H

Cl Cl Cl Cl Cl Cl $\xrightarrow{AlCl_3}$ Cl Cl Cl Cl Cl Cl Cl Cl Cl Cl Cl Cl

Br Br $\xrightarrow{CH_3Li}$

Tricyclo[4.1.0.0^{2,7}]heptane Tricyclo[4.1.0.0^{3,7}]heptane

Fig. 10-45. Some tricyclic and quadricyclic systems and their synthesis.

exists in many other bridged systems, for example, those shown in Fig. 10-37. As a result, the number of stereoisomers is always one-half of what it would be in the absence of this restriction. Thus the alcohol system corresponding to camphor (Fig. 10-47) presents two cis-bridged *dl* pairs, borneol and isoborneol, instead of the four *dl* pairs present in other systems with three asymmetric carbons. If the bridge were large enough, it ought to be possible to obtain the trans-bridged isomer also; this appears to have been achieved in the [4.4.1] system (Fig. 10-48).[66b]

CH_2 CH_3 CH_3 CH_3 H_3C CH_3 H_3C CH_3

Longifolene Cedrene

Fig. 10-46. Longifolene and cedrene.

Another restriction in bridged systems is given by the Bredt rule[67] which states that in small bridged systems one cannot, for steric reasons, have a double bond at the bridgehead position. Demonstrations of the Bredt rule are the failure of bromocamphor (Fig. 10-49) to eliminate hydrogen bromide

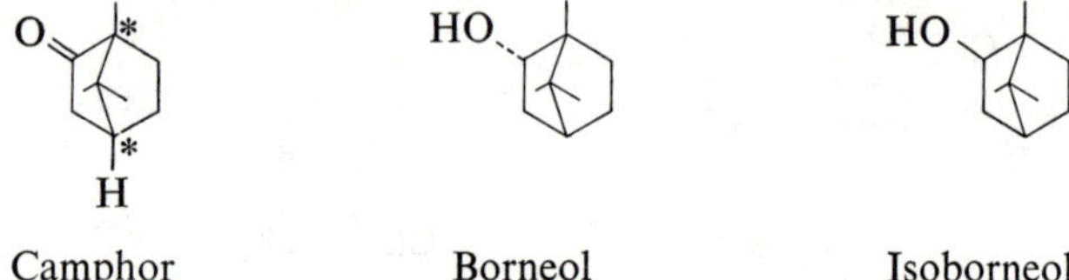

Camphor Borneol Isoborneol

Fig. 10-47. Camphor, borneol, and isoborneol.

in the presence of a base and the failure of the unsaturated analog of camphoric acid (Fig. 10-49) to form an anhydride. (Under forcing conditions, the anhydride forms with concomitant migration of the double bond to a non-bridgehead position.) Other interesting examples are shown in Fig. 10-50:

OH OH

H H

cis trans

Fig. 10-48. Cis- and trans-bridged bicyclic ring systems.

Bicyclo[2.2.2]octane-2,6-dione does not show the acidic properties of a 1,3-diketone[68] (the anion is incapable of resonance), and camphenonic and

[66b] W. G. Dauben and T. Westman, private communication; W. G. Dauben, Abstracts, Fourth Biennial Organic Symposium, Chemical Institute of Canada, Edmonton, Alberta, 1960.

[67] J. Bredt, *Ann.*, **437**, 1 (1924); cf. F. S. Fawcett, *Chem. Revs.*, **47**, 219 (1950).

[68] P. D. Bartlett and G. F. Woods, *J. Am. Chem. Soc.*, **62**, 2933 (1940).

ketopinic acids (Fig. 10-50), although they are β-keto acids, do not readily decarboxylate,[69] because they cannot achieve the required enolic transition state.

The limitations of the Bredt rule have been elegantly demonstrated by

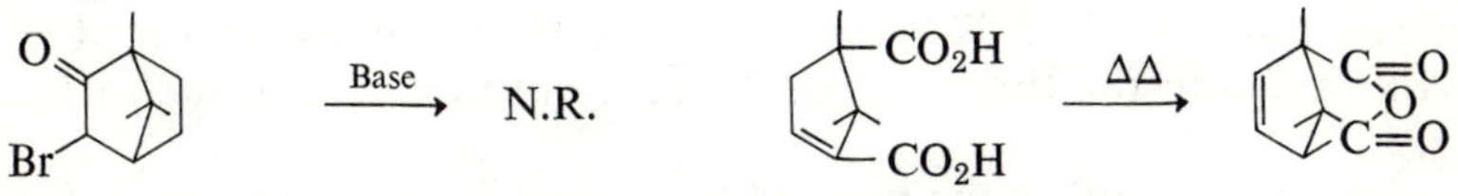

Base → N.R. ΔΔ →

Fig. 10-49. Examples of Bredt's rule.

Prelog and coworkers.[70] Ring closure of the diketone shown in Fig. 10-51 may give rise to either the fused-ring system A or the bridged-ring system B. When $n = 4$, only the fused system A is formed; when $n = 5$, the bridged system B is formed along with A; and when $n = 6$, only B results. Hence

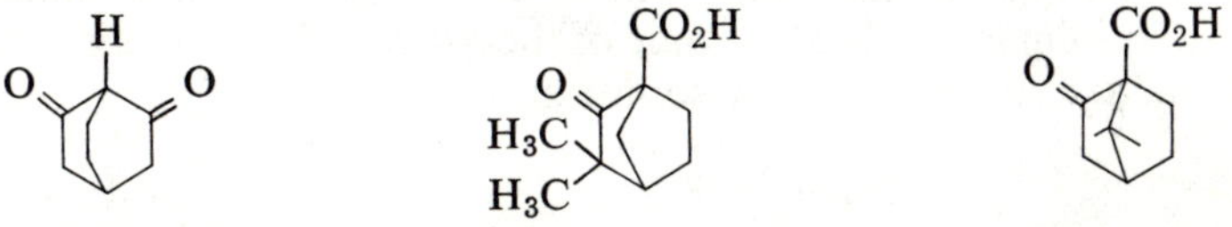

Bicyclo[2.2.2]octane-2,6-dione   Camphenonic acid   Ketopinic acid

Fig. 10-50. Further examples of the Bredt rule.

the Bredt rule no longer applies to the [5.3.1] system. It might be noted that this system contains a bridge with an eight-membered ring and that the eight-membered ring is also the smallest ring that can accommodate a trans double bond (Chap. 7).

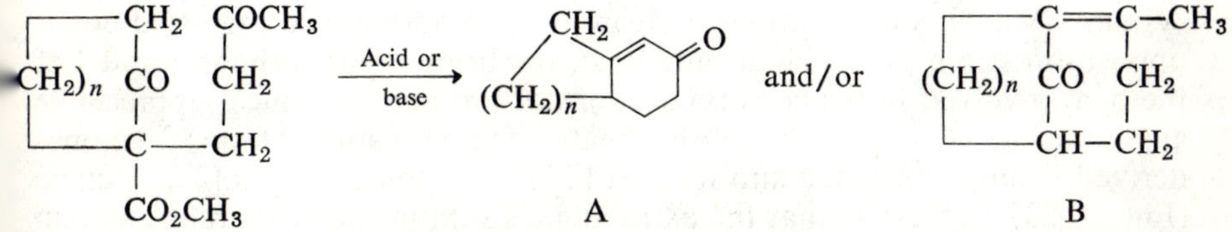

Acid or base → and/or

	A	B
$n=4$	65%	none
$n=5$	32%	14%
$n=6$	none	76%

Fig. 10-51. Limitations of Bredt's rule.

The stereochemical restrictions in decarboxylations of β-keto acids seem to be somewhat less severe. Whereas camphenonic acid and its isomer, ketopinic acid (Fig. 10-50), which contain the [2.2.1] system, as well as a similar

[69] J. Bredt, *J. prakt. Chem.* [2]**148**, 221 (1937).

[70] V. Prelog, L. Ruzicka, P. Barman, and L. Frenkiel, *Helv. Chim. Acta,* **31,** 92 (1948); cf. also V. Prelog, M. M. Wirth, and L. Ruzcika, *ibid.,* **29;** 1425 (1946), and Ref. 71.

β-keto acid containing the [2.2.2] system, fail to decarboxylate,[69] 2,6-diketobicyclo[3.3.1]nonane-1,3,5,7-tetracarboxylic acid 3,7-dimethyl ester (Fig. 10-52, *A*) can be decarboxylated by heating.[59] A similar acid (Fig. 10-52, *B*; $n = 3$), on the other hand, cannot be decarboxylated by heating in quinoline to 240°, although the next higher homolog ($n = 4$) is decarboxylated under these conditions.[71] The difference may well lie in the fact that in *A* the carbonyl group is in the three-carbon bridge, but in *B* it is in the one-carbon bridge.

It might be noted that the bridged benzene derivatives shown in Fig. 7-11 also do not follow Bredt's rule.

c. Some Further Reactions of the Norbornane System. Of the various bridged systems so far discussed, the norbornane (bicyclo[2.2.1]heptane) system is probably the most important because of its frequent occurrence in the bicyclic monoterpenes (cf. Fig. 10-37) and because of its easy accessibility by the Diels-Alder reaction starting with cyclopentadiene (Fig. 10-38). Some further reactions of this system will be considered in this section, especially since the system is involved in some of the Wagner-Meerwein rearrangements which make terpene chemistry so intriguing.

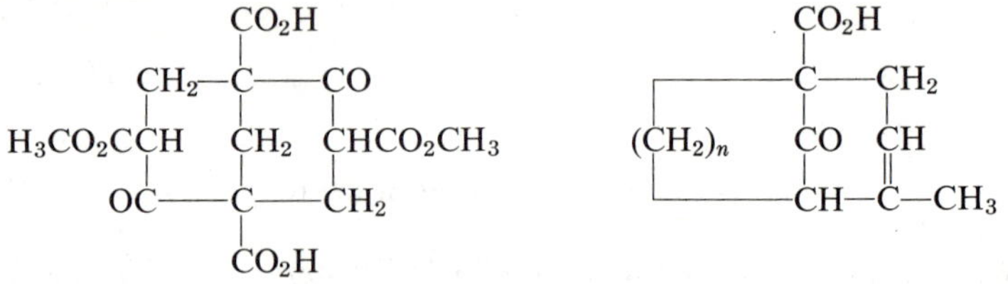

```
              CO2H                                   CO2H
               |                                      |
        CH2---C-----CO                      ┌--------C-----CH2
         |     |     |                      |        |      |
H3CO2CCH   CH2  CHCO2CH3              (CH2)n       CO     CH
         |     |     |                      |        |      ||
        OC----C-----CH2                     └--------CH----C---CH3
               |
              CO2H
               A                                      B
```

Fig. 10-52. Limitations of the Bredt rule; further examples.

Norbornane, despite its easy accessibility, is an appreciably strained system. Model considerations indicate that, in addition to the usual bond-opposition strain in a cyclohexane boat, the bridging of carbons 1 and 4 of the boat involved in the construction of norbornane introduces appreciable angle strain. Comparison of the heats of combustion of the 2-ketones derived from norbornane and its next higher homolog, bicyclo[2.2.2]octane (Fig. 10-53), suggests[72] that the extra angle strain in the norbornane system amounts to about 6.3 kcal./mole.† Norbornylene is even more strained, as

† Since the two ketones shown in Fig. 10-53 differ by a methylene group, their heats of combustion cannot be compared directly. Assuming that the bicyclooctane system II is free of angle strain, one subtracts from its heat of combustion 157.3 kcal./mole—this being the normal increment for an unstrained methylene group—to obtain the (hypothetical) heat of combustion of an angle-strain-free bicycloheptane system I. The difference between this and the actual heat of combustion of I is then ascribed to angle strain. Whether one is justified in using the 157.3-kcal. value for a methylene group, which is, after all, eclipsed, is questionable. If the proper increment were greater, the strain in the bicycloheptane system would come out to be even greater than it does.

[71] V. Prelog, P. Barman, and M. Zimmerman, *Helv. Chim. Acta,* **32,** 1284 (1949).

[72] K. Alder and G. Stein, *Ber.,* **67,** 613 (1934); G. Becker and W. A. Roth, *ibid.,* p. 627; J. Pirsch, *ibid.,* p. 1303.

evidenced by its high heat of hydrogenation[72, 73]—ca. 33 kcal./mole vs. a normal value of about 26 to 28 kcal./mole and a value of ca. 28 kcal./mole for bicyclo[2.2.2]octene—its tendency to undergo a reverse Diels-Alder reaction upon heating, its ready reaction with phenyl azide to give a dihydrotriazole,[72] its ready reduction to norbornane by means of lithium and ethylamine,[73a] and its ability to act as a dienophile in the Diels-Alder reaction. The last three points, of course, serve as evidence for an unusually reactive double bond. Norbornadiene (Fig. 10-54) is also quite strained, as evidenced by its heat of hydrogenation,[73] which averages 34 kcal./mole per double bond. Upon heating, it rearranges to tropylidene (Fig. 10-54);[74] the driving force in this process may be both loss of strain and gain of resonance energy.

I II

Fig. 10-53. Bicyclo[2.2.1]heptanone-2 and bicyclo[2.2.2]octanone-2.

Similar evidence for the strained nature of the norbornane system comes from the already mentioned rearrangement of the hydrogenated cyclopentadiene dimer to adamantane (Fig. 10-42).

Although the Diels-Alder synthesis of 2-substituted norbornanes usually produces the endo isomer by kinetic control (Fig. 10-38), this isomer is, in many cases, more sterically crowded than the exo form. Thus *exo*-dicyclopentadiene is formed by equilibration of the endo isomer upon heating† (Fig. 10-55).[75] By the same token, the exo side of the norbornane system is more accessible to a wide variety of reagents than is the endo side. This point is exemplified by the reduction of norcamphor (Fig. 10-56) to *endo*-norborneol (the hydrogen attacking the exo side of the ketone) by a wide variety of rea-

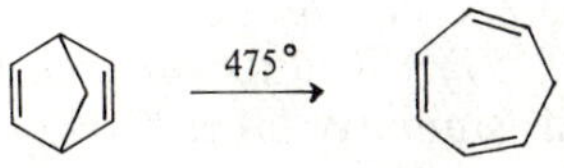

475°

Fig. 10-54. Norbornadiene-tropylidene rearrangement.

† However, just as the Alder rule (Sec. 10-3*a*) of endo addition lacks generality (Ref. 58*c*), so there are also cases where the exo-endo equilibrium favors the exo isomer only by a small amount or not at all. Thus for the cyclopentadiene–methyl acrylate adduct (*i*) the exo-endo equilibrium constant is close to unity: A. C. Cope, E. Ciganek, and N. A. Le Bel, *J. Am. Chem. Soc.*, **81**, 2799 (1959); J. A. Berson and D. A. Ben-Efraim, *ibid.*, **81**, 4083 (1959).

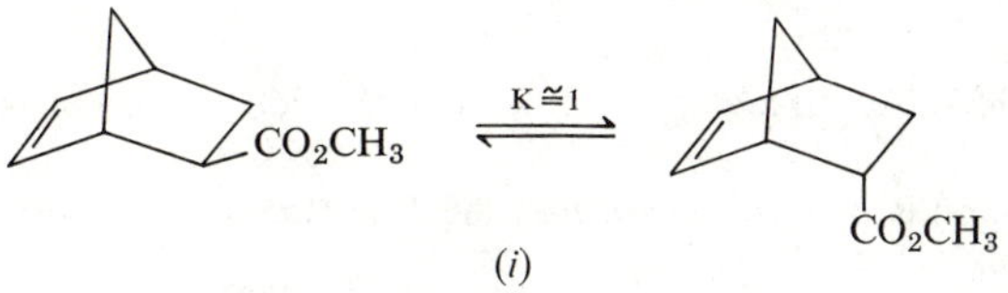

CO_2CH_3 $K \cong 1$ CO_2CH_3

(*i*)

[73] R. B. Turner, W. R. Meador, and R. E. Winkler, *J. Am. Chem. Soc.*, **79**, 4116 (1957).
[73a] J. G. Traynham, *J. Org. Chem.*, **25**, 833 (1960).
[74] Shell Chemical Corporation, New York, private communication (circular), June, 1956.
[75] K. Alder and G. Stein, *Ann.*, **504**, 216 (1933).

gents, including molecular hydrogen in the presence of a catalyst and lithium aluminum hydride.[76, 77] (The lithium aluminum hydride reduction involves steric approach control; cf. Fig. 8-54.) Another illustration (Fig. 10-56) is the oxidation of norbornylene to *exo*-norbornadiol.[78]

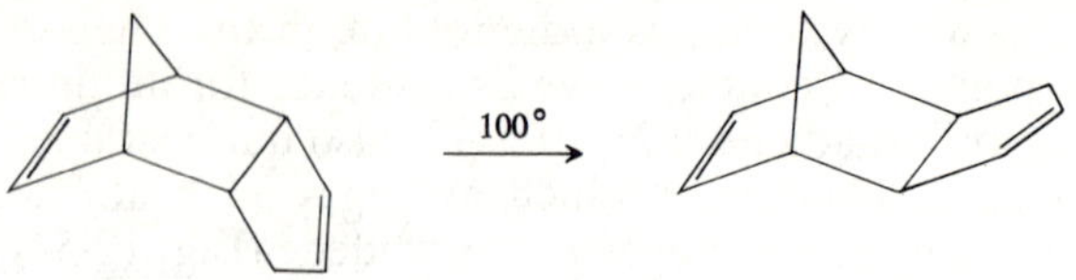

100°

Endo Exo

Fig. 10-55. Isomerization of *endo*-dicyclopentadiene.

A different steric situation prevails in the camphane system where the exo face of the molecule is strongly screened by the *gem*-dimethyl group at C_7. As a result, this system is generally approached by reagents from the endo side. Thus catalytic hydrogenation of camphor gives largely (over 90%)

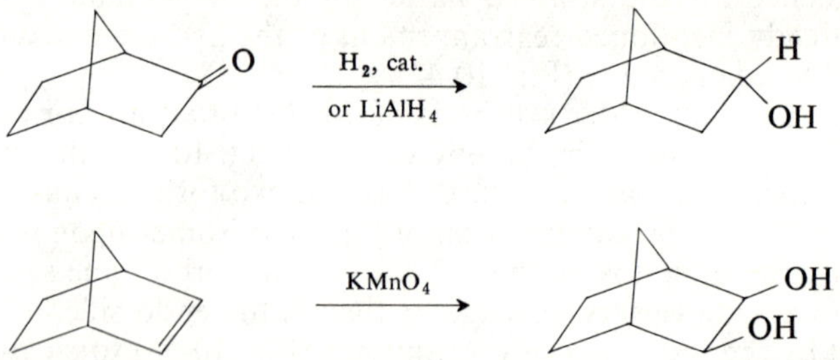

O H₂, cat. or LiAlH₄ H OH

KMnO₄ OH OH

Fig. 10-56. Approach of reagents on exo side of norbornane system.

isoborneol[79] which has the exo configuration (i.e., the hydrogen approaches from the endo side; Fig. 10-57). Reduction with lithium aluminum hydride (steric-approach-controlled) similarly gives mostly isoborneol (90%) with but a little borneol (10%).[76, 80]

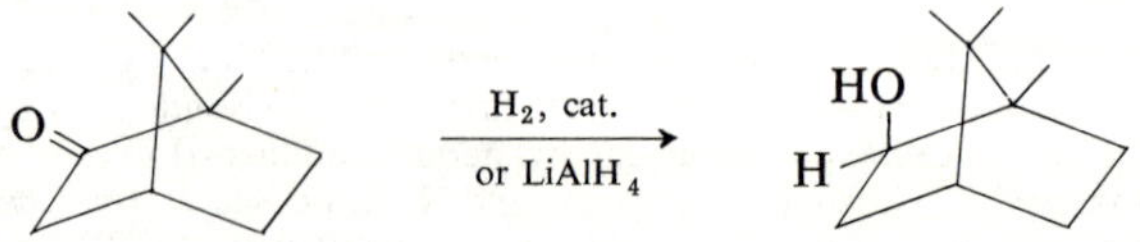

O H₂, cat. or LiAlH₄ HO H

Camphor Isoborneol

Fig. 10-57. Approach of reagents on endo side of bornane system.

[76] S. Beckmann and R. Mezger, *Chem. Ber.*, **89**, 2738 (1956); S. Beckmann, *Bull. soc. chim. France*, 1319 (1960).

[77] K. Alder, H. Wirtz, and H. Koppelberg, *Ann.*, **601**, 138 (1956).

[78] H. Kwart and W. G. Vosburgh, *J. Am. Chem. Soc.*, **76**, 5400 (1954).

[79] G. Vavon and P. Peignier, *Bull. soc. chim. France*, [4]**39**, 924 (1926).

[80] D. S. Noyce and D. B. Denney, *J. Am. Chem. Soc.*, **72**, 5743 (1950).

The last subject to be considered in this chapter is concerned with the well-known Wagner-Meerwein rearrangements in bicyclic monoterpenoids. When α-pinene is treated in ether solution with gaseous hydrogen chloride at −15°, an unstable addition product is formed which, upon warming, rearranges to bornyl chloride (Fig. 10-58).[81] The rearrangement has been depicted as proceeding through a succession of classical carbonium ions, but this picture cannot be correct, since the bornyl carbonium ion should give rise to a mixture of isobornyl and bornyl chloride.† From the stereochemical point of view, it would be attractive to postulate an attack of chloride at the bridgehead

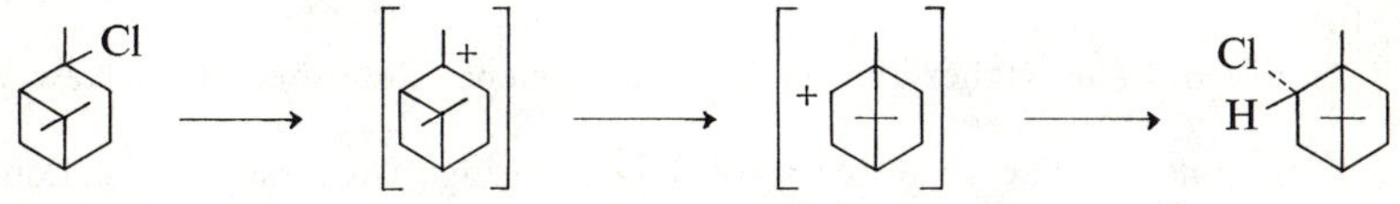

Cl Cl H + + A B

Fig. 10-58. α-Pinene hydrochloride–bornyl chloride rearrangement.

concerted with detachment of the bridge, for, as Fig. 10-58 shows, there is an inversion at the migration origin. (The chlorine attaches itself from the rear whereas the bridge was in front.) However, kinetic work[82] has shown that the rate of rearrangement is insensitive to added chloride ions, and therefore the concerted mechanism is untenable. Present evidence, discussed in detail elsewhere,[83] indicates that the rearrangement proceeds via a non-classical or mesomeric carbonium ion intermediate between A and B in Fig. 10-58, this ion being in equilibrium with the starting material and being converted slowly to the product in the rate-determining step which is acid-catalyzed.

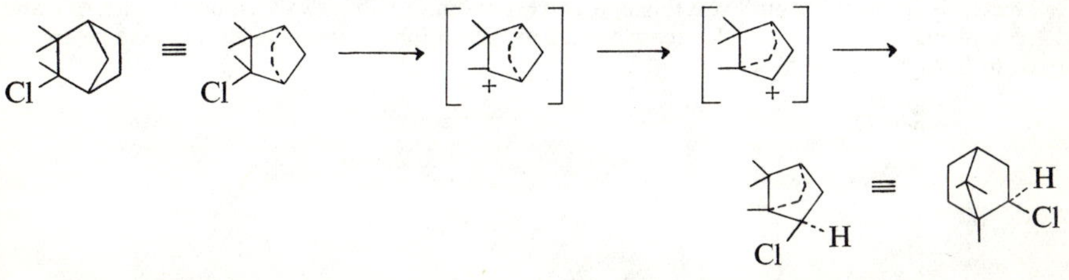

Cl ≡ Cl + + Cl H ≡ H Cl

Fig. 10-59. Camphene hydrochloride–isobornyl chloride rearrangement.

† The argument could be made that the carbonium ion gives rise exclusively to the thermodynamically more stable[81] bornyl chloride because of easier approach of the chloride ion from the unscreened endo side (*vide supra*). However, it is unlikely that this explanation could account for as high a stereospecificity as is actually observed. In any case, the argument could not apply to the rearrangement of camphene hydrochloride to *iso*bornyl chloride (*vide infra*).

[81] H. Meerwein and K. van Emster, *Ber.*, **53**, 1815 (1920); **55**, 2500 (1922).

[82] P. D. Bartlett and I. Pöckel, *J. Am. Chem. Soc.*, **60**, 1585 (1938).

[83] See E. S. Gould, "Mechanism and Structure in Organic Chemistry," Holt, Rinehart and Winston, New York, 1959, pp. 594–599; P. D. Bartlett, The Study of Organic Reaction Mechanisms, in H. Gilman, ed., "Organic Chemistry," vol. 3, John Wiley & Sons, Inc., New York, 1953, pp. 55–69.

Camphene also adds hydrogen chloride reversibly only under mild conditions.[81] Upon standing, the hydrochloride rearranges to isobornyl chloride† (Fig. 10-59); this rearrangement also involves inversion at the migration origin. The reader is advised to establish this point through the manipulation of models.

General References

[84] L. F. Fieser and M. Fieser, "Steroids," Reinhold Publishing Corporation, New York, 1959.

[85] P. de Mayo, "Mono- and Sesquiterpenoids," Interscience Publishers, Inc., New York, 1959.

[86] P. de Mayo, "The Higher Terpenoids," Interscience Publishers, Inc., New York, 1959.

[87] J. L. Simonsen, "The Terpenes," vol. II, Cambridge University Press, London, 1949.

[88] J. L. Simonsen and D. H. R. Barton, "The Terpenes," vol. III, Cambridge University Press, London, 1952.

[89] J. L. Simonsen and W. C. J. Ross, "The Terpenes," vols. IV and V, Cambridge University Press, London, 1957.

[90] K. W. Bentley, "The Alkaloids," Interscience Publishers, Inc., New York, 1957.

[91] R. H. F. Manske and H. L. Holmes, "The Alkaloids," vols. I–V, Academic Press, Inc., New York, 1950–1955.

[92] A. M. Patterson and L. T. Capell, "The Ring Index," Reinhold Publishing Corporation, New York, 1940.

[93] C. W. Shoppee, "Chemistry of the Steroids," Academic Press, Inc., New York, 1958.

[94] E. H. Rodd, ed., "Chemistry of Carbon Compounds," vol. IIA and B, Elsevier Publishing Company, Amsterdam, 1953.

---

† Actually an equilibrium is established corresponding to about 95% isobornyl chloride and 5% camphene hydrochloride. Under more drastic conditions, isobornyl chloride may be epimerized to bornyl chloride.[81]

Chapter 11

THE STEREOISOMERISM OF ALLENES AND RELATED COMPOUNDS

11-1. Enantiomerism in Allenes, Alkylidenecycloalkanes, and Spiranes

It was pointed out as early as 1875 by van't Hoff[1] that an unsymmetrically substituted allene should exist in two enantiomeric forms. A simple case is shown in Fig. 11-1; a necessary and sufficient condition for such an allene to be dissymmetric is that a $\neq$ b. The reason for the dissymmetry is that the groups a and b at one end of the system lie in a plane at right angles to those at the other end. If the doubly bonded carbon atoms are viewed as tetrahedra joined edge to edge—a view originally proposed by van't Hoff and recently reaffirmed in slightly modified form by Pauling[2]—the non-coplanarity of the two sets of groups follows directly from the geometry of the system

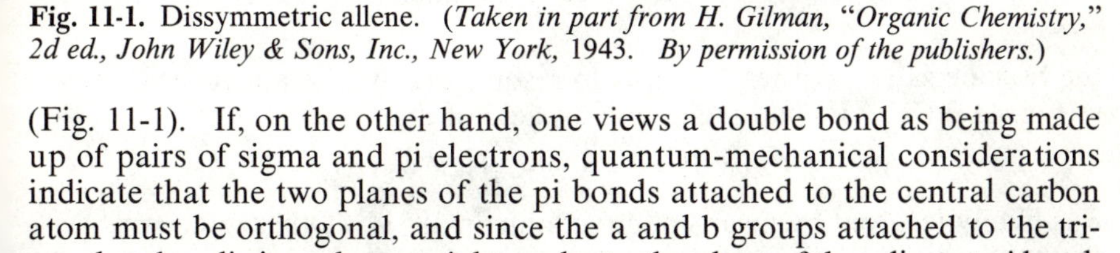

Fig. 11-1. Dissymmetric allene. (*Taken in part from H. Gilman, "Organic Chemistry," 2d ed., John Wiley & Sons, Inc., New York, 1943. By permission of the publishers.*)

(Fig. 11-1). If, on the other hand, one views a double bond as being made up of pairs of sigma and pi electrons, quantum-mechanical considerations indicate that the two planes of the pi bonds attached to the central carbon atom must be orthogonal, and since the a and b groups attached to the trigonal carbon lie in a plane at right angles to the plane of the adjacent pi bond, their planes are orthogonal to each other.

Note: All references above 23 are listed in the General References at the end of the chapter.

[1] J. van't Hoff, "La Chimie dans l' Espace," P. M. Bazendijk, Rotterdam, 1875, p. 29.

[2] L. Pauling, "The Nature of the Chemical Bond," 3d ed., Cornell University Press, Ithaca, N.Y., 1960, pp. 136–142.

Dissymmetric allenes have no dissymmetric atoms, being similar in this respect to dissymmetric biphenyls (Chap. 6) and the dissymmetric alkylidenecycloalkanes and spiranes to be discussed below. The allene type of isomerism is to be expected in any cumulene having an even number of cumulated double bonds, corresponding to the general formula $RR'C(=C)_nRR'$, where n is an even number greater than zero. Cumulenes having an odd number of double bonds (n odd), on the other hand, show cis-trans isomerism (Chap. 12).

The experimental realization of van't Hoff's prediction proved to be quite difficult, and 60 years elapsed before the first optically active allene was

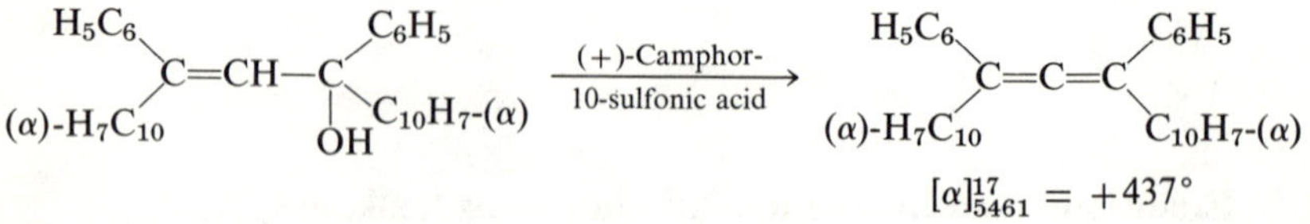

$$\underset{(\alpha)\text{-}H_7C_{10}}{\overset{H_5C_6}{>}}C{=}CH{-}\underset{OH}{C}\overset{C_6H_5}{\underset{C_{10}H_7\text{-}(\alpha)}{<}} \xrightarrow[\text{10-sulfonic acid}]{(+)\text{-Camphor-}} \underset{(\alpha)\text{-}H_7C_{10}}{\overset{H_5C_6}{>}}C{=}C{=}C\overset{C_6H_5}{\underset{C_{10}H_7\text{-}(\alpha)}{<}}$$

$$[\alpha]^{17}_{5461} = +437°$$

Fig. 11-2. Asymmetric synthesis of optically active allene.

obtained in the laboratory.[3] The route chosen was one of asymmetric synthesis: dehydration of 1,3-diphenyl-1,3-di-α-naphthyl-2-propen-1-ol (Fig. 11-2) with (+)-camphor-10-sulfonic acid gave (+)-1,3-diphenyl-1,3-di-α-naphthylallene (Fig. 11-2) in slight preponderance over its enantiomer. Fortunately, both the active allene (one part) and the *dl* compound formed concomitantly (20 parts) are high-melting solids, and the pure enantiomer could be separated from the racemate by fractional crystallization without excessive difficulty. The material has the high specific rotation $[\alpha]^{17}_{5461}$ +437° (benzene). Use of (−)-camphor-10-sulfonic acid gave the enantiomer of $[\alpha]^{17}_{5461}$ −438°. Shortly after this asymmetric synthesis was accomplished, the allenic acid shown in Fig. 11-3 ($R = CH_2CO_2H$) was resolved by crystallization of

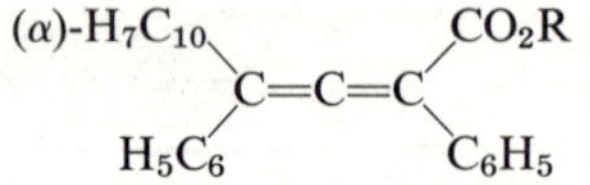

$$\underset{H_5C_6}{\overset{(\alpha)\text{-}H_7C_{10}}{>}}C{=}C{=}C\overset{CO_2R}{\underset{C_6H_5}{<}}$$

$R = CH_2CO_2H$ resolved
$R = H$ not resolved

Fig. 11-3. Allenic acids used in resolution experiments.

the brucine salt.[4] Earlier attempts to resolve the simpler allenic acid shown in Fig. 11-3 ($R = H$) had failed, but an acid of this type

$$CH_3CH{=}C{=}C(n\text{-}C_4H_9)CO_2H$$

was finally resolved by means of strychnine in 1951.[5] Other interesting syntheses of optically active allenes are the rearrangement of 1,3-diphenylpropyne, $PhC{\equiv}CCH_2Ph$, to active 1,3-diphenylallene, $PhCH{=}C{=}CHPh$, over an activated alumina catalyst impregnated with brucine (which gives the levorotatory

[3] P. Maitland and W. H. Mills, *J. Chem. Soc.*, 987 (1936); *Nature*, **135**, 994 (1935).
[4] E. P. Kohler, J. T. Walker, and M. Tishler, *J. Am. Chem. Soc.*, **57**, 1743 (1935).
[5] J. H. Wotiz and R. J. Palchak, *J. Am. Chem. Soc.*, **73**, 1971 (1951).

allene) or quinine (which gives the dextrorotatory enantiomer)[6] and the synthesis of 3-methyl-3-*t*-butyl-1-chloroallene, $CHCl{=}C{=}C(CH_3)C(CH_3)_3$, from optically active methyl-*t*-butylethynylcarbinol, $CH{\equiv}CC(CH_3)OHC(CH_3)_3$, and thionyl chloride.[7] This last example will be considered further in Sec. 11-3.

The antibiotic mycomycin,

$$HC{\equiv}C{-}C{\equiv}C{-}CH{=}C{=}CH{-}CH{=}CH{-}CH{=}CH{-}CH_2{-}CO_2H$$

is an allene derivative which occurs in nature in the optically active form.[7a]

Long before optically active allenes were known, the resolution of the similarly constituted alkylidenecycloalkane, 4-methylcyclohexylideneacetic acid (Fig. 11-4), was achieved in 1909.[8] In fact, this was the first molecule

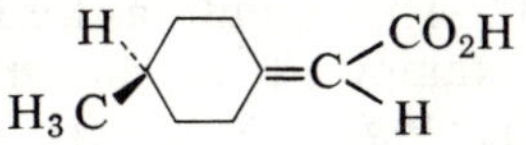

Fig. 11-4. 4-Methylcyclohexylideneacetic acid.

to be resolved which does not contain an asymmetric atom. Its asymmetry is due to the same cause as that of the allene shown in Fig. 11-1: The groups attached to the double bond extend in a plane at right angles to that defined by the groups attached at the 4 position of the ring. As a result, even when the ring is considered (for stereochemical purposes) in its average planar form, the molecule is devoid of a plane (or other element) of symmetry. A number of doubly bonded nitrogen derivatives—oximes, semicarbazones, and phenylhydrazone derivatives—shown in Fig. 11-5 have also been resolved.[9] Considering that ordinarily the syn and anti isomers of oximes (cf. Chap. 12) are not readily interconvertible, it is surprising that the compounds shown in Fig. 11-5 all show low optical stability and racemize on standing. Second-order asymmetric transformations (cf. Sec. 4-4*b*) may occur in the resolution of these compounds through their salts with alkaloids.

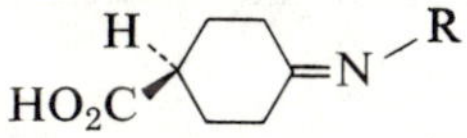

$R = OH, NHCONH_2, N(C_6H_5)COC_6H_5$

Fig. 11-5. Derivatives of 4-carboxycyclohexylidene imine.

When one replaces *both* the double bonds in an allene by rings, one obtains a bicyclic ring system in which the two rings have one atom in common.

[6] T. L. Jacobs and D. Dankner, *J. Org. Chem.*, **22**, 1424 (1957).

[7] S. R. Landor and R. Taylor-Smith, *Proc. Chem. Soc.*, 154 (1959).

[7a] W. D. Celmer and I. A. Solomons, *J. Am. Chem. Soc.*, **75**, 1372 (1953).

[8] W. H. Perkin, W. J. Pope, and O. Wallach, *Ann.*, **371**, 180 (1909); *J. Chem. Soc.*, **95**, 1789 (1909). A prior claim of the resolution of 4-methylcyclohexylideneacetic acid by W. Marckwald and R. Meth [*Ber.*, **39**, 1171 (1906)] is in error; these authors had 4-methylcyclohexeneacetic acid in hand.

[9] W. H. Mills and A. M. Bain, *J. Chem. Soc.*, **97**, 1866 (1910); **105**, 64 (1914). See also W. H. Mills and H. Schindler, *ibid.*, **123**, 312 (1923); W. H. Mills and B. C. Saunders, *ibid.*, 537 (1931).

Such bicyclic systems are called "spiranes." Although firmly entrenched, the name is somewhat inept, being derived from the Latin *spira* meaning pretzel, whereas the three-dimensional shape of a spirane (Fig. 11-6) does not resemble a pretzel at all. The two rings of the spirane, similar to the pi bonds of the allene, are orthogonal,† and the groups attached at the two ends define planes at right angles to the adjacent ring and to each other. Properly substituted

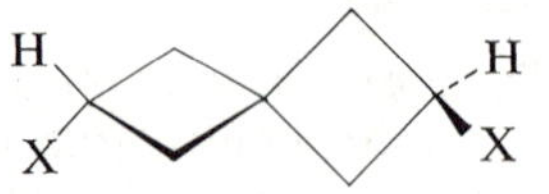

$X = CO_2H$ or NH_2

Fig. 11-6. Resolvable derivatives of spiro[3.3]heptane.

spiranes may be resolved.[10] The conditions for dissymmetry are analogous to those in allenes, simple examples being shown in Fig. 11-6.[11] The parent compound spiro[3.3]heptane (Fig. 11-6, X = H) is, of course, not resolvable because it has two planes of symmetry. These planes may be removed not only by attaching two unequal groups at the sides of the ring, as in the case of the allenes, but also by introducing suitable groups within the ring. An example of the latter kind[12] is shown in Fig. 11-7. The first spirane ever to

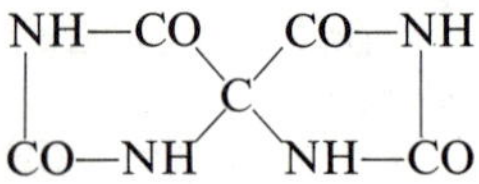

Fig. 11-7. 5,5′-Spirobihydantoin.

be resolved[13] was of this type (Fig. 11-8). Among other spiranes that have been obtained in optically active form, the quaternary ammonium salt shown in Fig. 11-9 is of interest,[14] since it suggests that quaternary nitrogen, like carbon, has tetrahedral configuration.

The compounds in Fig. 11-10 can be shown, by model studies,[10] to exist as three *dl* pairs. These molecules contain asymmetric carbon atoms as

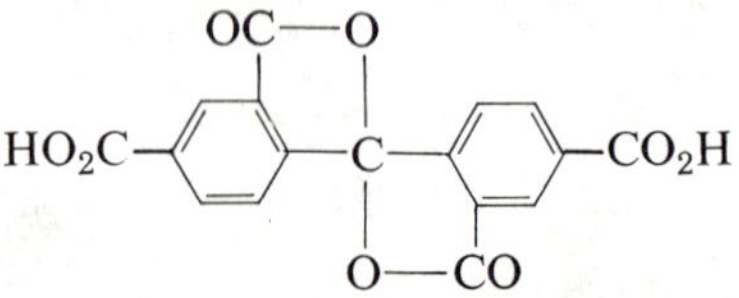

Fig. 11-8. First spirane resolved.[13]

† In the case of non-planar rings, such as cyclohexane, one should think in terms of a planar analog for the purpose of considering spirane isomerism.

[10] Cf. O. Aschan, *Ber.*, **35**, 3396 (1902).

[11] H. J. Backer and H. B. J. Schurink, *Rec. trav. chim.*, **50**, 921 (1931) (resolution was probably incomplete); S. E. Jensen and W. J. Pope, *J. Soc. Chem. Ind. (London)*, **51**, 316 (1932).

[12] W. J. Pope and J. B. Whitworth, *Proc. Roy. Soc. (London)*, **A 134**, 357 (1931).

[13] W. H. Mills and C. R. Nodder, *J. Chem. Soc.*, **117**, 1407 (1920).

[14] W. H. Mills and E. H. Warren, *J. Chem. Soc.*, **127**, 2507 (1925).

well as spirane asymmetry. Three distinct isomers were, in fact, isolated in both cases, although resolution of the presumed racemates was not attempted.[15]

H $CO_2C_2H_5$ $\overset{+}{N}$ Br^- H C_6H_5

Fig. 11-9. Resolved spirobispiperidinium salt.

11-2. Configurational Nomenclature[16]

The configurational nomenclature of the allenes, alkylidenecycloalkanes, and spiranes resembles that of the biphenyls (Sec. 6-4*c*). For purposes of assigning the configurational symbol, the compound to be named is drawn as

XH_2C-CH H_2C C CH_2 $CH-CH_2X$ O C=O O C=O $X = Br$ or H

Fig. 11-10. Spiranes containing asymmetric carbon atoms.

shown in Fig. 6-59. For example, the allene first obtained active by Maitland and Mills (Fig. 11-2) and represented in one of its enantiomeric forms in Fig. 11-11 is transformed as shown. Since near groups precede far groups (Sec. 6-4*c*), the vertically located substituents take precedence over the horizontal,

H_7C_{10} C_6H_5 C=C=C H_7C_{10} C_6H_5 ≡ $C_{10}H_7$ (1) (3) H_7C_{10} C_6H_5 (4) C_6H_5 (2)

Fig. 11-11. (R)-1,3-diphenyl-1,3-di-α-naphthylallene.

and, by virtue of the sequence rule (Sec. 5-2, page 92), naphthyl takes precedence over phenyl, since naphthyl has C—C bonds where phenyl has

[15] H. Leuchs and E. Gieseler, *Ber.*, **45**, 2114 (1912); H. Sutter and N. Wijkman, *Ann.*, **519**, 97 (1935).

[16] R. S. Cahn, C. K. Ingold, and V. Prelog, *Experientia*, **12**, 81 (1956).

only C—H. The substituents are therefore to be numbered as shown, and since substituents 1, 2, and 3 describe a right-handed screw, the configurational symbol is (R). Entirely analogous considerations apply to the spirane shown in Fig. 11-12.

HO_2C H ; CO_2H ; H ≡ CO_2H (1) ; (3) HO_2C ------ H (4) ; H (2)

Fig. 11-12. (R)-spirobicyclobutane-3,3′-dicarboxylic acid or (R)-spiro[3.3]heptane-3,6-dicarboxylic acid.

In the case of the alkylidenecyclohexane of Perkin, Pope, and Wallach (Fig. 11-13), the cyclohexane part is to be placed in front, since the groups attached to it jut out farther toward an external reference point than do the

HO_2C H ; C ; CH_3 ; H ≡ CH_3 (1) ; (3) HO_2C ------ H (4) ; H (2)

Fig. 11-13. (R)-4-methylcyclohexylideneacetic acid.

groups attached to the olefinic carbon. Hence the arrangement is as shown, and the enantiomer shown is (R). Similar considerations apply to the 4-carboxycyclohexanone oxime of Mills and Bain (Fig. 11-14). Here the nitrogen

HO ; N ; CO_2H ; H ≡ CO_2H (1) ; (3) HO ------ Zero (4) ; H (2)

Fig. 11-14. (R)-4-carboxycyclohexanone oxime.

has only one singly bonded substituent, and so the "missing" substituent is replaced by a bogus atom of atomic number zero which automatically falls lowest in the sequence rule. The last example to be considered is spiro-5,5-bihydantoin. Because of the symmetry of the molecule, it is immaterial which edge is placed in front. (This is also true of the compounds shown in Figs. 11-11 and 11-12.) In front, the proximate NH clearly takes precedence (by the sequence rule, since the atomic number of nitrogen is higher than that of carbon) over CO. In the rear, the distant NH and CO are to be considered, since these distant groups are closer to an external point located in the rear than the more proximate CO and NH groups. The situation is then as depicted in Fig. 11-15, and the symbol for the enantiomer shown is (R).

11-3. Correlation of Axial Dissymmetry and Centrodissymmetry

In most dissymmetric compounds one can discern one or more asymmetrically substituted atoms. Such atoms are sometimes called centers of asymmetry (a term which has been largely avoided in this book but is common in the literature), and the type of molecular dissymmetry which may be ascribed to the presence of asymmetric atoms is sometimes called "centroasymmetry"

(3) (4)
HN—CO
OC NH
C
OC NH
HN—CO
(1) (2)

≡

NH (1)
(3) NH - - - - - - - - - - CO (4)
CO (2)

Fig. 11-15. (R)-5,5′-spirobihydantoin.

or, more properly, "centrodissymmetry." In contrast, the dissymmetry of biphenyls, allenes, alkylidenecycloalkanes, and spiranes, which is not connected with the presence of asymmetric atoms, is sometimes called "axial asymmetry" or, better, "axial dissymmetry."† This is because one may look at such molecules in terms of an imaginary axis (running through the Ar—Ar bond in the biphenyls, through the C=C=C bond in the allenes, etc.) which is asymmetrically surrounded by the various substituents. Since the absolute configuration of centrodissymmetric molecules is known in many cases and ascertainable, in principle, in all cases, the configuration of an axially dissymmetric molecule could be established by suitable correlation with a centrodissymmetric molecule. It has already been shown in Sec. 6-4*c* how this was achieved in the case of the active biphenyls. In the present section, chemical correlations of axially dissymmetric and centrodissymmetric compounds will be discussed, and their possible usefulness in configurational assignment will be pointed out.

One such correlation[17] is shown in Fig. 11-16. Rearrangement of active

† A few active compounds, such as the ansa compounds and *p*-cyclophanes discussed in Chap. 6, fall into neither category.

[17] W. H. Perkin and W. J. Pope, *J. Chem. Soc.*, **99**, 1510 (1911).

4-methylcyclohexylideneacetic acid to 4-methylcyclohexeneacetic acid gives an inactive product, probably because a symmetrical carbonium ion intermediate is involved. Bromine addition to the active acid, however, gives a mixture of two active dibromides from which active alkylidenecycloalkanes can be regenerated by either dehydrobromination or debrominative decar-

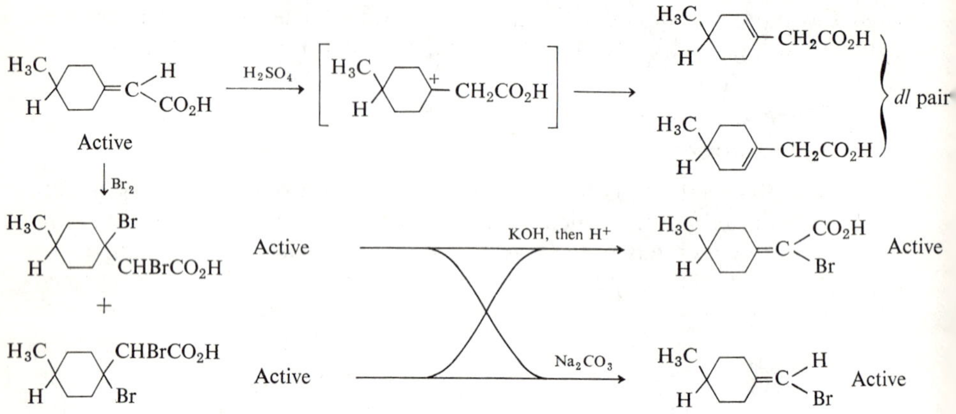

Fig. 11-16. Transformations of active 4-methylcyclohexylideneacetic acid.

boxylation (Fig. 11-16). Assuming bromine addition to be anti (cf. Chap. 12), one may assume the stereochemical correlation between the axial and centrodissymmetric compounds to be as shown in Fig. 11-17. Unfortunately it is not known which of the two dibromo acids is cis and which is trans with respect to the ring substituents (methyl and bromine), and neither is the con-

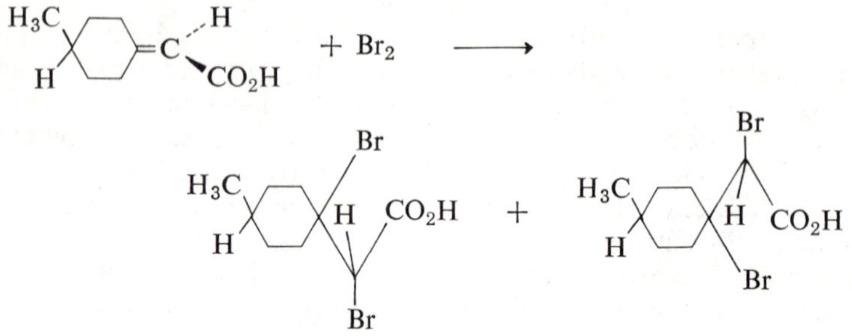

Fig. 11-17. Configurational correlation of 4-methylcyclohexylideneacetic acid with its bromine-addition products.

figuration of the asymmetric carbon (R—$\overset{*}{C}$HBr—CO_2H) in the dibromo acids known. Hence no configurational assignment to the 4-methylcyclohexylideneacetic acids is possible at this time. Similarly, the correlation of the configuration of (+)- or (−)-1,3-diphenylallene, PhCH=C=CHPh, with that of the quinine or brucine used in its synthesis from 1,3-diphenyl-1-

propyne, $PhC{\equiv}CCH_2Ph$ (*vide supra*), is not possible because the mechanism of the transformation is not clear. A more promising case for a configurational correlation would seem to be the synthesis of (+)-3-methyl-3-*t*-butyl-1-chloroallene from (+)-methylethynyl-*t*-butylcarbinol and thionyl chloride (Fig. 11-18).[7] The configuration of the (+) carbinol is not known but, on theoretical grounds to be explained in Chap. 14, is likely to be (S), as shown in Fig. 11-18.[17a] Stereospecific formation of the chloroallene from the carbinol could be either an S_N2' or an S_Ni' reaction. If it is S_N2', a double inversion should be involved,[18] whereas S_Ni' should go through a cyclic mechanism.[19]

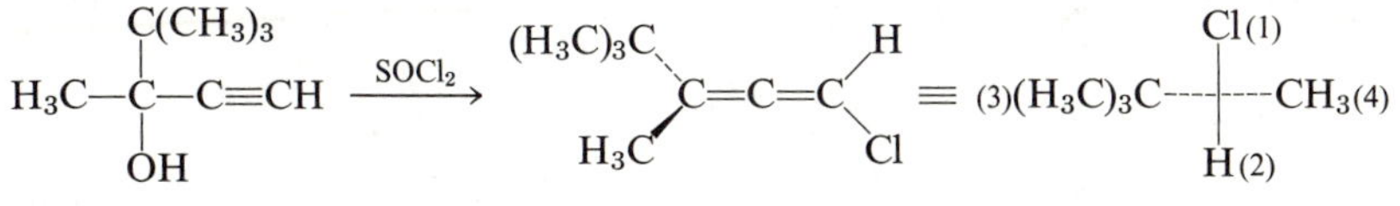

$$H_3C{-}C(C(CH_3)_3)(OH){-}C{\equiv}CH \xrightarrow{SOCl_2} (H_3C)_3C(H_3C)C{=}C{=}C(H)Cl \equiv$$

(3)$(H_3C)_3C$—[Cl(1), H(2)]—CH_3(4)

(+)-S (+) (R)

Fig. 11-18. Configurational correlation of (+)-methylethynyl-*t*-butylcarbinol and (+)-3-methyl-3-*t*-butyl-1-chloroallene.

In either case, the chlorine should approach from the side on which the hydroxyl group in the carbinol is located (see Fig. 11-19). The resultant (+)-chloroallene thus has the configuration shown in Fig. 11-18, i.e., the (R) configuration.[20]

Another such instance of configurational correlation is the pyrolysis of the dextrorotatory acetal of isobutyraldehyde and 3-butyn-2-ol to levorotatory 2,2-dimethyl-3,4-hexadienal (Fig. 11-20).[21] Here the configuration of the starting material is unambiguously known since it was made from (+)-3-butyn-2-ol and isobutyraldehyde. The carbinol has the (R) configuration

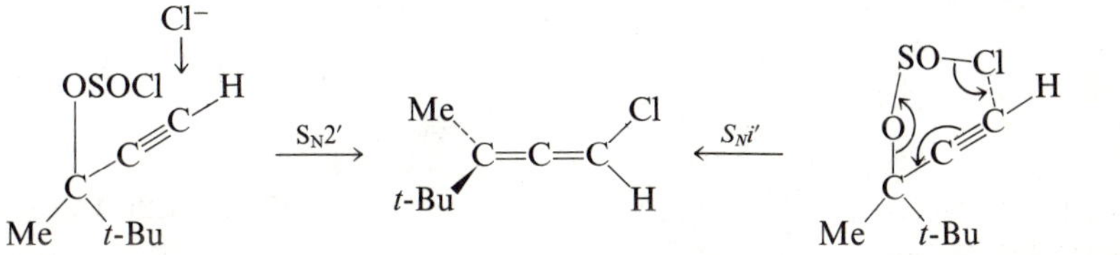

Cl⁻ → OSOCl, H, C≡C, C, Me, *t*-Bu $\xrightarrow{S_N2'}$ Me(*t*-Bu)C=C=C(Cl)H $\xleftarrow{S_Ni'}$ SO—Cl, O, H, C≡C, C, Me, *t*-Bu

Fig. 11-19. Conversion of (+)-methylethynyl-*t*-butylcarbinol to (+)-3-methyl-3-*t*-butyl-1-chloroallene.

because its enantiomer, (−)-3-butyn-2-ol, is converted by catalytic hydrogenation to (+)-2-butanol whose absolute configuration is known by correlation with tartaric acid (Chap. 5). The transformation of the acetal to the dienal (Fig. 11-20) also presents clear-cut stereochemistry, inasmuch as the Claisen rearrangement of the intermediate enol ether requires that the detach-

[17a] Now established: R. J. D. Evans and S. Landor, *Proc. Chem. Soc.,* 182, 1962.

[18] G. Stork and W. N. White, *J. Am. Chem. Soc.,* **78**, 4609 (1956).

[19] H. L. Goering, T. D. Nevitt, and E. F. Silversmith, *J. Am. Chem. Soc.,* **77**, 4042 (1955); F. F. Caserio, G. E. Dennis, R. H. DeWolfe, and W. G. Young, *ibid.,* **77**, 4182 (1955).

[20] E. L. Eliel, *Tetrahedron Letters,* no. 8, 16 (1960).

[21] E. R. H. Jones, J. D. Loder, and M. C. Whiting, *Proc. Chem. Soc.,* 180 (1960).

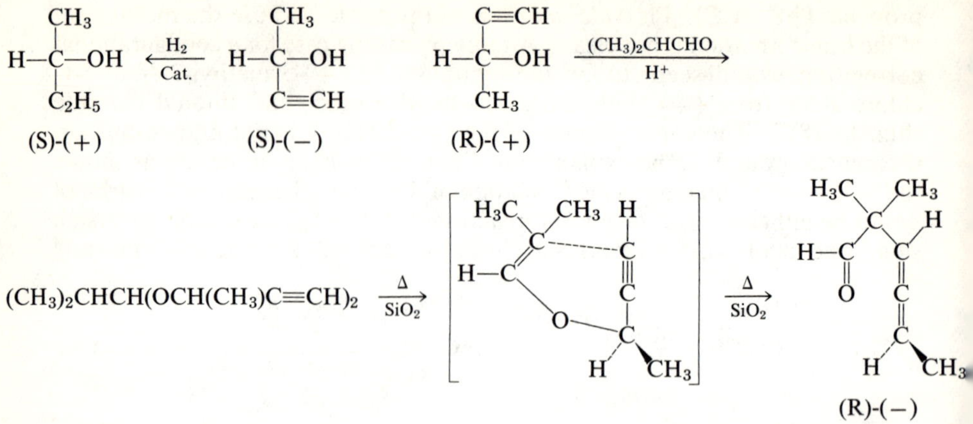

Fig. 11-20. Correlation of (+)-3-butyn-2-ol and (−)-2,2-dimethyl-3,4-hexadienal.

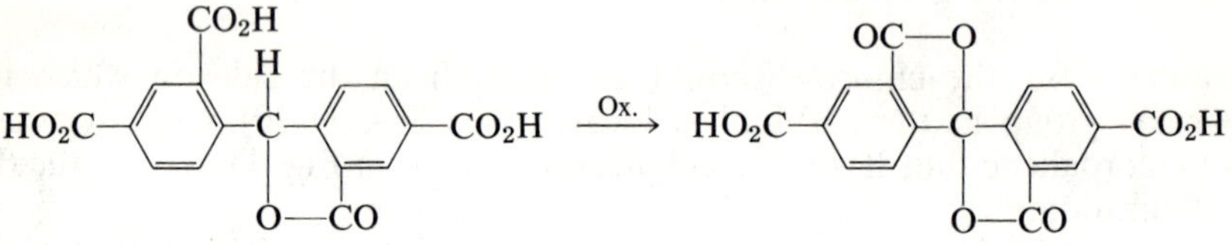

Fig. 11-21. Cyclization of an optically active lactone to a racemic spirane.

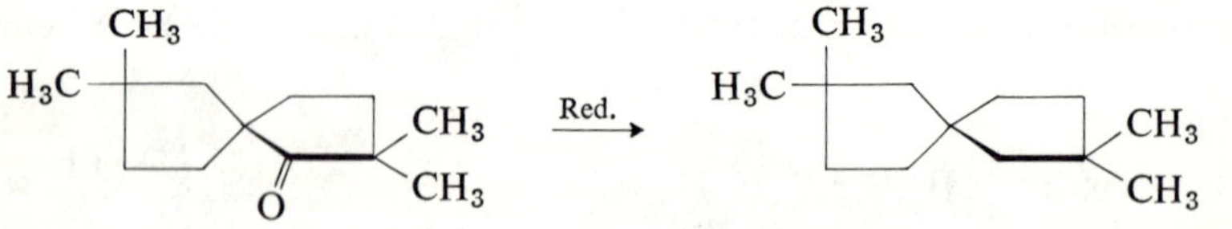

Fig. 11-22. Hypothetical configurational correlation of spirane with centrodissymmetric compound.

ment of the oxygen and attachment of the carbon occur on the same side of the allenic system. It may, therefore, be concluded that the configuration of (−)-2,2-dimethyl-3,4-hexadienal is as shown in Fig. 11-20, that is to say, R.

The configuration of a third allene, (+)-pentadienedioic acid,

$$HO_2CCH{=}C{=}CHCO_2H$$

has been established[22] by a Diels-Alder reaction with cyclopentadiene followed by ozonization of the resulting adduct to (+)-norcamphor (Fig. 10-56) of known configuration.[23]

[22] W. C. Agosta, *J. Am. Chem. Soc.*, **84**, 110 (1962). The original communication must be referred to for the details of this rather ingenious correlation.

[23] J. A. Berson, J. S. Walia, A. Remanick, S. Suzuki, P. Reynolds-Warnhoff, and D. Willner, *J. Am. Chem. Soc.*, **83**, 3986 (1961).

No reaction has as yet been described which would be suitable for the correlation of a spirane with a centrodissymmetric compound. Synthesis of the spirane shown in Fig. 11-8 by oxidative cyclization of a centrodissymmetric active lactone (Fig. 11-21) led to a racemic product.[24] It has been pointed out, however,[25] that, since the carbon atom common to the two rings in a spirane is tetracoordinate, spiranes, unlike other axially dissymmetric compounds, can be correlated with centrodissymmetric molecules by classical methods. A hypothetical example[25] is shown in Fig. 11-22; compound I could be related, in principle, to a dissymmetric tetraalkylmethane which in turn could be correlated with tartaric acid by methods described in Chap. 5.

General References

[26] R. Kuhn, Allene und Spirane, in K. Freudenberg, ed., "Stereochemie," Franz Deuticke, Leipzig, 1932, pp. 804–809.

[27] R. L. Shriner and R. Adams, Optical Isomerism, in H. Gilman, ed., "Organic Chemistry," vol. I, 2d ed., John Wiley & Sons, Inc., New York, 1943, pp. 337–343.

[28] "Nomenclature of Organic Chemistry, 1957," Butterworth & Co. (Publishers) Ltd., London, 1958, pp. 38–42, 69–70; *J. Am. Chem. Soc.*, **82**, 5560–5562, 5572–5573 (1960).

---

[24] W. H. Mills and C. R. Nodder, *J. Chem. Soc.*, **119**, 2094 (1921).

[25] K. Mislow, private communication.

Chapter 12

GEOMETRICAL ISOMERISM AND THE STEREOCHEMISTRY OF OLEFINS

12-1. Nature of Geometrical Isomerism

It has already been pointed out in Sec. 1-3 that structures containing double bonds or rings may exhibit geometrical isomerism. Geometrical isomerism in ring compounds was discussed in Chap. 7. This chapter will deal with geometrical isomerism in olefins.

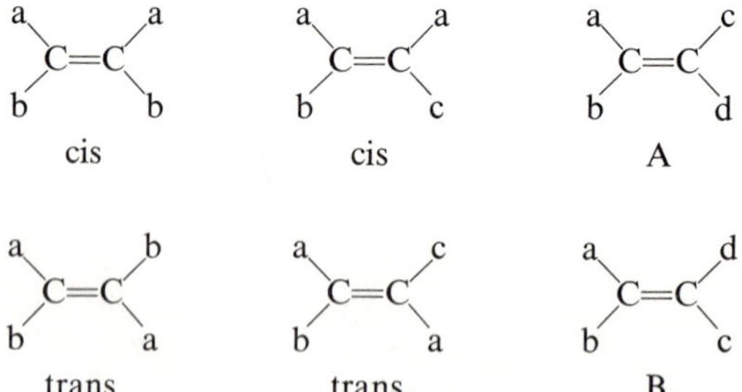

Fig. 12-1. Geometrical isomers.

The necessary and sufficient condition for geometrical isomerism to be found in an olefin abC=Ccd is similar to the condition for optical isomerism in the allene abC=C=Ccd (Chap. 11), namely, that $a \neq b$ and $c \neq d$. Three possible types are shown in Fig. 12-1. In the first two cases, where at least one group attached to the left carbon is equal to one attached to the right carbon, the geometrical isomers are distinguished by the prefixes "cis" and "trans," cis being used when the two equal groups are on the same side and trans when they are on opposite sides. This configurational nomenclature is not applicable to ethylenes in which all four substituents are different (A, B), and in fact no suitable general symbolism seems to have been developed to name such compounds. When the groups a and c are more alike than a and

Note: All references above 156 are listed in the General References at the end of the chapter.

d or b and c (e.g., if a is methyl, c ethyl, b hydrogen, and d halogen), A may be called the cis isomer and B the trans isomer.

Since, in general, the plane of the double bond is a plane of symmetry, geometrically isomeric olefins do not usually show optical activity. An exception occurs in compounds that contain asymmetric atoms (or other sources of dissymmetry) in addition to the double bond. A trivial instance of this has already been pointed out in Sec. 1-3. A more interesting case[1] is that of so-called "geometrical enantiomerism"† where the double bond leads to an increase in the number of optical isomers that are possible. This is illustrated in Fig. 12-2. In the absence of the double bond, a structure containing two like asymmetric atoms occurs as a *dl* pair and a meso form (Sec. 3-3). In the isomers shown in Fig. 12-2, where the A groups have the same configuration, the condition for geometrical isomerism is clearly not fulfilled, and there are no cis-trans isomers. When the two A groups have opposite configuration, it would appear at first sight that there should be two meso forms, one cis and one trans. Careful consideration discloses, however, that the two forms are not meso but are, in fact, dissymmetric and mirror images of each other. In other words, the structure represented in Fig. 12-2 exists in two diastereoisomeric *dl* pairs. An actual example will be considered below.

$A^R(A^R)C{=}C(a)(b)$ and $A^S(A^S)C{=}C(a)(b)$ — *dl* pair

$A^R(A^S)C{=}C(a)(b)$ and $A^S(A^R)C{=}C(a)(b)$ — *dl* pair

Fig. 12-2. Geometrical enantiomerism.

When there is more than one double bond fulfilling the condition for geometrical isomerism in a molecule, the number of geometrical isomers increases by a factor of 2 for each such double bond. Thus a structure with n different double bonds may exist in 2^n geometrically isomeric forms.‡ An example where $n = 2$ is shown[2] in Fig. 12-3.

Geometrical isomerism also occurs in cumulenes having an odd number of adjacent bonds. An example[3] is shown in Fig. 12-4. The isomers are not very stable configurationally, being readily interconverted by diffuse day-

† Called "geometric enantiomorphic isomerism" by R. E. Lyle and G. G. Lyle, *J. Org. Chem.*, **22**, 856 (1957).

‡ Unless two or more of the double bonds are identically substituted, in which case the number of isomers is diminished. Thus 2,4-hexadiene has only three geometrical isomers instead of the four predicted.

[1] Ref. 158, p. 240.

[2] L. Crombie, *J. Chem. Soc.*, 1007 (1955).

[3] R. Kuhn and D. Blum, *Chem. Ber.*, **92**, 1483 (1959).

light or by heating at 160°. Cumulenes with even numbers of double bonds show optical isomerism, as explained in Chap. 11.

A doubly bonded nitrogen may play the same part in geometrical isomerism as a doubly bonded carbon. Prominent examples are the oximes and other carbonyl derivatives for the case of C=N and the azo compounds for the case of N=N. Examples are shown in Fig. 12-5. Compounds containing C=N bonds are usually quite stable configurationally, which is surprising in view of the fact, to be discussed further in Chap. 13, that saturated tri-

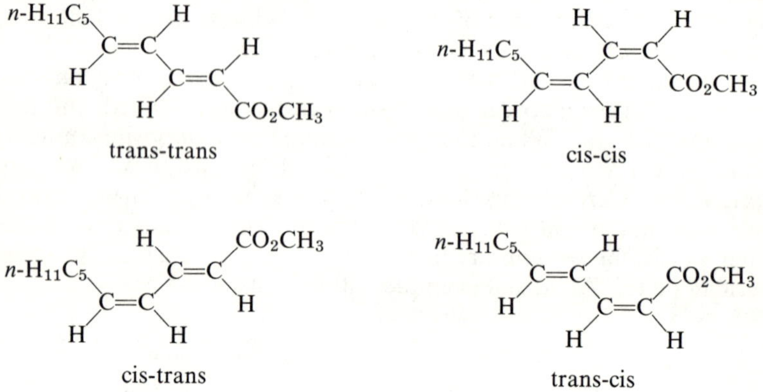

n-$H_{11}C_5$ H C=C H H C=C H CO_2CH_3
trans-trans

H H n-$H_{11}C_5$ C=C C=C CO_2CH_3 H H
cis-cis

H CO_2CH_3 n-$H_{11}C_5$ C=C C=C H H H
cis-trans

n-$H_{11}C_5$ H C=C CO_2CH_3 H C=C H H
trans-cis

Fig. 12-3. Geometrical isomerism in a diene.

covalent nitrogen compounds almost never occur as stable stereoisomers. *cis*-Azobenzene (Fig. 12-5)[4] is converted to the trans form by heating above its melting point of 71° or upon standing in solution.

The configuration of stereoisomeric oximes is usually denoted by the prefixes "syn" and "anti" instead of cis and trans. In an aldoxime, the syn isomer is the one in which the hydroxyl group of the oxime is on the side of the hydrogen of the aldehyde carbon. In a ketoxime, it is necessary to specify

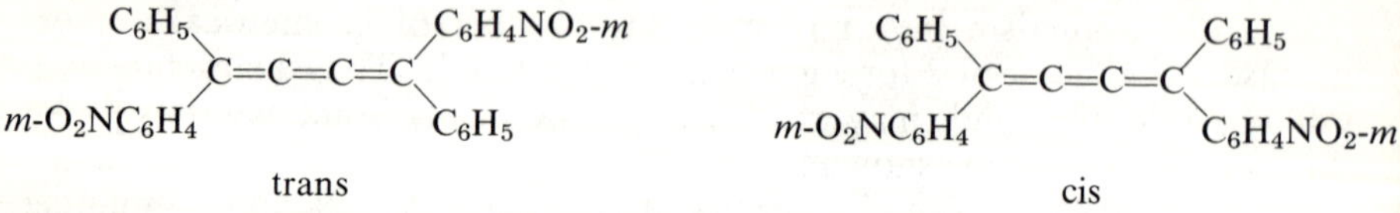

C_6H_5 $C_6H_4NO_2$-m C=C=C=C m-$O_2NC_6H_4$ C_6H_5
trans

C_6H_5 C_6H_5 C=C=C=C m-$O_2NC_6H_4$ $C_6H_4NO_2$-m
cis

Fig. 12-4. Geometrical isomerism in a 1,2,3-triene.

the group with respect to which the oxime OH group is syn. The nomenclature is illustrated in Fig. 12-5.

Geometrical enantiomerism (*vide supra*) may occur in an oxime of a ketone of the type A^R—CO—A^S, where A is a dissymmetric group. Thus, the compound shown in Fig. 12-6 has been resolved.[5] This particular example may, however, be considered alternatively as a case of alkylidenecycloalkane

[4] G. S. Hartley, *J. Chem. Soc.*, 633 (1938).
[5] R. E. Lyle and G. G. Lyle, *J. Org. Chem.*, **24,** 1679 (1959).

stereoisomerism, similar to the case shown in Fig. 11-5. Or, perhaps more logically, alkylidenecycloalkane stereoisomerism (Chap. 11) should be considered as a particular type of geometrical enantiomerism.

12-2. Determination of Configuration of Geometrical Isomers

Four methods are available for the determination of configuration of olefinic geometrical isomers. These may be compared to the methods of

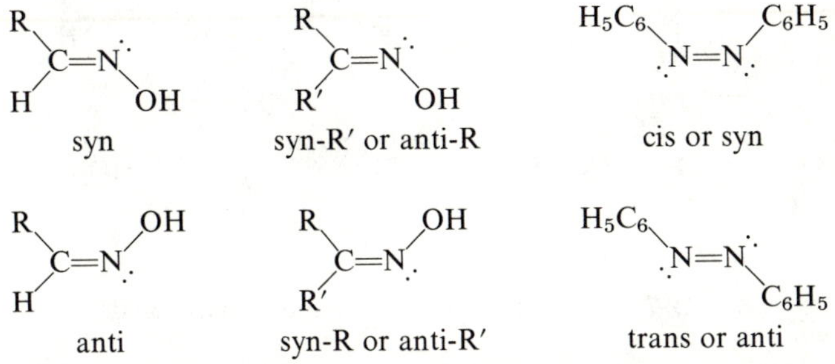

R, H: C=N, OH — syn
R, R′: C=N, OH — syn-R′ or anti-R
H_5C_6, C_6H_5: N=N — cis or syn

R, H: C=N, OH — anti
R, R′: C=N, OH — syn-R or anti-R′
H_5C_6, C_6H_5: N=N — trans or anti

Fig. 12-5. Geometrical isomerism due to C=N and N=N.

configurational assignment of enantiomers (Chap. 5) and especially to the methods used to assign configuration to geometrical isomers in rings (Sec. 7-1*b*). Two of the methods will be reviewed in this section, namely, configurational assignment by the formation of (or from) cyclic derivatives and configurational assignment by chemical correlation. Two other methods, those dependent on physical or spectral properties and those depending on the stereochemistry of olefin reactions, will be taken up later.

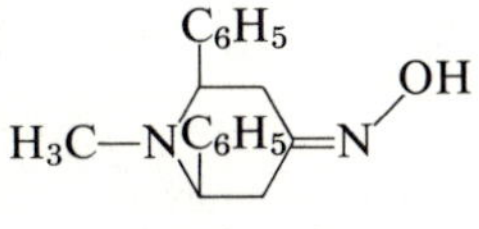

C_6H_5, OH
$H_3C—NC_6H_5=N$

Resolvable

Fig. 12-6. Example of geometrical enantiomerism.

The method of ring formation is illustrated by the behavior of maleic and fumaric acid, both of which have the butenedioic acid structure

$$HO_2CCH{=}CHCO_2H$$

Maleic acid readily forms a monomeric anhydride upon heating or treatment with dehydrating agents. Water regenerates the acid from its anhydride. Fumaric acid does not readily form an anhydride but upon vigorous heating is converted to the anhydride of maleic acid. It may be concluded that maleic acid is the cis isomer and fumaric acid is trans (Fig. 12-7). The same conclusion may be reached from the fact that maleic acid, and not fumaric acid, is obtained by oxidation of benzene or *p*-benzoquinone (Fig. 12-7), in both of which the configuration of the double bond is necessarily cis.† A similar argument allows one to assign the trans configuration to the stable *o*-hydroxycinnamic acid (coumaric acid) which is obtained from salicylaldehyde and acetic anhydride by the Perkin reaction. The cis isomer simultaneously formed cyclizes spontaneously and is isolated in the form of its lactone, coumarin. A somewhat greater resistance to cyclization is found in the *o*-aminocinnamic acids (Fig. 12-8). Here both isomers may be obtained by

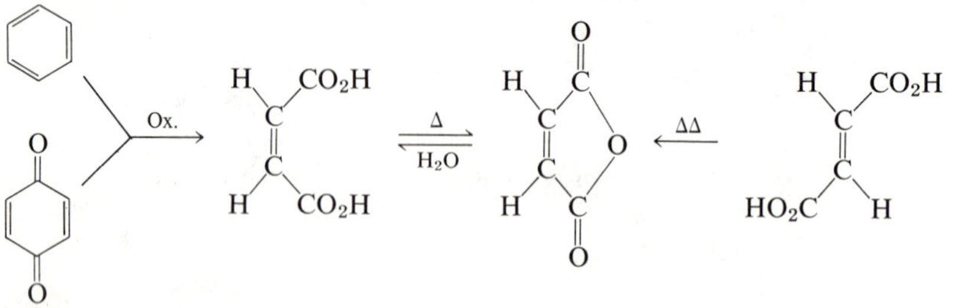

Fig. 12-7. Dehydration of maleic and fumaric acids.

reduction of the corresponding nitro acids,‡ but one is readily cyclized to carbostyril whereas the other is not.

Deamination of the isomer that is readily cyclized and that, therefore, has the cis configuration gives a cinnamic acid, m.p. 68°, which must be the cis isomer. It differs from the common cinnamic acid, m.p. 133°, which is obtained from the other isomer of *o*-aminocinnamic acid and to which the trans configuration is assigned (cf. Fig. 12-8).

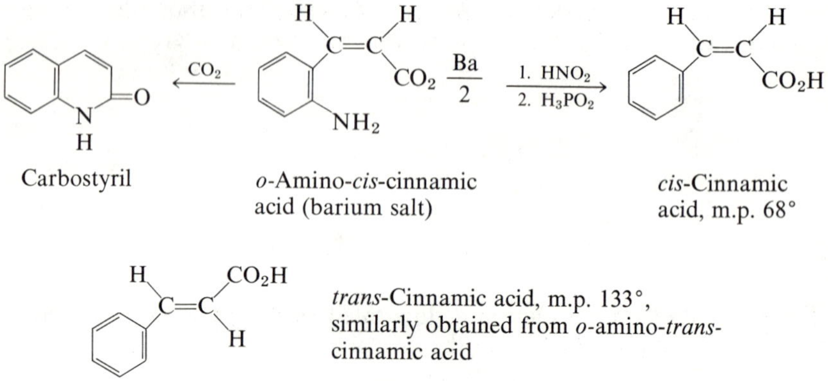

Carbostyril

o-Amino-*cis*-cinnamic acid (barium salt)

cis-Cinnamic acid, m.p. 68°

trans-Cinnamic acid, m.p. 133°, similarly obtained from *o*-amino-*trans*-cinnamic acid

Fig. 12-8. Configuration of cinnamic acid.[6]

† As pointed out in Chap. 7, no ring smaller than an eight-membered one can accommodate a trans double bond. The configuration in the five-membered ring present in maleic anhydride must thus also be cis.

‡ The cis acid may be isolated only in the form of its barium salt; spontaneous cyclization occurs upon acidification.

[6] R. Stoermer and P. Heymann, *Ber.*, **45**, 3099 (1912).

The configuration of oximes may be determined in similar fashion. The method of ring formation is illustrated by 2-chloro-5-nitrobenzaldoxime (Fig. 12-9).[7] Of the two isomers, one is readily cyclized by sodium hydroxide to an unstable indoxazole which spontaneously rearranges to a nitrile (Fig. 12-9). The other isomer is merely converted to a salt under these conditions. It is clear that the isomer that undergoes ring closure has the oxime hydroxyl group on the side of the ring and is therefore the anti isomer. Configura-

H C N OH Cl O₂N —NaOH→ [O₂N ... H C N O] —CO₂→ O₂N C≡N OH

H C N—OH Cl O₂N ⇌(NaOH / CO₂) salt

Fig. 12-9. Configuration of the 2-chloro-5-nitrobenzaldoximes by ring closure.

tional assignment by ring opening[8] is illustrated in Fig. 12-10. Benzil may form two monoximes, the syn-phenyl and the syn-benzoyl. One of these (which had previously been given the arbitrary configuration symbol β at a time when its configuration was unknown) is obtained by the oxidation of 3,4,5-triphenylisoxazole followed by hydrolysis.[8] It is clear (Fig. 12-10) that this must be the syn-benzoyl isomer and that therefore the other, or so-called

H_5C_6—C—C—C_6H_5 (N, C—C_6H_5, O) —O_3→ H_5C_6—C—C—C_6H_5 (N O, $OCOC_6H_5$) ⇌(NaOH / C_6H_5COCl) H_5C_6—C—C—C_6H_5 (N O, OH)

Benzil β-monoxime

↓ Beckmann rearrangement

$C_6H_5NHCOCOC_6H_5$

Fig. 12-10. Configuration of benzil β-monoxime by ring opening.

α isomer, is the syn-phenyl. It is of interest that the Beckmann rearrangement of the β-oxime gives the anilide of benzoylformic acid,[8]

$$C_6H_5COCONHC_6H_5$$

by phenyl migration, whereas the α-oxime gives compounds derived from dibenzoylimide, $C_6H_5CONHCOC_6H_5$, by benzoyl migration. Evidently the

[7] O. L. Brady and G. Bishop, *J. Chem. Soc.*, **127**, 1357 (1925).
[8] J. Meisenheimer, *Ber.*, **54**, 3206 (1921).

group that migrates is the one in the anti position (cf. Fig. 12-10 for the case of the β-oxime). This is quite general in the Beckmann rearrangement (cf. Sec. 5-4*f, ii*), and, in fact, the nature of the products of this rearrangement of a given oxime may be used to assign configuration to the oxime.[9]

A particularly interesting case of configurational assignment relates to the oximes of 1-acetyl-2-hydroxynaphthalene-3-carboxylic acids.[10] The anti-methyl oxime (Fig. 12-11) is resolvable because the hydroxyl group of the oxime interferes with that of the ring and produces restricted rotation (cf. Chap. 6). The syn-methyl oxime is not resolvable because rotation about the bond attached to the C_1 atom of the ring is not sufficiently restricted. The Beckmann rearrangement of these oximes proceeds in trans fashion, as is to be expected.

A second method for determination of configuration in the case of geometrical isomerism of ethylenes, as in the case of optical isomerism and in the case of geometrical isomerism of rings, is the method of chemical correlation.

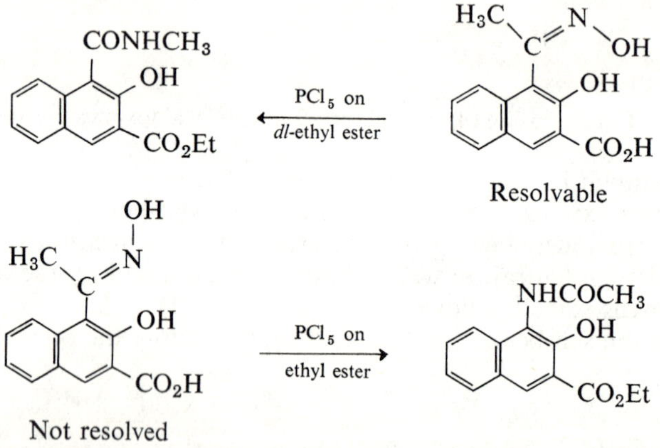

$CONHCH_3$ OH CO_2Et ← PCl_5 on *dl*-ethyl ester — H_3C C=N OH, OH, CO_2H Resolvable

H_3C C=N OH, OH, CO_2H Not resolved → PCl_5 on ethyl ester — $NHCOCH_3$ OH CO_2Et

Fig. 12-11. Oximes of 1-acetyl-2-hydroxynaphthalene-3-carboxylic acids.

Here a compound of unknown configuration is chemically related to one whose configuration is known. As in the case of ring compounds (Chap. 7), one has to guard against changes of configuration during the chemical transformations that are necessary. It is always best to carry out the correlation with *both* configurational isomers of the unknown and to make sure that they correlate with different isomers of the known.

An example of the correlative method is the transformation of γ,γ,γ-trichlorocrotonic acid, m.p. 114°, into fumaric acid by hydrolysis on the one hand and into crotonic acid, m.p. 72°, by reduction on the other,[11] as shown in Fig. 12-12. Since fumaric acid has the trans configuration (Fig. 12-7), the trichlorocrotonic acid, m.p. 114°, and the crotonic acid, m.p. 72°, must also be trans isomers. This assumes that hydrolysis of the trichlorocrotonic acid

[9] A. H. Blatt, *Chem. Revs.*, **12**, 215 (1933).
[10] J. Meisenheimer and O. Beisswenger, *Ann.*, **495**, 260 (1932).
[11] K. von Auwers and H. Wissebach, *Ber.*, **56**, 715 (1923).

does not involve a cis → trans isomerization nor that reduction of this acid involves a trans → cis isomerization. It would have been better if the correlation had been effected using *both* geometrical isomers of the trichloro acid, showing that the other isomer would give maleic acid upon hydrolysis and the low-melting crotonic acid, m.p. 15°, upon reduction, but unfortunately *cis*-γ,γ,γ-trichlorocrotonic acid is not known. It was shown, however, that maleic acid was not extensively converted to fumaric acid under the conditions of the acid hydrolysis of the trichlorocrotonic acid.†

Several of the physical methods of assigning configuration, e.g., from dipole-moment measurements; melting points; infrared, ultraviolet, or nuclear magnetic resonance spectra are also quite reliable. These will be discussed in the next section.

12-3. Physical Properties of Geometrical Isomers

a. Dipole Moments.[12, 157d, 162b] If in a compound of the type abC═Cab (Fig. 12-1) C—a has a strong bond moment but C—b has not, the cis isomer should have a considerable over-all dipole moment. On the other hand, in

H, CO_2H C═C HO_2C, H — Fumaric acid ← (H_2SO_4 / H_2O) — H, CCl_3 C═C HO_2C, H — γ,γ,γ,-Trichloro-crotonic acid — ((1) Zn,AcOH / (2) Na · Hg) → H, CH_3 C═C HO_2C, H — Crotonic acid, m.p. 72°

Fig. 12-12. Correlation of crotonic acid, m.p. 72°, with fumaric acid.

the centrosymmetrical trans isomer the bond moments are opposed, and the over-all moment is zero. Thus, *cis*-1,2-dichloro-, *cis*-1,2-dibromo-, and *cis*-1,2-diiodoethylene have dipole moments of 1.89, 1.35, and 0.75 D, respectively, whereas the moments of the corresponding trans isomers are zero. The difference is clearly useful in the assignment of configuration. One might expect that differences in dipole moments between geometrical isomers would also be found in molecules of the type abC═Cbc, where C—a and C—c have appreciable bond moments and C—b does not. Here one would not expect the dipole moment of the trans isomer to be zero (unless the bond moments of C—a and C—c are the same), but it should be smaller than the corresponding moment of the cis isomer. A case in point is 1-chloro-2-bromoethylene, CHCl═CHBr, the dipole moment being 1.55 D for the cis isomer and zero for the trans (the bond moments of C—Cl and C—Br

† A more precise use of the correlative method is the proof of the configuration of *cis*- and *trans*-1,4-dichloro-2-butene by lithium aluminum hydride reduction to *cis*- and *trans*-2-butene of known configuration (Sec. 12-3), described in Ref. 117.

[12] For sources of data, see L. G. Wesson, "Tables of Electric Dipole Moments," Technology Press, M. I. T., Cambridge, Mass., 1948; Landolt-Börnstein, "Zahlenwerte und Funktionen," vol. 1, pt. 3, 6th ed., Springer-Verlag, Berlin, Vienna, 1951.

are nearly identical). A contrary observation has, however, been reported[13] for 1-chloro-2-iodoethylene, $\mu_{cis} = 0.57$ D, $\mu_{trans} = 1.27$ D. This has been explained on grounds that the iodine may be electron-donating rather than electron-withdrawing: $\overset{+}{I}{=}CH{-}\overset{-}{C}H{-}Cl$; however, the explanation seems unlikely, since the larger of the two observed moments, 1.27 D, is smaller than the moment of vinyl chloride, $CH_2{=}CHCl$ (1.44 D). The fact that the 1.27-D moment is about half-way between that of *cis*-CHCl=CHCl and *cis*-CHI=CHI rather makes one believe that it should be assigned to the cis isomer of CHCl=CHI.† Dipole moments cannot easily be predicted for ethylenes with more complicated substituents in which rotation about the single bonds linking the substituents to the ethylene system leads to loss of cylindrical symmetry. Thus the dipole moment for diethyl maleate (2.54 D) is only slightly higher than that for its trans isomer diethyl fumarate (2.38 D), and the dipole moment for *cis*-2-butene-1,4-diol, $HOCH_2CH{=}CHCH_2OH$ (2.48 D), is nearly the same as that for the trans isomer (2.45 D).

When one of the substituents on the ethylenic system is electron-donating and the other is electron-withdrawing, the dipoles are additive rather than subtractive. Under these circumstances the trans isomer (in which the dipoles are parallel and therefore fully additive) has the higher moment. An example is 1-chloro-1-propene, $ClCH{=}CHCH_3$: cis, $\mu = 1.71$ D; trans, $\mu = 1.97$ D.

Relatively little work has been done on the dipole moments of simple olefinic hydrocarbons. The dipole moment of *trans*-2-butene is zero; that of cyclohexene, a cis olefin, is 0.75 D. *trans*-Cyclodecene has a moment of 0 to 0.15 D whereas the cis isomer has $\mu = 0.44$ D.[14] In the cycloöctene series, the cis isomer has a dipole moment of 0.43 D, but the strained trans isomer has the unusually large moment of 0.82 D.[15] These data will be discussed further below.

Geometrical isomerism about N=N bonds also manifests itself in differences in dipole moments. The moment of *cis*-azobenzene is 3.0 D whereas that of the trans isomer is zero.

b. Melting Points, Boiling Points, Densities, and Refractive Indices. Since the trans isomer has, in general, greater symmetry than the cis, it fits into the crystal lattice more easily and therefore usually has a higher melting point. Examples are maleic acid, m.p. 130°, and fumaric acid, m.p. 300°, their

† The dipole moment of *p*-chloroiodobenzene, 0.49 D, differs but slightly from the difference of the dipole moments of chlorobenzene (1.56 D) and iodobenzene (1.38 D). This also speaks against resonance forms with positive iodine being of major importance. The configuration of the chloroiodoethylenes has been assigned by H. Van de Walle and A. Henne [*Bull. soc. chim. Belges,* **34,** 399 (1925)] on the basis of the fact that one isomer eliminates hydrogen iodide with base about twice as fast as the other. Since this elimination should go more easily for trans substituents (see Sec. 12-5), the faster isomer should have the hydrogen and iodine trans, i.e., the chlorine and iodine cis. It would appear, however, that the difference in elimination rate was too small to make this assignment safe.

[13] J. Errera, *Physik. Z.,* **29,** 689 (1928).
[14] N. L. Allinger, *J. Am. Chem. Soc.,* **79,** 3443 (1957).
[15] N. L. Allinger, *J. Am. Chem. Soc.,* **80,** 1953 (1958).

methyl esters, m.p. −8.4° and +101°, respectively; cinnamic acid, cis, m.p. 68°, trans, m.p. 133°; crotonic acid, cis, m.p. 15°, trans, m.p. 72°; oleic acid, m.p. 16°, and its trans isomer elaidic acid, m.p. 51°; 1,2-dihaloethylenes, *cis*-dichloro, m.p. −80.5°, *trans*-dichloro, m.p. −50°, *cis*-dibromo, m.p. −53°, *trans*-dibromo, m.p. −6.5°, *cis*-diiodo, m.p. −14°, *trans*-diiodo, m.p. +72°; stilbene, cis, m.p. 1°, trans, m.p. 125°; 2-butene, cis, m.p. −139°, trans, m.p. −106°. There seem to be very few exceptions† to the rule that the cis isomer has the lower melting point. Along with this goes a higher solubility of the cis isomer, for example (in water), maleic acid, 78.8 g./100 ml.; fumaric acid, 0.7 g./100 ml.; *cis*-crotonic acid, 40.0 g./100 ml., *trans*-crotonic acid, 8.3 g./100 ml.; *cis*-cinnamic acid, 14.4 g./100 ml., *trans*-cinnamic acid, 0.1 g./100 ml. (all at 25°).

The relation of boiling point, density, and refractive index to configuration is not as direct as that of melting point and solubility. Since boiling point, density, and refractive index are inverse functions of molecular volume (cf. Sec. 8-4), the isomer that has the higher value for one of the three properties usually has the higher value for the other two also. As mentioned in Sec. 8-4, these properties can often be predicted on the basis of either the conformational rule or the dipole rule.[14, 16] Since cis and trans olefins almost always differ in dipole moment, the dipole rule should apply, i.e., the isomer of higher dipole moment should have the higher physical constants. Examples are shown in Table 12-1. The isomer of higher dipole moment is always listed first. (In some cases, the dipole moments are known from actual measurement; in others, their order has been guessed on the basis of what has been said in Sec. 12-3*a*.) It is found that boiling point, density, and refractive index usually correlate well with dipole moment,‡ although they do not correlate with configuration as such. Thus, whereas in the case of 1,2-dichlorethylene, 2-butene, diethyl butenedioate, and cyclodecene, the cis isomer has the higher constants, in the case of 1-chloro-1-propene, crotononitrile, and cyclooctene the trans isomer has the higher constants. It may also be seen that the conformational rule (correlation of higher constants with higher heat content) does not apply, inasmuch as *cis*-cyclodecene (cf. Table 7-3) and *cis*-1,2-dichloroethylene (cf. Sec. 12-4 below) both have lower enthalpy than their geometrical isomers and yet have higher physical constants.

c. Acid Strength.[162c, 165] The acid strengths of dibasic ethylenic acids are strongly dependent on configuration. This is a consequence of Bjerrum's law[17] which may be expressed in the form $pK_2 - pK_1 = 0.60 + 2.3N \cdot e^2/RT \cdot D \cdot r$, where pK_1 and pK_2 refer to the first and second dissociation constants of the dibasic acid, e is the electronic charge, N is Avogadro's number, R is the

† If the configurations of the 1-chloro-2-iodoethylenes have been assigned correctly (see, however, Sec. 12-3*a*), they are an exception in that the "cis" isomer melts at −36.4° and the "trans" isomer at −41°.

‡ CHCl=CHI again falls out of line (see above).

[16] A. E. van Arkel, *Rec. trav. chim.*, **51**, 1081 (1932); **52**, 1013 (1933); **53**, 246 (1934).

[17] N. Bjerrum. *Z. physik. Chem.*, **106**, 219 (1923).

gas constant, T is absolute temperature, D is the dielectric constant†, and r is the distance between the charges in the dianion. If this distance is large, $pK_2 - pK_1$ approaches 0.60 (= log 4); that is, K_1/K_2 becomes 4, which is simply the statistical ratio, considering that there is only one way of having the diacid or the dianion but there are two ways of getting the intermediate monoanion. In actual fact, the K_1/K_2 ratio always exceeds 4 because the two carboxyl groups are close enough together so that the electrostatic effect of one C=O dipole facilitates the departure of the first proton from the other carboxyl group, increasing K_1, whereas the electrostatic effect of the CO_2^- group in the monoanion retards the departure of the second proton, thus decreasing K_2. Since these electrostatic effects are greater, the closer together

Table 12-1
Illustrations of the Dipole Rule

Compound	b.p., °C. (760 mm.)	n_D^{20}	d_4^{20} g./ml.
CHCl=CHCl: cis	60.3	1.4486	1.2835
trans	48.4	1.4454	1.2565
CH_3CH=CHCl: trans	37.4	1.4054	0.935
cis	32.8	1.4055	0.9347
CHCl=CHI,[a] μ = 1.27 D	113	1.5715	2.1048 (15°)
μ = 0.57 D	117	1.5829	2.2080 (15°)
CH_3CH=$CHCH_3$: cis	3.7	1.3931 (−25°)	0.6213
trans	0.9	1.3848 (−25°)	0.6044
CH_3CH=CHCN: trans	122	1.4216	0.8239
cis	108	1.4182	0.8244
EtO_2CCH=$CHCO_2Et$: cis	223	1.4415	1.067
trans	218	1.4411	1.052
Cycloöctene: trans	75 (78 mm.)	1.4741 (25°)	0.8456 (25°)
cis	74–75 (84 mm.)	1.4684 (25°)	0.8448 (25°)
Cyclodecene: cis	194–195 (740 mm.)	1.4858	0.8760
trans	194–195 (740 mm.)	1.4821	0.8674

[a] Exception to the rule.

the two carboxyl groups are (this is merely a qualitative statement of Bjerrum's law), a cis dicarboxylic acid generally has a larger K_1, a smaller K_2, and a much larger K_1/K_2 ratio than its trans isomer. Maleic and fumaric acids provide a clear-cut example. For maleic acid (cis), K_1 is 1.3×10^{-2} and K_2 is 3.2×10^{-7}, K_1/K_2 being 40,000. For fumaric acid (trans), K_1 is 1.0×10^{-3} and K_2 is 3.2×10^{-5}, K_1/K_2 being 32.

More subtle differences between geometrical isomers are found in monobasic acids. The cis isomer is usually stronger but not by much. For exam-

† It might be thought that D is the dielectric constant of the medium in which the pK measurement is carried out. However, this assumption leads to some difficulty, inasmuch as part of the space between the charges of the anions is occupied by the molecule of the dibasic acid rather than by solvent. This molecule, being hydrocarbon-like, usually has a much lower dielectric constant than the solvents in which pK is measured. For a possible solution of this difficulty, see J. G. Kirkwood and F. H. Westheimer, *J. Chem. Phys.*, **6**, 506, 513 (1938).

ple, *cis*-crotonic acid has a pK_a of 4.44 whereas that of the trans isomer is 4.70; the ratio of the two acidity constants is less than 2. A slightly greater ratio (about 3) is found between *cis*-cinnamic acid, $pK_a = 3.96$, and *trans*-cinnamic acid, $pK_a = 4.44$. It is believed that these small differences are due to steric inhibition of resonance. Resonance of the type RCH=CH—C(OH)=O ↔ $R\overset{+}{C}H$—CH=C(OH)—$\overset{-}{O}$ is more important in the free acid than in its anion (where it would lead to the accumulation of two negative charges on the oxygen atoms of the carboxyl carbon). Therefore, the acid is more stabilized by this type of resonance than is the anion; i.e., the effect of the resonance is acid-weakening. When the resonance is inhibited (as by an ortho substituent in benzoic acid or a cis substituent in acrylic acid) the acid becomes stronger. The effect is greater in cinnamic acid than in crotonic acid, because in cinnamic acid the benzene ring is also involved in the resonance and the total effect of the resonance (and therefore of its inhibition) becomes larger.

d. Ultraviolet Spectra.[162a, 164] Simple olefins, such as 2-butene, show differences in absorption between cis and trans isomers in the ultraviolet below 200 mμ.[18] These differences are complex, however, and occur in a region that is not very easily accessible to experimental measurement.

More readily observable differences between cis and trans isomers are found in olefins of the type abC=Cbc, where the substituents a and c may enter into resonance interaction with the olefinic system. (The groups a and c may be the same.) The resonance interaction may be of the polyene type, C=C—C=C, of the styrene type, C=C—C_6H_5, or of the α,β-unsaturated carbonyl type, C=C—C=O. Examples are shown in Table 12-2. It is seen that the cis isomer in most cases has its absorption maximum at a slightly shorter wavelength and that its extinction coefficient is in all cases considerably smaller than that of the trans isomer. The major contributing factor to this situation seems to be steric inhibition of resonance. The *trans*-stilbene molecule, $C_6H_5CH=CHC_6H_5$, is planar and, therefore, resonance between the double bond and the (coplanar) phenyl rings is at a maximum. In *cis*-stilbene, on the other hand, the coplanar arrangement would be extremely crowded, and scale models indicate that both benzene rings must be turned at a considerable angle to the plane of the ethylenic double bond in order to avoid excessive van der Waals repulsions. Consequently, *cis*-stilbene has less resonance energy and is less stable than the trans isomer (see Sec. 12-4). The effect of this inhibition of resonance is even more important in the photo-excited state than in the ground state; as a result, the difference in energy levels between the excited state of *cis*-stilbene and the corresponding state of *trans*-stilbene is even greater than the corresponding difference in the ground states. It is clear from Fig. 12-13 that this results in a higher energy for the electronic transition and hence, since $\lambda = h \cdot c/\Delta E$, in a lower wavelength of absorption for *cis*-stilbene as compared with the trans isomer.

[18] E. P. Carr and H. Stücklen, *J. Am. Chem. Soc.*, **59**, 2138 (1937). See also T. H. Applewhite and R. A. Micheli, *Tetrahedron Letters*, 560 (1961).

Table 12-2
Ultraviolet Spectral Characteristics of Geometrical Isomers[a]

Compound	λ_{max}	ϵ_{max}
trans-Stilbene	2955	29,000
cis-Stilbene	2800	10,500
trans-α-Methylstilbene[b]	2700	20,100
cis-α-Methylstilbene[b]	2600	11,900
trans-1-Phenylbutadiene	2800	28,300
cis-1-Phenylbutadiene	2650	14,000
trans-Cinnamic acid	2950	27,000
cis-Cinnamic acid	2800	13,500
trans-Azobenzene	3190	20,000
cis-Azobenzene	3240	15,000
All-*trans*-β-carotene	4520	152,000
15,15′-*cis*-β-Carotene	4490	92,500
Dimethyl fumarate[c]	2140	34,000
Dimethyl maleate[c]	1980	26,000

[a] For source of data, see Ref. 164, pp. 129, 134. [b] Ref. 19. [c] Ref. 20.

It is interesting that the first four compounds shown in Table 12-2 show a second absorption band around 230 mμ (not shown in Table 12-2) which is more intense in the cis isomers than in the trans. It has been suggested that this absorption is due to an excited state in which one of the groups attached to the ethylene system is nearly coplanar with it whereas the other is almost completely at right angles and contributes nothing to the resonance. (This is in contrast to the excited state discussed earlier in which both groups are inclined

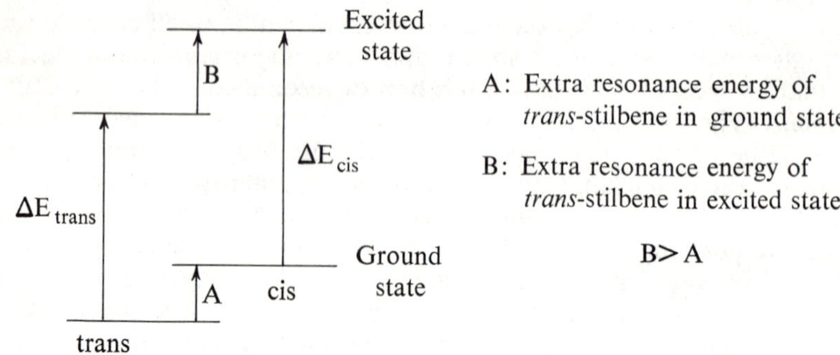

Fig. 12-13. Energy levels of ground states and excited states in stilbene.

at intermediate angles.) This excited state would be expected to be more important in the cis isomer.

Ultraviolet spectral studies have been of considerable importance in the

[19] D. J. Cram and F. A. Abd Elhafez, *J. Am. Chem. Soc.*, **74**, 5828 (1952).
[20] A. Wassermann and A. Smakula, *Z. physik. Chem.*, **A155**, 366 (1931).

assignment of geometrical configuration to the carotenoids[21] of which an example, β-carotene, is shown in Fig. 12-14.

In addition to the aforementioned configurational effects on electronic spectra, there are also important conformational effects. Butadiene is believed to exist in two stable (or metastable) conformations[22] which have been called "*s*-trans" and "*s*-cis."[23] These forms are shown in Fig. 12-15.†

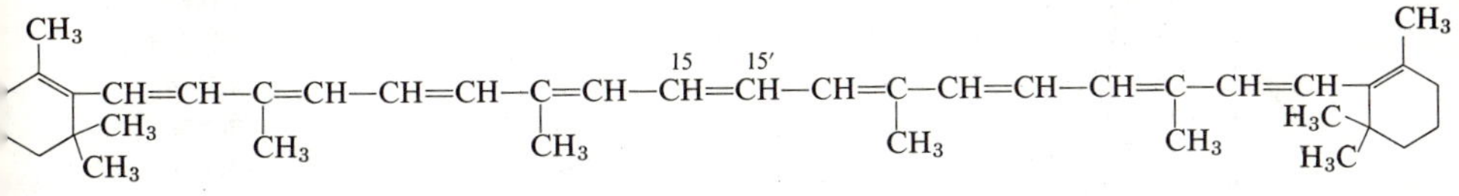

CH_3 —CH=CH—C=CH—CH=CH—C=CH—CH=CH—CH=C—CH=CH—CH=C—CH=CH— CH_3 (15, 15′; CH_3, CH_3, CH_3, CH_3, CH_3, CH_3, H_3C, H_3C)

Fig. 12-14. β-Carotene.

The difference in energy (cf. Chap. 6) between the two conformations is about 2.3 kcal./mole,[22] the *s*-trans conformation being the more stable.[24]‡ The energy barrier for conversion of the *s*-trans to the *s*-cis is about 4.9 kcal./mole.[22] Because of the predominance of the *s*-trans form, butadiene and most other acyclic dienes show the ultraviolet maximum corresponding to this form, with an extinction coefficient of ca. 20,000. On the other hand,

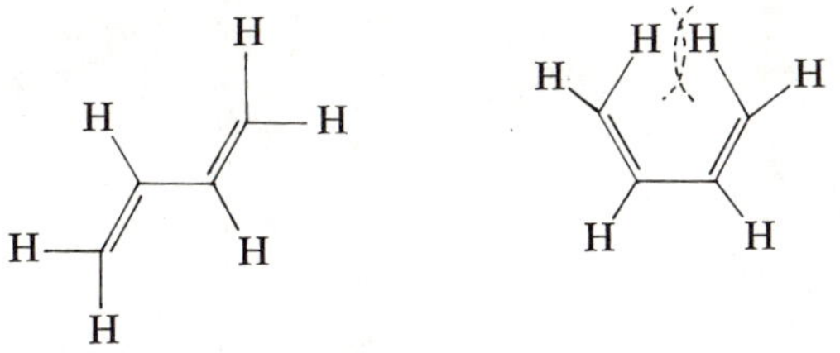

s-trans *s*-cis

Fig. 12-15. *s*-trans and *s*-cis forms of butadiene.

† It may be questioned whether the *s*-cis arrangement rather than a skew arrangement corresponds to the metastable energy minimum. In the *s*-cis arrangement the hydrogens at C_2 and C_3 are eclipsed, and two of the hydrogens at C_1 and C_4 are in bad steric interaction. As explained for biphenyls in Chap. 6, the diene system can be skewed away from the planar *s*-cis conformation by a considerable angle without there being an excessive loss of conjugation. It may be noted also that hexachlorobutadiene seems to exist in a non-coplanar rather than in the planar *s*-trans conformation for steric reasons: K. W. F. Kohlrausch and H. Wittek, *Ber.*, **75**, 227 (1942), and Ref. 24*c*.

‡ It might be noted, however, that only the *s*-cis form is capable of undergoing a Diels-Alder addition. Conjugated dienes which are constrained to the *s*-trans form do not undergo 1,4 addition with dienophiles.

[21] L. Zechmeister, *Chem. Revs.*, **34**, 267 (1944).

[22] J. G. Aston, G. Szasz, H. W. Woolley, and F. G. Brickwedde, *J. Chem. Phys.*, **14**, 67 (1946); R. G. Parr and R. A. Mulliken, *ibid.*, **18**, 1338 (1950). See also W. B. Smith and J. L. Massingill, *J. Am. Chem. Soc.*, **83**, 4301 (1961).

[23] R. S. Mulliken, *Revs. Modern Phys.*, **14**, 265 (1942); see also *J. Chem. Phys.*, **7**, 121 (1939).

[24] (*a*) V. Schomaker and L. Pauling, *J. Am. Chem. Soc.*, **61**, 1769 (1939); (*b*) K. Bradacs and L. Kahovec, *Z. physik. Chem.*, **B48**, 63 (1940); (*c*) G. J. Szasz and N. Sheppard, *Trans. Faraday Soc.*, **49**, 358 (1953).

Table 12-3
Extinction Coefficients of *s*-cis and *s*-trans 1,3-Dienes[a]

Compound	Conformation	ϵ_{max}
Butadiene	*s*-trans	21,000
1-Methylbutadiene	*s*-trans	23,000
2-Methylbutadiene	*s*-trans	24,000
1,1,4,4-Tetramethylbutadiene	*s*-trans	23,000
1,1,3-Trimethylbutadiene	[b]	8,500
Cyclopentadiene	*s*-cis	3,400
1,3-Cyclohexadiene	*s*-cis	8,000
1,2-Dimethylenecyclohexane	*s*-cis	10,000
Cholesta-3,5-diene[c]	*s*-trans	23,000
Cholesta-2,4-diene[c]	*s*-cis	6,500
Abietic acid[c]	*s*-trans	23,000
Levopimaric acid[c]	*s*-cis	7,100

[a] For sources of data see Ref. 25 and Ref. 164, p. 138.
[b] This diene may exist in an *s*-cis form, or alternatively a slightly skewed *s*-trans form may account for the low extinction.
[c] See Fig. 12-16 for formulas.

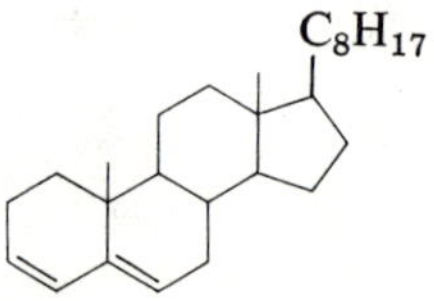

C_8H_{17} C_8H_{17}

Cholesta-3,5-diene 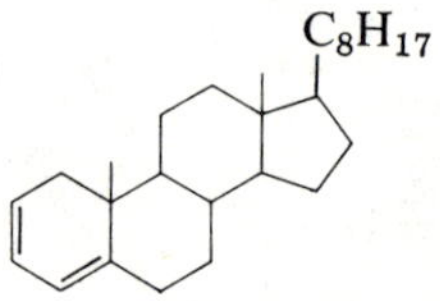 Cholesta-2,4-diene

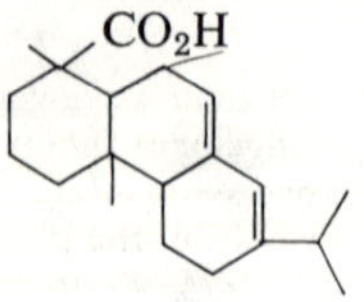

CO_2H CO_2H

Abietic acid 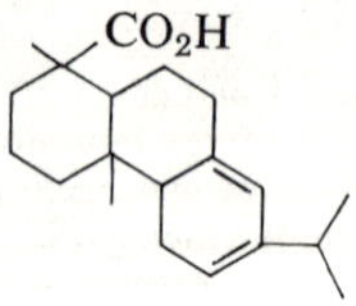 Levopimaric acid

Fig. 12-16. Structure of compounds in Table 12-3.

1,3-cyclohexadiene and other homocyclic conjugated dienes are necessarily *s*-cis. These compounds have much lower extinction coefficients ($<$10,000) in the ultraviolet, as shown in Table 12-3. Measurements of the extinction coefficient may evidently be used to distinguish "homoannular" conjugated

[25] (*a*) H. Booker, L. K. Evans, and A. E. Gillam, *J. Chem. Soc.,* 1453 (1940); (*b*) see also A. E. Gillam and E. S. Stern, "Electronic Absorption Spectroscopy in Organic Chemistry," Edward Arnold & Co., London, 1954; and Elsevier's "Encyclopedia of Organic Chemistry," Elsevier Publishing Co., Amsterdam.

dienes, such as cholesta-2,4-diene and levopimaric acid (Fig. 12-16), which are necessarily *s*-cis and "heteroannular" conjugated dienes, such as cholesta-3,5-diene and abietic acid (Fig. 12-16), which are necessarily *s*-trans. (In the homoannular dienes, the double bonds are in one and the same ring, whereas in the heteroannular dienes they are usually not.)

Table 12-4
Extinction Coefficients of *s*-cis and *s*-trans α, β-Unsaturated Ketones[a]

Compound	Conformation	ϵ_{max}
Mesityl oxide[b]	*s*-trans	14,000
1-Methylcyclohexene-3-one[b]	*s*-trans	12,950
Acetylcyclohexene[b]	*s*-trans	12,400
1-Acetyl-2-methylcyclohexene[b]	*s*-cis	6,890
Cholest-4-en-3-one[b]	*s*-trans	19,950
Cholest-4-en-3-ol-6-one[b]	*s*-cis	6,300

[a] For source of data see Refs. 25*b* and 26.
[b] Formulas in Fig. 12-17.

A similar situation is found in α,β-unsaturated ketones (Table 12-4). The normal *s*-trans conformation of the acyclic members is associated with an extinction coefficient in excess of 10,000. When steric factors force the molecule into the *s*-cis conformation, as with 2-methyl-1-acetylcyclohexene (Fig. 12-17), the extinction coefficient is much lower. In the case of the α,β-unsaturated ketones it should be noted that the homoannular representatives (such as cholest-4-en-3-one, Fig. 12-17) are *s*-trans and the heteroannular representatives (such as cholest-4-en-3-ol-6-one, Fig. 12-17) are *s*-cis. (Homoannular here means that the double bond and ketone function are in the same ring; heteroannular, that they are not.)

e. Infrared and Raman Spectra.[27, 157e, 164] Differences in infrared spectra of geometrical isomers are found in the 1650-cm.$^{-1}$ region (C=C stretching) and in the 970- and 690-cm.$^{-1}$ regions (=C—H out-of-plane vibration). For a molecular vibration to give rise to infrared absorption, it must produce a change in the dipole moment of the molecule. Clearly, such a change is not produced in the C=C stretching motion of *trans*-1,2-dichloroethylene, for the dipole moment of this molecule is zero (cf. Sec. 12-3*a*) and remains zero throughout the vibration. As a result, *trans*-1,2-dichloroethylene shows no double-bond stretching frequency in the infrared, although strong absorption due to this vibration is found in the Raman spectrum† at 1577 cm.$^{-1}$. On the other hand, *cis*-1,2-dichloroethylene shows strong infrared absorption at 1590

† In fact, in molecules of this type, *all* Raman-active vibrations are in principle infrared inactive, and vice versa.

[26] R. B. Turner and D. M. Voitle, *J. Am. Chem. Soc.*, **73**, 1403 (1951).
[27] N. Sheppard and D. M. Simpson, *Quart. Revs. (London)*, **6**, 1 (1952); L. J. Bellamy, "The Infrared Spectra of Complex Molecules," 2d ed., John Wiley & Sons, Inc., New York, 1958, chap. 3.

cm.$^{-1}$. Similar differences are shown between fumaric acid and maleic acid and between *trans*- and *cis*-3-hexene, $C_2H_5CH{=}CHC_2H_5$. The differences are less marked when the substitution on the ethylene is not symmetric, so that there is a small dipole moment even in the trans isomer. Thus, *trans*-2-hexene, $CH_3CH{=}CHC_3H_7$, shows a C=C stretching frequency at 1670 cm.$^{-1}$, although it is much less intense than the C=C stretching frequency of *cis*-2-hexene at 1656 cm.$^{-1}$. In compounds where the trans isomer has a strong dipole moment, such as crotonic acid (cf. Sec. 12-3*a*), there is also a strong C=C absorption.

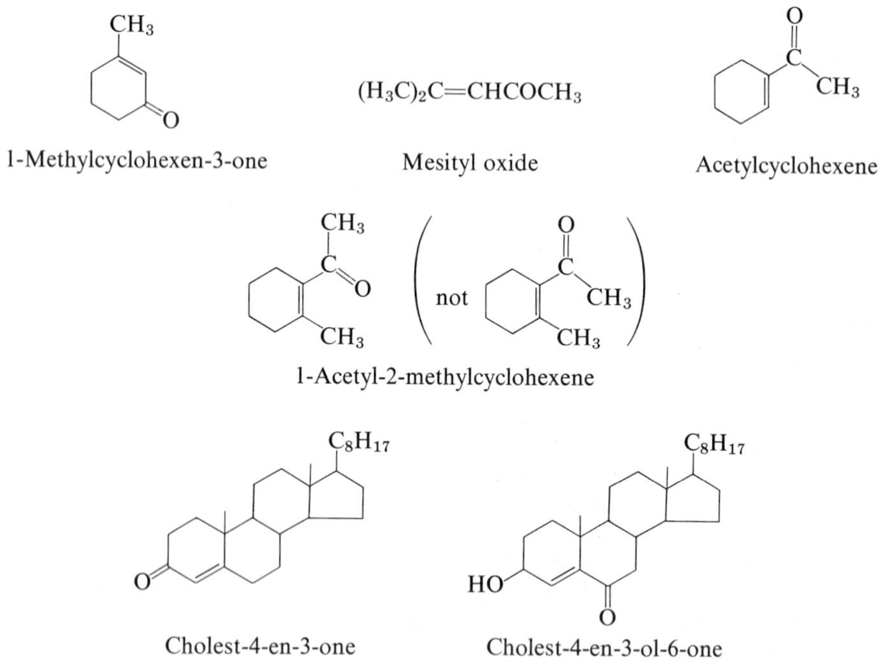

CH_3 O

$(H_3C)_2C{=}CHCOCH_3$ CH_3

1-Methylcyclohexen-3-one Mesityl oxide Acetylcyclohexene

CH_3 O CH_3 CH_3 CH_3 not

1-Acetyl-2-methylcyclohexene

C_8H_{17} C_8H_{17} O HO O

Cholest-4-en-3-one Cholest-4-en-3-ol-6-one

Fig. 12-17. Structure of compounds in Table 12-4.

There is also a small but measurable shift in the position of the C=C stretching band between cis and trans isomers. This shift can be measured in the Raman spectrum where both isomers have strong double-bond absorption. Thus cis olefins absorb at 1654 to 1657 cm.$^{-1}$ whereas trans olefins absorb at 1668 to 1671 cm.$^{-1}$.† Similar differences are found in the infrared in those instances where the trans isomer does show absorption. For example,[28] *cis*-propenylbenzene absorbs at 1653 cm.$^{-1}$ and the trans isomer at 1667 cm.$^{-1}$.

† There are other differences in the Raman spectra of cis and trans olefins; cf. W. Otting, "Der Raman-Effekt und seine analytische Anwendung," Springer-Verlag, Berlin, Vienna, 1952, p. 35; and J. H. Hibben, "The Raman Effect and Its Chemical Applications," Reinhold Publishing Corporation, New York, 1939, p. 170; also Ref. 157*e*.

[28] R. Y. Mixer, R. F. Heck, S. Winstein, and W. G. Young, *J. Am. Chem. Soc.*, **75**, 4094 (1953).

Whereas the C=C stretching band is helpful in distinguishing between cis and trans isomers in disubstituted olefins of the type aCH=CHb, it is of less general usefulness with trisubstituted olefins abC=CHc. When a and c are strongly polar groups and b is not, spectral differences between *cis*- and *trans*-abC=CHc are likely to be found. For example, *cis*-1,2-dichloro-1-propene, $CHCl=CClCH_3$, has a fairly intense absorption band at 1614 $cm.^{-1}$ whereas the trans isomer absorbs only weakly at 1615 $cm.^{-1}$. On the other hand, *cis*- and *trans*-3-methyl-2-pentene, $CH_3CH=C(CH_3)C_2H_5$, show practically no difference in the double-bond stretching region, both having medium-intensity bands at 1675 $cm.^{-1}$. Fortunately, isomers of this type can be distinguished by nuclear magnetic resonance spectroscopy (see below).

The C—H out-of-plane vibration is also often useful in distinguishing cis and trans 1,2-disubstituted olefins. The trans isomers almost invariably absorb at 895 to 990 $cm.^{-1}$. Examples are *trans*-2-butene, 964 $cm.^{-1}$; *trans*-1-chloro-1-propene, 930 $cm.^{-1}$; *trans*-1,2-dichloroethylene, 895 $cm.^{-1}$; crotonic acid, 912 $cm.^{-1}$. The cis isomers are usually free or almost free of absorption in this region and often show, instead, a band at 675 to 730 $cm.^{-1}$, e.g., *cis*-2-butene, 675 $cm.^{-1}$; *cis*-1-chloro-1-propene, 675 $cm.^{-1}$; *cis*-1,2-dichloroethylene, 697 $cm.^{-1}$.

The distinction between *s*-cis and *s*-trans α,β-unsaturated ketones (Sec. 12-3*d* above) may also be made on the basis of infrared spectra.[28a] The *s*-cis compounds show higher intensity of the C=C stretching vibration (near 1625 $cm.^{-1}$) and lower intensity of the C=O stretching frequency (near 1700 $cm.^{-1}$) than the *s*-trans compounds. Thus the intensity ratio of the C=O to the C=C band is of the order of 0.7 to 2.5 for the *s*-cis and of the order of 6 to 9 for the *s*-trans system.

f. Nuclear Magnetic Resonance Spectra. Inspection of the NMR spectra of monosubstituted olefins, $R—CH=CH_2$, indicates that there is an appreciable difference in coupling constant between the cis protons and the trans protons, the latter coupling much more strongly. Typical coupling constants[29] are 8 to 11 c.p.s. for the cis protons and 17 to 18 c.p.s. for the trans protons. Similar differences should be useful to distinguish cis and trans olefins of the type aCH=CHb. Strong coupling between the two ethylenic protons occurs only, however, if a and b are substantially different chemically; otherwise the two protons are equivalent or nearly equivalent and do not couple.

Differences in chemical shifts, rather than coupling constants, have been used[30] to differentiate cis and trans isomers of trisubstituted ethylenes of the type $CH_3aC=CHb$. The protons of the methyl group on carbon No. 1 are close enough to the cis substituent on carbon No. 2 that their observed chemical shift is influenced by the nature of this substituent (which may be either

[28a] R. L. Erskine and E. S. Waight, *J. Chem. Soc.*, 3425 (1960).

[29] J. A. Pople, W. G. Schneider, and H. J. Bernstein, "High-resolution Nuclear Magnetic Resonance," McGraw-Hill Book Company, Inc., New York, 1959, pp. 193 and 238–247; cf. M. Karplus, *J. Chem. Phys.*, **30,** 11 (1959).

[30] (*a*) L. M. Jackman and R. H. Wiley, *Proc. Chem. Soc.*, 196 (1958); *J. Chem. Soc.*, 2881, 2886 (1960); (*b*) D. Y. Curtin, H. Gruen, and B. A. Shoulders, *Chemistry & Industry (London),* 1205 (1958); (*c*) R. Morris, C. A. Vernon, and R. F. M. White, *Proc. Chem. Soc.*, 303 (1958). See also Ref. 152.

hydrogen or the atom or group b). An example is found in the methyl esters of the 2-methylbutenedioic acids (Fig. 12-18).

$$\begin{array}{ccc} H_3C & & H \\ & C{=}C & \\ H_3CO_2C & & CO_2CH_3 \end{array} \qquad \begin{array}{ccc} H_3C & & CO_2CH_3 \\ & C{=}C & \\ H_3CO_2C & & H \end{array}$$

Dimethyl citraconate δ_{CH_3} 190 c.p.s.

Dimethyl mesaconate δ_{CH_3} 184 c.p.s.

Fig. 12-18. Dimethyl citraconate and mesaconate.

In the methyl ester of the cis isomer, citraconic acid, the C-methyl group is close to the cis hydrogen, and the chemical shift of the methyl protons is 190 c.p.s. (at 40 mc., relative to the aromatic proton in toluene). In the methyl ester of the trans isomer, mesaconic acid, where the C-methyl group is close to the carbomethoxy group, the corresponding shift is only 184 c.p.s. The difference is clearly discernible and has been used to characterize the cis and trans isomers of dimethyl β-methylglutaconate,[30a]

$$H_3CO_2CCH_2C(CH_3){=}CHCO_2CH_3$$

NMR spectra have also found use in analyzing the geometrical isomerism of oximes,[31a] semicarbazones,[31b] and 2,4-dinitrophenylhydrazones.[31b]

g. X-ray and Electron Diffraction.[162d] Since the relative positions of the heavy atoms a and b in *cis*- and *trans*- aCH=CHb are different, X-ray diffraction or electron diffraction patterns of the isomers should indicate which configuration is at hand. Thus, a complete X-ray analysis of sorbic acid,[32] $CH_3CH{=}CH{-}CH{=}CHCO_2H$, indicates that this compound has the trans-trans configuration. Even an incomplete X-ray analysis may indicate whether the molecule has a center of symmetry and may thus permit assignment of configuration in compounds of the type aCH=CHa. Thus it has been confirmed[32a] that dimethyl fumarate has the trans configuration. The synthetic α,ω-diphenylpolyenes, $C_6H_5(CH{=}CH)_nC_6H_5$, prepared by Kuhn[33] have also been shown by X-ray diffraction to have the all-trans configuration.[34] The X-ray diffraction pattern of stretched rubber indicates[35] that it is the all-cis form of polyisoprene whereas gutta-percha (or balata) is the all-trans form (Fig. 12-19).

Electron diffraction may similarly be used to assign configuration to compounds that are readily vaporized, such as the 2-butenes[36] and the 1,2-dichloroethylenes.[37]

[31] (*a*) W. D. Phillips, *Ann. N. Y. Acad. Sci.*, **70**, 817 (1958); E. Lustig, *J. Phys. Chem.*, **65**, 491 (1961); (*b*) G. J. Karabatsos, J. D. Graham, and F. M. Vane, *J. Am. Chem. Soc.*, **84**, 37 (1962).
[32] K. Lonsdale, J. M. Robertson, and I. Woodward, *Proc. Roy. Soc. (London)*, **A178**, 43 (1941).
[32a] I. E. Knaggs and K. Lonsdale, *J. Chem. Soc.*, 417 (1942).
[33] R. Kuhn, *J. Chem. Soc.*, 605 (1938).
[34] J. Hengstenberg and R. Kuhn, *Z. Kryst. Mineral.*, **75**, 301 (1930); **76**, 174 (1930).
[35] C. W. Bunn, *Proc. Roy. Soc. (London)*, **A180**, 40 (1942).
[36] L. O. Brockway and P. C. Cross, *J. Am. Chem. Soc.*, **58**, 2407 (1936).
[37] L. O. Brockway, J. Y. Beach, and L. Pauling, *J. Am. Chem. Soc.*, **57**, 2693 (1935).

12-4. Relative Stability and Interconversion of Geometrical Isomers

a. Relative Stability of cis-trans Isomers. As pointed out in Chap. 7, "stability" may mean either thermochemical stability, measured by heat of formation, or thermodynamic stability, measured by free energy of formation. Thus the difference in stability between cis and trans isomers may be expressed either as a difference in their enthalpy or as a difference in their free energy. In most cases the isomer of lower enthalpy is also that of lower free energy. An exception to this is cyclododecene (Table 7-3) where the trans isomer has the higher enthalpy but (because of a favorable entropy) the lower free energy; such exceptions will probably prove to be very rare.

Thermochemical-stability differences between geometrical isomers may be measured as differences between their heats of combustion.[37a] Since the heats of combustion are large and their differences are small, this method is limited in accuracy. An alternative method is to determine the differences in heats of hydrogenation between the two isomers. The hydrogenation products of stereoisomeric cis and trans olefins are the same (except in the

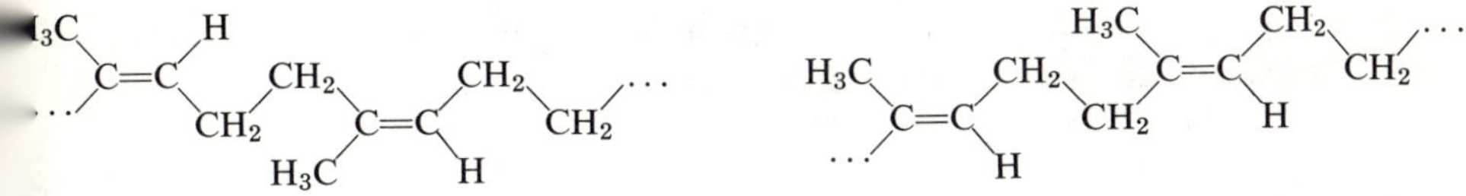

Rubber Gutta-percha

Fig. 12-19. Rubber and gutta-percha.

uncommon case of a tetrasubstituted ethylene, abC=Ccd, where none of the substituents is hydrogen; cf. Sec. 12-6*a*); therefore, the difference in heats of hydrogenation of the two isomers is equal to the difference in their heats of formation. If ΔH_{cis} denotes the heat of hydrogenation of the cis isomer, H_{cis} the heat of formation of the cis isomer, and H_P the heat of formation of the hydrogenation product, $\Delta H_{cis} = H_P - H_{cis}$. Similarly, for the trans isomer, $\Delta H_{trans} = H_P - H_{trans}$. Hence $\Delta H_{cis} - \Delta H_{trans} = H_{trans} - H_{cis}$.

For the isomeric 2-butenes, the heat of combustion of the cis isomer is 647.81 kcal./mole and that of the trans isomer is 646.81 kcal./mole, the difference being 1.0 kcal./mole.[38] As with most geometrical isomers, the cis has the higher heat content. The heats of hydrogenation of the 2-butenes are[39] 28.570 kcal./mole for the cis and 27.621 for the trans, the difference of 0.95 kcal./mole being in excellent agreement with the difference in the heats

[37a] For example, H. F. Bartolo and F. D. Rossini, *J. Phys. Chem.,* **64**, 1685 (1960); J. D. Rockenfeller and F. D. Rossini, *ibid.,* **65**, 267 (1961).

[38] F. D. Rossini, K. S. Pitzer, R. L. Arnett, R. M. Braun, and G. C. Pimentel, "Selected Values of Physical and Thermodynamic Properties of Hydrocarbons and Related Compounds," Carnegie Press, Pittsburgh, Pa., 1953, p. 475.

[39] G. B. Kistiakowsky, J. R. Ruhoff, H. A. Smith, and W. E. Vaughan, *J. Am. Chem. Soc.,* **57**, 876 (1935).

of combustion. Heats of hydrogenation of some other 1,2-dialkylethylenes, RCH=CHR′, have been measured[40] and their differences are as follows: methylisopropylethylene, 0.94 kcal./mole; methyl-*t*-butylethylene, 4.29 kcal./mole; di-*t*-butylethylene, 9.37 kcal./mole. Models indicate that in *cis*-2-butene (dimethylethylene) there is a barely perceptible steric interference of the methyl substituents. This interference may be responsible for the slight difference in enthalpy between *cis*- and *trans*-2-butene. In *cis*-methylisopropylethylene, the interference is hardly greater than in *cis*-2-butene, because the isopropyl group may be so rotated that its tertiary hydrogen "cogwheels" with the methyl group at the other end of the ethylenic system, as do the methyl hydrogens in *cis*-2-butene. The CH_3 groups on the isopropyl substituent are swung out of the way in this conformation. When one replaces the tertiary hydrogen in methylisopropylethylene, $CH_3CH{=}CHCH(CH_3)_2$, by a methyl group, one obtains methyl-*t*-butylethylene, $CH_3CH{=}CHC(CH_3)_3$, in which there is a moderately severe interaction in the cis isomer between the methyl and *t*-butyl groups; this evidences itself in a much bigger difference in enthalpy between the cis and trans isomers. Replacement of the methyl group on the other side by a *t*-butyl group leads to di-*t*-butylethylene,

$$(CH_3)_3CCH{=}CHC(CH_3)_3$$

in which there is now a very severe interaction between the two bulky *t*-butyl groups in the cis isomer, leading to a very large increase in enthalpy as one goes from the trans to the cis form.

Differences in heats of combustion of several pairs of geometrically isomeric higher di- and tri-alkylethylenes have also been measured.[37a]

Much less is known about enthalpy differences between geometrical isomers other than those of 1,2-dialkylethylenes. The heats of hydrogenation of *cis*- and *trans*-stilbene differ by 5.7 kcal./mole.[41] This rather large difference is not adequately explained by a steric effect. A model of *cis*-stilbene may be easily constructed without undue interference of the phenyl rings with each other, but to this end it is necessary to rotate the phenyl rings to a considerable extent out of the plane of the ethylenic double bond. As already explained in connection with ultraviolet spectra, this leads to inhibition of resonance and therefore diminished stability in the *cis*-stilbene, as compared with the trans isomer in which resonance of the conjugate system consisting of the double bond and the two phenyl rings is uninhibited.

The heats of combustion of the stilbenes have also been determined and are 1769.6 kcal./mole for the cis isomer and 1759.3 for the trans.[42] The difference is not, however, quantitatively significant because the trans isomer was burned in the solid state and the cis isomer as a liquid.

The difference in heats of hydrogenation between methyl *cis*- and *trans*-cinnamate is[41] 4.01 kcal./mole and that between diethyl maleate and diethyl fumarate 4.22 kcal./mole, the trans isomer possessing considerably greater thermochemical stability in both cases. In many other cases, greater stability

[40] R. B. Turner, D. E. Nettleton, and M. Perelman, *J. Am. Chem. Soc.*, **80**, 1430 (1958).
[41] R. B. Williams, *J. Am. Chem. Soc.*, **64**, 1395 (1942).
[42] J. Coops and G. J. Hoijtink, *Rec. trav. chim.*, **69**, 358 (1950).

of the trans isomer has been inferred from heat-of-combustion data which are cited in many textbooks. Unfortunately, most of these data are meaningless as far as thermochemical stability is concerned. In some cases (e.g., oleic acid and elaidic acid) a solid isomer was compared with a liquid isomer, and even where two solid isomers are compared (e.g., maleic acid and fumaric acid) much of the difference in heats of combustion may be due to different intercrystalline forces rather than due to inherent structural differences of the molecules. Thus the heat of combustion of *trans*-cinnamic acid,[43] 1040.9 kcal./mole, may be compared with that of any one of the three crystalline modifications of the cis isomer,[44] namely, that melting at 68° (1047.6 kcal./mole), that melting at 58° (1045.2 kcal./mole), and that melting at 42° (1043.8 kcal./mole). The respective differences in heats of combustion of 6.7, 4.3, and 2.9 kcal./mole are evidently devoid of any fundamental significance. Similarly, it has been pointed out[45] that the difference in heats of combustion of dimethyl maleate and dimethyl fumarate of 3.1 kcal./mole is reduced to an insignificant amount when the comparison is effected for the gaseous state.†

One of the few pairs of isomers whose heats of combustion have been carefully studied is *cis*- and *trans*-1,2-dichloroethylene. Here, contrary to the usual experience, the cis isomer is the more stable. The heats of combustion in the liquid state[46] are 261.07 kcal./mole for the cis isomer and 261.64 kcal./mole for the trans. These figures should be corrected for the known[47] heats of vaporization of 7.22 kcal./mole for the cis isomer and 6.90 for the trans, giving a difference in heats of combustion for the gaseous 1,2-dichloroethylenes of 0.25 kcal./mole, the cis isomer having the lower value. Although the difference is small, it is well documented by other data in the literature (*vide infra*).

The relative thermodynamic stability of geometrical isomers may be deduced from their free energy of formation, where it is known. Thus the difference in free energies of formation for the 2-butenes is 0.69 kcal./mole at 25° in favor of the trans isomer. It might be noted that this is less than the corresponding difference in enthalpy (0.95 kcal./mole), because *cis*-2-butene has a higher entropy by 1.04 cal./deg. mole than the trans isomer.‡

† It is not clear why the difference in the gas phase is so small, contrary to the difference in heats of hydrogenation of the ethyl esters in dilute solution. It should be pointed out that differences in heat of hydrogenation in solution have fundamental meaning only if the isomers compared have identical heats of solution in the solvent chosen. Fortunately, this is usually true, cf. ref. 37*a*.

‡ The higher entropy of *cis*-2-butene, despite the (slight) interference of the methyl groups, has been explained by W. G. Dauben and K. S. Pitzer in M. Newman, ed., "Steric Effects in Organic Chemistry," John Wiley & Sons, Inc., New York, 1956, p. 58. It is due to a lower energy barrier to rotation around the CH_3—C bonds in *cis*-2-butene as compared with *trans*-2-butene which, in turn, leads to more closely spaced torsional energy states in the cis isomer and thus to a higher entropy for the cis. The lower energy barrier for methyl rotation in cis- as

---

[43] P. Landrieu, F. Baylocq, and J. R. Johnson, *Bull. soc. chim. France,* [4]**45,** 36 (1929).
[44] F. Eisenlohr and A. Metzner, *Z. physik. Chem.,* **178,** 339 (1937).
[45] A. Wassermann, *Z. physik. Chem.,* **146,** 418 (1930).
[46] L. Smith, L. Bjellerup, S. Krook, and H. Westermark, *Acta Chem. Scand.,* **7,** 81 (1953).
[47] J. A. A. Ketelaar, P. F. van Velden, and P. Zalm, *Rec. trav. chim.,* **66,** 731 (1947).

For many pairs of geometrical isomers, free energies of formation (experimentally obtained from heat-of-combustion and entropy measurements) of the individual isomers are not available. Fortunately the two isomers can often be equilibrated chemically (see next section), and from the position of the equilibrium, their free-energy difference can be calculated directly. Thus, equilibration of the 2-butenes at 390°C. leads to a mixture containing 52.8% of the trans isomer;[49] thus the equilibrium constant is 52.8/47.2, or 1.12, and the free-energy difference at 390°C. is $-2.3RT \log K = -4.6 \times 663 \times 0.049$, or -0.15 kcal./mole. The calculated value $\Delta F = \Delta H - T\,\Delta S$ would be $-0.95 + 663 \times 0.00104$, or -0.26 kcal./mole; the agreement is moderately good. Equilibration of the stilbenes at 200° is said to give a mixture of 96% stilbene and 4% isostilbene,[50] corresponding to $\Delta F = 3.0$ kcal./mole. For this value to be compatible with the ΔH value of 5.7 kcal./mole (*vide supra*), isostilbene would have to have higher entropy than stilbene of no less than 5.7 cal./deg. mole. This may be an indication of a much greater rigidity of stilbene as compared with isostilbene, presumably because in stilbene torsional oscillation of the phenyl rings leads to a rapid drop in resonance stabilization. More accurate data would seem to be desirable to confirm this point.

A considerable amount of data on the isomerization of 1,2-dichloroethylene indicates that the cis isomer is favored at equilibrium. At 185°C. the equilibrium constant is 1.73, corresponding to 63.5% of the cis isomer.[51] From

---

compared with trans-l-substituted 1-propenes has been demonstrated experimentally by microwave spectroscopy in the 1-fluoro-1-propenes, $CH_3CH{=}CHF$.[48] The cause appears to be the following: The energy minimum for methyl rotation in $CH_3CH{=}CHX$ corresponds to the conformation in which the hydrogen atoms on the methyl group are staggered with respect to the hydrogen on the No. 2 carbon and eclipsed with respect to the C=C double bond. Models clearly show (cf. Fig. *i*) that in this conformation *in the cis isomer* there is steric repulsion between one of the methyl hydrogens and the X group. No such interaction occurs in the trans isomer, of course. The result of the steric interaction is a raising of the rotational energy minimum in the cis isomer as compared with the trans. Since the tops of the barriers are nearly the same, the effective height of the barrier (difference between top and bottom) is less for the cis than for the trans (Fig. *ii*).

In *cis*-2-butene the situation may be somewhat more complicated, inasmuch as the conformation in which both methyl groups have the position shown in Fig. *i* has very high steric repulsion and may not correspond to the actual minimum at all.

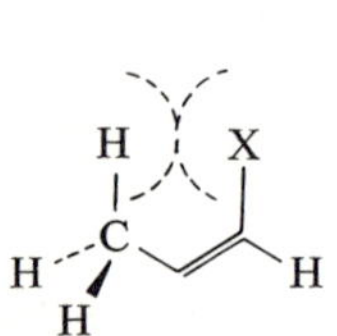

(*i*) Conformation of *cis*-$CH_3CH{=}CHX$.

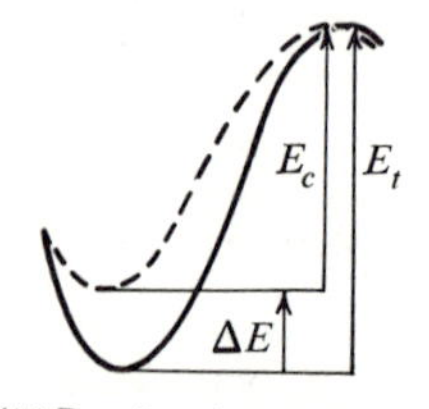

(*ii*) Barriers in *cis*- and *trans*-$CH_3CH{=}CHX$.

—— trans
- - - cis

E_t Barrier height in trans isomer
E_c Barrier height in cis isomer
ΔE H—X repulsion energy in stable conformation

---

[48] S. Siegel, *J. Chem. Phys.*, **27**, 989 (1957); R. A. Beaudet, Abstracts, Symposium on Molecular Structure and Spectroscopy, Ohio State University, Columbus, Ohio, 1960, p. 45.

[49] G. B. Kistiakowsky and W. R. Smith, *J. Am. Chem. Soc.*, **56**, 638 (1934).

[50] T. W. J. Taylor and A. R. Murray, *J. Chem. Soc.*, 2078 (1938).

[51] R. E. Wood and D. P. Stevenson, *J. Am. Chem. Soc.*, **63**, 1650 (1941).

the change in equilibrium constant with temperature, an enthalpy difference between the isomers of 0.45 to 0.72 kcal./mole has been calculated,[51-53] in fair agreement with the above-cited thermochemical value of 0.25 kcal./mole. The entropies of the two isomers are very nearly identical.[53] The greater stability of the cis isomer has been rationalized in terms of contributions of resonance forms of the type $\overset{+}{\text{Cl}}{=}\text{CH}{-}\overset{\delta-}{\text{CH}} \rightarrow \overset{\delta-}{\text{Cl}}$ which are more favored in the cis isomer because of the proximity of the two oppositely charged chlorines.[53] An alternative explanation is that the attractive London forces between the highly polarizable chlorine atoms in *cis*-1,2-dichloroethylene account for its greater stability. The London energy between the chlorine atoms (3A. apart) is 0.46 kcal./mole, just about enough to account for the experimental finding, provided that there is no compensating repulsive force of any kind.

In 1-bromo-1-propene, $BrCH{=}CHCH_3$, the cis isomer also predominates at equilibrium,[54] ΔF being -0.58 kcal./mole at 100°C. in the liquid phase[54a] and -0.85 kcal./mole in the gas phase.[54b] A number of related cases have been discussed in the literature.[54c]

The thermal isomerization of maleic acid is said to lead to nearly quantitative conversion to fumaric acid.[55]

b. Interconversion of Geometrical Isomers. The most straightforward, though not necessarily the best, way of interconverting geometrical isomers is by heating. In several instances, e.g., in the case of the stilbenes, 1,2-dichloroethylenes, and methyl cinnamates, the barrier is of the order of 40 to 45 kcal., the frequency factor of the interconversion (10^{11} to 10^{13}) being normal.[56] The energy barrier should be compared with that of 3 to 7 kcal. toward rotation about single bonds (Chap. 6). The energy situation is depicted in Fig. 12-20. Lower barriers may be found when the double bond in question is involved in resonance, as in dimethyl maleate,

$$H_3CO_2CCH{=}CHCO_2CH_3$$

dimethyl citraconate, $H_3CO_2CCH{=}C(CH_3)CO_2CH_3$, and polyenes, such as carotene (Fig. 12-14). It was claimed[57] at one time that there is a second mechanism involving low activation energies (of the order of 25 kcal.) and also low frequency factors (ca. 10^4). The suggestion has been made[58] that isomerizations of this type proceed via a triplet state rather than via the usual singlet transition state. Because of the repulsion of electrons of parallel spin, the triplet state (dashed line in Fig. 12-20) passes through an energy minimum

[52] A. R. Olson and W. Maroney, *J. Am. Chem. Soc.*, **56**, 1322 (1934); W. Maroney, *ibid.*, **57**, 2397 (1935).

[53] K. S. Pitzer and J. L. Hollenberg, *J. Am. Chem. Soc.*, **76**, 1493 (1954).

[54] (*a*) K. E. Harwell and L. F. Hatch, *J. Am. Chem. Soc.*, **77**, 1682 (1955); (*b*) P. S. Skell and R. G. Allen, *ibid.*, **80**, 5997 (1958); (*c*) H. G. Viehe, *Chem. Ber.*, **93**, 1697 (1960).

[55] B. Garre, *Z. anorg. Chem.*, **164**, 81 (1927).

[56] Cf. S. W. Benson, "Foundations of Chemical Kinetics," McGraw-Hill Book Company, Inc., New York, 1960, pp. 254–257; also Ref. 159, p. 106.

[57] G. B. Kistiakowsky and W. R. Smith, *J. Am. Chem. Soc.*, **57**, 269 (1935).

[58] J. L. Magee, W. Shand, and H. Eyring, *J. Am. Chem. Soc.*, **63**, 677 (1941). See also B. G. Gowenlock, *Quart. Revs. (London)*, **14**, 133 (1960).

when the two halves of the molecule are at right angles. This leads to a lowered activation energy, but there is also a lowered frequency factor because singlet-triplet transitions are highly forbidden. This explanation is no longer generally accepted, however, since in the case of the 2-butenes—which were originally considered the most remarkable instance of a low frequency factor in cis-trans isomerization—subsequent work[58a] has shown that the frequency factor is, in fact, normal. The other cases where the frequency factor is claimed to be low—notably the dimethyl maleate–dimethyl fumarate and dimethyl citraconate–dimethyl mesaconate (Fig. 12-18) equilibria—should perhaps be reinvestigated.

A wide variety of catalysts have been employed to bring about cis-trans isomerization.[59] Among these are found free radicals and free-radical generators, such as oxides of nitrogen and halogens in the presence of light; acids,

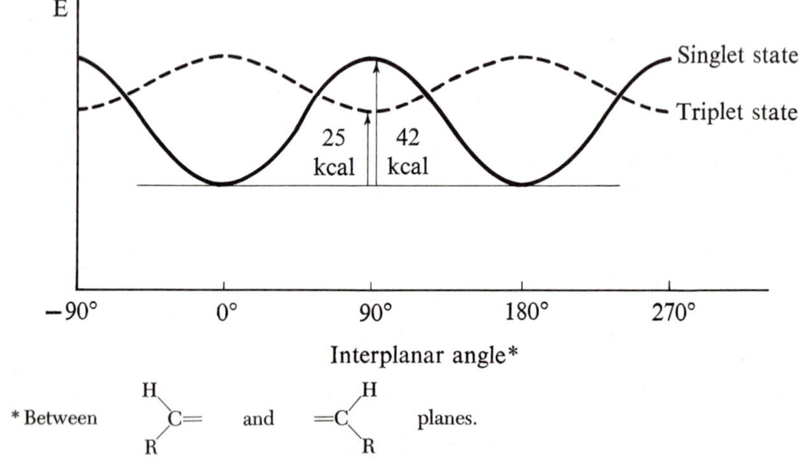

* Between H, R >C= and =C< H, R planes.

Fig. 12-20. Energy diagram for cis-trans isomerization.

such as halogen acids, sulfuric acid, and boron trifluoride; alkali metals, such as sodium; and hydrogenation-dehydrogenation catalysts and reagents, such as platinum and selenium. The mechanism of catalysis depends on the reagent employed. Radicals probably add to the double bond to form an adduct radical in which the two parts may then rotate with respect to each other. Subsequent departure of the addend radical produces the geometrical isomer of the starting material (Fig. 12-21). Light is often required to produce the catalyzing radical or atom, e.g., Br from Br_2, or I from I_2. The light may be of relatively long wavelength (the absorption maximum for bromine is at 4150 A., that for iodine near 5000 A.) and need not be absorbed by the double-bond system as such. The process is thus one of photosensitization and is clearly distinct from the photochemical isomerization which

[58a] W. F. Anderson, J. A. Bell, J. M. Diamond, and K. R. Wilson, *J. Am. Chem. Soc.,* **80,** 2384 (1958); B. S. Rabinovitch and K. W. Michel, *ibid.,* **81,** 5065 (1959).

[59] Cf. Ref. 159, p. 139.

will be discussed below.† It is of interest that the addition of an atom, such as iodine, to the double bond does not necessarily lead to isomerization; the atom may depart before the adduct radical has had time to rotate about the central bond.[60] This has been shown[60] by bringing about the isomerization of 1,2-diiodoethylene by means of radioactive iodine. Iodine exchange is found to be much faster than isomerization. As shown in Fig. 12-22, iodine

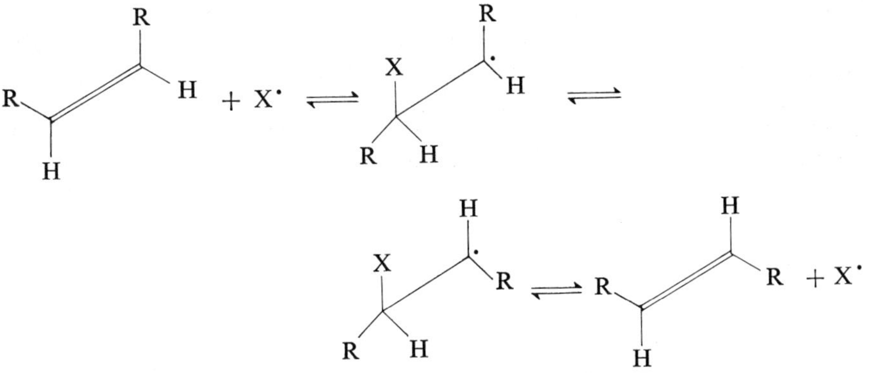

Fig. 12-21. Radical-catalyzed cis-trans isomerization.

exchange without isomerization may occur if the adduct radical loses an iodine atom before it has had time to rotate about the central bond.

Acid catalysts may act similarly by adding temporarily to the double bond. Thus boron trifluoride has been reported to isomerize *cis*-stilbene to *trans*-stilbene.[61] Acids are particularly effective in bringing about the isomerization of α,β-unsaturated carbonyl compounds, such as diethyl maleate,[62] perhaps via the mechanism shown in Fig. 12-23. In some cases, however, such

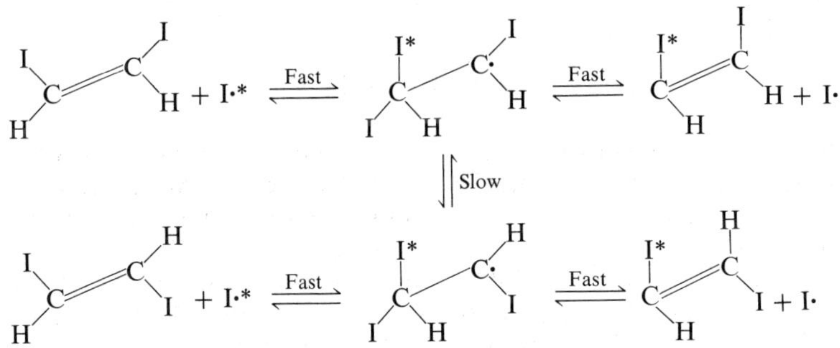

Fig. 12-22. Radical exchange vs. isomerization.

† Triplet states probably play an important part in photosensitized cis-trans isomerizations; cf. G. S. Hammond, P. A. Leermakers, and N. J. Turro, *J. Am. Chem. Soc.*, **83**, 2396 (1961).

[60] R. M. Noyes, R. G. Dickinson, and V. Schomaker, *J. Am. Chem. Soc.*, **67**, 1319 (1945).
[61] C. C. Price and M. Meister, *J. Am. Chem. Soc.*, **61**, 1595 (1939).
[62] W. I. Gilbert, J. Turkevich, and E. S. Wallis, *J. Org. Chem.*, **3**, 611 (1939).

as the isomerization of stilbene by hydrogen bromide,[63] the acid does not appear to be the true catalyst. The reaction fails in the absence of oxygen and light but proceeds readily in the presence of peroxide. It is stopped by radical inhibitors, such as hydroquinone. It seems that the hydrogen bromide here serves as a source of bromine atoms which bring about isomerization, as explained earlier. Hydrogen chloride is ineffective in this case.

Sodium or lithium converts *cis*-stilbene to the trans isomer.[64] This process may involve reversible addition of the metal to the double bond.

Dimethyl maleate is isomerized to dimethyl fumarate by means of secondary amines.[65] It has been suggested that hydrogen bonding of the amine hydrogen to the carbonyl oxygen of the ester is involved, i.e., that the process is similar to the one shown in Fig. 12-23. The alternative possibility that the amine, acting as a base, adds to the double bond to form a resonance-stabilized carbanion is ruled out by the observation that tertiary amines are ineffective as isomerization catalysts. It would seem, however, that reversible 1,4 addition of the amine (Fig. 12-24) provides a better explanation of the isomerization than simple hydrogen bonding.

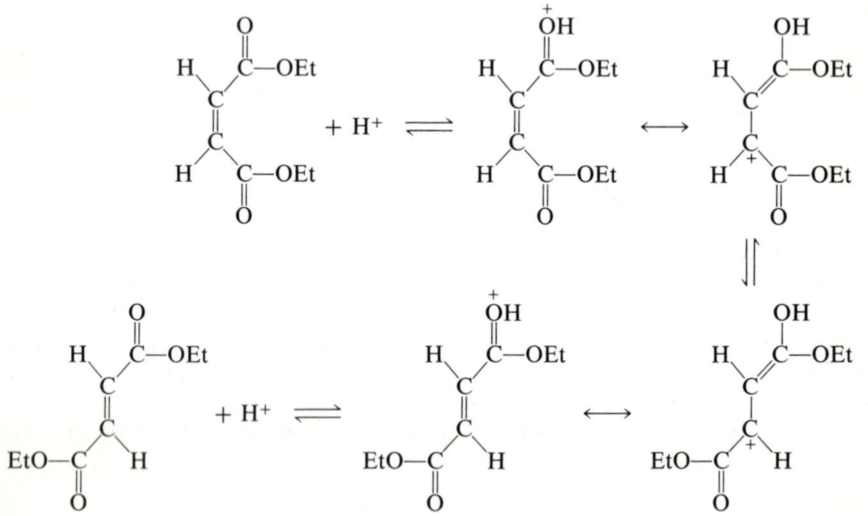

Fig. 12-23. Isomerization of diethyl maleate by acid.

Isomerization under the influence of hydrogenation-dehydrogenation catalysts or reagents, such as platinum[66] or selenium,[67] may involve reversible hydrogenation-dehydrogenation. A particularly facile isomerization is involved in the conversion of angelic acid (2-methyl-*cis*-crotonic acid) to the

[63] M. S. Kharasch, J. V. Mansfield, and F. R. Mayo, *J. Am. Chem. Soc.*, **59**, 1155 (1937).

[64] W. Schlenk and E. Bergmann, *Ann.*, **463**, 117 (1928).

[65] G. R. Clemo and S. B. Graham, *J. Chem. Soc.*, 213 (1930).

[66] R. C. Fuson, S. L. Scott, E. C. Horning, and C. H. McKeever, *J. Am. Chem. Soc.*, **62**, 2091 (1940).

[67] J. P. Kass and G. O. Burr, *J. Am. Chem. Soc.*, **61**, 1062 (1939).

chloride of tiglic acid (2-methyl-*trans*-crotonic acid) upon treatment with phosphorus oxychloride.[67a]

It has been suggested that the efficacy of some of the above-mentioned catalysts (e.g., NO, Na) is a result of their paramagnetic character.[67b]

An entirely different situation is found in the isomerization of geometrical isomers by photochemical means. This can be brought about only by light of a wavelength such that it is absorbed by the olefinic system (usually ultraviolet light). Since energy is absorbed, this process does not result in the establishment of thermal equilibrium. However, a steady state may be reached which generally corresponds to a predominance of the less stable isomer. Thus irradiation of maleic or fumaric acid with ultraviolet light leads to a mixture containing 75% of maleic acid,[68] and benzalacetophenone,

$$C_6H_5CH{=}CHCOC_6H_5$$

is converted to a mixture containing 74% of the cis and 26% of the trans isomer by irradiation with sunlight in a quartz vessel.[69] In the case of stilbene, the photostationary state corresponds to 68% of *cis*-stilbene and 32% *trans*-stilbene.[70]

$$\text{cis-}MeO_2C\text{–}CH{=}CH\text{–}CO_2Me + R_2NH \rightleftharpoons H(C(OH){=}OMe)C\text{–}C(H)(NR_2)CO_2Me \rightleftharpoons \text{trans-}MeO_2C\text{–}CH{=}CH\text{–}CO_2Me + HNR_2$$

Fig. 12-24. Base-catalyzed isomerization.

The mechanism of the photochemical isomerization has been explained as follows:[70, 71] The absorbed light raises the olefin molecule to an excited state whose energy content places it well above the higher barrier shown in Fig. 12-20. If the energy is not immediately lost by fluorescence, it may be degraded into vibrational and rotational energy, and a rotational motion about the central bond may be set up such that excited states of the cis and trans isomers are interconverted. Since the excited state of the cis isomer is higher in energy, the rotating molecule spends more time in this form than in the form corresponding to the excited state of the trans molecule.[71] This situation is similar to that found in a pendulum which spends most of its

[67a] S. M. Kupchan and A. Afonso, *J. Org. Chem.*, **25**, 2217 (1960); see also R. E. Buckles, G. V. Mock, and L. Locatell, *Chem. Revs.*, **55**, 659 (1955).

[67b] R. A. Harman and H. Eyring, *J. Chem. Phys.*, **10**, 557 (1942); H. McConnell, *ibid.*, **20**, 1043 (1952).

[68] A. Kailan, *Z. physik. Chem.*, **87**, 333 (1914).

[69] R. E. Lutz and R. H. Jordan, *J. Am. Chem. Soc.*, **72**, 4090 (1950).

[70] G. N. Lewis, T. T. Magel, and D. Lipkin, *J. Am. Chem. Soc.*, **62**, 2973 (1940).

[71] A. R. Olson, *Trans. Faraday Soc.*, **27**, 69 (1931); cf. G. K. Rollefson and M. Burton, "Photochemistry and the Mechanism of Chemical Reactions," Prentice-Hall, Inc., Englewood Cliffs, N.J., 1939, p. 252.

time in the positions near the ends of the swing, i.e., the positions of highest potential energy, because at these points its kinetic energy is lowest. Since, then, the rotating excited molecule spends more time in the configuration corresponding to the cis ground state, it has a greater chance to return to the cis ground state by further loss of energy than to the trans ground state. As a result the cis molecules predominate when the stationary state is reached.

There may be certain other points that have to be considered. Usually (cf. Sec. 12-3*d*) cis and trans isomers absorb light at slightly different wavelengths. If one irradiates with strictly monochromatic light, the isomer that absorbs the light should be quantitatively converted to the one that does not absorb. Even if the irradiation is with broad-spectrum ultraviolet light of nearly level intensity over the region of the absorption of the two geometrical isomers, one must take into account the fact that the trans isomer usually absorbs more strongly than the cis. As a result, even if the chances of the photoexcited state to drop back to the cis or trans ground state were equal, there should be some net trans → cis conversion.

A unidirectional (non-equilibrium) method of converting certain trans olefins to their nearly pure cis isomers in high over-all yield will be considered in the next chapter (Sec. 13-5). The method may also be used to convert cis olefins to the trans isomers.

12-5. Stereochemistry of Addition Reactions of Acetylenes and Their Reversal[72]

Addition of hydrogen to a non-terminal acetylene RC≡CR may lead to either the cis or the trans form of the olefin RCH=CHR.[73] When the hydrogenation is carried out catalytically over nickel or palladium catalysts (both of which allow the hydrogenation to be interrupted at the olefin stage), the predominant product (80 to 95%) is found to be the cis olefin.[74] Even hydrogenation of *cis-cis*-deca-2,8-diene-4,6-diyne,

$$CH_3CH{=}CHC{\equiv}CC{\equiv}CCH{=}CHCH_3$$

gives rise to all-*cis*-deca-2,4,6,8-tetraene, despite the very unfavorable steric crowding in the latter olefin.[75] A rare exception to cis hydrogenation is the acetylenic alcohol shown in Fig. 12-25; hydrogenation of this material over a palladium catalyst[76] produces the trans olefin if (and only if) carried out in the presence of a large amount of quinoline.†

The reason for cis hydrogenation is presumably that the hydrogenation is largely a one-stage process in which the two hydrogens must necessarily

† Semihydrogenation of butynediol-1,4 over palladium, which normally gives nearly exclusively *cis*-2-butenediol-1,4, gives a trans-rich product when carried out in the presence of alkali: F. J. McQuillin and W. O. Ord, *J. Chem. Soc.*, 2902 (1959).

[72] R. A. Raphael, "Acetylenic Compounds in Organic Synthesis," Academic Press, Inc., New York, 1955.

[73] Cf. K. N. Campbell and B. K. Campbell, *Chem. Revs.*, **31**, 77 (1942).

[74] M. Bourgel, *Compt. rend.*, **180**, 1753 (1925); *Bull. soc. chim. France*, [4]**45**, 1067 (1929); cf. Ref. 80.

[75] D. Holme, E. R. H. Jones, and M. C. Whiting, *Chemistry & Industry (London)*, 928 (1956).

[76] A. Mondon, *Ann.*, **577**, 181 (1952).

approach the triple bond from the same side (see the discussion of olefin hydrogenation in Sec. 12-6*a* below).

Chemical reduction of dialkylacetylenes with sodium and liquid ammonia in the presence of a hydrogen donor, such as water or alcohol, gives trans olefins. The exact course of this reaction does not seem to be clear. It is clear, however, that the outcome is controlled stereoelectronically rather than thermodynamically, for the reduction of cyclononyne[77] gives rise mainly to

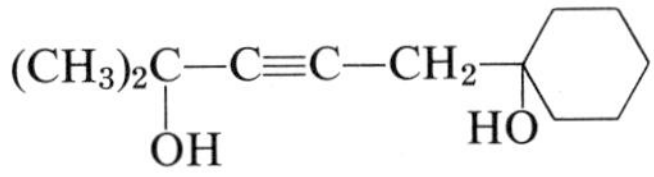

$(CH_3)_2C(OH)—C≡C—CH_2$—(1-hydroxycyclohexyl) [HO]

Fig. 12-25. Acetylene which is hydrogenated to trans olefin.

trans-cyclononene (79% of the total olefin[78]), despite the fact that this olefin is considerably less stable than its cis isomer (cf. Table 7-3).[78] Reduction of dialkylacetylenes to olefins by addition of diborane followed by treatment with acetic acid[79] leads to cis olefins (ca. 98% stereoselectivity). This process may involve a four-center cis addition followed by hydrolytic cleavage with retention of configuration (cf. Chap. 13), as shown in Fig. 12-26. On the other hand, reduction of acetylenic carbinols to allyl alcohols by means of

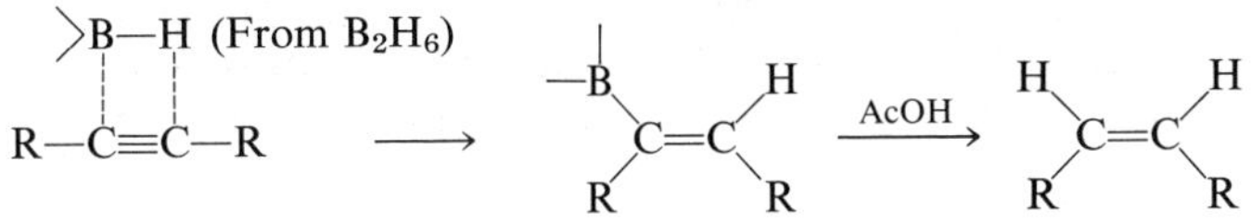

>B—H (From B_2H_6)

R—C≡C—R ⟶ (—B)(R)C=C(H)(R) —AcOH⟶ (H)(R)C=C(H)(R)

Fig. 12-26. Hydroboration of dialkylacetylene to cis olefin.

lithium aluminum hydride[80] produces the trans configuration at the double bond; an example[81] is shown in Fig. 12-27. The stereochemical course of this reduction appears reasonable in view of the postulated intermediate shown in Fig. 12-28. Cis addition is also observed in the addition of silicochloroform, $HSiCl_3$, to terminal acetylenes in the presence of a platinized charcoal catalyst[82] (cf. Fig. 12-51) and in the reaction of diphenylacetylene,

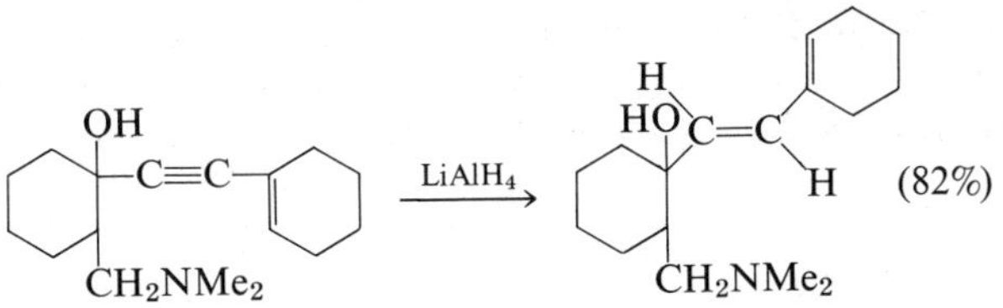

OH, —C≡C—, CH_2NMe_2 —$LiAlH_4$⟶ H, HO, C=C, H (82%), CH_2NMe_2

Fig. 12-27. Hydride reduction of acetylenic carbinol to allylic alcohol.

[77] A. T. Blomquist, L. H. Liu, and J. C. Bohrer, *J. Am. Chem. Soc.*, **74**, 3643 (1952).
[78] A. C. Cope, P. T. Moore, and W. R. Moore, *J. Am. Chem. Soc.*, **82**, 1744 (1960).
[79] H. C. Brown and G. Zweifel, *J. Am. Chem. Soc.*, **83**, 3834 (1961).
[80] R. A. Raphael, "Acetylenic Compounds in Organic Synthesis," Academic Press, Inc., New York, 1955, p. 30; cf. J. D. Chanley and H. Sobotka, *J. Am. Chem. Soc.*, **71**, 4140 (1949).
[81] R. A. Raphael and F. Sondheimer, *J. Chem. Soc.*, 3185 (1950).
[82] R. A. Benkeser and R. A. Hickner, *J. Am. Chem. Soc.*, **80**, 5298 (1958).

PhC≡CPh, with mercuric acetate,[83] which, after treatment with sodium chloride, gives α-acetoxy-β-chloromercuri-*cis*-stilbene,

cis-Ph(AcO)C=C(HgCl)Ph

Little work has been done with regard to electrophilic addition to dialkylacetylenes since the classical studies of Michael, who showed[84] that addition of bromine to acetylenedicarboxylic acid, $HO_2CC{\equiv}CCO_2H$, in aqueous

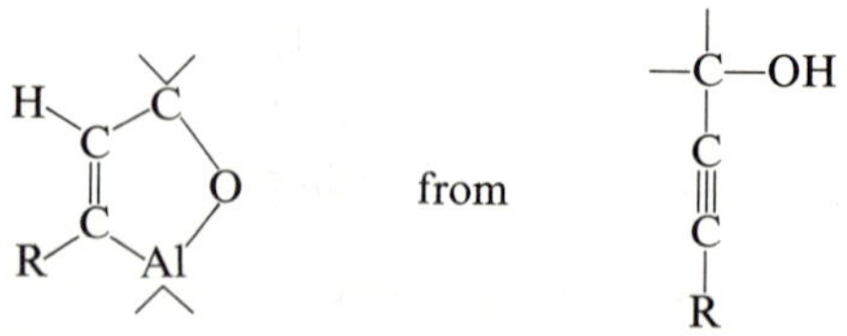

Fig. 12-28. Intermediate in hydride reduction of acetylenic carbinol.

solution gives dibromofumaric acid, *trans*-$HO_2CCBr{=}CBrCO_2H$, to the exent of 70% and the cis isomer, dibromomaleic acid, to the extent of 30%. Similar results were obtained when bromine was added to the corresponding diethyl ester in carbon tetrachloride solution. Clearly, trans addition predominates over cis addition, but the latter also occurs to an appreciable extent. Addition of hydrogen chloride to acetylenedicarboxylic acid, however, gives almost entirely the trans addition product chlorofumaric acid,[85]

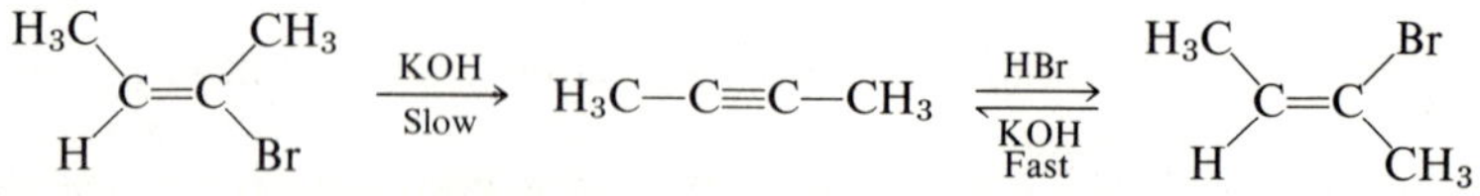

Fig. 12-29. Trans addition and elimination.

trans-$HO_2CCH{=}CClCO_2H$, and addition of hydrogen bromide to 2-butyne gives exclusively *trans*-2-bromo-2-butene (Fig. 12-29).[86] Elimination reactions of haloölefins to acetylenes similarly proceed best when the elements to be eliminated are located trans; thus *trans*-2-bromo-2-butene (Fig. 12-29) is converted to 2-butyne by potassium hydroxide several times faster than the corresponding cis isomer (Fig. 12-29).[86] Similarly, chlorofumaric acid loses hydrogen chloride in the presence of base about fifty times faster than does

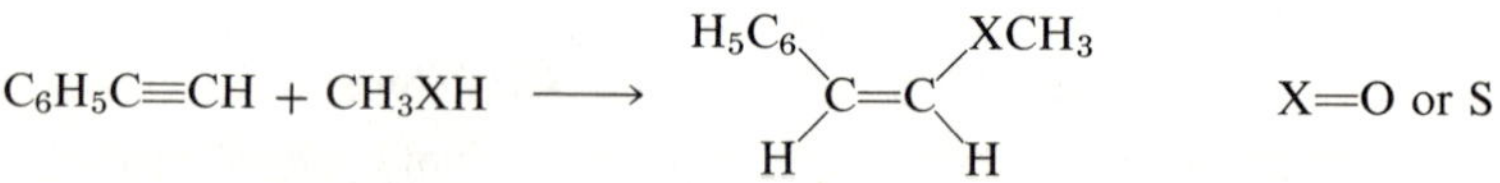

Fig. 12-30. Nucleophilic trans addition to phenylacetylene.

[83] G. Drefahl, G. Heublein, and A. Wintzer, *Angew. Chem.*, **70**, 166 (1958).
[84] A. Michael, *J. prakt. Chem.*, **46**, 210 (1892).
[85] A. Michael, *J. prakt. Chem.*, **52**, 321 (1895).
[86] M. Lépingle, *Bull. soc. chim. France,* [4]**39**, 741 (1926).

chloromaleic acid, the product in both cases being acetylenedicarboxylic acid.[87]

Strong bases, such as methoxide, and good nucleophiles, such as mercaptides and thiophenoxides, add to acetylenes in nucleophilic fashion.[166] The stereochemistry of such nucleophilic addition is trans.[88-90] Thus the addition of sodium methoxide in methanol to phenylacetylene (Fig. 12-30) gives[88] *cis*-β-methoxystyrene,† and similar addition of sodium methylmercaptide

$$H_3C{-}C{\equiv}C{-}CH_3 + p\text{-}H_3CC_6H_4SH \longrightarrow \underset{H}{\overset{H_3C}{>}}C{=}C\underset{CH_3}{\overset{SC_6H_4CH_3\text{-}p}{<}}$$

Fig. 12-31. Nucleophilic addition to 2-butyne.

(Fig. 12-30) gives methyl *cis*-β-styryl sulfide,[89] the cis isomers in both cases being formed by trans addition. The product obtained from sodium *p*-thiocresolate and 2-butyne, 2-*p*-tolylmercapto-2-butene, has also been assigned the trans structure on the basis of infrared evidence[89] and thus appears to involve trans addition (Fig. 12-31). It is believed[88, 89] that trans addition is a result of the electron pair contributed by the nucleophile and the electron pair dislodged from the triple bond tending to stay as far away from each other as possible. The addition may involve a carbanion of fleeting stability[88, 91] (cf. Chap. 13), or the addition of a proton may be concerted with the addition of the base,[91] as shown in Fig. 12-32.

12-6. Stereochemistry of Addition Reactions of Olefins

In Chap. 6 the stereochemistry of elimination reactions leading to olefins was taken up. This section deals with the reverse of this process, namely, addition reactions of olefins. Six different types of addition will be distin-

$$R{-}C{\equiv}C{-}R + R'S^- \longrightarrow \underset{\bar{:}}{\overset{R}{>}}C{=}C\underset{R}{\overset{SR'}{<}} \xrightarrow[\text{(solvent)}]{\text{EtOH}} \underset{H}{\overset{R}{>}}C{=}C\underset{R}{\overset{SR'}{<}}$$

$$\text{or via}\quad \underset{EtO-H}{\overset{R}{>}}C{=}C\underset{R}{\overset{SR'}{<}}$$

Fig. 12-32. Mechanism of nucleophilic addition.

† The configuration was assigned on the basis of a 937-cm.$^{-1}$ band in the infrared spectrum of the presumed trans isomer and a 726-cm.$^{-1}$ band in the spectrum of the cis; cf. Sec. 12-3*e*.

[87] A. Michael, *J. prakt. Chem.*, **52,** 308 (1895).
[88] S. I. Miller, *J. Am. Chem. Soc.*, **78,** 6091 (1956).
[89] W. E. Truce and J. A. Simms, *J. Am. Chem. Soc.*, **78,** 2756 (1956).
[90] F. Montanari and A. Negrini, *Gazz. chim. ital.*, **87,** 1061 (1957).
[91] W. E. Truce and R. F. Heine, *J. Am. Chem. Soc.*, **81,** 592 (1959).

guished: hydrogenation (catalytic or chemical), electrophilic addition, molecular addition other than hydrogenation, radical addition, addition of carbenes, and nucleophilic addition.

a. Hydrogenation.[92] Study of the stereochemistry of hydrogenation of olefins has, in the main, been confined to tetrasubstituted double bonds, including double bonds in rings. In principle, the stereochemistry of hydrogenation of disubstituted olefins could be studied, using deuterium. This has been done successfully[92a] with 2-butene using a chromium oxide gel catalyst; addition is cleanly cis. With the more common catalysts, there are complications due to H-D exchange. Catalytic hydrogenation of the 2,3-diphenylbutenes (Fig. 12-33) over palladium in acetic acid[93] is rather straightforward stereochemically in that the cis isomer gives *meso*-2,3-diphenylbutane (98%, containing 2% of the *dl* isomer), whereas the trans olefin gives (±)-2,3-diphenylbutane in equal predominance. Again hydrogenation is nearly exclusively cis. This result may be rationalized on the basis of a mechanism originally suggested by Polanyi:[94]

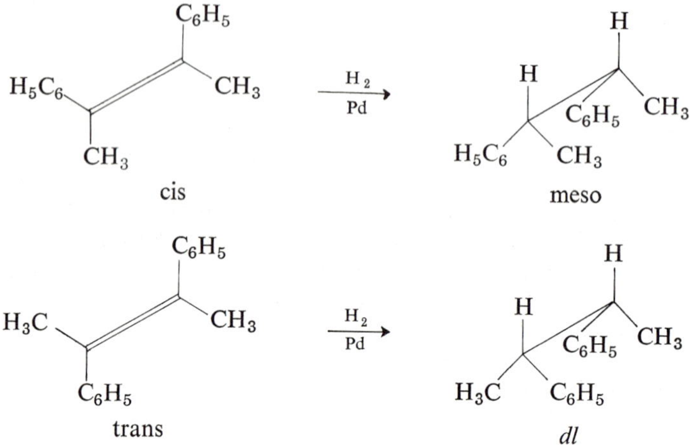

Fig. 12-33. Hydrogenation of the 2,3-diphenylbutenes.

(1) H_2 + catalyst ⇌ 2 H (hydrogen adsorbed on catalyst) [H bonded to * below]

(2) >C=C< + catalyst ⇌ >C(*)—C(*)<

(3) >C(*)—C(*)< + H(*) ⇌ >C(H)—C(*)<

(4) >C(H)—C(*)< + H(*) ⇌ >C(H)—C(H)<

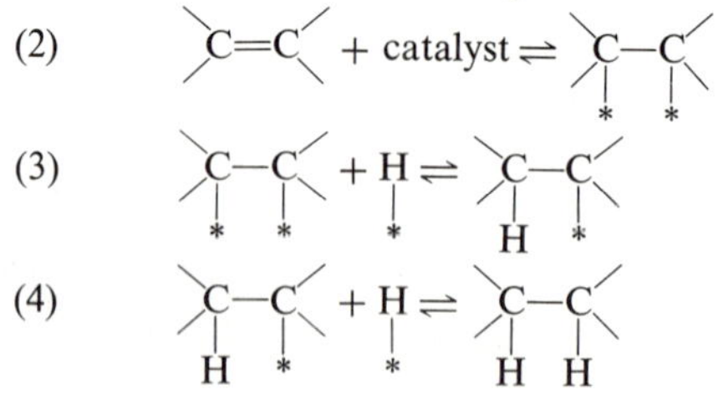

[92] R. L. Burwell, *Chem. Revs.*, **57**, 895 (1957).
[92a] M. Cardew and R. L. Burwell, *J. Am. Chem. Soc.*, **82**, 6289 (1960).
[93] F. von Wessely and H. Welleba, *Ber.*, **74**, 777 (1941).
[94] I. Horiuti and M. Polanyi, *Trans. Faraday Soc.*, **30**, 1164 (1934).

The olefin is anchored on the catalyst at two sites which are necessarily on the same side of the original double bond, and even though the addition of hydrogen is probably stepwise [i.e., steps (3) and (4) are discrete], as evidenced by the occurrence of H-D exchanges,[94] the above mechanism ensures cis hydrogenation as long as steps (2) and (3) are not appreciably reversible. However, there is increasing evidence that in many instances the picture is more complicated than that.[95] Thus hydrogenation of 1,2-dimethylcyclohexene[95a, b] and of 9,10-octalin[80, 95c] over platinum in acetic acid gives appreciable amounts of the trans product (Fig. 12-34). In the case of 1,2-dimethylcyclohexene, the amount of trans hydrogenation is pressure-dependent, increasing from about 5% at 300 atm. to 18% at 1 atm. With palladium as a catalyst, trans hydrogenation actually predominates to the extent of 3:1.[95d] Evidence has been presented[95b] that the formation of the trans-addition product involves reversal of the half-hydrogenated intermediate formed in step (3) above to an adsorbed form of 2,3-dimethylcyclohexene which is then desorbed by reversal of step (2) and subsequently re-adsorbed in such a way as to give a mixture of *cis*- and *trans*-1,2-dimethylcyclohexane (Fig. 12-35) upon hydrogenation.

CH_3 CH_3 $\xrightarrow[Pt]{H_2}$ CH_3 CH_3

82–95% cis
5–18% trans

$\xrightarrow[Pt]{H_2}$

50–66% cis
34–50% trans

Fig. 12-34. Hydrogenation of cyclic olefins.

Catalytic reduction of aromatic systems over platinum also involves preferential all-cis addition of hydrogen. The case of the perhydrophenanthrenes and of the diphenic acids has already been discussed in Chap. 10; *o*-, *m*-, and *p*-Xylene give the cis isomers of the corresponding dimethylcyclohexanes to the extent of 95, 86, and 74%.[95a] Catalytic reduction of alkylphenols may be an exception, for although one report states that *o*-cresol upon hydrogenation over platinum in acetic acid gives exclusively *cis*-2-methylcyclohexanol,[96] it was later claimed[97] that reduction of all three cresols under these conditions gives predominantly the *more stable* isomer, i.e. (cf. Chap. 8), the *trans*-2-, *trans*-4-, and *cis*-3-methylcyclohexanol. Reduction of *p*-cresol over ruthenium oxide, however, gives *cis*-4-methylcyclohexanol in slight predominance.[98] Reduction using Raney nickel at elevated temperatures and pressures leads to the more stable isomers, presumably by equilibration.[99] It is clear that the stereoselectivity in the hydrogenation of alkylphenols is not high, and the

[95] (*a*) S. Siegel and M. Dunkel, in "Advances in Catalysis," vol. 9, Academic Press, Inc., New York, 1957, pp. 15–24; (*b*) S. Siegel and G. V. Smith, *J. Am. Chem. Soc.*, **82,** 6082 (1960); (*c*) J. F. Sauvage, R. H. Baker, and A. S. Hussey, *ibid.*, **82,** 6090 (1960); (*d*) S. Siegel and G. V. Smith, *ibid.*, **82,** 6087 (1960). See also ref. 92*a*.

[96] R. H. Baker and R. D. Schuetz, *J. Am. Chem. Soc.*, **69,** 1250 (1947).

[97] R. J. Wicker, *J. Chem. Soc.*, 3299 (1957).

[98] E. L. Eliel and R. S. Ro, *J. Am. Chem. Soc.*, **79,** 5992 (1957).

[99] R. J. Wicker, *J. Chem. Soc.*, 2165 (1956).

predominant product is hard to predict. The situation is further complicated by the possibility that alkylcyclohexanones may be intermediates in some of these reductions. The hydrogenation of these ketones has already been discussed in Sec. 8-7.

It is of interest that from 1 to 11% racemization occurs in the catalytic hydrogenation of (−)-3-phenyl-1-butene, $CH_2{=}CHCH(Ph)CH_3$, to (−)-2-

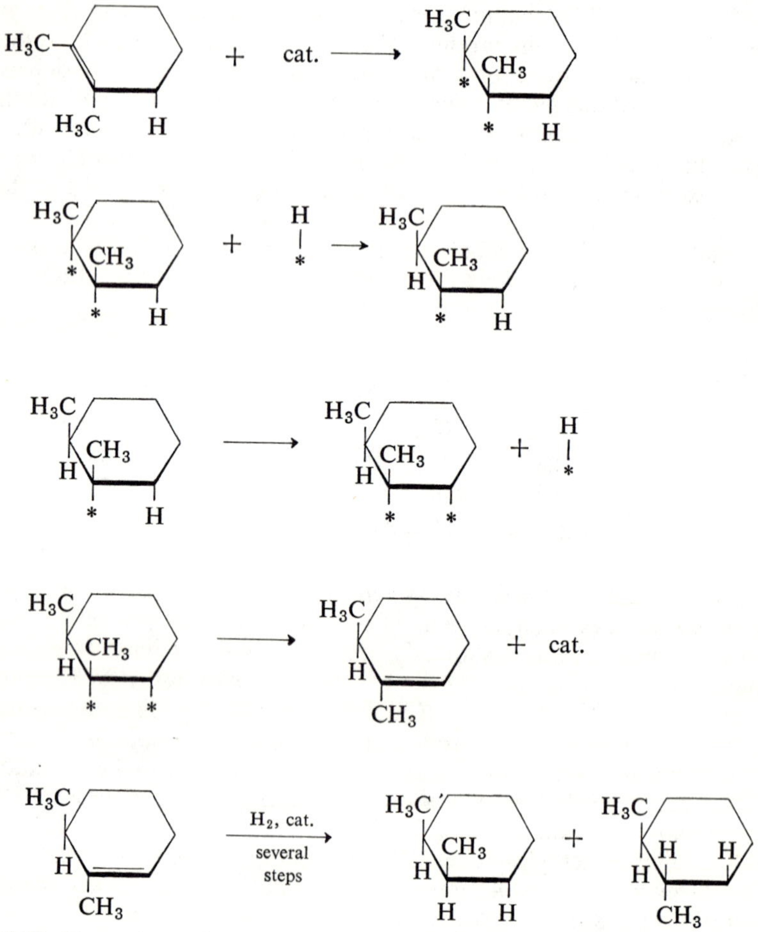

Fig. 12-35. Formation of *trans*-1,2-dimethylcyclohexane by hydrogenation of 1,2-dimethylcyclohexene.

phenylbutane, $CH_3CH_2CH(Ph)CH_3$, despite the fact that the asymmetric carbon is not directly involved in the hydrogenation and that the product is optically stable under the hydrogenation conditions.[100] Racemization has been shown not to involve exchange of the allylic hydrogen prior to reduc-

[100] D. J. Cram, *J. Am. Chem. Soc.*, **74**, 5518 (1952).

tion, but its exact mechanism has proved elusive.[101] Possibly there is some reversal of the half-hydrogenated state (cf. page 350) of 3-phenyl-1-butene to 3-phenyl-2-butene, $CH_3CH{=}C(Ph)CH_3$, necessarily inactive, which upon further hydrogenation would yield racemic 2-phenylbutane. Palladium catalysts, which cause most trans hydrogenation in the case of 1,2-dimethylcyclohexene (Fig. 12-35), also cause most racemization in the hydrogenation of (−)-3-phenyl-1-butene.

The only general chemical method for reduction of simple olefins seems to be hydroboration[102] followed by treatment with acetic acid. Since tetrasubstituted olefins can be hydroborated, the stereochemistry of the process can, in principle, be studied, but this has not yet been done.

Allylic alcohols can be reduced to saturated alcohols by means of lithium aluminum hydride followed by hydrolysis; thus cinnamyl alcohol,

$$C_6H_5CH{=}CHCH_2OH$$

gives hydrocinnamyl alcohol, $C_6H_5CH_2CH_2CH_2OH$. The steric course of this reduction in a tetrasubstituted olefin has not yet been defined.[103]

α,β-Unsaturated ketones are reduced to saturated ketones by lithium in ammonia containing a proton donor. It was originally believed[104] that this

R R′ O R″ R‴ —Li, NH₃ / EtOH→ R R′ O R″ H R‴

Fig. 12-36. Reduction of α,β-unsaturated ketones with lithium-ammonia-ethanol.

reduction led to the more stable epimer at the β-carbon atom, $\Delta^{1,10}$-2-octalone (Fig. 12-36, R = R′ = R″ = R‴ = H) thus giving the more stable *trans*-2-decalone. It has been found, however,[105] that the reduction gives trans decalones even when the corresponding cis isomers are more stable. Thus the reaction must be stereoelectronically rather than thermodynamically controlled, in analogy with the corresponding reduction of dialkylacetylenes mentioned earlier. The controlling factor is believed[105] to be the need of the orbital containing the electron pair at the β carbon to overlap with the pi electrons of the carbonyl group. Thus compounds A and B (Fig. 12-37) are both possible products of the reduction of an appropriate octalone but the alternative cis conformer C is not, since it does not fulfill the condition of overlap. If the carbanions shown in Fig. 12-37 are substantially tetrahedral (cf. Chap. 13) and protonation occurs without change of configuration at the carbanion site, then the formulations in Fig. 12-37 suggest that the reduction product will have configuration A or B but not C. If A is more stable than B, a trans-fused-ring system will be found in the reduction product, *even if A*

[101] W. A. Bonner and J. B. McKay, *J. Am. Chem. Soc.*, **82,** 5350 (1960).
[102] H. C. Brown and B. C. Subba Rao, *J. Am. Chem. Soc.*, **81,** 6428 (1959).
[103] Cf. J. F. Cavalla and J. F. McGhie, *J. Chem. Soc.*, 834 (1951).
[104] D. H. R. Barton and C. H. Robinson, *J. Chem. Soc.*, 3045 (1954).
[105] G. Stork and S. D. Darling, *J. Am. Chem. Soc.*, **82,** 1512 (1960).

is less stable than the alternative cis conformer C. In order to illustrate this point clearly, one must choose decalin systems in which the trans isomer is less stable than the cis. This will be the case if, in the formulations in Fig. 12-36, R and either R' or R''' are groups larger than hydrogen, for then there will be an appreciable 1-3 diaxial interaction in the trans decalone which would be absent in conformer C (Fig. 12-37) of the corresponding cis decalone. However, the same diaxial interactions appear in conformer B of the cis decalone. Thus, if C (Fig. 12-37) represented a possible transition state,

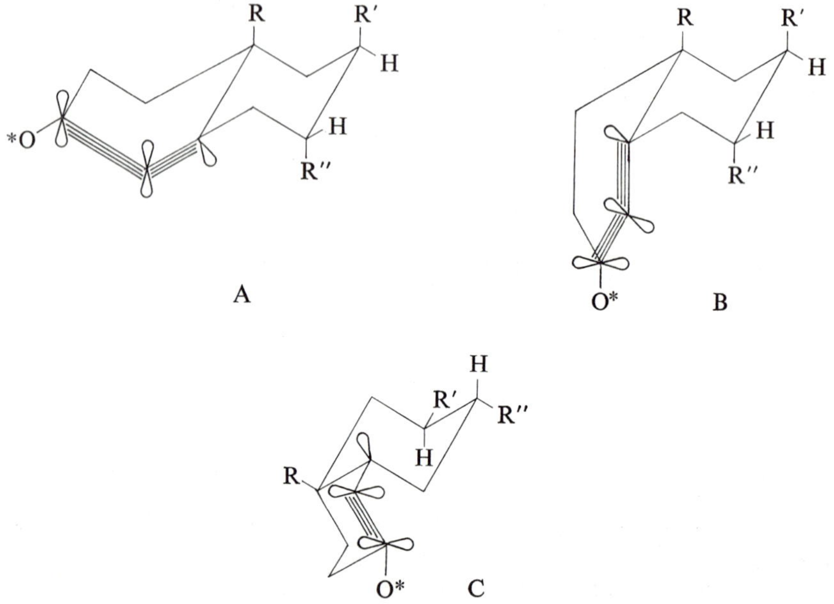

* The star stands for either an odd electron (if proton transfer occurs at the anion-radical stage, following transfer of one electron) or an O-metal bond (if proton transfer occurs following transfer of two electrons to the unsaturated ketone). While the former possibility appears more likely, the matter is as yet undecided. (Private communication from Professor Gilbert Stork.)

Fig. 12-37. Stereoelectronic control in reduction of unsaturated ketone. [*From G. Stork and S. D. Darling, J. Am. Chem. Soc.,* **82,** 1512 (1960). *By permission of the authors and editor.*]

the corresponding decalone should be formed. However, as already stated, transition state C is prohibited for stereoelectronic reasons, and between the alternative transition states A and B, A is preferred because of its equatorial-equatorial ring junction. In fact, chemical reduction of the octalones shown in Fig. 12-36 with R = CH_3 and either $R' = OCH_3$, $R''' = H$, and $R'' = H$ or CH_3, or $R' = H$ but $R''' = CH_3$ (with $R'' = i$-Pr) gave, in all cases, exclusively the trans-decalin system in the product, despite its lesser stability in these cases.

Cis decalones are obtained from octalones of the type shown in Fig. 12-36 upon catalytic hydrogenation over palladium on charcoal in the presence of hydrochloric acid.[106]

b. Electrophilic Addition. Many cases of electrophilic addition to double bonds are known. Addition of halogens, hydrogen halides, hypohalous acids, water in the presence of acids, sulfenyl halides, and formaldehyde in the presence of acid ("Prins reaction"), among others, is of this type. In most cases where the stereochemistry has been studied, such additions are found to be trans. Thus, addition of bromine to maleic acid gives (±)-2,3-dibromosuccinic acid, whereas addition of bromine to fumaric acid gives *meso*-2,3-dibromosuccinic acid.[107] Similar results are obtained with the *cis*- and *trans*-2-butenes.[108] The course of the addition as well as the postulated[109] intermediate bromonium ion is shown in Fig. 12-38. It may be seen that the steric course of the addition is anti as well as trans (cf. Chap. 6) and, as

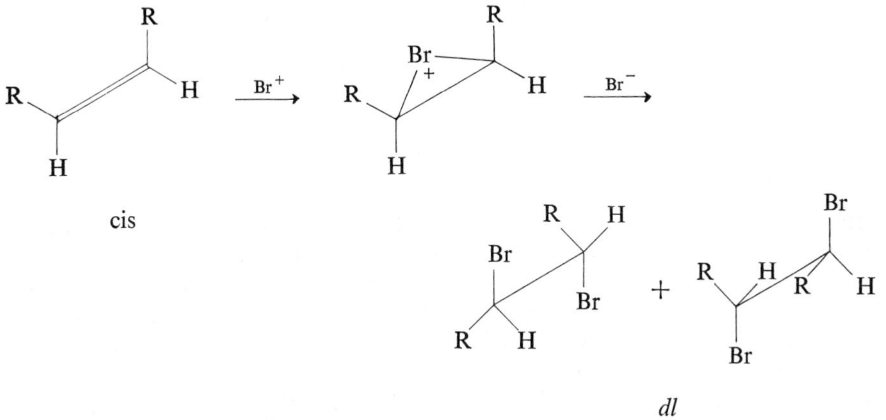

Fig. 12-38. Electrophilic trans addition to olefins.

already mentioned in Chap. 8, the addition in cyclic systems does lead to the expected *diaxial* trans isomers rather than to the more stable diequatorial trans isomers. Trans addition also occurs in the addition of hydrogen iodide to angelic and tiglic acid,[108] $CH_3CH{=}C(CH_3)CO_2H$; in the addition of 2,4-dinitrobenzenesulfenyl chloride, 2,4-$(O_2N)_2C_6H_3SCl$, to *cis*- and *trans*-2-butene;[110] in the addition of hypochlorous acid to cyclohexene; and in the addition of formaldehyde and acid to cyclohexene (Fig. 12-39).[111, 112] The

106 R. L. Augustine, *J. Org. Chem.*, **23**, 1853 (1958); R. L. Augustine and A. D. Broom, *ibid.*, **25**, 802 (1960).

107 B. Holmberg, *Svensk Kem. Tidskr.*, No. 5, 1911; *Chem. Abstr.*, **6**, 2072 (1912); A. McKenzie, *Proc. Chem. Soc.*, **27**, 150 (1911); *J. Chem. Soc.*, **101**, 1196 (1912).

108 W. G. Young, R. T. Dillon, and H. J. Lucas, *J. Am. Chem. Soc.*, **51**, 2528 (1929).

109 I. Roberts and G. E. Kimball, *J. Am. Chem. Soc.*, **59**, 947 (1937).

110 A. J. Havlik and N. Kharasch, *J. Am. Chem. Soc.*, **78**, 1207 (1956).

111 E. E. Smissman and R. A. Mode, *J. Am. Chem. Soc.*, **79**, 3447 (1957).

112 A. T. Blomquist and J. Wolinsky, *J. Am. Chem. Soc.*, **79**, 6025 (1957).

latter reaction, in a rigid system (Δ^2-*trans*-octalin), was shown to give the diaxial product (Fig. 12-39).[113]

Trans addition is also the usual course of the reaction of certain peracids, such as performic acid and peracetic acid in the presence of mineral acid with olefins.[114] In this case the first product is presumably the epoxide (which may be isolated when the oxidizing agent is perbenzoic or monoperphthalic acid or peracetic acid without added mineral acid used under mild conditions). Subsequent opening of the epoxide by rearward attack produces a

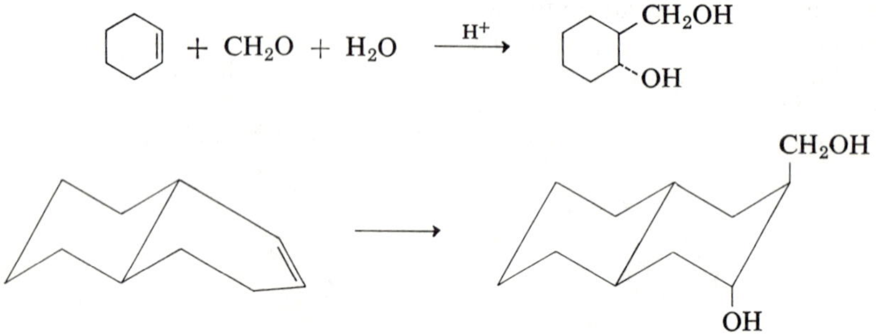

Fig. 12-39. Stereochemistry of the Prins reaction.

monoester of a glycol, the over-all stereochemical result being trans addition. Thus, reaction of cyclohexene with performic acid followed by hydrolysis gives *trans*-1,2-cyclohexanediol (Fig. 12-40).[114, 115] Exceptions to this stereochemical course have, however, been noted; thus the reaction of *cis*- and *trans*-*p*-methoxystilbene with perbenzoic acid in chloroform followed by hydrolysis gives the erythro and threo isomers of *p*-methoxyhydrobenzoin, respectively, as shown in Fig. 12-41.[116] In this case the intermediate epoxides are opened with retention of configuration. Other cases of this type have been discussed in Chap. 5 (page 102).

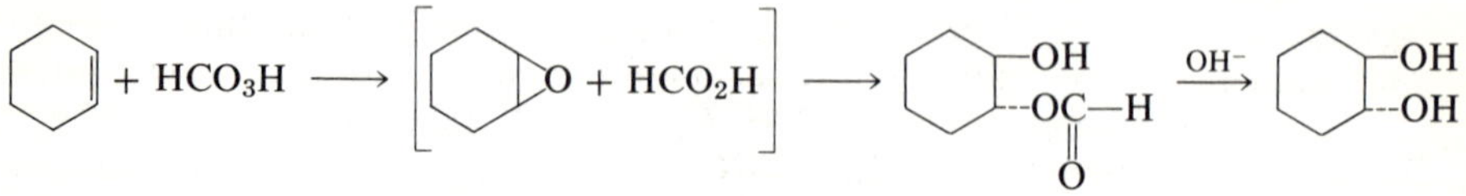

Fig. 12-40. Trans hydroxylation.

Less generality seems to attach to the steric course of ionic addition of reagents of the type HX. Hydrogen bromide adds trans to 1,2-dimethylcy-

[113] E. E. Smissman and D. T. Witiak, *J. Org. Chem.*, **25**, 471 (1960).

[114] D. Swern, in R. Adams, ed., "Organic Reactions," vol. 7, John Wiley & Sons, Inc., New York, 1953, pp. 378–433.

[115] J. English and J. D. Gregory, *J. Am. Chem. Soc.*, **69**, 2120 (1947); *Org. Syntheses*, Coll. Vol. III, p. 217.

[116] D. Y. Curtin, A. Bradley, and Y. G. Hendrickson, *J. Am. Chem. Soc.*, **78**, 4064 (1956). See also R. E. Parker and N. S. Isaacs, *Chem. Revs.*, **59**, 737 (1959).

clohexene,[116a] and hydrogen chloride adds trans to 1,2-dimethylcyclopentene.[116b] However, the nitric acid–catalyzed addition of water to 1,2-dimethylcyclohexene leads to a mixture of *cis*- and *trans*-1,2-dimethylcyclohexanol.[116c] Stereochemical anomalies in the addition of hydrogen chloride to 5-cholestene and 3-methyl-2-cholestene have already been discussed in Chap. 10.

The 1,4 addition of chlorine to butadiene has been shown[117] to give exclusively *trans*-1,4-dichloro-2-butene. This indicates that cyclic intermediates of the type shown in Fig. 12-42 are not involved.

CH_3O–C₆H₄–CH=CH–C₆H₅ —($PhCO_3H$)→ CH_3O–C₆H₄–CH(–O–)CH–C₆H₅ —((1) $PhCO_2H$, (2) OH^-)→ CH_3O–C₆H₄–CH(OH)–CH(OH)–C₆H₅

Fig. 12-41. Cis hydroxylation.

c. Molecular Addition. In addition to catalytic hydrogenation (Sec. 12-6*a*) there are a few other molecular-addition reactions which proceed in cis fashion, the best-known ones being the hydroxylation of olefins with osmium tetroxide and potassium permanganate. Oxidation of maleic acid with potassium permanganate gives *meso*-tartaric acid whereas similar oxidation of fumaric acid leads to (±)-tartaric acid[118] (Fig. 12-43).† Cyclohexene (cis) and permanganate give rise to *cis*-cyclohexanediol-1,2.[119] Similar stereochemical results are obtained in the oxidation of these compounds with osmium tetroxide in conjunction with sodium perchlorate[120] or hydrogen

CH=CH, CH_2 CH_2, $\overset{+}{Cl}$ HC—CH, CH_2 CH_2, Cl—Cl

Fig. 12-42. Cyclic intermediates in 1,4 addition that have been ruled out.

† These are probably the first reactions whose steric course was known, since the configurations of maleic and fumaric acids are readily derived from their relative ease of anhydride formation (cf. Sec. 12-2), and the configuration of the tartaric acids was known from the classical work of Pasteur (cf. Chap. 3). It is unfortunate that, compared with electrophilic-addition reactions (*vide supra*), these reactions are atypical in their stereochemical course; this misled Wislicenus to assume (in 1887) that all olefin addition is cis.

[116a] G. S. Hammond and T. D. Nevitt, *J. Am. Chem. Soc.*, **76**, 4121 (1954).
[116b] G. S. Hammond and C. H. Collins, *J. Am. Chem. Soc.*, **82**, 4323 (1960).
[116c] C. H. Collins and G. S. Hammond, *J. Org. Chem.*, **25**, 911 (1960).
[117] K. Mislow and H. M. Hellman, *J. Am. Chem. Soc.*, **73**, 244 (1951).
[118] A. Kekulé and R. Anschütz, *Ber.*, **13**, 2150 (1880); **14**, 713 (1881).
[119] J. Böeseken, *Rec. trav. chim.*, **47**, 683 (1928).
[120] K. A. Hofmann, O. Ehrhart, and O. Schneider, *Ber.*, **46**, 1657 (1913); J. Böeseken and J. van Giffin, *Rec. trav. chim.*, **39**, 183 (1920).

peroxide.[121] In the case of osmium tetroxide, it has been shown[122] that cyclic esters of the type shown in Fig. 12-44 are intermediates, and similar intermediates (Fig. 12-44) seem to be involved in the permanganate oxidation,[123] thus accounting for the cis additions.

Another method for producing cis glycols from olefins involves reaction with iodine and silver acetate in aqueous acetic acid.[124] This method apparently involves an ortho-acetate intermediate, as shown in Fig. 12-45.[125]

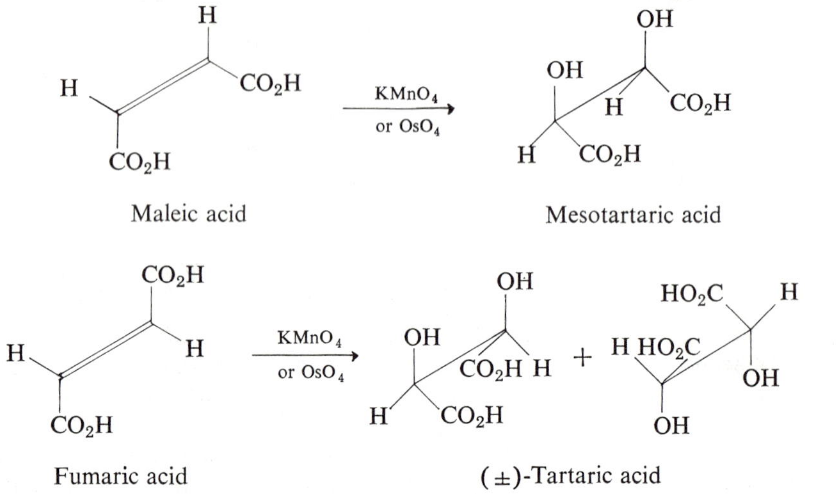

Maleic acid → ($KMnO_4$ or OsO_4) → Mesotartaric acid

Fumaric acid → ($KMnO_4$ or OsO_4) → (±)-Tartaric acid

Fig. 12-43. Conversion of maleic and fumaric acids to tartaric acid.

Whereas in a simple system, such as cyclohexene, the method shown in Fig. 12-45 leads to the cis glycol in the same way as potassium permanganate or osmium tetroxide, a difference becomes apparent in more complex systems where two diastereoisomeric cis glycols may exist. For example, in a $\Delta^{2,3}$ steroid such as 5α,22a-spirost-2-ene (partial formula shown in Fig. 12-45)[126] potassium permanganate or osmium tetroxide oxidation involves approach from the less hindered α side (cf. Chap. 10) to give the 2α,3α-diol. In the

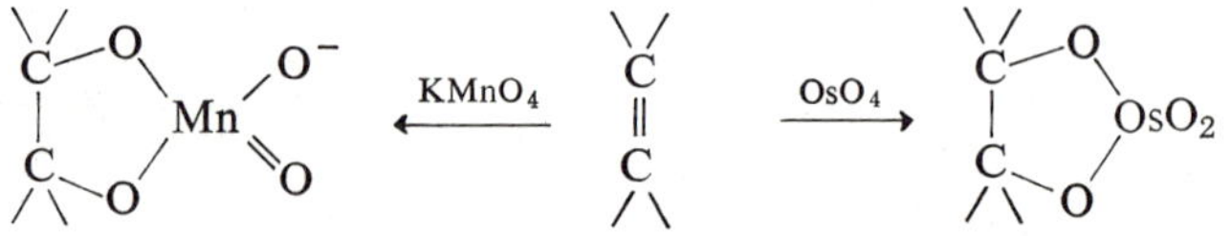

Fig. 12-44. Intermediates in oxidation of olefins to glycols.

[121] N. A. Milas and S. Sussman, *J. Am. Chem. Soc.*, **58**, 1302 (1936); **59**, 2345 (1937).
[122] R. Criegee et al., *Ann.*, **522**, 75 (1936); **550**, 99 (1942); *Angew. Chem.*, **51**, 519 (1938).
[123] K. B. Wiberg and K. A. Saegebarth, *J. Am. Chem. Soc.*, **79**, 2822 (1957).
[124] R. B. Woodward and F. V. Brutcher, *J. Am. Chem. Soc.*, **80**, 209 (1958).
[125] Cf. K. B. Wiberg and K. A. Saegebarth, *J. Am. Chem. Soc.*, **79**, 6256 (1957).
[126] C. Djerassi, L. B. High, T. T. Grossnickle, R. Ehrlich, J. A. Moore, and R. B. Scott, *Chemistry & Industry* (*London*), 474 (1955).

iodine–silver acetate–wet acetic acid method, however, the attack on the less hindered α side is by the original iodonium ion, and, as may be seen in Fig. 12-45, the final result is the formation of the $2\beta,3\beta$-diol in which the hydroxyl groups are on the more crowded side of the molecule.

As is to be expected on the basis of the mechanism,[125] water is necessary to produce the cis diols. The action of iodine and silver benzoate *in benzene*

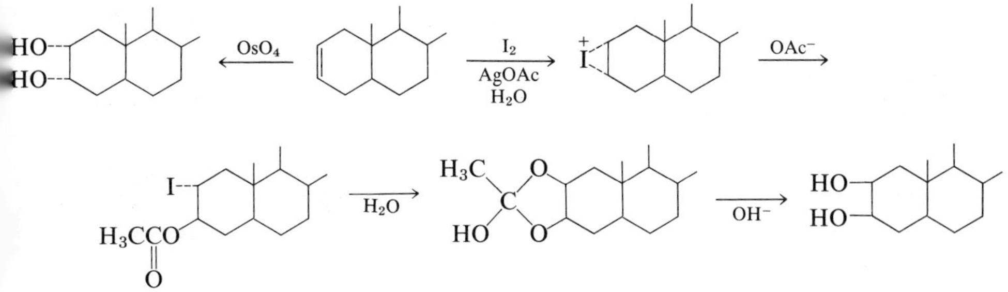

Fig. 12-45. Conversion of olefin to glycol by iodine–silver acetate–wet acetic acid.

on olefins gives dibenzoates by trans addition[127] (e.g., *meso*-hydrobenzoin dibenzoate, $C_6H_5CH(OCOC_6H_5)CH(OCOC_6H_5)C_6H_5$, from *trans*-stilbene), since in that case the intermediate ortho ester shown in Fig. 12-45 (benzoate instead of acetate) is opened by rearward attack of a benzoate ion with inversion.

Molecular cis addition is also observed in the reaction of cholesteryl benzoate with iodobenzene dichloride[128] which gives rise to $5\alpha,6\alpha$-dichlorocholestan-3β-yl benzoate (Fig. 12-46), the cis attack taking place on the less hindered α side of the molecule (cf. Chap. 10). It is not known how general

Fig. 12-46. Reaction of iodobenzene dichloride with cholesteryl benzoate.

[127] C. Prévost, *Compt. rend.*, **196**, 1129 (1933); **197**, 1661 (1933).

[128] C. J. Berg and E. S. Wallis, *J. Biol. Chem.*, **162**, 683 (1946); D. H. R. Barton and E. Miller, *J. Am. Chem. Soc.*, **72**, 370 (1950).

this reaction is. For example, acenaphthylene, upon treatment with iodobenzene dichloride, gives[129] *trans*-1,2-dichloroacenaphthene instead of the expected cis isomer, but then this system seems to behave altogether anomalously in that addition of chlorine gives the cis-dichloro compound.†

Cis addition of fluorine has been effected by means of lead tetrafluoride, prepared *in situ* from lead tetraacetate and hydrogen fluoride.[129a]

Another type of cis addition[129b] is the reduction of olefins to paraffins by means of diimide, HN=NH, generated *in situ* by oxidation of hydrazine, H_2NNH_2, or decarboxylation of azodiformic acid, $HO_2CN{=}NCO_2H$. Thus dimethylmaleic acid, $HO_2C(CH_3)C{=}C(CH_3)CO_2H$, is reduced to *meso*-2,3-dimethylsuccinic acid, *meso*-$HO_2CCH(CH_3)CH(CH_3)CO_2H$. The diimide is oxidized to nitrogen.

A molecular-addition reaction of particular interest is the already mentioned hydroboration reaction discovered by H. C. Brown.[130a] Oxidation of the olefin-diborane addition product by means of hydrogen peroxide leads to alcohols, the over-all result of hydroboration followed by oxidation being the addition of the elements of water to an olefin. Not only the direction of addition but also its stereochemistry is different from the earlier-mentioned acid-catalyzed addition of water to the double bond. Thus hydroboration of 1-methylcyclohexene[130b] (Fig. 12-47, top) gives pure *trans*-2-methylcyclohexanol, the product of cis addition in which the alcohol function ends up on the less substituted carbon.‡ In the case of steroids and terpenoids, addition is axial and from the less hindered face of the molecule (cf. Chap. 10); thus hydroboration followed by oxidation of cholest-4-ene gives cholestan-4-α-ol (Fig. 12-47, middle).[130c] Since hydroboration itself involves cis addition (cf. Fig. 12-26), the oxidation step must involve retention of configuration. This is reasonable if this step involves a molecular rearrangement of the type encountered in Sec. 5-4*f*, *ii*, in which the migrating group retains configuration. In the present case, the migration proceeds from boron to oxygen (Fig. 12-47, bottom).[130d]

† Cis addition to the 9,10 double bond of phenanthrene occurs with chlorine in the presence of lithium chloride: P. B. D. de la Mare and N. V. Klassen, *Chemistry & Industry (London)*, 498 (1960).

‡ Hydroboration has sometimes been considered "anti-Markovnikov addition," but this may be misleading, since the polarization of diborane is undoubtedly such that the boron is positive and the hydrogen negative. Thus, from the electrostatic point of view, addition is as expected, and whether it does or does not follow Markovnikov's rule depends on whether one uses the rule in its original form, or in a modified form, based on reagent polarity.

---

[129] S. J. Cristol, F. R. Stermitz, and P. S. Ramey, *J. Am. Chem. Soc.*, **78**, 4939 (1956).

[129a] A. Bowers, E. Denot, and R. Urquiza, *Tetrahedron Letters*, no. 20, 34 (1960).

[129b] E. J. Corey, D. J. Pasto, and W. L. Mock, *J. Am. Chem. Soc.*, **83**, 2957 (1961). For similar observations in the diimide reduction of acetylenes, cf. S. Hünig, H.-R. Müller, and W. Thier, *Tetrahedron Letters*, 353 (1961).

[130] (*a*) H. C. Brown, Organoboranes, in H. Zeiss, ed., "Organometallic Chemistry," Reinhold Publishing Corporation, New York, 1960, chap. 4; see especially pp. 166–169; (*b*) H. C. Brown and G. Zweifel, *J. Am. Chem. Soc.*, **83**, 2544 (1961); (*c*) S. Wolfe, M. Nussim, Y. Mazur, and F. Sondheimer, *J. Org. Chem.*, **24**, 1034 (1959); see also F. Sondheimer and S. Wolfe; *Can. J. Chem.*, **37**, 1870 (1959); M. Nussim and F. Sondheimer, *Chemistry & Industry* (*London*), 400 (1960); F. S. Alvarez and M. Arreguin, *ibid.*, 720 (1960); (*d*) A. G. Davies, D. G. Hare, and R. F. M. White, *ibid.*, 556 (1960); (*e*) H. C. Brown and G. Zweifel, *J. Am. Chem. Soc.*, **83**, 486 (1961).

Asymmetric synthesis of alcohols has been achieved by carrying out the hydroboration of an appropriate olefin with a dialkylborane of the type R^*_2BH, where R* is optically active.[130e] A suitable optically active reagent is obtained by adding diborane to α-pinene. Addition of this R^*_2BH reagent to *cis*-2-butene followed by hydrogen peroxide oxidation gives (−)-2-butanol of 87% optical purity.[130e] Since the α-pinene used in the experiment was probably no more than 90% optically pure, this means that the synthesis of the alcohol was *nearly completely asymmetric,* a most remarkable feat in view

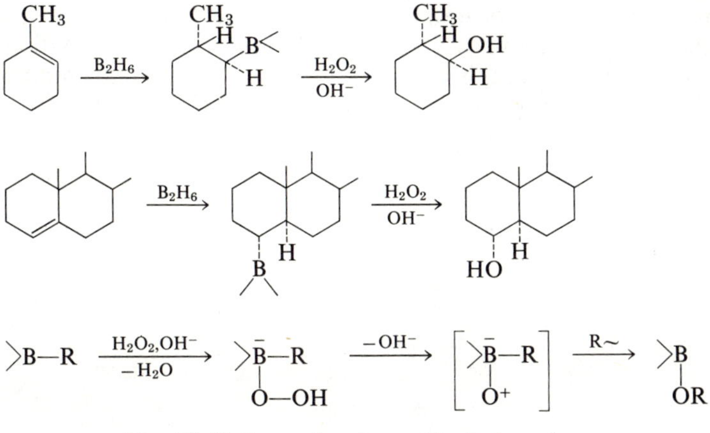

$$>B{-}R \xrightarrow[-H_2O]{H_2O_2, OH^-} >\bar{B}{-}R\ (O{-}OH) \xrightarrow{-OH^-} \left[>\bar{B}{-}R\ (O^+)\right] \xrightarrow{R\sim} >B{-}OR$$

Fig. 12-47. Stereochemistry of hydroboration.

of the general experience, cited in Sec. 4-4*d, ii,* that most other asymmetric syntheses (other than biochemical ones) are quite inefficient.

d. Free-radical Addition. Free radicals are known to add to multiple linkages, as evidenced, for example, by the fact that vinyl polymerization may be induced by radical catalysts. The stereochemistry of the addition was first elucidated in cyclohexane systems. Addition of hydrogen bromide to 1-methylcyclohexene,[131] 1-bromocyclohexene,[131, 132] and 1-chlorocyclohexene[132] under the influence of light or peroxide (which promote a free-

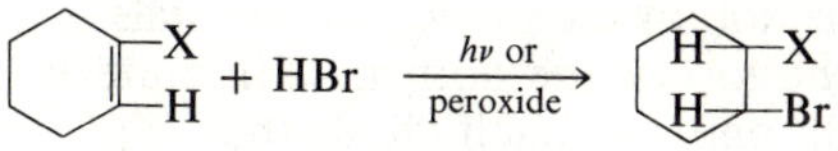

X = CH_3,Br,Cl Trans addition

Fig. 12-48. Free-radical addition of hydrogen bromide to 1-substituted cyclohexenes.

radical mechanism) proceeds in trans fashion to give cis-1,2-disubstituted cyclohexanes (Fig. 12-48). In the case of the halocyclohexenes, the stereoselectivity† of the addition was shown to exceed 99%.[132] A much lesser

† Regarding the precise meaning of stereoselectivity and stereospecificity, see Chap. 15.

[131] H. L. Goering, P. I. Abell, and B. F. Aycock, *J. Am. Chem. Soc.,* **74,** 3588 (1952).
[132] H. L. Goering and L. L. Sims, *J. Am. Chem. Soc.,* **77,** 3465 (1955).

degree of stereoselectivity (66 to 90% trans addition) was observed in the addition of RSH compounds, such as hydrogen sulfide, thiophenol, and thiolacetic acid, $HSCH_2CO_2H$, to 1-chlorocyclohexene[133] and 1-methylcyclohexene.[134]

Free-radical addition to acyclic olefins seems to be stereospecific† only under rather restricted conditions. Thus the light-induced addition of hydrogen bromide to the 2-bromo-2-butenes is stereospecifically trans if carried out at $-78°$.[134a] The cis olefin gives rise to *meso*-2,3-dibromobutane whereas the trans olefin yields (±)-2,3-dibromobutane (Fig. 12-49). However, if the addition is carried out at or near room temperature, both olefins give approximately the same mixture of ca. 25% *meso*-dibromide and 75% (±)-dibromide.‡ The rationalization[134a] of this result is as follows: Addition of a bromine atom to the olefins produces the intermediates shown in Fig. 12-50, A from the cis

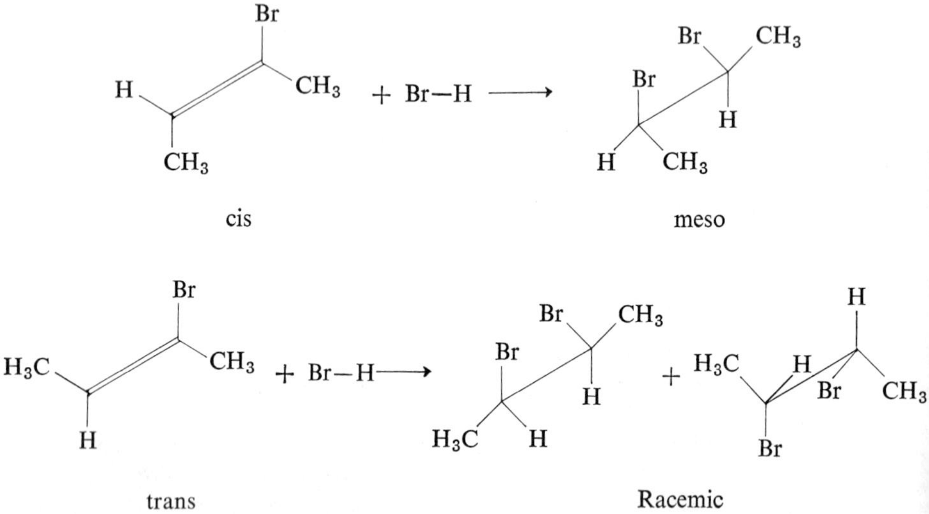

Fig. 12-49. Free-radical addition of hydrogen bromide to the 2-bromo-2-butenes.

olefin and B from the trans olefin. [It is assumed that the bromine atom approaches in the direction of the pi-electron orbitals of the olefin. Whether it approaches from the top or the bottom is immaterial, except that in the case of the trans olefin one approach yields the (+) form and the other the (−) form of the product. Since the two approaches are equally likely, a *dl* pair is actually produced.] If the chain transfer of a hydrogen atom from hydrogen bromide is faster than rotation of the intermediate radicals (Fig.

† Regarding the precise meaning of stereoselectivity and stereospecificity, see Chap. 15, p. 436.
‡ It is clear that even at room temperature the product composition is not thermodynamically controlled inasmuch as the *meso*-dibromide is more stable than the (±)-dibromide (cf. Chap. 6).

[133] H. L. Goering, D. I. Relyea, and D. W. Larsen, *J. Am. Chem. Soc.*, **78**, 348 (1956).
[134] F. G. Bordwell and W. A. Hewett, *J. Am. Chem. Soc.*, **79**, 3493 (1957).
[134a] H. L. Goering and D. W. Larsen, *J. Am. Chem. Soc.*, **81**, 5937 (1959).

12-50) around the C_2—C_3 bond and if it is assumed that the hydrogen atom approaches from the opposite side as the bromine atom, stereospecificity will be observed. But if rotation around the C_2—C_3 bond is faster than acquisition of hydrogen, then the intermediate radicals A and B will be interconverted, equilibrium lying on the side of the (less crowded) intermediate B. In this case an identical mixture of products is formed from the two starting olefins, the product formed via intermediate B predominating. Several possible reasons have been advanced as to why the hydrogen atom approaches from the side opposite that of the added bromine atom in A and B (Fig. 12-50). One is that the product can, in this way, be formed with a minimum eclipsing of bonds. A second one is that the olefin may already be complexed with

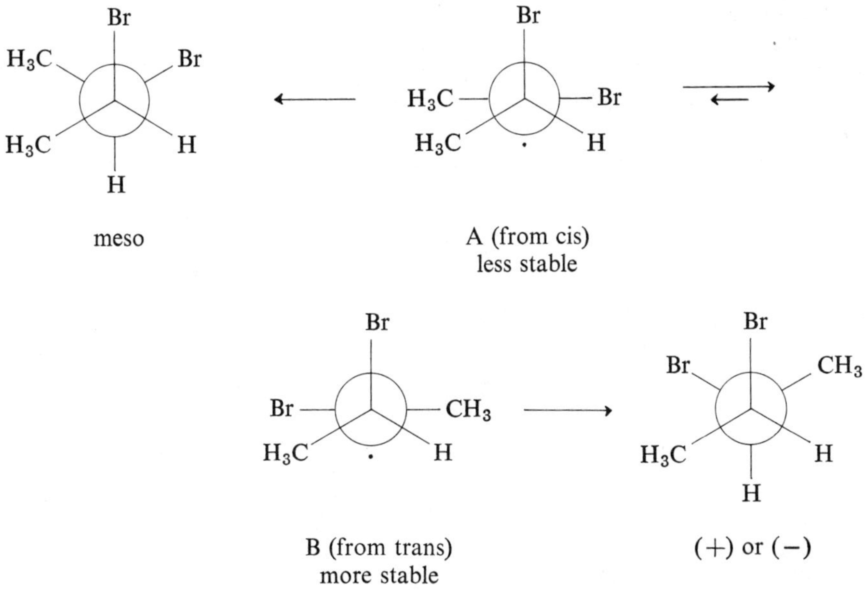

meso

A (from cis)
less stable

B (from trans)
more stable

(+) or (−)

Fig. 12-50. Intermediates in radical addition of hydrogen bromide to the 2-bromo-2-butenes.

hydrogen bromide when attacked by the bromine atom and that the bromine atom attacks from the opposite side of the complexed HBr molecule. A third possibility is that the radical, similar to a bromonium ion (Sec. 12-6*b*), receives a contribution from a bridged structure. Of the three explanations, the first seems to be most plausible in the light of present results.[135a–c]

It is clear that stereospecificity in this kind of addition depends on the competition between rate of chain transfer and rate of rotation about the C_2—C_3 bond. Thus, the radical additions to cyclohexenes (*vide supra*) are more cleanly trans than those to simple olefins because the barrier to rotation in cyclohexane (and thus, presumably, in the intermediate cyclohexyl radicals)

[135] (*a*) N. A. Le Bel, *J. Am. Chem. Soc.*, **82**, 623 (1960); (*b*) P. I. Abell and C. Chiao, *ibid.*, **82**, 3610 (1960); (*c*) N. P. Neureiter and F. G. Bordwell, *ibid.*, **82**, 5354 (1960).

is relatively high (cf. Sec. 8-1). The reaction of the 2-butenes with bromotrichloromethane[136] and with CH_3SD[137] is not stereospecific, presumably because the rate of chain transfer is relatively slow. The low stereoselectivity in the addition of RSH compounds to cyclohexenes is similarly explained. The stereoselectivity increases with increasing RSH concentration[133] and, in the case of the addition of CH_3SD to the 2-butenes, upon addition of DBr,[137] as would be expected on grounds that this increases the rate of chain transfer. (However, the stereospecific addition of CH_3SD to the 2-butenes in the presence of DBr may alternatively be interpreted as addition to an olefin—DBr pi complex.)

Trans addition has also been observed in the free-radical addition of silicochloroform[82] (Fig. 12-51) and hydrogen bromide[54b] in the liquid phase to alkylacetylenes. In the addition of hydrogen bromide in the gas phase to propyne, $CH_3—C{\equiv}CH$, complications result because the stereoisomeric 1-bromo-1-propenes, $CH_3CH{=}CHBr$, which are the products are readily interconverted under the conditions of the reaction. It should be mentioned that, in the radical addition of hydrogen bromide to the 2-bromo-2-butenes (Fig. 12-49) at room temperature, there is also some interconversion of the

$$\text{R(H)C=C(H)SiCl}_3 \xleftarrow{\text{peroxide}} \text{R—C}\equiv\text{CH} + \text{HSiCl}_3 \xrightarrow{\text{Pt/C}} \text{(R)(H)C=C(H)SiCl}_3$$

Trans addition — Cis addition

$$\text{R—C}\equiv\text{CH} \xrightarrow[h\nu]{\text{HBr}, -78^\circ} \text{R(H)C=C(H)Br}$$

Trans addition

Fig. 12-51. Addition of silicochloroform and of hydrogen bromide to acetylenes.

isomeric starting materials, presumably by reversion of the intermediates A and B (Fig. 12-50) to olefin by loss of bromine. This interconversion may complicate the interpretation of the reaction course.

It is not known whether radical addition to olefins in rigid six-membered-ring systems is trans-diequatorial or trans-diaxial. On the basis of the picture given above, it would be expected to be trans-diaxial.† It is, however, known that radical addition of a thiophenol to a bicyclic olefin will proceed exo.[138, 139] In this case the addition is cis, as shown in Fig. 12-52,[159] either because the hydrogen finds it easier to approach the intermediate radical from the less hindered exo side or because of reluctance of the chlorine and

† After this was written, Prof. Norman A. Le Bel, Wayne State University, kindly informed the author that he had observed diaxial addition of hydrogen bromide and of thiophenol to 2-chloro-4-*t*-butylcyclohexene under radical conditions.

---

[136] P. S. Skell and R. C. Woodworth, *J. Am. Chem. Soc.*, **77**, 4638 (1955).
[137] P. S. Skell and R. G. Allen, *J. Am. Chem. Soc.*, **82**, 1511 (1960).
[138] S. J. Cristol and G. D. Brindell, *J. Am. Chem. Soc.*, **76**, 5699 (1954).
[139] S. J. Cristol and R. P. Arganbright, *J. Am. Chem. Soc.*, **79**, 6039 (1957). See also Ref. 135*a*.

the thioether grouping to become eclipsed. In contrast, the addition of bromotrichloromethane to norbornylene apparently occurs in normal trans fashion.[140] Trans addition is also found in the reaction of cyclopentene and 1-methylcyclohexene with N_2O_4.[141]

e. Addition of Carbenes and Methylenes. This section will be concerned with the stereochemistry of the addition of divalent carbon compounds, $:CR_2$, to olefins. The two non-bonded electrons in $:CR_2$ may have either paired (antiparallel) or unpaired (parallel) spins. In the former case the compound is in a singlet state and is called a "carbene," whereas in the latter case it is in the triplet state and is called a "methylene." Whether a given divalent carbon species is a carbene or a methylene depends partly on the nature of the R groups in $:CR_2$. The species may be distinguished by the fact that toward a series of olefins a carbene shows electrophilic reactivity[142, 143] (which runs parallel to the reactivity in such typical electrophilic reactions as bromination

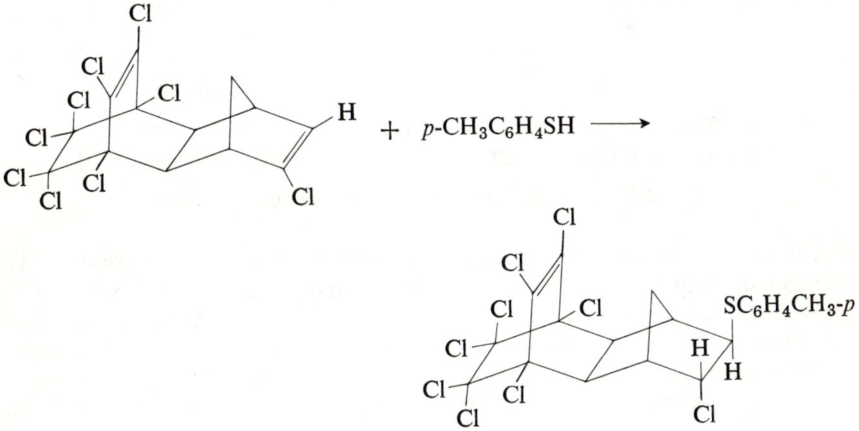

$$+\ p\text{-}CH_3C_6H_4SH \longrightarrow$$

Fig. 12-52. Exo-cis addition of radicals to bicyclic olefins.

and epoxidation), whereas a methylene shows the typical reactivity of a free-radical species.[144]

As far as the stereochemistry of addition is concerned, carbenes add to olefins stereospecifically, a cis olefin giving rise to a cis-1,2-disubstituted cyclopropane and a trans olefin giving a trans-1,2-disubstituted cyclopropane (Fig. 12-53).[145-148] On the other hand, methylenes, such as diphenylmeth-

[140] M. S. Kharasch and H. N. Friedlander, *J. Org. Chem.*, **14**, 239 (1949); cf. F. S. Fawcett, *Chem. Revs.*, **47**, 234 (1950).

[141] J. C. D. Brand and I. R. D. Stevens, *J. Chem. Soc.*, 629 (1958).

[142] P. S. Skell and A. Y. Garner, *J. Am. Chem. Soc.*, **78**, 5430 (1956).

[143] W. von E. Doering and W. A. Henderson, *J. Am. Chem. Soc.*, **80**, 5274 (1958).

[144] R. M. Etter, H. S. Skovronek, and P. S. Skell, *J. Am. Chem. Soc.*, **81**, 1008 (1959).

[145] W. von E. Doering and P. LaFlamme, *J. Am. Chem. Soc.*, **78**, 5447 (1956).

[146] R. C. Woodworth and P. S. Skell, *J. Am. Chem. Soc.*, **81**, 3383 (1959).

[147] P. S. Skell and A. Y. Garner, *J. Am. Chem. Soc.*, **78**, 3409 (1956).

[148] H. M. Frey, *J. Am. Chem. Soc.*, **80**, 5005 (1958).

ylene and propargylene (Fig. 12-54), add non-specifically to the 2-butenes, converting either geometrical isomer to a mixture of cis- and trans-1,2-dimethylsubstituted cyclopropanes.[144, 149] Presumably the addition of a carbene to an olefin is a concerted process, much as the addition of OH^+ in epoxide formation, which is also stereospecific, producing a cis epoxide from a cis olefin and a trans epoxide from a trans olefin. In contrast, the addition

H_3C, CH_3, H, H: C=C + : CR_2 ⟶ H_3C, CH_3, H, H, R R (cyclopropane)

H_3C, H, H, CH_3: C=C + : CR_2 ⟶ H_3C, H, H, CH_3, R R (cyclopropane)

R = Br (: CBr_2 generated from $CHBr_3$ and *t*-BuOK)
R = H (: CH_2 generated by photolysis of CH_2N_2 or $CH_2{=}C{=}O$)

Fig. 12-53. Stereospecific addition of carbenes to olefins.

of a methylene to an olefin must be stepwise, for the following reason: The electron spins of the two electrons in the methylene are parallel, whereas the spins of the two electrons in the pi orbital of the olefin are antiparallel. Thus, when one of the methylene electrons bonds with one of the pi electrons to form the first bond, the other methylene electron and the second pi electron have parallel spins and are not able to bond until one of the spins is inverted. This

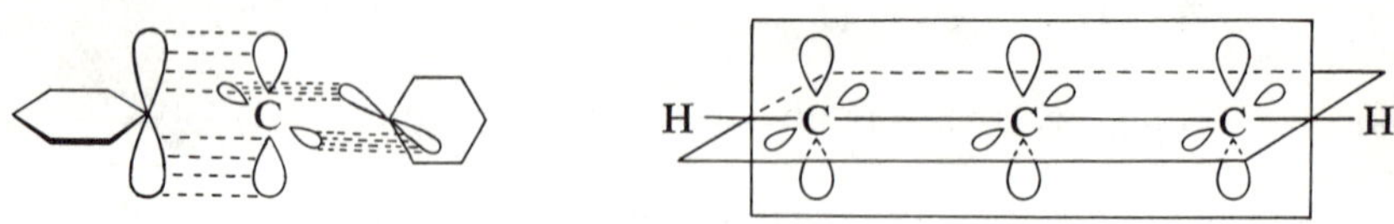

Diphenylmethylene Propargylene

Fig. 12-54. Diphenylmethylene and propargylene. [*From R. M. Etter, H. S. Skovronek, and P. S. Skell, J. Am. Chem. Soc.,* **81,** 1008 (1959); *and P. S. Skell and J. Klebe, ibid.,* **82,** 247 (1960). *By permission of the authors and editor.*]

process takes time, and the addition of the methylenes is therefore stepwise, proceeding via an intermediate which lives long enough to rotate around the central bond and lose steric identity (Fig. 12-55). Equilibration of the intermediates is not, however, complete, as evidenced by the fact that, although both cyclopropanes are formed from both 2-butenes, their ratio depends on the starting material as well as on the methylene that adds. The reason why

[149] P. S. Skell and J. Klebe, *J. Am. Chem. Soc.,* **82,** 247 (1960).

diphenylmethylene and propargylene, unlike most other $:CR_2$ species, are biradicals is clear from Fig. 12-54: The unpaired electrons in these species are effectively stabilized by pi-orbital overlap.

The stereospecificity of the addition of methylene (obtained by photolysis of diazomethane, CH_2N_2) to the 2-butenes is reduced when the reaction is carried out in a large excess of nitrogen as a diluent.[149a] This has been taken to mean that under these conditions CH_2 is converted from the less stable

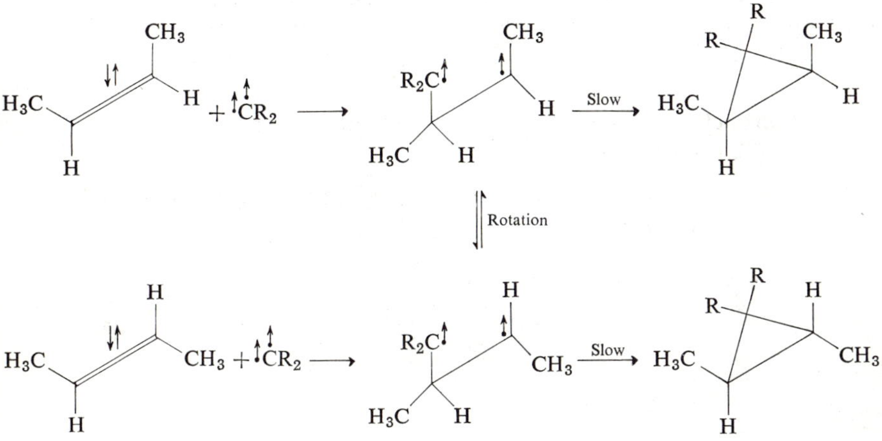

Fig. 12-55. Non-stereospecific addition of methylenes to olefins.

(presumably excited) singlet state to the more stable triplet (biradical) state.

f. Nucleophilic Addition. Except under very stringent conditions, nucleophilic reagents add to double bonds only when they are activated by electron-withdrawing groups. The Michael addition[150] to systems such as α,β-unsaturated carbonyl compounds, —C=C—C=O, is a prominent example of nucleophilic addition. When the α and β carbons are both substituted, diastereoisomeric products may result. An example is the addition of malonic

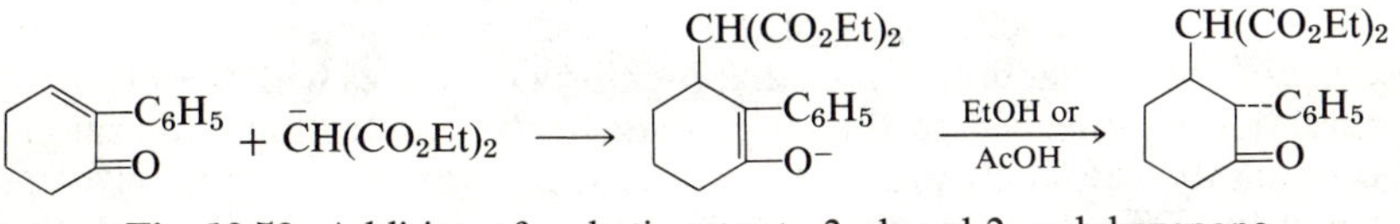

Fig. 12-56. Addition of malonic ester to 2-phenyl-2-cyclohexenone.

ester to 2-phenyl-2-cyclohexenone (Fig. 12-56); the trans product results exclusively.[151] However, since the addition is probably 1,4 and proceeds via the enolate, the over-all stereochemistry of the process probably reflects the

[149a] F. A. L. Anet, R. F. W. Bader, and A. M. Van der Auwera, *J. Am. Chem. Soc.*, **82**, 3217 (1960); H. M. Frey, *ibid.*, **82**, 5947 (1960).

[150] E. D. Bergmann, D. Ginsburg, and R. Pappo, in R. Adams, ed., "Organic Reactions," vol. 10, John Wiley & Sons, Inc., New York, 1959.

[151] W. E. Bachmann and E. J. Fornefeld, *J. Am. Chem. Soc.*, **72**, 5529 (1950); cf. D. Ginsburg and R. Pappo, *J. Chem. Soc.*, 938 (1951), and R. A. Abramovitch and J. M. Muchowski, *Can. J. Chem.*, **38**, 557 (1960).

stereochemistry of ketonization rather than the stereochemistry of nucleophilic addition. Ketonization may either be thermodynamically controlled (giving the more stable trans product) or involve equatorial approach of a proton (cf. Chap. 8); in the present case the thermodynamically controlled product seems to result.

Another stereochemical problem concerns the addition of nucleophiles to octalones of the type shown in Fig. 12-37. By analogy with what was said (Sec. 12-6*a*) about chemical reduction of such systems, one would predict that a nucleophile would approach the system from the axial side. This has apparently not been tested.

12-7. Stereochemistry of Nucleophilic Displacement in Olefinic Halides

For reasons to be mentioned below, nucleophilic-displacement reactions of the type RR′C=CR″X + Y⁻ → RR′C=CR″Y + X⁻ are related to the

Fig. 12-57. Stereochemistry of nucleophilic displacement in vinyl halide.

elimination and addition reactions of olefins discussed in the previous two sections. In general, olefinic halides are sluggish in nucleophilic-displacement reactions, and special structural features are usually required to observe such reactions. The stereochemistry of the above displacement reaction was first observed[152] in the reaction of the ethyl β-chlorocrotonates with thiophenoxide or ethyl mercaptide. With the latter reagent the cis starting material gives predominantly (85%) cis product, and the trans starting material gives predominantly (92%) the trans product, as shown in Fig. 12-57. With ethoxide and phenoxide, on the other hand, the reaction was non-stereospecific in that both starting materials gave the same product. Predominant retention of configuration has also been observed[153] in the reaction of the 1-bromo-2-*p*-nitrophenylethylenes with iodide and of β-halovinyl aryl sulfones,

$$ArSO_2CH{=}CHX$$

with various thiophenols in the presence of sodium bicarbonate.[154] A pos-

[152] D. E. Jones, R. O. Morris, C. A. Vernon, and R. F. M. White, *J. Chem. Soc.*, 2349 (1960).
[153] S. I. Miller and P. K. Yonan, *J. Am. Chem. Soc.*, **79**, 5931 (1957).
[154] F. Montanari, *Gazz. chim. ital.*, **86**, 747 (1956).

sible path explaining retention is shown in Fig. 12-58. The group Y^- adds to the olefinic system, and X^- departs before the intermediate carbanion has had time to rotate about the central bond. The situation is somewhat similar to that depicted for radical addition to olefins (Fig. 12-50 and especially Fig. 12-22). The fact that the radicals shown in Figs. 12-50 and 12-22 are probably planar at the tricovalent carbon whereas the carbanion (Fig. 12-58) is probably tetrahedral is likely to be of little consequence, inasmuch as the two tetrahedral forms of the carbanion are rapidly interconverted by molecular vibration (cf. Chap. 13).

Two other paths of apparent nucleophilic displacement in vinyl halides have been recognized,[153, 155, 156] namely, elimination followed by addition and addition followed by elimination. These paths lack stereospecificity in the sense that two geometrically isomeric starting materials give one and the same

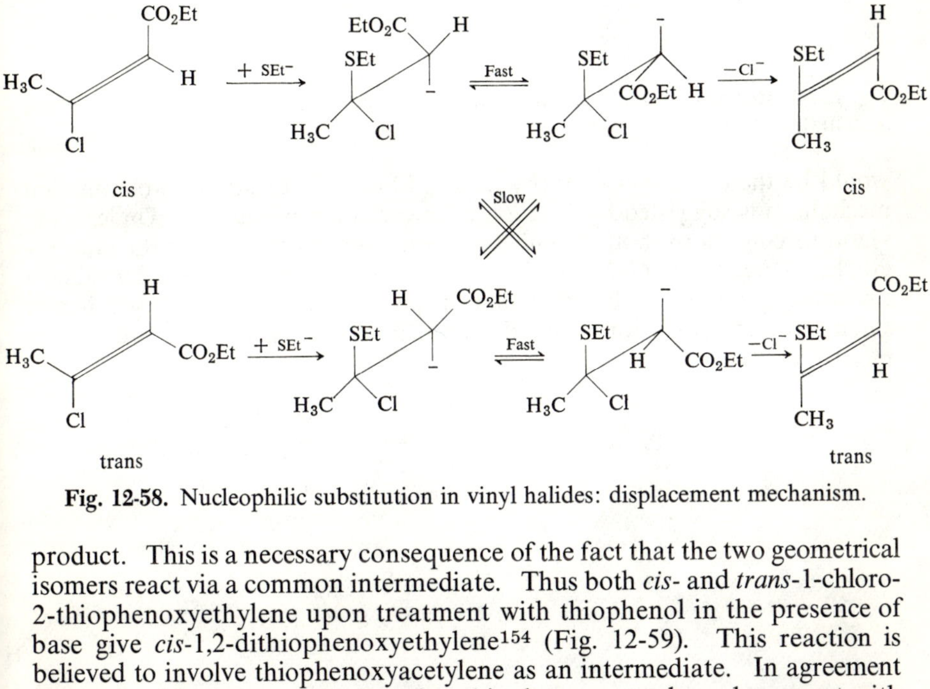

Fig. 12-58. Nucleophilic substitution in vinyl halides: displacement mechanism.

product. This is a necessary consequence of the fact that the two geometrical isomers react via a common intermediate. Thus both *cis*- and *trans*-1-chloro-2-thiophenoxyethylene upon treatment with thiophenol in the presence of base give *cis*-1,2-dithiophenoxyethylene[154] (Fig. 12-59). This reaction is believed to involve thiophenoxyacetylene as an intermediate. In agreement with this assumption is the fact that thiophenoxyacetylene does react with thiophenol to give *cis*-1,2-dithiophenoxyethylene (trans nucleophilic addition; cf. Sec. 12-5) and that *cis*-1-chloro-2-thiophenoxyethylene undergoes reaction with thiophenol and base faster than the trans isomer (the former undergoes trans elimination whereas the latter requires cis elimination; cf. Fig. 12-29). *cis*-1,2-Dichloroethylene also reacts with thiophenol and base to give the ultimate product shown in Fig. 12-59.[156a] In this case the first step is trans

[155] W. E. Truce, M. M. Boudakian, R. F. Heine, and R. J. McManimie, *J. Am. Chem. Soc.*, **78,** 2743 (1956).

[156] W. E. Truce and M. M. Boudakian, *J. Am. Chem. Soc.*, **78,** 2748 (1956).

[156a] W. E. Parham and J. Heberling, *J. Am. Chem. Soc.*, **77,** 1175 (1955).

elimination of HCl to give chloroacetylene followed by trans addition of thiophenol to give the cis starting material shown in Fig. 12-59. *trans*-1,2-Dichloroethylene is inert because it does not eliminate hydrogen chloride under the conditions of the reaction.[155]

When 1,1-dichloroethylene is treated with *p*-thiocresol and base under mild conditions, the product is *trans*-1-chloro-2-*p*-thiocresoxyethylene (*B*, Fig. 12-60). This is evidently not the product of elimination-addition which

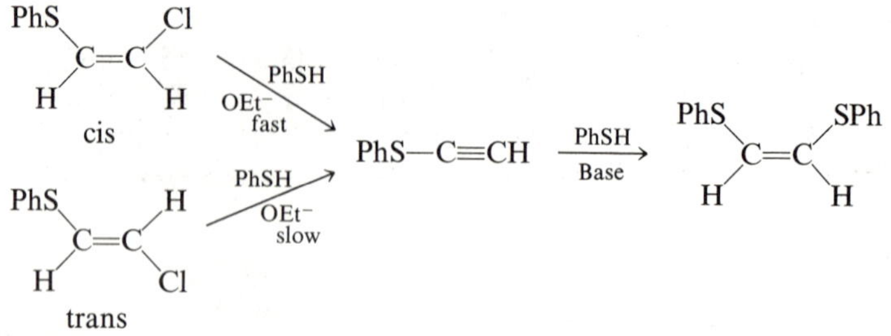

PhS(H)C=C(Cl)H (cis) —PhSH, OEt⁻, fast→ PhS—C≡CH —PhSH, Base→ PhS(H)C=C(SPh)H

PhS(H)C=C(H)Cl (trans) —PhSH, OEt⁻, slow→ PhS—C≡CH

Fig. 12-59. Reaction of the 1-chloro-2-thiophenoxyethylenes with thiophenol and base (elimination-addition).

would be the cis isomer of B (by trans addition). An addition-elimination mechanism is suggested by isolation of the addition product A. Under more vigorous conditions, both A and B are converted to *cis*-1,2-di-*p*-thiocresoxyethylene (Fig. 12-60, *C*) by an elimination-addition mechanism. The almost exclusive formation of B rather than its cis isomer is a result of the greater stability of the trans form and the transition state leading toward it ("cis effect"; cf. Chap. 6).

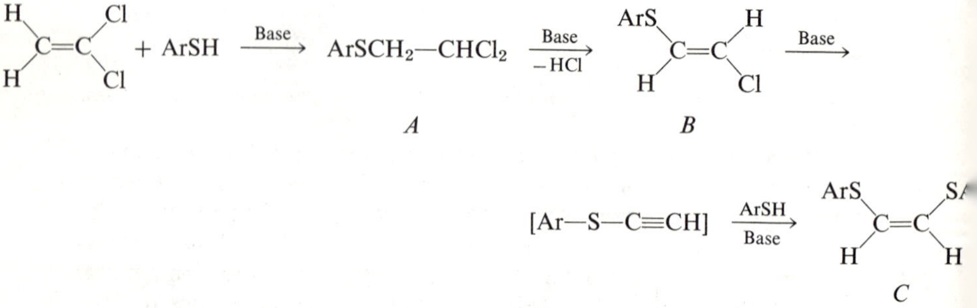

$H_2C{=}CCl_2$ + ArSH —Base→ $ArSCH_2{-}CHCl_2$ (*A*) —Base, −HCl→ ArS(H)C=C(H)Cl (*B*) —Base→

[Ar—S—C≡CH] —ArSH, Base→ ArS(H)C=C(SAr)H (*C*)

Fig. 12-60. Reaction of 1,1-dichloroethylene with *p*-thiocresol (addition-elimination in first stage).

General References

[157] K. Freudenberg, ed., "Stereochemie," Franz Deuticke, Leipzig, 1932: (*a*) F. Ebel, Die Isomerieerscheinungen, pp. 641–659; (*b*) A. Wassermann, Physikalische Eigenschaften und Konfiguration geometrisch isomerer Äthylenkörper, pp. 721–780; (*c*) R. Kuhn, Cis-trans Umlagerung der Äthylenkörper, pp. 913–920; (*d*) K. L. Wolf and O. Fuchs, Sterischer Bau und elektrische Eigenschaften, pp. 255–258; (*e*) A.

Dadieu, Ramaneffekt und Stereochemie, pp. 187–188; (*f*) S. Goldschmidt, Cis-trans isomere Athylenverbindungen und deren Reaktionsgeschwindigkeit, pp. 508–523.

[158] C. S. Marvel, Geometric Isomerism, in H. Gilman, ed., "Organic Chemistry," John Wiley & Sons, Inc., New York, 1943, pp. 444–477.

[159] L. Crombie, Geometrical Isomerism about Carbon-Carbon Double Bonds, *Quart. Revs. (London),* **6,** 101 (1952).

[160] G. Egloff, G. Hulla, and V. I. Komarewsky, "Isomerization of Pure Hydrocarbons," Reinhold Publishing Corporation, New York, 1942.

[161] C. C. Price, "Mechanism of Reactions at Carbon-Carbon Double Bonds," Interscience Publishers, Inc., New York, 1946.

[162] E. A. Braude and F. C. Nachod, eds., "Determination of Organic Structures by Physical Methods," Academic Press, Inc., New York, 1955: (*a*) E. A. Braude, Ultraviolet and Visible Light Absorption, chap. 4; (*b*) L. E. Sutton, Dipole Moments, chap. 9; (*c*) H. C. Brown, D. H. McDaniel, and O. Häfliger, Dissociation Constants, chap. 14; (*d*) J. Karle and I. L. Karle, Electron Diffraction, chap. 10.

[163] L. L. Ingraham, Steric Effects on Certain Physical Properties, in M. S. Newman, ed., "Steric Effects in Organic Chemistry," John Wiley & Sons, Inc., New York, 1956, chap. 11.

[164] E. A. Braude and E. S. Waight, The Relationships between the Stereochemistry and Spectroscopic Properties of Organic Compounds, in W. Klyne, ed., "Progress in Stereochemistry," vol. 1, Butterworth & Co. (Publishers) Ltd., London, 1954, chap. 4.

[165] V. Gold, Steric Effects and Acid-base Equilibria, in P. B. D. de la Mare and W. Klyne, eds., "Progress in Stereochemistry," vol. 3, Academic Press, Inc., New York, 1962, chap. 5.

[166] W. E. Truce, Nucleophilic Reactions of Thiols with Acetylenes and Chloroethylenes, in N. Kharasch, ed., "Organic Sulfur Compounds," vol. 1, Pergamon Press, Ltd., London, 1961.

Chapter 13

THE STEREOCHEMISTRY OF TRICOVALENT CARBON

13-1. Introduction

Carbon has four valence electrons and ordinarily forms four covalent bonds involving four bond orbitals containing two electrons each. When one of the orbitals is not engaged in a bond, the carbon atom is tricovalent. The non-bonding orbital may be empty, in which case one is dealing with a carbonium ion, $RR'R''C^+$; it may contain one electron, in which case a carbon radical $RR'R''C^\cdot$ is at hand; or it may contain two non-bonding electrons, giving rise to a carbanion $RR'R''C:^-$. Tricovalent carbon compounds are usually of very fleeting stability and are found only as short-lived reaction intermediates, except in a few rare cases where the non-bonding orbital overlaps strongly with other pi orbitals in the molecule, as, for example, in the triphenylmethyl cation, radical, and anion. Nevertheless, the stereochemistry of tricovalent carbon intermediates is of interest, since, though short-lived, they are quite common and since their stereochemistry is often reflected in the stereochemistry of the reaction products obtained via these intermediates.

13-2. Carbonium Ions[1]

When six bonding electrons are arranged about a carbon atom, their most stable electronic configuration is such that they partake of the maximum amount of *s* character in their orbitals. Therefore, the orbitals in a carbonium ion $RR'R''C^+$ have sp^2 hybridization, and the ion is planar,† with valence angles of 120°, and incapable of enantiomerism (unless one or more of the attached R groups is dissymmetric). Reactions of optically active compounds $RR'R''CX$ which proceed via "free" carbonium ions $RR'R''C^+$ therefore necessarily lead to racemic products.

Note: All references above 85 are listed in the General References at the end of the chapter.

† A theoretical estimate by G. E. Kimball (see Ref. 12, footnote 4) suggests that a tetrahedral carbonium ion (sp^3) is less stable than the corresponding planar form (sp^2) by 24 kcal. Trimethylborane—which is isoelectronic with the *t*-butyl cation—is planar: H. A. Levy and L. O. Brockway, *J. Am. Chem. Soc.*, **59**, 2085 (1937).

[1] See Ref. 90, pp. 148–153.

The statement just made must be considered as axiomatic, for, since most organic reactions proceed in solution, the carbonium-ion intermediates are almost never free. If the reaction proceeds in a well-ionizing solvent, the carbonium ion is apt to be associated with solvent molecules (solvation), whereas in a poorly ionizing solvent it is apt to be associated with its gegenion in the form of an ion pair (e.g., triphenylmethyl chloride, Ph_3CCl, would give rise to the ion pair $Ph_3C^+Cl^-$). Therefore, one can rarely observe a free carbonium ion, and, in fact, the degree to which the ion $RR'R''C^+$ is free has sometimes been deduced from the extent of racemization in the conversion of its precursor RR'R"CX to some product RR'R"CY.† The circuitous nature of the argument is evident and can be justified only because the answer obtained is usually consistent with our body of knowledge about the stability of ions as a function of structure, solvent, and other factors.

The most convincing evidence for the planarity of carbonium ions comes from cases [1a, 2] where the ion is stable enough to persist for a considerable time. Thus, when phenyl-*p*-biphenylyl-α-naphthylmethylthioglycolic acid (Fig. 13-1) is treated with perchloric or sulfuric acid or mercuric chloride, a deep blue "halochromic salt" containing presumably the phenyl-*p*-biphenylyl-α-naphthyl carbonium ion is formed in solution. Upon quenching with water, this is converted to completely racemic phenyl-*p*-biphenylyl-α-naphthylcarbinol.[1a] Since the halochromic salts have been quite conclusively shown to be salts of carbonium ions, the experiment demonstrates either that the ion itself is racemic or that complete racemization takes place in its formation or in its reaction with water, a less plausible alternative. There is, to be sure, some question as to how far this case may be generalized since triaryl carbonium ions exist in a number of resonance forms in which the central carbon atom becomes doubly bonded. One such form is shown in Fig. 13-1. This special feature may account for the inability of these particular ions to be optically active.

In many instances, carbonium ions are believed to be intermediates in solvolysis or other nucleophilic displacement reactions.[3] Such reactions (called "S_N1 reactions") would be expected to occur with complete racemization if the starting material is optically active due to asymmetry at the carbon that becomes the carbonium ion. Experience, however, does not bear this out. For example, α-phenethyl chloride, $C_6H_5CHClCH_3$, is hydrolyzed in aqueous acetone to the alcohol $C_6H_5CHOHCH_3$ at a rate that is independent of the concentration of added alkali.[4] The rate-determining step is, there-

† Although commonly drawn, this conclusion now appears to be quite unsafe inasmuch as the carbonium-ion partner in an ion pair may also racemize without any actual separation of the ion pair taking place. See, for example, S. Winstein and J. S. Gall, *Tetrahedron Letters*, no. 2, 31 (1960); S. Winstein, A. Ledwith, and M. Hojo, *ibid.*, 341 (1961).

[1a] E. S. Wallis, *J. Am. Chem. Soc.*, **53**, 2253 (1931); M. Gomberg and W. E. Gordon, *ibid.*, **57**, 119 (1935).

[2] E. S. Wallis and F. H. Adams, *J. Am. Chem. Soc.*, **55**, 3838 (1933).

[3] See (*a*) C. K. Ingold, "Structure and Mechanism in Organic Chemistry," Cornell University Press, Ithaca, N.Y., 1953, chap. 7; (*b*) A. Streitwieser, *Chem. Revs.*, **56**, 571 (1956).

[4] E. D. Hughes, C. K. Ingold, and A. D. Scott, *J. Chem. Soc.*, 1201 (1937).

fore, believed to be the formation of the carbonium ion $C_6H_5\overset{+}{C}HCH_3$ rather than bimolecular displacement of chloride by hydroxide or water. Nevertheless, the product alcohol is only 89 to 96% racemized, the remaining 4 to 11% corresponding to inversion of configuration.† Even in a solvolysis reaction involving a tertiary benzylic carbonium-ion intermediate, the methanolysis of 2-phenylbutyl acid phthalate (Fig. 13-2, A), there is some preservation of activity, the extent of racemization being 88 to 93%.[5] When the carbonium ions involved are purely aliphatic, the degree of racemization is considerably less, for example, 34% in the hydrolysis of 2-bromoöctane[6] and

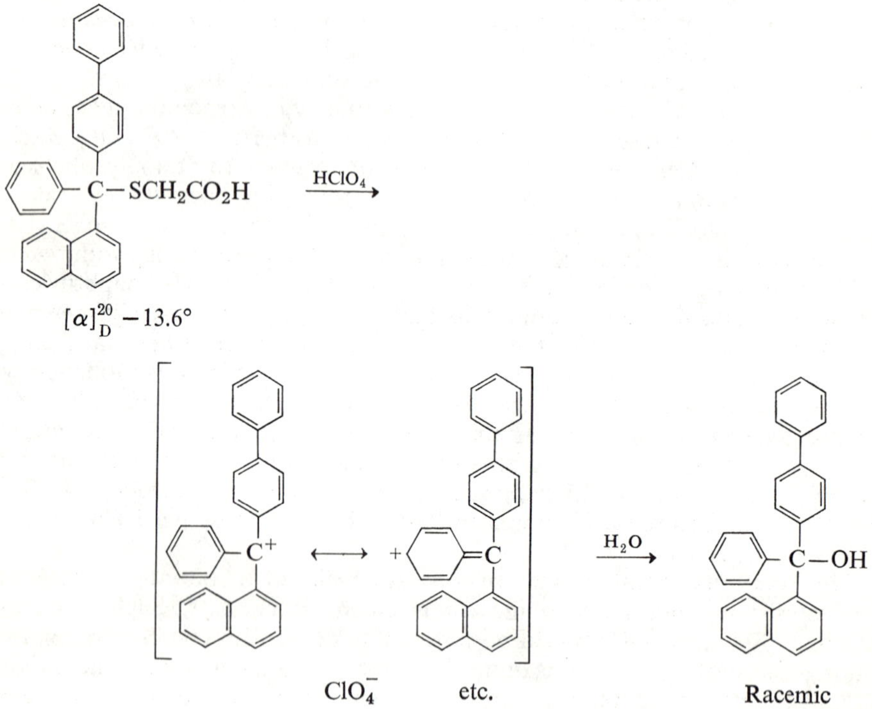

$C-SCH_2CO_2H \xrightarrow{HClO_4}$

$[\alpha]^{20}_D$ −13.6°

$C^+ \longleftrightarrow {}^+C_6H_5{=}C$ $\xrightarrow{H_2O}$ $C-OH$

ClO_4^- etc. Racemic

Fig. 13-1. Racemization via stable carbonium ion.

46% in the methanolysis of 3,5-dimethyl-3-hexyl phthalate (Fig. 13-2, B).[7] In the reaction of optically active 1-butylamine-1-*d*, $CH_3CH_2CH_2CHDNH_2$, with nitrous acid in acetic acid medium—which presumably involves the

† Using the corrected value of $[\alpha]^{25}_D = 109°$ of R. L. Burwell, A. D. Shields, and H. Hart [*J. Am. Chem. Soc.*, **76**, 908 (1954)] for the rotation of $C_6H_5CHClCH_3$.

[5] H. H. Zeiss, *J. Am. Chem. Soc.*, **75**, 3154 (1953).

[6] E. D. Hughes, C. K. Ingold, and S. Masterman, *J. Chem. Soc.*, 1196 (1937). The calculation of the extent of racemization depends somewhat on the uncertain value for the specific rotation of 2-bromoöctane; cf. N. Kornblum, L. Fishbein, and R. A. Smiley, *J. Am. Chem. Soc.*, **77**, 6261 (1955).

[7] W. v. E. Doering and H. H. Zeiss, *J. Am. Chem. Soc.*, **75**, 4733 (1953).

n-butyl cation as an intermediate inasmuch as there is much rearrangement to sec-butyl derivatives—the 1-butyl-1-*d* acetate, $CH_3CH_2CH_2CHDOAc$, formed is only 24 to 40% racemized.[8]

In all these cases there may be some suspicion that the reaction does not actually proceed through a carbonium ion but involves a nucleophilic displacement of water, methanol, or acetic acid on the halide, phthalate, or diazonium ion, RN_2^+. In most of the instances cited, however, such an alternative is unlikely. For example, in the hydrolysis of α-phenethyl chloride, if rearward attack by water is involved in the conversion to the alcohol, it is not clear why the hydroxide ion should neither accelerate the reaction nor enter into the kinetics. An alternative explanation for the absence of complete racemization is called for.

The salient point appears to be that, although a free carbonium ion is planar and, therefore, inactive, a solvated carbonium ion may be dissymmetric and therefore optically active, provided that the solvent shell is arranged about the ion in dissymmetric fashion. In the formation of a carbonium ion

$$o\text{-}C_6H_4(CO_2H)CO_2-C(CH_3)(R)-CH_2CH_3 \xrightarrow[\circlearrowleft]{CH_3OH} CH_3CH_2-C(CH_3)(R)-OCH_3$$

A: $[\alpha]_D^{24}$ + 4.44° $\qquad$ $[\alpha]_D^{23}$ + 0.20°

B: $[\alpha]_D^{22}$ +6.1° $\qquad$ $[\alpha]_D^{23}$ + 2.53°

A: $R{=}C_6H_5$

B: $R{=}(CH_3)_2CHCH_2$

Fig. 13-2. Solvolysis of 2-phenylbutyl and 3,5-dimethyl-3-hexyl acid phthalates.

$RR'R''C^+$ from a molecule $RR'R''CX$, the solvent shell is likely to be arranged dissymmetrically at the beginning, inasmuch as solvent molecules find it easier to approach the ion from the side opposite to the departing ion (or molecule) $X^{(-)}$ than from the side from which X departs. Thus, although the *bare* carbonium ion is planar, the *solvated* carbonium ion is asymmetric. If the solvent (or another species) now attaches itself covalently to the carbonium ion before the ion becomes symmetrically solvated, the resulting product may still be optically active. In most cases the covalent attachment occurs from the side opposite to the one from which X had departed, especially if the covalent bonding involves solvent molecules, and the stereochemical course is therefore predominantly inversion. It may happen, however, that the covalent bonding occurs preferentially from the side from which X departed, and in this case the result is predominant retention of configuration. This is especially likely when the covalent bond involves a part of the leaving group X, for example, in the decomposition of a chlorosulfite, R—OSOCl, where the departing group X is —OSOCl and the newly bonding group is Cl

[8] A. Streitwieser and W. D. Schaeffer, *J. Am. Chem. Soc.*, **79**, 2888 (1957).

(formed from —OSOCl by loss of SO_2), the product being the chloride R—Cl. Processes of this type are called "S_Ni reactions."[9]

The more stable the intermediate carbonium ion is, the more chance it has to survive until it becomes symmetrically solvated, i.e., optically inactive. Covalent attachment of a solvent molecule or other entity after solvation has become symmetric leads to racemic products, since approach to either face of the carbonium ion is equally likely. This state of affairs is approached in those solvolyses which were mentioned earlier that involve benzylic carbonium-ion intermediates of the type Ar—$\overset{+}{C}$RR′, as this ion is stabilized by resonance.

The evidence presented so far for the planar geometry of the carbonium ion may not strike the reader as very convincing, except perhaps in the case of benzylic ions; with these, resonance of the type shown in Fig. 13-1 might account for the planarity. Fortunately, there is another set of data which indicates that, if a carbonium ion is sterically prevented from becoming planar, it either will not form at all or will form only at a greatly reduced rate. These data are concerned with carbonium ions at bridgeheads of bicyclic compounds, such as the ions which would be formed from the compounds shown in Fig. 13-3 (*A, C, D*) by loss of halogen. Bartlett and Knox showed[10] that apocamphyl chloride (Fig. 13-3, *A*) does not give a precipitate of silver chloride when

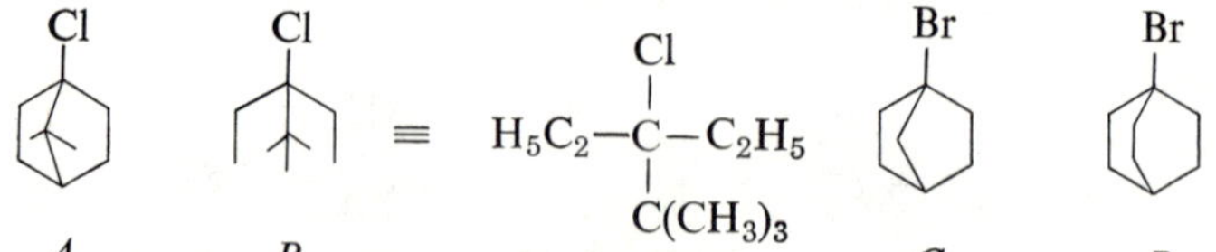

Cl Cl Cl Br Br

≡ H_5C_2—C—C_2H_5

$C(CH_3)_3$

A *B* *C* *D*

Fig. 13-3. Unreactive bridgehead halides (*A, C, D*).

boiled with aqueous alcoholic silver nitrate for 48 hrs. In contrast, *t*-butyl chloride and the chloride *B* (Fig. 13-3), the open-chain analog of *A*, react very rapidly with silver nitrate in the cold. The difference in rate between *A* and *B* is by a factor of well over 10^9. It was later found[11] that apocamphyl chloride (*A*) does react with aqueous silver nitrate when heated in an autoclave at 205°, but the organic products were unidentified and the reaction does not appear to be a simple solvolysis.

Partial solvolysis takes place when 1-bromonorbornane (Fig. 13-3, *C*) is heated with aqueous silver nitrate at 150° for 48 hrs.[12] The difference between *C* and *A* may result from the well-known greater propensity of bromine, compared with chlorine, to depart as an anion. 1-Bromobicyclo[2.2.2]octane (Fig. 13-3, *D*) solvolyzes in aqueous ethanol in the presence of silver nitrate at room temperature in 4 hrs.[12] In the absence of silver nitrate in aqueous dioxane at 100°, the solvolysis rate of *D* is about $1/10^6$ of that of *t*-butyl

[9] Cf. Ref. 87, pp. 79–83.

[10] P. D. Bartlett and L. H. Knox, *J. Am. Chem. Soc.*, **61**, 3184 (1939).

[11] W. v. E. Doering and E. F. Schoenewaldt, *J. Am. Chem. Soc.*, **73**, 2333 (1951).

[12] W. v. E. Doering, M. Levitz, A. Sayigh, M. Sprecher, and W. P. Whelan, *J. Am. Chem. Soc.*, **75**, 1008 (1953).

bromide.[12] The compounds shown in Fig. 13-4 have also been successfully solvolyzed,[13, 14] 1-bromoadamantane being remarkably reactive,[13] although still less so than *t*-butyl bromide by a factor of over 800.[13b]

The finding that bicyclooöctyl halides are solvolyzed more readily than bicycloheptyl halides is consistent with the hypothesis that carbonium ions are intermediates in these solvolyses and that the preferred geometry of these (and other) carbonium ions is planar. To the extent that the ion is distorted from planarity, it becomes less stable and the rate of solvolysis decreases. The bicycloheptyl cations derived from *A* and *C*, Fig. 13-3, have scant possibility of becoming even approximately planar without excessive (and therefore energetically very expensive) distortion of the valence angles. In fact, the normal C—C—X valence angle in the starting halides is probably greater than the normal tetrahedral angle, and the ion would tend to be distorted *away* from planarity. In the bicyclooöctyl compounds (Figs. 13-3, *D*, and 13-4), the normal C—C—Br valence angle is tetrahedral, and in the corresponding carbonium ions it is possible to distort the valence angle somewhat toward the point where a planar geometry at the No. 1 carbon would result. Although this distortion is undoubtedly not complete, it may proceed part

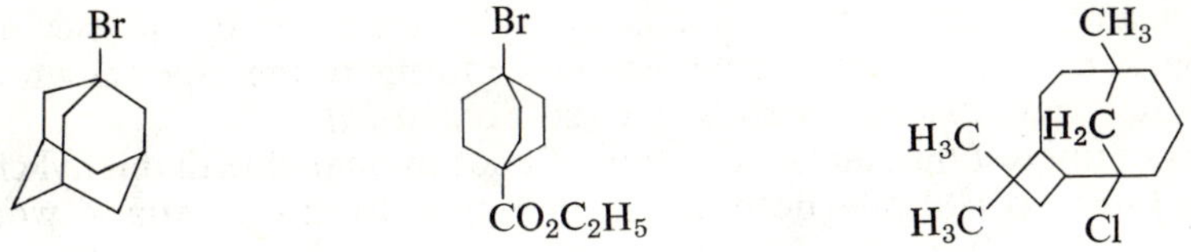

1-Bromoadamantane | 1-Bromo-4-carbethoxy-[2.2.2]bicyclooctane | β-Caryophyllene chloride

Fig. 13-4. Bridgehead halides that have been solvolyzed.

way without excessive expenditure of energy, and to the extent that the carbonium ion at least approaches planarity, it is less unstable than the bicycloheptyl cation. Along the lines of this argument, even greater stability of the ion (and therefore a higher solvolysis rate) would be expected in still larger bicyclic ring systems where the bridgehead can be deformed more readily toward planarity. In fact, β-caryophyllene chloride (Fig. 13-4), which contains the [4.3.1] system, may be solvolyzed to the corresponding acetate in boiling glacial acetic acid, a relatively poorly ionizing solvent.[15]

Of particular interest in this connection are the 1-substituted "triptycenes" in Fig. 13-5. Despite their analogy with triphenylmethyl systems (which yield the triphenylcarbonium ion with great readiness), these systems show absolutely no tendency to form the corresponding carbonium ion.[16] No

[13] (*a*) H. Stetter, M. Schwarz, and H. H. Hirschhorn, *Chem. Ber.*, **92**, 1629 (1959); H. Stetter, J. Mayer, M. Schwarz, and K. Wulff, *ibid.*, **93**, 226 (1960). (*b*) P. v. R. Schleyer and R. D. Nicholas, *J. Am. Chem. Soc.*, **83**, 2700 (1961).

[14] J. D. Roberts, W. T. Moreland, and W. Frazer, *J. Am. Chem. Soc.*, **75**, 637 (1953).

[15] G. G. Henderson, J. M. Robertson, and C. A. Kerr, *J. Chem. Soc.*, 62 (1926).

[16] P. D. Bartlett and E. S. Lewis, *J. Am. Chem. Soc.*, **72**, 1005 (1950).

evidence of formation of a halochromic salt occurs when the bromide (Fig. 13-5, X = Br) is dissolved in liquid sulfur dioxide[16] or when the alcohol (X = OH) is dissolved in concentrated sulfuric acid.[17] The iodide (X = I) produces no silver iodide after 48-hr. boiling with aqueous alcoholic silver nitrate[17] and is therefore less reactive than *t*-butyl bromide by at least a factor of 10^{15}. The much lesser reactivity of the triptycyl system (Fig. 13-5) as compared with the bicyclo[2.2.2]octyl system (Figs. 13-3, *D*, and 13-4) may be due in part to the greater rigidity of the former (which makes it harder for the bridgehead carbonium ion to approach planarity) and in part to the

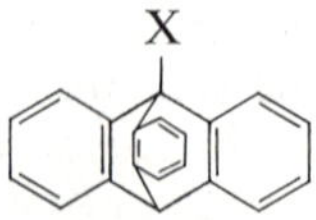

X = OH, Br, I

Fig. 13-5. 1-Substituted triptycenes.

unfavorable inductive effect (electron withdrawing) of the phenyl groups. The resonance overlap with the aromatic pi electrons which makes the triphenylcarbonium ion so stable is prohibited in the triptycyl cation because the vacant *p* orbital of the ion is at right angles to the pi electrons of the benzene ring. If one wrote resonance forms of the type shown in Fig. 13-1, they would violate Bredt's rule (cf. Sec. 10-3*b*).

It is of interest that the S_Ni reaction of apocamphanol with thionyl chloride (Fig. 13-6) fails also,[10] despite the fact that front-side attack would be

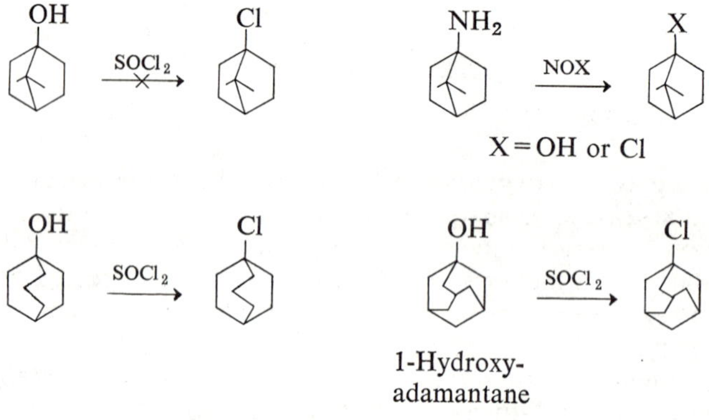

OH —$SOCl_2$ (✗)→ Cl NH_2 —NOX→ X

X = OH or Cl

OH —$SOCl_2$→ Cl   OH —$SOCl_2$→ Cl

1-Hydroxy-adamantane

Fig. 13-6. S_Ni reactions at bridgeheads.

involved in the decomposition of the chlorosulfite. As mentioned before, this reaction probably involves an ion pair,[18] and, in the case of apocamphanol, is prevented by the high energy content of the apocamphyl cation. In contrast, the reaction of apocamphylamine (Fig. 13-6) with nitrous acid (X = OH) or nitrosyl chloride (X = Cl) succeeds in giving apocamphanol and apocam-

[17] P. D. Bartlett and F. D. Greene, *J. Am. Chem. Soc.*, **76**, 1088 (1954).

[18] D. J. Cram, *J. Am. Chem. Soc.*, **75**, 332 (1953); see also E. S. Lewis and C. E. Boozer, *ibid.*, **74**, 308 (1952); **75**, 3182 (1953).

phyl chloride, respectively.[10, 19] Here the extra driving force provided by the departure of the stable nitrogen molecule presumably overcomes the handicap of the instability in the carbonium ion. Larger-ring bridgehead alcohols, such as bicyclo[3.2.2]-1-nonanol[19a] and 1-hydroxyadamantane[13a] (Fig. 13-6), do react with thionyl chloride to give the corresponding chlorides.

Parenthetically it might be mentioned that the apocamphyl system (Fig. 13-3, *A*) and triptycyl system (Fig. 13-5) as well as the compounds shown in Fig. 13-4 are, of course, completely incapable of undergoing bimolecular nucleophilic displacement (S_N2) reactions, since there is no possibility of approaching the bridgehead atom from the rear or of inverting its configuration. Similarly, elimination reactions (E_2 or E_1) fail because the products of such reactions would violate Bredt's rule (Sec. 10-3*b*). Thus 1-chloroapocamphane does not react with boiling aqueous alcoholic potassium hydroxide, nor does apocamphyl *p*-toluenesulfonate react with lithium iodide in acetone.[10] 1-Bromotriptycene does not react with either boiling alcoholic sodium hydroxide or sodium sulfide,[16] and β-caryophyllene chloride (Fig. 13-4) does not react with sodium ethoxide.[15]†

Yet another line of evidence for the preference of carbonium ions to be planar comes from studies of the solvolysis rates of highly branched tertiary halides. If in a halide RR′R″CX the alkyl groups are very bulky, they may crowd each other sterically. This leads to a steric repulsion or strain in the molecule which has been termed[20] "B-strain." If, indeed, the carbonium ion $RR'R''C^+$ is planar, then, in going from the halide RR′R″CX to the carbonium ion $RR'R''C^+$, the R—C—R angle will increase from 109°28′ to 120°, and the steric compression of the alkyl groups will be diminished. This diminution of strain in the carbonium ion (and the transition state leading to it which presumably resembles the ion[21]) leads to steric assistance of the reaction and reflects itself in an increased solvolysis rate of the more highly branched halides as compared with, say, *t*-butyl halide. Such an effect is, indeed, observed.[22–25] For example,[22b] dimethylneopentylcarbinyl chloride

† All bridgehead halides are, of necessity, tertiary halides and as such would not undergo S_N2 reactions readily anyway. However, S_N2 reactions of *simple* tertiary halides with good nucleophiles are known, e.g., the reaction of *t*-butyl chloride with sodium thiophenoxide reported by P. B. D. de la Mare and C. A. Vernon, *J. Chem. Soc.*, 41 (1956).

---

[19] A number of other examples have been recorded; cf. Ref. 90, p. 151.

[19a] C. A. Grob, M. Ohta, E. Renk, and A. Weiss, *Helv. Chim. Acta*, **41**, 1191 (1958).

[20] H. C. Brown, H. Bartholomay, and M. D. Taylor, *J. Am. Chem. Soc.*, **66**, 435 (1944). The original application of the B-strain concept to tertiary amines is no longer tenable, as the valence angle in trimethylamine is less than the tetrahedral angle; cf. page 384.

[21] G. S. Hammond, *J. Am. Chem. Soc.*, **77**, 334 (1955).

[22] (*a*) H. C. Brown and R. S. Fletcher, *J. Am. Chem. Soc.*, **71**, 1845 (1949); (*b*) H. C. Brown and H. L. Berneis, *ibid.*, **75**, 10 (1953); (*c*) H. C. Brown and Y. Okamoto, *ibid.*, **77**, 3619 (1955); (*d*) H. C. Brown and I. Moritani, *ibid.*, **77**, 3623 (1955).

[23] (*a*) P. D. Bartlett and M. S. Swain, *J. Am. Chem. Soc.*, **77**, 2801 (1955); (*b*) P. D. Bartlett and M. Stiles, *ibid.*, **77**, 2806 (1955).

[24] F. Brown, T. D. Davies, I. Dostrovsky, O. J. Evans, and E. D. Hughes, *Nature*, **167**, 987 (1951).

[25] Summarized in Ref. 87, pp. 70–73. See also H. C. Brown, *Record Chem. Progr. (Kresge-Hooker Sci. Lib.)*, **14**, 83 (1953).

(Fig. 13-7, A) solvolyzes twenty-one times as fast as *t*-butyl chloride, and dineopentylmethylcarbinyl chloride (Fig. 13-7, B) solvolyzes 580 times as fast.†

It is of interest that the difficulty of achieving rearward solvation of the highly branched halide during the departure of the halide anion does not keep these halides from solvolyzing rapidly. Such solvation can, therefore, play but a minor part in determining the activation energy of solvolysis.

13-3. Free Radicals[26]

In a free radical RR′R″C·, seven valence electrons are arranged about the carbon atom, i.e., one more than in the carbonium ion. There are two possible arrangements: sp^2 for the bonding electrons, with the odd electron occupying a pure *p* orbital, or sp^3 for all electrons, including the odd one. The former arrangement would correspond to planar geometry, the latter to tetrahedral geometry (or pyramidal geometry, if one disregards the odd electron). In the former arrangement there are six electrons each having ⅓ *s* character; in the latter, seven electrons, each having ¼ *s* character. Thus there is more over-all *s* character in the bonds in the sp^2 arrangement, and it might be expected that the radical prefers the planar arrangement. More refined theo-

```
       CH3     CH3                      CH3     CH3     CH3
       |       |                        |       |       |
  CH3—C—CH2—C—CH3                  CH3—C—CH2—C—CH2—C—CH3
       |       |                        |       |       |
       Cl      CH3                      CH3     Cl      CH3

            A                                   B
```

Fig. 13-7. Highly branched tertiary chlorides.

retical arguments[27] tend to suggest, nevertheless, that the radical is pyramidal. Perhaps more important than this rather tentative conclusion is the realization that the energy difference between a planar and a tetrahedral radical is considerably less than that between a planar and tetrahedral carbonium ion (*vide supra*).

Experimental studies of the methyl radical using ultraviolet spectroscopy[28]

† E. D. Hughes, C. K. Ingold, and V. J. Shiner [*J. Chem. Soc.*, 3827 (1953)] have objected to the steric explanation of the high solvolysis rates of highly branched tertiary halides, preferring to explain the acceleration in terms of inductive, hyperconjugative, and anchimeric effects. While such effects may contribute to the high rates in a number of cases, they do not seem to suffice. Inductive and hyperconjugative effects of different alkyl groups do not vary greatly and can hardly account for the large increases in rate. Anchimeric effects may indeed be important in the hydrolysis of halides such as tri-*t*-butylcarbinyl chloride,[23a, 24] which give rise exclusively to rearranged products, but are not likely to be important with halides such as dimethylneopentylcarbinyl chloride whose hydrolysis gives dimethylneopentylcarbinol in 80% yield.[22b]

[26] See Ref. 87, pp. 140–142 and 149, and Ref. 90, pp. 155–157.

[27] J. W. Linnett and A. J. Poe, *Trans. Faraday Soc.*, **47**, 1039 (1951); A. D. Walsh, *J. Chem. Soc.*, 2296 (1953).

[28] G. Herzberg and J. Shoosmith, *Can. J. Phys.*, **34**, 523 (1956); cf. G. Herzberg, *Ann. Rev. Phys. Chem.*, **9**, 327 (1958).

and electron paramagnetic resonance[29] indicate that this radical is planar or nearly planar.

If a radical is planar, reactions leading to or proceeding via radicals should involve complete loss of optical activity, provided that the carbon atom that becomes tricovalent in the product or intermediate is the one which was originally dissymmetric. A number of examples have been studied. Optically active phenyl-*p*-biphenyl-α-naphthylmethylthioglycolic acid (Fig. 13-1) is racemized by treatment with triphenylmethyl (a free radical),[2] presumably because phenyl-*p*-biphenyl-α-naphthylmethyl radicals are formed.† Furthermore, attempts to resolve phenyl-*p*-biphenylyl-α-naphthylmethyl radicals by chromatography on several optically active adsorbents[30] all failed.‡

Chlorination at the tertiary carbon of optically active 2-methyl-1-chlorobutane (Fig. 13-8) gives rise to racemic 1,2-dichloro-2-methylbutane.[31] This reaction presumably proceeds via the tertiary radical shown in Fig. 13-8. Also shown in this figure is the dimerization of optically active 2-phenylbu-

$$\underset{\text{Active}}{\underset{|\atop CH_3}{CH_3CH_2CHCH_2Cl}} \xrightarrow{Cl_2} \underset{\text{Racemic}}{\underset{|\atop CH_3}{CH_3CH_2CClCH_2Cl}} \quad \text{via} \quad \underset{|\atop CH_3}{CH_3CH_2\dot{C}CH_2Cl}$$

$$\underset{|\atop C_6H_5}{CH_3CH_2CHCH_3} \xrightarrow{Ac_2O_2} \underset{\text{meso + racemic}}{\begin{array}{c} C_6H_5 \\ | \\ CH_3CH_2CCH_3 \\ | \\ CH_3CH_2CCH_3 \\ | \\ C_6H_5 \end{array}} \quad \text{via} \quad \underset{|\atop C_6H_5}{CH_3CH_2\dot{C}CH_3}$$

Fig. 13-8. Racemization in reactions proceeding via free radicals.

tane to a mixture of *meso-* and *dl*-3,4-diphenyl-3,4-dimethylhexane which presumably proceeds via the 2-phenyl-2-butyl radical.[32] The bromination of optically active 1-phenylethane-1-*d*, $PhCHDCH_3$, to 1-phenyl-1-bromoethane-1-*d*,¶ $PhCDBrCH_3$, by means of *N*-bromosuccinimide also yields a racemic product, presumably via the radical $Ph\dot{C}DCH_3$.[33]

† Although the formation of these radicals in the reaction has not been demonstrated with certainty, their existence was inferred from the appearance of a reddish-brown color which replaces the yellow color of the triphenylmethyl radical upon prolonged reaction. The color is discharged when oxygen is admitted to the reaction vessel.

‡ Negative evidence of this kind is, of course, never conclusive. As stressed in Chap. 4, resolution is a difficult art, and failure may always be ascribed to technical causes rather than to lack of resolvability of the substrate.

¶ Because of the isotope effect, this product predominates over the product $PhCHBrCH_3$ in which deuterium rather than hydrogen is replaced by bromine.

[29] T. Cole, H. O. Pritchard, N. R. Davidson, and H. M. McConnell, *Mol. Phys.*, **1**, 406 (1958); see also M. Karplus, *J. Chem. Phys.*, **30**, 15 (1959).

[30] G. Karagounis, *Helv. Chim. Acta*, **32**, 1840 (1949).

[31] H. C. Brown, M. S. Kharasch, and T. H. Chao, *J. Am. Chem. Soc.*, **62**, 3435 (1940).

[32] E. L. Eliel, P. H. Wilken, F. T. Fang, and S. H. Wilen, *J. Am. Chem. Soc.*, **80**, 3303 (1958).

[33] H. J. Dauben and L. L. McCoy, *J. Am. Chem. Soc.*, **81**, 5404 (1959).

It must be realized that, although the loss of stereochemical identity is consistent with a planar structure for the radical, it does not demand such a structure. A pyramidal (and, therefore, dissymmetric) radical which is interconverted, through molecular vibrations, with its enantiomer at a rate much faster than the rate at which it reacts to form products is consistent with the experimental facts.† However, more detailed evidence regarding the preferred geometry of radicals may be obtained from the ease with which radicals are formed when they are either constrained to be planar or prevented from becoming planar. The former case is illustrated in Fig. 13-9. The doubly bridged triphenylmethyl radical is believed to have some compul-

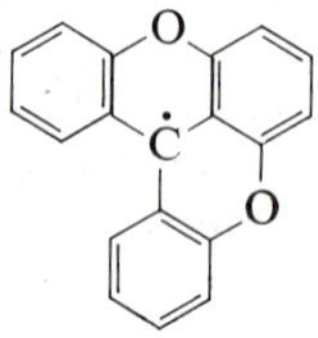

Fig. 13-9. Radical constrained to planar form.

sion to be planar on account of the oxygen bridges.[34] Nevertheless, the radical forms easily, for the corresponding hexaphenylethane is highly dissociated.[34] While this would speak for a preferred planar structure for the radical, bridgehead radicals have also been prepared which cannot be planar. Thus decomposition of apocamphoyl peroxide[35] in carbon tetrachloride gives, among other products, diapocamphyl (9%), the coupling product of two apocamphyl radicals (Fig. 13-10), and 1-chloroapocamphane (36%), the product of attack of the apocamphyl radical on the solvent. Similarly, the decomposition of triptoyl peroxide (Fig. 13-11)[17] in benzene gives triptycene (45%, by abstraction of a hydrogen by the triptycyl radical from the benzene solvent) and a high-melting hydrocarbon which may be ditriptycyl;

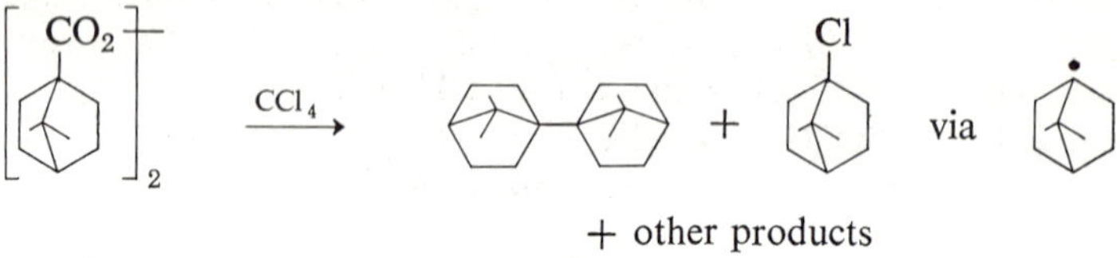

+ other products

Fig. 13-10. Decomposition of apocamphoyl peroxide.

in the presence of iodine, 1-iodotriptycene is formed.[17] While bridgehead radicals are thus capable of formation, they appear to be considerably less stable than other tertiary alkyl radicals. This is evidenced by the fact[17] that apocamphoyl and triptoyl peroxide decompose at a rate only slightly greater than that of acetyl peroxide, whereas other peroxides of the type

$$R_3CCO_2O_2CCR_3$$

† This point will be discussed further in connection with carbanions (Sec. 13-4).

[34] O. Neunhoffer and H. Haase, *Chem. Ber.*, **91**, 1801 (1958).
[35] M. S. Kharasch, F. Engelmann, and W. H. Urry, *J. Am. Chem. Soc.*, **65**, 2428 (1943).

undergo a rapid concerted decomposition into $2R_3C\cdot + 2CO_2$.[17] Furthermore, the instability of the apocamphyl and triptycyl radicals is evidenced by their great avidity to attack the solvent (chlorine abstraction from carbon tetrachloride and even hydrogen abstraction from benzene; the latter is quite unusual). The triptycyl radical shows, of course, none of the stability of the triphenylmethyl radicals, since the resonance stabilization of the latter is totally lacking. This is further emphasized by the failure of triptycene to

CO_2 ... $]_2$ $\xrightarrow{C_6H_6}$ (triptycene) via (triptycyl radical)

+ other products

Fig. 13-11. Decomposition of triptoyl peroxide.

undergo free radical chlorination by sulfuryl chloride in the presence of benzoyl peroxide.[36]

In view of all the evidence so far cited that radicals do not hold their configuration and are probably planar, it is of considerable interest that some radical reactions do give optically active products. These reactions are decompositions of optically active peroxides $\overset{*}{R}CO_2O_2C\overset{*}{R}$ to esters $\overset{*}{R}CO_2\overset{*}{R}$ in which the alcohol part as well as the acid part of the ester retains activity.

Acid (active) + alcohol (active)

↑ H_2O

$R{-}CH(CH_3){-}C({=}O){-}O{-}O{-}C({=}O){-}CH(CH_3){-}R \longrightarrow R{-}CH(CH_3){-}C({=}O){-}O{-}CH(CH_3){-}R$ via (cyclic transition state: $R{-}CH(CH_3){-}C$, O—O, C=O, O, CH(CH_3)—R)

Active → Active

$R = C_2H_5$[37]
$R = C_6H_5$[38]
$R = C_6H_5CH_2$[39]

$+ R{-}CH(CH_3){-}CH(CH_3){-}R$
active ($R = C_6H_5$ only)

+ other products

or

$[R{-}CH(CH_3){-}CO_2\cdot \quad \cdot O_2C{-}CH(CH_3){-}R]$

↓

$[R{-}CH(CH_3){-}CO_2\cdot \quad \cdot CH(CH_3){-}R + CO_2]$

Fig. 13-12. Partial retention of activity in formation of esters from peroxides.

Examples[37-39] are shown in Fig. 13-12. Two explanations have been advanced to account for the partial retention of activity in the alcohol part of the ester. According to one,[37] the ester is formed by a cyclic mechanism. This expla-

[36] P. D. Bartlett, M. J. Ryan, and S. G. Cohen, *J. Am. Chem. Soc.,* **64,** 2649 (1942).
[37] M. S. Kharasch, J. Kuderna, and W. Nudenberg, *J. Org. Chem.,* **19,** 1283 (1954).
[38] F. D. Greene, *J. Am. Chem. Soc.,* **77,** 4869 (1955). The hydratropyl peroxide was not isolated but decomposed *in situ* after formation from the chloride and sodium peroxide.
[39] D. F. DeTar and C. Weis, *J. Am. Chem. Soc.,* **79,** 3045 (1957).

nation is supported by the finding[40] (through isotopic labeling) that the two oxygen atoms of the peroxide do not become equivalent in the ester. The other explanation[38, 39] invokes recombination of an $RCO_2\cdot$ and $R\cdot$ radical (the latter formed by decarboxylation of another $RCO_2\cdot$, a process which is very rapid[41]) by collision before the radicals have the opportunity to diffuse away from each other to assume random orientation in solution.† Here the $RCO_2\cdot$ radical recombines with $R\cdot$ preferentially from the same side from which it separated originally, and partial retention of configuration results. This explanation is in better agreement with the fact that retention is not complete, with the fact[38] that hydratropyl peroxide (Fig. 13-12, R = C_6H_5) gives not only optically active hydratropyl hydratropate but also partly active 2,3-diphenylbutane, $C_6H_5CH(CH_3)CH(CH_3)C_6H_5$, and with other studies[42] on the decomposition of peroxides to esters.

A third possible explanation, namely, that activity in the alcohol part of the ester is produced by the activity in the $\overset{*}{R}CO_2\cdot$ part by a process of asymmetric induction (Sec. 4-4*d, ii*), is extremely unlikely, for the energy of activation for the recombination of the radicals is low or nil, so that there could not possibly be a big difference in energy of activation between combination of (R)-$RCO_2\cdot$ and (R)-$R\cdot$ on the one hand and (R)-$RCO_2\cdot$ and (S)-$R\cdot$ on the other, and yet the activity in the alcohol part of the ester is found to be high.

Although the ester formed from active β-phenylisobutyryl peroxide (Fig. 13-12) is optically active in the alcohol part, the simultaneously formed products of attack on solvent carbon tetrachloride, namely, RCl and $RCCl_3$ (R = 1-phenyl-2-propyl), are totally racemic,[39] indicating once again that when the radical $R\cdot$ becomes free, it loses its configurational identity.

13-4. Carbanions[43]

A carbanion $RR'R''C:^-$ almost certainly has the sp^3 configuration and is, therefore, tetrahedral and should be capable of giving optically active products when it is generated from optically active precursors. While this view is correct (*vide infra*), it does not tell the whole story. Tertiary amines $RR'R''N:$ which are isoelectronic‡ and presumably isosteric‡ with carbanions are also tetrahedral (or pyramidal, if the lone pair is disregarded), with a valence angle in trimethylamine of 108.7°.[44] Yet such amines have never been resolved even at low temperatures.[45] The explanation[45a] seems to be that tertiary

† This process is commonly called "cage recombination" or recombination in a "solvent cage" or is said to be due to a "cage effect."

‡ Meaning that they have the same number and kind of valence electrons and the same geometry.

---

40 M. A. Greenbaum, Ph.D. Thesis, Yale University, 1957; cf. H. H. Lau and H. Hart, *J. Am. Chem. Soc.*, **81**, 4897 (1959), footnote 29.

41 For example, see J. R. Nash, W. H. Hamill, and R. R. Williams, *J. Phys. Chem.*, **60**, 823 (1956).

42 D. F. DeTar and C. Weis, *J. Am. Chem. Soc.*, **79**, 3041 (1957).

43 Ref. 87, pp. 152–156 and 163.

44 D. R. Lide and D. E. Mann, *J. Chem. Phys.*, **28**, 572 (1958).

45 T. D. Stewart and C. Allen, *J. Am. Chem. Soc.*, **54**, 4027 (1932); see also Ref. 86, pp. 402–413.

45a First suggested by J. Meisenheimer, L. Angermann, O. Finn, and E. Vieweg, *Ber.*, **57**, 1747 (1924).

amines, and presumably also carbanions, may be converted into their enantiomers by a molecular vibration of the type shown in Fig. 13-13. Such interconversion takes place at a very high rate,[46] and, therefore, tertiary amines, while dissymmetric, exist as a mixture of rapidly interconverted (and therefore inseparable) mirror-image forms. This view has been confirmed in the case of *N*-ethylethyleneimine (Fig. 13-14) which at room temperature shows two kinds of ring proton resonances in the NMR spectrum due to hydrogens cis to the ethyl group and trans to it.[47] At 120° the inversion rate increases to the point where only one average resonance is seen in the spectrum. Coalescence of the two lines occurs at 110°, and at this temperature the mean lifetime of the enantiomers may be calculated to be ca. 0.015 sec.

R, R′, R″ N: ⇌ :N R, R′, R″

Fig. 13-13. Interconversion of amine enantiomers.

The mean lifetime of the enantiomeric forms of cyclic imines with larger rings (such as *N*-alkyl-trimethyleneimines, -pyrrolidines, and -piperidines) is even considerably shorter than that.[47] In the case of *N*-ethylethyleneimine it was estimated[47] that resolution could not be successful at temperatures above −50°.

Returning to the carbon analogs, it thus appears highly unlikely that a stable carbanion RR′R″C:$^-$ could ever be isolated in optically active form, even though the ion is dissymmetric. Experimentally, no activity was preserved in the carbanion derived by treatment of the optically active benzoxanthene derivative shown in Fig. 13-15 with sodium in liquid ammonia,[48] despite earlier claims to the contrary.[2] Carbanions derived from nitro compounds

H, H, H, H, N, CH_2-CH_3 ⇌ H, H, H, H, N, CH_2-CH_3

Fig. 13-14. *N*-Ethylethyleneimine.

of the type $RR'\bar{C}-NO_2 \longleftrightarrow RR'C=NO_2^-$ also lose the activity of their $RR'CHNO_2$ progenitors.[48a]

Where carbanions are short-lived reaction intermediates, their stereochemical fate depends on how long they persist. It was shown earlier (Fig. 4-5) that in the case of a strongly resonance-stabilized carbanion, such as that derived from phenyl *sec*-butyl ketone, $C_6H_5COCH(CH_3)C_2H_5$, each act of

[46] R. S. Berry, *J. Chem. Phys.*, **32**, 933 (1960).
[47] A. T. Bottini and J. D. Roberts, *J. Am. Chem. Soc.*, **80**, 5203 (1958).
[48] G. Wittig, F. Vidal, and E. Bohnert, *Chem. Ber.*, **83**, 359 (1950).
[48a] N. Kornblum, N. N. Lichtin, J. T. Patton, and D. C. Iffland, *J. Am. Chem. Soc.*, **69**, 307 (1947); N. Kornblum, J. T. Patton, and J. B. Nordmann, *ibid.*, **70**, 746 (1948).

hydrogen exchange (presumably via a carbanion intermediate) produces complete racemization.[48b] The same does not, however, seem to be true for carbanions that are less stabilized by resonance.† Thus the hydrogen-deuterium exchange reaction of optically active α-phenethyl-α-*d* methyl ether,

$$C_6H_5CD(CH_3)OCH_3$$

with potassium *t*-butoxide in *t*-butyl alcohol, $(CH_3)_3COH$, proceeds with at least 94% retention of configuration in each act of exchange (although the same reaction in dimethyl sulfoxide, a more ionizing solvent, produces total racemization).[49a] Similarly, high retention is observed in the exchange‡ of optically active ethylbenzene-α-*d*, $C_6H_5CHDCH_3$, with lithium cyclohexylamide in cyclohexylamine.[49b]

A number of similar cases where carbanions are formed by cleavage of carbon-carbon bonds has been studied by Cram and coworkers.[50, 51] Representative examples are shown in Fig. 13-16. The hydrogen necessary to protonate the carbanion was supplied by solvent in some cases, but in others no protonated solvent was present and the proton must have been abstracted internally.

H_5C_6 S–CH_2–CO_2H $[\alpha]_D$ −37.6° $\xrightarrow{Na/NH_3}$ H_5C_6 (carbanion) $\xrightarrow{NH_4Cl}$ H_5C_6 H Racemic

Fig. 13-15. Stable, optically inactive carbanion.

The most significant outcome of this work is the observation that the reactions may involve either predominant retention or inversion of configuration or extensive or complete racemization. Which stereochemical course is observed seems to depend mainly on the solvent. In dimethyl sulfoxide, a non-protonated, polar, and well-ionizing solvent, racemization is extensive. In solvents of low dielectric constant, such as benzene, dioxane, or *t*-butyl

† N. L. Allinger, R. B. Hermann, and C. Djerassi [*J. Org. Chem.*, **25**, 922 (1960)] have suggested an analogy of the carbanion with nitrogen in this case: Dialkylanilines are pyramidal in structure but the strongly resonance-stabilized amides are planar. See, however, Ref. 49*a*.

‡ In this case one observes the exchange of hydrogen by hydrogen, since the exchange of deuterium by hydrogen gives inactive ethylbenzene. The rate of hydrogen-hydrogen exchange, in turn, is calculated from the observed rate of deuterium-hydrogen exchange and the known isotope effect of the reaction.

[48b] See also D. J. Cram, B. Rickborn, C. A. Kingsbury, and P. Haberfield, *J. Am. Chem. Soc.*, **83**, 3678 (1961).

[49] (*a*) D. J. Cram, C. A. Kingsbury, and B. Rickborn, *J. Am. Chem. Soc.*, **83**, 3688 (1961); (*b*) A. Streitwieser, D. E. Van Sickle, and L. Reif, *ibid.*, **84**, 258 (1962).

[50] D. J. Cram and coworkers, *J. Am. Chem. Soc.*, **81**, 5740, 5750, 5754, 5760, 5767, 5785 (1959).

[51] D. J. Cram, J. L. Mateos, F. Hauck, A. Langemann, K. R. Kopecky, W. D. Nielsen, and J. Allinger, *J. Am. Chem. Soc.*, **81**, 5774 (1959).

alcohol, there is extensive retention. In protonated solvents of high dielectric constant, such as ethylene gylcol and diethylene gylcol, there is predominant inversion. There is also dependence of the steric course of the reaction on the nature of the gegenion (cation)—whether lithium, potassium, or tetramethylammonium. The effects have been interpreted in detail in terms of the structure, solvation, and degree of separation of the ion-pair intermediates involved in these reactions.[51] From the stereochemical point of view, it is interesting that carbanion reactions resemble carbonium-ion reactions (cf. Sec. 13-2) in that they may lead to retention or inversion of configuration as well as racemization. This suggests that maintenance of activity in the case of the carbanion intermediates in the above-discussed reactions may be a result not of maintenance of a given tetrahedral configuration of the anion

$$C_6H_5-C(CH_3)(C_2H_5)-C(OH)(R_2)-R_1 \xrightarrow{C_6H_5\bar{N}HK^+} C_6H_5-C(CH_3)(C_2H_5)-C(O^-K^+)(R_2)-R_1 \longrightarrow$$

$$R_1\underset{\displaystyle O}{\overset{}{C}}R_2 + \left[C_6H_5-C(CH_3)(C_2H_5)^-K^+\right] \xrightarrow{C_6H_5NH_2} C_6H_5-C(CH_3)(C_2H_5)-H$$

e.g., $R_1 = R_2 = CH_3$, 83% retention

$$C_6H_5-C(CH_3)(OCH_3)-\underset{\| \atop O}{C}-C_6H_5 \xrightarrow[\text{Dioxane}]{t\text{-BuO}^-K^+} t\text{-BuO}\underset{\| \atop O}{C}C_6H_5 + \left[C_6H_5-C(CH_3)(OCH_3)^-K^+\right] \xrightarrow{*} C_6H_5-C(CH_3)(OCH_3)-H$$

61% retention

* Internal proton abstraction.

Fig. 13-16. Stereochemistry of carbanionic cleavage reaction.

but rather of unsymmetrical solvation of an anion which oscillates between the two possible pyramidal configurations.† In this respect, there may be a close analogy between carbanions and carbonium ions.[51] Specifically, the retention mechanism (approach of an internal proton to the front side of the carbanion in the ion-pair intermediate) resembles the S_Ni mechanism in nucleophilic substitution (approach of the anion to the front side of the carbonium ion in the ion-pair intermediate). The racemization mechanism, involving symmetric solvation of the carbanion in a well-ionizing solvent,

† The best evidence that unsymmetric solvation rather than dissymmetry as such is responsible for preservation of activity in carbanions comes from the observation cited earlier that even benzylic carbanions are capable of preserving optical activity. These anions are presumably planar because of resonance, as evidenced by their deep color (Ref. 49*a*).

resembles the racemizing carbonium-ion mechanism (extreme S_N1) which appears to involve symmetric solvation of a relatively stable carbonium ion. The inversion mechanism (approach of an external proton to the rear face of the carbanion) resembles the partial inversion normally observed in S_N1 reactions (preferential covalent binding of the solvent at the rear face of the carbonium ion). Finally, the complete retention of configuration in certain electrophilic displacement reactions or organometallics to be discussed below may correspond to the complete inversion of configuration observed in extreme S_N2 displacements. The reason why bimolecular electrophilic displacement (S_E2) involves retention whereas bimolecular nucleophilic displacement (S_N2) involves inversion appears to be that the former involves attack of the electrophile on the bonding electrons whereas the latter involves attack of the nucleophile on the atom that undergoes displacement.

Most other studies of carbanions have involved organometallic compounds. The situation here is complicated by the fact that the carbon-metal bond in such compounds is certainly partly covalent and only partly ionic. An additional complication exists in that the formation of organometallic reagents may involve radical intermediates which may be responsible for racemization when it occurs. No acyclic Grignard reagent has ever been obtained which can maintain dissymmetry at the site of attachment of the metal.[52] For example,[53] the treatment of (+)-2-bromoöctane, $CH_3CHBrC_6H_{13}$, with magnesium gave an inactive Grignard reagent† which upon treatment with acetone led to inactive 2-octyldimethylcarbinol, $(CH_3)_2COHCH(CH_3)C_6H_{13}$.‡ Conversion of either *cis*- or *trans*-3-methylcyclohexyl chloride to the Grignard reagent followed by carbonation gives, from either starting material, a mixture of 26% *trans*-3-methylcyclohexanecarboxylic acid and 74% *cis*-3-methylcyclohexanecarboxylic acid.[54] Organo*lithium* compounds may, however, show fleeting optical stability. Thus[55] (−)-2-iodoöctane, $CH_3CHIC_6H_{13}$, and 2-butyllithium, $CH_3CHLiC_2H_5$, in ether–petroleum ether at −70° gave 2-octyllithium, $CH_3CHLiC_6H_{13}$, which was evidently dissymmetric, for upon carbonation it yielded (−)-2-methyloctanoic acid, $CH_3CH(CO_2H)C_6H_{13}$. The over-all steric course was retention of configuration (20%) accompanied by racemization (80%). When the 2-octyllithium solution was warmed to 0°C. for 20 min. before carbonation, the 2-methyloctanoic acid obtained was totally racemic.[55] Greater optical stability is observed when dissymmetric alkyllithium compounds are prepared in *ether-free* pentane.[56] 2-Butyllithium

† Actually the Grignard solution was faintly active, but this was ascribed to unchanged bromide, since enough active bromide was isolated after hydrolysis of the Grignard reagent to account for the observed activity of the Grignard solution.

‡ However, as mentioned in Sec. 4-4*d, i, sec*-butylmagnesium chloride may give rise to optically active products when dissolved in (+)-2,3-dimethoxybutane.

---

[52] Cf. R. H. Pickard and J. Kenyon, *J. Chem. Soc.,* **99,** 65 (1911). The case of Grignard reagents in dissymmetric solvents has been taken up in Chap. 4. Regarding the cyclopropyl Grignard reagent, see page 393.

[53] C. W. Porter, *J. Am. Chem. Soc.,* **57,** 1436 (1935). See, however, F. C. Whitmore and B. J. Harriman, *ibid.,* **60,** 2821 (1938).

[54] H. L. Goering and F. H. McCarron, *J. Am. Chem. Soc.,* **80,** 2287 (1958).

[55] R. L. Letsinger, *J. Am. Chem. Soc.,* **72,** 4842 (1950).

[56] D. Y. Curtin and W. J. Koehl, *Chemistry & Industry (London),* 262 (1960).

prepared from 2-octyllithium and (−)-di-2-butylmercury (*vide infra*) in pentane at −5 to −10° upon carbonation (after a time lapse of 30 min.) gave 2-methylbutyric acid, $CH_3CH(CO_2H)C_2H_5$, with 47 to 56% retention of configuration.[56]

It would appear[56] that the effect of ether solvent in lessening the optical stability of dissymmetric lithium compounds is through increasing the carbanionic character of the organometallic reagent and decreasing its covalent character. It is not surprising, therefore, that when one goes to more covalent organometallic compounds, such as organomercury compounds, one finds that they can be resolved by classical methods and are quite optically stable. Thus 2-butylmercuric halides, $CH_3CH(HgX)CH_2CH_3$ (X = halide), may be resolved by conversion to the corresponding mandelates (X = active $C_6H_5CHOHCO_2^-$).[57,58] After purification of the diastereoisomeric salts, the 2-butylmercuric bromide (X = Br) regenerated by treatment of the mandelate with sodium bromide is optically active and quite stable. While this finding has perhaps little bearing on the optical stability of carbanions, it is worth noting here that reaction of (−)-2-butylmercuric bromide with bromine[58] to give 2-bromobutane, $CH_3CHBrC_2H_5$, proceeds with considerable preservation of optical activity, especially if carried out at low temperatures (−65°). The exact degree of stereospecificity and the configurational relation of starting material and product cannot be ascertained in this case, because the configuration and maximum rotation of the organomercury compound are unknown. However, in the related case of the reaction of 4-methylcyclohexylmercuric bromide,[59] where configurations could be assigned spectroscopically,† it was found that reaction with bromine under polar conditions (in acetic acid or pyridine) gave complete or nearly complete *retention* of configuration, as shown in Fig. 13-17.‡ Retention of configuration is also observed in the reaction of di-2-butylmercury with mercuric bromide shown in Fig. 13-18.[60] (The reader is advised to deduce from the rotations indicated that retention of configuration occurs in the displacement.) These reactions probably proceed by a bimolecular electrophilic-displacement mechanism, as originally proposed by Winstein, Traylor, and Garner.[61]

† By preparing the 1-deutero-1-bromomercuri-4-methylcyclohexanes and observing that in one the C—D stretching frequency corresponded to axial deuterium and in the other to equatorial deuterium (cf. Table 8-5). The isomer with the axial deuterium must have equatorial —HgBr and that with equatorial deuterium axial —HgBr. If it is assumed that the methyl group in both isomers is equatorial, the former isomer is trans and the latter cis: F. R. Jensen and L. H. Gale, *J. Am. Chem. Soc.*, **82**, 145 (1960).

‡ In a non-polar solvent (carbon tetrachloride) there is complete loss of stereochemical identity, a mixture of *cis*- and *trans*-4-methylcyclohexyl bromide being formed. Under these conditions the reaction is believed to proceed by a non-stereospecific radical path; cf. Sec. 13-3.

[57] H. B. Charman, E. D. Hughes, and C. K. Ingold, *J. Chem. Soc.*, 2523 (1959).

[58] F. R. Jensen, L. D. Whipple, D. K. Wedegaertner, and J. A. Landgrebe, *J. Am. Chem. Soc.*, **82**, 2466 (1960).

[59] F. R. Jensen and L. H. Gale, *J. Am. Chem. Soc.*, **82**, 148 (1960).

[60] (*a*) H. B. Charman, E. D. Hughes, and C. K. Ingold, *J. Chem. Soc.*, 2530 (1959); (*b*) F. R. Jensen, *J. Am. Chem. Soc.*, **82**, 2469 (1960).

[61] S. Winstein, T. G. Traylor, and C. S. Garner, *J. Am. Chem. Soc.*, **77**, 3741 (1955); see also G. F. Wright, *Can. J. Chem.*, **30**, 268 (1952).

All the evidence so far presented indicates that non-conjugated carbanions are tetrahedral (or pyramidal) rather than planar. It is, therefore, not surprising that the formation of carbanions at bridgehead positions does not meet with any special difficulty.[62] For example, the trisulfone shown in Fig.

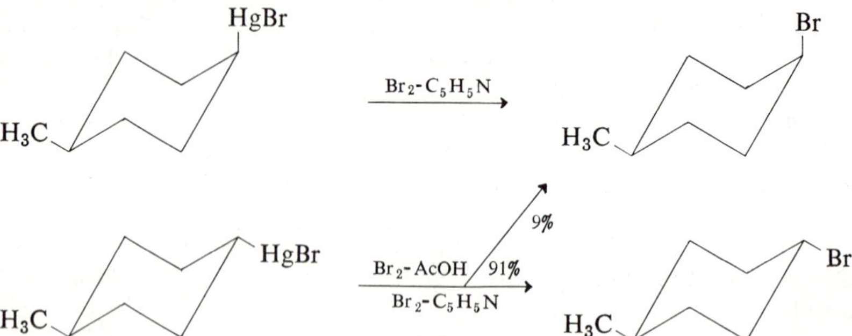

Fig. 13-17. Reaction of the 4-methylcyclohexylmercuric bromides with bromine in polar solvents.

13-19 shows the expected high acidity.[63] [This case differs from that of the bicyclic dicarbonyl compounds discussed in Chap. 10 (Fig. 10-50) in that the anion shown in Fig. 13-19 cannot be involved in overlap of the lone pair with the *p* electrons of sulfur. Either the acid-strengthening effect of the sulfone

$$(-)\text{-}CH_3\text{—}\underset{\underset{HgBr}{|}}{CH}\text{—}CH_2CH_3 + (\pm)\text{-}CH_3\text{—}\underset{\underset{MgBr}{|}}{CH}\text{—}CH_2CH_3 \longrightarrow$$

$[\alpha]_D^{22} - 6.49°$

$$\underbrace{\begin{matrix} CH_3\text{—}CH\text{—}CH_2CH_3 \\ | \\ Hg \\ | \\ CH_3\text{—}CH\text{—}CH_2CH_3 \end{matrix} + \begin{matrix} CH_3\text{—}CH\text{—}CH_2CH_3 \\ | \\ Hg \\ | \\ CH_3CH_2\text{—}CH\text{—}CH_3 \end{matrix}}_{[\alpha]_D^{22} - 5.52°} \xrightarrow{HgBr_2}$$

$$\underbrace{(-)\text{-}CH_3\text{—}\underset{\underset{HgBr}{|}}{CH}\text{—}CH_2CH_3 + (\pm)\text{-}CH_3\text{—}\underset{\underset{HgBr}{|}}{CH}\text{—}CH_2CH_3}_{[\alpha]_D^{22} -3.36°}$$

Fig. 13-18. Formation of active di-2-butylmercury and its reaction with mercuric bromide.[60b]

group is purely electrostatic, or, if there is electron overlap, it must be with the *d* electrons of the sulfur. Such overlap may not be governed by Bredt's rule.[64]]

[62] See Ref. 87, pp. 153–154.

[63] W. v. E. Doering and L. K. Levy, *J. Am. Chem. Soc.*, **77**, 509 (1955).

[64] See also H. E. Zimmerman and B. S. Thyagarajan, *J. Am. Chem. Soc.*, **82**, 2505 (1960); E. J. Corey and E. T. Kaiser, *ibid.*, **83**, 490 (1961); D. J. Cram, D. A. Scott, and W. D. Nielsen, *J. Am. Chem. Soc.*, **83**, 3696 (1961).

The reaction of 1-bromotriptycene (Fig. 13-20) with butyllithium to give 1-lithiotriptycene[65a]—which, in turn, reacts normally with benzophenone to give 1-triptycyldiphenylcarbinol[65b]—has also been cited[62] to illustrate ready formation of carbanions at bridgehead positions. It should, however, be considered that the transition state leading to the bridgehead lithium com-

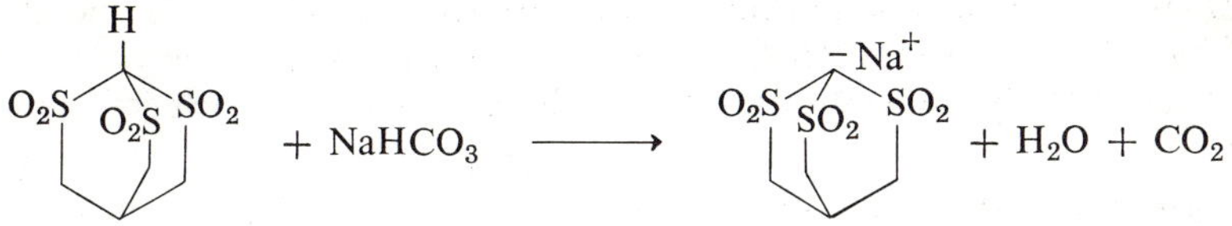

$$+ NaHCO_3 \longrightarrow + H_2O + CO_2$$

Fig. 13-19. Acidic proton at bridgehead position.

pound may be four-membered with little carbanion character (Fig. 13-20) and that the lithium compound itself may be largely covalent. Unlike triphenylmethyllithium, 1-lithiotriptycene is colorless. This might be related to its highly covalent character; on the other hand, even if the 1-triptycyl carbanion *is* formed, it would, according to Bredt's rule, not be expected to show

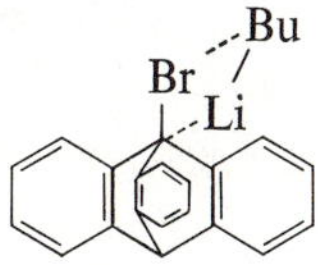

Fig. 13-20. Possible transition states in formation of 1-lithiotriptycene.

the benzylic resonance which is presumably responsible for the red color of triphenylmethyllithium.

A bridgehead carbanion cannot undergo the oscillatory motion shown in Fig. 13-13. Thus such an anion should be able to hold its configuration indefinitely. This has not been tested experimentally, but the point has been

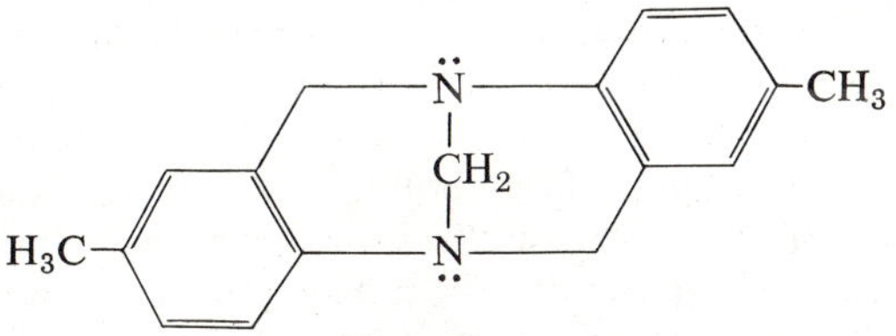

$[\alpha]_D^{17} = 287°$

Fig. 13-21. Tröger's base.

established in an analogous nitrogen compound: Tröger's base (Fig. 13-21).[66] Tröger's base can be resolved by chromatography on lactose hydrate and is optically stable in a neutral medium. (In an acid medium, racemization

[65] (*a*) G. Wittig and U. Schöllkopf, *Tetrahedron,* **3,** 91 (1958); (*b*) G. Wittig and W. Tochtermann, unpublished; see Ref. 90, footnote 49.

[66] V. Prelog and P. Wieland, *Helv. Chim. Acta,* **27,** 1127 (1944).

occurs, presumably because formaldehyde or a methylene-imonium intermediate, $>\overset{+}{N}{=}CH_2\ HN<$, is formed in reversible fashion from the methylene group between the two nitrogen atoms.)

Greater configurational stability than that of saturated carbanions is found in olefinic carbanions of the type $RR'C{=}\bar{C}R''$. These anions are the carbon analogs of nitrogen compounds of the type $RR'{=}NR''$. It was pointed out in Chap. 12 that oximes $RR'C{=}NOH$ and similar compounds are capable of existing as fairly stable cis and trans isomers. This would seem to indicate that an oscillation of the type $>C{=}N{\diagup}^{OH} \rightleftharpoons >C{=}N{\diagdown}_{OH}$ involving the linear transition state $>C{=}N{-}OH$ occurs less easily (i.e., requires higher activation energy) than the process shown in Fig. 13-13.

A number of organolithium compounds of the type $RR'C{=}CR''Li$ have been prepared in distinct cis and trans forms which are quite stable configurationally.[67] For example, *cis*-1-bromo-1-propene (Fig. 13-22) may be converted to an organolithium reagent by treatment with lithium metal. Subsequent addition of benzaldehyde gives *cis*-1-phenyl-2-buten-1-ol. In the same reaction sequence *trans*-1-bromo-1-propene gives *trans*-1-phenyl-2-buten-1-ol (Fig. 13-22). The two stereoisomeric lithium reagents are not interconverted to any great extent (less than 8%) by boiling in ether solution for about 3 hrs. The configuration of the 1-propenyllithium compounds shown in Fig. 13-22 has been assigned on the basis of infrared spectral evidence,[68] and it is thus assured that both steps—the formation of the alkenyllithium reagents and their further reactions—proceed with retention of configuration, rather than with inversion. The configurational stability of the different vinyllithium compounds that have been studied varies with their structure. For example, *cis*- or *trans*-1,2-diphenyl-2-(*p*-chlorophenyl)vinyllithium,

$$\underset{\text{cis}}{\left(H_3C\right)(H)C{=}C(Br)(H)} \xrightarrow[\text{ether, }\Delta]{\text{Li}} \left[(H_3C)(H)C{=}C(Li)(H)\right] \xrightarrow{C_6H_5CHO,\ 100\%} \underset{\text{cis}}{(H_3C)(H)C{=}C(CHOHC_6H_5)(H)}$$

$$\underset{\text{trans}}{\left(H_3C\right)(H)C{=}C(H)(Br)} \xrightarrow[\text{ether, }\Delta]{\text{Li}} \left[(H_3C)(H)C{=}C(H)(Li)\right] \xrightarrow{C_6H_5CHO} \underset{\text{trans}}{(H_3C)(H)C{=}C(H)(CHOHC_6H_5)}\ (92\%;\ 8\%\ \text{cis})$$

Fig. 13-22. Stereospecificity in preparation and reaction of vinyllithium compounds.

[67] D. Y. Curtin and J. W. Crump, *J. Am. Chem. Soc.*, **80**, 1922 (1958); see also earlier references there cited.

[68] N. L. Allinger and R. B. Hermann, *J. Org. Chem.*, **26**, 1040 (1961); see also A. N. Nesmeyanov and A. E. Borisov, *Tetrahedron*, **1**, 158 (1957), and earlier papers by A. N. Nesmeyanov and coworkers there cited.

$$p\text{-}ClC_6H_4(C_6H_5)C{=}C(C_6H_5)Li$$

prepared from the corresponding bromides by exchange with ethereal butyllithium at -50 to $-35°$ reacts stereospecifically with carbon dioxide,[69] methanol,[69, 70] formaldehyde,[70] and methyl iodide.[70] However, when the exchange reaction with butyllithium is carried out at 0 or 25° (30-min. reaction time) before the vinyl lithium reagent is decomposed with methanol, mixtures of *cis*- and *trans*-1,2-diphenyl-2-(*p*-chlorophenyl)ethylenes,

$$p\text{-}ClC_6H_4(C_6H_5)C{=}CHC_6H_5$$

are obtained.[70] At $+25°$ the two isomeric lithium reagents seem to be in equilibrium, since they give approximately the same product mixture. Similarly, *cis*-1,2-diphenylvinyllithium, *cis*-$C_6H_5CH{=}CLiC_6H_5$, is readily converted into the trans isomer above $-40°$ in benzene-ether solution.[68, 71]

Cyclopropyl carbanions (or, more properly stated, the corresponding organolithium reagents, since it is not known to what extent these reagents are ionic) seem to have optical stability intermediate between that of the ions

H_5C_6, H_5C_6 / CH_3, Br (cyclopropane) $\xrightarrow{C_4H_9Li}$ H_5C_6, H_5C_6 / CH_3, Li $\xrightarrow{CH_3OH}$ H_5C_6, H_5C_6 / CH_3, H

A, $[\alpha]_D^{26}$ $+106°$ B, $[\alpha]_D^{26}$ $-78°$

A $\xrightarrow{Mg}$ H_5C_6, H_5C_6 / CH_3, MgBr $\xrightarrow{H_2O}$ ditto, $[\alpha]_D^{23}$ $-13.6°$

Fig. 13-23. Retention of configuration in cyclopropyllithium derivative and cyclopropyl Grignard reagent.

corresponding to saturated organolithium reagents and those corresponding to the more stable vinyllithium reagents. Treatment of (+)-2,2-diphenyl-1-bromo-1-methylcyclopropane (A, Fig. 13-23) with butyllithium in ether–petroleum ether–benzene below 6° followed by decomposition with methanol gave (−)-1-methyl-2,2-diphenylcyclopropane (B) with 80% retention of configuration (Fig. 13-23).[72a] Even the Grignard reagent derived from A gives rise to optically active B upon hydrolysis, although the optical yield is only about 11% (Fig. 13-23).[72b] Analogous results are obtained when the organolithium or Grignard reagents shown in Fig. 13-23 are carbonated.[72b] The greater extent of racemization of B when prepared via a Grignard reagent (as compared with an alkyllithium reagent) may reflect racemization occurring during the preparation of the RMgBr reagent, possibly involving R· radicals.[72b]

[69] D. Y. Curtin and E. E. Harris, *J. Am. Chem. Soc.*, **73**, 2716 (1951).

[70] D. Y. Curtin, H. W. Johnson, and E. G. Steiner, *J. Am. Chem. Soc.*, **77**, 4566 (1955).

[71] However, D. Y. Curtin and E. E. Harris, *J. Am. Chem. Soc.*, **73**, 4519 (1951), find the 1,2-diphenylvinyllithium isomers stable in ether up to at least $-21°$.

[72] (*a*) H. M. Walborsky and F. J. Impastato, *J. Am. Chem. Soc.*, **81**, 5835 (1959); (*b*) H. M. Walborsky and A. E. Young, *ibid.*, **83**, 2595 (1961).

13-5. Stereochemistry of Hydrogenolysis

This section deals with hydrogenolysis reactions of the type

$$RR'R''CX \rightarrow RR'R''CH$$

brought about either by hydrogen gas in the presence of catalysts or by metal-acid combinations. The placing of these reactions in a chapter dealing with tricovalent carbon may be questioned, since the mechanism of the reductions is not yet clear. Nevertheless, it is worth summarizing the known facts, even though a consistent pattern may not emerge.

Chemical reduction of (−)-2-phenyl-2-chloropropionic acid,

$$C_6H_5C(CH_3)ClCO_2H$$

with zinc in acetic acid[73] gives (−)-hydratropic acid, $C_6H_5CH(CH_3)CO_2H$,

trans $\xrightarrow[CHCl_3]{Cl_2,\ SbCl_5}$ erythro $\xrightarrow[C_3H_7OH]{KOH}$ cis $\xrightarrow[NH_3]{Na}$ cis

Fig. 13-24. Conversion of trans to cis olefins.

with inversion of configuration.[74] Reduction of chloroölefins of the type RCH=CClR′ with sodium in liquid ammonia to olefins RCH=CHR′, on the other hand, proceeds with retention of geometrical configuration.[75] This reduction forms part of an interesting reaction sequence by means of which trans olefins have been converted to their cis isomers, and vice versa;[75] the entire sequence is shown in Fig. 13-24.

Chemical reduction of bridgehead halides, e.g., reduction of 4-chlorotricyclene (Fig. 13-25) by sodium and ethanol, proceeds normally.[75a]

The stereochemistry of hydrogenolysis by Raney nickel has been studied

[73] E. Ott and K. Krämer, *Ber.*, **68**, 1655 (1935). For a similar case, see M. Gut and M. Uskoković, *J. Org. Chem.*, **25**, 792 (1960).

[74] Cf. E. L. Eliel and J. P. Freeman, *J. Am. Chem. Soc.*, **74**, 923 (1952).

[75] M. C. Hoff, K. W. Greenlee, and C. E. Boord, *J. Am. Chem. Soc.*, **73**, 3329 (1951).

[75a] H. Meerwein and R. Wortmann, *Ann.*, **435**, 190 (1924); regarding the structure of the chloride, cf. J. Houben and E. Pfankuch, *Ann.*, **501**, 219 (1933).

largely by William Bonner and coworkers.[76-80] Examples are listed in Fig. 13-26. In three instances there was retention of configuration with some racemization, in two instances complete racemization, and in one instance inversion with some racemization. However, in the last case the configurational assignment of the starting material (sulfone) rests entirely upon an application of the rule of shift (Sec. 5-4*d*) and must be considered tentative.

The stereochemistry of hydrogen-deuterium exchange has been studied in similar fashion. Thus when (+)-2-phenylpropionamide,

$$C_6H_5CH(CH_3)CONH_2$$

is treated with deuterated Raney nickel,† the hydrogen atom on the asymmetric carbon atom (along with a considerable portion of the methyl and phenyl hydrogens) is exchanged by deuterium, but the amide largely retains its configuration.[81] On the other hand, partial exchange of the α-hydrogen (along with some of the methyl hydrogen) in lactic acid, $CH_3CHOHCO_2H$, by deuterium from deuterium oxide over a platinum catalyst proceeds with inversion of configuration.[82]

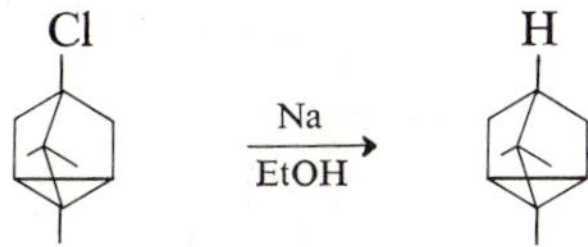

Fig. 13-25. Reduction of 4-chlorotricyclene.

13-6. Stereochemistry of Radiolysis

Because of increasing interest in radiation chemistry, a few stereochemical aspects of this subject may be mentioned here. Exposure of crystalline D-(+)-glucose to recoil tritium leads to incorporation of tritium in the molecule without formation of any D-galactose.[83] Apparently H-T exchange proceeds without change of configuration (at least there is no change at C_4). Similar exposure of crystalline L-(+)-alanine, $CH_3CH(NH_2)CO_2H$, leads to the α-tritiated species $CH_3CT(NH_2)CO_2H$, among others, again without change of configuration.[84] Gamma irradiation of (−)-mandelic acid partly destroys the material and, in part, racemizes it without altering it chemically.[85] Un-

† In all these cases, the hydrogen (or deuterium) adsorbed on the Raney nickel brings about the hydrogenolysis, the nickel acting as a catalyst.

[76] W. A. Bonner, *J. Am. Chem. Soc.,* **74,** 1034 (1952).
[77] W. A. Bonner, *J. Am. Chem. Soc.,* **74,** 5089 (1952).
[78] W. A. Bonner, J. A. Zderic, and G. A. Casaletto, *J. Am. Chem. Soc.,* **74,** 5086 (1952).
[79] W. A. Bonner and J. A. Zderic, *J. Am. Chem. Soc.,* **78,** 3218 (1956).
[80] W. A. Bonner and T. W. Greenlee, *J. Am. Chem. Soc.,* **81,** 3337 (1959); cf. D. J. Cram, K. R. Kopecky, F. Hauck, and A. Langemann, *ibid.,* p. 5754, footnote 4a.
[81] W. A. Bonner and J. A. Zderic, *J. Am. Chem. Soc.,* **78,** 4369 (1956).
[82] J. Bell, K. A. Macdonald, and R. I. Reed, *J. Chem. Soc.,* 3459 (1953).
[83] H. Keller and F. S. Rowland, *J. Phys. Chem.,* **62,** 1373 (1958).
[84] J. G. Kay, R. P. Malsan, and F. S. Rowland, *J. Am. Chem. Soc.,* **81,** 5050 (1959).
[85] P. Y. Feng and S. W. Tobey, *J. Phys. Chem.,* **63,** 759 (1959).

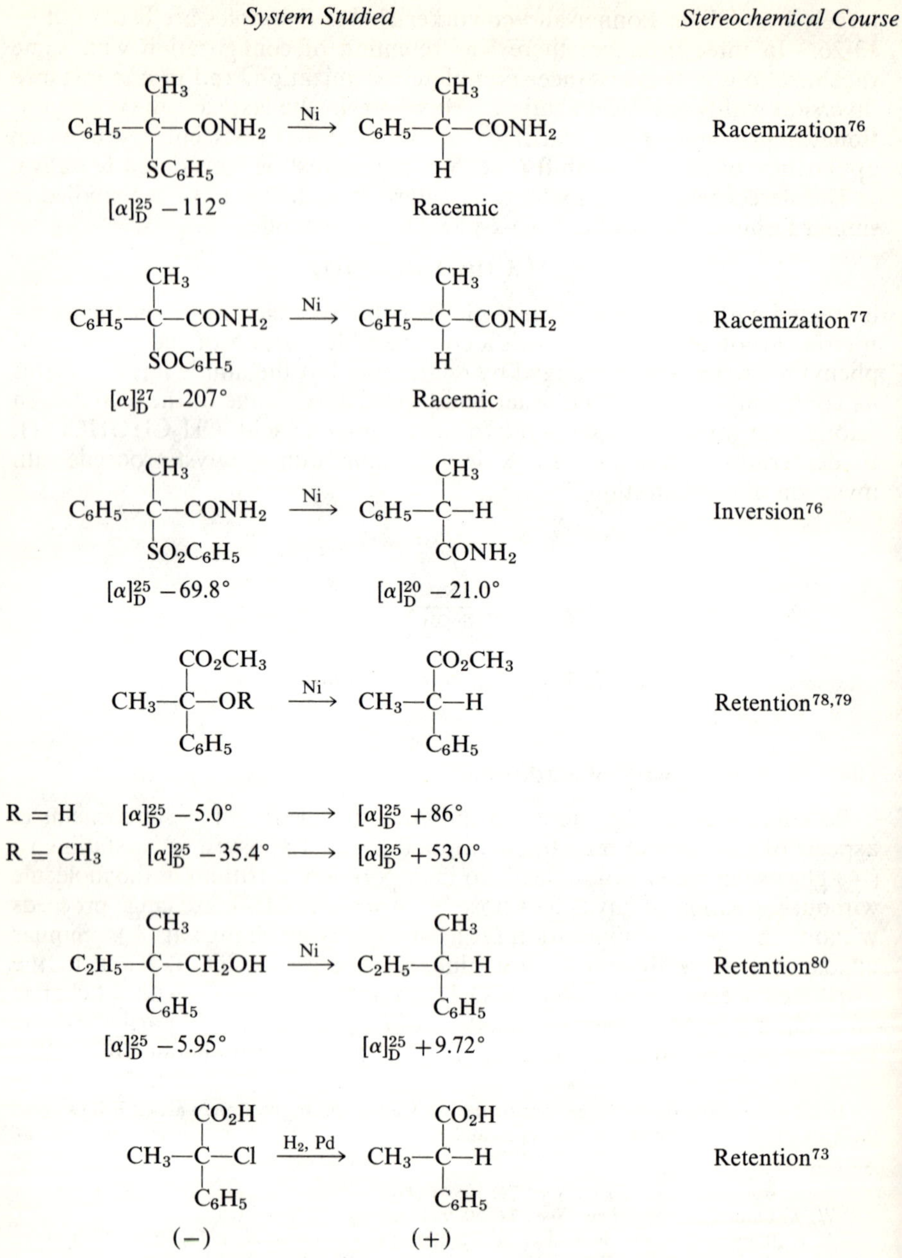

System Studied			*Stereochemical Course*
C_6H_5—C(CH_3)(SC_6H_5)—$CONH_2$ $[\alpha]_D^{25}$ −112°	$\xrightarrow{Ni}$	C_6H_5—C(CH_3)(H)—$CONH_2$ Racemic	Racemization[76]
C_6H_5—C(CH_3)(SOC_6H_5)—$CONH_2$ $[\alpha]_D^{27}$ −207°	$\xrightarrow{Ni}$	C_6H_5—C(CH_3)(H)—$CONH_2$ Racemic	Racemization[77]
C_6H_5—C(CH_3)($SO_2C_6H_5$)—$CONH_2$ $[\alpha]_D^{25}$ −69.8°	$\xrightarrow{Ni}$	C_6H_5—C(CH_3)($CONH_2$)—H $[\alpha]_D^{20}$ −21.0°	Inversion[76]
CH_3—C(CO_2CH_3)(C_6H_5)—OR	$\xrightarrow{Ni}$	CH_3—C(CO_2CH_3)(C_6H_5)—H	Retention[78,79]
R = H $[\alpha]_D^{25}$ −5.0°	⟶	$[\alpha]_D^{25}$ +86°	
R = CH_3 $[\alpha]_D^{25}$ −35.4°	⟶	$[\alpha]_D^{25}$ +53.0°	
C_2H_5—C(CH_3)(C_6H_5)—CH_2OH $[\alpha]_D^{25}$ −5.95°	$\xrightarrow{Ni}$	C_2H_5—C(CH_3)(C_6H_5)—H $[\alpha]_D^{25}$ +9.72°	Retention[80]
CH_3—C(CO_2H)(C_6H_5)—Cl (−)	$\xrightarrow{H_2,\ Pd}$	CH_3—C(CO_2H)(C_6H_5)—H (+)	Retention[73]

Fig. 13-26. Stereochemistry of hydrogenolysis.

fortunately, in this experiment, the recovered mandelic acid was recrystallized, with the resulting danger of preferential crystallization of either the enantiomer or the racemate. Thus the measured rotation of the crystalline material may not correspond to that of the mandelic acid originally recovered in the irradiation.

General References

[86] R. L. Shriner and R. Adams, Optical Isomerism, in H. Gilman, ed., "Organic Chemistry," John Wiley & Sons, Inc., New York, 1943, pp. 383–400.

[87] E. L. Eliel, Substitution at Saturated Carbon Atoms, in M. S. Newman, ed., "Steric Effects in Organic Chemistry," John Wiley & Sons, Inc., New York, 1956, chap. 2.

[88] J. E. Leffler, "The Reactive Intermediates of Organic Chemistry," Interscience Publishers, Inc., New York, 1956.

[89] D. E. Applequist and J. D. Roberts, Displacement Reactions at Bridgeheads of Bridged Polycarbocyclic Systems, *Chem. Revs.*, **54**, 1065 (1954).

[90] U. Schöllkopf, Substitutions-Reaktionen am Brückenkopf bicyclischer Verbindungen, *Angew. Chem.*, **72**, 147 (1960).

Chapter 14

OPTICAL ROTATION AND OPTICAL ROTATORY DISPERSION

14-1. Relation between Rotation and Configuration

Much has been said in this book about molecular dissymmetry, and many instances have been cited where optical rotation is used as a tool to probe molecular geometry. However, little mention has been made in either a qualitative or a quantitative way of the fundamental relation between optical rotation and molecular dissymmetry, i.e., about how dissymmetry gives rise to optical activity, and to what extent. There are historical reasons for this, for although optical rotation was discovered at the beginning of the nineteenth century and became a powerful tool in the hands of chemists by the end of that century (see Chap. 1), a fundamental understanding of the phenomenon lagged far behind and received only intermittent attention on the part of a few physicists. Over a hundred years elapsed before the molecular basis of optical rotatory power began to be elaborated by Gray, de Malleman, Born, and Boys,[1] and successful theories of optical rotation have only recently been proposed by the schools of W. Kuhn,[2] J. G. Kirkwood,[3] and H. Eyring.[4] These theories have successfully predicted the configuration of (R)-(−)-2-butanol,[2, 4b] (R)-(+)-2-epoxybutane,[3b] and (R)-(+)-1,2-dichloropropane.[3b] The configuration of (S)-(+)-ethylbenzene-*α*-*d* (Chap. 3) has also been correctly calculated.[4f] Moreover, the theories correctly predict the approximate magnitude of the rotation.† The application of the abovementioned theories of rotation to more complex molecules does not, however, at the present time appear promising.[4e, g] Considerably greater expectations

Note: All references above 49 are listed in the General References at the end of the chapter.

† Since the theories involve certain simplifications, they are not totally reliable. One of the original calculations of the (+)-2-butanol configuration gave the wrong answer because of an error in sign (Ref. 3*a*), and an early calculation for a deuterium compound of type RR′CHD [P. Fréon, *J. phys. radium,* [8]**1,** 374 (1940)] predicted zero rotation. See also page 168.

---

[1] See Ref. 51 for a discussion of the early theories.

[2] For example, W. Kuhn, *Z. physik. Chem.,* **B31,** 23 (1935); see also Refs. 50 and 54.

[3] (*a*) J. G. Kirkwood, *J. Chem. Phys.,* **5,** 479 (1937); (*b*) W. W. Wood, W. Fickett, and J. G. Kirkwood, *ibid.,* **20,** 561 (1952).

[4] (*a*) E. U. Condon, W. Altar, and H. Eyring, *J. Chem. Phys.,* **5,** 753 (1937); (*b*) E. Gorin, J. Walter, and H. Eyring, *ibid.,* **6,** 824 (1938); (*c*) E. U. Condon, *Revs. Modern Phys.,* **9,** 433 (1937); (*d*) C. G. LeFèvre and R. J. W. LeFèvre, *Chemistry & Industry (London),* 1283 (1959); (*e*) W. Kauzmann, "Quantum Chemistry," Academic Press, Inc., New York, 1957, pp. 616–635, 703–723. See also Ref. 52.

[4f] W. Fickett, *J. Am. Chem. Soc.,* **74,** 4204 (1952).

[4g] W. Kauzmann, F. B. Clough, and I. Tobias, *Tetrahedron,* **13,** 57 (1961).

may be held for theoretical calculations of optical-rotatory-dispersion curves (Sec. 14–2). Successful predictions of such curves on the basis of absorption data for molecules as complicated as 12-ketocholanic acid have been made,[5a] and many interesting developments in this area may well be expected in the near future.[5b-d] The treatment is, however, quite involved mathematically.

In the following pages will be presented, in greatly simplified form, some of the basic principles of optical rotation. This will be followed (page 401) by a recently developed semi-empirical approach to the calculation of optical rotations[6] which has the virtue of being relatively simple as well as giving the reader an insight into the origin of the rotation phenomenon.

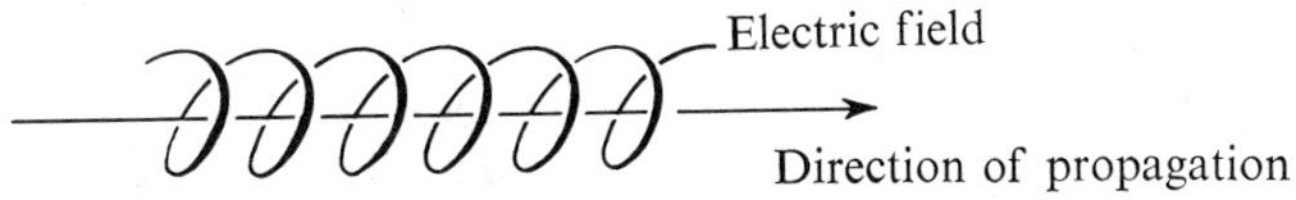

Fig. 14-1. Right circularly polarized light wave.

A wave of plane-polarized light may be considered to be made up of two types of "circularly polarized light": one right circularly polarized wave and one left circularly polarized wave. A "circularly polarized wave" is one whose plane of polarization rotates continuously and in the same sense around the axis of propagation of the wave. Thus the electric field of a right circularly polarized wave may be described as a right-handed screw or helix twisting around the direction of propagation (Fig. 14-1), whereas a left circularly polarized wave describes a left-handed screw. Figure 14-2 shows how the electric vector of a right circularly polarized wave ($\mathbf{E}_R$) and that of a

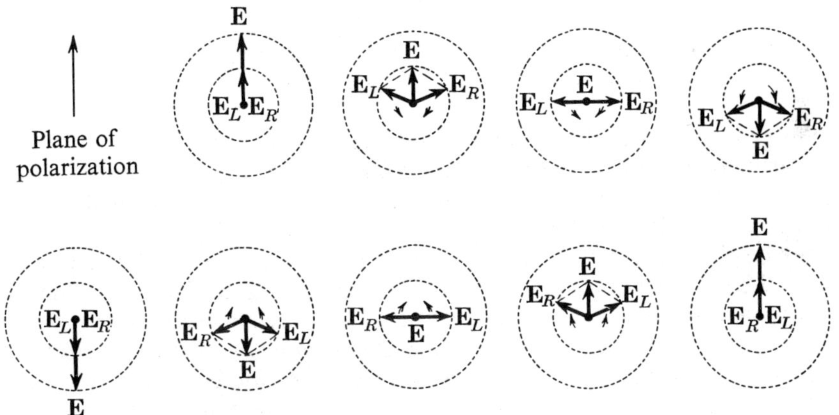

Fig. 14-2. Right and left circularly polarized electric field vectors combining to give plane-polarized field vector. (*From C. Djerassi, "Optical Rotatory Dispersion," copyright,* 1960, *McGraw-Hill Book Company, Inc.*)

[5] (*a*) A. Moscowitz, chap. 12 in Ref. 55; (*b*) W. Moffitt and A. Moscowitz, *J. Chem. Phys.*, **30**, 648 (1959); (*c*) A. Moscowitz, *Tetrahedron*, **13**, 48 (1961); (*d*) W. Kuhn, *ibid.*, **13**, 1 (1961). See also Ref. 40*b*.

[6] (*a*) J. H. Brewster, *J. Am. Chem. Soc.*, **81**, 5475 (1959); (*b*) *id.*, *ibid.*, p. 5483; (*c*) *id.*, *ibid.*, p. 5493; (*d*) *Tetrahedron*, **13**, 106 (1961). See also Ref. 4*g*.

left circularly polarized wave ($\mathbf{E}_L$) combine to give the vector of a plane-polarized wave (**E**) which, starting out with a maximum value, decreases to zero and then to a minimum and grows again to zero and back to the maximum. (The plane-polarized wave travels away from the observer along an axis passing through the centers of the circles in Fig. 14-2.) Actually, circularly polarized light is not a figment of the physicist's imagination but may actually be produced by passing plane-polarized light through a specially cut glass prism known as "Fresnel's rhomb." For our present purpose, however, we resolve the plane-polarized light into two beams of oppositely circularly polarized light merely as a matter of convenience in understanding optical rotation. The salient feature of an optically active substance is that it is "circularly birefringent," i.e., that it has unequal refractive indices for right and left circularly polarized light. Since the velocity of light in a medium is given by $v = c/n$, where c is the velocity of light *in vacuo* and n is the refractive index of the medium, the result of circular birefringence is an unequal

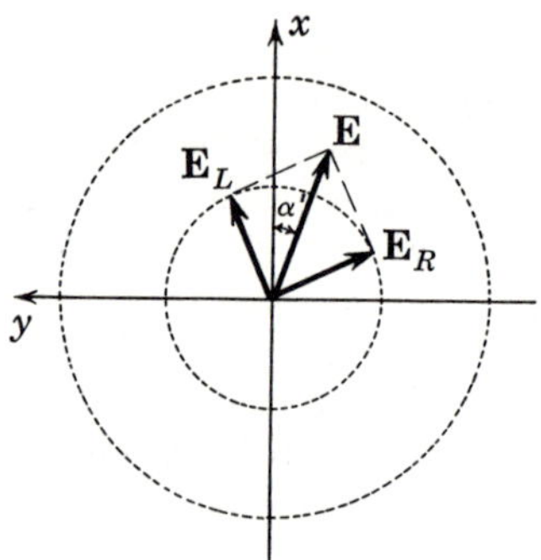

Fig. 14-3. Optical rotation as a result of circular birefringence. (*From C. Djerassi, "Optical Rotatory Dispersion," copyright,* 1960, *McGraw-Hill Book Company, Inc.*)

rate of propagation of the left and right circularly polarized rays. If the right circularly polarized ray $\mathbf{E}_R$ travels faster than the left circularly polarized one $\mathbf{E}_L$, the result will be as shown in Fig. 14-3. The resultant wave will still pulsate in a plane, i.e., be plane-polarized, but the plane of polarization will no longer be the x plane but will make an angle α' with the x plane; in other words, the circularly birefringent medium has rotated the plane of polarization by an angle α'. If the right circularly polarized wave travels faster, α' is positive and the medium is dextrorotatory, whereas if the left circularly polarized wave is faster, α' is negative and the medium is levorotatory. The angle of rotation per unit path length is[7] $\alpha = (n_L - n_R)\pi/\lambda$, where n_L and n_R are the indices of refraction for left and right circularly polarized light, respectively, and λ is the vacuum wavelength of the light.

Our task now becomes the correlation of circular birefringence with molecular dissymmetry. Since refractive index is related to polarizability, it is logical that circular birefringence (and thus optical activity) should be

[7] A. Fresnel, *Ann. chim. et phys.*, [2]**28**, 147 (1825).

related to some dissymmetry of polarizability in the molecule. According to Brewster,[6] "a center of optical activity can usefully be described as an asymmetric screw pattern of polarizability." Such a screw pattern may originate in either of two ways or a combination thereof.[8] The first way is illustrated by a compound such as chlorobromoiodomethane, CHClBrI, in which the dissymmetry of polarizability is caused strictly by the difference in polarizability of the groups attached to the asymmetric atom. This is called "atomic asymmetry." The second way is illustrated by methylethylpropylmethane (3-methylhexane), $CH_3CH_2CH(CH_3)CH_2CH_2CH_3$, in which the dissymmetry of polarizability is caused by the spatial arrangement (conformation) of the groups in the molecule. This is called "conformational asymmetry" and usually makes a larger contribution to molecular rotation than does atomic asymmetry. It is the neglect of conformational asymmetry which made an earlier empirical attempt to correlate rotation and configuration[9] of rather limited applicability.†

a. Atomic Asymmetry.[6a] Without going into the details of derivation, it may be stated that in a system Cabcd which shows strictly atomic asymmetry (i.e., in which the groups either have no conformational alternatives—such as halogen atoms—or are conformationally symmetrical—as a methyl

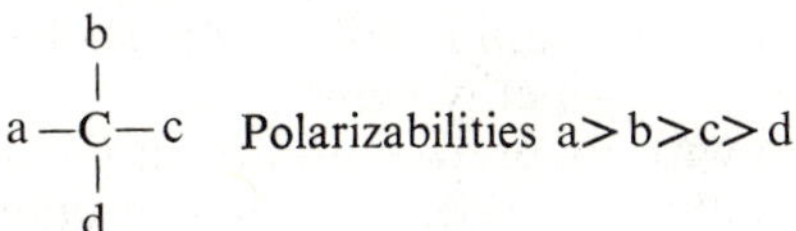

Fig. 14-4. Atomic asymmetry: dextrorotatory configuration.

group) the configuration shown (as a Fischer projection) in Fig. 14-4 is dextrorotatory when the polarizabilities of the attached atoms are a > b > c > d. The polarizabilities of the four groups are derived from the atomic refraction[10] of the atoms attached to the asymmetric carbon. In cases of doubt, some empirical assignments of refractions were made; thus the values for C=C and C≡C are one-half the value of the group refractivities, and the vinyl (and phenyl) groups were deemed to precede carboxyl (although the attached atoms have identical refractivities) because the bond refractivity of C=C is greater than that of C=O. The order thus obtained (refraction of attached atom in parentheses) is I (13.954), Br (8.741), SH (7.729), Cl (5.844), C≡C (3.580; value for whole group, 7.159), CN (3.580; value for whole group, 5.459), C=C (3.379; value for whole group, 6.757), C_6H_5 (same as C=C), CO_2H (3.379; value for whole group, 4.680), CH_3 (2.591), NH_2 (2.382), OH (1.518), H (1.028), D (1.004), F (0.81).

† A third type of asymmetry, "permolecular asymmetry," accounts for the optical activity of allenes, spiranes, etc. (Chap. 11). The importance of this type of asymmetry in centrodissymmetric compounds remains to be assessed (Ref. 6*a*).

[8] D. H. Whiffen, *Chemistry & Industry, (London),* 964 (1956).

[9] (*a*) R. E. Marker, *J. Am. Chem. Soc.,* **58,** 976 (1936); cf. (*b*) J. A. Mills, *Chemistry & Industry (London),* 218 (1953).

[10] A. I. Vogel, *J. Chem. Soc.,* 1833 (1948).

By way of application, we may predict the rotation of (S)-hydratropic acid and of (S)-α-phenethyl chloride (configurations shown in Fig. 14-5). The usual Fischer projections of these compounds are rewritten so as to conform as closely as possible to the pattern shown in Fig. 14-4, and the substituents are labeled a, b, c, and d in order of decreasing polarizability (see listing above). It is clear that (S)-hydratropic acid has the configuration symbolized in Fig. 14-4 and is therefore dextrorotatory, whereas (S)-α-phenethyl chloride has the opposite configuration from that shown in Fig. 14-4 and is therefore levorotatory. Both these predictions prove to be correct. As a third example we may consider (−)-ethylbenzene-α-*d*.[11] Since the polarizability sequence is $C_6H_5 > CH_3 > H > D$ (see above) and since the levorotatory isomer

$$\begin{array}{c} CO_2H \\ | \\ H_3C—C—H \\ | \\ C_6H_5 \end{array} \equiv \begin{array}{c} CO_2H(b) \\ | \\ (a)H_5C_6—C—CH_3(c) \\ | \\ H \\ (d) \end{array} \qquad \begin{array}{c} CH_3 \\ | \\ H—C—Cl \\ | \\ C_6H_5 \end{array} \equiv \begin{array}{c} C_6H_5(b) \\ | \\ (a)Cl—C—H(d) \\ | \\ CH_3(c) \end{array}$$

Fig. 14-5. Application of Brewster's treatment to (S)-hydratropic acid and (S)-α-phenethyl chloride.

should have a configuration enantiomeric to that shown in Fig. 14-4, the predicted configuration[12] for (−)-ethylbenzene-α-*d* is that shown in Fig. 14-6. This is in accordance with its preparation from (−)-α-phenethyl chloride (Fig. 14-5) and lithium aluminum deuteride,[11] a reaction now known (cf. Sec. 3-1) to involve inversion of configuration.

Some doubt might attach to the prediction of configuration of carboxylic acids in the manner indicated, since the CO_2H group has conformational alternatives by virtue of rotation around the bond linking it to the asymmetric carbon. This does not seem to produce serious difficulty, however, unless there is another substituent with which the carboxyl group may form

$$\begin{array}{c} CH_3 \\ | \\ H_5C_6—C—D \\ | \\ H \end{array} \equiv \begin{array}{c} CH_3 \\ | \\ D—C—H \\ | \\ C_6H_5 \end{array}$$

Fig. 14-6. (−)-Ethylbenzene-α-*d*.

hydrogen bonds, as in a hydroxy acid or amino acid. An example is lactic acid (Fig. 14-7). According to the rule embodied in Fig. 14-4, the enantiomer shown in Fig. 14-7 should be dextrorotatory (polarizability sequence $CO_2H > CH_3 > OH > H$) whereas, in fact, it is levorotatory in water (although its salts and esters are dextrorotatory). This anomaly is ascribed to the existence of the hydrogen-bonded form shown in Fig. 14-7 which, according to the rules applicable to cyclic compounds (see Sec. 14-1*d* below), ought to be levorotatory. In some instances (e.g., with mandelic acid, $PhCHOHCO_2H$) both atomic polar-

[11] E. L. Eliel, *J. Am. Chem. Soc.*, **71**, 3970 (1949).
[12] Cf. J. H. Brewster, *Tetrahedron Letters*, nos. 20, 23 (1959).

izability and consideration of the cyclic structure predict the same sense of rotation, and in these cases there is no ambiguity. In yet other cases, e.g., atrolactic acid, $PhCMeOHCO_2H$, the ambiguity disappears if the assumption is made[12] that hydroxyl and amino, *if attached to a benzylic carbon atom,* rank ahead of carbon substituents.†

b. Conformational Asymmetry.[6a] In 3-methylhexane, $C_2H_5CH(CH_3)C_3H_7$, the three atoms attached to the asymmetric carbon are identical and have the same polarizability. Therefore, the substance should have no atomic asymmetry. Nevertheless, it displays appreciable rotation: $[M]_D = 9.9°$, due to

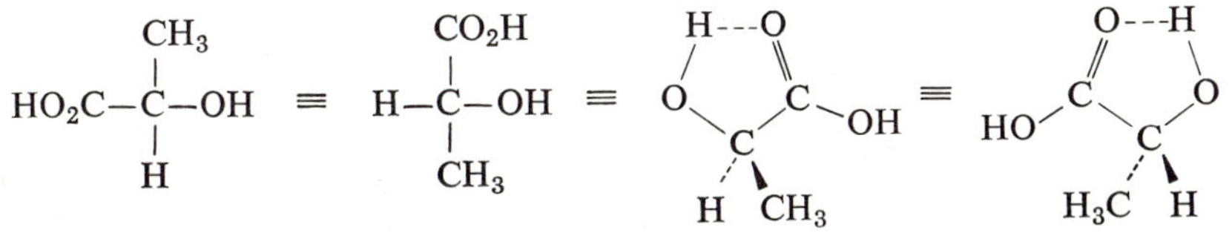

Fig. 14-7. (R)-(−)-Lactic acid, $[M]_D$ −3° (H_2O).

the *conformational asymmetry* of the alkyl groups in question. The simplest unit of conformational asymmetry is a skewed chain of four atoms as shown (in a Newman projection) in Fig. 14-8. It is found that units I and II make a dextrorotatory contribution to $[M]_D$ which is $\Delta[M] = k \cdot A \cdot A'$, where k is a constant which happens to be the same for I and II and A and A' are functions of the polarizabilities of the attached atoms a and a′. III and IV make a corresponding levorotatory contribution. In terms of accessible parameters, it is found that $\Delta[M] = 160\sqrt{R_a R_a'}$, where R_a and R_a' are the atomic refractions of the attached atoms a and a′ as given in Sec. 14-1*a*.

Fig. 14-8. Units of conformational asymmetry.

In looking at actual molecules, one must, of course, consider full or complete conformations of the type shown in Fig. 14-9. These may be considered as made up of six units of the type shown in Fig. 14-8, and taking into account the proper direction of the skew turns one has, for the staggered conformation, $[M] = \Sigma\Delta[M] = k(A \cdot A' - A' \cdot \mathrm{H} + \mathrm{H} \cdot \mathrm{H} - \mathrm{H} \cdot \mathrm{H} + \mathrm{H} \cdot \mathrm{H} - A \cdot \mathrm{H})$‡ $= k(A \cdot A' - A' \cdot \mathrm{H} + \mathrm{H} \cdot \mathrm{H} - A \cdot \mathrm{H})$; this may be written symboli-

† This assumption was originally made in order to predict correctly the rotation of C_6H_5CHDOH and $C_6H_5CHDNH_2$.

‡ Summing the terms in clockwise direction and considering that $A \cdot A' = A' \cdot A$.

cally $[M] = k(A - \text{H})(A' - \text{H})$. The rotation for the eclipsed conformation turns out to be the same; here one has to go around the axis twice instead of once to sum up all the terms. The product $k(A - \text{H})(A' - \text{H})$ may be either evaluated empirically from a few suitable compounds of known rotation or evaluated from atomic refractivities, for $[M] = k(A \cdot A' - A' \cdot \text{H} + \text{H} \cdot \text{H} - A \cdot \text{H}) = 160R_{\text{a}}^{1/2}R_{\text{a}'}^{1/2} - 160R_{\text{a}'}^{1/2}R_{\text{H}}^{1/2} + 160R_{\text{H}}^{1/2}R_{\text{H}}^{1/2} - 160R_{\text{a}}^{1/2}R_{\text{H}}^{1/2} =$

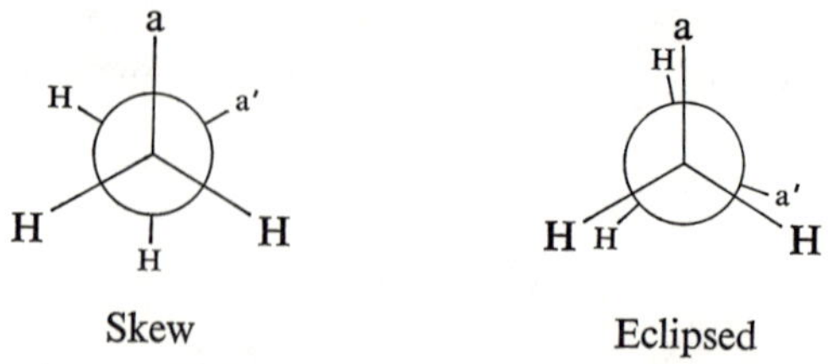

Skew Eclipsed

Fig. 14-9. Complete conformations corresponding to units of conformational asymmetry.

$160(R_{\text{a}}^{1/2} - R_{\text{H}}^{1/2})(R_{\text{a}'}^{1/2} - R_{\text{H}}^{1/2})$. It is convenient to list a series of such products where a is an alkyl carbon and a′ is some other atom or group; these may be written $k(\text{C} - \text{H})(\text{X} - \text{H})$. In Table 14-1 are listed the empirical and calculated values of this product for a number of atoms and groups X.

We may now see how these data are used to predict molecular rotation. The problem is complicated by the fact that chains of four atoms may exist in three stable conformations (cf. Chap. 6) and even more conformations

Table 14-1
Conformational Rotatory Powers

Atom or group	Empirical value	Calculated value
I	250	268
Br	180	192
SH	—	174
Cl	170	139
CN	160	87
C_6H_5	140	82
CO_2H	90	82
CH_3	60	60
NH_2	55	53
OH	50	23
H	0	0
F	—	−10

need to be considered for longer chains. In order to keep the conformational accounting simple, Brewster makes the following three assumptions:

1. Doubly skewed conformations (Fig. 14-10, *A*) contribute negligibly.
2. Conformations of the type shown in Fig. 14-10, *B*, contribute negligibly; these resemble 1,3-diaxially substituted cyclohexanes where the energetic situation is quite unfavorable (Chap. 8).
3. All other stable conformations contribute equally.

We may now apply these principles to 2-chlorobutane (Fig. 14-11). On the basis of rule 1 above, conformation C may be disregarded, and on the basis of rule 3 the other two conformational isomers, A and B, weigh equally. For A we have $[M]_A = k(\mathrm{C\cdot H - H\cdot Cl + Cl\cdot C - C\cdot H + H\cdot H - H\cdot C}) = k(\mathrm{Cl - H})(\mathrm{C - H})$. For B, $[M]_B = k(\mathrm{C\cdot H - H\cdot Cl + Cl\cdot H - H\cdot H + H\cdot C - C\cdot C}) = -k(\mathrm{C - H})(\mathrm{C - H})$. Since both weigh equally, $[M] = \frac{1}{2}([M]_A + [M]_B)$. From Table 14-1, using empirical values, $[M]_A = 170$ and $[M]_B =$

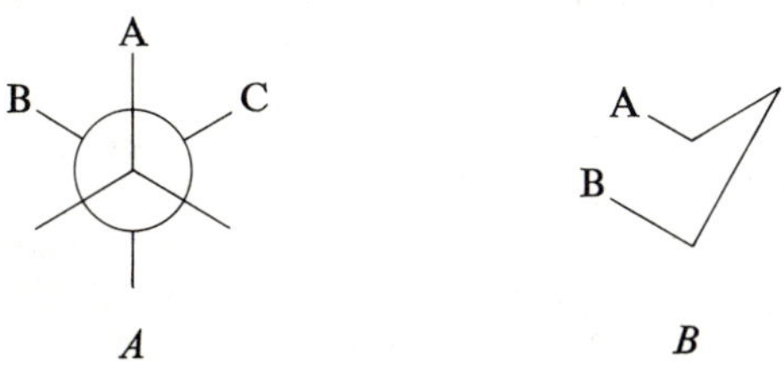

Fig. 14-10. Conformations which are neglected.

-60. Hence the predicted molecular rotation for (S)-2-chlorobutane (Fig. 14-11) is $\frac{1}{2}(170 - 60) = +55°$. The upper limit of the observed rotation[13] is 36.° Thus the prediction is correct in sign and order of magnitude.

When this treatment is applied to molecules with conformational alternatives about more than one bond—such as 3-methylhexane, where there are three bonds to be considered—the process of summing up the contributions

CH_3, H, H, H, Cl, CH_3 — *A*

CH_3, H_3C, H, H, Cl, H — *B*

CH_3, H, CH_3, H, Cl, H — *C*

$\equiv$ Cl—C—H (with CH_2CH_3 above, CH_3 below) $\equiv$ H—C—Cl (with CH_3 above, CH_2CH_3 below)

Fig. 14-11. Conformations of 2-chlorobutane.

to the molecular rotation becomes quite complex and time-consuming. For a treatment of such cases, the reader is referred to the original paper.[6a] Considerable success has been achieved in the prediction of the configuration and magnitude of rotation of a number of compounds of type $RR'CHCH_3$

[13] R. L. Letsinger, L. G. Maury, and R. L. Burwell, *J. Am. Chem. Soc.*, **73**, 2373 (1951).

and RR′CHOH, where R and R′ are flexible chains. Compounds of the type RR′CR″R‴ are predicted to have nearly zero rotation and the one case known—methylethyl-*n*-propylisobutylmethane—confirms this prediction.[14]

With the simpler compounds CH_3CHXR, where R is a flexible chain and X a rigid group (exemplified above by 2-chlorobutane), good agreement is also obtained except in cases where X is OH or NH_2. Some cases where both atomic and conformational asymmetry come into play—of the general type RCHXY, where R is a flexible chain and X and Y are rigid groups—have also been considered successfully,[6a] but other such cases have as yet resisted prediction.

c. Saturated Cyclic Compounds.[6b] The principles of conformational asymmetry also apply to saturated five- and six-membered cyclic compounds. Since only one or at most two conformations need be considered, the treatment is simpler than for acyclic compounds. Cyclohexane is considered in the chair form, cyclopentane in the average planar form or (in the D ring of steroids) in a puckered form.

In cyclohexane itself (Newman notation, Fig. 14-12), although individual carbon atoms show conformational asymmetry, the asymmetry of successive

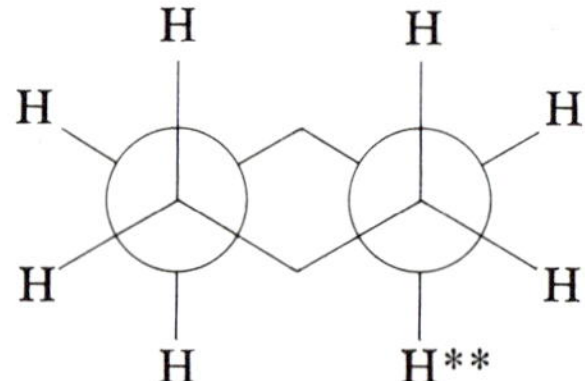

Fig. 14-12. Newman projection of cyclohexane.

ring atoms is equal and opposite and the net result around the ring is zero rotation. Introduction of an equatorial substituent (e.g., at the starred position, Fig. 14-12) adds no asymmetry, since it simply replaces a $+H \cdot H - H \cdot H$ interaction by $+H \cdot X - H \cdot X$ (X being the equatorial substituent). An axial substituent (doubly starred position) changes the rotational contribution of the two-carbon segment on the right from $+k(C - H)^2$ to $+k(C - X)(C - H)$, but since a change equal in magnitude and opposite in direction occurs at the next carbon to the rear, the net rotation is still zero. The conclusion here reached is that, in a cyclohexane ring, a substituent located in a segment $—CH_2—CHX—CH_2—$ makes no contribution to the rotation.[15] That this conclusion is approximately correct may be gleaned from the small rotation of the *cis*- and *trans*-3-methylcyclohexanols ($[M]_D = 4.6$ and $8.3°$, respectively) and the fact that the addition of a 3-substituent in cholestane, $[M]_D = 93°$, does not usually produce vast changes in molecular rotation.†

† This is true only as a rough first approximation. Complications are introduced by second-order atomic polarizability differences, e.g., between $—CH_2CH_2—$ and $—CH_2CH(CH_3)—$ in the 3-methylcyclohexanols, and possibly by asymmetry effects caused by ring distortion.

[14] A. Streitwieser and T. R. Thomson, *J. Am. Chem. Soc.*, **77**, 3921 (1955).

[15] Cf. W. M. Stokes and W. Bergmann, *J. Org. Chem.*, **17**, 1194 (1952); W. Klyne and W. M. Stokes, *J. Chem. Soc.*, 1979 (1954).

The situation is different when two substituents are next to each other, as in 2-methylcyclohexanol. In this case one has to consider the rotatory contribution along three bonds—the 1,2 bond and the bonds linking the 1 and 2 carbons to the rest of the ring (6-1 and 2-3)—and add this contribution (if any) to that of the unsubstituted ring system. For a diequatorially substituted trans isomer the situation looks as in Fig. 14-13, and the rotatory contribution of the 1,2 bond is $k(-C \cdot H + C \cdot C - C \cdot H + O \cdot H - C \cdot O + C \cdot H)$. This takes the place of $k(-C \cdot H + C \cdot C - C \cdot H + H \cdot H - C \cdot H + C \cdot H)$ before the second equatorial substituent was introduced, the difference being

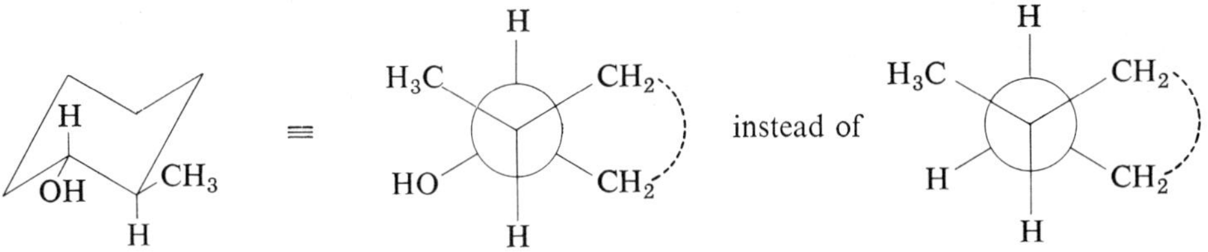

H CH₃ OH H ≡ H H₃C CH₂ HO CH₂ H instead of H H₃C CH₂ H CH₂ H

(Disubstituted) (Monosubstituted)

Fig. 14-13. *trans*-2-Methylcyclohexanol segment.

$k(O \cdot H - C \cdot O - H \cdot H + C \cdot H)$ or $-k(C - H)(O - H)$.† This then is the extra contribution of the 1,2-diequatorial substituents, since otherwise, as explained above, the introduction of the equatorial substituents in the ring makes no contribution. From the listing given earlier (Sec. 14-1*b*), the contribution $-k(C - H)(O - H)$ for equatorial methyl and hydroxyl in the configuration shown in Fig. 14-13 is $-50°$, and this is the predicted $[M]_D$ for *trans*-2-methylcyclohexanol, in good agreement with the experimental value of $-44.2°$. The *trans*-2-methylcyclohexanol segment shown also occurs in 4α-cholestanol (partial formula, Fig. 14-14), and the predicted molecular

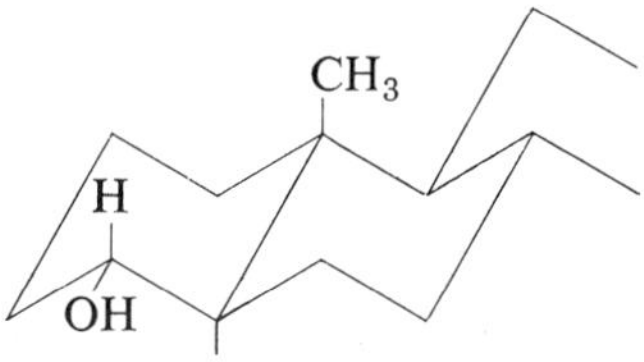

CH₃ H OH

Fig. 14-14. A/B rings in 4α-cholestanol.

rotation for this isomer would thus be $93° - 50°$, or $43°$ ($93°$ being the molecular rotation of the unsubstituted parent hydrocarbon cholestane). The observed rotation is $18°$, and the agreement is none too good. This point will be returned to below.

† The general result is $-k(A - H)(B - H)$ for the configuration shown in (*i*).

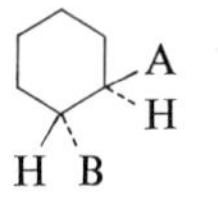

A H H B

(*i*)

The rotational contribution of a 1,2-trans-diaxially substituted ring segment is zero, since each axial substituent is effectively isolated from the other.

Finally we must consider the contribution of a 1,2-cis-disubstituted segment (*e*,*a*). In this case, two bonds contribute toward the rotation, the 1,2 bond and the bond next to the axially substituted ring carbon. The situation is depicted in Fig. 14-15 for the isomer shown of *cis*-2-methylcyclohexanol. We shall deal with this case in general terms, calling the methyl substituent A and the hydroxyl substituent B. The extra rotatory contribution of the 1,2 bond upon introducing the axial substituent is $k(\mathrm{A\cdot B - B\cdot C - A\cdot H + C\cdot H})$ and of the 2,3 bond $k(-\mathrm{B\cdot H + B\cdot C + H\cdot H - C\cdot H})$, whence the total contribution of the segment is $k(\mathrm{A\cdot B - A\cdot H - B\cdot H + H\cdot H}) = k(\mathrm{A - H})(\mathrm{B - H})$, i.e., numerically the same as that of the diequatorial segment. We may apply this to 4β-cholestanol (as in Fig. 14-14, but axial

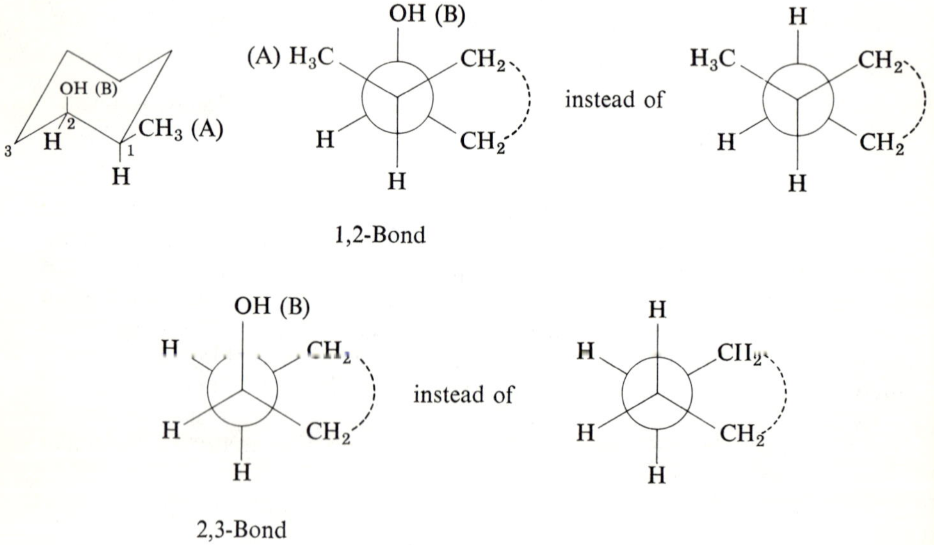

1,2-Bond

2,3-Bond

Fig. 14-15. *cis*-2-Methylcyclohexanol segment.

hydroxyl) whose rotation should be greater than that of cholestane by $k(\mathrm{C - H})(\mathrm{O - H})$ or 50°, giving a calculated $[M]_\mathrm{D}$ of 93 + 50, or 143°. The observed value is 115°. It is interesting that, while neither the value for the 4α- nor that for the 4β-hydroxycholestane is in good agreement with the calculated value, their *difference* (−97°) is in good agreement with the calculated difference (−100°). Brewster[6b] suggests that this may be due to a deformation of the cholestane ring system from the perfect chair conformations.

Returning now to *cis*-2-methylcyclohexanol, the situation here is complicated because the molecule exists in two conformations. Inspection of Fig. 14-16 indicates that the predicted molecular rotations of the two conformers are equal and opposite. This is because in the expression

$$[M] = k(\mathrm{A - H})(\mathrm{B - H})$$

the magnitude of the rotation is independent of which group is A and which B; changing the conformation of the molecule reverses the rotatory contribution of the two groups and, therefore, changes the sign of the predicted molecular rotation. The predicted rotation for the first conformation is $+50°$ and for the second $-50°$; the observed rotation of $+15°$ is in agreement with the expected predominance of the conformation in which the

Note $+$C·O term

Note $-$C·O term

Fig. 14-16. Conformers of (+)-2-methylcyclohexanol.

smaller (cf. Table 8-6) hydroxyl group is axial and the larger methyl group equatorial.

For a consideration of cyclopentane and tetrahydropyran derivatives Brewster's original paper should be referred to;[6b] an empirical treatment of sugars and cyclitols has also been presented by Whiffen.[8]

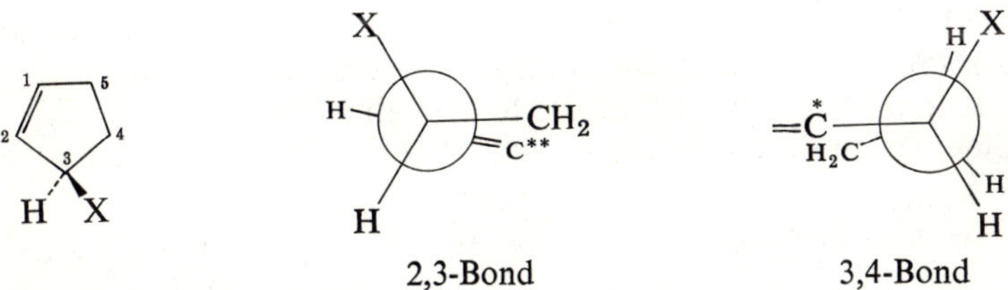

Fig. 14-17. Bonds contributing to rotation of 3-substituted cyclopentene.

d. Endocyclic Olefins.[6c] In a 3-substituted cyclopentene (Fig. 14-17) there are two bonds that contribute to rotation, 2-3 and 3-4, shown in Fig. 14-17. The contribution of the former is $k(\text{X} \cdot \text{C}^{**} - \text{C}^{**} \cdot \text{H} + \text{H} \cdot \text{H} - \text{H} \cdot \text{X}) = k(\text{C}^{**} - \text{H})(\text{X} - \text{H})$ and of the latter $k(\text{X} \cdot \text{H} - \text{H} \cdot \text{C}^{*} + \text{C}^{*} \cdot \text{H} - \text{H} \cdot \text{H} + \text{H} \cdot \text{C} - \text{C} \cdot \text{X}) = -k(\text{C} - \text{H})(\text{X} - \text{H})$. Provided that X is more polarizable than hydrogen (which is true for all substituents except fluorine), the first conformation is dextrorotatory and the second levorotatory. Since, however, $(C^{**} - H) > (C - H)$,† the dextrorotatory conformation is stronger, and the molecule in Fig. 14-17 is dextrorotatory in the configuration shown.‡ Experimentally, this has been found to be true for 3-methylcyclopentene ($X = CH_3$) and for chaulmoogric acid [$X = -(CH_2)_{12}CO_2H$], among other compounds. The rotation for 3-ethylcyclopentene, $[M]_D = 118°$, is greater than that for 3-methylcyclopentene, $[M]_D = 64°$. This may be predicted on the basis of Fig. 14-18; the contribution of the first conformation is $k(-C^* \cdot H + C \cdot H - C \cdot H + H \cdot H - H \cdot R - R \cdot C^*) = k(C^* - H)(R - H) = 140°$ and that of the second conformation $k(-C^* \cdot H + C \cdot H - C \cdot R + R \cdot H - H \cdot H + H \cdot C^*) = -k(C - H)(R - H) = -60°$. Assuming, as a fair approximation, that both conformations contribute equally, the molecular rotation should be ½(140 − 60), or 40°, higher for $R = CH_3$ (3-ethylcyclopentene) than for R = H (3-methylcyclopentene), the experimental value being 54°.

3-Substituted cyclohexenes may be considered qualitatively similar to the 3-substituted cyclopentenes, and one thus arrives at the conclusion that the configuration shown in Fig. 14-19 corresponds to the more dextrorotatory epimer (or the dextrorotatory isomer, if the asymmetric carbon shown is the

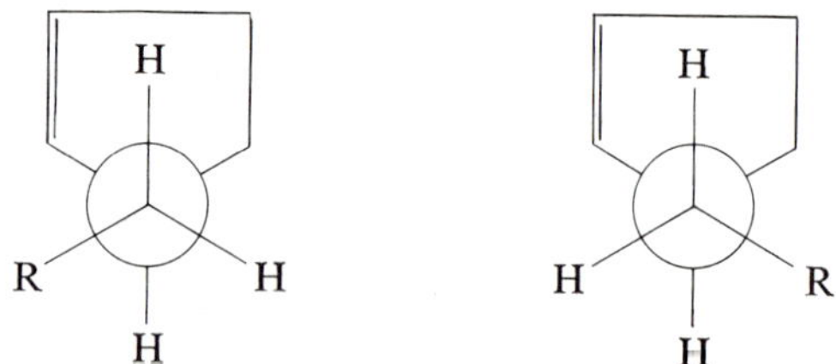

Fig. 14-18. Conformations of 3-ethylcyclopentene ($R = CH_3$).

only one) for all substituents except fluorine. This statement is a generalization of an empirical rule first proposed by Mills.[16] When X = alkyl, the rotatory contribution is ca. +130°.[6c]

4-Substituted cyclohexenes will now be considered. If the substituent is equatorial, the asymmetric carbon itself (Fig. 14-20) makes no rotatory contribution (Sec. 14-1*b*), and the rotation is due to the cyclohexene system as such. It must be realized that the half-chair of cyclohexene (Fig. 8-49) is dissymmetric, and the only reason why it shows no optical activity is that it

† This may be gleaned from the listing in Sec. 14-1*b*, where it appears that

$$k(C - H)(C_6H_5 - H) > k(C - H)(CH_3 - H)$$

and therefore $(C_6H_5 - H) > (CH_3 - H)$. The ethylenic group is assumed to have the same polarizability as phenyl and any alkyl group the same as methyl.

‡ Analysis of atomic asymmetry of the 3-substituted cyclopentene shown in Fig. 14-17 in terms of the diagram in Fig. 14-4 (left as an exercise for the reader) shows that it is dextrorotatory also when X is more polarizable than vinyl or less polarizable than methyl (but more so than hydrogen). For X = vinyl, phenyl, or methyl there is no atomic asymmetry. Only for $X = CO_2H$ is the atomic-asymmetry contribution levorotatory and contrary to the conformational contribution. However, the conformational contribution prevails in this case also.

[16] J. A. Mills, *J. Chem. Soc.*, 4976 (1952).

inverts between two conformations that are mirror images of each other. If one of the two conformations is preferred because of a substituent that wants to be equatorial, the dissymmetry of the ring comes to the fore. The rotatory contribution of this system seems to be ca. 160°, this being the molecular rotation of a variety of 1,4-disubstituted cyclohexenes, regardless of the nature of the substituent.† We see then that the configuration shown in Fig. 14-20 is dextrorotatory, or, if part of a larger asymmetric system, that it will be more dextrorotatory than its epimer.‡ This rule has also been stated empirically by Mills.[16]

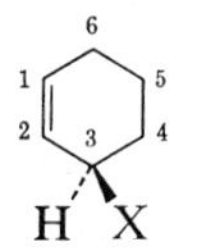

6 1 5 2 3 4
H X

Fig. 14-19. More dextrorotatory 3-alkylcyclohexene.

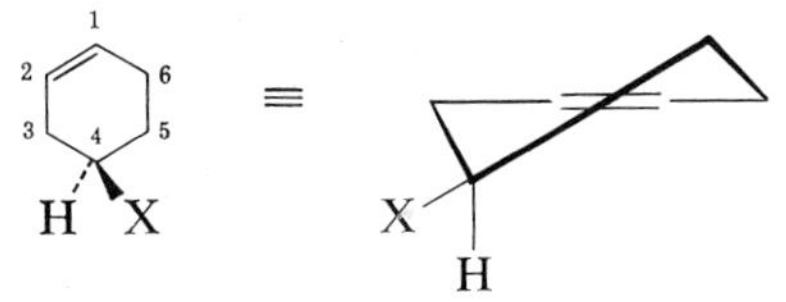

1 2 6 3 4 5
H X ≡ X H

Fig. 14-20. 4-Substituted cyclohexene, $[M]_D = 160°$.

As an example of the application of these concepts we shall consider the *trans*- and *cis*-2-menthenes (Fig. 14-21). To place the isopropyl group in the pseudoequatorial position, one must use the opposite half-chair configuration from that shown in Fig. 14-20, giving a rotatory contribution of $-160°$. In *trans*-2-menthene the isopropyl and methyl groups each are of the configuration shown in Fig. 14-19 and contribute 130° each to the rotation; in *cis*-2-

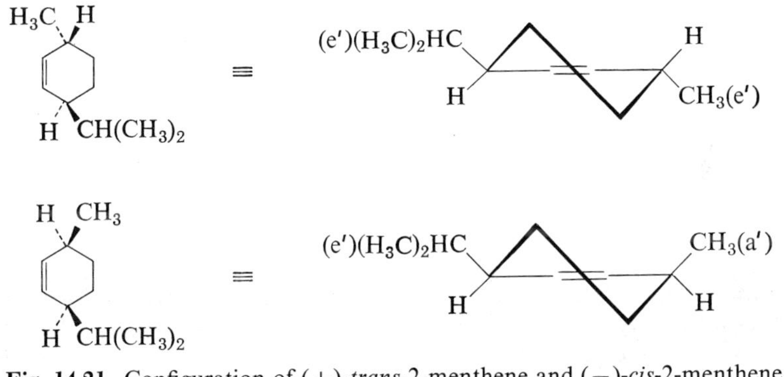

H_3C H
H $CH(CH_3)_2$
≡
(e′)$(H_3C)_2HC$ H H CH_3(e′)

H CH_3
H $CH(CH_3)_2$
≡
(e′)$(H_3C)_2HC$ CH_3(a′) H H

Fig. 14-21. Configuration of (+)-*trans*-2-menthene and (−)-*cis*-2-menthene.

menthene, their configurations are opposed and cancel each other. Finally, there is a conformational contribution of the isopropyl group which is seen, from Fig. 14-22, to be $k(-C^* \cdot H + C \cdot H - C \cdot C + C^* \cdot C) = k(C^* - C)(C - H)$. Since this value is not available, one transforms this as follows:

$$k(C^* - C)(C - H) = k(C^* - H - C + H)(C - H)$$
$$= k(C^* - H)(C - H) - k(C - H)^2 = 140 - 60 = 80°$$

† It could not have been foreseen a priori that the 1-substituent makes no contribution either; cf. Ref. 6*c*.

‡ Caution is indicated when axial substituents are present at C_4; cf. Ref. 6*c*.

The complete predicted molecular rotation for *trans*-2-menthene (configuration shown) is then $-160 + 260 + 80 = 180°$; the observed is 185°. For *cis*-2-menthene (configuration shown) one calculates $-160 + 80 = -80°$; the observed is $-64°$. Agreement here, as in a number of other cases,[6c] is very good.

e. Conclusion. At least two successful predictions of configuration have so far been made on the basis of Brewster's treatment: one in the case of a pair of synthetic quercitols (cyclohexanepentols),[16a] the other in the case of the alkaloid anagyrine.[4d, 16b] There has also been an application to norbornane derivatives[16c] which revealed several instances where the configuration found differed from that predicted. Some (but not all) the discordant cases involve molecules with phenyl substituents at the asymmetric carbon atom.

At the present time it appears that Brewster's approach (which applies only to rotation at the D line) is promising for compounds devoid of absorption bands in the near ultraviolet, such as saturated hydrocarbons, olefins, alcohols, amines, and halides. It now appears not to apply, in general, to compounds which are chromophoric in the near ultraviolet and give rise to Cotton effects (Sec. 14-2), such as ketones and phenyl derivatives.[17] Fortunately, these are

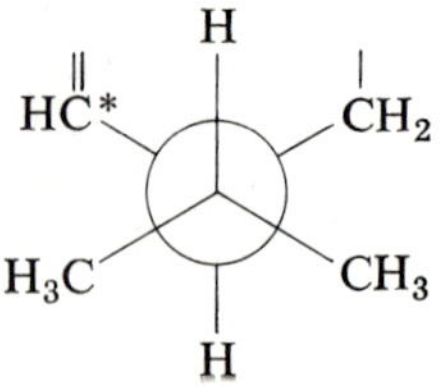

Fig. 14-22. Isopropyl conformation in the menthenes.

precisely the types of compounds which lend themselves to the experimental determination of configuration by optical rotatory dispersion, to be discussed in the next section.†

14-2. Optical Rotatory Dispersion [55-58]

a. Terminology. Just as dissymmetric media generally have different indices of refraction for right and left circularly polarized light (Sec. 14-1), their absorption coefficients for the two kinds of circularly polarized light differ also. As a result, the two vectors $\mathbf{E}_L$ and $\mathbf{E}_R$ in Fig. 14-2 are unequal in length,

† It must be emphasized that rotation at the D line may be the summation of the long-wavelength tailing of several Cotton effects, and an ultimate understanding of such rotation is probably dependent on a complete knowledge of all these effects.

---

[16] (*a*) G. E. McCasland, S. Furuta, L. F. Johnson, and J. N. Shoolery, *J. Am. Chem. Soc.*, **83**, 2335 (1961); (*b*) S. Okuda, K. Tsuda, and H. Kataoka, *Chemistry & Industry (London)*, 1115 (1961); (*c*) J. A. Berson, J. S. Walia, A. Remanick, S. Suzuki, P. Reynolds-Warnhoff, and D. Willner, *J. Am. Chem. Soc.*, **83**, 3986 (1961).

[17] J. H. Brewster, personal communication.

and their resultant, instead of increasing and decreasing in a given plane (the plane of polarization), actually sweeps out in an elongated ellipse. The plane-polarized light passing through a dissymmetric medium therefore becomes elliptically polarized, the major axis of the ellipse taking the place of the plane of polarization.† This phenomenon is known as "circular dichroism." Circular dichroism is not easy to measure in the ultraviolet—the only region where it can be important for colorless compounds—but the combination of circular dichroism and circular birefringence (Sec. 14-1), known as the "Cotton effect,"[17a] may be studied by observing the change of optical rotation with wavelength, the so-called "optical rotatory dispersion."

If one plots optical rotation vs. wavelength—and such plots may now be obtained quite readily down to about 220 mμ by means of the photoelectric spectropolarimeters mentioned in Chap. 2‡—one obtains curves of the type

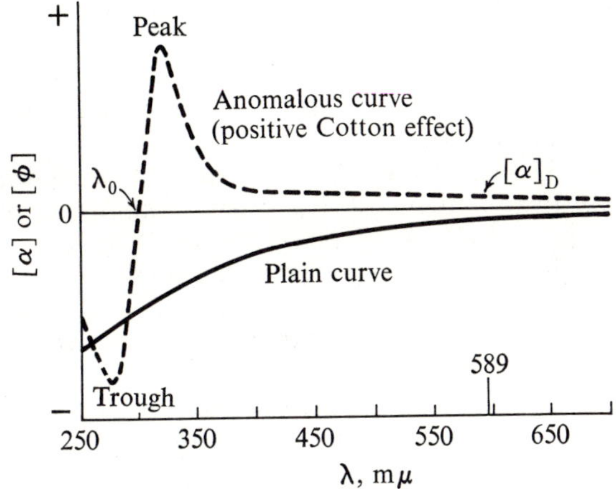

Fig. 14-23. Rotatory-dispersion curves. (*From C. Djerassi, "Optical Rotatory Dispersion," copyright,* 1960, *McGraw-Hill Book Company, Inc.*)

shown in Fig. 14-23. For compounds that absorb only in the far ultraviolet—meaning, for present purposes, well below 220 mμ—no Cotton effect is observed, precisely because the effect occurs only near an absorption maximum. There is, however, an increase in optical rotation as the wavelength decreases. The resulting plot is shown as the solid curve in Fig. 14-23; such a plot is called by Djerassi a "plain curve."¶ On the other hand, if the com-

† Except in the vicinity of a light absorption band, the ellipse is extremely long and thin, so that for experimental purposes it does not differ appreciably from a plane.

‡ Professor Djerassi has informed the author that, under unusually favorable circumstances, measurements can now be extended almost down to 210 mμ.

¶ To be distinguished from a "normal curve." A normal curve is defined as one devoid of inflections, maxima or minima, and crossings of the zero-rotation axis. A "plain curve" lacks inflections and extrema but may cross the zero axis.

[17a] A. Cotton, *Ann. chim. et phys.*, [7]8, 347 (1896); S. Mitchell, "The Cotton Effect," G. Bell & Sons, Ltd., London, 1933.

pound has an absorption maximum in or near the region of the measurement, an "anomalous curve" results (dashed line in Fig. 14-23). Such an anomalous curve has a high point called "peak" and a low point called "trough" (the terms maximum and minimum are avoided to prevent confusion with ultraviolet absorption maxima and minima). If, approaching the region of the Cotton effect from long wavelengths, one passes first through a peak and then through a trough, the Cotton effect is called "positive" but if the trough is reached first and then (at lower wavelength) the peak, one speaks of a "negative" Cotton-effect curve. The dashed curve in Fig. 14-23 shows a positive Cotton effect.

The point at which the anomalous curve crosses the zero axis of rotation (λ_0 in Fig. 14-23) corresponds in wavelength very closely to an ultraviolet absorption maximum. Contemplation of Fig. 14-23 shows that the plain curve (solid line) may well be the long-wavelength end of a negative Cotton-effect curve, inasmuch as it is well known that almost all organic compounds absorb light if one goes to short enough wavelengths (180 to 200 mμ). The nature of the curve (plain or anomalous) is therefore dependent on the wavelength region that one observes, and, as already mentioned, the present lower limit of observation is 220 mμ.

It has long been known that a plain curve of the type shown in Fig. 14-23 can often be described mathematically by an equation first proposed by P. Drude,[18] namely, $[\alpha] = K/(\lambda^2 - \lambda_0^2)$ (known as the "Drude equation"), where λ is the wavelength at which the specific rotation $[\alpha]$ is observed and K and λ_0 are empirical constants. λ_0 corresponds to a wavelength close to an ultraviolet absorption maximum. It is clear that, if the Drude equation held down to wavelength λ_0, the specific rotation would become infinite, and this is clearly not true. Cotton-effect curves thus do not obey the simple Drude equation,† and two- or more-term equations of the type

$$[\alpha] = K_1/(\lambda^2 - \lambda_0^2) + K_2/(\lambda^2 - \lambda_1^2)$$

have been proposed. Such equations appear to be of limited usefulness, and at the time of this writing an empirical approach to Cotton-effect curves—in which one measures, plots, and compares the curves for a number of related compounds—seems to prove more fruitful than attempts to fit such curves by mathematical equations and relate the parameters in these equations to structural factors (see, however, page 399).

b. Plain Curves. Although, as will be seen in the following sections, by far the most rewarding results from rotatory-dispersion measurements are obtained when a Cotton effect is observed; even plain curves can yield helpful information. In general, specific rotations near the lower end of the accessible region ($[\alpha]_{250}$) are much greater than corresponding rotations for the same compound at the sodium D line ($[\alpha]_{589}$), factors of 10 between the

† Curves that obey the one-term Drude equation have been called "simple"; others, "complex." These terms will not be used here.

[18] P. Drude, "Lehrbuch der Optik," 2d ed., S. Hirzel Verlag, Leipzig, 1906. See also W. Heller, *J. Phys. Chem.*, **62,** 1569 (1958).

two rotations being quite common.† As a result, measurements of rotation can be effected with less substance at the lower wavelengths, small changes in concentration can be followed more readily,‡ and differences in specific rotation between different substances may be enhanced. (For example, the rotations of 4-cholestene and 5-cholestene which differ by less than 10% at 589 mμ differ by nearly a factor of 2 at 300 mμ.)

Another application of rotatory dispersion which is useful even where plain curves are encountered is in deciding whether or not a given compound is optically active. Thus neither the alkaloid pilocereine[19a] (Fig. 14-24) nor the degradation product of vitamin B_{12} shown in Fig. 14-25[19b] shows appreciable optical activity at the sodium D line. Pilocereine remains inactive at all accessible wavelengths and is, therefore, certain to be a racemate. The compound shown in Fig. 14-25, on the other hand, has $[M]_{365}$ $-165°$ and is clearly not racemic.

Plain curves may be of importance in making configurational assignments in cases where the rotatory-dispersion curve crosses the zero axis of rotation. An illustration[20] is provided by the *o*-, *m*-, and *p*-iodophenyl ethers of lactic acid, $IC_6H_4OCH(CH_3)CO_2H$. It might be thought that, regardless of the

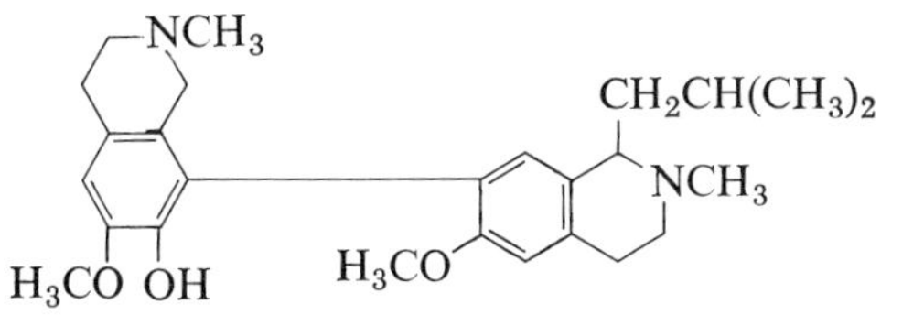

NCH_3 $CH_2CH(CH_3)_2$ NCH_3 H_3CO OH H_3CO

Fig. 14-24. Pilocereine.

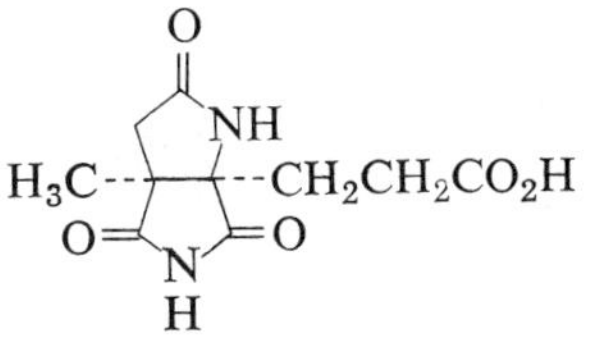

O NH H_3C $CH_2CH_2CO_2H$ O N O H

Fig. 14-25. Degradation product of vitamin B_{12}.

position of the iodine in the ring, the derivatives of like rotation would have the same configuration, and, in fact, identical configuration was demonstrated for the dextrorotatory meta and para isomers by the quasi-racemate method (Sec. 5-4*c*). However, the dispersion curves shown in Fig. 14-26 clearly show that the dextrorotatory *m* and *p* isomers and the *levo*rotatory (at the D line) *o* isomer have the same configuration. The inversion of sign at the D line is caused by the fact that the dispersion curve of the ortho isomer (but not those of the meta and para isomers) intersects the zero axis of rotation. Below about 310 mμ all three isomers of the same configuration have the same

† This is true a fortiori in the case of anomalous dispersion curves. Because of the steep rise of the dispersion curve near a peak (and the steep drop near a trough), rotations in the wavelength near these "extrema" may be 100 to 1000 times greater than rotations at the D line.

‡ This may be of value when optical rotation is used as an analytical tool (e.g., in ascertaining the composition of a binary mixture) or when it is used to follow reaction kinetics. See, for example, N. L. Allinger, R. B. Hermann, and C. Djerassi, *J. Org. Chem.*, **25**, 922 (1960).

[19] (*a*) C. Djerassi, S. K. Figdor, J. M. Bobbitt, and F. X. Markley, *J. Am. Chem. Soc.*, **79**, 2203 (1957); (*b*) V. M. Clark, A. W. Johnson, I. O. Sutherland, and A. R. Todd, *J. Chem. Soc.*, 2383 (1958).

[20] B. Sjoberg, *Arkiv Kemi*, **15**, 451 (1960).

sign of rotation. This example indicates that caution is necessary when one tries to draw conclusions as to molecular configuration from rotational data at the D line only (e.g., Secs. 5-4*d* and 14-1).

Plain curves have also played a part in the determination of conformation of polypeptides.[21] In solvents which discourage the formation of intramolecular hydrogen bonds (e.g., in dichloroacetic acid, which hydrogen-bonds intermolecularly to the polypeptide), these peptides have a random conformation of the polymer chain and display plain dispersion curves which obey the Drude equation. On the other hand, in solvents, such as chloroform which permit intramolecular hydrogen bonding within the polypeptide, the molecules coil up in helical form (cf. Fig. 15-10), and a more complex type

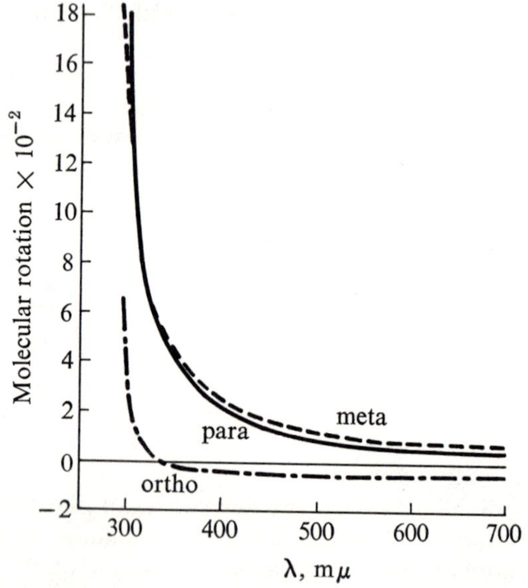

Fig. 14-26. Rotatory-dispersion curves of the α-(*p*-iodophenoxy)-propionic acids. (*From C. Djerassi, "Optical Rotatory Dispersion," copyright,* 1960, *McGraw-Hill Book Company, Inc.*)

of rotatory dispersion results. Rotatory dispersion thus throws light on the extent of helix formation in a polypeptide.

c. Rotatory Dispersion of Ketones. By far the most extensive measurements of optical rotatory dispersion, especially in the laboratories of C. Djerassi, have been made on saturated† ketones. The prinicipal reason for this has been one of experimental simplicity: The carbonyl chromophore absorbs in a readily accessible region of the ultraviolet (280 to 290 mμ) and

† By "saturated" here is meant that there is absence of unsaturation in the vicinity of the carbonyl function.

[21] E. R. Blout, Ref. 55, chap. 17. See also, for example, R. H. Karlson, K. S. Norland, G. D. Fasman, and E. R. Blout, *J. Am. Chem. Soc.*, **82**, 2268 (1960). See also Ref. 50.

the absorption is not strong enough to interfere with measurement of the optical rotation right through the region of the absorption band. As a result, it is usually possible to obtain complete Cotton-effect curves for saturated ketones. An example is shown in Fig. 14-27 which depicts the dispersion curve of 3-cholestanone.[22] The peak (*P*) and trough (*T*) are clearly visible, and one may define the amplitude (*a*) and breadth (*b*) of such a curve. The point where the curve crosses the zero-rotation axis (at ca. 285 mμ) coincides closely with the ultraviolet absorption maximum. In an abbreviated description of such a curve one indicates the concentration (*c*, 0.10 g./100 ml.), temperature (29 to 31°),† and a few salient points on the curve—$[\alpha]_{700} + 37°$, $[\alpha]_{589} + 55°$, $[\alpha]_{307} + 959°$ (peak),‡ $[\alpha]_{267} - 740°$ (trough),‡ $[\alpha]_{245} - 362°$. The

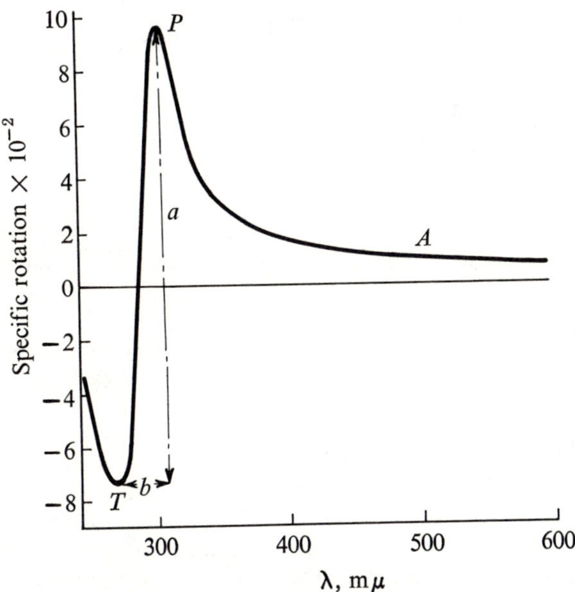

Fig. 14-27. Rotatory-dispersion curve of 3-cholestanone. (*From C. Djerassi, "Optical Rotatory Dispersion," copyright,* 1960, *McGraw-Hill Book Company, Inc.*)

dispersion curve of 3-cholestanone is a "single, positive Cotton-effect curve" because there is a single UV absorption band (due to the carbonyl function) which causes the Cotton effect. When there are two or more absorption bands, multiple Cotton-effect curves result which are more complex (e.g., Fig. 14-37).

† Accurate control of temperature is often not necessary in rotatory-dispersion measurements, because the shape of the curves (as distinct from the exact value of the rotation) is usually quite insensitive to temperature.

‡ The words "peak" and "trough" are usually not included in the description of the curve but are implied in the data.

[22] C. Djerassi, W. Closson, and A. E. Lippman, *J. Am. Chem. Soc.*, **78**, 3163 (1956).

Another helpful factor in the study of rotatory dispersion of saturated ketones has been the ready availability of a wealth of suitable optically active substrates among the ketosteroids as well as terpenoid ketones.

i. Structure. One of the applications of rotatory-dispersion studies is as a tool in structure elucidation, especially in the location of functional groups.

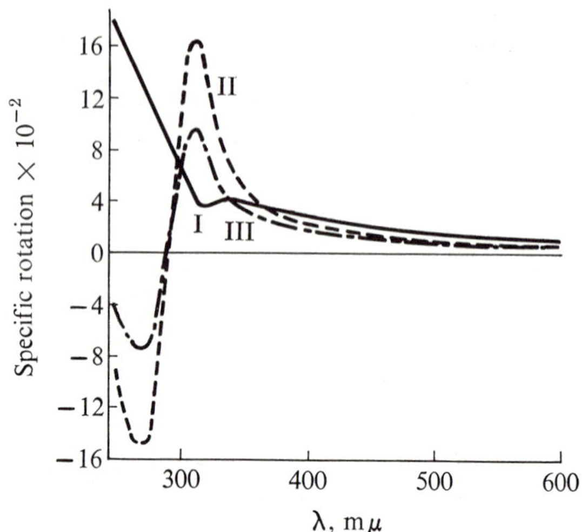

Fig. 14-28. Rotatory dispersion of cholestan-1-one (I), -2-one (II), and -3-one (III). (*From C. Djerassi, "Optical Rotatory Dispersion," copyright,* 1960, *McGraw-Hill Book Company, Inc.*)

Figure 14-28 illustrates both the applicability and limitations of the method. The dispersion curve of the 1-ketosteroid is clearly quite distinct from that of the 2 and 3 isomers, but the curves for the latter are so similar as not to be useful as diagnostic tools.† The rotatory dispersion of the ring A ketones is relatively insensitive to changes in the C and D rings, and that of a ring D

Fig. 14-29. Androstan-17-one.

ketone is insensitive to changes in the A ring. Thus the substitution of hydroxy (OH) or mercapto (SH) groups at 3α or 3β in androstan-17-one (Fig. 14-29) shifts the peak and trough of the dispersion curve by less than 4 mμ and the amplitude by less than 900° (in a total of about 14,000°).

† See Sec. 14-2*g* for a means to distinguish these ketones by dispersion measurements.

An example of the application of these principles is in the location of one of the hydroxyl groups in rubijervine (Fig. 14-30, A).[23] Oxidation gave a ketone, rubijervone-12 (Fig. 14-30, B), whose dispersion curve showed it to be a 12-ketosteroid; hence one of the hydroxyl groups in rubijervine is at C_{12}.

It is of interest that, whereas the introduction of a single methyl group at C_2 or C_4 in cholestanone or the introduction of two methyl groups at C_2 does not change the sign of the Cotton-effect curve in 3-cholestanone, introduction of a *gem*-dimethyl group at C_4, as in 4,4-dimethylcholestanone (Fig. 14-31, I) or the triterpene 3-lanostanone (Fig. 14-31, II), leads to an inversion

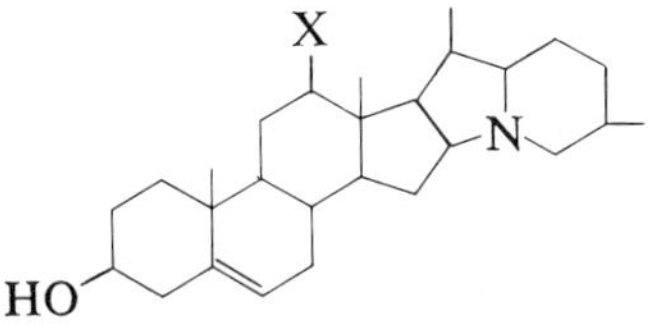

X

N

HO

A, X = —OH

B, X = =O

Fig. 14-30. Rubijervine (A) and rubijervone-12 (B).

of the Cotton effect,[24] as shown in Fig. 14-32. A similar change is observed in going from 10-methyl-2-decalone to 1,1,10-trimethyl-2-decalone.

An entirely different type of structural change which reflects itself in the nature of the dispersion curve is change of ring size in a cycloalkanone.[25] (R)-(+)-3-methylcyclopentanone and (R)-(+)-3-methylcyclohexanone show positive Cotton-effect curves, but (R)-(−)-3-methylcycloheptanone has a negative curve, and the negative Cotton effect persists in (R)-3-methylcyclononanone and (R)-3-methylcyclopentadecanone. The inversion of the curve in going from the six- to the seven-membered ring is in accordance with the octant rule (Sec. 14-2*e*), provided that the seven-membered ketone is not in

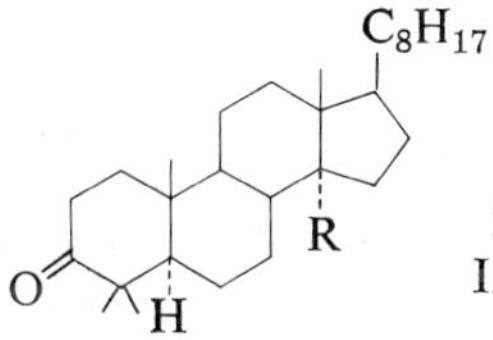

C_8H_{17}

R

O

H

I, R = H

II, R = CH_3

Fig. 14-31. 4,4-Dimethyl-3-cholestanone (I) and 3-lanostanone (II).

the chair conformation. This agrees with calculations for cycloheptanone of the type discussed in Sec. 9-2.[26]

It is worth noting that five-membered-ring ketones show a Cotton effect of the same sign as six-membered ones of like configuration. The amplitude of the curve is, however, about five times as great for the five-membered ketones as for the six-membered. This is true of the 8-methylhydrindan-1-ones (and

[23] S. W. Pelletier and D. M. Locke, *J. Am. Chem. Soc.,* **79,** 4531 (1957).

[24] C. Djerassi, O. Halpern, V. Halpern, and B. Riniker, *J. Am. Chem. Soc.,* **80,** 4001 (1958).

[25] C. Djerassi and G. W. Krakower, *J. Am. Chem. Soc.,* **81,** 237 (1959).

[26] N. L. Allinger, *J. Am. Chem. Soc.,* **81,** 5727 (1959).

-2-ones) in comparison with the corresponding methyldecalones (Fig. 14-33) as well as of the monocyclic ketones mentioned earlier; it is also true of 17-ketosteroids (e.g., Fig. 14-29) compared with 3-ketosteroids.

ii. Configuration. Just as structure affects the shape of the dispersion curve, so does configuration in the vicinity of the chromophore. An example is

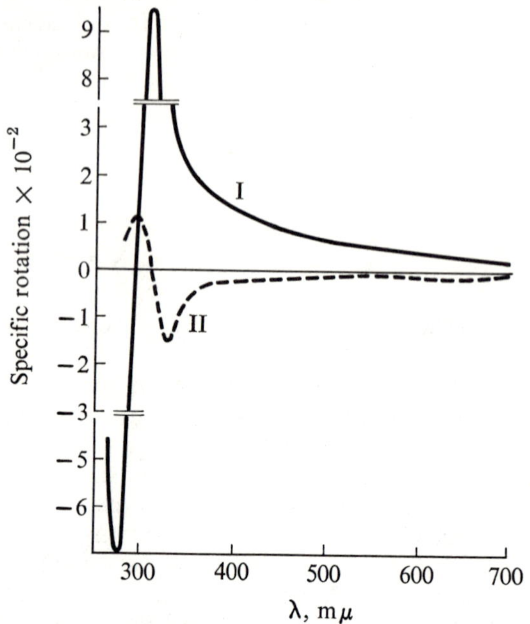

Fig. 14-32. Rotatory-dispersion curves of cholestan-3-one (I) and 4,4-dimethylcholestan-3-one (II). (The double line indicates a discontinuity in scale.) (*From C. Djerassi, "Optical Rotatory Dispersion," copyright,* 1960, *McGraw-Hill Book Company, Inc.*)

provided by androstan-17β-ol-3-one (A/B trans) and the corresponding 5β isomer (A/B cis) whose curves are shown in Fig. 14-34. Clearly, where the position of the ketone chromophore is known, the dispersion curve can be used to decide on configuration in the vicinity of the keto group. On the other hand,

H_3C O H_3C O

H   H

Fig. 14-33. *trans*-8-Methylhydrindan-1-one and *trans*-9-methyl-1-decalone.

there is little effect of changes of configuration distant from the chromophore; thus the curves for pregnan-3α-ol-20-one acetate (Fig. 14-35, *A*) and 5α-pregnan-3β-ol-20-one acetate (Fig. 14-35, *B*) are quite similar [whereas the curve for 17α-pregnan-3α-ol-20-one acetate (Fig. 14-35, *C*) is entirely different]. An application of the principle is shown in the elucidation of the stereochemistry of the hydrogenation of 19-nortestosterone (Fig. 14-36) with chemical reduc-

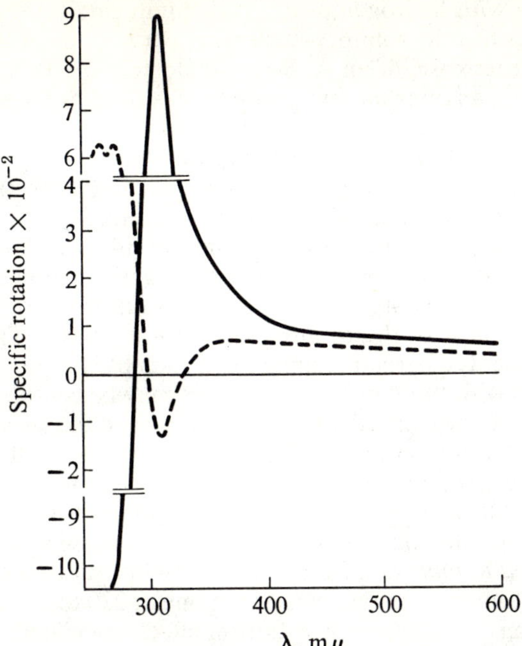

Fig. 14-34. Rotatory-dispersion curves of the androstan-17β-ol-3-ones (solid, A/B trans; dashed, A/B cis. The double line indicates a discontinuity in scale). (*From C. Djerassi, "Optical Rotatory Dispersion," copyright,* 1960, *McGraw-Hill Book Company, Inc.*)

A *B* *C*

Fig. 14-35. Pregnan-3α-ol-20-one acetate (*A*), 5α-pregnan-3β-ol-20-one acetate (*B*), and 17α-pregnan-3α-ol-20-one acetate (*C*).

Fig. 14-36. 19-Nortestosterone.

ing agents[27] and with hydrogen over a ruthenium catalyst.[28] In the former case, the product has the rotatory-dispersion curve of an A/B-trans 3-ketosteroid; in the latter case, of an A/B-cis 3-ketosteroid. It is of interest that the absence of the angular methyl group at C_{10} does not seem to affect the rotatory-dispersion curve of the 3-ketosteroids.

iii. Conformation. In a mobile cyclohexane system (cf. Sec. 8-6), e.g., in a monocyclic cyclohexane derivative or in a cis-decalin derivative (Fig. 10-21) optical rotatory dispersion depends on the preferred conformation of the molecule. This point will be elaborated in Sec. 14-2*d* (α-haloketones) and Sec. 14-2*e* (the octant rule) below. For the moment, it should be noted that conformation, as well as structure and configuration, is of importance in determining the shape of the rotatory-dispersion curve. Thus the above-described assignments of structure or configuration by comparison of rotatory-dispersion curves with those of known analogs may be considered reasonably safe only if the unknown and the known analog have the same conformation in the vicinity of the chromophore. For example, the original configurational assignment of the sesquiterpenoid eremophilone[29] had to be reversed[30] when it was found that the analogy made to the 4-ketosteroids was invalid because of conformational differences between the compounds originally compared.

iv. Unsaturated Ketones and Diketones. α,β-Unsaturated ketones present two ultraviolet absorption maxima, a strong one at 220 to 260 mμ and a weak one above 300 mμ. Since there is a Cotton effect associated with each such band and the effects are additive, a rather complex rotatory-dispersion curve is apt to result. A curve of this type is called a "multiple Cotton-effect curve," exemplified in Fig. 14-37 which shows the rotatory dispersion of testosterone. The slight negative Cotton effect is associated with the long-wavelength band. Recent extension[30a] of the measurements to shorter wavelengths has put in evidence the much stronger Cotton effect due to the short-wavelength band. In testosterone, this is positive; its onset may be seen in Fig. 14-37.

Although rotatory-dispersion curves of α,β-unsaturated ketones have been studied quite extensively, space limitations prevent further consideration here. For a discussion of such curves as well as a consideration of compounds containing more than one ketone function, the reader is referred to the already cited book by Djerassi.[55]

d. The Axial Haloketone Rule. It was mentioned in Sec. 8-7 that introduction of an axial halogen next to a carbonyl group in a cyclohexanone affects the infrared absorption frequency of the carbonyl group and that introduction of either an equatorial or an axial α-halogen leads to a shift in the ultraviolet maximum of the ketone chromophore (Table 8-8). An equatorial bromine, for example, produces a hypsochromic shift (i.e., to shorter wavelength) of about 5 mμ, and an axial bromine produces a bathochromic shift

[27] A. Bowers, H. J. Ringold, and E. Denot, *J. Am. Chem. Soc.,* **80,** 6115 (1958).

[28] R. T. Rapala and E. Farkas, *J. Am. Chem. Soc.,* **80,** 1008 (1958).

[29] C. Djerassi, R. Riniker, and B. Riniker, *J. Am. Chem. Soc.,* **78,** 6362 (1956).

[30] C. Djerassi, R. Mauli, and L. H. Zalkow, *J. Am. Chem. Soc.,* **81,** 3424 (1959); L. H. Zalkow, F. X. Markley, and C. Djerassi, *ibid.,* **81,** 2914 (1959).

[30a] C. Djerassi, R. Records, E. Bunnenberg, K. Mislow, and A. Moscowitz, *J. Am. Chem. Soc.* **84,** 870 (1962).

(i.e., to longer wavelength) of 28 mμ. It is to be expected that corresponding shifts should be observed in the rotatory-dispersion peaks (or troughs) of α-halocyclohexanones as compared with those of their halogen-free precursors, and this is indeed found.[31] Similar shifts occur with α-hydroxy- and α-acetoxycyclohexanones.

In addition to measuring the shift in the extrema (peaks or troughs) produced by the above-mentioned α substituents, one may concern oneself with changes in the shape of the dispersion curve. Equatorial substituents have little effect on the curve whereas an axial acetoxy group increases the amplitude at least twofold. Similar increases occur with axial halogen. Of greater importance is the fact, however, that axial α-halogens may invert the sign of the Cotton-effect curve. Empirically it has been found in a number of cases studied that the inversion of the Cotton-effect curve (or lack thereof) upon

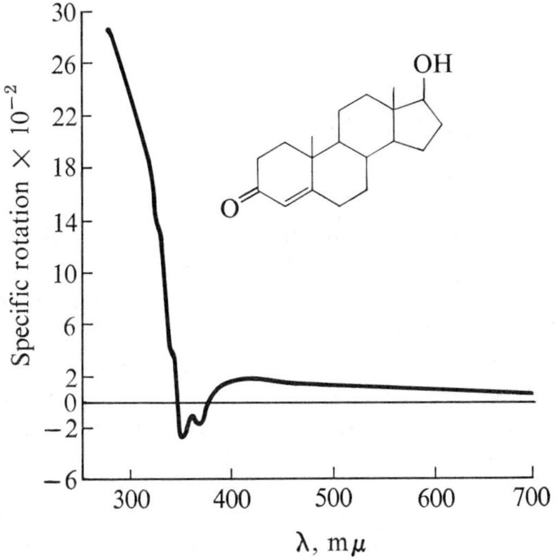

Fig. 14-37. Rotatory dispersion of testosterone. *(Adapted from C. Djerassi, "Optical Rotatory Dispersion," copyright, 1960, McGraw-Hill Book Company, Inc.)*

introduction of an axial α-halogen depends on the position, configuration, and conformation of the haloketone. If one knows two of these factors and observes the rotatory-dispersion curve, the third factor can be deduced. The following paragraphs will illustrate this point.

The "axial haloketone rule" states the matter as follows:[32] The effect of introducing an axial chlorine, bromine, or iodine atom next to the keto group of a cyclohexanone may affect the sign of the Cotton effect of the parent ketone. The Cotton effect of the α-halocyclohexanone may be predicted by viewing along the O═C axis in a model so placed that the carbonyl group occupies the head of the chair (or boat) closest to the observer. If the halo-

[31] C. Djerassi, J. Osiecki, R. Riniker, and B. Riniker, *J. Am. Chem. Soc.,* **80,** 1216 (1958).
[32] C. Djerassi and W. Klyne, *J. Am. Chem. Soc.,* **79,** 1506 (1957); also Ref. 55, p. 122.

gen is now on the left of the line of view† (Fig. 14-38*a*), the compound will exhibit a negative Cotton effect but if it is on the right (Fig. 14-38*b*), a positive Cotton effect will be observed.‡

The rule may be used in four ways.

1. If there is doubt whether an α-halogen is equatorial or axial, reversal of the Cotton effect upon halogenation proves axial location of the halogen. The converse is obviously not true: The halogen could be axial without reversing the Cotton effect, but even in this case it should be readily distinguishable from equatorial halogen through the bathochromic shift of the extrema and the increased amplitude.[33]

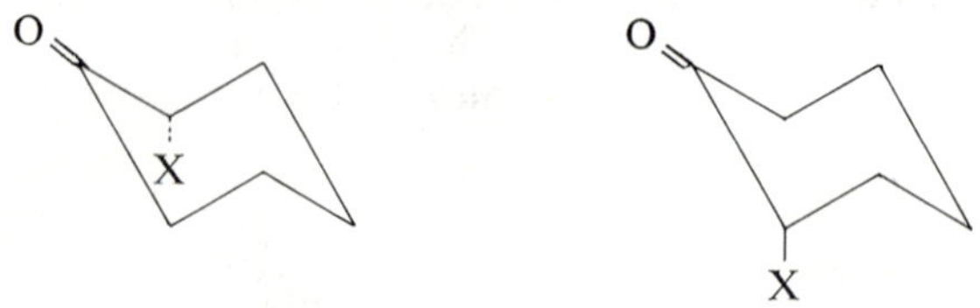

(*a*) Negative Cotton effect (*b*) Positive Cotton effect

Fig. 14-38. Axial haloketone rule.

2. If configuration and conformation of the parent ketone are known and the α-halogen is axial (demonstrated as explained above, or by infrared or ultraviolet measurements), its location may be deduced from the sign of the Cotton effect of the haloketone. For example, 2,3-seco-cholestan-6-one-2,3-dioic acid (Fig. 14-39) upon bromination gives an axial α-bromo derivative which, a priori, could be either the 5- or the 7-bromide. Since the Cotton effect for this bromide is negative, the haloketone rule implies that it is the 5 isomer.

C_8H_{17}
CH_3
HO_2CH_2C
HO_2CH_2C
O
7
H

Fig. 14-39. 2,3-seco-Cholestan-6-one-2,3-dioic acid.

3. If the conformation of the ring system is fixed and the location of the axial α-halogen is known, the absolute configuration of the α-haloketone (and

† Actually a plane passing through the carbonyl group and the No. 4 carbon and thus bisecting the chair or boat.

‡ α-Fluoroketones do not follow the axial haloketone rule. This is consistent with the fact that fluorine, unique among the halogens, has a lower atomic refraction than hydrogen (page 401) and with the expected negative conformational rotatory power of fluorine (Table 14-1).

[33] C. Djerassi and J. Staunton, *J. Am. Chem. Soc.*, **83**, 736 (1961).

thus of the parent ketone) may be deduced. An example[33] is provided by (−)-*trans*-1-decalone (Fig. 14-40). This ketone may, a priori, have either configuration A or B in Fig. 14-40. Now bromination of *trans*-1-decalone gives, among other products, the axial 2-bromoketone. (That the product is not the equatorial 2-bromoketone may be deduced from infrared as well as rotatory-dispersion data as discussed above; that it is not a 9-bromoketone follows from the fact that its reduction returns pure *trans*-1-decalone whereas the 9-bromoketones—which are also formed in the bromination—upon reduction give the expected mixtures of *cis*- and *trans*-1-decalone.[34]) When (−)-*trans*-1-decalone is brominated, the product is (+)-2β-bromo-*trans*-1-decalone (C or D, Fig. 14-40) which has a strong positive Cotton effect. According to the haloketone rule (cf. Fig. 14-38), this means that the dextrorotatory bromoketone has configuration C rather than configuration D (Fig. 14-40) and therefore its precursor, (−)-*trans*-1-decalone, must have configuration A rather than configuration B. In this particular instance, the conclusion as to the configuration of the parent ketone A could also have been reached by a direct comparison of its dispersion curve with that of 10-methyldecalone of

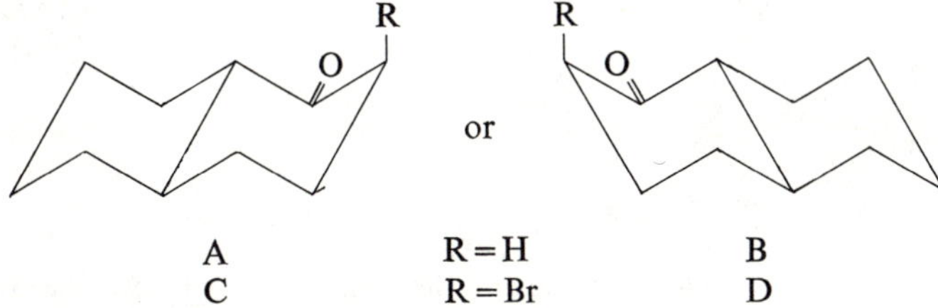

Fig. 14-40. Enantiomeric configurations of *trans*-1-decalone and its axial 2-bromo derivative.

known configuration[33] (cf. Sec. 14-2*f*), but the use of the haloketone rule provides independent corroboration of the configurational assignment, in addition to illustrating a rather general method.

4. If the α-haloketone is axial and the position of the halogen and the configuration of the molecule are known, the conformation of the ring may be deduced. For example, chlorination of (+)-3-methylcyclohexanone (absolute configuration as shown in Fig. 14-41) gives a mixture from which a pure crystalline isomer was isolated and shown, by chemical means, to be 2-chloro-5-methylcyclohexanone.[35] Two configurational isomers are possible, and each may exist in one or the other of two conformations, as shown in Fig. 14-41. The spectral properties of the crystalline chloroketone indicate that in octane solution it has axial halogen and the Cotton effect in octane is negative; this is compatible with configuration A but not with B. Hence the chloroketone is the trans isomer. When the rotatory-dispersion curve is measured in methanol, the Cotton effect is found to have become positive. This means that the trans isomer has changed from the diaxial conformation

[34] H. E. Zimmerman and A. Mais, *J. Am. Chem. Soc.*, **81**, 3644 (1959).

[35] C. Djerassi, L. E. Geller, and E. J. Eisenbraun, *J. Org. Chem.*, **25**, 1 (1960). See also Ref. 33.

A to the diequatorial conformation A′, presumably because in methanol the dipole repulsion between the carbonyl group and the adjacent equatorial halogen is not as serious as in octane with its much lower dielectric constant (cf. Sec. 8-7). Such solvent effects have also been found in 2-bromocyclohexanone.[36]

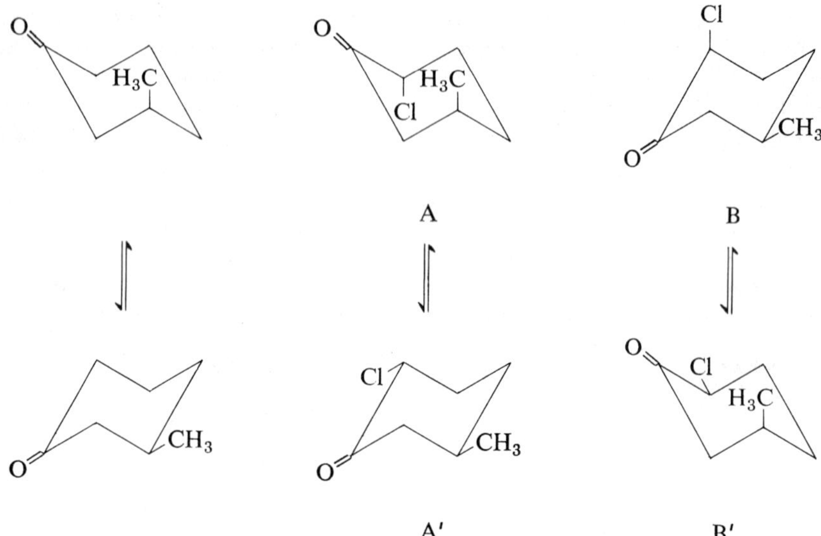

Fig. 14-41. 3-Methylcyclohexanone and the 2-chloro-5-methylcyclohexanones.

It was mentioned above that configurational assignments based on the haloketone rule may be vitiated if the ketone ring exists in the boat form. Such boat forms are not as unlikely as it may seem, inasmuch as the difference between the chair and boat forms of cyclohexanone has been estimated[26] to be only half as much (2.7 vs. 5.6 kcal./mole) as the difference between the chair and boat forms of cyclohexane, and other unfavorable conformational

Fig. 14-42. 2α-Methylcholestan-3-one.

interactions in a substituted cyclohexanone may well force such a molecule to assume the boat conformation (cf. the footnote on page 291). An example appears to have been found in the bromination of 2α-methylcholestan-3-one (Fig. 14-42) at the 2 position to give a 2-bromo-2-methylcholestan-3-one.

[36] J. Allinger and N. L. Allinger, *Tetrahedron,* **2,** 64 (1958).

Infrared and ultraviolet spectral measurements[37] indicate that the bromine in this compound is axial, and thus it appeared at first that the bromination product was 2β-bromo-2α-methylcholestan-3-one whose formation would have involved the usual axial attack of bromine in a kinetically controlled step (Sec. 8-7). However, when the rotatory-dispersion curve of the bromoketone was recorded, it was found[38] that instead of the expected strong positive Cotton effect it shows a negative one. Two suggestions are possible, namely, either that the haloketone rule does not hold in this case or else that the bromination product is not the 2β-bromo-2α-methyl isomer. The former suggestion must be rejected, for 2β-bromo-2α-methyl-3-ketosteroids have been prepared[38] and they do, indeed, show the expected strong positive Cotton effect. It, therefore, appears that the above-discussed bromination product is 2α-bromo-2β-methylcholestan-3-one and that its anomalous Cotton effect is due to the existence of the A ring in the boat form (Fig. 14-43).† It is clear that if the 2α-bromo-2β-methyl compound existed as a chair (Fig. 14-43), it would have two very unfavorable interactions: an axial 1,3-dimethyl interaction and an equatorial bromoketone interaction. The reason why bromination gives the 2α-bromo, instead of the 2β-bromo, isomer seems to be that it is difficult for the bromine to come in axially over the methyl

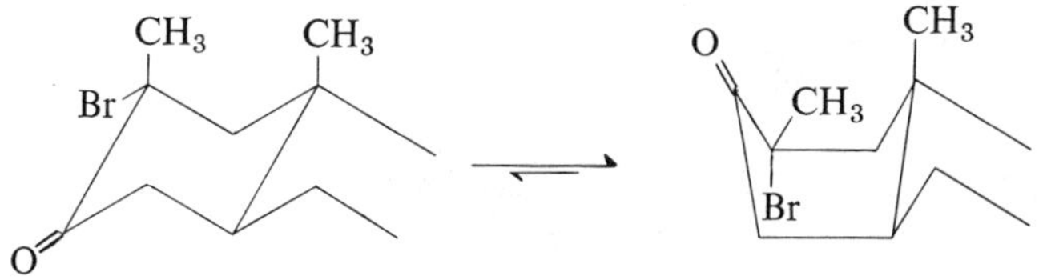

CH3 CH3 Br O ⇌ O CH3 CH3 Br

Fig. 14-43. 2α-Bromo-2β-methylcholestan-3-one (partial formula).

group at C_{10}, for in a 2-methyl-19-norsteroid (methyl group at C_{10} absent) bromination of the 3-ketone gives the expected axial 2-bromo compound.[39] Since axial attack on the chair form of 2α-methylcholestan-3-one (Fig. 14-42) to form the 2β bromo derivative is sterically so unfavorable, the small number of molecules of the starting material in the boat form may, instead, react by axial attack of bromine to give the 2α-bromo-2β-methyl isomer, as observed. The alternative would be equatorial attack on the chair form of the starting material.‡

e. The Octant Rule.[40] The octant rule is an empirical rule which permits

† In the chair form the bromine is equatorial, and the Cotton effect should be unchanged from that of the parent ketone; this is not the case.

‡ See Ref. 39 for a detailed discussion of these alternatives.

[37] Y. Mazur and F. Sondheimer, *J. Am. Chem. Soc.*, **80**, 5220 (1958).

[38] C. Djerassi, N. Finch, and R. Mauli, *J. Am. Chem. Soc.*, **81**, 4997 (1959).

[39] R. Villotti, H. J. Ringold, and C. Djerassi, *J. Am. Chem. Soc.*, **82**, 5693 (1960); C. Djerassi, N. Finch, R. C. Cookson, and C. W. Bird, *ibid.*, **82**, 5488 (1960).

[40] (*a*) W. Moffitt, R. B. Woodward, A. Moscowitz, W. Klyne, and C. Djerassi, *J. Am. Chem. Soc.*, **83**, 4013 (1961); (*b*) see also K. Mislow, M. A. W. Glass, A. Moscowitz, and C. Djerassi, *ibid.*, **83**, 2771 (1961).

one to deduce the sign of the Cotton effect for a considerable number of compounds from their structure, configuration, and conformation.† The compound to be considered is oriented in a three-dimensional coordinate system as shown in Fig. 14-44. Instead of speaking about the coordinate axes, we shall consider the three orthogonal planes whose intersection defines the coordinate axes and we shall orient the molecule with respect to these planes as follows: One plane (A) is put through the carbonyl group and the No. 4 carbon. This plane is vertical in the diagram. A second, horizontal plane (B) passes through the No. 1 carbon atom, and the cyclohexanone is so tilted that it also passes through No. 2 and No. 6 (which for purposes of the octant rule are designated R2 and L2). The third plane (C) passes approximately midway through the carbon-oxygen bond and at right angles to it. The mid-point of the C═O bond is the origin of the coordinate system. The

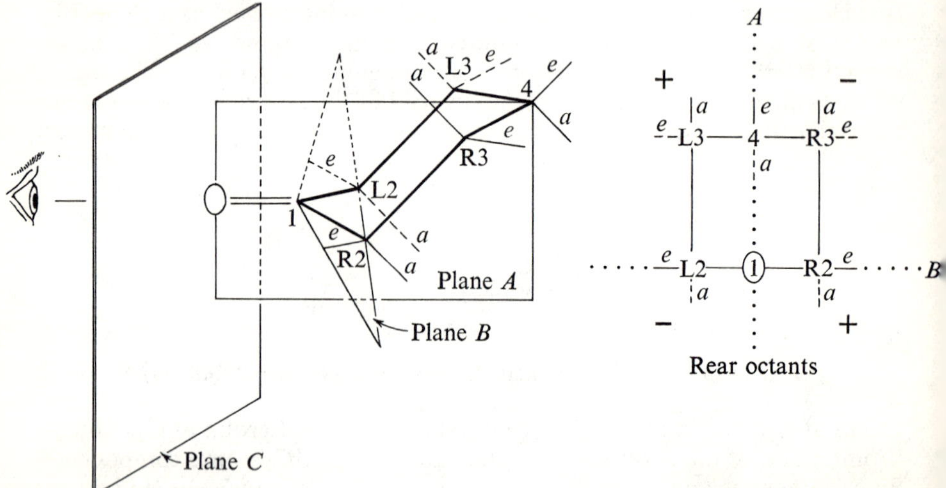

Fig. 14-44. Octant rule. (*From C. Djerassi, "Optical Rotatory Dispersion," copyright,* 1960, *McGraw-Hill Book Company, Inc.*)

rotatory dispersion of a molecule depends on which of the eight octants (into which space is divided by the coordinate planes) its substituents (or the majority of its substituents) fall into. Since only rarely do substituents bend over the carbonyl group toward the oxygen and beyond, the four octants nearest the observer are usually vacant. The octant rule states that substituents lying in the coordinate planes make no appreciable contribution to the rotatory dispersion. Substituents in the far-lower-right and far-upper-left octant make a positive contribution to the Cotton effect; substituents in the far-upper-right or far-lower-left octant make a negative contribution. In those rare cases where substituents are found in the near octants, their contribution is opposite to that of substituents in the far octants.

Further inspection shows that equatorial substituents at R2 and L2 make

† As explained in the previous section, a more useful application of such a rule is to deduce configuration or conformation when the other variables are known.

little contribution to the rotatory dispersion because they are nearly contained in the *B* plane.† Axial substituents at R2 (lower right) and all substituents at L3 (upper left) make a positive contribution. Axial substituents at L2 (lower left) and all substituents at R3 (upper right) make a negative contribution.‡ Substituents at C_4 make no contribution because they lie in the *A* plane. It is customary in the usual cases where the front octants are empty to simplify the octant diagram in a planar representation, as shown in Fig. 14-45.

Two applications of the rule must suffice to illustrate its usefulness.

i. Configuration. Determination of configuration may be illustrated by (+)-*trans*-10-methyl-2-decalone (Fig. 14-46). If this molecule has the configuration shown, carbons 8, 7, and 6, being equatorially attached to C_9, are in the upper left octant whereas carbon 5 and the angular methyl group are in the *A* plane and make no contribution. No substituents are attached at C_1 or C_3. Since the only substituents outside the coordinate plane are thus

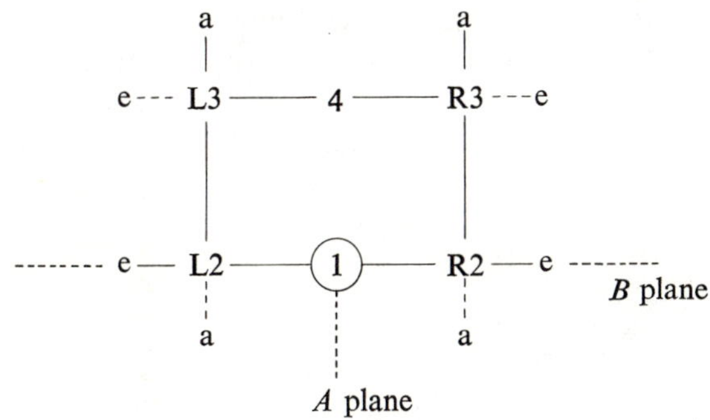

Fig. 14-45. Planar octant diagram.

in the upper left octant, the Cotton effect for the molecule should be positive. Since, in fact, the Cotton effect for (+)-*trans*-10-methyl-2-decalone is positive, the absolute configuration shown in Fig. 14-46 is the correct one.[41]

† The finding, mentioned in Sec. 14-2*c*, that 4,4-dimethylcholestan-3-one (Fig. 14-31) has a negative Cotton effect whereas 4β-methylcholestan-3-one (which differs only by the absence of the equatorial 4α-methyl group) has a positive one is clearly an exception to the rule that equatorial L2 substituents do not influence the sign of the Cotton effect. However, the case of 4β-methylcholestan-3-one is a rather delicately poised one to begin with, inasmuch as the molecule has an axial L2 substituent, which should make a negative contribution to the Cotton effect but whose contribution is apparently more than offset by the positive contribution of the L3 substituent (carbon 6 of the B ring). See also J. S. E. Holker and W. B. Whalley, *Proc. Chem. Soc.*, 464 (1961); N. L. Allinger and M. A. DaRooge, *Tetrahedron Letters*, 676 (1961); J.-M. Lehn, J. Levisalles, and G. Ourisson, *ibid.*, 682 (1961).

‡ It may be seen that the axial haloketone rule (Fig. 14-38) is a special case of the octant rule with the additional proviso that the effect of the axial halogen next to the ketone function overrides the effect of all other substituents, because of the high atomic refractivity of chlorine, bromine, and iodine (page 401). A similar situation occurs with β,γ-unsaturated ketones.[40b]

---

[41] C. Djerassi, R. Riniker, and B. Riniker, *J. Am. Chem. Soc.*, **79**, 1506 (1957). Actually, the configuration of the dextrorotatory decalone was known, and the example thus serves as a check on the octant rule. The rule has been checked in a number of other cases of ketones of known absolute configuration.

ii. Conformation. Here again we shall choose a rather simple illustration where the answer is known in advance and thus serves to check on the reasonableness of the octant rule.[42] (+)-3-Methylcyclohexanone is known[43] to have the R configuration shown in Fig. 14-47. If one orients the ketone appropriately for application of the octant rule, one obtains either the equatorial representation with methyl on the upper left (substituent at L3) or the axial representation with methyl on the upper right (at R3) (see Fig. 14-45). There are no other substituents and therefore the equatorial conformer should have a positive Cotton effect and the axial conformer a negative Cotton effect. The fact that the Cotton effect is indeed positive[25, 44] agrees with the conclusion, to be expected on conformational grounds (Chap. 8), that the equatorial isomer predominates.

f. Absolute Configuration. One of the important problems in natural-product chemistry is the determination of the absolute configuration of a new optically active product derived from natural sources. A number of methods for this purpose have been described in Chap. 5. Many of these methods are tedious and time-consuming. It is fortunate, therefore, that optical

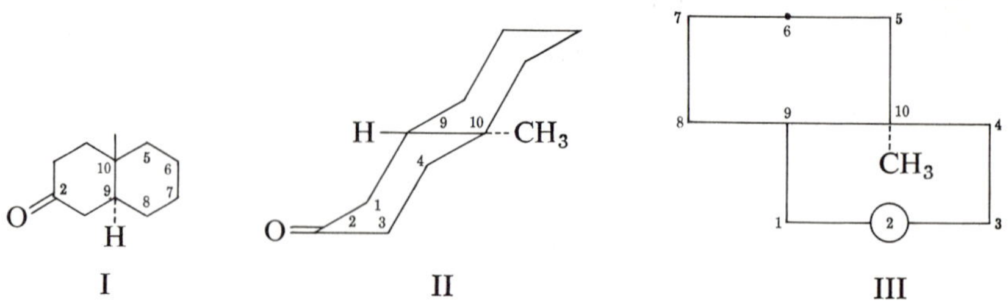

Fig. 14-46. (+)-*trans*-10-Methyl-2-decalone. (*From C. Djerassi, "Optical Rotatory Dispersion," copyright,* 1960, *McGraw-Hill Book Company, Inc.*)

rotatory dispersion provides a rather facile method of determining absolute configuration, although it is not quite as reliable as some of the other methods described earlier.† Two ways of determining absolute configuration, one based on the axial haloketone rule and one based on the more general octant rule, have already been described in Sec. 14-2*d* and *e*. A third method is based on the postulate that, since groups distant from the chromophore have relatively little effect on rotatory dispersion, it is legitimate to compare two chemically quite different species, *provided that their structure is similar in the vicinity of the chromophore.* An example is provided by the diterpene cafestol which occurs in the coffee bean. Degradation of cafestol (Fig. 14-48, *A*)

† Djerassi (Ref. 55, p. 135) has pointed out that no method of determining configuration based on measurements of rotation at a single wavelength (e.g., Sec. 5-4*d*) can be any more reliable than the rotatory-dispersion method that compares rotations over a range of wavelengths. However, methods of configurational assignment on the basis of rotation are not as reliable anyway as some of the other methods cited in Chap. 5.

[42] For further cases, see Ref. 55, chap 13.

[43] Cf. E. J. Eisenbraun and S. M. McElvain, *J. Am. Chem. Soc.*, **77**, 3383 (1955).

[44] H. S. French and M. Naps, *J. Am. Chem. Soc.*, **58**, 2303 (1936).

gave ketone *B* (Fig. 14-48) whose rotatory-dispersion curve is very nearly the reflection across the zero-rotation axis of the dispersion curve for 4α-ethylcholestan-3-one (Fig. 14-48, *C*). The curves are shown in Fig. 14-49. Since *B* and *C* (Fig. 14-48) are structurally similar in the vicinity of the carbonyl chromophore, it was concluded[45] that they are of opposite configuration, as drawn in Fig. 14-48.

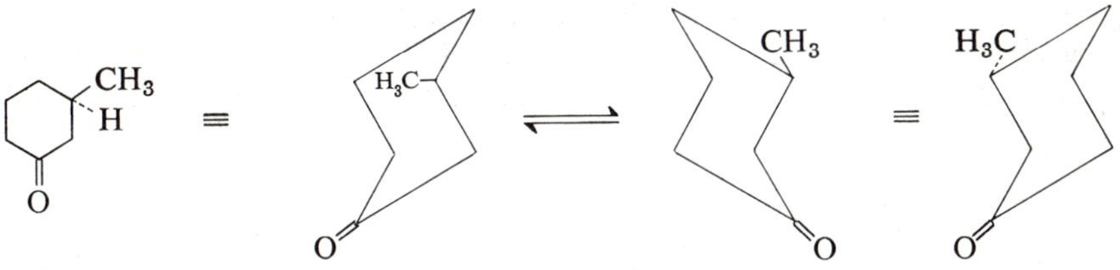

Fig. 14-47. (R)-(+)-3-Methylcyclohexanone.

It should be noted that the determination of "absolute configuration" by optical rotatory dispersion is not an "absolute" method in the sense that the X-ray diffraction method discussed in Sec. 5-3 is.† Rather, it is a quite powerful correlative method in which a compound of unknown configuration is compared with a series of compounds of known configuration.

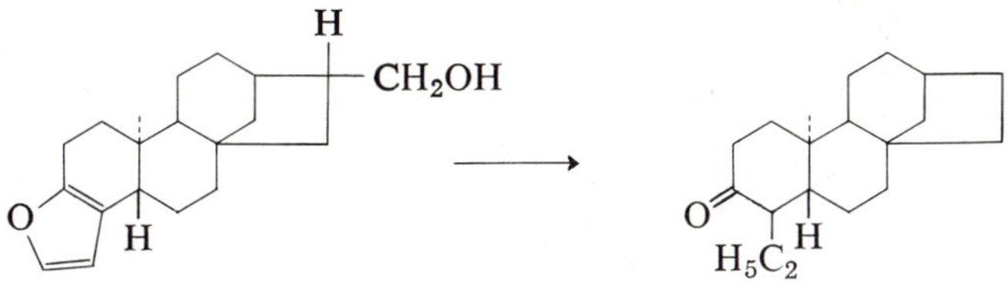

Fig. 14-48. Elucidation of configuration of cafestol.

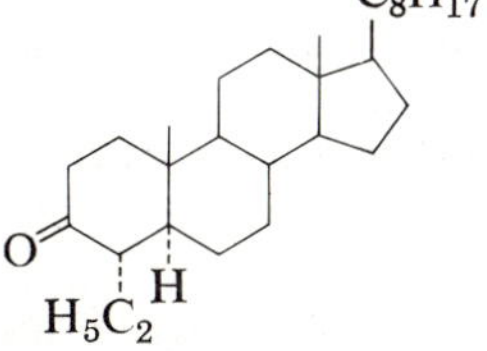

A particularly interesting and far-reaching example is the correlation, by optical rotatory dispersion, of a centrodissymmetric compound with an axially dissymmetric biphenyl[46] (cf. Sec. 6-4*c*).

† At least until such time as a complete theoretical prediction is possible.

[45] C. Djerassi, M. Cais, and L. A. Mitscher, *J. Am. Chem. Soc.*, **81**, 2386 (1959).
[46] K. Mislow and C. Djerassi, *J. Am. Chem. Soc.*, **82**, 5247 (1960).

g. Ketal Formation.[47] When a ketone is dissolved in methanol and a drop of hydrochloric acid is added, an equilibrium is established in which the ketone is converted to a greater or lesser extent to a dimethyl ketal. Although the ketal can be isolated only after removal of the acidic catalyst, its formation in solution can be readily observed, since it goes hand in hand with disappearance of the carbonyl chromophore and its attendant Cotton effect. The degree to which the Cotton effect is reduced under these circumstances gives a quantitative measure of the extent of ketal formation. For example, addition of HCl to a solution of (+)-3-methylcyclohexanone in methanol reduces the Cotton effect by 93%, but in ethanol only a 33% reduction occurs,

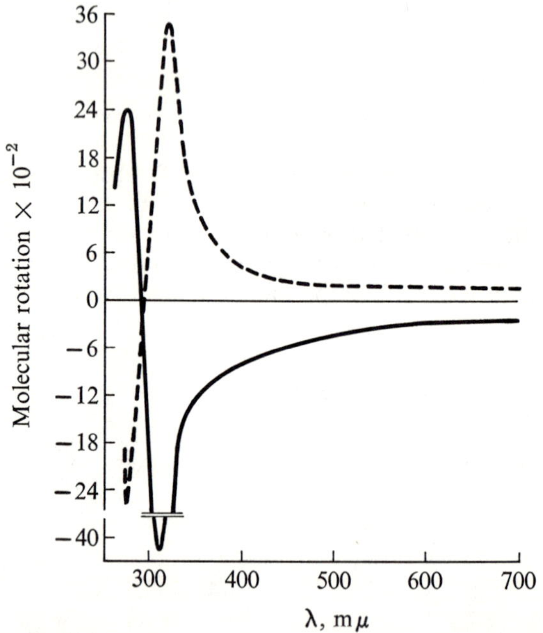

Fig. 14-49. Rotatory-dispersion curves of ketone from cafestol (solid line) and 4α-ethylcholestan-3-one (dashed line). (The double line indicates a discontinuity in scale.) (*From C. Djerassi, "Optical Rotatory Dispersion," copyright,* 1960, *McGraw-Hill Book Company, Inc.*)

and in isopropyl alcohol there is no reduction. As expected, ketal formation is strongly affected by steric factors. Similarly, the Cotton-effect curves of (+)-3-methylcyclopentanone and (−)-3-methylcycloheptanone in methanol show only a 24 and 21% reduction, respectively, upon addition of acid. The lesser amount of ketal formation here as compared with the six-membered ketone is in agreement with the prediction from I-strain (Sec. 9-4).

A useful application of ketal formation is found in the differentiation of 2- and 3-ketosteroids. It was said in Sec. 14-2*c* that the dispersion curves of

[47] C. Djerassi, L. A. Mitscher, and B. J. Mitscher, *J. Am. Chem. Soc.,* **81,** 947 (1959). However, ketals rather than hemiketals are generally involved; cf. D. G. Kubler and L. E. Sweeney, *J. Org. Chem.,* **25,** 1437 (1960).

these isomers are not distinctive. However, whereas the 3-ketosteroid experiences a considerable reduction (to about 20% of the original) of its Cotton effect in the presence of methanol and acid, only a trifling reduction is obtained in the case of the 2-ketosteroid. This is to be expected, since in the ketal of the 2-ketosteroid the —OR group must be axial and therefore experiences a 1,3-diaxial interaction with the angular methyl group at C_{10}. This interaction (which is, of course, absent in the 3-ketosteroid derivative) shifts the equilibrium away from the ketal back to the free ketone.

h. Other Chromophores. Although the discussion in this section has been limited to consideration of ketones, some work, though less extensive, has been carried out on other chromophores.[48] Of particular interest is the fact that "non-chromophoric groups," i.e., groups that have plain rotatory-dispersion curves because their absorption maximum is too far in the ultraviolet, can sometimes be converted to "chromophoric groups" with Cotton-effect curves by suitable derivatization. Thus α-phenethylamines show a Cotton effect in the form of their phthalimides,[49a] α-substituted carboxylic acids show Cotton effects after conversion to their acylthiourea derivatives,[49b] and olefins show Cotton effects upon conversion to osmate esters (Fig. 12-44) by means of osmium tetroxide.[49c] The sign of the Cotton effect of such derivatives may give an indication of the configuration of the parent compound.[49b, d] This extension of the optical-rotatory-dispersion method of assigning configuration evidently provides a fruitful field for further work; it is but an example of the many exciting developments to be expected in this active field in the near future.

General References

[50] B. H. Levedahl and T. W. James, eds., Proceedings of a Conference on Rotatory Dispersion, *Tetrahedron,* **13,** 1–240 (1961).

[51] T. M. Lowry, "Optical Rotatory Power," Longmans, Green & Co., Inc., New York, 1935.

[52] W. J. Kauzmann, J. E. Walter, and H. Eyring, Theories of Optical Rotatory Power, *Chem. Revs.,* **26,** 339 (1940).

[53] E. Hückel, Die Theorien der Natürlichen Optischen Aktivität Gasförmiger und Flüssiger Isotroper Stoffe, *Z. Electrochem.,* **50,** 13 (1944).

[54] W. Kuhn, Optical Rotatory Power, *Ann. Rev. Phys. Chem.,* **9,** 417 (1958).

[55] C. Djerassi, "Optical Rotatory Dispersion," McGraw-Hill Book Company, Inc., New York, 1960.

[56] W. Klyne, Optical Rotatory Dispersion and the Study of Organic Structures, in R. A. Raphael, E. C. Taylor, and H. Wynberg, eds., "Advances in Organic Chemistry," vol. 1, Interscience Publishers, Inc., New York, 1960, pp. 239–348.

[57] C. Djerassi, Some Recent Applications of Optical Rotatory Dispersion Studies to Organic Chemical Problems, *Record Chem. Progr. (Kresge-Hooker Sci. Lib.),* **20,** 101 (1959).

[58] W. Klyne and A. C. Parker, Optical Rotatory Dispersion, in A. Weissberger, ed., "Physical Methods of Organic Chemistry," vol. I, pt. 3, 3d ed., Interscience Publishers, Inc., New York, 1960, pp. 2335–2385.

---

[48] For example, Ref. 55, chaps. 14 and 15; G. G. Lyle, *J. Org. Chem.,* **25,** 1779 (1960).

[49] (*a*) J. H. Brewster and S. F. Osman, *J. Am. Chem. Soc.,* **82,** 5754 (1960); (*b*) C. Djerassi and K. Undheim, *ibid.,* **82,** 5756 (1960); (*c*) E. Bunnenberg and C. Djerassi, *ibid.,* **82,** 5953 (1960); (*d*) B. Sjoberg, A. Fredga, and C. Djerassi, *ibid.,* **81,** 5002 (1959).

Chapter 15

STEREOSELECTIVE SYNTHESIS AND STEREOREGULATED POLYMERIZATION

Stereoselectivity is of great importance in the design of syntheses of natural products. It has also become extremely important in the synthesis of certain polymers whose carefully regulated molecular architecture endows them with special, often highly desirable physical properties. Despite the evident practical significance of these topics, they can be taken up only briefly within the framework of a textbook. Stereoselective synthesis—like synthesis altogether—is still an art which cannot be taught rigorously. It is possible to set down certain principles and to illustrate how they have been used, but the application of these principles to new problems still requires originality and ingenuity in the highest degree. Stereoregulated polymerization is an art even more empirical. Although there is now a fairly clear picture as to what a stereoregulated polymer is, there is still much controversy on how it is formed, and as yet there seems to be little basis for prediction as to whether a new catalyst system will produce a highly stereoregulated polymer or not.

15-1. Stereoselective Synthesis

By "stereoselective synthesis" is meant a synthesis that produces one diastereoisomer (or diastereoisomeric *dl* pair) of a given structure in considerable predominance over all the other possible diastereoisomers (or diastereoisomeric *dl* pairs) of the same structure. Stereoselective synthesis is *not* usually concerned with the synthesis of optically active molecules. Usually (although not invariably) syntheses of natural products are carried out with racemic starting materials or intermediates, and resolution is carried out only at or near the end. Stereoselective synthesis is concerned with synthesizing the right diastereoisomeric *dl* pair, not with synthesizing the naturally occurring enantiomer of this pair. The latter may be obtained by one of the standard methods of resolution (cf. Sec. 4-4) once the appropriate *dl* pair is at hand.†

Note: All references above 29 are listed in the General References at the end of the chapter.

† It was explained in Chap. 4 (e.g., Fig. 4-35) that if in a synthesis of A—B from A (where A and B are asymmetric carbon atoms) (R)-A gives, say, (R)-A—(R)-B in preference over (R)-A—(S)-B, then (±)-A will give the diastereoisomeric *dl* pair (R)-A—(R)-B-plus-(S)-A—(S)-B in predominance over (R)-A—(S)-B-plus-(S)-A—(R)-B.

Perhaps the best way to illustrate the point is by means of a simple example: the synthesis of (−)-ephedrine and (+)-ψ-ephedrine. These diastereoisomers are shown in Fig. 15-1. (−)-Ephedrine is a pharmacologically active substance isolated from the Chinese drug mahuang; it is valuable, for example, in the control of asthma. Its diastereoisomer, (+)-ψ-ephedrine (which also occurs naturally), is less valuable pharmacologically. To obtain (−)-ephedrine synthetically, one looks for a method that will give (±)-ephedrine (which can subsequently be resolved by means of active mandelic acid) in preference over (±)-ψ-ephedrine. A logical synthetic precursor is (±)-α-methylaminopropiophenone (Fig. 15-2) which is readily obtained from pro-

(−)-Ephedrine	(+)-ψ-Ephedrine
CH_3	CH_3
H—C—$NHCH_3$	H—C—$NHCH_3$
H—C—OH	HO—C—H
C_6H_5	C_6H_5

Fig. 15-1. The diastereoisomeric ephedrines.

piophenone by bromination followed by reaction with methylamine. Reduction of this ketone with sodium amalgam gives, however, a mixture of (±)-ephedrine and (±)-ψ-ephedrine, with the latter predominating slightly;[1] this synthesis is thus largely non-stereoselective and not suitable for the preparation of (±)-ephedrine. On the other hand, the catalytic reduction of α-methylaminopropiophenone hydrochloride with hydrogen over platinum gives almost exclusively (±)-ephedrine which may be readily isolated from the reduction products in over 90% yield.[2] This, then, is a stereoselective synthesis of ephedrine. The reactions are summarized in Fig. 15-2.

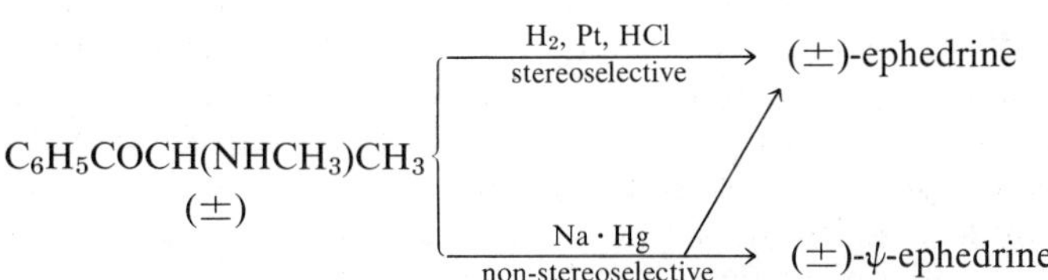

Fig. 15-2. Reduction of α-methylaminopropiophenone.

The particular stereoselective synthesis here described is one based on "kinetic control" of the reaction products. Ephedrine is formed in predominance over ψ-ephedrine in the catalytic hydrogenation because it is formed at a faster rate. In other instances, stereoselective synthesis may be based on "thermodynamic control." This means that both isomers or mainly the undesired isomer may be formed originally, but then, by a process of equilibration, are converted to the desired isomer. An example will be given below. In the case of ephedrine this method is not applicable, for although ephedrine

[1] A. Eberhard, *Arch. Pharm.*, **253**, 62 (1915); **258**, 97 (1920).
[2] J. F. Hyde, E. Browning, and R. Adams, *J. Am. Chem. Soc.*, **50**, 2287 (1928).

and ψ-ephedrine can be interconverted by heating with 25% aqueous hydrochloric acid in a sealed tube, the position of equilibrium corresponds to about 60% ψ-ephedrine and 40% ephedrine, neither isomer being obtained in anything resembling a state of purity in the process.[3]

Another stereoselective synthesis of ephedrine is by the reductive amination of phenylacetylcarbinol, $C_6H_5CHOHCOCH_3$, with methylamine and hydrogen over platinum. It is of interest that this method has been applied to (−)-phenylacetylcarbinol, in which case it gives directly the natural (−)-ephedrine.[4,5] (−)-Phenylacetylcarbinol may be synthesized either from (−)-mandelic acid[4] or by a fermentative mixed acetoin condensation between benzaldehyde and acetaldehyde in the presence of the enzyme carboligase.[6] Although this constitutes a very ingenious route to the natural, active (−)-ephedrine, it is emphasized again that the use of active starting material or the obtention of an active product is *not* necessarily implied in the concept of stereoselective synthesis. The synthesis of (±)-ephedrine to the near exclusion of (±)-ψ-ephedrine from (±)-phenylacetylcarbinol is also a stereoselective synthesis.

Before other examples of stereoselectivity are considered, a comment on nomenclature is in order. Many authors have used (and are still using) the term "stereospecific synthesis" where "stereoselective synthesis" is used here. In this book, however, the terms are used in the sense proposed by H. Zimmerman and coworkers.[7] Thus "stereospecific" as applied to a reaction means that stereoisomerically different starting materials give rise to stereoisomerically different products. Examples are the formation of *meso*-2,3-dibromobutane by addition of bromine to *trans*-2-butene as contrasted to the formation of the *dl*-dibromide from the cis olefin (Fig. 12-38) and the reverse of this process in the iodine-induced elimination (Fig. 6-17). "Stereoselective," in contrast, simply means that, of two (or more) possible stereoisomeric products in a reaction, one is produced in predominance (usually in great predominance)† over all the others. An example here is the predominant formation of *trans*-stilbene (rather than *cis*-stilbene) by elimination of hydrogen chloride from desyl chloride, $C_6H_5CHClCH_2C_6H_5$ (Fig. 6-32). It might be noted that all stereospecific processes are stereoselective, but not all stereoselective processes are stereospecific. For example, the low-temperature free-radical addition of hydrogen bromide to the 2-bromo-2-butenes (page 362) is both stereospecific and stereoselective in that the cis olefin gives *meso*-2,3-dibromobutane whereas the trans olefin gives the *dl*-dibromide. At higher temperature the reaction is still stereoselective (formation of 75% *dl*

† The term stereoselective is capable of semiquantitative and quantitative modulation, such as "highly stereoselective," "weakly stereoselective," "90% stereoselective," etc.

[3] H. Emde, *Helv. Chim. Acta*, **12**, 377 (1929).

[4] K. Freudenberg, E. Schöffel, and E. Braun, *J. Am. Chem. Soc.*, **54**, 234 (1932).

[5] Cf. also G. Hildebrandt and W. Klavehn, German Patent 548,459 (1930); *Chem. Abstr.*, **26**, 3623 (1932).

[6] C. Neuberg and J. Hirsch, *Biochem. Z.*, **115**, 282 (1921).

[7] H. E. Zimmerman, L. Singer, and B. S. Thyagarajan, *J. Am. Chem. Soc.*, **81**, 108 (1959), footnote 16.

isomer to 25% meso isomer) but no longer stereospecific; both olefins now give the same product mixture.

In natural-product synthesis, it is evidently high stereoselectivity (in the above-defined sense) which is important, not stereospecificity.†

Many natural products contain numerous asymmetric carbon atoms. Cholestanol (Fig. 10-29), for example, contains nine, so that there are 2^9, or 512, stereoisomers forming 256 diastereoisomeric *dl* pairs. In a stereoselective synthesis of such a compound one has the choice of either introducing the asymmetric centers one by one, attempting in each step to obtain the desired diastereoisomer (i.e., the one that has the correct relative configuration of the asymmetric atoms) in predominance, or attempting to devise syntheses which introduce several asymmetric atoms at one time stereoselectively in such a way as to give mainly the desired diastereoisomer. The latter course, while more rapid, is also more difficult; thus the introduction of four new asymmetric atoms at a time in a compound that already has one (or more) asymmetric atoms may produce as many as 16 diastereoisomers of which only one is useful. Also, when several asymmetric atoms are introduced in one step of a synthesis, it is often difficult, if not impossible, to make a firm prediction as to which stereoisomer will result in predominance. Thus, even if one does obtain the desired diastereoisomer (i.e., the naturally occurring isomer or one that can be elaborated by further transformations into the desired natural product), the synthesis does not serve to confirm the (presumably known) stereochemistry of the natural product. To the extent that one of the objectives of synthesis is to confirm the stereochemistry as well as the structure of the compound synthesized, the above is a somewhat serious shortcoming.

The ingenious synthesis of (±)-epiandrosterone‡ (Fig. 15-3) by W. S. Johnson and coworkers[8] illustrates a case where no less than five asymmetric carbon atoms are introduced stereoselectively in one step. The tetracyclic starting material (±)-I is readily available from 5-methoxy-2-tetralone. It contains only one asymmetric carbon. Upon reduction with lithium in liquid ammonia–ethanol–dioxane, compound II containing seven asymmetric carbon atoms was obtained along with III. The six asymmetric atoms that are common to II and epiandrosterone all had the correct relative configuration. (The seventh asymmetric atom in II is readily epimerizable in the basic reaction mixture, and its stereochemistry would undoubtedly be such that the

† It should be mentioned that in other places "stereospecificity" has been used to denote high stereoselectivity, e.g., F. G. Bordwell and P. S. Landis, *J. Am. Chem. Soc.*, **80**, 6383 (1958), footnote 7. To this author, the use of one technical term to denote a semiquantitative modulation of another term seems undesirable, since it is not clear how "high" stereoselectivity must be before it is called stereospecificity. (Bordwell and Landis proposed 98^+%, but this seems rather arbitrary.) It should not be concealed, however, that the present terminology is not perfect either. For example, the addition of bromine to cyclohexene to give *trans*-1,2-dibromocyclohexane must be called stereoselective (only *cis*-cyclohexene is available for the experiment) whereas the analogous addition of bromine to *cis*-2-butene is called stereospecific.

‡(+)-Epiandrosterone is found in human urine as a steroid metabolite. It can be converted synthetically to testosterone, one of the major steroidal hormones.

---

[8] W. S. Johnson, B. Bannister, and R. Pappo, *J. Am. Chem. Soc.*, **78**, 6331 (1956).

C/D ring fusion in II is trans.) Hydrogenation of II to IV followed by methylation of the methine carbon next to the carbonyl group (after protection of the active methylene group by furfural) gives a mixture of the two possible C_{13} epimers of which the one having the stereochemistry of epiandrosterone actually constitutes only the minor product. This step of the synthesis is thus not stereoselective, but fortunately the desired isomer may be readily isolated from the mixture by crystallization and chromatography. The product (V) contains all the asymmetric atoms of epiandrosterone, and

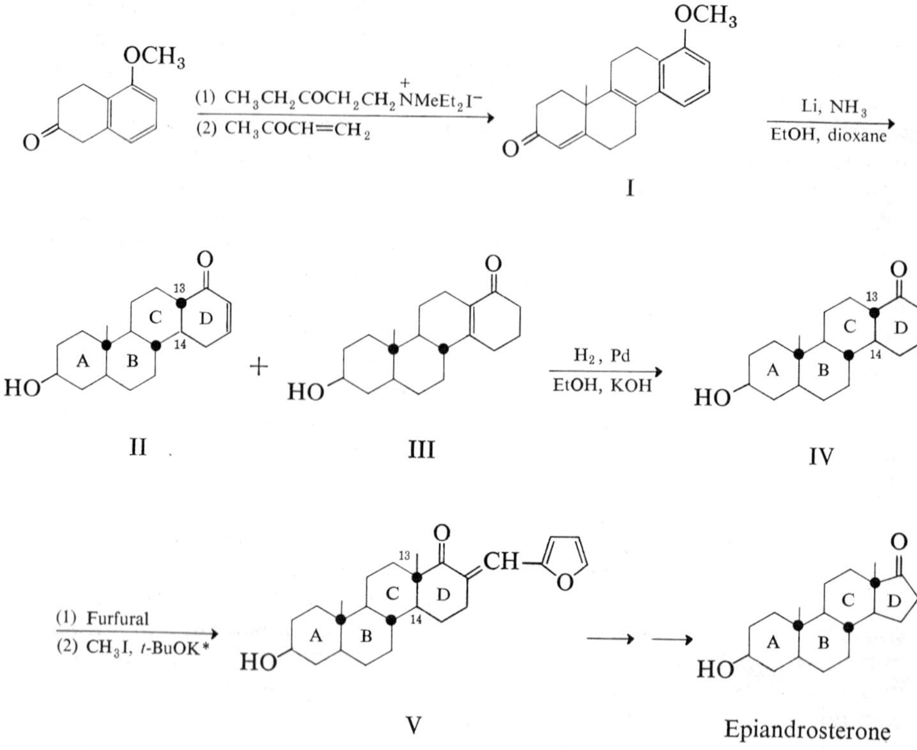

(1) $CH_3CH_2COCH_2CH_2\overset{+}{N}MeEt_2I^-$
(2) $CH_3COCH{=}CH_2$

I

Li, NH_3
EtOH, dioxane

II + III

H_2, Pd
EtOH, KOH

IV

(1) Furfural
(2) CH_3I, *t*-BuOK*

V

Epiandrosterone

*3-Hydroxy group temporarily protected by dihydropyran

Fig. 15-3. Synthesis of (±)-epiandrosterone.

the remaining steps of the synthesis serve to contract ring D to a five-membered one without any stereochemical change.

It is of interest that, although III has two fewer asymmetric carbon atoms than II, its hydrogenation over palladium-charcoal in ethanol containing potassium hydroxide gave the same saturated ketone (IV) as the hydrogenation of II. In this hydrogenation, then, asymmetry at C_{14} is introduced stereoselectively, the reason for this being, presumably, that the molecule is adsorbed on the catalyst from the less-hindered α side (cf. Chap. 10). The fact that hydrogenation appears to occur in trans fashion should not be considered anomalous, for since the hydrogenation medium is basic, C_{13} will be epimerized to the more stable configuration even if hydrogenation is originally cis.

It is sometimes stated that Johnson's stereoselective steroid synthesis is based on thermodynamic control, inasmuch as the reason for the correct stereochemistry of the product II (Fig. 3-15) obtained seems to be that the four rings are joined in the stereochemically most stable form. This classification is not, however, correct, for with the exception of C_{13} (whose configuration in II is irrelevant anyway) none of the asymmetric atoms in the product II are epimerizable after the product is formed. The synthesis must, therefore, be considered one of kinetic stereochemical control. This is especially true now that it is known (cf. Fig. 12-37) that the reason for the formation of a trans junction of rings in lithium–liquid ammonia–alcohol reductions is a stereoelectronic rather than a thermodynamic one.

True thermodynamic stereochemical control is rare and occurs at best in one or two steps of a given stereoselective synthesis. An example is the synthesis of yohimbine[9] shown in Fig. 15-4. The starting material for this synthesis is the Diels-Alder addition product of butadiene and quinone which, upon reduction with zinc and acetic acid, gives *cis*-Δ^6-octalin-1,4-dione (Fig. 15-4, I). This material is subjected to a glycidic ester synthesis with ethyl chloroacetate and potassium *t*-butoxide. In the course of this reaction, the *cis*-octalin system of I is epimerized to the (desired) *trans*-octalin system in the glycidic ester II by what is evidently a thermodynamically controlled process. Saponification and decarboxylation of the glycidic ester II leads to an aldehyde (III) in which the ring carbon C_1 is readily epimerizable. Either during formation of this aldehyde or during its subsequent oxidation to the acid IV by means of alkaline silver oxide, the exocyclic carbon is placed in the equatorial position by what appears once again to be a process of thermodynamic control. The acid IV thus has the most stable all-trans stereochemistry. Condensation with tryptamine followed by osmium tetroxide hydroxylation gives compound V which is reduced to VI by hydrogen over platinum. The reduction step is subject to kinetic stereochemical control, inasmuch as the hydrogen approaches from the more accessible equatorial side (cf. Chap. 8) to give the less stable axial alcohol. Compound VI now has the four asymmetric carbon atoms of the E ring of yohimbine in the correct relative stereochemical disposition. The fifth asymmetric atom (C_3 in yohimbine) appears in the course of the glycol cleavage of VI followed by a Pictet-Spengler ring closure to VII. Interestingly enough, this carbon atom is disposed in the less stable configuration (ring C axially fused at C_3), for upon further elaboration of VII the product is *dl*-pseudoyohimbine (VIII) rather than *dl*-yohimbine (in which ring fusion at C_3 is equatorial). Resolution of VIII gave (±)-pseudoyohimbine, one of the known naturally occurring alkaloids of the yohimbine family. Although epimerization at C_3 of pseudoyohimbine derivatives to the thermodynamically more stable yohimbine derivatives has been reported in

[9] E. E. van Tamelen, M. Shamma, A. W. Burgstahler, J. Wolinsky, R. Tamm, and P. E. Aldrich, *J. Am. Chem. Soc.*, **80**, 5006 (1958); cf. *Ann. Rep. on Prog. Chem. (Chem. Soc. London)*, **55**, 307 (1958).

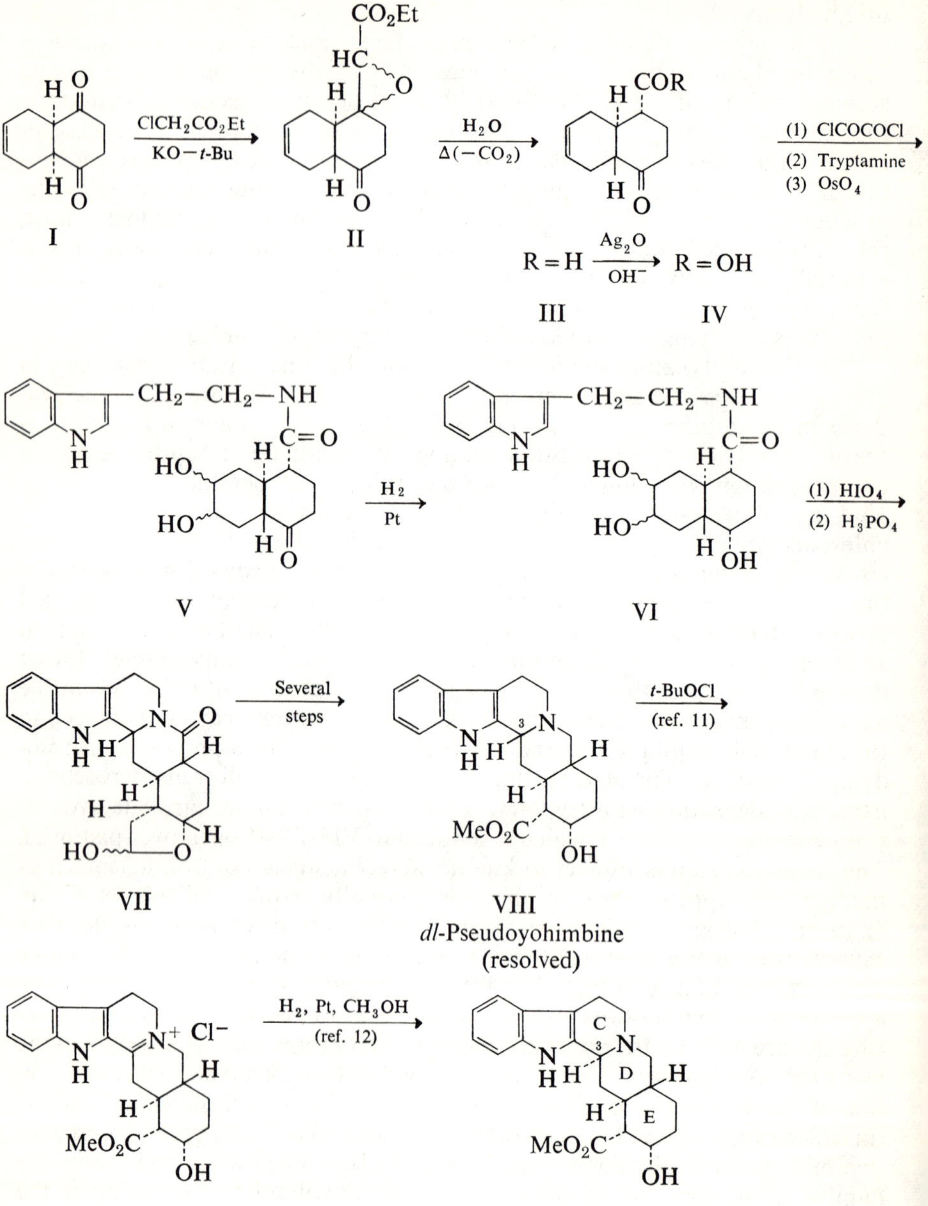

I $\xrightarrow[\text{KO}-t\text{-Bu}]{\text{ClCH}_2\text{CO}_2\text{Et}}$ II $\xrightarrow[\Delta(-\text{CO}_2)]{\text{H}_2\text{O}}$ III (R = H) $\xrightarrow[\text{OH}^-]{\text{Ag}_2\text{O}}$ IV (R = OH) $\xrightarrow{}$ (1) ClCOCOCl (2) Tryptamine (3) OsO_4

V $\xrightarrow[\text{Pt}]{\text{H}_2}$ VI $\xrightarrow{}$ (1) HIO_4 (2) H_3PO_4

VII $\xrightarrow{\text{Several steps}}$ VIII *dl*-Pseudoyohimbine (resolved) $\xrightarrow[\text{(ref. 11)}]{t\text{-BuOCl}}$

$\xrightarrow[\text{(ref. 12)}]{\text{H}_2,\ \text{Pt},\ \text{CH}_3\text{OH}}$ Yohimbine

Yohimbine

Fig. 15-4. Total synthesis of yohimbine.

the literature,[10] such epimerization requires conditions too drastic to be applied to pseudoyohimbine itself. A smooth conversion of pseudoyohimbine to yohimbine is, however, available through an oxidation[11]-reduction[12] sequence, as shown in Fig. 15-4. The reduction step, if carried out with hydrogen over platinum in methanol in the absence of base, produces yohimbine (in preference to pseudoyohimbine) in highly stereoselective fashion. This conversion, of course, is an example of kinetic stereochemical control.

The last stereoselective synthesis to be presented here is the elegant total synthesis of reserpine (Fig. 15-5) by Woodward and coworkers.[13] Reserpine has six asymmetric carbons: C_3, C_{15}, C_{16}, C_{17}, C_{18}, and C_{20}. Its synthesis, summarized in Fig. 15-7, illustrates a number of interesting points in stereochemistry, including an ingenious new principle which may be called "conformational stereochemical control." As explained in Chap. 8, in cyclohexane systems, which are so common in natural products, equatorial and axial groups show differential reactivity and stability. This difference may sometimes be exploited to place a given substituent in the desired stereochemical configuration; e.g., in compound IV, Fig. 15-4, the carboxyl

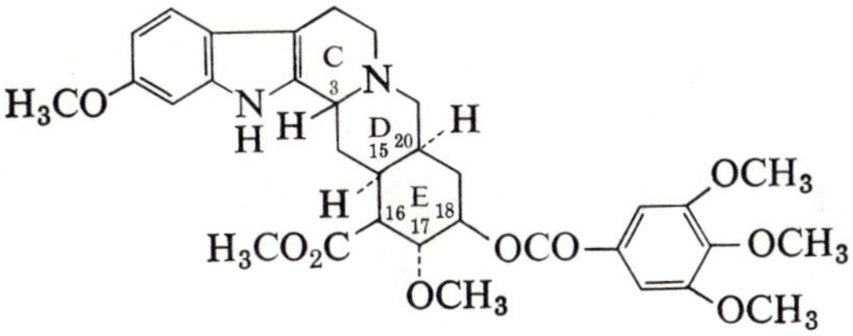

Fig. 15-5. Reserpine.

group ends up in the desired configuration (in which it is equatorial). However, the situation may be such that kinetic or thermodynamic control (or both) operates in such a way that the substituent is placed in the *undesired* configuration. Here a remedy may be available, if the cyclohexane system is mobile in nature, i.e., if one of the possible chair forms may be converted into the other (cf. Sec. 8-6). The same operation which placed the substituent under consideration in the undesired configuration in the original chair conformation will place it in the desired conformation in the *inverted* chair conformation. A simple though hypothetical case is depicted in Fig. 15-6. Substituent X is trans to the carboxyl group in compound A. It cannot be shifted to the cis position by equilibration, since it is already in the more stable equatorial position. However, by formation of the lactone B, the ring is forced into the inverted chair conformation *in which substituent X has become axial.* If X is now equilibrated by a suitable chemical process, it will

[10] M. M. Janot, R. Goutarel, and M. Amin, *Compt. rend.,* **230,** 2041 (1950).
[11] W. O. Godtfredsen and S. Vangedal, *Acta Chem. Scand.,* **10,** 1414 (1956).
[12] F. L. Weisenborn and P. A. Diassi, *J. Am. Chem. Soc.,* **78,** 2022 (1956).
[13] R. B. Woodward, F. E. Bader, H. Bickel, A. J. Frey, and R. W. Kierstead, *Tetrahedron,* **2,** 1 (1958).

end up in the more stable equatorial position, *which is now cis to the carboxyl group* (compound C). Subsequent hydrolysis of the lactone C gives the hydroxyacid D in which X has the opposite configuration it had in A.

The above example illustrates "conformational thermodynamic control," but, as will be seen below, there is also the kinetic analog, "conformational kinetic control."

The synthesis of reserpine (Fig. 15-7) begins with a Diels-Alder condensation of quinone and *trans*-vinylacrylic acid to give the potential D and E rings of the alkaloid in the desired cis fusion. For mechanistic reasons, the carboxyl group is also cis to the ring fusion in the intermediate I. Reduction of I with sodium borohydride proceeds selectively to give the keto alcohol II, epoxidation of which leads to III. If the epoxide in III were now opened with acetic acid, one would, by the principle of diaxial ring opening (Chap. 8), expect to obtain IV in which the asymmetric atoms at the potential C_{17}

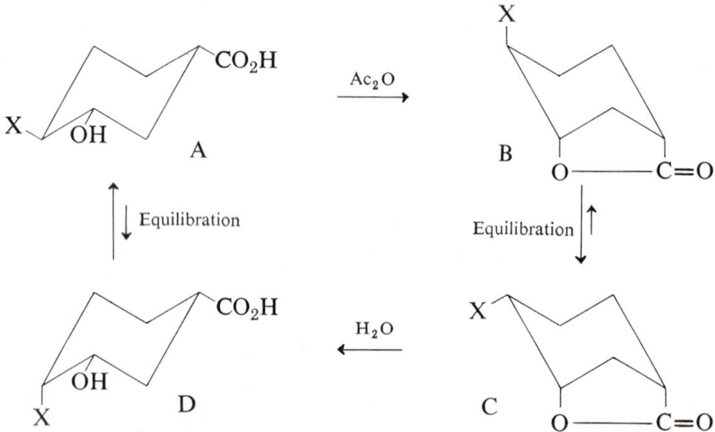

Fig. 15-6. Conformational stereochemical control.

and C_{18} positions have the undesired configuration relative to the carboxyl group at C_{16}. Therefore III was converted to the lactone V. Now, while in III the carboxyl group is in the more stable equatorial position and, therefore, the epoxide ring has the conformation as labeled in Fig. 15-7, formula III,† in V lactone formation forces the carboxyl group to be axial, so that it can join with the hydroxyl group. (The junction of ring D to ring E next to the hydroxyl group must also be axial.) Thus lactone formation forces an inversion of the chair and switches the equatorial and axial nature of the epoxide linkages. Diaxial opening of the epoxide V now gives the desired stereoisomer VI. The transformation of III to VI via V is clearly an example of conformational kinetic control as explained above.

† This is an oversimplification, inasmuch as the epoxide ring produces distortion of the chair (Chap. 8). However, from the way in which III is labeled, it is easy to see that diaxial ring opening will produce IV.

Although VI has the asymmetric centers of reserpine at the potential C_{15}, C_{16}, C_{17}, C_{18}, and C_{20} positions all in the correct relative stereochemical configuration, and could presumably have been elaborated to the alkaloid, the actual synthesis took another track. Treatment of V with aluminum isopropoxide led to reduction of the carbonyl function, relactonization to a γ-lactone, attack of the free alkoxide upon the epoxide, and finally elimination of water (all in one step) to give VII. Addition of methanol to VII gave VIII, the methanol approaching the double bond from the less-hindered "convex face" of the *cis*-decalin system, i.e., from the side of the angular hydrogen atoms.† Moreover, the five-membered lactone ring in VIII would be expected to end up in the thermodynamically more stable cis fusion.

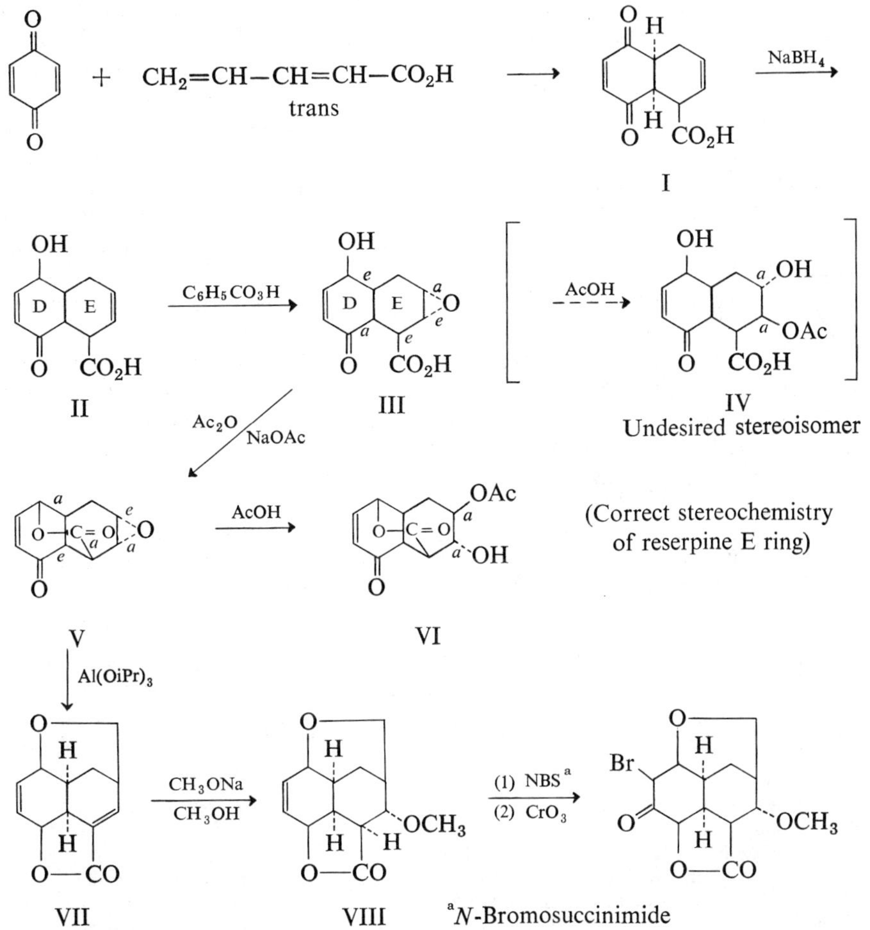

Fig. 15-7a. Synthesis of reserpine (1).

† It is a general proposition that in *cis*-decalin the "convex face" (side of the angular hydrogens) is more accessible than the opposite or "concave face."

Hence compound VIII again has all the substituents of the E ring in reserpine in correct relative stereochemistry.† It was elaborated, as shown, into IX which was condensed with 6-methoxytryptamine and further elaborated via X into methyl (±)-isoreserpate (XI). In the formation of XI, the last asymmetric carbon at C_3 assumes the thermodynamically more stable configuration, with the ring juncture equatorial and the hydrogen axial. Unfortunately, this is the opposite of the natural configuration in reserpine. To correct this situation, conformational stereochemical control was again resorted to, this time of the thermodynamic kind. Compound XI was converted to the 16-18 lactone XII, forcing the carboxyl group at C_{16} and the hydroxyl group at C_{18}, which were originally equatorial, to become axial; in the process the entire system of three chairs of the C, D, and E rings of reserpine is inverted. This changes the conformation of all the substituents of the E ring as well as of the D ring (see the appropriate letters in Fig. 15-7, XI, XII). In particular, the C_2—C_3 linkage, which was equatorial to ring D in X, now becomes axial to ring D in XII. This linkage therefore now has the unstable conformation and, upon equilibration with pivalic acid, changes to the more stable equatorial conformation (XIII). Compound XIII has the correct stereochemistry of reserpine and is converted to the latter by treatment with methanol (leading to breakage of the lactone ring and spontaneous reconversion to the more stable chair conformation) followed by esterification with 3,4,5-trimethoxybenzoyl chloride.

An alternative conversion[12] of the isoreserpine (XI) to the reserpine system makes use of the fact that alkaloids of the yohimbe type are oxidized at C_3 by mercuric acetate if the hydrogen at C_3 is axial but not if it is equatorial.‡ Thus reserpine (Fig. 15-5) is unaffected by this oxidant, but methyl isoreserpate 18-acetate (XI, Fig. 15-7) is oxidized to a salt of the dehydrocompound X. Reduction of X with zinc and acetic acid (in lieu of sodium borohydride) gives directly methyl reserpate 18-acetate.¶

† An alternative and more convenient route to VIII starts[13] with the methyl ester of I—from quinone and methyl vinylacrylate—which is reduced with aluminum isopropoxide to *i*, converted by bromine to *ii*, which in turn gives VIII on treatment with methanolic sodium methoxide.

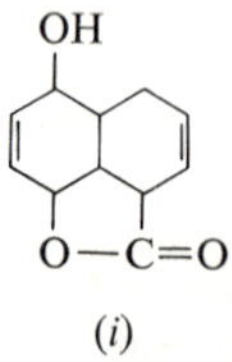

(*i*) 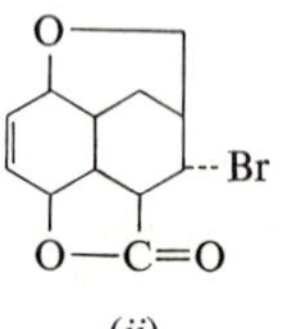 (*ii*)

‡ This is probably true of quinolizidine ring systems in general: cf. N. J. Leonard, A. S. Hay, R. W. Fulmer, and V. W. Gash, *J. Am. Chem. Soc.*, **77**, 439 (1955).

¶ Reduction of X to methyl reserpate 18-acetate followed by hydrolysis and esterification with 3,4,5-trimethoxybenzoyl chloride to give reserpine evidently provides a short-cut in the synthesis of the alkaloid. Of more general interest, however, is the possibility of going from a yohimbe alkaloid with the more stable configuration at C_3 to one with the less stable configuration by the combination of mercuric acetate oxidation and zinc–acetic acid reduction. Surprisingly, hydrogen-over-platinum reduction of dehydrocompounds of type X leads to the more stable configuration at C_3, just as sodium borohydride does (see also Fig. 15-4).

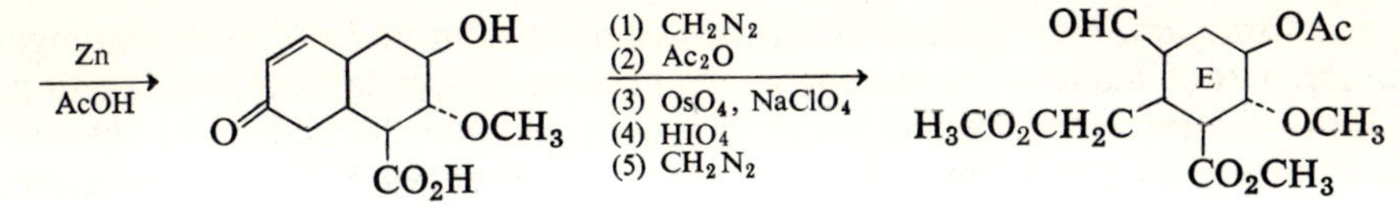

Zn / AcOH → (OH, OCH₃ bicyclic enone, CO_2H)

(1) CH_2N_2
(2) Ac_2O
(3) OsO_4, $NaClO_4$
(4) HIO_4
(5) CH_2N_2

→ IX (OHC, OAc, E, $H_3CO_2CH_2C$, OCH_3, CO_2CH_3)

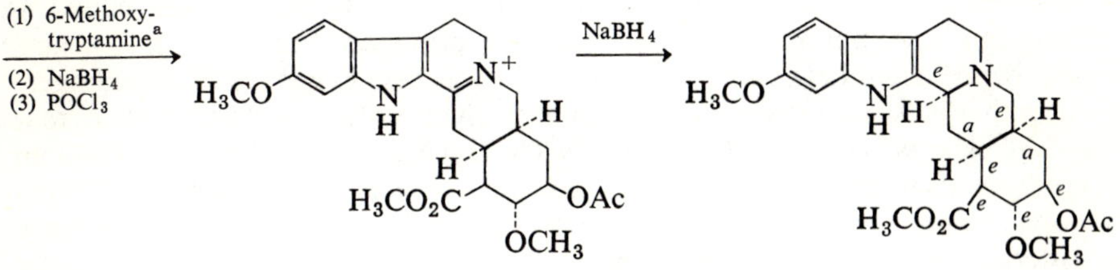

(1) 6-Methoxy-tryptamine[a]
(2) $NaBH_4$
(3) $POCl_3$

→ X (H_3CO, N, H, N^+, H, H, H_3CO_2C, OAc, OCH_3)

$NaBH_4$ → XI

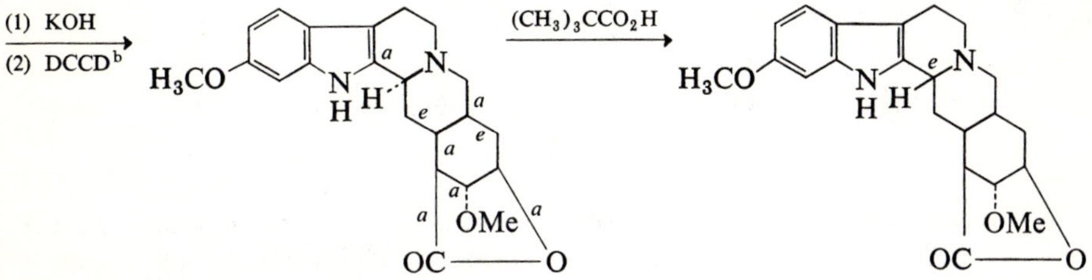

(1) KOH
(2) DCCD[b]

→ XII

$(CH_3)_3CCO_2H$ → XIII

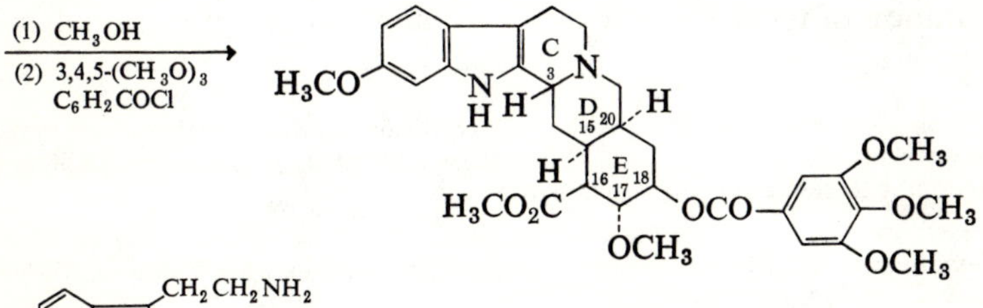

(1) CH_3OH
(2) 3,4,5-$(CH_3O)_3C_6H_2COCl$

→ (reserpine: H_3CO, C, D, E, 3, 15, 16, 17, 18, 20, H_3CO_2C, OCH_3, OCO, OCH_3, OCH_3, OCH_3)

[a] H_3CO–indole–$CH_2CH_2NH_2$

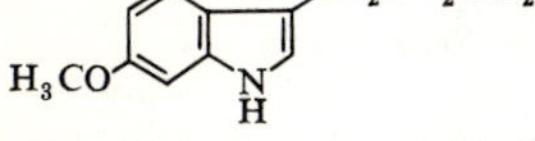

[b] N,N'-Dicyclohexylcarbodiimide

Fig. 15-7b. Synthesis of reserpine (2).

In summary, stereoselective synthesis requires, in addition to much ingenuity, a knowledge of the stereochemical course of a wide variety of reactions and of the position of equilibrium in a variety of acyclic and cyclic systems. A good deal of such information has been presented earlier in this book: Secs. 4-4, 6-3, 8-2, 8-3, 8-5, 8-6, 10-3, 12-4, 12-5, and 12-6.

15-2. Stereoregulated Polymerization[30-32]

When a monosubstituted ethylene, $RCH{=}CH_2$ (sometimes called an "α-olefin"), is polymerized, the resulting polymer (Fig. 15-8) has a large number of asymmetric carbon atoms.† It was first suggested by Staudinger[14] that the lack of crystallinity of many such polymers (e.g., polystyrene, $R = C_6H_5$; polyvinyl acetate, $R = OCOCH_3$) is due to the irregular (random) configuration of these asymmetric atoms. When later it was found[15] that polystyrene fractions obtained by polymerization at different temperatures differed somewhat in physical properties, Huggins[16] attributed such differences to different relative configuration of the asymmetric atoms. Much more clear-cut differences of this nature were subsequently found by Schildknecht

```
          R          R          H          R          H
          |          |          |          |          |
~~~CH2—C*—CH2—C*—CH2—C*—CH2—C*—CH2—C*~~~
          |          |          |          |          |
          H          H          R          H          R
```

Fig. 15-8. Polymer of monosubstituted ethylene.

and coworkers in polymers of isobutyl vinyl ether and methyl vinyl ether.[17] When isobutyl vinyl ether is treated with boron trifluoride gas in pentane at ca. −40°, very rapid polymerization (so-called "flash polymerization") occurs, and the resulting polymer is tacky and rubber-like. However, if the polymerization is carried out slowly by boron trifluoride etherate at −60 to −80°, the polymer is more brittle and possesses crystallinity, as evidenced by X-ray pattern. Schildknecht correctly attributed the difference to random configuration of the asymmetric carbon atoms in the rubbery polymer con-

† A carbon atom near the center of the polymer chain sees only negligible *structural* differences between the two ends of the chain, but there are usually *configurational* differences and such an atom is therefore properly asymmetric.

[14] H. Staudinger, "Die Hochmolekularen Organischen Verbindungen," Springer-Verlag, Berlin, Vienna, 1932, pp. 114, 165.

[15] T. Alfrey, A. Bartovics, and H. Mark, *J. Am. Chem. Soc.*, **65**, 2319 (1943).

[16] M. L. Huggins, *J. Am. Chem. Soc.*, **66**, 1991 (1944).

[17] C. E. Schildknecht, S. T. Gross, H. R. Davidson, J. M. Lambert, and A. O. Zoss, *Ind. Eng. Chem.*, **40**, 2104 (1948); C. E. Schildknecht, S. T. Gross, and A. O. Zoss, *ibid.*, **41**, 1998 (1949).

trasted to ordered configuration in the crystalline polymer, although the nature of the orderly arrangement was not clear at the time.†

The real breakthrough in the field of stereoregulated polymerization came in 1955 when G. Natta and coworkers in Milan published[18] a series of detailed X-ray analyses of polymers of monosubstituted olefins obtained with different polymerization catalysts. These X-ray analyses show the exact relative configuration of the asymmetric atoms in the olefin polymer. Natta distinguishes three types of arrangements, shown in Fig. 15-9: "isotactic polymers," in which all asymmetric carbon atoms have the same configuration;‡ "syndiotactic polymers," in which successive asymmetric atoms alternate in configuration;† and "atactic polymers," in which the configuration of the asymmetric

† Schildknecht believed that his crystalline polymer was syndiotactic (see below) because in this configuration there appeared to be least crowding of the isobutoxy groups. However, later work by Natta indicates that this crystalline polyvinyl isobutyl ether is isotactic.

‡ There is some question about the meaning of the statement that two successive asymmetric carbon atoms in a polymer chain have the same configuration, inasmuch as there is a structural change in going from one carbon to the next, and as pointed out in Chap. 5, the concept of "like configuration" is not well defined when applied to atoms bearing unlike substituents. We shall define the configuration of a chain link —CH_2^a—CHX—CH_2^b— in such a way that the CH_2^a group always points to one and the same end of the chain, regardless of the position of the link in the chain. In this sense, the configurations of all the carbons in the isotactic species in Fig. 15-9 are the same. The situation here is different from that in a small molecule, such as *meso*-tartaric acid. Alternatively, one may define an isotactic arrangement as one in which all substituents are on the same side of a sequence of segments with the backbone chain placed in its planar zigzag conformation (Fig. *i*). A syndiotactic arrangement is one in which the substituents are on alternate sides of such a sequence (Fig. *ii*). It must be kept in mind that, whereas Fig. *i* correctly represents the configuration of an isotactic chain, the chain usually does not exist in the conformation shown, because of the large steric interaction of the substituents (similar to a 1,3-diaxial interaction in cyclohexane). The actual conformation is spiral-shaped, similar to the α helix of a polypeptide (Fig. 15-10), without the hydrogen bonds, of course.

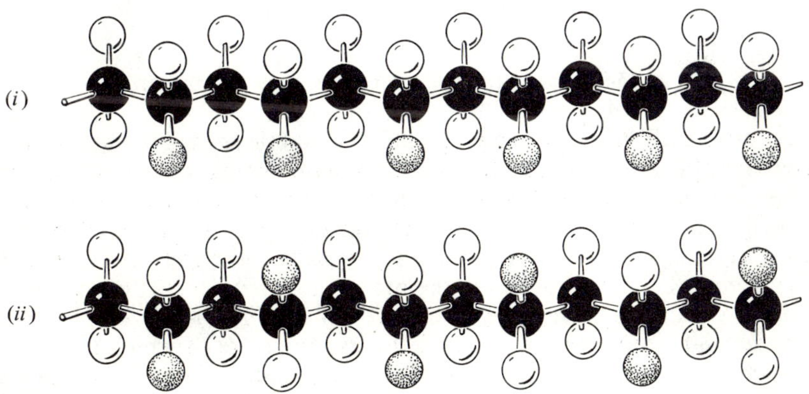

[18] G. Natta, P. Pino, P. Corradini, F. Danusso, E. Mantica, G. Mazzanti, and G. Moraglio, *J. Am. Chem. Soc.*, **77**, 1708 (1955); G. Natta and P. Corradini, *J. Polymer Sci.*, **20**, 251 (1956); G. Natta, *ibid.*, **16**, 143 (1955); G. Natta, *Makromol. Chem.*, **16**, 213 (1955).

atoms is random. The "stereoregulated" (i.e., isotactic or syndiotactic) polymers have higher densities, higher melting points, and lower solubility than the atactic polymers and frequently may be drawn into fibers, all properties which make them quite valuable technologically.

In trying to understand the formation of stereoregulated polymers, one must keep in mind that the addition of a monomer unit of type $RCH{=}CH_2$ to a growing polymer made up of such units involves the creation of a new asymmetric atom in a compound which already has a number of asymmetric atoms. The two possible arrangements of the new asymmetric atom give rise to two diastereoisomers of the growing polymer chain as a whole. As was already explained in Chap. 4, such diastereoisomers do not usually form at equal rates. Normally one might expect that the new asymmetric carbon either would prefer to go in with the same configuration as the adjacent

Isotactic	Syndiotactic	Atactic
⌇	⌇	⌇
H—C—R	H—C—R	H—C—R
CH_2	CH_2	CH_2
H—C—R	R—C—H	R—C—H
CH_2	CH_2	CH_2
H—C—R	H—C—R	H—C—R
CH_2	CH_2	CH_2
H—C—R	R—C—H	H—C—R
CH_2	CH_2	CH_2
⌇	⌇	R—C—H
		CH_2
		⌇

Fig. 15-9. Configurationally different types of polymers.

asymmetric carbon in the growing chain (isotactic arrangement) or would prefer the opposite configuration (syndiotactic arrangement) but not both for the same monomer and catalyst. In fact, it is quite unlikely that there is complete randomness, i.e., that neither arrangement is preferred over the other. In other words, one is unlikely to obtain a completely atactic polymer in the stereochemical sense.† On the other hand, neither is it likely that one obtains a completely syndiotactic or a completely isotactic polymer chain, unless special circumstances prevail, the exact nature of which has not yet been defined in general terms. Empirically, stereoregulated polymers seem to be obtained with heterogeneous catalysts of the Ziegler type (aluminum

† B. Coleman [*J. Polymer Sci.,* **31,** 155 (1958)] has pointed this out and has emphasized that isotactic, syndiotactic, and atactic arrangements, rather than occupying separate conceptual compartments, are the two extremes and the mid-point, respectively, of a graded range.

alkoxide–titanium chloride combinations) as well as with solid chromium trioxide on silica-alumina and molybdenum-trioxide-on-alumina catalysts discovered in the laboratories of the Phillips Petroleum Company and the Standard Oil Company of Indiana, respectively. Polymerization on such solid surfaces may well be stereoregulated by a preferred relative arrangement of the adsorbed monomer and the growing polymer chain. However, in the case of methacrylate monomers, stereoregulated polymerization has been brought about also by apparently homogeneous catalysts such as alkyllithium[19] and Grignard reagents.[20] It has been suggested[21] that, in solvents in which polymer chains are coiled in the shape of a helix, one configuration of an adding monomer unit fits much better into the growing polymer chain than the other, thus accounting for stereoregularity.

The above brief discussion has been limited to monosubstituted olefins, where the only fully regular arrangements are the isotactic and the syndiotactic ones. A larger number of regular arrangements is possible in polymers of 1,2-disubstituted olefins,[22] RCH=CHR′, and in alternating copolymers of two different monosubstituted olefins,[23] $RCH{=}CH_2$ and $R'CH{=}CH_2$.

Stereoregulated polymers from inactive monomers are not optically active. Inspection of Fig. 15-9 discloses that, if the ends of the polymer chain are disregarded, a plane of symmetry may be passed through each carbon atom in the isotactic polymer and through each second (i.e., each tertiary) carbon atom in the syndiotactic chain. But even if one does consider the ends of the chain, one must keep in mind that, for each polymer chain having a given configuration at the end of the chain, one will be formed which has the opposite configuration, because the first asymmetric carbon atom in the chain has an equal chance of being (R) or (S) (unless the initiator or polymerization catalyst itself is dissymmetric).

Instead of being entirely isotactic, or entirely syndiotactic, or completely atactic, a polymer may have chain segments of considerable length which are isotactic alternating with segments which are atactic. Similarly, syndiotactic and atactic or syndiotactic and isotactic segments may occur in the same polymer. Also, an isotactic segment of one configuration may be followed by an isotactic segment of opposite configuration, etc. Polymers of these types are called "stereoblock polymers."

An interesting way of synthesizing an isotactic macromolecule has been devised by Price and Osgan[24] who polymerized (+)-propylene oxide with potassium hydroxide. In this case the feature of stereoregularity is already built into the monomer, inasmuch as all the monomer units have the same

[19] T. G. Fox, B. S. Garrett, W. E. Goode, S. Gratch, J. F. Kincaid, A. Spell, and J. D. Stroupe, *J. Am. Chem. Soc.*, **80**, 1768 (1958).

[20] R. G. J. Miller, B. Mills, P. A. Small, A. Turner-Jones, and D. G. M. Wood, *Chemistry & Industry (London)*, 1323 (1958).

[21] M. Szwarc, *Chemistry & Industry (London)*, 1589 (1958).

[22] C. L. Arcus, *J. Chem. Soc.*, 2801 (1955).

[23] C. L. Arcus, *J. Chem. Soc.*, 1189 (1957).

[24] C. C. Price and M. Osgan, *J. Am. Chem. Soc.*, **78**, 4787 (1956).

configuration at the asymmetric carbon. Since the nature of the propagation step

$$RO^- + \overset{\;\;O}{CH_2CHCH_3} \rightarrow ROCH_2CH(CH_3)O^-$$

is such that the configuration of the asymmetric carbon of the epoxide is not affected, the asymmetric atoms in the polymer chain must also have identical configuration. In fact, the polymer, in addition to being optically active, is a crystalline solid, whereas the inactive polymer obtained by polymerizing (±)-propylene oxide with potassium hydroxide, while of approximately the same chain length, is a liquid.[24] This difference in properties provides further clear-cut evidence that the high crystallinity of isotactic polymers is indeed due to their configurational regularity, as postulated by Natta and supported by the already mentioned X-ray crystallographic studies.

When (+)-propylene oxide is polymerized by ferric chloride, the resulting polymer may be separated into two fractions, one partly crystalline (isotactic) and one partly amorphous (atactic).[24] The optical activity of the crystalline fraction is about eight times as high as that of the amorphorus fraction and is about equal to that of the polymer obtained with KOH. In acid-catalyzed polymerization of an epoxide, a carbonium-ion-like intermediate of the type $Cl_3Fe^-OCH_2CH(CH_3)^+$ may be involved, and if this intermediate is free, the asymmetric carbon is racemized and an atactic, inactive polymer results.† On the other hand, if the intermediate is not free and retains configuration during the polymerization, an isotactic polymer is again obtained. A mechanism accounting for maintenance of asymmetry in the monomer units and explaining why, once the polymer starts out active, it will continue to grow by the addition of non-racemizing monomer units has been proposed.[25] The polymerization of (±)-propylene oxide with ferric chloride also produces a mixture of atactic and isotactic polymer, and the properties of the isotactic fraction are essentially the same as those of the isotactic fraction obtained from the active monomer, except for the absence of optical activity.[24] Here again the isotactic chains must contain monomer units all of the same configuration—or at least sizable stereoblocks made up of monomer units of one configuration—but since there is an equal chance of starting a polymer chain (or a stereoblock) from a (+)-oxide or a (−)-oxide molecule, there will be equal numbers of chains (or stereoblocks) of opposite configuration and the resulting polymer will be racemic.‡

† The observed slight activity may be due to incomplete fractionation of the atactic and isotactic portions of the polymer, or it may be caused by small isotactic stereoblocks in the atactic polymer.

‡ Inspection of the projection formula (*i*) of an isotactic polypropylene oxide chain shows that, unlike an isotactic polypropylene chain (Fig. 15-9), it does not have any element of symmetry.

```
      |
      O
      |
      CH2
      |
  H—C—CH3      (i)
      |
      O
      |
      CH2
      |
  H—C—CH3
      |
```

[25] M. Osgan and C. C. Price, *J. Polymer Sci.*, **34**, 153 (1959); E. J. Corey, *Tetrahedron Letters*, no. 2, 1 (1959).

The formation of isotactic polymers from optically active monomers has an interesting analogy in the formation of polymeric materials in nature. For example, as already mentioned in Chap. 5, proteins are made up, as far as it is known, exclusively from L-amino acids. Proteins may, therefore, be considered as isotactic polymers. It was also mentioned in Chap. 5 that monomeric D-amino acids occur in lower organisms, and the question was raised as to why they do not seem to be found in protein hydrolysates. The answer may be that the polymerization of amino acids in the living organism, as the polymerization of propylene oxide catalyzed by ferric chloride, occurs under conditions that favor formation of an isotactic polymer, and if the L-amino acids are in substantial excess over the D-amino acids in the available mixture of monomers, the few D-amino acids that are present may be "rejected" in the formation of the protein. On the other hand, the concentration of the D-amino acids might be too small to allow them to polymerize by themselves to form a protein enantiomeric to the natural one. The stereoregulating influence in the polymerization of α-amino acids may be the formation of an α helix (Fig. 15-10) stabilized by hydrogen bonding,[26] in which the isotactic arrangement may be favored over an atactic arrangement, by virtue of the better steric fit of the R groups in the recurring —CHR—CO—NH— unit (cf. Ref. 18). Experimentally it has been found[27] that, in the polymerization of γ-benzyl-L-glutamate-*N*-carboxyanhydride (Fig. 15-11) initiated by sodium methoxide in dioxane solution, addition of the D enantiomer greatly reduces both the polymerization rate and the degree of polymerization. This is ascribed[27] to difficulties encountered when the growing α helix made up of L-amino acid monomers tries to incorporate a D-amino acid fragment. In fact, the drop in rate is so large as to suggest not merely a rejection of the unnatural monomer unit but a partial blocking of the polymerization process by the encounter of the growing chain with the unnatural monomer.

In diene polymerization, there are avenues of stereoregulation in addition to those already mentioned. Thus, butadiene, in addition to giving isotactic and syndiotactic 1,2-olefin polymers, $—CH_2—CH(CH{=}CH_2)—$, may also give rise to a 1,4 polymer in which the repeating unit is

$$—CH_2—CH{=}CH—CH_2—$$

Clearly, there can be cis-trans isomerism about the double bond, and the polymer may be all-cis, all-trans, or mixed. By use of proper catalysts, stereoregulated all-cis and all-trans 1,4 polymers may indeed be obtained, and catalysts of this type have been used[28] to polymerize isoprene to natural rubber which is the all-cis polymer (Fig. 12-19). 1,4 Polymerization which also generates asymmetric carbon atoms along the chain occurs with monomers of the type RCH=CH—CH=CHR′ to give polymers in which the recurring unit is $—\overset{*}{C}HR—CH{=}CH—\overset{*}{C}HR'—$. Depending on the configu-

[26] See L. Pauling, "Nature of the Chemical Bond," 3d ed., Cornell University Press, Ithaca, N.Y., 1960, pp. 498–502.

[27] M. Idelson and E. R. Blout, *J. Am. Chem. Soc.*, **80**, 2387 (1958); cf. P. Doty and R. D. Lundberg, *ibid.*, **78**, 4810 (1956).

[28] F. W. Stavely et al., *Ind. Eng. Chem.*, **48**, 778 (1956); S. E. Horne et al., *ibid.*, **48**, 784 (1956).

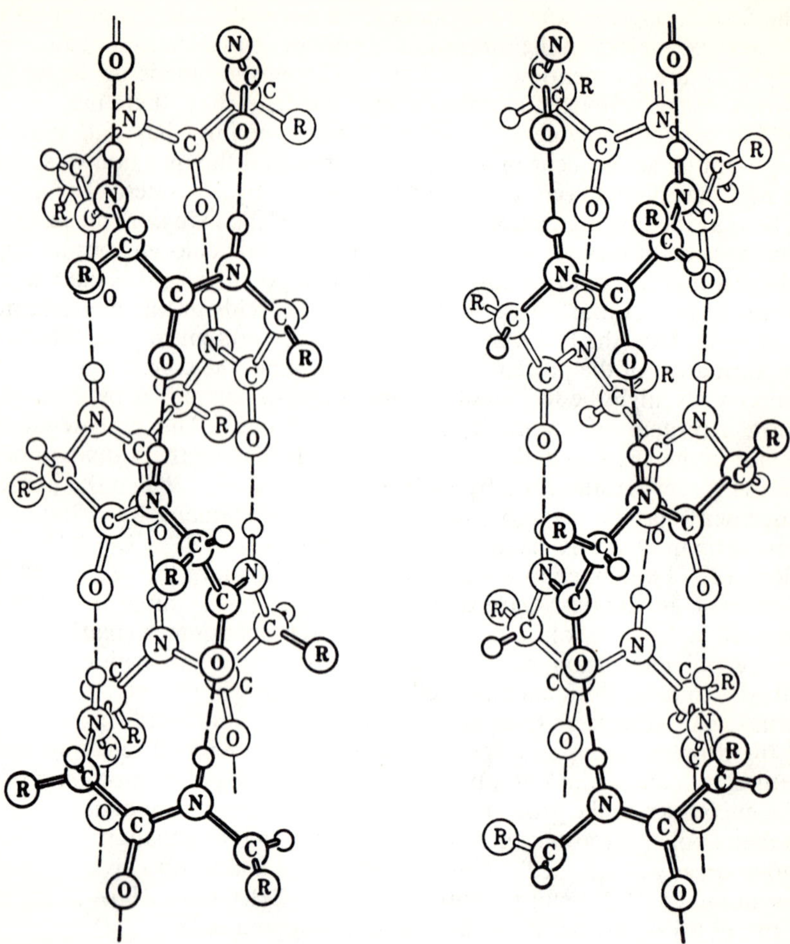

Fig. 15-10. α Helix of polypeptide. [*From R. B. Corey and L. Pauling, Rend. ist. lombardo sci.,* P1, **89,** 10 (1955). *By permission of the authors.*]

$$\begin{array}{l} C_6H_5CH_2OCOCH_2CH_2CH\!-\!CO \\ \qquad\qquad\qquad\qquad\quad | \qquad\quad \backslash \\ \qquad\qquad\qquad\qquad\quad | \qquad\quad\ O \\ \qquad\qquad\qquad\qquad\quad | \qquad\quad / \\ \qquad\qquad\qquad\qquad\ NH\!-\!CO \end{array}$$

Fig. 15-11. γ-Benzyl-L-glutamate-*N*-carboxyanhydride (a Leuchs anhydride).

ration of the starred carbon atoms, the polymer may exist in threo and erythro as well as cis and trans forms. Such polymers have been termed "tritactic" and have been obtained in optically active form by the use of optically active polymerization catalysts.[29] Many other types of stereoregulated polymers will undoubtedly be prepared in the future.

[29] *Chem. Eng. News,* Jan. 16, 1961, p. 44.

General References

[30] N. G. Gaylord and H. F. Mark, "Linear and Stereoregular Addition Polymers," Interscience Publishers, Inc., New York, 1959.

[31] C. L. Arcus, The Stereochemistry of Addition Polymers, in P. B. D. de la Mare and W. Klyne, eds., "Progress in Stereochemistry," vol. 3, Academic Press, Inc., New York, 1962, chap. 7.

[32] G. Natta, Stereospecifische Katalysen und isotaktische Polymere, *Angew. Chem.*, **68**, 393 (1956).

NAME INDEX

The letter *n.* following a page reference indicates a text footnote, the letter *t.* a table footnote.

SUBJECT INDEX

Relationship between Percentage of More Stable Isomer at Equilibrium, Equilibrium Constant *K*, and Standard Free-energy Difference ΔF^0 at 25.00°C. and 80.00°C. for an Equilibrium of Isomers: A $\rightleftharpoons$ B

% Stabler Isomer	K	ΔF^0_{25} cal./mole	ΔF^0_{80} cal./mole
50	1.00	0	0
55	1.22	119	141
60	1.50	240	285
65	1.86	367	434
70	2.33	502	595
75	3.00	651	771
80	4.00	821	973
85	5.67	1028	1217
90	9.00	1302	1542
95	19.00	1745	2066
98	49.00	2306	2731
99	99.00	2723	3225
99.9	999.0	4092	4847
99.99	9999	5457	6464